Lutz F. Tietze, Theophil Eicher,
Ulf Diederichsen, Andreas Speicher

Reactions and Syntheses

in the Organic Chemistry Laboratory

BICENTENNIAL
BICENTENNIAL
1807
⟨⊗⟩WILEY
2007
BICENTENNIAL
BICENTENNIAL

WILEY-VCH Verlag GmbH & Co. KGaA

The Authors

Prof. Dr. Dr. h.c. Lutz F. Tietze
Institute of Organic and Biomolecular Chemistry
Georg-August-University
Tammannstr. 2
37077 Göttingen
Germany

Prof. Dr. Dr. h.c. Theophil Eicher
Saarland University
FR 8.1 – Organic Chemistry
66123 Saarbrücken
Germany

Prof. Dr. Ulf Diederichsen
Institute of Organic and Biomolecular Chemistry
Georg-August-University
Tammannstr. 2
37077 Göttingen
Germany

PD Dr. Andreas Speicher
Saarland University
FR 8.1 – Organic Chemistry
66123 Saarbrücken
Germany

All books published by Wiley-VCH are carefully produced. Nevertheless, authors, editors, and publisher do not warrant the information contained in these books, including this book, to be free of errors. Readers are advised to keep in mind that statements, data, illustrations, procedural details or other items may inadvertently be inaccurate.

Library of Congress Card No.:
applied for

British Library Cataloguing-in-Publication Data
A catalogue record for this book is available from the British Library.

**Bibliographic information published by
the Deutsche Nationalbibliothek**
Die Deutsche Nationalbibliothek lists this publication in the Deutsche Nationalbibliografie; detailed bibliographic data are available in the Internet at <http://dnb.d-nb.de>.

© 2007 WILEY-VCH Verlag GmbH & Co. KGaA, Weinheim

Printed in the Federal Republic of Germany

Printed on acid-free paper

Printing Strauss GmbH, Mörlenbach
Binding Litges & Dopf Buchbinderei GmbH, Heppenheim
Wiley Bicentennial Logo Richard J. Pacifico

ISBN: 978-3-527-31223-8

The three research groups in Göttingen and Saabrücken participated in equal parts in the preparation of the experimental sections of the syntheses presented in this book. The following collaborators were engaged in checking, testing and elaborating the selected literature procedures:

The *Tietze* group

Dirk Spiegl, Deshan Liu, Florian Stecker, Sabine Schacht, Christian Brazel, Niels Böhnke, Prof. Dr. Dr. h.c. L. F. Tietze, Dr. Julia Zinngrebe, Florian Lotz, Heiko Schuster, Dr. Francisco Colunga, Dr. Stephan Hettstedt, Dr. Xiong Chen, Thomas Redert

The *Speicher* group

Timo Backes   Sabrina Bleif   Mandy Döring   Matthias Groh

Luisa Gilmore   Judith Holz   Sandra Kern   Boris Weidenhof

The *Diederichsen* group

Katja Bensmann   Stefan Cortekar   Matthias Decke   Nicola Diezemann

Katharina Fejfar   Ansgar Fitzner   Juliane Gräfe   Daniel Heinrich

Nicole Hildebrandt   Nadine Jede   Andrea Küsel   André Nadler

Marian Pitulescu   Ruzica Ranevski   Anmol Kumar Ray   Philipp Schneggenburger

Ratika Srivastava   Thorsten Stafforst   Angelina Weiß   Brigitte Worbs

Contents

Reactions and Syntheses in the Organic Chemistry Laboratory. L. F. Tietze, Th. Eicher, U. Diederichsen, A. Speicher
Copyright © 2007 WILEY-VCH Verlag GmbH & Co. KGaA, Weinheim
ISBN: 978-3-527-31223-8

3 Heterocyclic compounds

Abbreviations and Symbols

General abbreviations and symbols

g	gram	$[\alpha]_D$	specific rotation
mg	milligram	*ee*	enantiomeric excess
L	liter	*ds*	diastereoselectivity
mL	milliliter	TLC	thin-layer chromatography
mol	mole	HPLC	high-performance liquid chromatography
mmol	millimole	ca.	approximately
min	minute(s)	ref.	literature reference
h	hour(s)	p.	page
d	day(s)	ed.	edition
°C	degrees Celsius	Ed(s).	editor(s)
%	percent	cf.	compare
mp	melting point	dec.	decomposition
bp	boiling point	M_r	relative mass
n_D^{20}	refractive index at Na D line (at 20°C)	rt	room temperature

Spectroscopic abbreviations

IR	infrared spectrum
$\tilde{\nu}$	wave number (in cm^{-1})
^{1}H NMR	proton nuclear magnetic resonance spectrum
^{13}C NMR	^{13}C nuclear magnetic resonance spectrum
δ (ppm)	chemical shift relative to tetramethylsilane ($\delta_{TMS} = 0$)
s	singlet
d	doublet
dd	doublet of doublets
t	triplet
dt	doublet of triplets
q	quartet
quint	quintet
sext	sextet
sept	septet
m	multiplet
br	broad
Hz	Hertz
J	coupling constant
UV/VIS	ultraviolet/visible spectrum
nm	nanometer
λ_{max} (log ε)	wavelength of the absorption maximum (molar extinction coefficient)

Reactions and Syntheses in the Organic Chemistry Laboratory. L. F. Tietze, Th. Eicher, U. Diederichsen, A. Speicher
Copyright © 2007 WILEY-VCH Verlag GmbH & Co. KGaA, Weinheim
ISBN: 978-3-527-31223-8

Abbreviations for substituents...

Ac	–COCH$_3$	acetyl	*i*Bu	–CH$_2$CH(CH$_3$)$_2$	*iso*-butyl
Ar		aryl	*s*Bu	–CH(CH$_3$)CH$_2$CH$_3$	*sec*-butyl
Me	–CH$_3$	methyl	*t*Bu	–C(CH$_3$)$_3$	*tert*-butyl
Et	–CH$_2$CH$_3$	ethyl	Mes	–SO$_2$CH$_3$	methanesulfonyl
Pr	–CH$_2$CH$_2$CH$_3$	propyl	Ph	–C$_6$H$_5$	phenyl
*i*Pr	–CH(CH$_3$)$_2$	*iso*-propyl	Tf	–SO$_2$CF$_3$	trifluoromethanesulfonyl
*n*Bu	–(CH$_2$)$_3$CH$_3$	*n*-butyl	Tos	–SO$_2$C$_6$H$_4$CH$_3$	*p*-toluenesulfonyl

...and commonly used compounds...

AIBN	azoisobutyronitrile
DABCO	1,4-diazabicyclo[2.2.2]octane
DBN	1,5-diazabicyclo[4.3.0]non-5-ene
DBU	1,8-diazabicyclo[4.3.0]undec-7-ene
DCC	dicyclohexylcarbodiimide
DDQ	dichlorodicyano-*p*-benzoquinone
DIBAL	diisobutylaluminum hydride
Diglyme	diethylene glycol dimethyl ether
DME	dimethoxyethane
DMF	*N,N*-dimethylformamide
DMPU	1,3-dimethyl-3,4,5,6-tetrahydro-2(1*H*)-pyrimidone
DMSO	dimethyl sulfoxide
Et$_2$O	diethyl ether
EtOH	ethanol
LAH	lithium aluminum hydride
MeOH	methanol
NBS	*N*-bromosuccinimide
NCS	*N*-chlorosuccinimide
PPA	polyphosphoric acid
TBAF	tetra-*n*-butylammonium fluoride
TFA	trifluoroacetic acid
THF	tetrahydrofuran

...and retrosynthesis

disc	bond disconnection
FGI	functional group interconversion
FGA	functional group addition

Preface

(1) Background

The book "Reactions and Syntheses in the Organic Chemistry Laboratory" was first published in German in 1981, with a second edition in 1991, and was translated into Japanese in 1984 (2nd edition 1995), English in 1989, Chinese in 1999, Russian in 2000, and Korean in 2002. The intention was

- to associate classes of compounds and functionalities with reaction types and mechanisms,

- to offer a great number of reliable preparative procedures of general importance,

- to show usefulness and robustness of the offered procedures for the synthesis of selected interesting compounds of relevance in biology, pharmacy, and medicine.

Since the last German edition, many new preparative procedures have been developed showing high chemo-, regio-, diastereo-, and enantioselectivity, which frequently approach the selectivity of enzymatic transformations with the advantage of a lower substrate specificity. In addition, new methods such as combinatorial chemistry, solid-phase chemistry, high-pressure chemistry, and the use of microwaves for heating have been introduced. Moreover, the efficiency of a synthesis, which can be defined as the increase in complexity per transformation, the avoidance of toxic reagents as well as solvents, and the preservation of resources are important issues in modern preparative organic chemistry. Significant developments in the last years have been realized in transition metal catalysis, organocatalysis, and domino reactions. This progress has been impressively documented in "Classics of Total Synthesis" [1], "Organic Synthesis Highlights" [2], and "Domino Reactions in Organic Synthesis" [3].

As a consequence, we now present the book „Reactions and Syntheses in the Organic Chemistry Laboratory"

(a) in a new form with respect to its concept and organization,

(b) extensively renewed with respect to its content.

- Basic units as well as main objectives are *syntheses* (up to multi-step syntheses with > 5 steps) of interesting and instructive target molecules from various fields of Organic Chemistry. Each synthesis is centred around one ore more methods and reactions principles of general synthetic relevance.

- As before, the users of the new book are provided with carefully elaborated experiments, which are described in preparative and analytical detail. However, experiments and syntheses are accompanied throughout in concentrated form by the required general, theoretical and mechanistic background and explanations. Special attention is given to retrosynthetic analysis and alternative approaches of synthesis for a given target molecule.

- To allow the inclusion of a representative and qualified spectrum of contemporary synthetic methods, more than 70 % of the contents of the former book have been replaced by more recent and more relevant experimental examples. The remaining (elder) syntheses have been „updated" with respect to description of their general background.

Considering the various types of potential users of the book in the past, there has been a definite and broad acceptance among chemists and pharmacists on a more advanced level, besides graduate students and researchers at universities and in industry. From these considerations, the following consequences have emerged for the 3rd edition:

- General laboratory information, such as safety, first aid, performance of chemical reactions, instrumentation and standard apparatus, isolation and purification of products, has been omitted. Methods for the formation and transformation of basic functional groups in organic compounds, regarded as being important for the elementary education level in organic laboratory practice, are not described. These topics are comprehensively covered in other qualified textbooks [4–6].

- The deletion of these elementary aspects of organic chemistry has allowed us to describe more of the advanced synthetic methods and to include mechanistic aspects as well as to incorporate total syntheses and retrosynthetic analyses.

(2) Organization of the book and directions for its use

The book is divided into four chapters with several subchapters:

Chapter **1**   C–C Bond formation

Chapter **2**   Oxidation and reduction

Chapter **3**   Heterocyclic compounds

Chapter **4**   Selected natural products

The subchapters (e.g. **1.1**, **1.2**, etc.) contain the different procedures and syntheses specified at the beginning of the section and in the Table of *Contents* (cf. p. II) and are organized as follows:

(a) In the general part (a) the *structural formula* of the target molecule and the *topics* of the presented synthesis (important for rapid information!) are given, which is followed by *introductory information* on the target molecule, *retrosynthesis* [7], and *planning of the synthesis* (possibilities, strategies, and synthetic alternatives; considerations on practicability for laboratory use).

(b) In part (b) the *synthesis of the target molecule* and the synthetic steps performed in the experimental part are described. This is accompanied by information about the mechanism(s), the stereochemical outcome, and the selectivity of the transformations (specific reaction principles). Finally, the number of steps performed and the yields obtained are summarized. In general, section (b) contains a complete *scheme of the synthesis* performed.

(c) In section (c) individual *experimental procedures* are described.

Each procedure has the following structure:

- An identification number, which characterizes the prepared compound according to chapter, subchapter, and synthesis (e.g. **1.1.1.1**, **1.1.1.2**, etc.); the identification number carries one or more asterisks (*, **, ***) according to the degree of difficulty of the procedure.

- Literature reference(s) for the prepared compound.

- A formula equation, which gives structures of reactants and products, and their relative molecular masses. In general, apparatus is not discussed in detail; however, in special cases, information about specialized equipment (photochemical, high-pressure, microwave, etc.) is given.

- Throughout, the procedures are presented in two parts. The first, describing the reaction, often includes additional notes about purification and characteristics of the substrates, such as toxicity and safety remarks. The second describes the work-up, isolation, and purification of the product, along with criteria of purity (mp, bp, n_D, TLC/R_f, $[\alpha]_D$), notes about characteristics of the product, and other crucial experimental details.

- Characterization of the product by spectral data (IR, UV/VIS, ^{1}H and ^{13}C NMR, MS). In selected cases, the preparation of derivatives together with their instrumental and chemical analysis is given.

(d) The presentation of each synthesis is concluded by a compilation of the *literature references* cited in the sections (a)–(c). They cover the primary literature on the synthesis, its steps and its topics, and refer to important collective articles, reviews, and textbooks of advanced organic chemistry [8].

[1] K. C. Nicolaou, E. J. Sorensen, *Classics in Total Synthesis*, VCH, Weinheim, **1997**; K. C. Nicolaou, S. A. Snyder, *Classics in Total Synthesis II*, Wiley-VCH, Weinheim, **2003**.

[2] *Organic Synthesis Highlights I–V* (Eds.: J. Mulzer, H.-J. Altenbach, M. Braun, K. Krohn, H.-U. Reissig, H. Waldmann, H.-G. Schmalz, Th. Wirth), VCH/Wiley-VCH, Weinheim, **1991-2003**.

[3] L. F. Tietze, G. Brasche, K. M. Gericke, *Domino Reactions in Organic Synthesis*, Wiley-VCH, Weinheim, **2006**.

[4] *Organikum*, 21st ed., Wiley-VCH, Weinheim, **2001**.

[5] S. Hünig, P. Kreitmeier, G. Märkl, J. Sauer, *Arbeitsmethoden in der Organischen Chemie* (mit Einführungs-praktikum), Verlag Lehmanns, Berlin, **2006**.

[6] R. C. Larock, *Comprehensive Organic Transformations* (A Guide to Functional Group Preparations), 2nd ed. Wiley-VCH, Weinheim, **1999**.

[7] Retrosynthesis is oriented toward the concepts and terminology of S. Warren, *Organic Synthesis – The Disconnection Approach*, John Wiley & Sons, New York, **1982**; S. Warren, *Designing Organic Syntheses*, John Wiley & Sons, New York, **1978**; E. J. Corey, X.-M. Cheng, *The Logic of Chemical Synthesis*, John Wiley & Sons, New York, **1989**.

[8] For example: M. B. Smith, J. March, *March's Advanced Organic Chemistry*, 6th ed., John Wiley & Sons, Inc., New York, **2007**; F. A. Carey, R. J. Sundberg, *Organische Chemie*, VCH, Weinheim, **1995**; G. Quinkert, E. Egert, Ch. Griesinger, *Aspekte der Organischen Chemie*, VCH, Weinheim, since **1995**; R. Brückner, *Reaktionsmechanismen* (Organische Reaktionen, Stereochemie, moderne Synthesemethoden), 3rd ed., Spektrum Akademischer Verlag, Heidelberg, **2004**; E. L. Eliel, S. H. Wilen, M. P. Doyle, *Basic Organic Stereochemistry*, John Wiley & Sons, New York, **2001**; J.-H. Fuhrhop, G. Li, *Organic Synthesis*, 3rd ed., Wiley-VCH, Weinheim, **2003**; P. J. Kocieński, *Protecting Groups*, 3rd ed., Georg Thieme Verlag, Stuttgart, **2005**; G. Helmchen, R. W. Hoffmann, J. Mulzer, E. Schaumann, *Houben-Weyl, Methods of Organic Chemistry, Stereoselective Synthesis*, Vol. E21, 4th ed., Georg Thieme Verlag, Stuttgart, **1996**; S. Hauptmann, G. Mann, *Stereochemie*, Spektrum Akademischer Verlag, Heidelberg, **1996**.

Acknowledgements

The authors are indebted, above all, to the collaborators of the groups in Göttingen and Saarbrücken, who performed the laboratory work for testing the suitability of the selected synthetic examples; they are presented in pictures directly following the title page of this book.

L. F. T. is especially indebted to Christian Brazel, Katja Grube, Tom Kinzel, Dirk Spiegl and Florian Stecker for their outstanding valuable contributions in the preparation of some parts of the manuscript.

We are also thankful to the Fonds der Chemischen Industrie for generous financial support.

Furthermore, the authors are grateful to Prof. Dr. U. Kazmaier, Institute of Organic Chemistry, University of the Saarland, Prof. Dr. P. Knochel, Institute of Organic Chemistry Lutwig-Maximilian University, München, Prof. Dr. J. A. Wisner, University of Western Ontario, London Ontario, Canada and Dr. L. Kattner, Fa. Endotherm, Saarbrücken, for making available experimental procedures from their research field, and to Prof. Dr. R. Schmidt for helpful suggestions. T. E. thanks Prof. Drs. H. Becker, J. Jauch, U. Kazmaier, and G. Wenz for providing collegial hospitality and support during the preparation of this book.

Special thanks are due to Dr. E. Maase, Dr. R. Kirsten, Dr. S. Pauly and Dr. M. Köhl from the staff of the editorial office of Wiley-VCH for their efficient assistance and cooperativity.

Göttingen and Saarbrücken, August 2007

Lutz F. Tietze, Theophil Eicher, Ulf Diederichsen, Andreas Speicher

1   C–C Bond formation

Introduction

C–C Bond formations are essential for the construction of the backbone of any organic compound in this book. Their mechanistic description is used as a general tool for their classification. Thus, in chapters **1.1–1.8**, the focus is on transformations in which nucleophilic, electrophilic, radical, and pericyclic reactions, as well as reactions mediated by organometallics and transition metal compounds, play the decisive role.

Chapter **1.1** deals with *nucleophilic addition to the carbonyl group of aldehydes, ketones, and derivatives of carboxylic acids (esters, anhydrides, etc.), addition to acceptor-substituted olefins (Michael addition),* and *carbonyl olefination.*

Examples are

- the classical addition of organometallics to aldehydes or ketones, such as the addition of Grignard compounds (**1.1.1**), allylsilanes (**1.1.2**, conducted according to the Tietze protocol in enantioselective fashion), and the Ti-induced geminal dimethylation of C=O groups (**1.1.6**);

- the desymmetrization of achiral anhydrides by Grignard compounds in the presence of a chiral catalyst (**1.1.3**);

- conjugate addition to auxiliary-substituted α,ß-unsaturated carbonyl systems (**1.1.4**) and trapping of an enolate formed by 1,4-addition to an acceptor-substituted olefin according to the Baylis–Hillman concept (**1.1.5**);

- carbonyl olefination, such as by the Wittig reaction (**1.1.6**) and the Lombardo reaction (**1.1.7**).

Chapter **1.2** deals with *alkylation reactions of aldehydes/ketones or carboxylic acids and β-dicarbonyl compounds* at their α- and γ-positions, respectively.

Examples are

- the auxiliary-based stereoselective α-alkylation of ketones according to the SAMP method of Enders (**1.2.1**);

- the auxiliary-based stereoselective α-alkylation of chiral oxazolidinones according to the Evans methodology (**1.2.2**);

- the γ-alkylation of acetoacetate (**1.2.3**).

Reactions and Syntheses in the Organic Chemistry Laboratory. L. F. Tietze, Th. Eicher, U. Diederichsen, A. Speicher
Copyright © 2007 WILEY-VCH Verlag GmbH & Co. KGaA, Weinheim
ISBN: 978-3-527-31223-8

Chapter **1.3** deals with *reactions of the aldol and Mannich type.*

Examples are

- the synthesis of olivetol (**1.3.1**), involving aldol addition/condensation, Michael addition/Claisen condensation (cf. **1.4**), and a domino process for aromatization;

- the synthesis of the Wiechert–Hajos ketone (**1.3.2**) by Michael addition and (*S*)-proline-catalyzed enantioselective aldol addition/condensation (organocatalytic Robinson annelation);

- the diastereoselective aminoalkylation of an enamine via an *N*-acyl imino ester (**1.3.3**);

- the enantioselective Mukaiyama aldol reaction (**1.3.4**) of an aldehyde with a silylenol ether catalyzed by a chiral (acyloxy)borane, as well as the synthesis of the required chiral CAB ligand from (*S,S*)-tartaric acid;

- the enantioselective synthesis of an oxocyclohexane carboxylic ester (**1.3.5**) by a domino-Michael-aldol process using Jørgensen's catalyst as well as its synthesis from *L*-phenylalanine.

Chapter **1.4** deals with *electrophilic and nucleophilic acylation* reactions.

Examples are

- the synthesis of a chiral 1-alkylated 2-oxocyclopentane-1-carboxylate (**1.4.1**) by Dieckmann cyclization of an adipate, chemoenzymatic enantioselective reduction of a β-ketoester, α-alkylation of a chiral β-hydroxyester, and re-oxidation;

- the synthesis of both enantiomers of an α-hydroxyester (**1.4.2**) by an ECP protocol starting with succinylation of an arene by *O*-acetylated (*S*)-malic anhydride;

- the synthesis of (*S*)-naproxene (**1.4.3**) starting with a thermodynamically controlled Friedel–Crafts acylation of a naphthalene derivative and an enzymatic hydrolysis of an ester moiety;

- addition of an α-metalated (*O*-silyl)cyanohydrin as an acyl anion equivalent to an enone (**1.4.4**) according to the Hünig protocol.

Chapter **1.5** deals with *reactions of alkenes* proceeding *via carbenium ions.*

Examples are

- the synthesis of piperine (**1.5.1**) by a Lewis acid-catalyzed reaction of acetals with ethyl vinyl ether;

- the synthesis of cicloxilic acid (**1.5.2**) by way of a stereoselective Prins reaction;

- the synthesis of β-ionone (**1.5.3**) with formation of the cyclohexene ring by a cationic cyclization.

Chapter **1.6** deals with *transition-metal-catalyzed reactions* and is focused on some representative transformations using Pd and Cu as catalysts, such as:

- the Heck reaction, applied for a novel ligand-free stilbene synthesis (**1.6.1**);

- the Suzuki–Miyaura cross-coupling reaction for the synthesis of a functionalized biaryl (**1.6.2**);

- the Sonogashira cross-coupling reaction, used for diarylalkyne formation (**1.6.3**);

- a domino cuprate reaction of a β-mesyloxy alkenone, applied for the synthesis of 3,3-dimethylcyclohexanone (**1.6.4**).

Other transition-metal-mediated reactions are described in **1.1.6/1.1.7/2.1.1** (Ti), **2.1.2** (Mn), **3.3.2** (Ru), **4.2.2** (Rh).

Chapter **1.7** deals with *pericyclic reactions* (cycloadditions, electrocyclic transformations, sigmatropic reactions).

Examples are

- the synthesis of tranylcypromine (**1.7.1**) with cyclopropane formation by [1+2]-addition of carbenes to alkenes;

- the synthesis of [10]annulene (**1.7.2**) using a [1+2]-cycloaddition and an electrocyclic ring-opening (norcaradiene–cycloheptatriene rearrangement);

- the synthesis of a heptalene derivative (**1.7.3**) by [2+2]-cycloaddition of ADE to azulene followed by electrocyclic ring-opening;

- the synthesis of bishomocubane (**1.7.4**) by [4+2]-cycloaddition of 1,3-cyclohexadiene to a 1,4-quinone followed by intramolecular photochemical [2+2]-cycloaddition;

- the synthesis of α-terpineol (**1.7.5**) by the use of a Diels–Alder reaction as the key step, which is performed (a) for *rac*-α-terpineol and (b) for (+)-α-terpineol by an auxiliary-assisted Evans process (cf. **1.2.2**);

- generation of a bicyclo[2.2.2]octene derivative (**1.7.6**) by oxidation of an arene with a hypervalent iodine compound and trapping of the intermediate cyclohexadienone by [4+2]-cycloaddition.

The hetero-Diels–Alder reaction of α,β-unsaturated carbonyl compounds to electron-rich olefins ([4+2]-cycloaddition with inverse electron demand) is another type of pericyclic reaction and is described in **4.1.1**.

Chapter **1.8** deals with some basic *radical reactions*.

Examples are

- the radical addition of CCl_4 to an olefin (telomerization) combined with a Claisen [3,3]-sigmatropic rearrangement (**1.8.1**);

- the radical addition of arenes to activated olefins (Meerwein arylation) and the photochemical cyclization of *cis*-stilbenes to phenanthrenes (**1.8.2**).

1.1 Nucleophilic addition to aldehydes, ketones, carboxylic acid derivatives (esters, anhydrides, α,β-unsaturated carbonyl compounds)

1.1.1 (E)-4-Acetoxy-2-methyl-2-butenal

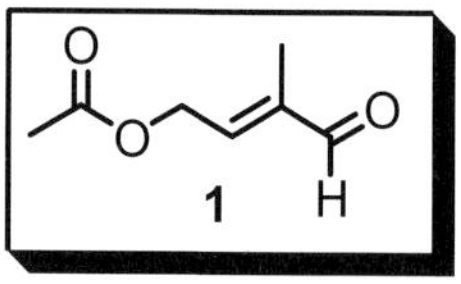

Topics:
- Preparation of a C_5-building block for vitamin A synthesis
- Allylic alcohols from ketones and vinyl Grignard compounds
- Acetylation of an allyl alcohol with allylic inversion
- Kornblum oxidation R–CH$_2$–X → R–CH=O

(a) General

(E)-2-Methyl-2-butenal bearing an acetoxy group at the 4-position can be regarded as a functional isoprene unit and is used as a C_5-building block for the synthesis of terpenes by carbonyl olefination [1]. Thus, in the classical industrial vitamin A synthesis of BASF (cf. **4.2.6**), (E)-4-acetoxy-2-methyl-2-butenal (**1**) is combined with the C_{15}-ylide **2** in a Wittig reaction to give vitamin A acetate **3**:

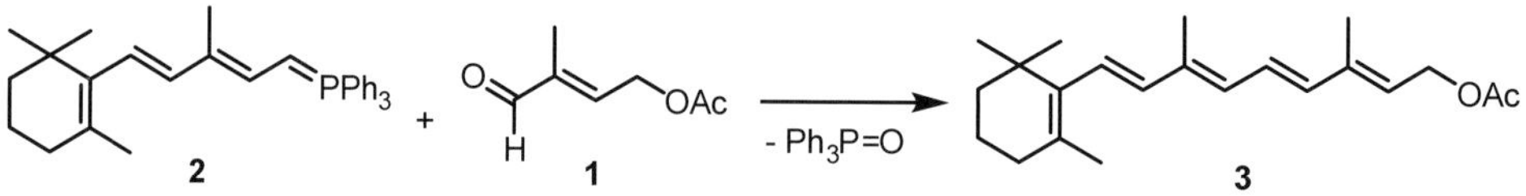

Retrosynthesis of the target molecule **1** can be conducted in two directions (**A/B**) via the intermediates **4/5** and further by allylic inversions to allyl alcohols **6/7**. These should result from the acetone derivatives **8/9** either by addition of allyl metals or by ethynylation followed by partial hydrogenation of the primarily formed acetylenic alcohols (approaches **I/II**). Both approaches **I/II** have been described in the literature.

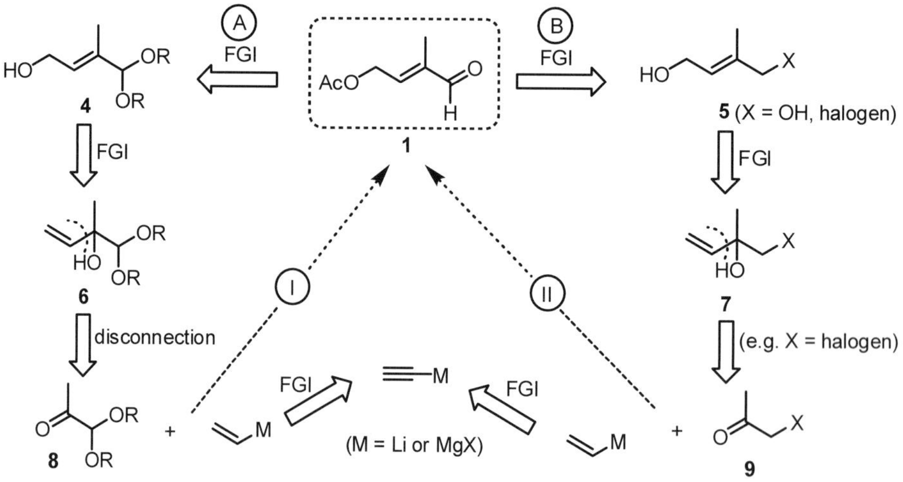

Approach **I** corresponds to a former industrial synthesis of **1** by BASF [2] starting with oxidation (nitrosation in the presence of methanol) of acetone to give methylglyoxal dimethyl acetal (**10**). This is followed by ethynylation with acetylene, partial hydrogenation, and acetylation (**10** → **11** → **12** → **13**). The synthesis is completed by a Cu(II)-catalyzed allylic inversion and acid hydrolysis of the acetal function (**13** → **1**). Alternatively, oxygenation of the dienol acetate **15** with O_2 in glacial acetic acid in the presence of a Pd/Cu catalyst leads to the allyl-inverted acylal **14**. Hydrolysis of the latter gives **1** [3, 4]; **15** can be obtained from the readily available tiglic aldehyde (**16**) and isopropenyl acetate:

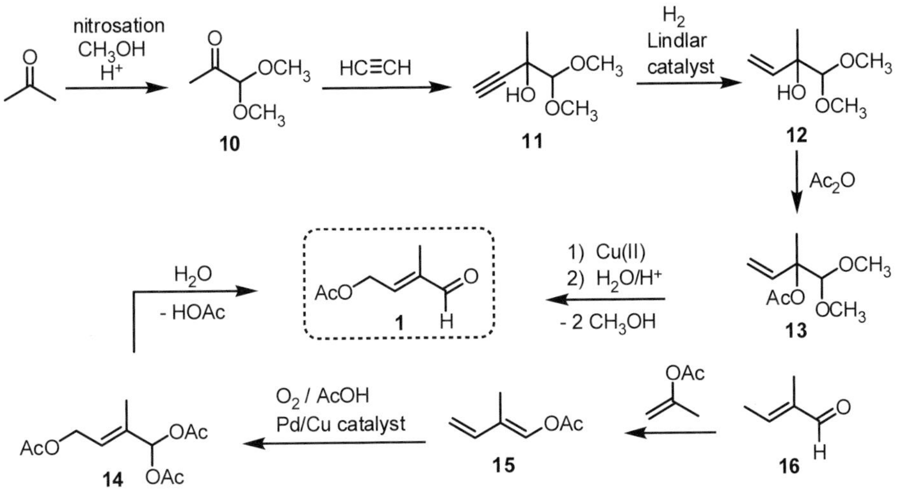

Approach **II** is the basis of a laboratory synthesis of **1** [5], which is described in detail in section (b).

More recently, two other processes have been introduced for industrial syntheses of **1** [6] starting from (a) 3-formyl crotonate **17** and (b) 3,4-epoxy-1-butene (**20**), respectively. In (b), the key step is a regioselective Rh-catalyzed hydrocarbonylation (→ **18**) of the diacetate **19**, obtained by ring-opening of **20** with acetic anhydride.

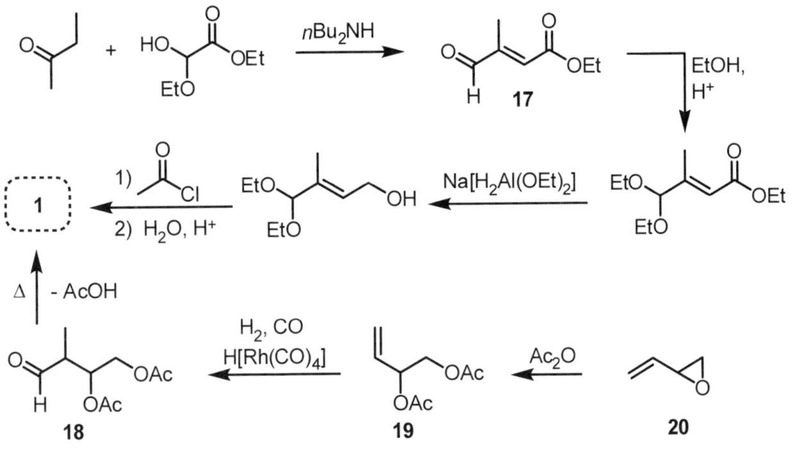

Likewise, isoprene monoepoxide (**21**) undergoes ring opening with subsequent oxidative chlorination upon reaction with CuCl$_2$/LiCl. The product is (*E*)-4-chloro-2-methyl-2-butenal (**22**), which yields **1** upon substitution of chlorine by acetate [7]:

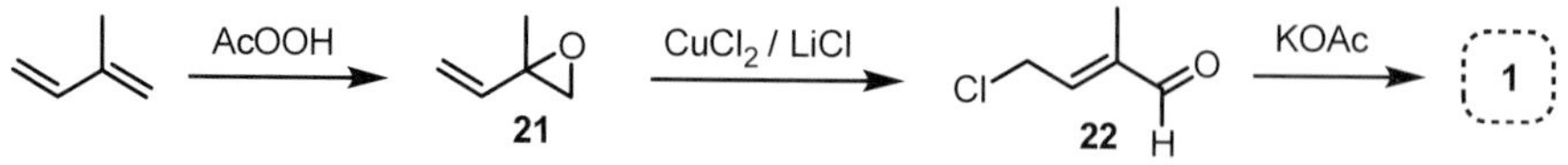

A more complex synthesis of **1** [8] is initiated by ene-type chlorination [9] of prenyl benzyl ether (**23**) with hypochlorite. In this reaction, the double bond is regioselectively transposed to the *gem*-dimethyl position to give **24**, in which the allylic chlorine can be substituted by dimethylamine (→ **25**). The benzyl ether moiety is replaced by acetate and the formed allylamine **27** is oxidized with peracetic acid to afford exclusively the (*Z*)-configured allyloxyamine **28**. This transformation involves a [2,3]-sigmatropic rearrangement of the primarily formed *N*-oxide **26**. *N*-Alkylation of **28** with CH$_3$I followed by thermal Hofmann-like elimination of (CH$_3$)$_3$N finally provides **1** via **29**:

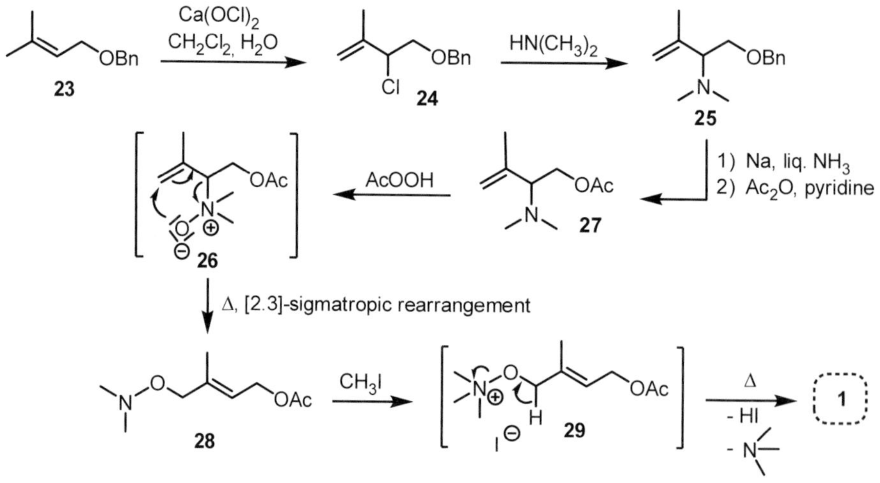

(b) Synthesis of 1

The synthesis of **1** starts with the addition of vinyl magnesium bromide to chloroacetone (**30**) to afford the isoprene chlorohydrin (**31**). For the formation and handling of vinyl Grignard compounds, the use of tetrahydrofuran as solvent is crucial [10]. When the tertiary alcohol **31** is treated with acetic anhydride in the presence of *p*-toluenesulfonic acid, the product is not the tertiary acetate **32** but the thermodynamically more stable primary acetate **33**, resulting from an allylic inversion involving an allylic cation formed from **31** or a Cope rearrangement of **32**.

For the final step of the synthesis, the primary chloride in **33** is converted into the aldehyde group of **1** by means of Kornblum oxidation with dimethyl sulfoxide. The disadvantage of the Kornblum oxidation (in particular, odor of (CH$_3$)$_2$S!) can be avoided by the use of *N*-ethylmorpholine *N*-oxide (**34**), which cleanly oxidizes primary allyl chlorides to the corresponding aldehydes [11, 12].

Thus, the target molecule **1** is obtained in a three-step sequence in an overall yield of 48% (based on **30**).

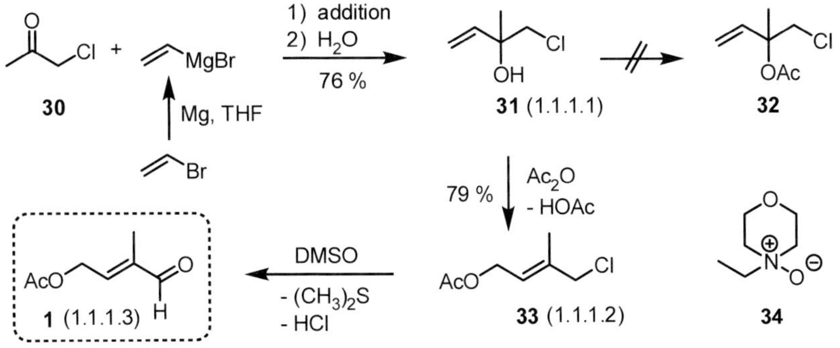

(c) Experimental procedures for the synthesis of 1

1.1.1 **   1-Chloro-2-methyl-3-buten-2-ol (isoprene chlorohydrin) [5]

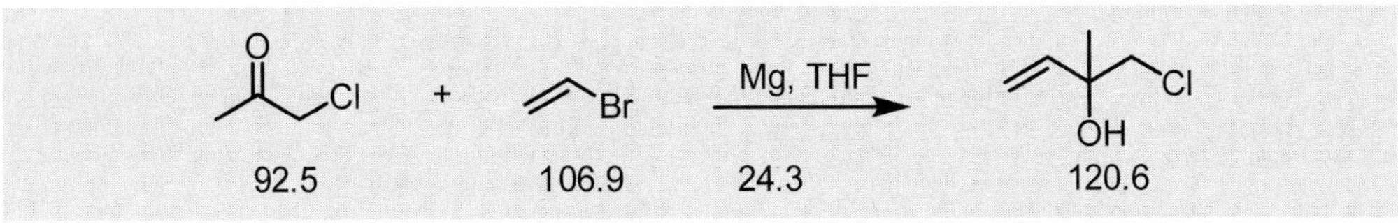

Magnesium turnings (7.30 g, 0.30 mol) are added to anhydrous THF (70 mL) under N_2 and a small amount of ethyl bromide (1 g) is added to start the reaction. Vinyl bromide (32.0 g, 0.30 mol) in THF (90 mL) is then added dropwise with stirring at such a rate that the temperature never exceeds 40 °C (ca. 90 min). Stirring is continued for 30 min, the dark-grey solution is cooled to 0 °C, and a solution of chloroacetone (18.5 g, 0.20 mol) (note) in THF is added dropwise over 45 min. Stirring is continued at room temperature for 1 h.

The adduct is hydrolyzed by the dropwise addition of ice-cold saturated aqueous NH_4Cl solution (100 mL) at 0 °C. The phases are separated and the aqueous phase is extracted with Et_2O (2 × 100 mL). The combined organic phases are washed with 2% $NaHCO_3$ solution (100 mL) and H_2O (100 mL), and dried over Na_2SO_4. The solvent is evaporated and the residue is fractionally distilled to give a colorless oil. The yield is 18.3 g (76%), bp_{17} 48–49 °C, n_D^{20} = 1.4608.

IR (film): $\tilde{\nu}$ (cm^{-1}) = 3420 (br, OH), 3080 (vinyl CH), 1640 (C=C).

^{1}H NMR (CDCl$_3$): δ (ppm) = 6.20–5.05 (12 lines of H_2C=CH system; 5.55–5.05 AB part; 6.20–5.75 X-part, J_{AB} = 1.8 Hz, J_{AX} = 16.2 Hz, J_{BX} = 10.3 Hz), 3.54 (s, 2 H, CH$_2$), 2.60 (s, 1 H, OH; exchangeable)], 1.41 (s, 3 H, CH$_3$).

Note: Chloroacetone (lachrymator!) is distilled (bp_{760} 118–119 °C) through a short packed column before use.

1.1.1.2   *   (*E*)-1-Acetoxy-4-chloro-3-methyl-2-butene [5]

A solution of *p*-toluenesulfonic acid monohydrate (2.54 g, 13.4 mmol) in glacial acetic acid (60 mL) is added dropwise to a stirred solution of isoprene chlorohydrin (**1.1.1.1**, 15.3 g, 127 mmol) in acetic anhydride (20 mL) and glacial acetic acid (60 mL) at 15 °C over a period of 15 min. The temperature of the bath is raised to 55 °C and stirring is continued for 24 h.

The solution is cooled and carefully poured into a mixture of 10% NaOH (800 mL) and ice (200 g). The resulting mixture is extracted with Et_2O (3 × 100 mL), and the combined ethereal phases are dried over Na_2SO_4 and concentrated *in vacuo*. The residue is fractionally distilled to give the product as a colorless oil; 16.3 g (79%), bp_{10} 91–93 °C, n_D^{20} = 1.4658.

> **IR** (film): $\tilde{v}$ (cm^{-1}) = 1740 (C=O), 1235, 1025, 685.
>
> **^{1}H NMR** (CCl$_4$): δ (ppm) = 5.68 (t$_{br}$, *J* = 7 Hz, 1 H, =CH), 4.57 (d, *J* = 7 Hz, 2 H, OCH$_2$), 4.09, 3.99 (s, total 2 H, ratio 1:6, Z-CH$_2$Cl/E-CH$_2$Cl), 2.02 (s, 3 H, OCOCH$_3$), 1.85 (s$_{br}$, 3 H, =C–CH$_3$).

Note: The ^{1}H NMR spectrum indicates a 6:1 mixture of the *E*/*Z* stereoisomers of the product.

1.1.1.3   *   (*E*)-Acetoxy-2-methyl-2-butenal [5]

K$_2$HPO$_4$ (19.9 g, 114 mmol), KH$_2$PO$_4$ (4.14 g, 30.0 mmol), and NaBr (1.20 g, 11.6 mmol) are suspended in a stirred solution of allyl chloride **1.1.1.2** (16.1 g, 99.5 mmol) in anhydrous DMSO (120 mL). The mixture is heated to 80 °C and stirred for 24 h (Hood! Formation of dimethyl sulfide!).

The mixture is then cooled and poured into H$_2$O (400 mL) and CCl$_4$ (200 mL) (Caution!). The phases are separated, the aqueous phase is extracted with CCl$_4$ (100 mL), and the combined organic layers are dried over Na$_2$SO$_4$. The solvent is evaporated *in vacuo* and the yellow residue is fractionally distilled to give the acetoxy aldehyde as a colorless oil; 11.2 g (80%), bp$_2$ 66–72 °C, n_D^{20} = 1.4647 (note).

> **IR** (film): $\tilde{v}$ (cm^{-1}) = 2720 (aldehyde CH), 1735 (C=O acetoxy), 1690 (C=O), 1645.
>
> **^{1}H NMR** (CDCl$_3$): δ (ppm) = 9.55 (s, 1 H, aldehyde CH; Z-isomer: δ = 10.23), 6.52 (tq, *J* = 6.0, 1.0 Hz, 1 H, =CH), 4.93 (dq, *J* = 6.0, 1.0 Hz, 2 H, OCH$_2$), 2.12 (s, 3 H, OCOCH$_3$), 1.81 (dt, *J* = 1.0, 1.0 Hz, 3 H, = C–CH$_3$).

Note: If smaller amounts of starting material are used, column chromatography (silica gel, 0.06–0.2 mm, eluent *n*-hexane/Et$_2$O, 9:1) is recommended as the purification procedure.

[1]   H. Pommer, A. Nürrenbach, *Pure Appl. Chem.* **1975**, *43*, 527.

[2]   W. Reif, H. Grassner, *Chemie-Ing. Techn.* **1973**, *45*, 646.

[3]   R. H. Fischer, H. Krapf, J. Paust, *Angew. Chem.* **1988**, *100*, 301; *Angew. Chem. Int. Ed. Engl.* **1988**, *27*, 285.

[4]   Y. Tanabe, *Hydrocarbon Process.* **1981**, *60*, 187.

[5]   a) J. Huet, H. Bouget, J. Sauleau, *C. R. Acad. Sci., Ser. C.* **1970**, *271*, 430; b) J. H. Babler, M. J. Coghlan, M. Feng, P. Fries, *J. Org. Chem.* **1979**, *44*, 1716.

[6]   Ullmann's *Encyclopedia of Industrial Chemistry*, 6th ed., Vol. 38, p. 119, Wiley-VCH, Weinheim, **2003**.

[7]   G. Eletti-Bianchi, F. Centini, L. Re, *J. Org. Chem.* **1976**, *41*, 1648.

[8]   S. Inoue, N. Iwase, O. Hiyamoto, K. Sato, *Chem. Lett.* **1986**, 2035.

[9]   S. Suzuki, T. Onichi, Y. Fujita, J. Otera, *Synth. Commun.* **1985**, *15*, 1123.

[10] H. Normant, *Adv. Org. Chem.* **1960**, *2*, 1.

[11] S. Suzuki, T. Onishi, Y. Fujita, M. Misawa, J. Otera, *Bull. Chem. Soc. Jpn.* **1986**, *59*, 3287; for **1**, a yield of 88% is reported.

[12] Sulfoxides anchored on ionic liquids are reported to represent non-volatile and odorless reagents for Swern oxidation: X. Me, T. H. Chan, *Tetrahedron* **2006**, *62*, 3389.

1.1.2   (*S*)-2,3-Dimethyl-hex-5-en-3-ol

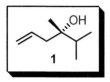

1

Topics:
- Stereoselective allylation of ketones
- Synthesis of enantiopure tertiary allyl alcohols
- Domino reactions

(a) General

Allylations of aldehydes and ketones using allyl silanes, allyl stannanes, or allyl boronates are very important methods for the preparation of homoallylic alcohols [1]. The reactions can also be performed in an enantioselective way, with high asymmetric induction if aldehydes are used as substrates. Otherwise, the stereoselective allylation of ketones such as alkyl methyl ketones is much more difficult and so far only one method exists for this transformation, which was developed by Tietze and co-workers [2, 3].

The following asymmetric allylation of alkyl methyl ketones can be regarded as a three-component domino reaction, in which a ketone **1**, allyl trimethylsilane (**2**) as allylating reagent, and a chiral silyl ether **3** as *O*-alkylating component react to give the homoallylic ether **5**. A catalytic amount of triflic acid is needed for the initiation of this multicomponent process.

$$R^1 = \text{Alkyl} \qquad R^2 = \text{Me}$$

The allylation proceeds via an intermediate carboxenium ion **4**, which is attacked by the allyl silane **2**. If a benzyl trimethylsilyl ether **3** is used, the formed product **5** can be cleaved reductively to give the corresponding homoallylic tertiary alcohol **6**.

Chiral silyl ethers such as **3a** [4, 5] and **3b** [6] give only a very low selectivity of 1.8:1, whereas an excellent facial differentiation could be achieved with the norpseudoephedrine and mandelic acid derivatives **3c** [2] and **3d** [3], respectively.

3a   **3b**   **3c**   **3d**

(b) Synthesis of 1

First, the chiral silyl ether **9** is synthesized, which is derived from mandelic acid (**7**) and used in the asymmetric allylation of 3-methyl-2-butanone (**10**) [3]. Mandelic acid (**7**) is activated using acetyl chloride and reacted with ammonia to give the mandelic acid amide (**8**). This is reduced with borane-tetrahydrofuran complex to afford the corresponding amine. Formation of the trifluoroacetamide by employing ethyl trifluoroacetate and silylation of the free hydroxy group with chlorotrimethylsilane in the presence of triethylamine leads to the chiral silyl ether **9**.

7 **8** (1.1.2.1) **9** (1.1.2.2)

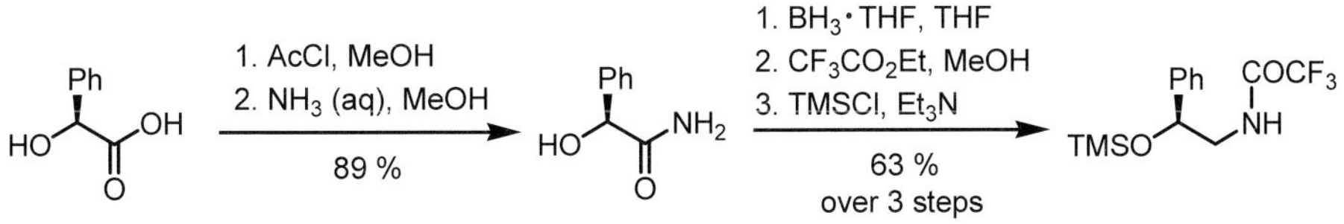

Second, 3-methyl-2-butanone (**10**) is allylated with allyl trimethylsilane (**2**) in the presence of **9** and a catalytic amount of triflic acid in CH_2Cl_2 at –78 °C. The formed benzyl ether **11** is reductively cleaved employing lithium and 4,4'-di-(*tert*-butyl)-1,1'-biphenyl (DBBP) (**12**) to give (*S*)-2,3-dimethyl-hex-5-en-3-ol (**13**). Recent computational investigations employing the B3LYP/G–31+G(d) level of theory in dichloromethane solution indicate that the reaction proceeds via an S_N1-type mechanism with an open transition state for the addition of the allyl silane [7].

2 **10** **9**

11 (1.1.2.3) **13** (1.1.2.5)

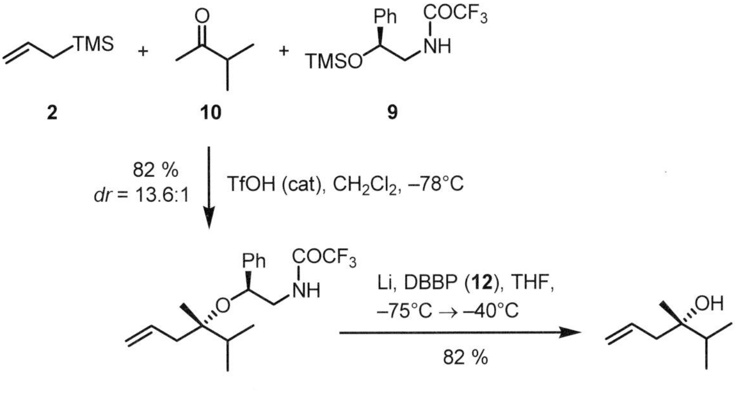

DBBP (**12**) can be easily prepared from biphenyl (**14**) by Friedels–Crafts alkylation using 2-chloro-2-methylpropane (**15**) and aluminum trichloride.

DBBP

14 **15** **12** (1.1.2.4)

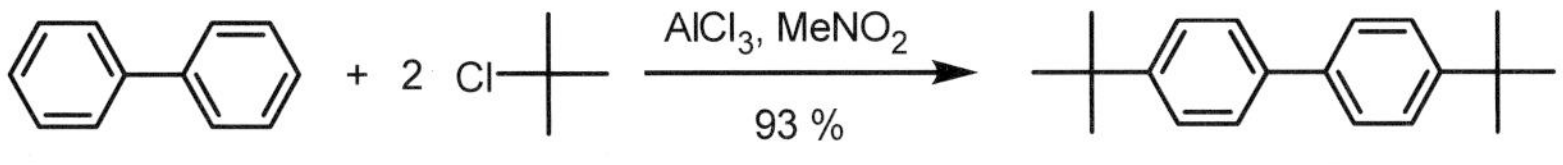

(c) Experimental procedures for the synthesis of 1

1.1.2.1 * (S)-Mandelic acid amide [3]

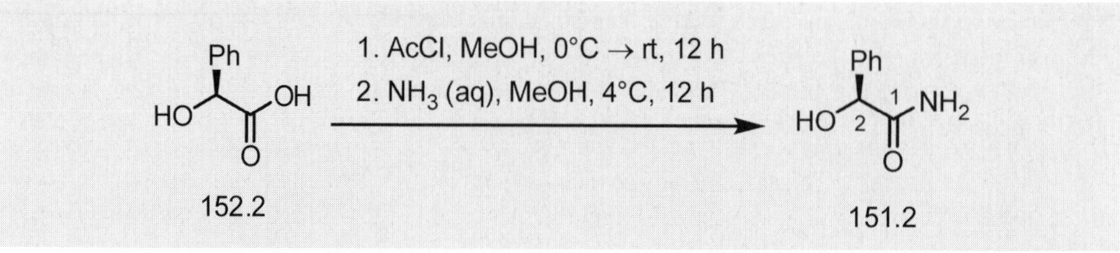

A 1 L oven-dried two-necked round-bottomed flask, equipped with a magnetic stirring bar and a three-way stopcock attached to an argon balloon, is charged with (S)-mandelic acid (25.0 g, 164 mmol) and anhydrous MeOH (670 mL). After cooling the mixture to 0 °C with an ice-water bath, acetyl chloride (31.5 mL, 443 mmol) is added dropwise from a dropping funnel at 0 °C with stirring. Stirring is continued for 12 h at room temperature.

Thereafter, the solvent is evaporated under reduced pressure ($T < 40$ °C), the residue is redissolved in MeOH (125 mL), and ice-cold saturated aqueous NH_3 solution (310 mL) is added with stirring. The flask is stored in a refrigerator (4 °C) overnight.

After evaporation of the solvent, the residue is recrystallized (EtOH/EtOAc) to afford colorless crystals; 22.0 g (89%), mp 120–121 °C, $[\alpha]_D^{20} = +78.0$ ($c = 1.7$, acetone).

> **IR** (NaCl): $\tilde{\nu}$ (cm^{-1}) = 3357, 3187, 2927, 1681, 1495, 1452, 1422, 1295, 1190, 1102, 1056, 928, 895, 859, 763, 710.
>
> **^{1}H NMR** (300 MHz, [D$_6$]DMSO): δ (ppm) = 7.43–7.22 (m, 6 H, Ph-H, NH), 7.13 (s, 1 H, NH), 5.95 (d, 1 H, J = 4.7 Hz, 2-H), 4.83 (d, 1 H, J = 4.7 Hz, OH).
>
> **^{13}C NMR** (50 MHz, [D$_6$]DMSO): δ (ppm) = 174.5 (C-1), 141.3 (Ph-C), 127.8 (2 × Ph-CH), 127.2 (Ph-CH), 126.4 (2 × Ph-CH), 73.4 (C-2).
>
> **MS** (EI, 70 eV): m/z (%) = 151 (10) [M]$^+$, 107 (100) [M – CONH$_2$]$^+$.

1.1.2.2 ** (S)-2,2,2-Trifluoro-N-(2-phenyl-2-trimethylsilanyloxyethyl)acetamide [3]

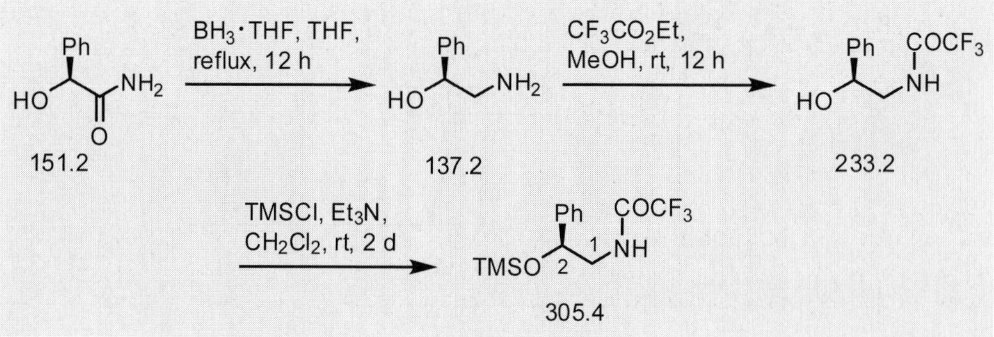

An oven-dried 500 mL three-necked round-bottomed flask, equipped with a magnetic stirring bar, an addition funnel, a stopper, and a reflux condenser with a three-way stopcock attached to an argon balloon, is charged with the amide **1.1.2.1** (21.6 g, 143 mmol). A solution of borane-tetrahydrofuran complex in THF (1 M, 300 mL, 300 mmol) is added by means of the addition funnel (Caution: gas formation!). The mixture is heated to reflux for 12 h and then cooled to room temperature, whereupon MeOH (60 mL) is carefully added (Caution: gas formation!) through the addition funnel. After evaporation of the solvent under reduced pressure, the crude product is dissolved in anhydrous MeOH (185 mL). Ethyl trifluoroacetate (14.4 mL, 143 mmol) is added dropwise and the mixture is stirred for 12 h at room temperature. The solvent is then removed under reduced pressure and the residue is taken up in CH_2Cl_2 (230 mL). NEt_3 (37.9 mL, 274 mmol) is then added with stirring, the mixture is cooled to 0 °C in an ice-water bath, chlorotrimethylsilane (17.4 mL, 143 mmol) is added, and stirring is continued for 2 d at room temperature.

The reaction is quenched with H_2O (150 mL), and the crude product is extracted with CH_2Cl_2 (3 × 250 mL). The combined organic layers are washed with brine (200 mL), dried over $MgSO_4$, and concentrated in vacuo. The residue is purified by column chromatography on silica gel using *n*-pentane/*tert*-butyl methyl ether (5:1) as eluent to afford the acetamide as a colorless oil; 27.5 g (63 % over three steps), $[\alpha]_D^{20} = +53.5$ ($c = 1.0$, $CHCl_3$).

IR (NaCl): $\tilde{v}$ (cm^{-1}) = 3317, 3090, 3032, 2959, 1708, 1556, 1494, 1454, 1366, 1254, 1166, 1027, 963, 842, 756.

^{1}H NMR (300 MHz, $CDCl_3$): δ (ppm) = 7.31–7.18 (m, 5 H, Ph-H), 6.60 (s, 1 H, NH), 4.74 (dd, 1 H, J = 8.4, 3.8 Hz, 1-H), 3.66 (ddd, 1 H, J = 13.5, 7.5, 3.8 Hz, 2-H$_a$), 3.19 (ddd, 1 H, J = 13.5, 8.4, 4.5 Hz, 2-H$_b$), –0.02 (s, 9 H, Si(CH$_3$)$_3$).

^{13}C NMR (50 MHz, $CDCl_3$): δ (ppm) = 140.9 (Ph-C), 128.5 (2 × Ph-CH), 128.1 (Ph-CH), 125.8 (2 × Ph-CH), 72.7 (C-1), 47.4 (C-2).

MS (DCI, NH_3, 200 eV): m/z (%) = 629 (2) $[2M + NH_4]^+$, 340 (50) $[M + 2NH_4]^+$, 323 (100) $[M + NH_4]^+$.

1.1.2.3   ***   (4*S*,1'*S*)-4,5-Dimethyl-4-(1'-phenyl-2'-trifluoroacetamido-1'-ethoxy)-1-hexene [3]

An oven-dried 500 mL two-necked round-bottomed flask, equipped with a magnetic stirring bar, a stopper, and a three-way stopcock attached to an argon balloon, is charged with **1.1.2.2** (13.8 g, 40.2 mmol), 3-methyl-2-butanone (6.92 g, 80.4 mmol), and CH_2Cl_2 (250 mL). After cooling to −78 °C, allyltrimethylsilane (12.8 mL, 80.4 mmol) is added by means of a syringe at −78 °C, and then precooled trifluoromethanesulfonic acid (750 μL, 8.40 mmol) is added. The reaction mixture is stirred for 6 h at −78 °C.

The reaction is then quenched with NEt$_3$ (18 mL) and MeOH (200 mL). The product is extracted with CH$_2$Cl$_2$ (4 × 250 mL), and the combined extracts are dried over Na$_2$SO$_4$ and concentrated. The residue is purified by column chromatography on silica gel using *n*-pentane/Et$_2$O (9:1) as eluent to afford a 13.6:1 mixture of the diastereomeric homoallylic ethers as a colorless oil; 11.3 g (82%), $R_f = 0.77$ (*n*-pentane/Et$_2$O, 5:1).

---

^{1}H NMR (300 MHz, CDCl$_3$): δ (ppm) = 7.39–7.21 (m, 5 H, Ph-H), 6.69 (s$_{br}$, 1 H, N-H), 5.84 (ddt, J = 17.0, 10.2, 7.2 Hz, 1 H, 2-H), 5.13 (d, J = 17.0 Hz, 3 H, 1-H$_Z$), 5.12 (d, J = 10.2 Hz, 1 H, 1-H$_E$), 4.66 (dd, J = 8.2, 4.3 Hz, 1 H, 1'-H), 3.60 (ddd, J = 13.6, 7.2, 4.3 Hz, 1 H, 2'-H$_b$), 3.29 (ddd, J = 13.6, 8.2, 4.5 Hz, 1 H, 2'-H$_a$), 2.32 (m$_c$, 2 H, 3-H$_2$), 1.74 (sept, J = 6.9 Hz, 1 H, 5-H), 0.91 (d, J = 6.9 Hz, 3 H, 6-H$_{3/b}$), 0.84 (s, 3 H, 4-CH$_3$), 0.80 (d, J = 6.9 Hz, 3 H, 6-H$_{3/a}$).

^{13}C NMR (50.3 MHz, CDCl$_3$): δ (ppm) = 156.95 (q, $^2J_{CF}$ = 38 Hz, C=O), 141.95 (Ph-C), 133.89 (C-2), 128.48, 127.88, 126.30 (Ph-C), 118.08 (C-1), 115.84 (q, $^1J_{CF}$ = 288 Hz, CF$_3$), 80.99 (C-4), 71.61 (C-1'), 46.87 (C-2'), 40.25 (C-3), 35.04 (C-5), 21.10 (4-CH$_3$), 17.15 (C-6).

1.1.2.4   **   4,4'-Di-(*tert*-butyl)-1,1'-biphenyl (DBBP) [3]

A dry 250 mL three-necked round-bottomed flask, equipped with a magnetic stirring bar, a reflux condenser, an addition funnel, and a stopper, is first charged with nitromethane (75 mL) and biphenyl (13.2 g, 85.0 mmol) under an argon atmosphere. Anhydrous aluminum trichloride (3.00 g, 22.5 mmol) is then added, whereupon the color of the solution changes to deep violet. A solution of 2-chloro-2-methylpropane (17.4 g, 190 mmol) in nitromethane (20 mL) is then added dropwise to the stirred mixture over 30 min. At the end of the addition, the reaction starts with a vigorous evolution of HCl gas (Caution!) and stirring is continued for 14 h.

The reaction mixture is then poured onto crushed ice in a 500 mL beaker and allowed to warm to room temperature. It is extracted with nitromethane/*n*-pentane (1:1; 3 × 50 mL), and the combined organic phases are washed with brine (50 mL), dried over MgSO$_4$, and filtered. The yellowish filtrate is treated with silica gel (100 g) to provide a colorless solution, which is concentrated. The resulting white powder is recrystallized from nitromethane to give white needles of the biphenyl; 21.0 g (93%), mp 128–129 °C.

---

^{1}H NMR (300 MHz, CDCl$_3$): δ (ppm) = 7.55–7.44 (m, 8 H, Ar-H), 1.37 (s, 18 H, 2 × C(CH$_3$)$_3$).

^{13}C NMR (50.3 MHz, CDCl$_3$): δ (ppm) = 139 (Ar-C), 150 (Ar-C), 127 (Ar-CH), 126 (Ar-CH), 34.5 (2 × $\underline{C}$(CH$_3$)$_3$), 31.4 (2 × C($\underline{C}$H$_3$)$_3$).

MS (EI, 70 eV): *m/z* (%) = 266 (41) [*M*]$^+$, 251 (100) [*M* – CH$_3$]$^+$.

1.1.2.5   ***   (S)-2,3-Dimethyl-hex-5-en-3-ol [3]

343.2   128.2

An oven-dried 1 L two-necked round-bottomed flask, equipped with a magnetic stirring bar, a stopper, and a three-way stopcock attached to a balloon filled with argon, is charged with degassed THF (220 mL). **1.1.2.4** (6.88 g, 25.8 mmol) and lithium granules (max. Ø 1 mm) are added and the mixture is vigorously stirred at room temperature until the solution shows a deep-blue colour. The mixture is stirred for an additional 2 h at 0 °C and then cooled to –78 °C. A solution of the homoallylic ether **1.1.2.3** (11.1 g, 32.3 mmol) in anhydrous THF (50 mL) is then added dropwise. After stirring for 1 h at –78 °C, the mixture is allowed to warm to –45 °C and stirred for an additional 2 h. The reaction is then quenched by adding solid NH_4Cl at –45 °C (until the black solution turns yellow) and the mixture is allowed to warm to room temperature. The solid residue is filtered off and then a half-saturated NH_4Cl solution (120 mL) is added. The aqueous layer is extracted with Et_2O (3 × 100 mL) and the combined organic layers are washed with brine (60 mL), dried over Na_2SO_4, and concentrated. The residue is purified by column chromatography on silica gel using *n*-pentane to elute DBBP; then, a gradient of *n*-pentane/Et_2O from 15:1 to 3:1 as eluent affords the tertiary alcohol as a colorless oil; 3.40 g (82%), $[\alpha]_D^{20}$ = +14.2 (*c* = 1.0, CHCl$_3$), R_f = 0.43 (*n*-pentane/Et_2O, 5:1).

IR (NaCl): $\tilde{\nu}$ (cm^{-1}) = 3380, 3314, 3082, 2976, 2936, 2882, 1448, 1418, 1380, 1276, 1192, 1088, 1052, 912, 882.

^{1}H NMR (200 MHz, CDCl$_3$): δ (ppm) = 5.89 (m$_c$, 1 H, 5-H), 5.22–5.07 (m, 2 H, 6-H$_2$), 2.23 (d, *J* = 7.0 Hz, 2 H, 4-H$_2$), 1.69 (sept, *J* = 7.0 Hz, 1 H, 2-H), 1.40 (s$_{br}$, 1 H, OH), 1.08 (s, 3 H, 3-CH$_3$), 0.94, 0.90 (2 × d, *J* = 7.0 Hz, 6 H, 1-H$_3$, 2-CH$_3$).

^{13}C NMR (50 MHz, CDCl$_3$): δ (ppm) = 134.1 (C-5), 118.6 (C-6), 74.25 (C-3), 44.12 (C-4), 36.84 (C-2), 22.83 (3-CH$_3$), 17.53, 16.90 (C-1, 2-CH$_3$).

MS (EI, 70 eV): *m/z* (%) = 87 (84) [$M - C_3H_5$]$^+$, 69 (34) [C_5H_9]$^+$, 43 (100) [C_3H_7]$^+$.

---

[1]   a) S. E. Denmark, J. Fu, *Chem. Rev.* **2003**, *103*, 2763; b) I. Fleming, A. Barbero, D. Walter, *Chem. Rev.* **1997**, *97*, 2063; c) J. A. Marshall, *Chem. Rev.* **1996**, *96*, 31; d) C. E. Masse, J. S. Panek, *Chem. Rev.* **1995**, *95*, 1293; e) Y. Yamamoto, N. Asao, *Chem. Rev.* **1993**, *93*, 2207; f) I. Fleming, *Org. React.* **1989**, *37*, 57.

[2]   a) L. F. Tietze, L. Völkel, C. Wulff, B. Weigand, C. Bittner, P. McGrath, K. Johnson, M. Schäfer, *Chem. Eur. J.* **2001**, *7*, 1304; b) L. F. Tietze, B. Weigand, L. Völkel, C. Wulff, C. Bittner, *Chem. Eur. J.* **2001**, *7*, 161; c) L. F. Tietze, K. Schiemann, C. Wegner, C. Wulff, *Chem. Eur. J.* **1998**, *4*, 1862; d) L. F. Tietze, C. Wegner, C. Wulff, *Eur. J. Org. Chem.* **1998**, 1639; e) L. F. Tietze, K. Schiemann, C. Wegner, *J. Am. Chem. Soc.* **1995**, *117*, 5851.

[3]   L. F. Tietze, S. Hölsken, J. Adrio, T. Kinzel, C. Wegner, *Synthesis* **2004**, *13*, 2236.

[4]   a) T. Mukaiyama, M. Ohshima, N. Miyoshi, *Chem. Lett.* **1987**, 1121; b) T. Mukaiyama, H. Nagaoka, M. Murakami, M. Ohshima, *Chem. Lett.* **1985**, 977; c) J. Kato, N. Iwasawa, T. Mukaiyama, *Chem. Lett.* **1985**, 743.

[5]   A. Mekhalfia, I. E. Markó, *Tetrahedron Lett.* **1991**, *32*, 4779.

[6]   L. F. Tietze, C. Wegner, C. Wulff, *Synlett* **1996**, 471.

[7]   L. F. Tietze, T. Kinzel, S. Schmatz, *J. Am. Chem. Soc.* **2006**, *128*, 11483.

1.1.3   (*S*)-5-Oxo-3,5-diphenylpentanoic acid methyl ester

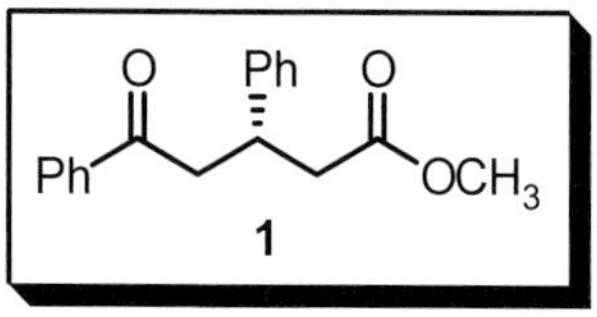

Topics:
- Knoevenagel condensation, Michael addition
- "Acid cleavage" of acetoacetate, anhydride formation
- Enantioselective asymmetric desymmetrization of a cyclic *meso*-anhydride by a Grignard compound in the presence of (–)-sparteine as a stereocontrolling agent
- Determination of enantiomeric excess by HPLC on a chiral phase

(a) General

The desymmetrization of *meso* and other prochiral compounds represents an important approach in asymmetric synthesis [1]. The desymmetrization of five- and six-membered cyclic anhydrides of the *meso* type (such as **2** and **4**) by ring-opening attack of nucleophiles at one of the enantiotopic carbonyl groups is broadly possible (1) by chiral alcohols or amines, (2) by achiral alcohols in combination with enzymes or other organocatalysts [2]. Clearly, the catalytic variant is by far the more attractive route for reasons of atom economy and preparative efficiency [3], e.g.:

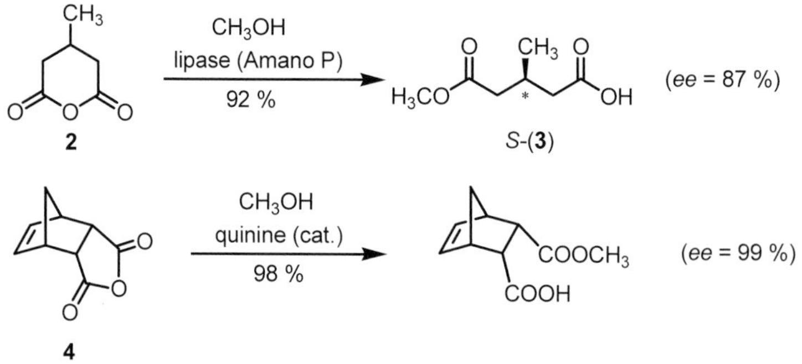

Only a few examples of the desymmetrization of cyclic anhydrides with carbon nucleophiles have been described [2]. Recently, asymmetric ring-opening reactions of 3-substituted glutaric anhydrides (like **2**) were found [4] to occur with Grignard compounds in the presence of (–)-sparteine as a chiral complexing ligand system, which has been shown [5] to be a very versatile organocatalyst [8] for asymmetric stereocontrol in reactions of organolithium compounds.

As described in section (b), a simple synthesis of the chiral target molecule **1** [6] can be achieved by application of the above protocol.

(b) Synthesis of 1

3-Phenylglutaric anhydride (**5**) is reacted with phenylmagnesium bromide in toluene at –78 °C in the presence of 1.3 equivalents of (–)-sparteine (**6**) as a stereocontrolling agent to give the δ-keto acid **7** in 78% yield and with high enantioselectivity (*ee* = 96%) [4]. The enantiomeric purity of **7** may be determined by HPLC on a chiral phase of the methyl ester **1**, easily accessible from the acid **7** by *O*-alkylation of its potassium salt with methyl iodide.

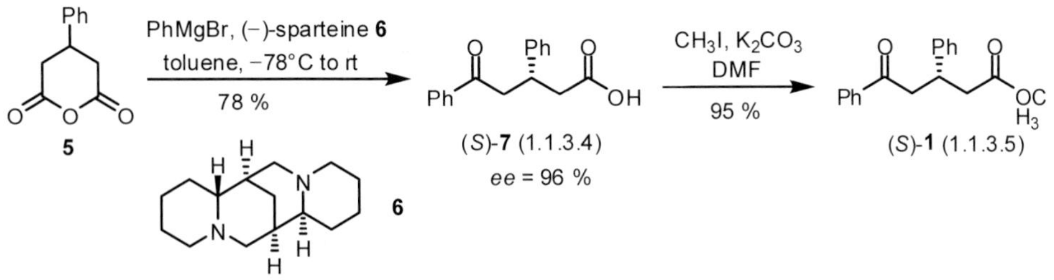

The *meso*-anhydride **5** is conventionally prepared [7] starting from benzaldehyde and two moles of ethyl acetoacetate. The product **9** of this base-catalyzed three-component reaction results from a domino process involving Knoevenagel condensation of the first molecule of acetoacetate with benzaldehyde followed by Michael addition of the second acetoacetate to the primary condensation product **8** in the presence of piperidine:

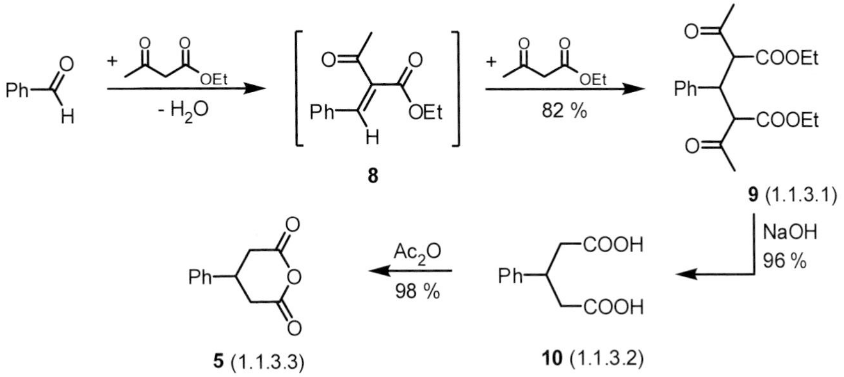

The diester **9** undergoes a two-fold "acid cleavage" of the acetoacetate moieties upon treatment with NaOH in EtOH, which results in the loss of two molecules of acetate and saponification of the carboxylic acid ester moieties to give 3-phenylglutaric acid (**10**). Finally, the diacid **10** is transformed to the cyclic anhydride **5** using acetic anhydride.

Thus, the target molecule **1** is obtained in a four-step sequence with an overall yield of 60% (based on benzaldehyde).

(c) Experimental procedures for the synthesis of 1

1.1.3.1   *   2,4-Diacetyl-3-phenyl-pentanedioic acid diethyl ester [7]

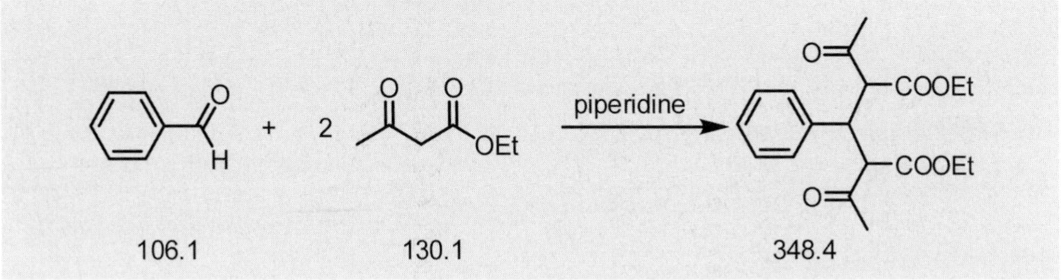

Ethyl acetoacetate (75.5 g, 0.58 mol, note 1) and benzaldehyde (28.5 g, 0.27 mol) are dissolved in anhydrous EtOH (160 mL). Piperidine (4 mL) is added and the solution is heated to reflux for 10 min and then stirred for 12 h at room temperature. The product begins to crystallize after 2–3 h (note 2).

The slurry is cooled to –20 °C (MeOH/dry ice), and the product is collected by filtration, washed with EtOH (–20 °C) until the washings remain colorless, and dried in vacuo. One obtains 77.2 g (82%) of colorless crystals; mp 150–152 °C (note 3).

IR (KBr): $\tilde{v}$ (cm^{-1}) = 3515 (enol OH), 2980 (C–H), 1740/1720 (C=O), 1500, 1470, 1380, 1190, 830.

^{1}H NMR (CDCl$_3$): very complex because of enolization and different E/Z geometries.

Notes: (1) Ethyl acetoacetate is freshly distilled prior to use (bp$_{15}$ 76–77 °C).

(2) If the slurry becomes too viscous for stirring, it is diluted with additional EtOH (50–100 mL).

(3) The product is sufficiently pure according to TLC (SiO$_2$/Et$_2$O). It may be recrystallized from EtOH, which increases the mp to 154–155 °C.

1.1.3.2   *   3-Phenylglutaric acid [7]

The diester **1.1.3.1** (69.7 g, 0.20 mol) is added to a mixture of 50% aqueous NaOH (200 mL) and EtOH (200 mL) and the resulting mixture is heated to reflux for 3 h.

The slurry obtained is diluted with H$_2$O (200 mL), concentrated to dryness in vacuo, and the residue is taken up in H$_2$O (400 mL). The resulting solution is acidified to pH 2 using concentrated HCl and the phenylglutaric acid is extracted with Et$_2$O (3 × 150 mL). The combined extracts are dried (Na$_2$SO$_4$) and concentrated to a volume of 100–120 mL. Cooling to –20 °C results in crystallization of the product, which is collected by filtration, washed with a small amount of Et$_2$O (–20 °C), and dried. The yield is 40.0 g (96%) as colorless crystals, mp 145–147 °C; the product is sufficiently pure for further use.

IR (KBr): $\tilde{v}$ (cm^{-1}) = 3200–2500 (OH, associated), 1720/1705 (C=O), 1495, 1450, 1420, 820.

^{1}H NMR ([D$_6$]acetone): δ (ppm) = 10.88 (s$_{br}$, 2 H, COOH), 7.25 (m$_c$, 5 H), 3.66 (quintet, J = 6.7 Hz, 1 H), 2.72 (d, J = 6.7 Hz, 4 H).

1.1.3.3   *   3-Phenylglutaric anhydride [7]

208.2   190.2

The dicarboxylic acid **1.1.3.2** (26.0 g, 0.125 mol) in acetic anhydride (140 mL) is heated under reflux for 2 h.

The reaction mixture is then concentrated to dryness. The residue is suspended in Et$_2$O (100 mL), filtered off, washed with Et$_2$O, and dried in vacuo. The yield is 23.3 g (98%) as colorless crystals, mp 103–105 °C; the product is sufficiently pure for further use.

IR (solid): $\tilde{v}$ (cm^{-1}) = 3034, 2980, 1809, 1751, 1243, 1172, 1066, 953, 763, 702, 605, 591.

^{1}H NMR (CDCl$_3$): δ (ppm) = 7.40 (t, J = 7.4 Hz, 2 H), 7.34 (t, J = 7.6 Hz, 1 H), 7.20 (d, J = 7.25 Hz, 2 H, Ar–H), 3.42 (m$_c$, 1 H, PhC–H), 3.11 (dd, J = 17.3/4.4 Hz, 2 H) and 2.89 (dd, J = 17.3/11.4 Hz, 2 H, CO–CH$_2$).

^{13}C NMR (CDCl$_3$): δ (ppm) = 165.84, 139.10, 129.40, 128.17, 126.25, 37.15, 34.11.

1.1.3.4   ***   (S)-5-Oxo-3,5-diphenylpentanoic acid [4]

190.2   PhBr: 157.0,   234.4   268.3
                         Mg: 24.3

Approximately one-fifth of a solution of bromobenzene (1.02 g, 6.50 mmol) in anhydrous Et$_2$O (10 mL) is added to magnesium turnings (158.0 mg, 6.50 mmol) under nitrogen. Once the Grignard reaction has started, the remainder of the bromobenzene solution is added dropwise. The solution is then heated to reflux until all of the magnesium turnings have reacted.

The Grignard solution is added dropwise to a solution of (–)-sparteine (1.52 g, 6.50 mmol) in anhydrous toluene (25 mL) at room temperature under a nitrogen atmosphere. The solution is stirred for 3 h and then cooled to –78 °C. A solution of 3-phenylglutaric anhydride **1.1.3.3** (951 mg, 5.00 mmol) in toluene (10 mL) is added dropwise to the Grignard/(–)-sparteine solution at –78 °C. The mixture is stirred at this temperature for an additional 3 h before being allowed to warm to room temperature.

The solution is quenched with 2 M NaOH (50 mL), stirred thoroughly, and extracted with Et$_2$O (3 × 20 mL). The aqueous layer is separated and acidified with concentrated HCl under ice-cooling. The carboxylic acid precipitates and is collected by suction filtration, washed with H$_2$O (2 × 20 mL), and air-dried to give 1.05 g (78%) of colorless crystals, mp 126–127 °C.

IR (solid): $\tilde{v}$ (cm^{-1}) = 3030 (Ph-H), 1734 (C=O, acid), 1698 (C=O, ketone), 1682, 1595, 1578.

^{1}H NMR ([D$_6$]DMSO): δ (ppm) = 12.06 (s$_{br}$, 1 H, COOH), 7.91 (d, J = 7.3 Hz, 2 H, ArH), 7.60 (t, J = 7.6 Hz, 1 H, ArH), 7.48 (t, J = 7.6 Hz, 2 H, ArH), 7.26 (d, J = 7.0 Hz, 2 H, ArH), 7.23 (t, J = 7.6 Hz, 2 H, ArH), 7.13 (t, J = 7.3 Hz, 1 H, ArH), 3.66 (quintet, J = 7.9 Hz, 1 H, PhCH), 3.44 (dd, J = 17.1, 7.9 Hz, 1 H) and 3.37 (dd, J = 17.1, 6.3 Hz, 1 H, C<u>H</u>$_2$CO$_2$H), 2.69 (dd, J = 15.8, 6.3 Hz, 1 H) and 2.56 (dd, J = 15.8, 8.5 Hz, 1 H, PhCOC<u>H</u>$_2$).

^{13}C NMR ([D$_6$]DMSO): δ (ppm) = 198.40, 172.85, 143.83, 136.68, 133.07, 128.61, 128.10, 127.84, 127.48, 126.20, 43.96, 40.37, 37.21.

1.1.3.5   *   (*S*)-5-Oxo-3,5-diphenylpentanoic acid methyl ester [4]

Methyl iodide (2.84 g, 20.0 mmol) is added dropwise with stirring to a solution of the carboxylic acid **1.1.3.4** (268 mg, 1.00 mmol), anhydrous K$_2$CO$_3$ (207 mg, 1.50 mmol), and DMF (5 mL). The resulting mixture is stirred overnight at room temperature.

It is then quenched with 10% K$_2$CO$_3$ solution (10 mL). The product is extracted with Et$_2$O (5 × 10 mL) and the combined ethereal extracts are washed with brine (3 × 10 mL) and dried over Na$_2$SO$_4$. The solvent is removed *in vacuo* and the product is allowed to crystallize; one obtains 268 mg (96%) as colorless crystals, mp 82 °C.

IR (solid): $\tilde{v}$ (cm^{-1}) = 2970, 2870, 1735 (C=O, ester), 1680 (C=O, ketone), 1596, 1578, 1496.

^{1}H NMR (CDCl$_3$): δ (ppm) = 7.91 (d, J = 6.9 Hz, 2 H, ArH), 7.53 (t, J = 7.6 Hz, 1 H, ArH), 7.43 (t, J = 7.9 Hz, 2 H, ArH), 7.26 (m, 3 H, ArH), 7.19 (m, 2 H, ArH), 3.88 (quintet, J = 7.3 Hz, 1 H, PhC<u>H</u>), 3.58 (s, 3 H, OCH$_3$), 3.39 (dd, J = 16.7, 6.9 Hz, 1 H) and 3.33 (dd, J = 16.7, 6.9 Hz, 1 H, C<u>H</u>$_2$CO$_2$Me), 2.82 (dd, J = 15.3, 7.3 Hz, 1 H) and 2.69 (dd, J = 15.3, 7.3 Hz, 1 H, PhCOC<u>H</u>$_2$).

^{13}C NMR (CDCl$_3$): δ (ppm) = 198.13, 172.29, 143.36, 136.95, 133.09, 128.62, 128.07, 127.33, 127.18, 126.82, 51.55, 44.55, 40.58, 37.53.

The enantiomeric ratio of 98:2 (96% *ee*) may be determined by HPLC on a Daicel Chiralcel OD-H column (4.6 × 250 mm; isopropanol/*n*-hexane (20:80), 0.5 mL min^{-1}; UV 254 nm, baseline separation). A reference sample of racemic 5-oxo-3,5-diphenylpentanoic acid methyl ester may be obtained by performing the reaction **1.1.3.4** in the absence of sparteine and then generating the methyl ester according to the procedure described above.

[1] S. R. Magnuson, *Tetrahedron* **1995**, *51*, 2167.

[2] A. C. Spivey, B. J. Andrews, *Angew. Chem.* **2001**, *113*, 3227; *Angew. Chem. Int. Ed.* **2001**, *40*, 3131.

[3] Although the ring-opening of *meso* cyclic anhydrides by enantiomerically pure alcohols, amines, and other chiral nucleophiles is highly diastereoselective, the preparative value is limited by the amounts of chiral nucleophile required and by the necessity of further reactions for removal of the chiral auxiliary (see ref. [2]).

[4] R. Shintani, G. C. Fu, *Angew. Chem.* **2002**, *114*, 1099; *Angew. Chem. Int. Ed.* **2002**, *41*, 1057.

[5] a) D. Hoppe, T. Hense, *Angew. Chem.* **1997**, *109*, 2376; *Angew. Chem. Int. Ed. Engl.* **1997**, *36*, 2282; b) P. A. Beak et al., *Acc. Chem. Res.* **2000**, *33*, 715.

[6] For comparison, see: A. Diaz-Ortiz, E. Diez-Barra, A. de la Hoz, P. Prieto, A. Moreno, *J. Chem. Soc. Perkin Trans. 1* **1996**, *259*.

[7] In analogy to Th. Eicher, H. J. Roth, *Synthese, Gewinnung und Charakterisierung von Arzneistoffen*, p. 80, Georg Thieme Verlag, Stuttgart, **1986**.

[8] A. Berkessel, H. Kröger, *Asymmetric Organocatalysis*, Wiley-VCH, Weinheim, **2005**.

1.1.4   (S)-3-Phenylheptanoic acid

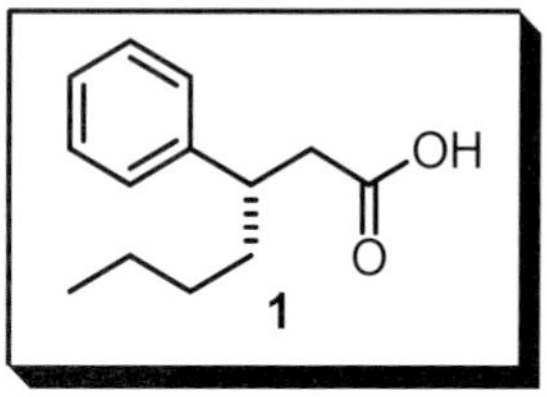

1

Topics: • (–)-Ephedrine as a chiral auxiliary
•   Chemoselective *N*-acylation of (–)-ephedrine
•   Enantioselective conjugate addition of a Grignard compound to an α,β-unsaturated carboxamide
•   Hydrolytic cleavage of a carboxamide, regeneration of the chiral auxiliary

(a) General

Nucleophilic addition to a C=C double bond bearing an electron-withdrawing functionality (such as COR, COOR, CONR$_2$, C≡N, etc.) is often referred to as conjugate addition (Michael addition, 1,4-addition). Generally, Michael addition follows a two-step scheme,

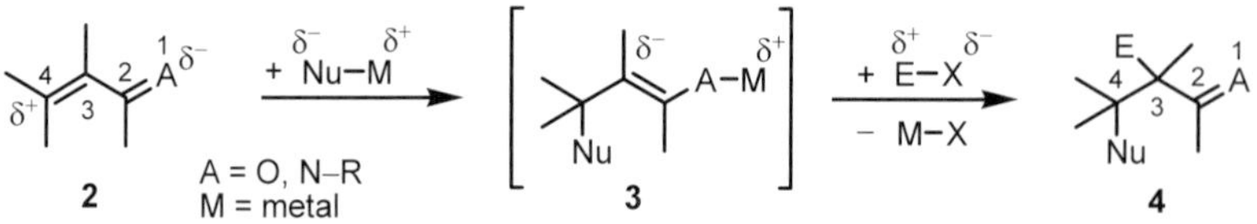

A = O, N–R
M = metal

2   **3**   **4**

in which a nucleophilic reagent (mainly an organometallic) adds to a conjugated system **2** in a 1,4-mode to give an enolate-like intermediate **3**, which is "trapped" by attack of an electrophilic reagent (proton acid, alkylating reagent, etc.) thus creating two new stereogenic sp^3 centers (positions 3/4) in the final addition product **4** [1].

Michael addition can be conducted in an asymmetric fashion in three ways, namely (1) by attachment of the conjugated system to a chiral auxiliary, (2) by the use of a chiral nucleophilic reagent, and (3) by performing the reaction in the presence of a chiral catalyst or with a chiral ligand on the Nu-M system. All of these types of asymmetric Michael addition are covered by literature examples.

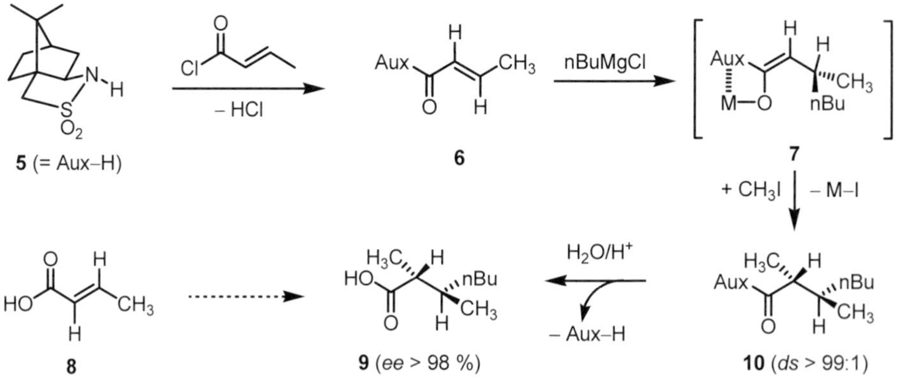

5 (= Aux–H)   **6**   **7**

+ CH$_3$I │ – M–I

8   **9** (*ee* > 98 %)   **10** (*ds* > 99:1)

(1): The camphor-derived sultam auxiliary **5** (Oppolzer auxiliary [2]) can be used in Michael additions of the crotonic acid system [3] (**8 → 9**) to control both conjugate addition of an organometallic (**6 → 7**) and *in situ* trapping by subsequent alkylation of the primary adduct (**7 → 10**). The observed high stereoselectivity is due to coordination of the metal by the donor groups CO and SO$_2$ in the substrate **6**.

Representative of this category is the synthesis of the target molecule **1**, which is discussed in section (b).

(2): The lithium amide of the readily available chiral secondary amine **11** (representing a chiral ammonia synthon) can undergo a highly diastereoselective 1,4-addition to α,β-unsaturated esters. This procedure is used for the synthesis of almost enantiopure (S)-β-amino esters, e.g. β-tyrosine methyl ester **13** from (4-benzyloxy)cinnamic ester **12** (**12** → **14** → **13**) ([4], cf. **4.4.2**):

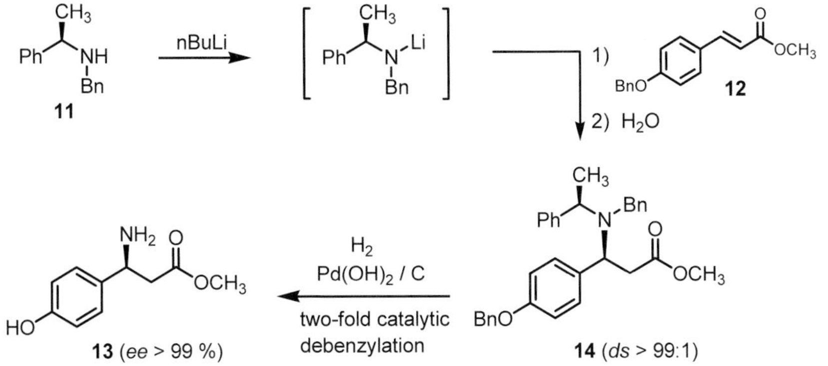

13 (ee > 99 %)

14 (ds > 99:1)

two-fold catalytic debenzylation

(3): Malonates undergo enantioselective conjugate addition to a variety of acyclic enones in the presence of a chiral phenylalanine-based imidazolidine organocatalyst [5] (cf. **1.3.5**), e.g.:

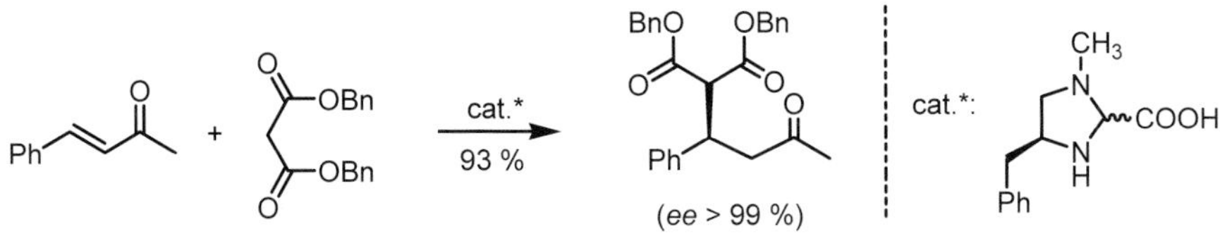

(ee > 99 %)

Malonates also add highly stereoselectively to cyclic enones under catalysis by chiral Ru amido complexes [6], e.g.:

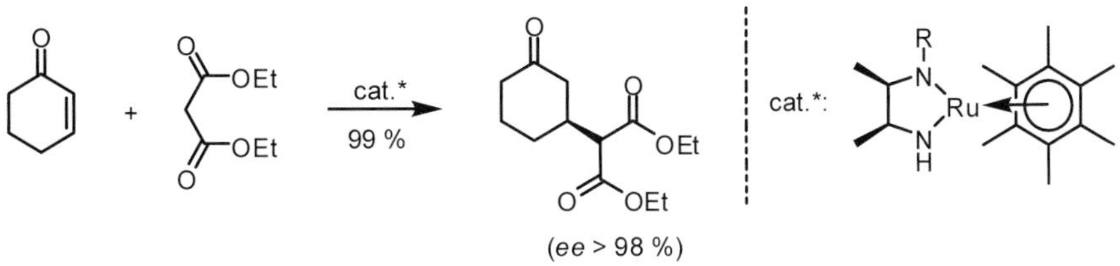

(ee > 98 %)

Interestingly, arylboronic acids are effective as nucleophilic partners in asymmetric Michael additions to cyclic [7] and acyclic enone systems [8], which are catalyzed by Rh complexes bearing axial chiral phosphoramidite ligands and which proceed with high enantioselectivities, e.g.:

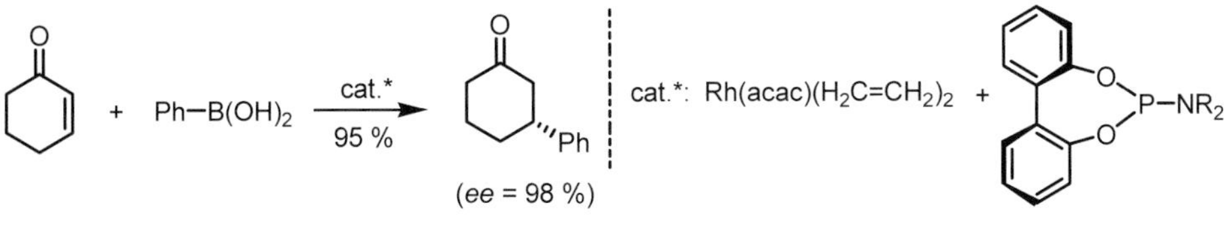

(ee = 98 %)

(b) Synthesis of 1

First, a cinnamic acid unit has to be connected to (–)-ephedrine (15) as chiral auxiliary. With cinnamoyl chloride in the presence of "proton sponge" (16) as a base, N-acylation of the auxiliary occurs chemoselectively to give the desired product 17. Then, n-butyl magnesium bromide (in an excess of six equivalents) is reacted with the α,β-unsaturated amide 17 at –40 to –20 °C in diethyl ether to yield the Michael adduct 18 with a diastereoselectivity of 99:1 (ref. [9]: >99:1):

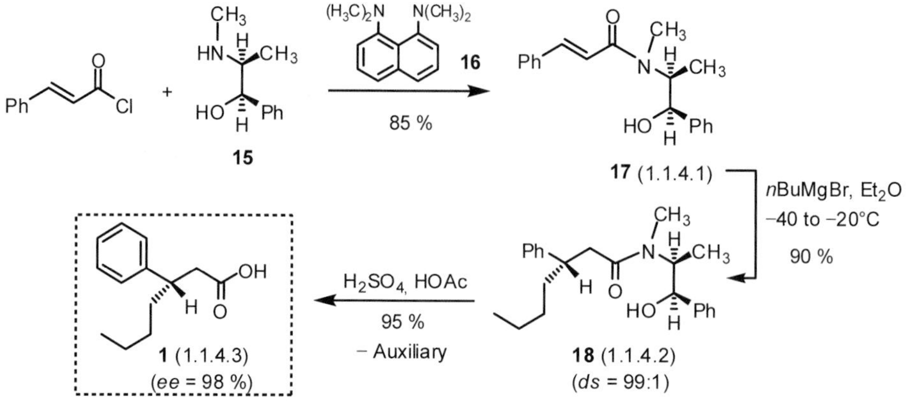

Finally, the auxiliary is removed by acid hydrolysis of the adduct 18 to provide the chiral acid 1 in nearly enantiopure (S)-form ($ee \approx 98\%$) together with (–)-ephedrine in quantitative yield ("regenerative" use of a chiral auxiliary, cf. 1.7.5).

The high diastereoselectivity of the described Michael reaction is probably due to initial formation of the rigid Mg-chelate complex 19:

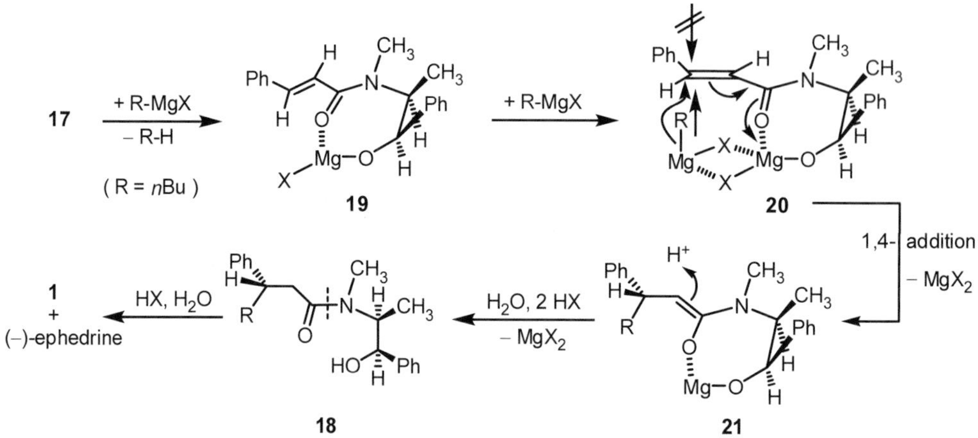

A second molecule of the Grignard reagent then coordinates to the Mg of the chelate complex, inducing attack on the enone system from the sterically less hindered (lower) side of the chelate (opposite to the methyl and phenyl groups at the stereogenic centers of the chiral auxiliary) as depicted in 20. Protonation of the resulting enolate 21 and removal of the Mg complement the Michael addition to yield amide 18, which is cleaved upon acidic hydrolysis to give the chiral acid 1 and the auxiliary 15.

In this way, the target molecule **1** is formed in a three-step sequence with an overall yield of 53% (based on cinnamoyl chloride).

(c) Experimental procedures for the synthesis of 1

1.1.4.1   *   (E)-N-((1S,2R)-2-Hydroxy-1-methyl-2-phenylethyl)-N-methyl-3-phenyl-acrylamide [9]

166.3	165.2	214.3	295.4

A solution of cinnamoyl chloride (1.67 g, 10.0 mmol) in anhydrous THF (10 mL) is added dropwise to a stirred solution of (−)-ephedrine (1.65 g, 10.0 mmol) and proton sponge (2.14 g, 10.0 mmol) in anhydrous THF (15 mL) under a nitrogen atmosphere. The reaction mixture is stirred for 1 h at room temperature.

The precipitate formed is then removed by vacuum filtration and washed with Et$_2$O. The filtrate is washed with HCl (1 N, 2 × 10 mL) and brine (3 × 10 mL). The organic layer is dried (MgSO$_4$) and the solvents are removed under vacuum to give 2.50 g (85%) of a light-yellow resin-like solid, sufficiently pure according to TLC (SiO$_2$/Et$_2$O: R_f = 0.42); $[\alpha]_D^{20}$ = −150 (c = 1.00, CHCl$_3$).

IR (solid): $\tilde{v}$ (cm^{-1}) = 3334 (OH), 3059, 3026, 2977, 1642 (amide C=O), 1578.

^{1}H NMR (CDCl$_3$): δ (ppm) = 7.71 (d, J = 15.5 Hz, 1 H, PhC$\underline{H}$=CHCO), 7.52 (d, J = 2.5 Hz, 2 H, Ar–H), 7.31–7.40 (m, 8 H, Ar–H), 6.79 (d, J = 15.5 Hz, 1 H, PhCH=C$\underline{H}$CO), 4.94 (d, not resolved, 1 H, HOC$\underline{H}$Ph), 4.55 (m$_c$, 1 H, NCH), 4.51 (s$_{br}$, 1 H, OH), 2.85 (s, 3 H, NCH$_3$), 1.26 (d, J = 6.9 Hz, 3 H, NCHC$\underline{H}_3$).

^{13}C NMR (CDCl$_3$): δ (ppm) = 168.43, 143.27, 141.80, 135.20, 129.78, 128.83, 128.21, 127.89, 127.53, 126.36, 117.91, 67.96, 58.76, 33.46, 12.11.

1.1.4.2   *   (S)-3-Phenyl-heptanoic acid ((1S,2R)-2-hydroxy-1-methyl-2-phenyl-ethyl) methyl amide** [9]

295.4	nBuBr: 137.0	353.5
	Mg: 24.3	

A solution of nBuMgBr in anhydrous Et$_2$O (30 mL), prepared from n-butyl bromide (4.11 g, 30.0 mmol) and magnesium turnings (730 mg, 30.0 mmol), is added at −78 °C to a solution of

N-cinnamoylephedrine **1.1.4.1** (1.48 g, 5.00 mmol) in Et$_2$O (70 mL) under a nitrogen atmosphere. The solution is then stirred at –40 to –20 °C with monitoring by TLC (SiO$_2$/Et$_2$O) until the reaction is complete (~48 h).

A phosphate buffer solution (50 mL, pH 7) is added to the reaction mixture, which is then allowed to warm to room temperature. The insoluble material is filtered off through Celite and washed with EtOAc (30 mL). The layers are separated and the aqueous layer is extracted with EtOAc (30 mL). The combined extracts are dried (MgSO$_4$) and concentrated in vacuo. The resulting crude Michael adduct is purified by chromatography (SiO$_2$/Et$_2$O; R_f = 0.60); yield 1.60 g (90%) as a light-yellow resin-like solid.

IR (solid): $\tilde{\nu}$ (cm^{-1}) = 3365 (O–H), 3027, 2955, 2928, 2858, 1611 (amide C=O).

^{1}H NMR (CDCl$_3$): δ (ppm) = 7.60–7.35 (m, 10 H, ArH), 4.73 (d, not resolved, 1 H, PhC*H*OH), 4.41 (m$_c$, 1 H, NCH), 4.08 (s$_{br}$, 1 H, OH), 3.15 (m$_c$, 1 H, PhCH), 2.49 (s, 3 H, NCH$_3$), 2.57 and 2.48 (dd, J = 14.6, 7.6 Hz, 2 H, PhCHC*H$_2$*CO), 1.75–1.60 (m, 2 H, CH$_2$), 1.11–1.31 (m, 4 H, CH$_2$–CH$_2$), 1.08 (d, J = 7.0 Hz, 3 H, NCHC*H$_3$*), 0.83 (t, J = 7.3 Hz, 3 H, CH$_3$).

^{13}C NMR (CDCl$_3$): δ (ppm) = 174.00, 144.62, 141.54, 128.42, 128.05, 127.68, 127.33, 126.39, 126.25, 77.17, 58.05, 42.61, 41.45, 35.69, 33.36, 29.70, 22.64, 13.97, 12.00.

1.1.4.3   *   (*S*)-3-Phenylheptanoic acid [9]

353.5   206.3

The Michael adduct **1.1.4.2** (1.41 g, 4.00 mmol) is dissolved in acetic acid (10 mL) and sulfuric acid (6 N, 20 mL) and then refluxed for 3 h.

The mixture is extracted with Et$_2$O (3 × 20 mL) and the combined extracts are washed with brine (2 × 20 mL). The carboxylic acid is extracted with NaOH (1 M, 2 × 10 mL). The aqueous phase is washed with Et$_2$O (20 mL), then acidified with concentrated HCl and extracted with Et$_2$O (3 × 10 mL). The combined organic phases are dried (MgSO$_4$) and concentrated *in vacuo* to yield 785 mg (95%) of a colorless oil; $[\alpha]_D^{20}$ = +33.6 (c = 4.0, benzene). The enantiopurity is 98% based on the optical rotation [9].

IR (solid): $\tilde{\nu}$ (cm^{-1}) = 3029, 2957, 2927, 2858, 1706 (C=O).

^{1}H NMR (CDCl$_3$): δ (ppm) = 7.31–7.24 (m, 2 H, ArH), 7.22–7.14 (m, 3 H, ArH), 3.06 (m$_c$, 1 H, PhCH), 2.65 (dd, J = 15.5, 7.1 Hz, 1 H, COCH$_2$), 2.59 (dd, J = 15.5, 7.9 Hz, 1 H, COCH$_2$), 1.57–1.67 (m, 2 H, CH$_2$), 1.08–1.30 (m, 4 H, CH$_3$–C*H$_2$*–C*H$_2$*), 0.82 (t, J = 7.3 Hz, 3 H, CH$_3$).

^{13}C NMR (CDCl$_3$): δ (ppm) = 178.28, 143.97, 128.45, 127.43, 126.47, 41.84, 41.49, 35.93, 29.46, 22.55, 13.90.

[1] For an instructive introduction to asymmetric conjugate addition, see G. Procter, *Asymmetric Synthesis*, p. 127, Oxford University Press, **1999**.

[2] W. Oppolzer, R. J. Mills, W. Pachinger, T. Stevenson, *Helv. Chim. Acta* **1986**, *69*, 154; see also ref. [1].

[3] W. Oppolzer, G. Poli, *Tetrahedron Lett.* **1986**, *27*, 4717.

[4] S. G. Davies, O. Ichihara, *Tetrahedron: Asymmetry* **1991**, *2*, 183. This work was extended: S. G. Davies, S. W. Epstein, O. Ishihara, A. Smith, *Synlett* **2001**, 1599; in addition, the intermediately formed enolates were subjected to electrophilic oxygenation with (−)-CSO (camphor-sultam oxaziridine) to give α-hydroxy-β-amino esters.

[5] N. Halland, P. S. Aburel, K. A. Joergensen, *Angew. Chem.* **2003**, *115*, 685; *Angew. Chem. Int. Ed.* **2003**, *42*, 661; compare: a) F.-Y. Zhang, E. J. Corey, *Org. Lett.* **2000**, *2*, 1097; b) H. Ito, T. Nagahara, K. Ishihara, S. Saito, H. Yamamoto, *Angew. Chem.* **2004**, *116*, 1012; *Angew. Chem. Int. Ed.* **2004**, *43*, 994; c) H. Huang, E. N. Jacobsen, *J. Am. Chem. Soc.* **2006**, *128*, 7170 (enantioselective conjugate addition to nitroalkenes).

[6] M. Watanabe, K. Murata, T. Ikarya, *J. Am. Chem. Soc.* **2003**, *125*, 7508; see also: a) Y. Xu, K. Ohori, T. Ohshima, M. Shibasaki, *Tetrahedron* **2002**, *58*, 2585; b) J. G. Boiteau, A. J. Minnaard, B. L. Feringa, *J. Org. Chem.* **2003**, *68*, 9481.

[7] Y. Iguchi, R. Itooka, N. Miyaura, *Synlett* **2003**, 1040.

[8] O. Meyer, J.-M. Becht, G. Helmchen, *Synlett* **2003**, 1539; *Synthesis* **2003**, 2805.

[9] T. Mukaiyama, N. Iwasawa, *Chem. Lett.* **1981**, 913.

1.1.5   Ethyl 8-chloro-4-methylnaphthalene-2-carboxylate

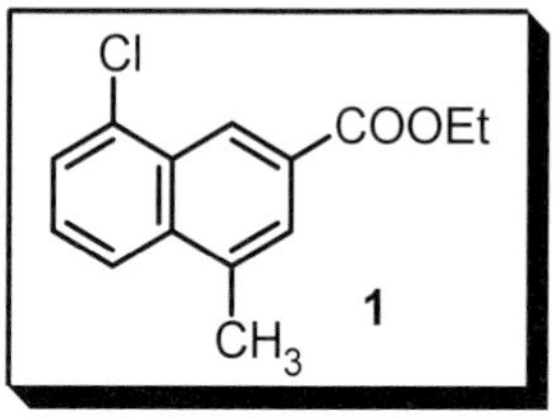

Topics:
- Baylis–Hillman reaction of an aryl aldehyde with an acrylate in the presence of DABCO
- Acetylation of an OH function
- Transformation of an acetylated Baylis–Hillman adduct to a functionalized naphthalene system by reaction with nitroalkane/base (domino process)

(a) General

The addition of carbonyl compounds, mainly aldehydes or aldimines, to acceptor-substituted alkenes (e.g., acrylates, acrylonitrile, enones, etc.) induced by tertiary amines (or phosphines), preferentially DABCO, is known as the Baylis–Hillman reaction [1, 2]:

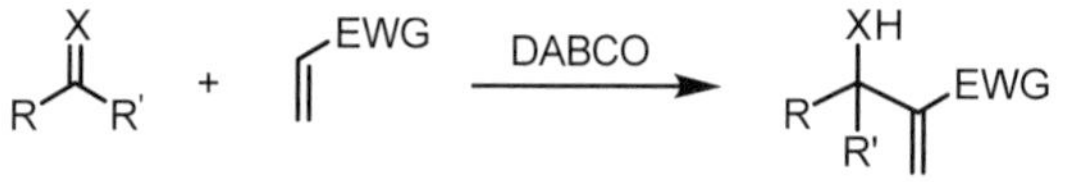

$X = O, NR^1$
EWG = electron withdrawing group
$(COOR^2, CN, COR^2$ etc.$)$

The generally accepted mechanism for the Baylis–Hillman process is illustrated by the reaction of an aldehyde with acrylate under the catalytic influence of DABCO:

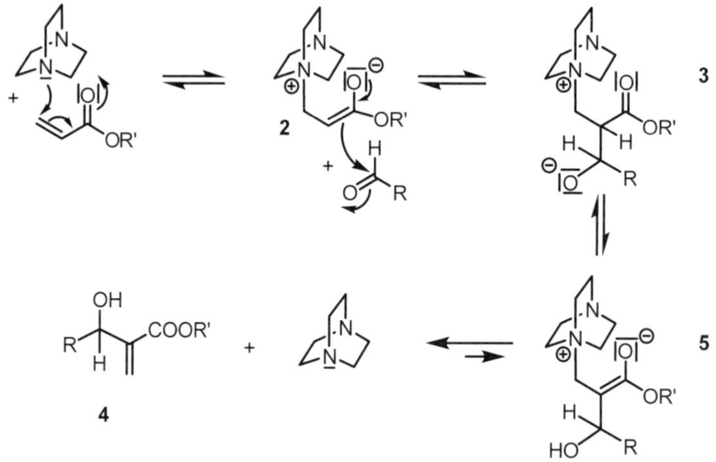

The first step involves a Michael-type addition of DABCO to the acrylate to produce the zwitterionic enolate betaine **2**, which then adds as a nucleophile to the aldehyde carbonyl group in an aldol-like fashion to give the zwitterion **3**. Subsequent proton transfer (**3 → 5**) and release of the tertiary amine complete the catalytic cycle and provide the Baylis–Hillman adduct **4**.

The product **4** contains three different functionalities and is therefore capable of undergoing several different transformations [2]. Moreover, if the electrophilic component in the Baylis–Hillman reaction carries additional functionalities, domino reactions [3] can be induced, which lead to the formation of carbocyclic and heterocyclic compounds [4], as illustrated by the following examples (1)–(3):

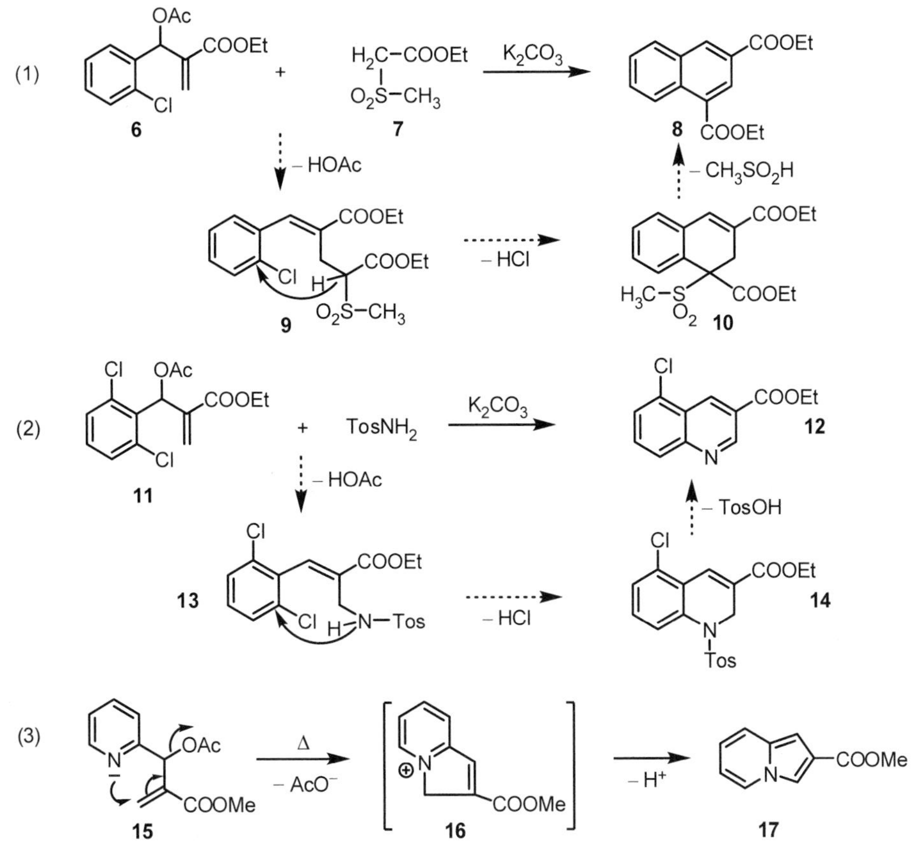

(1)

OAc
COOEt
Cl
6

+

$H_2C-COOEt$
O_2S-CH_3
7

K_2CO_3

COOEt
8   COOEt

$- HOAc$

COOEt
COOEt
Cl H
O_2S-CH_3
9

$- HCl$

COOEt
COOEt
H_3C-S COOEt
O_2 **10**

$- CH_3SO_2H$

(2)

Cl OAc
COOEt
Cl
11

+ $TosNH_2$

K_2CO_3

Cl
COOEt
N
12

$- HOAc$

Cl
COOEt
Cl H N$-$Tos
13

$- HCl$

Cl
COOEt
N
Tos
14

$- TosOH$

(3)

OAc
N
COOMe
15

Δ
$- AcO^-$

$\oplus$N
COOMe
16

$- H^+$

N
COOMe
17

In (1), the acetylated Baylis–Hillman adduct **6** (obtained from 2-chlorobenzaldehyde and ethyl acrylate with subsequent acetylation) reacts with the sulfone **7** in the presence of a base. The product is the naphthalene derivative **8**, which is formed via S_N' attack (→ **9**), intramolecular S_NAr reaction (→ **10**), and finally elimination of sulfinic acid [5].

In (2), the acetylated Baylis–Hillman adduct **11** (obtained as above from 2,6-dichlorobenzaldehyde) is reacted with p-toluenesulfonamide/base to give the quinoline derivative **12** in a sequence analogous to (1) (**11** → **13** → **14** → **12**) [6].

In (3), the acetylated Baylis–Hillman adduct **15** (obtained from pyridine-2-aldehyde and methyl acrylate with subsequent acetylation) is transformed into **17** by thermolysis. The process involves an S_N' substitution of the allylic acetate moiety in **15** to give the indolizinium ion **16** followed by deprotonation [7a].

As a further example, the synthesis of **1** is described in detail in (b) [8, 9].

(b) Synthesis of 1

The synthesis of **1** [9] starts with the reaction of 2,6-dichlorobenzaldehyde (**18**) and ethyl acrylate in the presence of DABCO, which provides the Baylis–Hillman adduct **19**. As in most cases, this DABCO-initiated Baylis–Hillman process requires a long reaction time of 5 d for completion, probably in this case due to steric hindrance in the o,o'-disubstituted benzaldehyde **18**. In other cases, the reaction time may be decreased to ca. 12 h by the use of a catalytic system consisting of DABCO, triethanolamine, and the Lewis acid La(OTf)$_3$ [10]. Such acceleration, however, could not be observed for the transformation **18** → **19**. Nevertheless, triethanolamine was used as solvent which proved to be more effective even than octanol [10c]. In the next step, the adduct **19** is acetylated using acetic anhydride to give the acetate **11**:

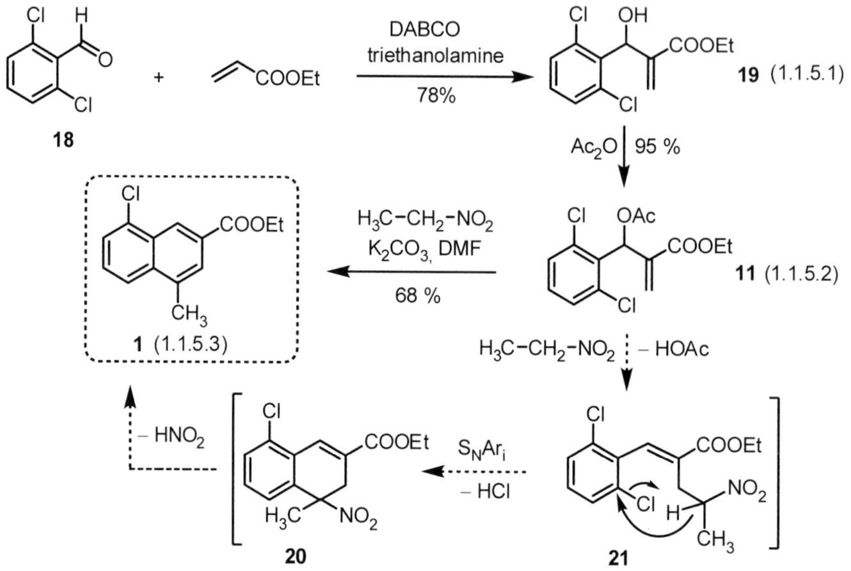

The acetate **11** reacts smoothly with nitroethane in *N,N*-dimethylformamide in the presence of K$_2$CO$_3$ as a base to provide the target molecule **1** in good yield (68%). Again, the functionalized naphthalene system is formed by way of a three-step domino process. First, the nitronate anion from nitroethane displaces acetate in **11** in an S_N'-like fashion to give the cinnamic ester **21**; second, one of the arene *ortho*-halogens (activated by the α,β-unsaturated ester moiety) is substituted by the nitronate in **21** in an intramolecular S_NAr reaction to afford **20**; third, aromatization of the 1,2-dihydronaphthalene intermediate **20** takes place by base-induced elimination of HNO$_2$.

Thus, the naphthalene-2-carboxylic ester **1** is obtained in a three-step sequence with an overall yield of 50% (based on aldehyde **18**).

(c) Experimental procedures for the synthesis of 1

1.1.5.1   *   2-[(2,6-Dichlorophenyl)hydroxymethyl]-acrylic acid ethyl ester [10]

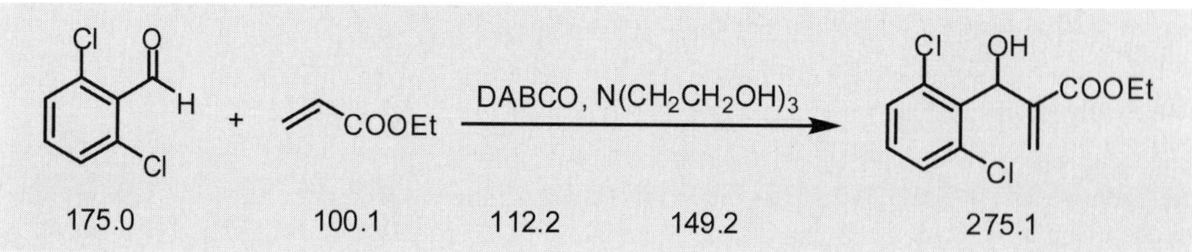

| 175.0 | 100.1 | 112.2 | 149.2 | 275.1 |

In contrast, the synthesis of **1** according to approach II (from retrosynthesis according to **B**) was found to be superior to all other alternatives [5] and is described in the following section with experimental details. This access via benzosuberone **12** requires the construction of the *gem*-dimethyl moiety from a carbonyl group, a transformation elegantly accomplished by use of the titanium reagent $(CH_3)_2TiCl_2$ [6].

4-Methyl-3-nitroacetophenone (**21**) is subjected to a Wittig reaction with the commercially available C_4-phosphonium salt **22** in the presence of KO*t*Bu as base. The carbonyl olefination results in the formation of the unsaturated ester **23** (obtained as a mixture of *E/Z* isomers). Hydrogenation of the C=C double bond and the nitro group in **23** using Pd/C in ethanol provides the amino ester **24**. The primary aromatic amine function in **24** is then transformed into a phenolic OH group by the classical two-step process of diazotization with aqueous HNO_2 and nucleophilic substitution of the diazonium group by hydroxide in methanol. In this process, the ester function is also hydrolyzed to give the carboxylic acid **26**. Ring-closure to the benzosuberone **25** by intramolecular Friedel–Crafts acylation is then achieved by treatment with polyphosphoric acid. After methylation of the phenolic OH group using dimethyl sulfate/NaOH (**25 → 12**), geminal dimethylation at the C=O group of **12** is accomplished with $(CH_3)_2TiCl_2$ at –30 °C to give the benzocycloheptene **18**. Finally, the methyl ether function in **18** is cleaved with BBr_3 to give the *ar*-himachalan **1** in a linear seven-step sequence with an overall yield of 18% (based on **21**).

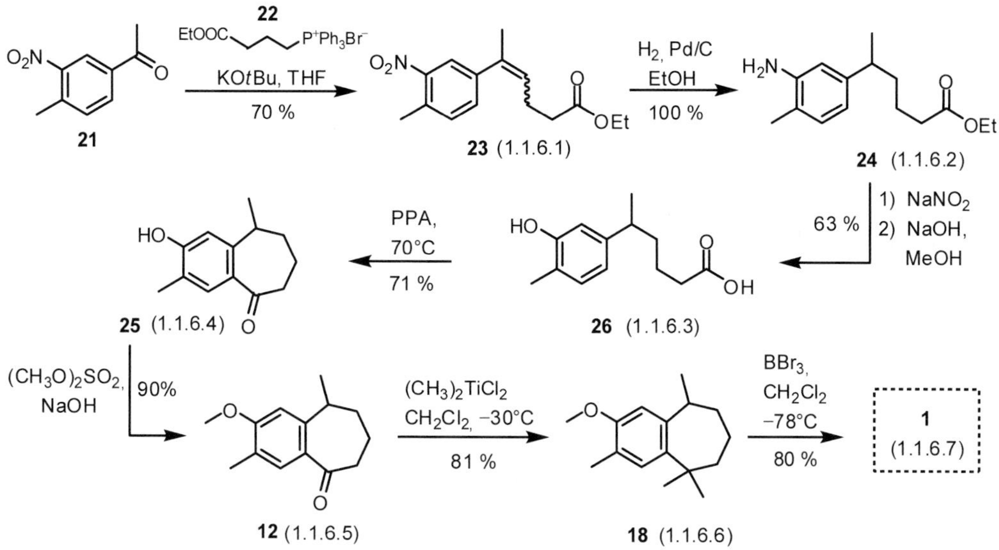

For the geminal dimethylation of **12**, two equivalents of $(CH_3)_2TiCl_2$ per carbonyl group are required. This leads to the following mechanism: (1) methyl transfer from titanium to the carbonyl carbon atom by nucleophilic addition of Ti-CH_3 to C=O, (2) methyl migration within the ion-pair ate-complex **29**. As the driving force, the large difference in Δ_H Ti–O versus Δ_H Ti–C (480 vs. 250 kJ mol^{-1}) can be assumed [6, 7].

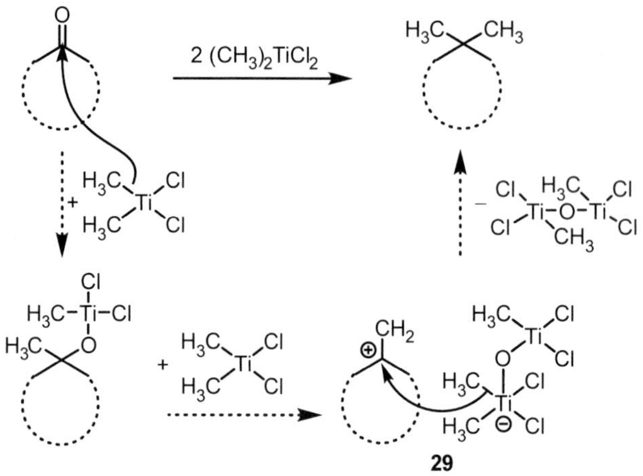

29

(c)   Experimental procedures for the synthesis of 1

1.1.6.1   *   Ethyl 5-(4-methyl-3-nitrophenyl)-4-hexenoate [5]

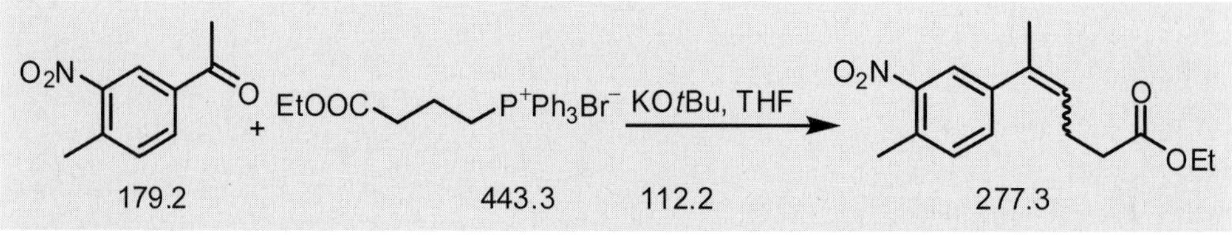

179.2   443.3   112.2   277.3

(3-Carbethoxypropyl)triphenylphosphonium bromide (42.8 g, 94.0 mmol) is added to a stirred solution of KOtBu (10.0 g, 90.0 mmol) in anhydrous THF (100 mL) and stirring is continued for 1.5 h. A solution of 4-methyl-3-nitroacetophenone (11.2 g, 72.0 mmol) in THF (100 mL) is then added dropwise with stirring. When the addition is complete, the dark mixture is heated to reflux for 12 h.

The reaction mixture is cooled to room temperature, poured into H$_2$O (500 mL), and extracted with Et$_2$O (4 × 250 mL). The combined ethereal phases are washed with H$_2$O (5 × 200 mL) and dried over MgSO$_4$. The solvent is removed, and the crude oily product is dissolved in the minimum volume of CH$_2$Cl$_2$ and purified (a) by rapid filtration through SiO$_2$ (eluent: CH$_2$Cl$_2$), (b) by chromatography on SiO$_2$ (eluent: Et$_2$O/petroleum ether, 1:6). The product (14.0 g, 70%, 2:1 mixture of E/Z stereoisomers) is used directly in the next step.

IR (film): $\tilde{v}$ (cm^{-1}) = 1770 (C=O), 1655 (C=C).

^{1}H NMR (CDCl$_3$): δ (ppm) = 7.95 (d, J = 1.8 Hz, 1 H, Ar-H, Z), 7.79 (d, J = 1.3 Hz, 1 H, Ar-H, E), 7.50 (dd, J = 8.0/1.8 Hz, 1 H, Ar-H, Z), 7.35–7.29 (m, 2 H, Ar-H, E), 7.26 (d, J = 8.0 Hz, 1 H, Ar-H, Z), 5.82 (m$_c$, 1 H, =CH, Z), 5.52 (m$_c$, 1 H, =CH, E), 4.15 (q, J = 7.1 Hz, 2 H, OCH$_2$, Z), 4.10 (q, J = 7.1 Hz, 2 H, OCH$_2$, E), 2.59 (s, 3H, Ar-CH$_3$, E), 2.57 (s, 3H, Ar-CH$_3$, Z), 2.55–2.44 and 2.36–2.25 (m, 4 H, CH$_2$–CH$_2$, E and Z), 2.06 (d, J = 1.3 Hz, 3 H, =C–CH$_3$, Z), 2.03 (d, J = 1.3 Hz, 3 H, =C–CH$_3$, E), 1.28 (t, J = 7.1 Hz, 3H, Z), 1.23 (t, J = 7.1 Hz, 3 H, E).

MS (CI, CH$_4$, 120 eV): m/z (%) = 277 (76) [M]$^+$.

1.1.6.2   *   Ethyl 5-(3-amino-4-methylphenyl)hexanoate [5]

277.3   249.4

5% Pd/C catalyst (ca. 0.5 g) is added to a solution of the unsaturated ester **1.1.6.1** (10.0 g, 36.0 mmol) in EtOH (200 mL). Hydrogenation is carried out in a hydrogenation apparatus for 12 h under a hydrogen pressure of 2.5 bar.

The catalyst is then filtered off and rinsed with EtOH. The EtOH solution is concentrated *in vacuo*. The product (9.00 g, 100%) is obtained as a faintly yellow oil, which is homogeneous according to TLC and is used in the next step without further purification.

IR (film): $\tilde{v}$ (cm^{-1}) = 3455, 3370 (NH$_2$), 1740 (C=O).

^{1}H NMR (CDCl$_3$): δ (ppm) = 6.95 (d, J = 7.5 Hz, 1 H, Ar-H), 6.52 (d, J = 7.5 Hz, 1 H, Ar-H), 6.49 (s$_{br}$, 1 H, Ar-H), 4.09 (q, J = 7.1 Hz, 2 H, OCH$_2$), 3.57 (s$_{br}$, 2 H, NH$_2$), 2.59–2.54 (m, 1 H, C$\underline{H}$–CH$_3$), 2.26–2.22 (m, 2 H, CH$_2$), 2.12 (s, 3 H, Ar-CH$_3$), 1.59–1.48 (m, 4 H, (CH$_2$)$_2$), 1.24 (t, J = 7.1 Hz, 3H, CH$_3$), 1.19 (d, J = 6.6 Hz, 3 H, CH–C$\underline{H}_3$).

MS (CI, CH$_4$, 120 eV): *m/z* (%) = 249 (3) [*M*]$^+$.

1.1.6.3   *   5-(3-Hydroxy-4-methylphenyl)hexanoic acid [5]

249.4   222.3

The amino ester **1.1.6.2** (8.00 g, 32.0 mmol) is stirred with HCl (5 M, 20 mL). When most of the ester has dissolved, the reaction mixture is cooled in an ice bath and a solution of NaNO$_2$ (2.5 M, 13 mL, 32.5 mmol) is added with stirring at such a rate that the internal temperature does not exceed 5 °C. More NaNO$_2$ solution is added until the I$_2$/starch test for free HNO$_2$ is positive (ca. 15 min after the last addition); the excess of HNO$_2$ is then destroyed by the addition of urea. The solution of the diazonium salt thus obtained is heated to 100 °C (water bath) until the -evolution of N$_2$ ceases.

After cooling to room temperature, the resulting two-phase system is extracted with Et$_2$O (3 × 50 mL), the combined extracts are dried over MgSO$_4$, and the solvent is removed *in vacuo*. The residue (ester of **1.1.6.3**) is dissolved in a solution of NaOH (5.12 g, 128 mmol) in MeOH (100 mL) and stirred at room temperature for 12 h.

The solvent is then removed *in vacuo*, the residue is dissolved in H$_2$O (100 mL), and the (alkaline) solution is washed with Et$_2$O (3 × 50 mL). The organic extracts are discarded, and the aqueous phase is brought to pH 1 by the addition of concentrated HCl (stirring!) and extracted with Et$_2$O (3 × 50 mL). The combined ethereal extracts are dried (MgSO$_4$), the solvent is removed, and the residue is purified by chromatography on SiO$_2$ (eluent: Et$_2$O/petroleum ether, 3:2). The product is obtained as an orange solid, 4.50 g (63%), mp 96–97 °C.

IR (film): $\tilde{v}$ (cm^{-1}) = 3450 (OH), 1720 (C=O).

^{1}H NMR (CDCl$_3$): δ (ppm) = 7.01 (d, J = 7.5 Hz, 1 H, Ar-H), 6.65 (d, J = 7.5 Hz, 1 H, Ar-H), 6.58 (s, 1 H, Ar-H), 2.61–2.56 (m, 1 H, C$\underline{H}$–CH$_3$), 2.31–2.28 (m, 2 H, CH$_2$), 2.19 (s, 3 H, Ar-CH$_3$), 1.57–1.49 (m, 4 H, (CH$_2$)$_2$), 1.19 (d, J = 6.6 Hz, 3 H, CH–C$\underline{H}_3$).

MS (CI, CH$_4$, 120 eV): m/z (%) = 222 (76) $[M]^+$.

1.1.6.4   *   2-Hydroxy-3,9-dimethyl-6,7,8,9-tetrahydro-5*H*-benzo[*a*]cyclohepten-5-one [5]

222.3 → PPA, 70°C → 204.3

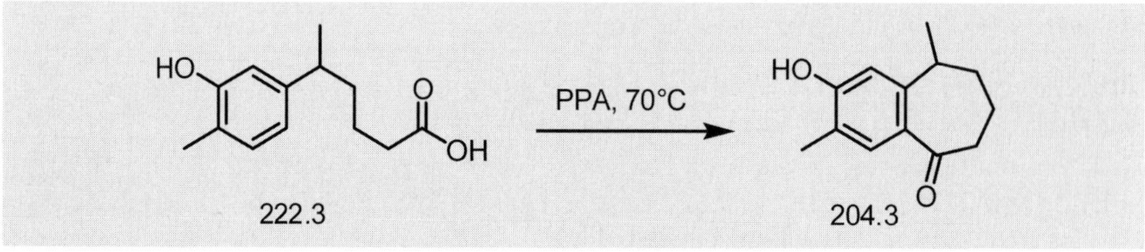

The finely powdered carboxylic acid **1.1.6.3** (2.00 g, 9.00 mmol) is suspended in polyphosphoric acid (20 mL, 85% P$_4$O$_{10}$). The resulting orange suspension is heated to 70 °C for 2 h with intense stirring.

The dark-red reaction mixture is then poured into H$_2$O (50 mL) and extracted with Et$_2$O (3 × 25 mL). The combined ethereal extracts are dried (MgSO$_4$), the solvent is removed *in vacuo*, and the residue is purified by column chromatography (SiO$_2$, eluent: Et$_2$O/petroleum ether, 1:2). The product is obtained in the form of colorless crystals, 1.30 g (71%), mp 142–143 °C.

IR (film): $\tilde{v}$ (cm^{-1}) = 3115 (OH), 1670 (C=O).

^{1}H NMR ([D$_6$]DMSO): δ (ppm) = 9.90 (s, 1 H, OH), 7.29, 6.74 (s, each 1 H, Ar-H), 3.09–3.02 (m$_c$, 1 H, C$\underline{H}$–CH$_3$), 2.58–2.51 (m, 2 H, CH$_2$), 2.11 (s, 3 H, Ar-CH$_3$), 1.92–1.76 (m, 2 H, CH$_2$), 1.48–1.34 (m, 2 H, CH$_2$), 1.28 (d, J = 6.6 Hz, 3 H, CH–C$\underline{H}_3$).

MS (EI, 70 eV): m/z (%) = 204 $[M]^+$.

1.1.6.5   *   2-Methoxy-3,9-dimethyl-6,7,8,9-tetrahydro-5*H*-benzo[*a*]cyclohepten-5-one [5]

204.3 → (CH$_3$O)$_2$SO$_2$, NaOH → 218.3

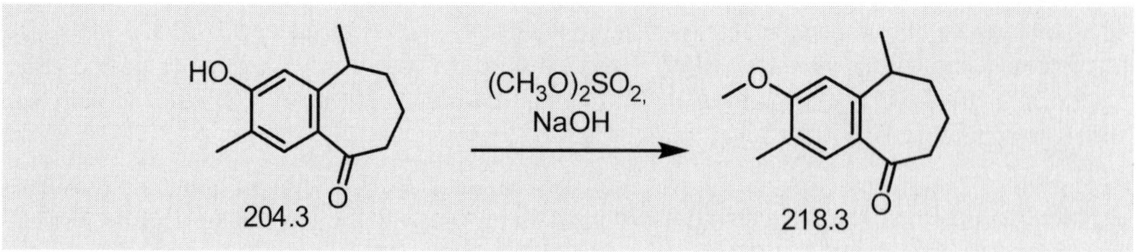

The hydroxybenzosuberone **1.1.6.4** (1.00 g, 5.0 mmol) is added over a period of 5 min to a stirred solution of NaOH (200 mg, 5.00 mmol) in H_2O (2.0 mL). Dimethyl sulfate (0.63 g, 5.0 mmol, 500 μL) is then added and stirring is continued for 30 min at room temperature; more $(CH_3O)_2SO_2$ (same amount as before) is then added and stirring is continued for 1 h at room temperature and for 30 min at 100 °C (water bath).

The reaction mixture is then cooled to room temperature, diluted with H_2O (10 mL), and extracted with Et_2O (3 × 20 mL). The combined extracts are dried (MgSO$_4$), the solvent is removed *in vacuo*, and the oily residue is purified by chromatography on SiO_2 (eluent: CH_2Cl_2). The product is obtained as a faintly yellow solid, 0.96 g (90%), mp 61–62 °C.

IR (film): $\tilde{v}$ (cm^{-1}) = 1690 (C=O).

¹H NMR (CDCl₃): δ (ppm) = 7.44, 6.69 (s, each 1 H, Ar-H), 3.88 (s, 3 H, OCH₃), 3.13 (m$_c$, 1 H, C$\underline{H}$–CH₃), 2.74–2.68, 2.61–2.53 (m, each 1 H, CH₂), 2.19 (s, 3 H, Ar-CH₃), 1.98–1.83, 1.66–1.49 (m, each 2 H, CH₂), 1.39 (d, J = 7.0 Hz, 3 H, CH–C$\underline{H}$₃).

MS (EI, 70 eV): *m/z* (%) = 218 (77) [M]$^+$.

1.1.6.6   **   **2-Methoxy-3,5,5,9-tetramethyl-6,7,8,9-tetrahydro-5*H*-benzo[*a*]cycloheptene [5]

Under an N_2 atmosphere, a solution of dimethylzinc in toluene (2 M, 2.5 mL, 5.00 mmol) is added dropwise to a stirred solution of titanium tetrachloride (0.96 g, 5.00 mmol) in anhydrous CH_2Cl_2 (25 mL) at such a rate that an internal temperature of –30 °C is maintained. After 15 min, a solution of the ketone **1.1.6.5** (0.50 g, 2.30 mmol) in CH_2Cl_2 (1.0 mL) is added dropwise at –30 °C. During the addition, the brown color of the reaction mixture changes to an intense dark brown. The mixture is allowed to warm to room temperature and is then heated under reflux for 12 h.

The reaction mixture is poured into H_2O (50 mL) and extracted with CH_2Cl_2 (3 × 20 mL). The extracts are combined, washed successively with H_2O (100 mL) and saturated NaHCO₃ solution (100 mL), and dried over MgSO₄. The solvent is removed and the residue is purified by rapid filtration through silica gel (eluent: CH_2Cl_2). The product is obtained as a colorless oil, which is homogeneous according to TLC; 0.43 g (81%).

---

¹H NMR (CDCl₃): δ (ppm) = 7.12, 6.71 (s, each 1 H, Ar-H), 3.81 (s, 3 H, OCH₃), 3.27 (m$_c$, 1 H, C$\underline{H}$–CH₃), 2.18 (s, 3 H, Ar-CH₃), 1.79–1.74, 1.65–1.52 (m, each 3 H, aliph. H), 1.39, 1.31 [s, each 3 H, C(CH₃)₂], 1.36 (d, J = 7.1 Hz, 3 H, CH–C$\underline{H}$₃).

MS (CI, CH₄, 120 eV): *m/z* (%) = 232 (66) [M]$^+$.

1.1.6.7   *   2-Hydroxy-3,5,5,9-tetramethyl-6,7,8,9-tetrahydro-5*H*-benzo[*a*]cycloheptene [5]

232.4   218.3

A 1.0 M solution of boron tribromide (4.0 mL, 8.0 mmol) in CH_2Cl_2 is added to a stirred solution of the methoxy compound **1.1.6.6** (0.42 g, 1.48 mmol) in anhydrous CH_2Cl_2 (40 mL) at –78 °C. The reaction mixture is allowed to warm to room temperature over 12 h.

H_2O (50 mL) is then added, the organic phase is separated, the aqueous phase is extracted with CH_2Cl_2 (3 × 20 mL), and the combined organic phases are dried ($MgSO_4$). The solvent is removed and the crude product is purified by column chromatography (SiO_2, eluent: CH_2Cl_2) to give 0.25 g (80%) of the hydroxy-*ar*-himachalan as a yellowish oil, which is pure according to TLC.

IR (film): $\tilde{v}$ (cm^{-1}) = 3370 (OH).

^{1}H NMR (CDCl$_3$): δ (ppm) = 7.10, 6.65 (s, each 1 H, Ar-H), 4.57 (s$_{br}$, 1 H, OH), 3.22 (m$_c$, 1 H, C$\underline{H}$–CH$_3$), 2.21 (s, 3 H, Ar-CH$_3$), 1.83–1.49 (m, 6 H, (CH$_2$)$_3$), 1.39 and 1.30 (s, each 3 H, C(CH$_3$)$_2$), 1.29 (d, J = 7.1 Hz, 3 H, CH–C$\underline{H}_3$).

MS (CI, CH$_4$, 120 eV): *m/z* (%) = 218 (478) [*M*]$^+$.

Note: The ^{1}H NMR spectrum is identical to that of the natural product according to ref. [4].

[1]   a) T. C. Joseph, S. Dev, *Tetrahedron* **1968**, *24*, 3809; b) R. C. Pandey, S. Dev, *Tetrahedron* **1968**, *24*, 3829.

[2]   F. Bohlmann, M. Lonitz, *Chem. Ber.* **1983**, *111*, 843.

[3]   H. Becker, A. Schmidt, unpublished results; A. Schmidt, Ph.D. thesis, Saarbrücken, **1996**.

[4]   G. U. Devi, G. S. K. Rao, *Indian J. Chem.* **1976**, *14B*, 162.

[5]   Th. Eicher, A. Schmitz, unpublished results; A. Schmitz, Ph.D. thesis, Saarbrücken, **2000**.

[6]   N. Krause, *Metallorganische Chemie*, p. 166, Spektrum Akademischer Verlag, Heidelberg, **1996**; for an alternative Al-organic method for the conversion of a carbonyl group to a geminal dimethyl functionality, see: C. U. Kim. P. F. Misco, B. Y. Luh, M. M. Mansuri, *Tetrahedron Lett.* **1994**, *35*, 3017.

[7]   For the use of titanium complexes with chiral ligands as valuable tools in enantioselective synthesis, see: D. J. Ramon, M. Yus, *Chem. Rev.* **2006**, *106*, 2126.

1.1.7   Methylenecyclododecane

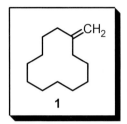

1

Topics: • Carbonyl olefination by the Lombardo reaction

(a) General

The term "carbonyl olefination" covers a series of C–C bond-forming reactions, which allow the chemo- and stereoselective formation of an olefinic double bond at carbonyl groups of aldehydes, ketones, esters, amides, etc., according to the following general scheme:

$$R^1_{R^2}{=}O \;+\; \overset{\oplus\, \text{or}}{\underset{R^4}{\overset{\delta\oplus}{X}{-}\overset{R^3}{\underset{}{C}}{}^{\ominus}}} \longrightarrow \underset{R^2}{\overset{R^1}{}}{=}\underset{R^4}{\overset{R^3}{}} \;+\; \overset{\text{neutral or}}{\overset{\ominus}{X}{=}O}$$

2   **3**

As reactands for carbonyl olefinations, α-carbanionic species of type **2** are required, which attack as nucleophiles at the carbonyl C-atom. They also exhibit electrophilic properties at the structural unit X for attack at the carbonyl oxygen, which is removed as an X=O moiety **3** in the olefination process. If X carries a positive charge, **2** represents a betaine or ylide. Several types of reactands **2** containing P-, Si-, and metal centers in the X part have been developed as carbonyl olefinating reagents.

(1) The classical procedure for carbonyl olefination is the Wittig reaction [1], using phosphoranes **5**. The phosphoranes **5** (easily obtainable by deprotonation of the corresponding α-CH phosphonium salts **4**) react with aldehydes or ketones to give alkenes and phosphine oxide:

$$\underset{X^{\ominus}}{\overset{\oplus}{Ph_3P}}{-}\underset{R^2}{\overset{R^1}{C}}{-}H \quad \xrightarrow[-\,HX]{\text{base}} \quad \left[\begin{array}{c} \overset{\oplus}{Ph_3P}{-}\overset{R^1}{\underset{R^2}{C}}{}^{\ominus} \\ \updownarrow \\ Ph_3P{=}\underset{R^2}{\overset{R^1}{}} \end{array} \right] \quad \xrightarrow{\underset{R^4}{\overset{R^3}{}}{=}O} \quad \left[\underset{R^2\;R^3}{\overset{O{-}PPh_3}{R^1{-}\underset{}{|}{-}R^4}} \right] \xrightarrow{-\,Ph_3P{=}O} \underset{R^2}{\overset{R^1}{}}{=}\underset{R^4}{\overset{R^3}{}}$$

4   **5**   **6**

The mechanism and stereochemistry of the Wittig reaction have been thoroughly investigated [2, 3]. Intermediates are the oxaphosphetanes **6** (from non-stabilized P-ylides), which collapse thermally by elimination of phosphine oxide. As a simplified rule, resonance-stabilized P-ylides give rise to (*E*)-alkenes, whilst non-stabilized P-ylides preferentially lead to (*Z*)-alkenes. The high value of the Wittig reaction is documented by a very large number of applications in the synthesis of alkenes (cf. **1.1.6, 3.5.1, 4.2.6**).

(2) The Horner reaction is a prominent example of the principle of "PO-activated olefination" [1], in which α-carbanions **8** (R = OR') from phosphonates react with aldehydes and ketones; phosphonamides and phosphine oxides behave analogously:

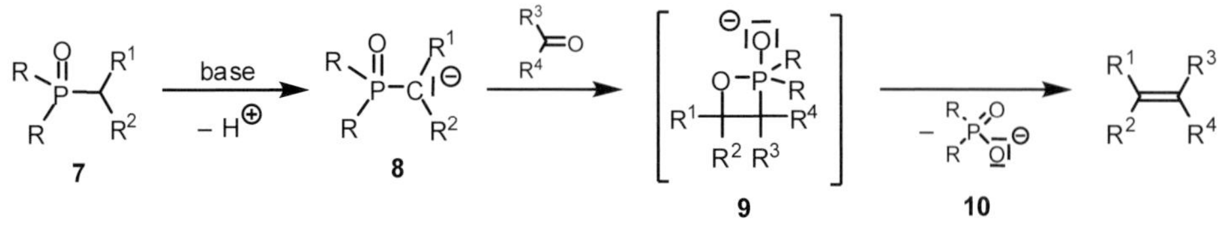

The mechanism of the Horner reaction is comparable to that of the Wittig reaction [2, 3]. Oxaphosphetanes **9** can be assumed as primary intermediates, which are cleaved by olefin formation and elimination of phosphate **10** (R = OR'). The main advantages of the Horner reaction lie in the facts that

- the reactivity of the α-carbanions **8** often proves to be superior to that of the corresponding P-ylides **5**, thus allowing olefination of carbonyl substrates not or less susceptible to the Wittig method, e.g. cyclohexanone;

- the phosphates **10** (R = OR') formed in the olefination process are water-soluble, thus considerably improving the isolation and purification procedures for the olefinic products.

Phosphonates **7** (R = OR') are obtained from phosphites and halogenoalkanes by the Arbusov reaction (cf. **4.2.6**).

(3) In the Peterson olefination (sometimes called the sila-Wittig reaction) [4], α-lithiated trialkylsilanes **12** – obtained from tetraalkylsilanes of type **11** by metalation with LDA or *n*BuLi – react with aldehydes or ketones:

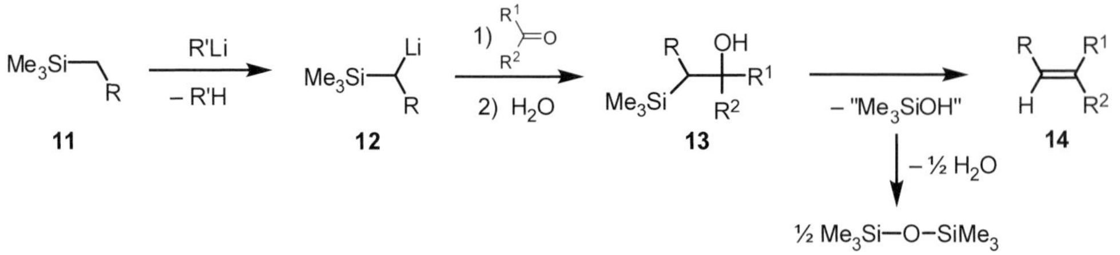

Initially, a β-hydroxysilane **13** is obtained, which is transformed into the olefin **14** by elimination of a silanol $Me_3Si–OH$ (finally appearing as $Me_3Si–O–SiMe_3$). The stereochemistry of olefin formation **13** → **14** can often be controlled by whether an acid or a base is used for the silanol elimination. Use of an acid generally leads to an *anti*-elimination (transition state **5**) whereas use of base leads to a *syn*-elimination (transition state **16**); in this way, the stereoselective formation of either (*Z*)- or (*E*)-alkenes **14** can be achieved.

This concept has been verified in several modifications [1]. Particularly successful outcomes have been achieved with the enantiomers of the hydrazine derived from (S)- or (R)-prolinol methyl ether (SAMP/RAMP, **9/10**), both of which are readily available [2] and thus allow the preparation of both enantiomers of an α-alkyl ketone [3].

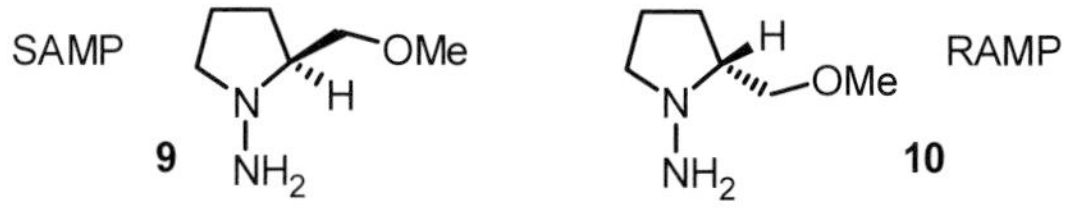

Here, the SAMP method (Enders method) is used to prepare the ketone **1** with high enantiopurity [4].

(b) Synthesis of 1

First, the SAMP auxiliary is condensed with diethyl ketone (**2**) to give the chiral hydrazone **11**:

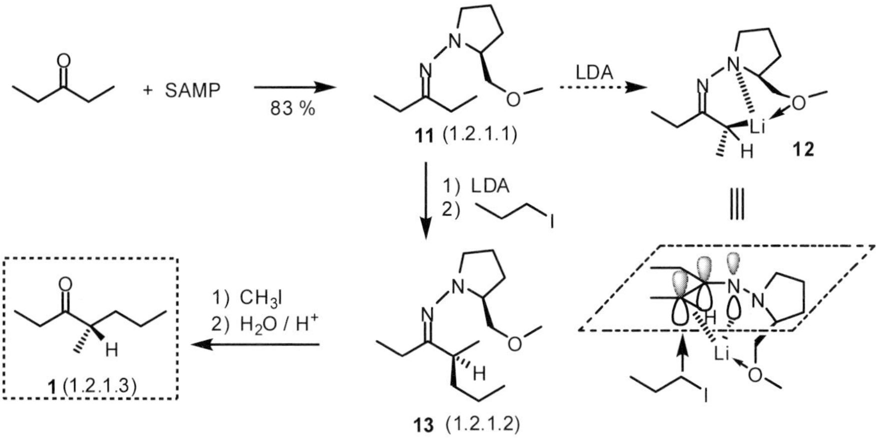

The hydrazone **11** is metalated with LDA and the formed azaenolate **12** is reacted with n-propyl iodide to provide the α-alkylated hydrazone **13** as a single diastereomer. The alkylation process presumably follows an S_E2 mechanism with retention of configuration, as visualized in the transition state **12**. Li chelation of the methoxy group is obviously responsible for the high degree of diastereoselection observed.

The hydrazone **13** is cleaved by alkylation with CH_3I to give the corresponding hydrazonium salt, which is cleaved by acid hydrolysis in an aqueous two-phase system to give the α-alkylated ketone **1** with 99% ee. This mild method of hydrazone cleavage has the drawback that the chiral auxiliary cannot be recovered ("sacrificial" vs. "regenerative" use of a chiral auxiliary [5]).

Ozonolysis is another method that has been introduced for the cleavage of SAMP hydrazones; it leads directly to the α-alkylated carbonyl source and a SAMP-derived nitrosamine [6].

(c) Experimental procedures for the synthesis of 1

1.2.1.1 ** Preparation of the chiral hydrazone [4]

| 86.1 | 130.2 | 198.3 |

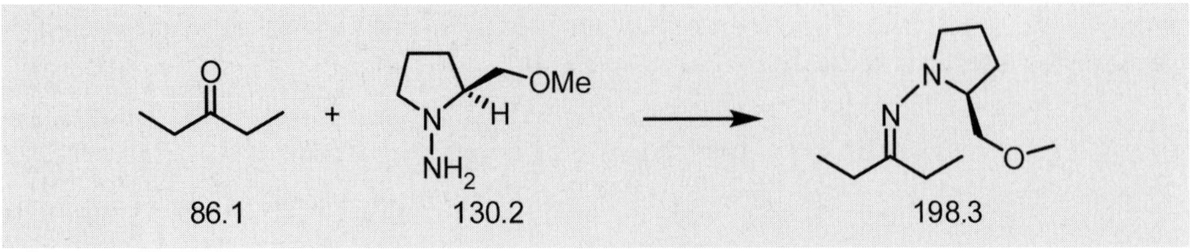

(–)-(S)-1-Amino-2-(methoxymethyl)pyrrolidine [2] (2.60 g, 20.0 mmol) and 3-pentanone (distilled, bp_{760} 101–102 °C; 1.89 g, 22.0 mmol, ≈2.32 mL) are stirred at 60 °C for 20 h in a 25 mL single-necked flask.

The reaction mixture is then diluted with anhydrous CH_2Cl_2 (20 mL) and dried over Na_2SO_4. The solvent is evaporated and the residue is distilled from a deactivated Kugelrohr (trimethylchlorosilane is distilled from the apparatus at atmospheric pressure). Any remaining solvent in the distillate is evaporated *in vacuo* to leave a colorless oil; 3.29 g (83%), $bp_{0.04}$ 46 °C (oven temperature 50–55 °C).

1.2.1.2 *** Diastereoselective alkylation of the chiral hydrazone [4]

1) Li-$N[CH(CH_3)_2]_2$
2) n-C_3H_7I

1) iPr_2NH: 101.2
2) 170.0

| 198.3 | | 240.4 |

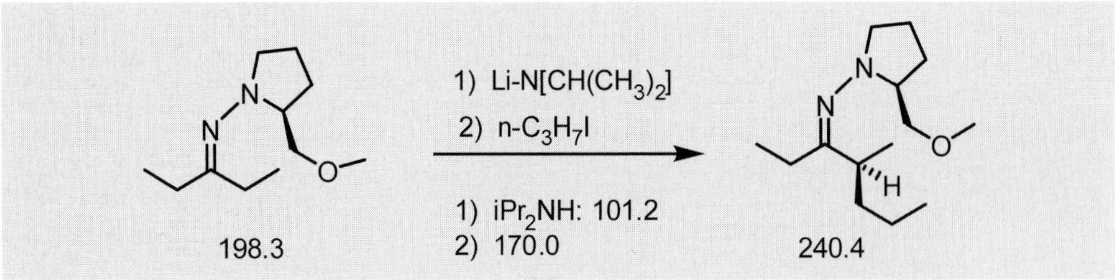

A solution of diisopropylamine (distilled from CaH_2, bp_{760} 84 °C; 0.84 g, 8.30 mmol, ≈ 1.17 mL) in anhydrous diethyl ether (40 mL) is prepared under nitrogen at –78 °C in an oven-dried, 100 mL two-necked flask fitted with a septum. *n*-Butyllithium in *n*-hexane (1.6 M, 5.2 mL) is added by means of a syringe (cannula) and the solution is stirred for 10 min at –78 °C. The solution is then warmed to 0 °C over ca. 30 min and the hydrazone prepared in **1.2.1.1** (1.52 g, 7.70 mmol) is slowly added dropwise from a syringe. Stirring is continued at 0 °C for 10 h. The solution is then cooled to –110 °C (petroleum ether/N_2), propyl iodide (distilled, bp_{760} 102 °C; 1.47 g, 8.65 mmol, ~0.84 mL) is added dropwise over 10 min through a cannula (the cannula is cooled during the addition), and stirring is continued for 1 h.

The mixture is warmed to room temperature, diluted with CH_2Cl_2 (40 mL), filtered, and the solvent is evaporated. The residue is used immediately for the next step.

1.2.1.3   ***   Cleavage of the alkylated hydrazone [4]

240.4   141.9   128.2

The crude product prepared in **1.2.1.2** and methyl iodide (3.54 g, 25.0 mmol, ~1.56 mL; Caution: carcinogenic) are heated under reflux. Excess methyl iodide is evaporated *in vacuo*. The formed hydrazonium iodide (green-brown oil) is stirred vigorously with *n*-pentane (60 mL)/HCl (6 M, 40 mL) for 60 min.

The organic phase is separated and the aqueous phase is extracted with *n*-pentane (2 × 50 mL). The combined organic layers are washed with brine and dried over Na_2SO_4. The solvent is evaporated and the green residue is distilled from a Kugelrohr (pretreated with trimethylchlorosilane); colorless liquid, 475 mg (48% based on SAMP), bp$_{110}$ 140 °C (oven temperature), $[\alpha]_D^{20}$ = +16.5 (*c* = 1.2, *n*-hexane) (note).

IR (film): $\tilde{v}$ (cm^{-1}) = 1710 (C=O), 740.

^{1}H NMR (CDCl$_3$): δ (ppm) = 2.45 (m, 3 H, CH$_2$CO + CHCO), 1.8–0.7 (m, 13 H, CH$_2$ + CH$_3$).

Note: Observed *ee* = 87% (*S*); reported [3]: $[\alpha]_D^{20}$ = +22.1 (*c* = 1.0, *n*-hexane), *ee* = 99.5% (*S*).

[1]   G. Procter, *Asymmetric Synthesis*, p. 58, Oxford University Press, Oxford, **1999**.

[2]   SAMP and RAMP are commercially available. SAMP can be prepared starting from (*S*)-proline according to L. F. Tietze, Th. Eicher, *Reaktionen und Synthesen im organisch-chemischen Praktikum und Forschungslaboratorium*, p. 443, Georg Thieme Verlag, Stuttgart, **1991**.

[3]   a) A simple example is the enantioselective preparation of both enantiomers of the defense substance of the "daddy longleg" cranefly: D. Enders, U. Braus, *Liebigs Ann. Chem.* **1983**, 1439. For other applications of the SAMP methodology, see: b) D. Enders, K. Klatt, *Synthesis* **1996**, 1403; c) J. L. Vicario, A. Job, M. Wolberg, M. Mueller, D. Enders, *Org. Lett.* **2000**, *6*, 1023; d) D. Enders, M. Voith, *Synlett* **2002**, 29; e) D. Enders, M. Boudou, J. Gries, *New Methods Asymm. Synth. Nitrogen Heterocycl.* **2005**, 1.

[4]   a) D. Enders, H. Eichenauer, *Chem. Ber.* **1979**, *112*, 2933; b) D. Enders, H. Eichenauer, *Angew. Chem.* **1979**, *91*, 425; *Angew. Chem. Int. Ed. Engl.* **1979**, *18*, 397; c) for a review on asymmetric synthesis with the RAMP/SAMP auxiliaries, see ref. [3b].

[5]   For a review on the principles and recent applications of chiral auxiliaries, see: Y. Guas, F. Glorius, *Synthesis* **2006**, 1899.

[6]   D. Enders, H. Kipphardt, P. Fey, *Organic Synthesis* **1987**, *65*, 183.

1.2.2 (*S*)-2-Isopropylhex-4-yn-1-ol

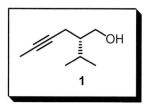

1

Topics:
- Enantioselective α-alkylation of an alkanoic acid by application of the Evans methodology
- Synthesis of a chiral oxazolidinone (Evans auxiliary) from (*S*)-valine
- Synthesis of 1-bromobut-2-yne (alkylating reagent)
- *N*-Acylation and enantioselective α-alkylation of a chiral *N*-acyloxazolidinone
- Reductive removal of the Evans auxiliary, transformation of the *N*-acyl moiety into the corresponding primary alcohol

(a) General

The target molecule **1** was required as a building block in the context of a multi-step natural product synthesis [1]. Retrosynthesis of **1** leads to **2** and, furthermore, to isovaleric acid (**3**) and the propargylic halide **4** as substrates.

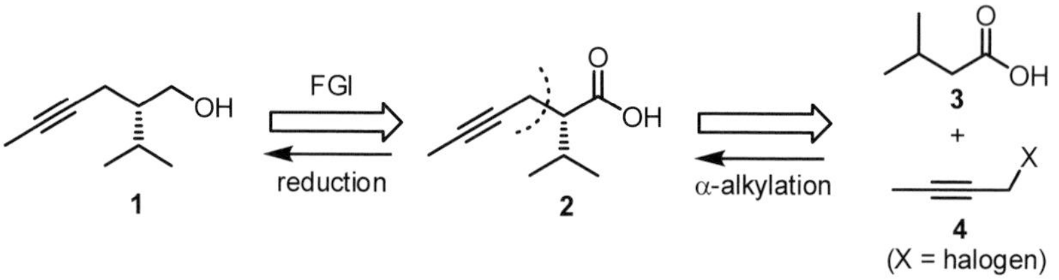

Accordingly, the synthesis consists of an α-alkylation of the acid **3** followed by reduction of the carboxyl group of the formed product **2** to provide the primary alcohol **1**. The main objective for a stereochemically concise synthesis of **1** is stereoselective α-alkylation of an alkanoic acid R–CH₂–CO₂H (**5**). For stereodifferentiation in the alkylation process, carboxylic acid derivatives **6** can be employed, in which a chiral auxiliary is introduced at the acyl C-atom (cf. **1.2.1**) [2]. The chiral auxiliary influences the configuration of the enolates **7a/7b** (*Z* or *E*, formed by deprotonation of **6**) and the facial selectivity of the alkylation (*re* or *si*) according to the following scheme:

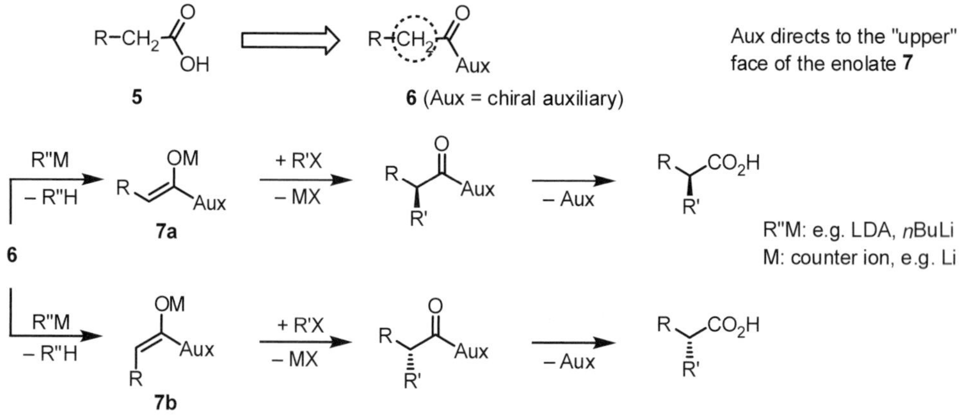

Aux directs to the "upper" face of the enolate **7**

R"M: e.g. LDA, *n*BuLi
M: counter ion, e.g. Li

For high stereoselectivity in the enolate formation and the alkylation, formation of a chelate by coordination of the metal ion to an appropriate functionality of the chiral auxiliary is necessary. Widely used chiral carboxylic acid derivatives are **8** and **11**, containing an oxazolidinone as the chiral auxiliary (Evans auxiliaries) [3].

Deprotonation of **8** and **11** with LDA produces the chelated enolates **9** and **12**, respectively, with a Z-selectivity of >99:1, which can then be α-alkylated with alkyl halides (only reactive alkyl halides such as methyl, benzyl, allyl, and propargyl can be used) with very high levels of diastereoselectivity to give the products **10** and **13**:

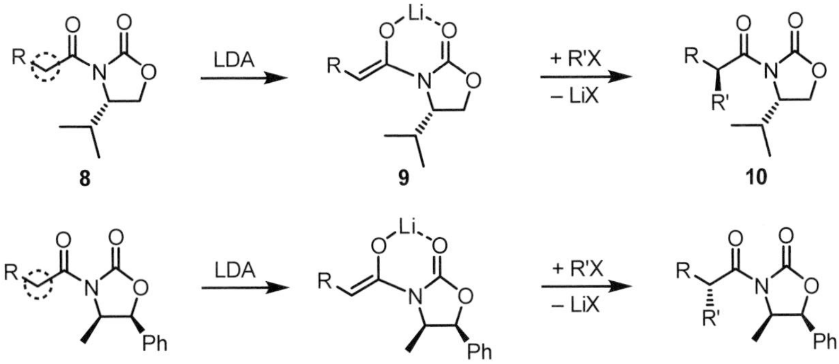

It should be noted that the enolates **9** and **12** can also be used in aldol reactions. In these transformations, using a (Z)-enolate a *syn*-product is obtained via a closed transition state, whereas with an (E)-enolate the *anti*-product is formed predominately. By adding one mole of a Lewis acid, the stereochemical outcome is reversed since under these conditions an open transition structure is preferred.

Removal of the chiral auxiliary from the α-alkylated N-acyl oxazolidinones **10** and **13** may be achieved by hydrolysis, alcoholysis or reduction, as illustrated for **10**.

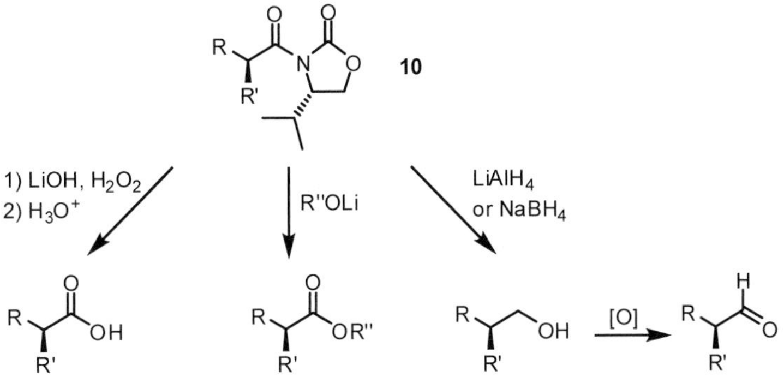

In this way, almost enantiopure α-alkylated carboxylic acids, esters, primary alcohols, and aldehydes can be obtained.

The chiral oxazolidinones **15** and **17**, as parts of the chiral *N*-acyl derivatives **8** and **11**, are prepared from readily available 1,2-amino alcohols such as *L*-valinol (**14**), formed from *L*-valine, and norephedrine (**16**) by reaction with diethyl carbonate. The acylation of **15** and **17** to give **8** and **11**, respectively, is accomplished by deprotonation with *n*BuLi or LDA followed by reaction of the anion with an acid chloride:

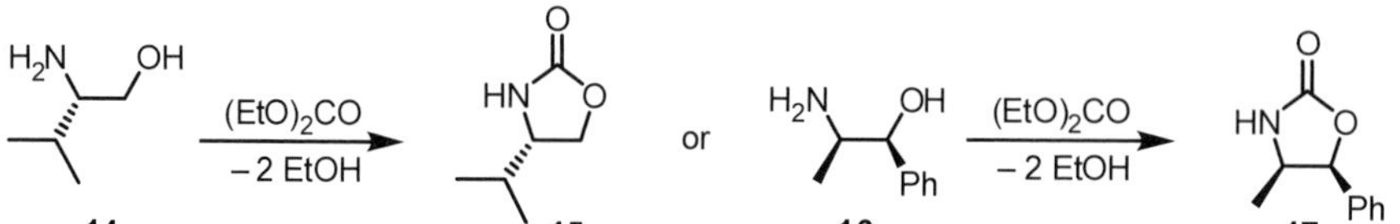

Due to the complementary outcomes of their α-alkylations, the systems **8** and **11** allow the preparation of both enantiomers of an (α-alkyl)alkanoic acid R–CHR'–CO$_2$H.

In section (b), the synthesis of the target molecule **1** from a chiral α-alkylated isovaleric acid (**2**) is presented, which is accessible by application of the auxiliary **15** and the Evans methodology.

(b) Synthesis of 1

The synthesis of **1** is convergent and is divided into three parts. First, the auxiliary **15** is prepared from (*S*)-valine; second, the propargylic halide **22** is synthesized from propargyl alcohol (**23**); third, the auxiliary **15** is acylated, then the diastereoselective α-alkylation with **22** is performed, and finally the auxiliary is removed reductively.

(1) (*S*)-Valine (**18**) is reduced with LiAlH$_4$ in THF to give (*S*)-valinol (**14**), which is transformed into the chiral oxazolidinone **15** by cyclocondensation with diethyl carbonate in the presence of K$_2$CO$_3$ [4, 5].

(2) Propargylic alcohol (**23**) is converted to its tetrahydropyranyl ether **21** by reaction with dihydropyran in the presence of concentrated HCl. The THP ether **21** is deprotonated at the terminal acetylene function and the formed acetylide is methylated *in situ* with CH$_3$I to give **19**. The THP ether in **19** is cleaved by treatment with phosphorus tribromide in the presence of pyridine in diethyl ether, thereby generating 1-bromo-2-butyne (**22**) required as alkylating agent [6, 7].

(3) The oxazolidinone **15** is *N*-deprotonated using *n*BuLi in THF and acylated with isovaleroyl chloride (both at −78 °C) to give the *N*-isovaleroyloxazolidinone **20** in almost quantitative yield. α-Alkylation of **20** is achieved in THF by deprotonation at the CH$_2$ group with LDA to yield the enolate **9**, R = (CH$_3$)$_2$CH. Subsequent reaction with the propargylic halide **22** in the presence of DMPU at −78 °C cleanly affords the alkylation product **24** as a single diastereomer (93% yield, *ds* = 150:1).

Finally, treatment of **24** with LiAlH$_4$ in THF at −78 °C leads to reductive removal of the auxiliary with formation of the chiral acetylenic alcohol **1** [6, 7].

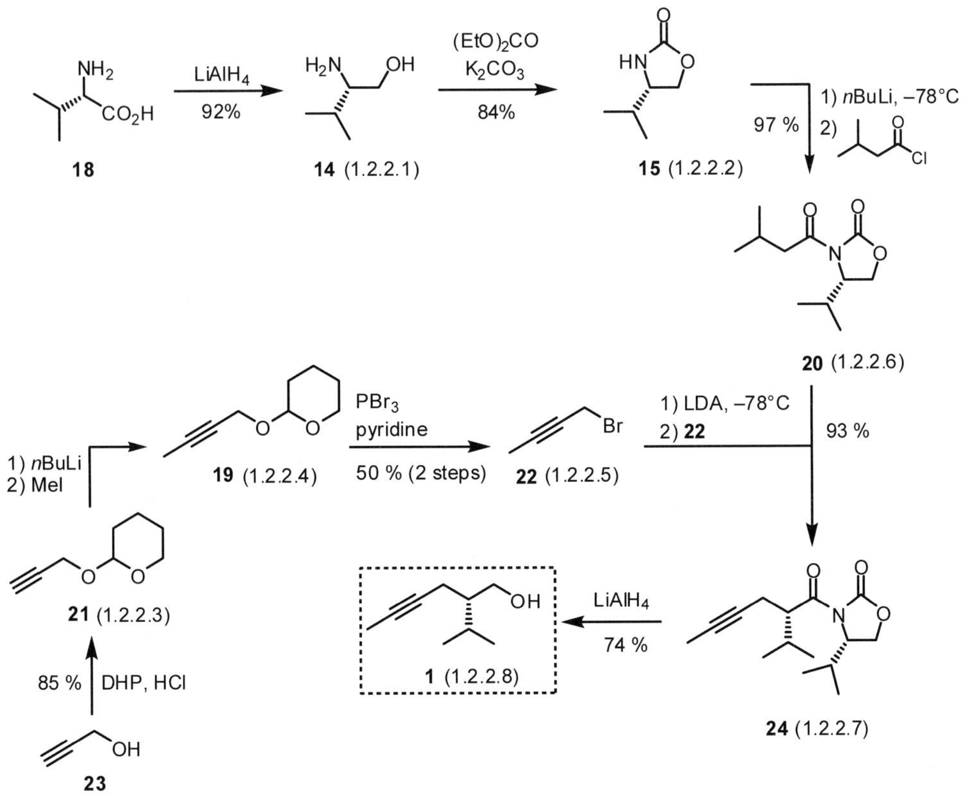

(c) Experimental procedures for the synthesis of 1

1.2.2.1   **   (2S)-2-Amino-3-methylbutan-1-ol (L-valinol) [5]

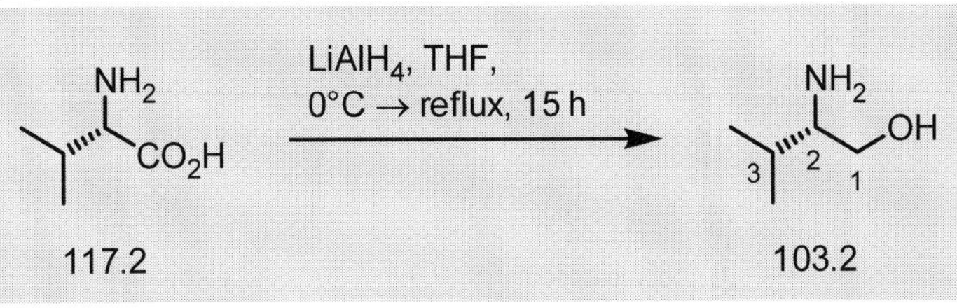

In a flame-dried three-necked round-bottomed flask equipped with a reflux condenser and a mechanical stirrer, LiAlH$_4$ (25.8 g, 0.68 mol) is suspended in anhydrous THF (300 mL) and cooled to 0 °C (nitrogen atmosphere). L-Valine (40.0 g, 0.34 mol) is carefully added in 1 g portions under vigorous stirring. The reaction mixture is heated to reflux for 15 h.

It is then cooled to 0 °C and ice-cold water (40 mL) is carefully added (dropwise at the beginning). The grey-white aluminum salts are filtered off, suspended in a THF/H$_2$O mixture (4:1, 200 mL), stirred for 30 min, and then this mixture is also filtered. The process is repeated once more. The combined filtrates are concentrated *in vacuo*, the residue is dissolved in CHCl$_3$ (200 mL), and the mixture is refluxed in a Dean–Stark apparatus. The solvent is evaporated *in vacuo* and the residue is distilled under reduced pressure to afford *L*-valinol as a colorless solid; 32.2 g (92%), mp 55–56 °C, bp_{16} 85–86 °C, $[\alpha]_D^{20} = +25.7$ ($c = 1$, $CHCl_3$).

---

^{1}H NMR (200 MHz, $CDCl_3$): δ (ppm) = 3.66 (dd, $J = 12.0$, 3.5 Hz, 1 H, 1-H_b), 3.33 (dd, $J = 8.0$, 12.0 Hz, 1 H, 1-H_a), 2.58 (ddd, $J = 8.0$, 6.5, 3.5 Hz, 1 H, 2-H), 2.30 (s_{br}, 3 H, NH_2, OH), 1.01 (dsept, $J = 7.0$, 6.5 Hz, 1 H, 3-H), 0.91 (d, $J = 7.0$ Hz, 6 H, 2 × CH_3).

1.2.2.2   **   (4*S*)-Isopropyloxazolidin-2-one [5]

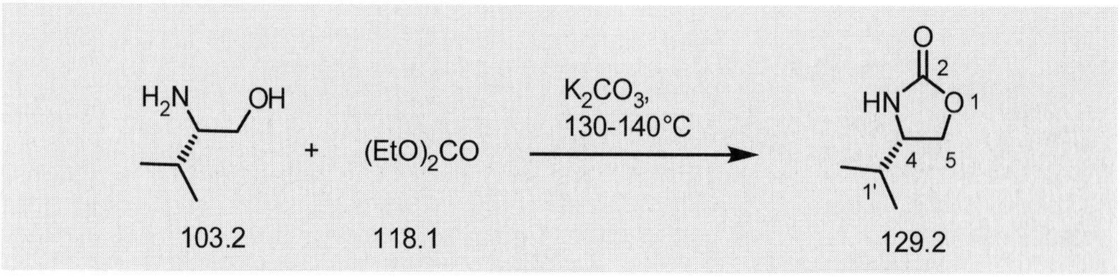

In a micro distillation apparatus equipped with a Vigreux column (30 cm), and an internal thermometer, a mixture of *L*-valinol **1.2.2.1** (31.0 g, 0.30 mol), diethyl carbonate (76.8 g, 0.65 mol), and anhydrous K_2CO_3 (4.13 g, 0.03 mol) is slowly heated to 130–140 °C. EtOH is distilled off in the course of the reaction (internal temperature should not exceed 100 °C, temperature at the top of the Vigreux column should not exceed 85 °C). After EtOH formation has ceased, the mixture is heated for another 30 min.

It is then cooled to room temperature, diluted with CH_2Cl_2 (200 mL), and filtered. The filtrate is washed with saturated $NaHCO_3$ solution (2 × 50 mL), dried over $MgSO_4$, and concentrated under reduced pressure, and the residue is crystallized to afford the oxazolidinone; 32.7 g (84%), mp 74–75 °C, $[\alpha]_D^{20} = -19.2$ ($c = 1.24$, EtOH).

---

^{1}H NMR (200 MHz, $CDCl_3$): δ (ppm)= 7.14 (s_{br}, 1 H, NH), 4.47 (dd, $J = 9.0$, 8.5 Hz, 1 H, 5-H_b), 4.12 (dd, $J = 9.0$, 6.0 Hz, 1 H, 5-H_a), 3.64 (ddd, $J = 8.5$, 6.5, 6.0 Hz, 1 H, 4-H), 1.76 (dsept, $J = 7.0$, 6.5 Hz, 1 H, 1'-H), 0.97 (d, $J = 7.0$ Hz, 3 H, CH_3), 0.90 (d, $J = 7.0$ Hz, 3 H, CH_3).

1.2.2.3   *   2-(2-Propynyloxy)tetrahydropyran [6,7]

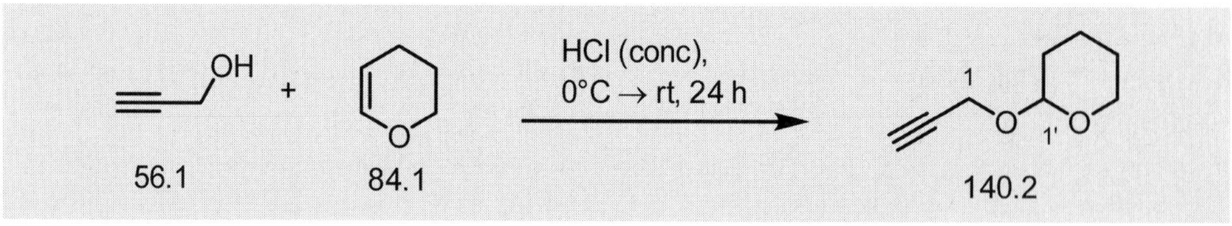

Concentrated HCl (1 µL) is added to a stirred mixture of propargyl alcohol (28.1 g, 500 mmol) and 3,4-dihydro-2H-pyran (DHP) (43.7 g, 520 mmol) at 0 °C. Stirring is continued for 24 h at room temperature.

KOH (900 mg) is then added and the mixture is stirred for another 15 min. Fractional distillation using a Vigreux column affords the tetrahydropyran as a colorless liquid; 59.6 g (85%), bp_{20} 72–80 °C.

---

^{1}H NMR (200 MHz, CDCl$_3$): δ (ppm) = 4.78 (t, J = 3.2 Hz, 1 H, 1'-H), 4.26 (dq, J = 15.2, 2.3 Hz, 1 H, 1-H$_b$), 4.09 (dq, J = 15.2, 2.3 Hz, 1 H, 1-H$_a$), 3.90–3.75 (m, 1 H, 5'-H$_b$), 3.52–3.43 (m, 1 H, 5'-H$_a$), 2.02 (t, J = 2.5 Hz, 1 H, 3-H), 1.88–1.40 (m, 6 H, 2'-H$_2$, 3'-H$_2$, 4'-H$_2$).

1.2.2.4   **   2-(2-Butynyloxy)tetrahydropyran [6, 7]

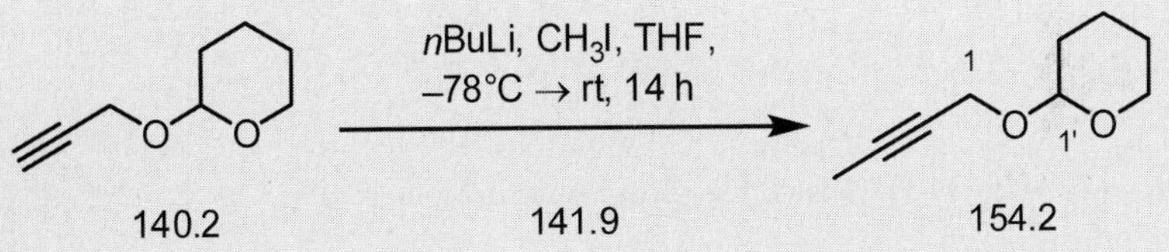

140.2   141.9   154.2

n-Butyllithium in n-hexane (2.5 M, 172 mL, 429 mmol) is added dropwise over 1 h to a stirred solution of the tetrahydropyran **1.2.2.3** (50.0 g, 357 mmol) in THF (600 mL) at –78 °C. Stirring is continued at –78 °C for 5 h, then methyl iodide (152 g, 1.07 mol, 66.9 mL) (Caution: carcinogenic!) is added and the solution is allowed to warm to room temperature over 14 h.

The reaction is quenched by the addition of H$_2$O (20 mL) and the solvent is evaporated under reduced pressure. The brown, oily crude product is dissolved in benzene (150 mL) and this solution is concentrated in vacuo to remove the remaining H$_2$O by azeotropic distillation. The residue is fractionally distilled to afford the product as a colorless oil; 50.4 g (91%), bp_{10} 75–80 °C.

---

^{1}H NMR (200 MHz, CDCl$_3$): δ (ppm) = 4.76 (t, J = 3.2 Hz, 1 H, 1'-H), 4.27 (dq, J = 15.2, 2.3 Hz, 1 H, 1-H$_b$), 4.14 (dq, J = 15.2, 2.3 Hz, 1 H, 1-H$_a$), 3.86–3.73 (m, 1 H, 5'-H$_b$), 3.54–3.42 (m, 1 H, 5'-H$_a$), 1.90–1.42 (m, 6 H, 2'-H$_2$, 3'-H$_2$, 4'-H$_2$), 1.81 (t, J = 2.3 Hz, 3 H, 4-H$_3$).

1.2.2.5   **   1-Bromo-2-butyne [6, 7]

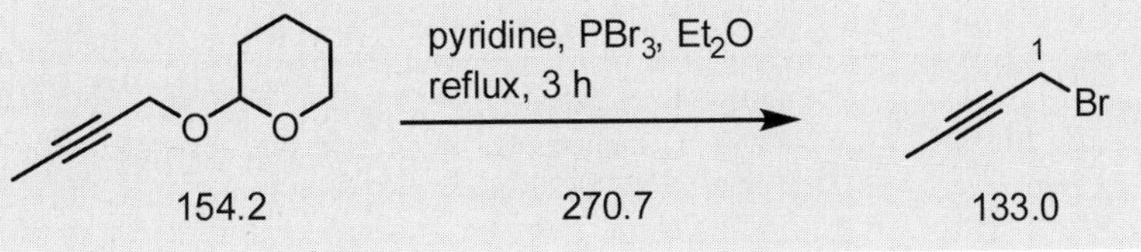

154.2   270.7   133.0

Phosphorus tribromide (37.6 g, 139 mmol, 13.1 mL) is added dropwise to a solution of the tetrahydropyran **1.2.2.4** (42.9 g, 278 mmol) and pyridine (0.2 mL) in Et$_2$O (25 mL) and the mixture is heated under reflux for 3 h.

The reaction is quenched with H$_2$O (50 mL), the organic layer is separated, and the aqueous layer is extracted with Et$_2$O (2 × 150 mL). The combined organic layers are washed with saturated NaHCO$_3$ solution (2 × 150 mL), dried over Na$_2$SO$_4$, and concentrated *in vacuo*. The residue is fractionally distilled with a Vigreux column to afford 1-bromo-2-butyne as a colorless liquid; 21.6 g (58%), bp$_{43}$ 38–43 °C, R_f = 0.72 (*n*-pentane/MeO*t*Bu = 5:1).

---

^{1}H NMR (200 MHz, CDCl$_3$): δ (ppm) = 3.91 (q, J = 2.5 Hz, 2 H, 1-H$_2$), 1.89 (t, J = 2.5 Hz, 3 H, 4-H$_3$).

MS (EI): *m/z* (%) = 135 (60) [M + 2H]$^+$, 133 (60) [M]$^+$.

1.2.2.6   **   (*S*)-4-Isopropyl-3-isovaleroyl-oxazolidin-2-one [6, 7]

n-Butyllithium in *n*-hexane (2.6 M, 78.2 mL, 203 mmol) is added dropwise with stirring to a solution of the oxazolidinone **1.2.2.2** (25.0 g, 194 mmol) in anhydrous THF (800 mL) at –78 °C and stirring is continued for 30 min. Isovaleroyl chloride (25.7 g, 213 mmol, 26.2 mL) is then added dropwise and the reaction mixture is stirred at –78 °C for 20 min and at 0 °C for 30 min.

The reaction is quenched by the addition of K$_2$CO$_3$ solution (1 M, 150 mL) and the solvents are removed *in vacuo*. After the addition of H$_2$O (500 mL), the layers are separated and the aqueous layer is extracted with MeO*t*Bu (4 × 400 mL). The combined organic layers are washed with brine (2 × 100 mL), dried over Na$_2$SO$_4$, and concentrated under reduced pressure. The residue is fractionally distilled to afford the product as a colorless liquid; 41.3 g (100%), bp$_{0.008}$ 70–85 °C, R_f = 0.47 (*n*-pentane/Et$_2$O, 1:1).

---

^{1}H NMR (200 MHz, CDCl$_3$): δ (ppm) = 4.38 (m$_c$, 1 H, 4'-H), 4.26–4.08 (m, 2 H, 5'-H$_2$), 2.88 (dd, J = 15.9, 7.2 Hz, 1 H, 2-H$_b$), 2.64 (dd, J = 15.9, 7.2 Hz, 1 H, 2-H$_a$), 2.32 (dsept, J = 7.3, 4.0 Hz, 1 H, *i*Pr-CH), 2.13 (non, J = 7.2 Hz, 1 H, 3-H), 1.00–0.76 (m, 12 H, 4 × CH$_3$).

1.2.2.7 *** (S,S)-4-Isopropyl-3-(2-isopropyl-hex-4-ynoyl)-oxazolidin-2-one [6, 7]

213.3 + 133.0 → 265.4

LDA, DMPU, THF, –78°C → rt, 14 h

n-Butyllithium in *n*-hexane (2.8 M, 41.5 mL, 116 mmol) is added dropwise to a stirred solution of diisopropylamine (12.8 g, 126 mmol, 17.8 mL) in anhydrous THF (250 mL) at –78 °C. Stirring is continued at 0 °C for 45 min and then the mixture is cooled to –78 °C once more. A solution of the oxazolidinone **1.2.2.6** (22.5 g, 105 mmol) in THF (25 mL) is added dropwise with stirring over 1 h at –78 °C and the resulting mixture is stirred for a further 2 h. DMPU (32 mL) and the freshly prepared bromide **1.2.2.5** (20.0 g, 150 mmol) are then added over 1 h. The reaction mixture is allowed to slowly warm to room temperature and is stirred for a further 14 h.

Saturated aqueous NH_4Cl solution (100 mL) is then added, the organic layer is separated, and the aqueous layer is extracted with Et_2O (3 × 150 mL). The combined organic layers are successively washed with ice-cold HCl (1 M, 100 mL), saturated $NaHCO_3$ solution (100 mL), and brine (100 mL), and dried over Na_2SO_4. After removal of the solvents under reduced pressure, the residue is purified by column chromatography on silica gel (*n*-pentane/MeO*t*Bu, 20:1 → *n*-pentane/MeO*t*Bu, 3:1) to afford the alkylation product as a colorless oil; 26.0 g (93%), *ds* = 150:1, $[\alpha]_D^{20}$ = +63.6 (*c* = 0.5, $CHCl_3$), R_f = 0.41 (*n*-pentane/MeO*t*Bu, 5:1).

IR (NaCl): $\tilde{v}$ (cm^{-1}) = 3376, 2964, 2924, 2876, 1780, 1698, 1468, 1432, 1388.

UV (CH_3CN): λ_{max} (nm) (lg ε) = 206.0 (3.5222).

^{1}H NMR (200 MHz, $CDCl_3$): δ (ppm) = 4.52 (ddd, *J* = 7.6, 3.8, 3.8 Hz, 1 H, 2-H), 4.34–4.16 (m, 2 H, 5'-H$_2$), 3.93 (ddd, *J* = 9.5, 7.0, 5.0 Hz, 1 H, 4'-H), 2.61–2.32 (m, 3 H, 3-H$_2$, 2-H), 1.98 (oct, *J* = 7.0 Hz, 1 H, 4'-H), 1.71 (t, *J* = 2.8 Hz, 3 H, 6-H$_3$), 0.95 (d, *J* = 6.8 Hz, 3 H, *i*Pr-CH$_3$), 0.94 (d, *J* = 7.0 Hz, 6 H, 2 × *i*Pr-CH$_3$), 0.92 (d, *J* = 7.0 Hz, 3 H, *i*Pr-CH$_3$).

^{13}C NMR (50.3 MHz, $CDCl_3$): δ (ppm) = 174.8 (C-1), 153.6 (C-2'), 76.36 (C-4), 76.23 (C-5), 62.89 (C-5'), 58.47 (C-4'), 48.23 (C-2), 29.76 (4'-*i*Pr-CH), 28.32 (2-*i*Pr-CH), 20.55 (2-*i*Pr-CH$_3$), 19.00 (C-3), 18.87 (2-*i*Pr-CH$_3$), 17.75 (4'-*i*Pr-CH$_3$), 14.37 (4'-*i*Pr-CH$_3$), 3.37 (C-6).

MS (DCI, NH_3, 200 eV): *m/z* (%) = 549 (40) $[2M+NH_4]^+$, 283 (100) $[M+NH_4]^+$.

1.2.2.8   **   (S)-2-Isopropylhex-4-yn-1-ol [6, 7]

265.4   140.2

A suspension of LiAlH$_4$ (5.74 g, 151 mmol) in THF (66 mL) is added dropwise to a stirred solution of the oxazolidinone **1.2.2.7** (20.0 g, 75.4 mmol) in anhydrous THF (300 mL) at –78 °C and stirring is continued for 20 h.

The reaction is then quenched by the dropwise addition of H$_2$O (6 mL) and the mixture is allowed to warm to room temperature, whereupon 15% NaOH solution (6 mL) and H$_2$O (20 mL) are added. The precipitate formed is filtered off, washed with THF, and extracted with Et$_2$O using a Soxhlet apparatus for 14 h. The solvent is then evaporated under reduced pressure, the residue is redissolved in benzene, and this solution is concentrated once more under reduced pressure to remove small amounts of H$_2$O by azeotropic distillation. The residue is fractionally distilled with a Vigreux column to afford the alcohol as a colorless liquid; 7.8 g (74%), bp$_{0.5}$ 48–49 °C, $[\alpha]_D^{20} = -3.0$ ($c = 0.5$, CHCl$_3$), $R_f = 0.27$ (*n*-pentane/MeO*t*Bu, 5:1).

IR (NaCl): $\tilde{\nu}$ (cm^{-1}) = 3346, 2960, 2922, 2876, 1388, 1368, 1072, 1040.

^{1}H NMR (200 MHz, CDCl$_3$): δ (ppm) = 3.79–3.59 (m, 2 H, 1-H$_2$), 2.37–2.07 (m, 2 H, 3-H$_2$), 1.89 (s, 1 H, OH), 1.78 (m$_c$, 1 H, *i*Pr-CH), 1.75 (t, J = 2.6 Hz, 3 H, 6-H$_3$), 1.45 (m$_c$, 1 H, 2-H), 0.91 (d, J = 6.5 Hz, 3 H, *i*Pr-CH$_3$), 0.88 (d, J = 6.5 Hz, 3 H, *i*Pr-CH$_3$).

^{13}C NMR (50.3 MHz, CDCl$_3$): δ (ppm) = 77.70 (C-4), 76.79 (C-5), 63.79 (C-1), 46.14 (C-2), 27.78 (*i*Pr-CH), 19.92 (*i*Pr-CH$_3$), 19.73 (*i*Pr-CH$_3$), 18.24 (C-3), 3.40 (C-6).

MS (EI, 70 eV): *m/z* (%) = 140 (2) [*M*]$^+$, 125 (31) [*M* – CH$_3$]$^+$, 97 (100) [*M* – *i*Pr]$^+$, 53 (20) [CH$_3$CCCH$_2$]$^+$.

---

[1]   L. F. Tietze et al., unpublished results.

[2]   An instructive review on α-alkylation of enolates is given in: G. Procter, *Asymmetric Synthesis*, p. 41, Oxford University Press, Oxford, **1999**.

[3]   a) D. A. Evans, *Asymmetric Synthesis* (Ed.: J. D. Morrison), Vol. 3, p. 1, Academic Press, New York, **1984**; b) D. A. Evans et al., *Pure Appl. Chem.* **1981**, *53*, 1109; c) *Houben-Weyl*, 4th ed., Vol. 21a, p. 883, Georg Thieme Verlag, Stuttgart, **1995**.

[4]   D. A. Evans, M. D. Ennis, D. J. Mathre, *J. Am. Chem. Soc.* **1982**, *104*, 1737.

[5]   C. Schneider, *Ph.D. Thesis*, University of Göttingen, **1992**.

[6]   C. Bittner, *Ph.D. Thesis*, University of Göttingen, **2002**.

[7]   S. Hölsken, *Ph.D. Thesis*, University of Göttingen, **2004**.

1.2.3   3-Oxo-5-phenylpentanoic acid methylester

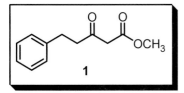

1

Topic: • γ-Alkylation of acetoacetate

(a) General

Among the numerous methods for the synthesis of β-ketoesters [1], the elongation of acetoacetate by γ-alkylation is the most relevant for β-ketoesters of structural type **1**.

In acetoacetate, the α-CH$_2$ group shows far stronger C–H acidity than the γ-CH$_3$ group (ΔpK_a (α vs. γ) ca. 10). As a consequence, attack of electrophiles can be regioselectively directed either to the α-position through formation of the monoanion **2** or to the γ-position through formation of the (ambident) dianion **3**.

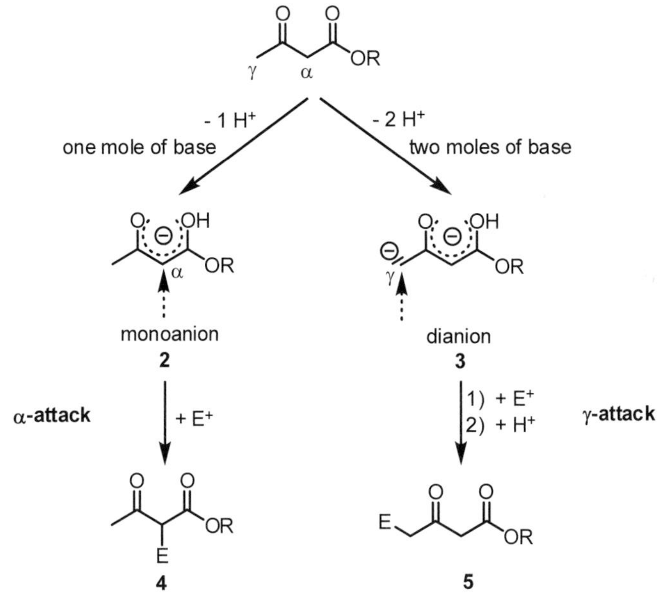

$$O \qquad O$$
$$\gamma \qquad \text{OR}$$
$$\alpha$$

- 1 H$^+$ - 2 H$^+$
one mole of base   two moles of base

monoanion
2

dianion
3

α-attack | + E$^+$

1) + E$^+$
2) + H$^+$ γ-attack

O   O
OR
E
4

O   O
E   OR
5

Accordingly, with an alkyl halide as attacking electrophile, acetoacetate is transformed to the product **4** of α-alkylation using one equivalent of base, whereas with two equivalents of a sufficiently strong base product **5** is obtained, since the γ-CH$_2$ group in the dianion is of higher electron density than the (delocalized) α-CH group [2].

A useful and preparatively versatile alternative for the synthesis of γ-substituted acetoacetates like **1** is the C$_2$-chain elongation of aldehydes with ethyl diazoacetate catalyzed by tin(II) chloride [3]:

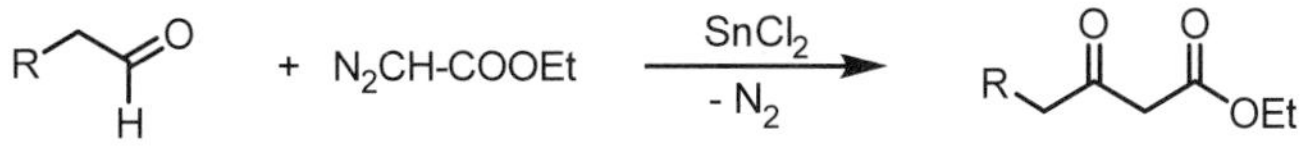

$$R \overset{O}{\underset{H}{\diagup}} \quad + \quad N_2CH\text{-}COOEt \quad \xrightarrow[- N_2]{SnCl_2} \quad R \overset{O \quad O}{\diagdown} OEt$$

(b) Synthesis of 1

If methyl acetoacetate is reacted with benzyl chloride in the presence of NaOCH$_3$ in anhydrous methanol, the "classical" α-alkylation of acetoacetate occurs via formation of the α-monoanion **6** and its nucleophilic attack at the benzyl halide to give methyl 2-benzyl-3-oxobutanoate (**7**) in 80% yield [4]:

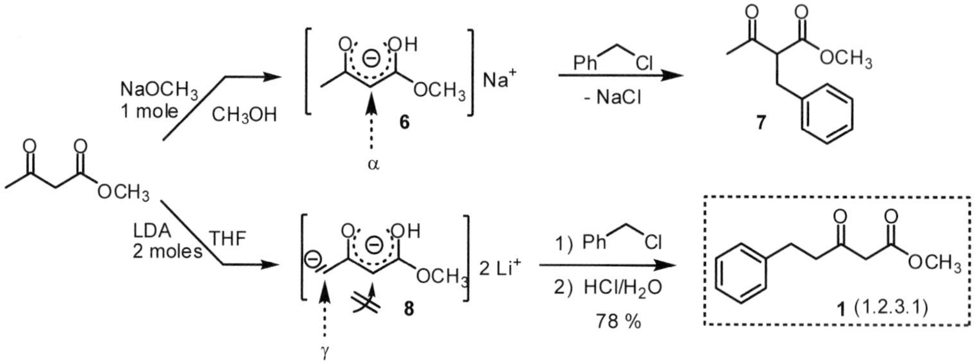

If methyl acetoacetate is reacted with two equivalents of LDA in THF at 0 °C and subsequently with benzyl chloride, **1** is obtained after work-up with aqueous HCl in 78% yield. Initially, the acetoacetate dianion **8** is formed, which undergoes regioselective γ-alkylation with the benzyl halide in an S_N process [5].

(c) Experimental procedure for the synthesis of 1

1.2.3.1   **   3-Oxo-5-phenylpentanoic acid methyl ester [5]

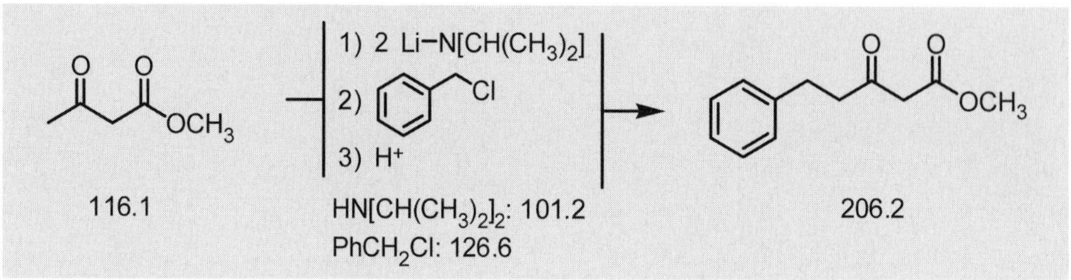

In a 250 mL two-necked flask, fitted with a septum, an inert gas attachment (N$_2$), and a magnetic stirrer, diisopropylamine (5.15 g, 50.0 mmol, ca. 7.13 mL) (note 1) is added by means of a syringe to anhydrous THF (100 mL). n-Butyllithium in n-hexane (1.6 M, 32.5 mL, 52.0 mmol) is slowly added dropwise with stirring at 0 °C. After 20 min, methyl acetoacetate (2.80 g, 24.0 mmol, ca. 2.60 mL) is added dropwise and stirring is continued for 20 min at 0 °C (formation of the dianion). Finally, benzyl chloride (3.04 g, 24.0 mmol, ca. 2.76 mL) is added dropwise and stirring is continued for an additional 20 min at 0 °C.

A mixture of concentrated HCl (10 mL), H_2O (25 mL), and Et_2O (75 mL) is added to the reaction mixture. The organic phase is separated, the aqueous phase is extracted with Et_2O (2 × 50 mL), and the combined organic phases are washed with saturated $NaHCO_3$ and NaCl solutions. After drying ($MgSO_4$), the solvent is evaporated and the residue is distilled *in vacuo* to give the product as a colorless oil; 3.67 g (78%), $bp_{0.2}$ 116–117 °C; n_D^{20} = 1.5293.

IR (film): $\tilde{\nu}$ (cm^{-1}) = 3080, 3060, 3030, 1745, 1715 (C=O), 1600, 1495.

^{1}H NMR (CDCl$_3$): δ (ppm) = 7.22 (s, 5 H, Ar-H), 3.67 (s, 3 H, COOCH$_3$), 3.39 (s, 2 H, 2-CH$_2$), 2.86 (s, 4 H, 4- and 5-CH$_2$) (note 2).

Notes: (1) Diisopropylamine is distilled over CaH_2 before use; bp_{760} 83–84 °C.

(2) In the ^{1}H NMR spectrum, the 4- and 5-CH$_2$ signals happen to coincide and thus appear as a singlet. In addition, small peaks due to the enol form of **1** are observed.

[1]   R. C. Larock, *Comprehensive Organic Transformations*, 2nd ed., p. 1528, Wiley-VCH, New York, **1999**, and literature cited therein.

[2]   M. B. Smith, J. March, *March's Advanced Organic Chemistry*, 5th ed., p. 458, John Wiley & Sons, Inc., New York, **2001**; see also ref. [1], p. 1539.

[3]   C. R. Holmquist, E. J. Roskamp, *J. Org. Chem.* **1989**, *54*, 3258. By application of this method to hydrocinnamic aldehyde, the ethyl ester of **1** is formed in 86% yield; for the mechanism, see ref. [2], p. 1407.

[4]   In analogy to *Organikum*, 21st ed., p. 608, Wiley-VCH, Weinheim, **2001**.

[5]   T. M. Harris, C. M. Harris, *Org. React.* **1969**, *17*, 155; *Tetrahedron* **1977**, *33*, 2159.

1.3   Reactions of the aldol and Mannich type

1.3.1   Olivetol

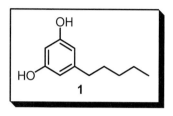

1

Topics:
- Synthesis of a phenolic natural product (intermediate of cannabinoid biosynthesis
- Inter- and intramolecular aldol reaction
- Domino process: Michael addition/intramolecular Claisen condensation
- Ester cleavage and decarboxylation, aromatization of a cyclic 1,3-diketone

(a) General

Olivetol (**1**) and the corresponding carboxylic acid **2** are intermediates in the biosynthesis of cannabinoids. They are likely to be formed from hexanoic acid, acetyl-CoA, and malonyl-CoA via the polyketide **5** [1]. After cyclocondensation, aromatization, and, in the case of **1**, decarboxylation, the resorcinol derivatives **1** and **2** are prenylated by geranyl pyrophosphate to give cannibigerol (**3**) and cannabigerolic acid (**4**). **3** is the biogenetic precursor of tetrahydrocannabinol (**6**), one of the active ingredients of cannabis. More recently, olivetol has been shown to be a building block in lichen constituents [2].

1 (R = H)
2 (R = COOH)

cannabinoids

3 (R = H)
4 (R = COOH)

5

6

Retrosynthesis of olivetol **1** can be performed in two ways. In **A**, oxidation of the pentyl side chain leads to the oxo compound **7**, which can be transformed to α-resorcylic acid **8**, an inexpensive substrate with the correct arrangement of phenolic hydroxyl groups, and a halobutane. A metalorganic species (X = Li or Mg) must be prepared from the halobutane for the reaction with **8**. Syntheses following a concept based on **A** have been reported in the literature [3].

Retrosynthetic analysis according to **B** leads to hexanal, acetone, and malonate via the intermediates **10** and **11**. This approach has been influenced by biosynthetic considerations and might be called a biomimetic synthesis. Thus, in the synthesis of natural products, knowledge of the biosynthesis can be very helpful.

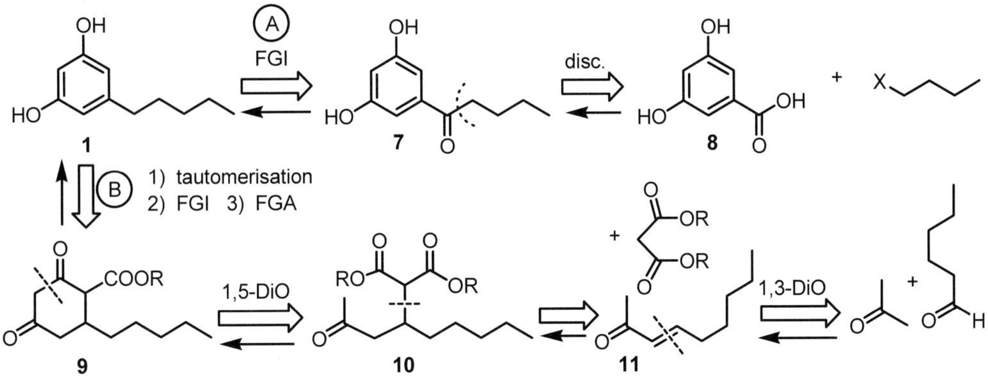

(b) Synthesis of 1

For the synthesis of **1** according to retroanalysis **B**, the base-induced aldol addition of acetone to hexanal (**12**) is used to give the β-hydroxy ketone **13**, acid-catalyzed elimination of water from which yields 3-nonen-2-one (**11**). This α,β-unsaturated ketone is reacted with dimethyl malonate in the presence of NaOCH$_3$ to yield the cyclic β-ketoester **9** in a base-catalyzed domino process consisting of a 1,4-addition of malonate to the enone (→ **10**) followed by intramolecular Claisen condensation to afford the β-ketoester **9**. Reaction of **9** with bromine in DMF yields olivetol (**1**) in a reaction cascade that includes bromination, elimination of HBr, tautomerization, ester cleavage, and decarboxylation:

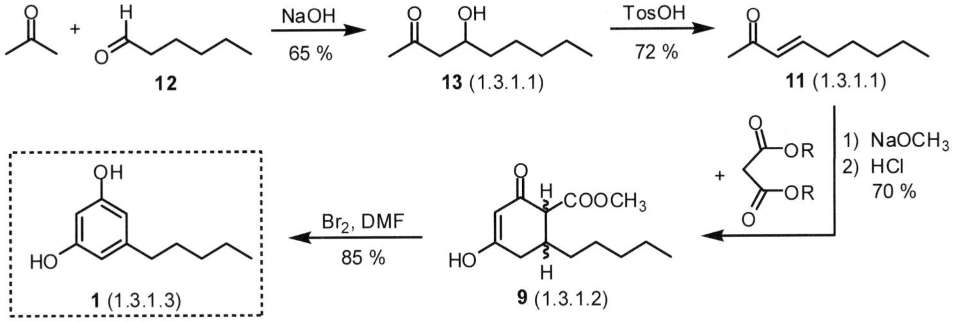

(c) Experimental procedures for the synthesis of 1

1.3.1.1   *   3-Nonen-2-one [5]

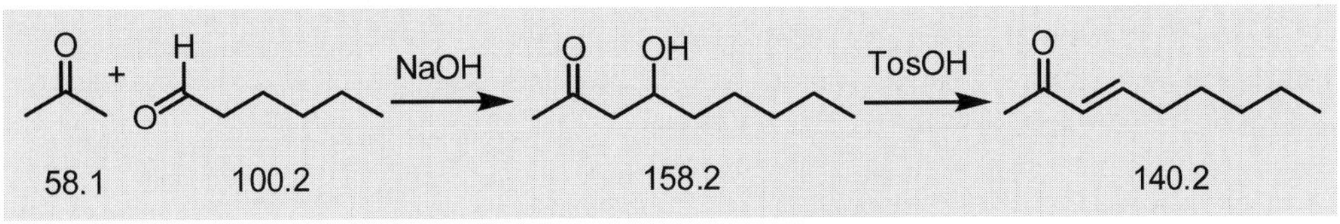

A solution of hexanal (distilled, bp$_{760}$ 131–132 °C; 100 g, 1.00 mol) in acetone (175 mL) is added dropwise to a mixture of acetone (175 mL) and sodium hydroxide (2.5 N, 50 mL) at 10–15 °C over 2.5 h. Stirring is continued at room temperature for 1 h.

The mixture is neutralized (pH 7) with ice-cold HCl (6 N), concentrated to a volume of ca. 150 mL, and extracted three times with Et$_2$O (total 200 mL). The combined extracts are washed with saturated NaHCO$_3$ and NaCl solutions and dried over Na$_2$SO$_4$. Evaporation of the solvent and distillation of the residue *in vacuo* gives 103 g (65%) of the aldol adduct, bp$_{10}$ 108–109 °C.

IR (film): $\tilde{v}$ (cm^{-1}) = 3450 (OH), 1710 (C=O).

A mixture of the aldol adduct (95.0 g, 0.60 mol), *p*-toluenesulfonic acid (300 mg), and anhydrous Na$_2$SO$_4$ (40 g) in benzene (200 mL, Caution!) is heated under reflux for 1 h.

Na$_2$SO$_4$ is removed by filtration and the organic phase is washed with saturated NaHCO$_3$ solution and with brine, and dried over Na$_2$SO$_4$. The benzene is evaporated and the residue is distilled over a small Vigreux column; yield 60.6 g (72%), bp$_{10}$ 88–89 °C.

IR (film): $\tilde{v}$ (cm^{-1}) = 2970, 2940, 2880, 2870 (CH), 1685 (C=O), 1630 (C=C), 1360.

^{1}H NMR (CDCl$_3$): δ (ppm) = 6.83 (dt, *J* = 16.0, 7.0 Hz, 1 H, HC=C–C=O), 6.06 (dt, *J* = 16.0, 1.5 Hz, 1 H, C=CH–C=O), 2.20 (s, 3 H, CO–CH$_3$), 2.5–2.0 (m, 2 H, CH$_2$–C=C), 1.7–1.1 (m, 6 H, CH$_2$), 0.90 (t, *J* = 5.0 Hz, 3 H, CH$_3$).

1.3.1.2 * Methyl 2-hydroxy-4-oxo-6-pentylcyclohex-2-ene-1-carboxylate [4]

140.2   132.1   240.3

Sodium (Caution! 8.10 g, 0.35 mol) is added in portions to anhydrous MeOH (200 mL, hood! H$_2$ evolution!). After the sodium has completely dissolved, dimethyl malonate (52.8 g, 0.40 mol) is added with stirring and the mixture is heated to 60 °C. 3-Nonen-2-one **1.3.1.1** (42.1 g, 0.30 mol) is added dropwise over 30 min and the solution is heated under reflux for 3 h.

The solvent is then evaporated *in vacuo* and the residue is dissolved in H$_2$O (250 mL). The aqueous solution is washed with CHCl$_3$ (3 × 50 mL, discard) and then acidified with concentrated HCl to pH 3–4. The precipitated oily solid is dissolved in CHCl$_3$ (200 mL), and the aqueous phase is extracted with CHCl$_3$ (3 × 50 mL). The combined organic phases are washed with brine, dried over MgSO$_4$, and concentrated. The residue is recrystallized from a mixture of *n*-hexane/diisopropyl ether/isopropanol, 25:13:2 (150 mL), to give colorless needles; 50.5 g (70%), mp 98–100 °C.

IR (KBr): $\tilde{v}$ (cm^{-1}) = 3300–2200 (OH), 1740, 1605 (C=O), 1510, 1440, 1415, 1360.

^{1}H NMR (CDCl$_3$): δ (ppm) = 9.26 (s, 1 H, OH), 5.50 (s, 1 H, C=CH), 3.83 (s, 1.5 H, OCH$_3$), 3.75 (s, 1.5 H, OCH$_3$), 3.7–2.8 (m, 1 H, CH–CO$_2$Me), 2.8–2.3 (m, 2 H, CH$_2$–C=O), 1.5–1.1 (m, 9 H, CH + CH$_2$), 0.90 (t, *J* = 6 Hz, 3 H, CH$_3$).

1.3.1.3   *   1,3-Dihydroxy-5-pentylbenzene (olivetol) [4]

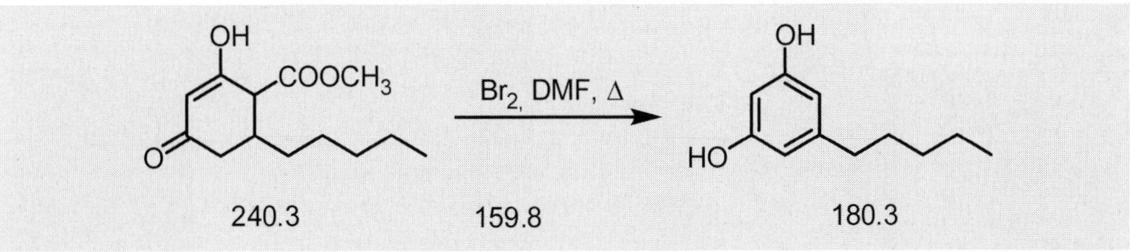

240.3   159.8   180.3

Bromine (Caution! Hood! 7.50 g, 47.0 mmol, ~2.40 mL) is added dropwise to a stirred solution of the methyl ester **1.3.1.2** (12.0 g, 50.0 mmol) in anhydrous dimethylformamide (30 mL) at 0 °C over a period of 90 min. The solution is slowly heated to 160 °C and is stirred for 10 h at the same temperature. Evolution of carbon dioxide starts at ca. 100 °C.

The solution is cooled and concentrated *in vacuo* (ca. 100 °C/10 mbar). The residue is dissolved in H_2O (30 mL) and extracted three times with Et_2O (total 105 mL). The combined ethereal extracts are washed sequentially with H_2O (30 mL), 10% aqueous $NaHCO_3$ (2 × 20 mL), HOAc (2 M, 2 × 20 mL), and twice with brine, dried over Na_2SO_4, and concentrated. The residue is distilled in a Kugelrohr to give a light-pink viscous oil that solidifies at 4 °C. It may be recrystallized from Et_2O; yield 7.22 g (85%), $bp_{0.01}$ 150 °C (oven temperature), mp 85–86 °C.

IR (film): $\tilde{v}$ (cm^{-1}) = 3600–2000 (OH), 1610 (C=C, arom.), 1590.

^{1}H NMR (CDCl₃): δ (ppm) = 6.85 (s, 2 H, OH), 6.25 (s_{br}, 3 H, arom. H), 2.33 (t, J = 7 Hz, 2 H, Ph–CH_2), 1.5–1.0 (m, 6 H, CH_2), 0.83 (t, J = 6 Hz, 3 H, CH_3).

---

[1]   P. Nuhn, *Naturstoffchemie*, 3rd ed., p. 61, S. Hirzel Verlag, Stuttgart, **1997**.

[2]   A. G. Gonzales et al., *Z. Naturforsch.* **1991**, *46c*, 12.

[3]   a) S. A. Barker, R. L. Settine, *Org. Prep. Proc. Int.* **1979**, *11*, 87; b) A. A. Durrani, J. H. P. Tyman, *J. Chem. Soc. Perkin Trans. I* **1980**, 1658.

[4]   A. Focella, S. Teitel, A. Brossi, *J. Org. Chem.* **1977**, *42*, 3456.

[5]   J. G. Tishchenko, L. S. Stanishevskii, *J. Gen. Chem. USSR* **1963**, *33*, 134.

1.3.2   (+)-(7a*S*)-7,7a-Dihydro-7a-methyl-1,5(6*H*)-indanedione

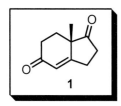

1

Topics: • Acylation of succinic acid

• Michael addition

• Asymmetric intramolecular aldol condensation with a chiral catalyst, enantioselective organocatalysis, Eder–Sauer–Wiechert–Hajos–Parrish reaction

• Asymmetric Robinson annulation

(a) General

The hydrindene **1** is an important building block in numerous syntheses of steroids and of other natural products [1]. Its preparation is one of the first outstanding examples of the importance of enantioselective organocatalysis, which has gained general acceptance in recent years.

The retrosynthesis of **1** follows a retro-Robinson annulation, which consists of a retro-aldol reaction and a retro-Michael addition to give 2-methyl-cyclopentane-1,3-dione (**4**) and methyl vinyl ketone (**3**) as starting materials.

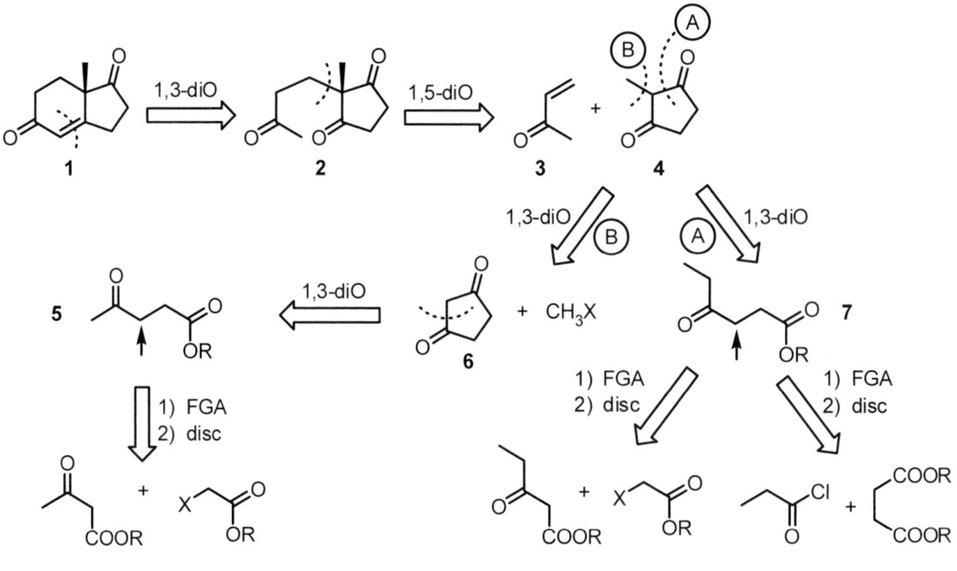

The dione **4** can either be obtained from acetoacetate and haloacetate via a γ-ketoester and cyclopenta-1,3-dione (**6**) according to pathway **B** or from the γ-ketoester **7** according to pathway **A**. For the preparation of **7**, again two different approaches could be used. According to the proposed retrosynthetic analysis, the synthesis of **1** has been achieved in an enantioselective way by 1,4-addition of 2-methylcyclopentane-1,3-dione (**4**) to methyl vinyl ketone to give the Michael adduct **2** and subsequent asymmetric intramolecular aldol condensation of **2** in the presence of (*S*)-proline as organocatalyst to give **1** in high chemical yield and excellent enantiopurity [2].

(b) Synthesis of 1

The procedure presented here was developed by Eder, Sauer, and Wiechert [3], as well as by Hajos and Parrish [4]. Cyclization of the Michael adduct **2** initially provides the *cis*-aldol adduct **8** as a single diastereomer in 88% yield and with 84% *ee*; subsequently, **8** is subjected to acid-catalyzed H_2O-elimination with TosOH in benzene to give the desired product **1** in 81% yield:

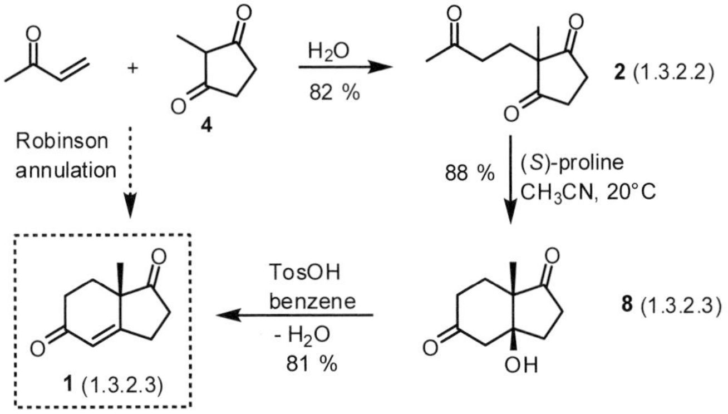

The role of proline as chiral organocatalyst can be interpreted in terms of a mechanism [5] based on initial enamine formation between (S)-proline and the carbonyl group in the side-chain of **2**. Subsequent ring-closure of the enamine **9** by addition to one of the remaining C=O groups leads to the iminium carboxylate betaine **11**. A transition state **10** with hydrogen-bond differentiation between the two diastereotopic C=O groups may account for the high stereoselectivity of the ring-closure reaction (**9 → 11**). The catalytic cycle is terminated by hydrolysis of **11** to yield the aldol adduct **8** with regeneration of the catalyst:

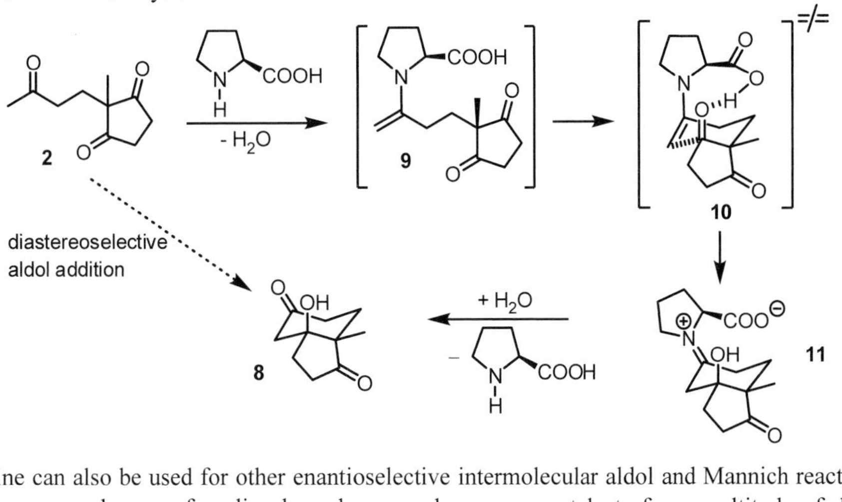

(S)-Proline can also be used for other enantioselective intermolecular aldol and Mannich reactions [5, 6]. Moreover, analogues of proline have been used as organocatalysts for a multitude of different reactions [7].

For the synthesis of 2-methyl-cyclopentane-1,3-dione (**4**) an efficient one-step procedure [8] is used, which consists of the acylation of succinic acid with propionyl chloride in the presence of $AlCl_3$ according to the retrosynthetic pathway **A** [9].

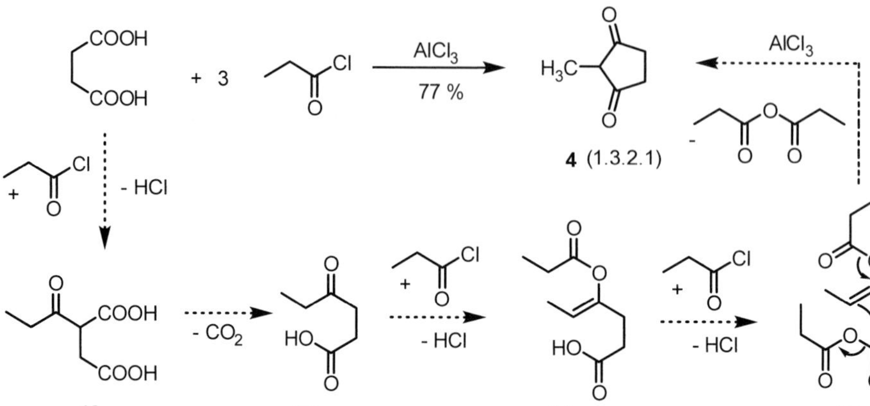

4 (1.3.2.1)

12   **13**   **14**   **15**

Since 3 equivalents of the acid chloride are required, a domino process is likely to occur, which involves α-acylation of succinic acid (→ **12**), decarboxylation of the β-keto acid **12** (→ **13**), and acylation of its enol (→ **14**); finally, activation of the remaining carboxyl function by formation of a mixed anhydride (or chloride) (→ **15**) and Claisen-like cyclization of the acylenol functionality in **15** lead to the dione **4**.

(c)   Experimental procedures for the synthesis of 1

1.3.2.1   *   2-Methylcyclopentane-1,3-dione [8]

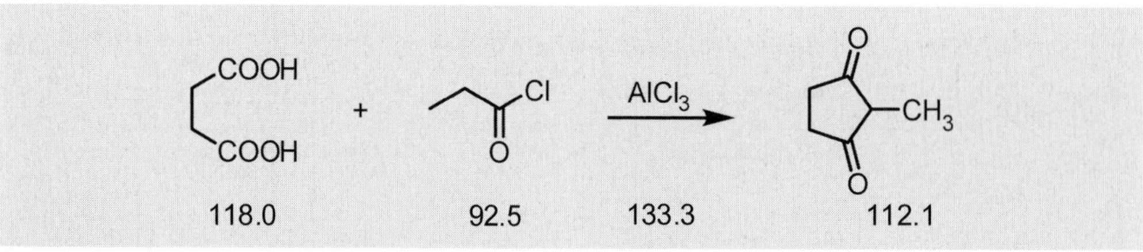

118.0   92.5   133.3   112.1

Finely powdered succinic acid (5.90 g, 0.50 mol) is added in small portions to a solution of anhydrous aluminum chloride (200 g, 1.50 mol) in anhydrous nitromethane (200 mL), causing vigorous gas evolution (HCl! Hood!). When HCl evolution has ceased, propionyl chloride (139 g, 1.50 mol) is added and the mixture is heated to 80 °C for 3 h. A red solution results.

The solution is cooled and poured onto ice (400 g). The mixture is maintained at –10 °C for 15 h, allowing the product to crystallize. The solid is collected by filtration, washed with 10% NaCl solution and toluene (each 200 mL), and recrystallized from H_2O after treatment with activated charcoal to give colorless prisms; 43.0 g (77%), mp 214–216 °C.

IR (KBr): $\tilde{\nu}$ (cm^{-1}) = 3200–2600 (assoc. OH), 1590.

^{1}H NMR ([D$_4$]MeOH): δ (ppm) = 4.84 (s, OH), 2.44 [s, CH$_2$–CH$_2$; keto form], 2.9–2.25 (m, CH$_2$–CH$_2$; enol form], 1.54 (s, CH$_3$); keto-enol tautomeric mixture.

1.3.2.2 * 2-Methyl-2-(3-oxobutyl)-cyclopentane-1,3-dione [3, 4]

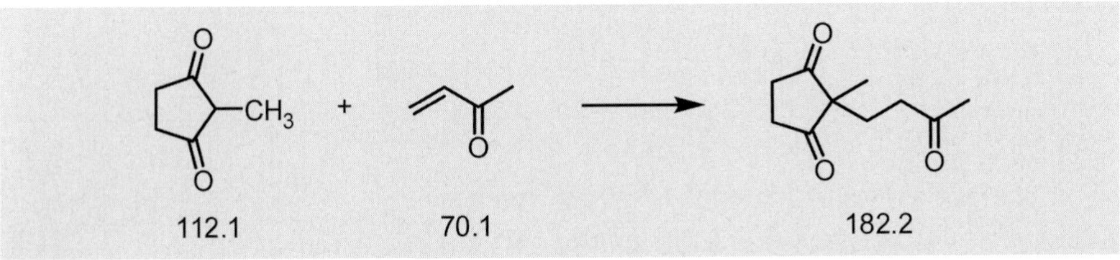

112.1   70.1   182.2

Methyl vinyl ketone (14.0 g, 200 mmol, ~16.2 mL) is added in one portion to a suspension of 2-methyl-1,3-cyclopentanedione **1.3.2.1** (11.2 g, 100 mmol) in H_2O (25 mL) and the mixture is stirred for 5 d at room temperature under a nitrogen atmosphere.

The clear, red-brown solution is then extracted with toluene (3 × 25 mL). The combined extracts are dried over $MgSO_4$ and stirred for 2 h at room temperature with activated charcoal. The charcoal is removed by filtration and washed with hot toluene (50 mL). The combined filtrates are concentrated and the residue is fractionally distilled *in vacuo* to give a colorless oil; 15.0 g (82%), $bp_{0.1}$ 108–110 °C.

IR (film): $\tilde{v}$ (cm^{-1}) = 2970, 2930, 2875 (CH), 1765, 1720 (C=O), 1450, 1420, 1370, 1170.

^{1}H NMR (CDCl$_3$): δ (ppm) = 2.79 (s, 4 H, cyclopentane-H), 2.60–1.65 (m, 4 H, CO–CH$_2$–CH$_2$), 2.09 (s, 3 H, CH$_3$–CO), 1.10 (s, 3 H, CH$_3$).

1.3.2.3 * (+)-(7aS)-7,7a-Dihydro-7a-methyl-1,5(6H)-indanedione [3, 4]

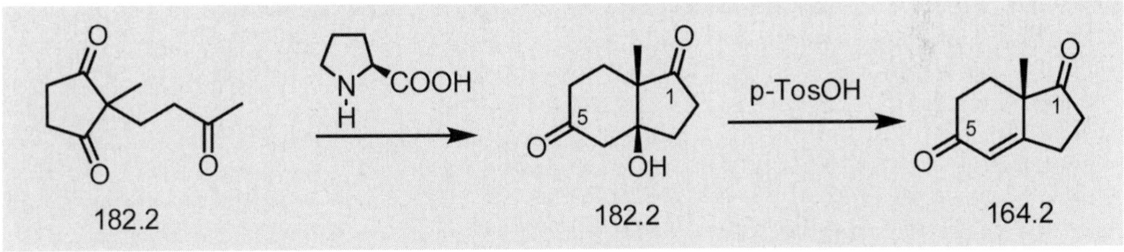

182.2   182.2   164.2

(1) (+)-(3aS,7aS)-3a,4,7,7a-Tetrahydro-3a-hydroxy-7a-methyl-1,5(6H)-indanedione

A solution of the triketone **1.3.2.2** (5.60 g, 30.7 mmol) and (–)-(S)-proline (3.54 g, 30.7 mmol) in acetonitrile (40 mL) is stirred at room temperature for 6 d under nitrogen (balloon). The initially light-yellow solution becomes dark brown to black.

Proline is collected by filtration and washed with a small amount of acetonitrile. The filtrate is concentrated *in vacuo*, the dark-brown residue is dissolved in EtOAc (100 mL), and this solution is filtered through silica gel (10 g). The silica gel is rinsed with additional EtOAc (150 mL) and the combined filtrates are concentrated *in vacuo*. A light-brown residue is obtained, which solidifies after 14 h at –20 °C. Recrystallization from Et$_2$O gives light-yellow crystals. The yield is 4.90 g (88 %), mp 119–120 °C, $[\alpha]_D^{20}$ = +60 (c = 0.5, CHCl$_3$).

> **IR** (film): $\tilde{\nu}$ (cm^{-1}) = 3470 (OH), 1740 (5-ring C=O), 1710 (6-ring C=O), 1305, 1270, 1065.
>
> **^{1}H NMR** (CDCl$_3$): δ (ppm) = 2.84 (s, 1 H, OH), 2.63 (s, 2 H, CO–C$\underline{H}_2$–CHOH), 2.6–1.65 (m, 8 H, CH$_2$), 1.21 (s, 3 H, CH$_3$).

(2) (+)-(7aS)-7,7a-Dihydro-7a-methyl-1,5(6H)-indanedione

A mixture of the hydroxy ketone prepared in step (1) (3.64 g, 20.0 mmol), anhydrous p-toluene-sulfonic acid (25 mg, 0.15 mmol), and molecular sieves (4 Å, 5 g) in anhydrous benzene (30 mL, Caution!) is heated under reflux for 30 min.

The mixture is then cooled, NaHCO$_3$ solution (1 M, 2 mL) is added, and the phases are separated. The organic phase is dried over MgSO$_4$ and concentrated *in vacuo*. The residue is a yellow oil, which solidifies in 14 h at –20 °C. The product is washed with ice-cold Et$_2$O and recrystallized from Et$_2$O/ n-pentane; 2.66 g (81%), mp 64–65 °C, $[\alpha]_D^{20}$ = +362 (c = 0.1, benzene). The compound is almost enantiopure with > 98% ee.

> **IR** (KBr): $\tilde{\nu}$ (cm^{-1}) = 3045 (CH, olef.), 1745 (C=O), 1660 (C=C), 1455, 1355, 1235, 1150, 1065.
>
> **^{1}H NMR** (CDCl$_3$): δ (ppm) = 5.95 (m, 1 H, C=CH), 2.95–1.75 (m, 8 H, CH$_2$), 1.30 (s, 3 H, CH$_3$).

[1] For example, taxol: S. J. Danishefsky, J. J. Masters, W. B. Young, J. T. Link, L. B. Snyder, T. V. Magee, D. K. Jung, R. C. A. Isaacs, W. G. Bornmann, C. A. Alaimo, C. A. Coburn, M. J. Di Grandi, *J. Am. Chem. Soc.* **1996**, *118*, 2843.

[2] R. A. Micheli et al., *J. Org. Chem.* **1975**, *40*, 675.

[3] U. Eder, G. Sauer, R. Wiechert, *Angew. Chem.* **1971**, *83*, 492; *Angew. Chem. Int. Ed. Engl.* **1971**, *10*, 496.

[4] Z. G. Hajos, D. R Parrish, *J. Org. Chem.* **1974**, *39*, 1612.

[5] a) L. Hoang, S. Bahmanyar, K. N. Houk, B. List, *J. Am. Chem. Soc.* **2003**, *125*, 16; b) F. R. Clemente, K. N. Houk, *Angew. Chem.* **2004**, *116*, 5890; *Angew. Chem. Int. Ed.* **2004**, *43*, 5766.

[6] a) P. J. Dalko, L. Moisan, *Angew. Chem.* **2004**, *116*, 5248; *Angew. Chem. Int. Ed.* **2004**, *43*, 5138; b) J. Seayad, B. List, *Org. Biomol. Chem.* **2005**, *3*, 719. c) Amino acids as organocatalysts in carbohydrate synthesis: U. Kazmaier, *Angew. Chem.* **2005**, *117*, 2224; *Angew. Chem. Int. Ed.* **2005**, *44*, 2186; d) organo-catalyzed direct aldol reactions of aldehydes with ketones: L.-Z. Gong et al., *J. Am. Chem. Soc.* **2005**, *127*, 9285; e) A. Cordova et al., *Chem. Commun.* **2005**, 3586.

[7] a) B. Westermann, *Nachrichten aus der Chemie* **2003**, *51*, 802; b) A. Berkessel, H. Gröger, *Asymmetric Organocatalysis*, Wiley-VCH, Weinheim, **2005**.

[8] H. Schick, G. Lehmann, G. Hilgetag, *Chem. Ber.* **1969**, *102*, 3238.

[9] For comparison, see the multistep synthesis of **1** in *Org. Synth.*: J. P. John, S. Swaminathan, P. S.Venkataraman, *Org. Synth. Coll. Vol. IV*, **1963**, 840.

1.3.3   Cyclohexyl 2-benzoylamino-2-(2'-oxocyclohexyl) acetate

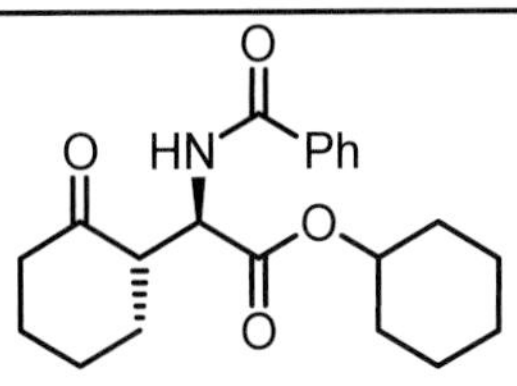

1

Topics: • Diastereoselective aminoalkylation of an enamine by an N-acyl imino ester (modified aldol reaction)

• Esterification of an N-acyl-α-amino acid

• Formation of an N-acyl imino ester by α-halogenation/dehydrohalogenation

(a) General

In the directed aldol reaction [1], equivalents of enolates **2**, e.g. α-lithiated imines **3** (Wittig aldol reaction), silyl enol ethers **4** (Mukaiyama aldol reaction, cf. **1.3.4**), or enamines **5**, react with aldehydes or ketones to give the product **6** by an aldol addition and/or **7** by an aldol condensation:

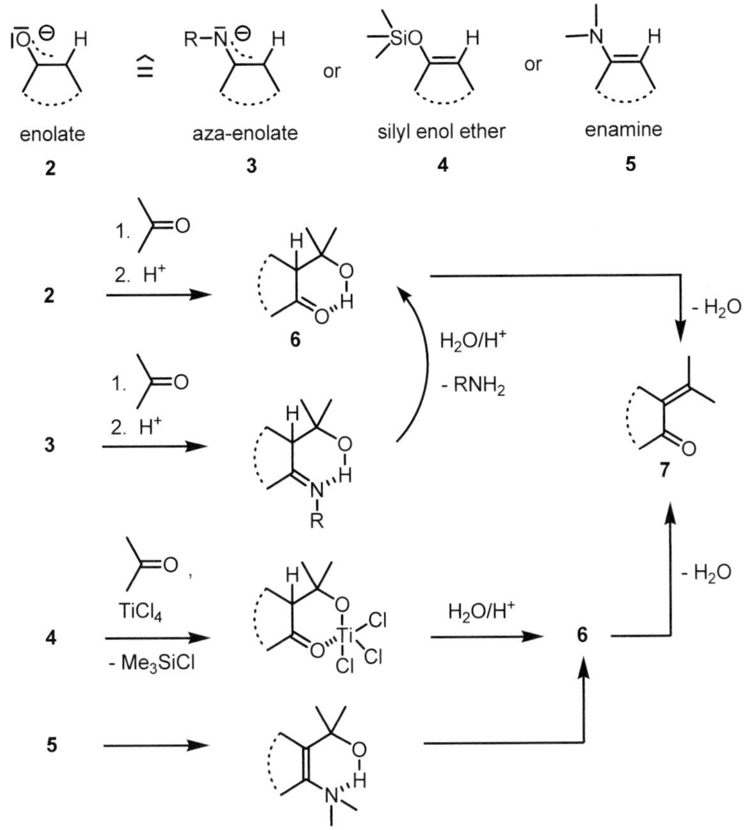

enolate   aza-enolate   silyl enol ether   enamine

2   **3**   **4**   **5**

Enamines of cycloalkanones are easily accessible and can undergo aldol condensation with aldehydes under equilibrium conditions with azeotropic removal of H₂O and subsequent acid hydrolysis [2]:

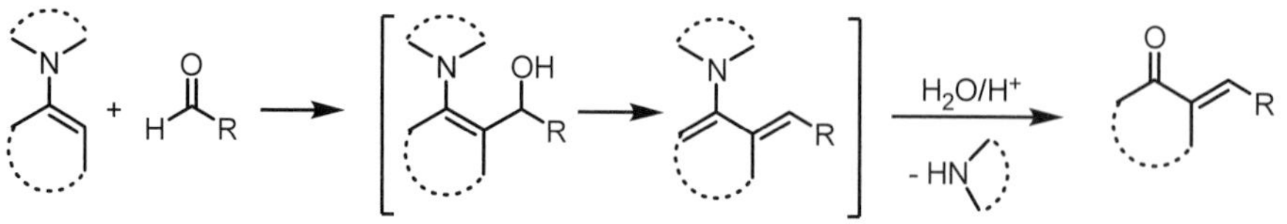

When enamines of this type are reacted with acyl iminoacetates **8** as electrophilic substrates, an aza-analogous aldol addition takes place to give *N*-acyl-γ-keto-α-aminoesters **9** [3], as exemplified in section (b).

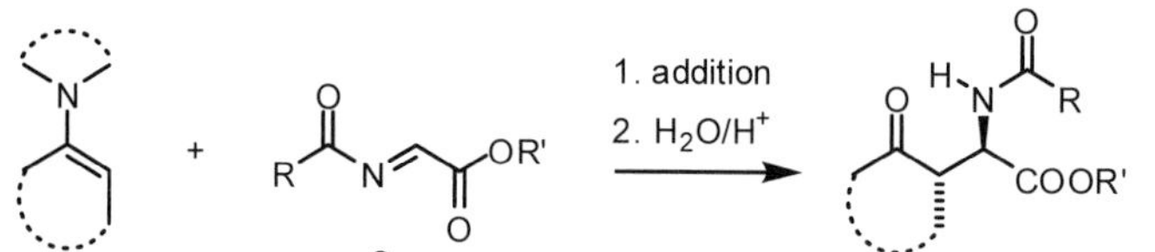

The relative configuration of the products **9** is *anti* (X-ray). The high diastereoselectivity (>96% *de*) in this aza-modified aldol process is consistent with a hetero-Diels–Alder-like transition state **10** for the formation of an intermediate **11**, which may undergo ring-opening either to the zwitterion **12** or the enamine **13**. After acid hydrolysis, the *anti* product **9** is obtained [3]:

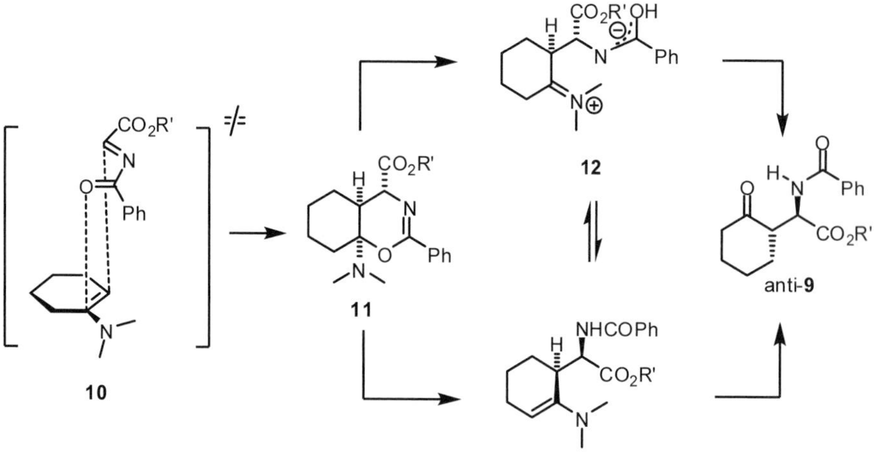

For the formation of enantioenriched products, chiral esters and chiral enamines can be used. Following the concept of double stereodifferentiation [4], the (+) and (–)-menthyl esters of **8** (R = Ph) are reacted with the chiral enamine **14** derived from (*S*)-proline. Using the (+)-menthyl ester **15**, reaction proceeds in quantitative yield and with complete diastereo- and enantioselectivity (*de* = *ee* >99%) and gives the pure compound **16** with (1'*S*,2*R*)-configuration at the newly formed stereogenic centers ("matched" case), while the (–)-menthyl ester (**8**, R = Ph) leads to a product of type **9** with *de* >98% and *ee* = 45% ("mismatched" case [5]).

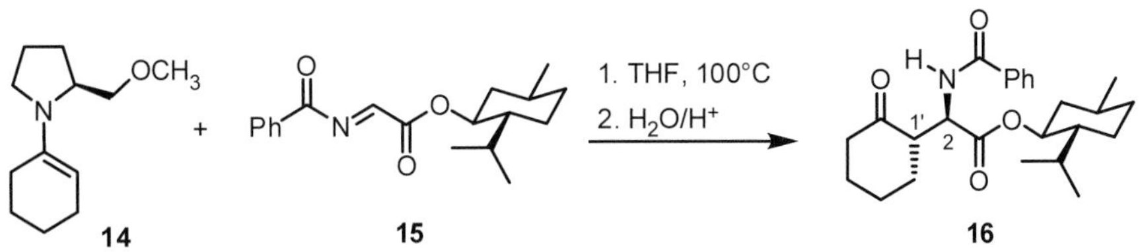

It should be noted that acyl iminomalonates **17** represent interesting electrophilic building blocks and can be used for the synthesis of α-amino acids [6] by reaction with Grignard compounds followed by hydrolysis and decarboxylation:

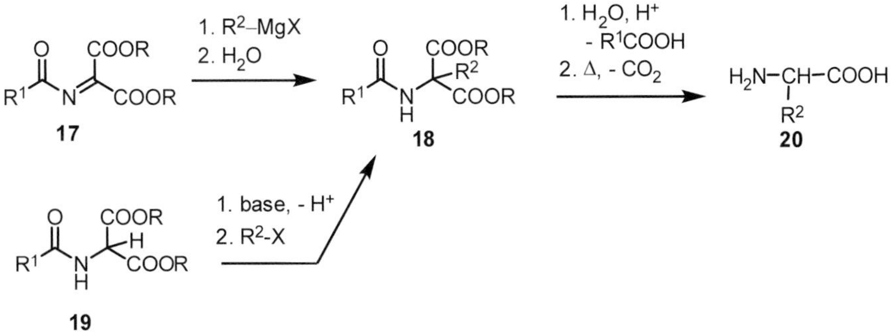

This mode of formation of α-amino acids is an alternative to a method [7] in which acyl aminomalonates **19** are alkylated in the presence of a base to give **18**. Acyl iminomalonates **17** represent the "umpoled" version [8] of the acyl amidomalonate anion **21**:

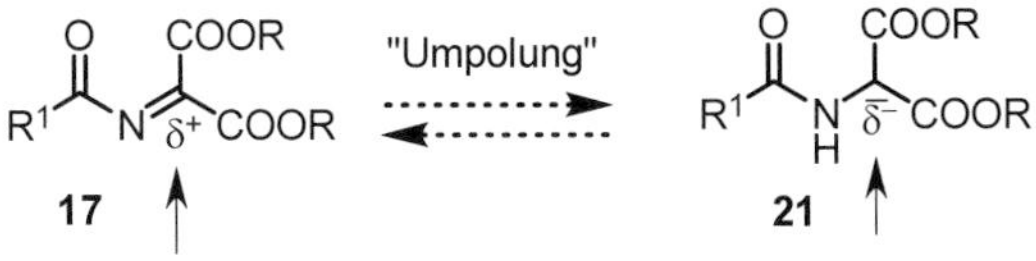

(b) Synthesis of 1

Commercially available hippuric acid (**22**) is subjected to azeotropic esterification with cyclohexanol in the presence of TosOH in toluene. Photobromination of the cyclohexyl ester **23** with Br_2 in CCl_4 occurs at the α-position to the COOR group and affords the bromo ester **24**. In the concluding steps, the bromo ester **24** is transformed into the benzoyl iminoacetate **25**, which, without isolation, leads to **1** by reaction with the enamine **26** followed by hydrolysis. For this reaction, a solution of the bromo ester **24** in THF is treated at −78 °C firstly with triethylamine and then with the enamine morpholinocyclohexene (**26**) [9]. The reaction presumably proceeds via the intermediates **27** and **28**. After hydrolysis of the reaction mixture at pH 4–5, **1** is isolated in 79% yield and with *ds* >98:2, thus documenting the high level of stereoselectivity of the aza-modified aldol process. In the described process, **1** is obtained as a racemic mixture since neither a chiral enamine nor a chiral ester is used.

Interestingly, the diastereoselectivity seen with the corresponding methyl or ethyl esters is significantly more temperature-dependent compared to that with the larger cyclohexyl ester **23** used here. The methyl and ethyl esters give high diastereoselectivity in the reaction with the enamine **26** only at −100 °C (*ds* >98:2), whereas at −78 °C a *ds* of only around 85:15 is achieved [3].

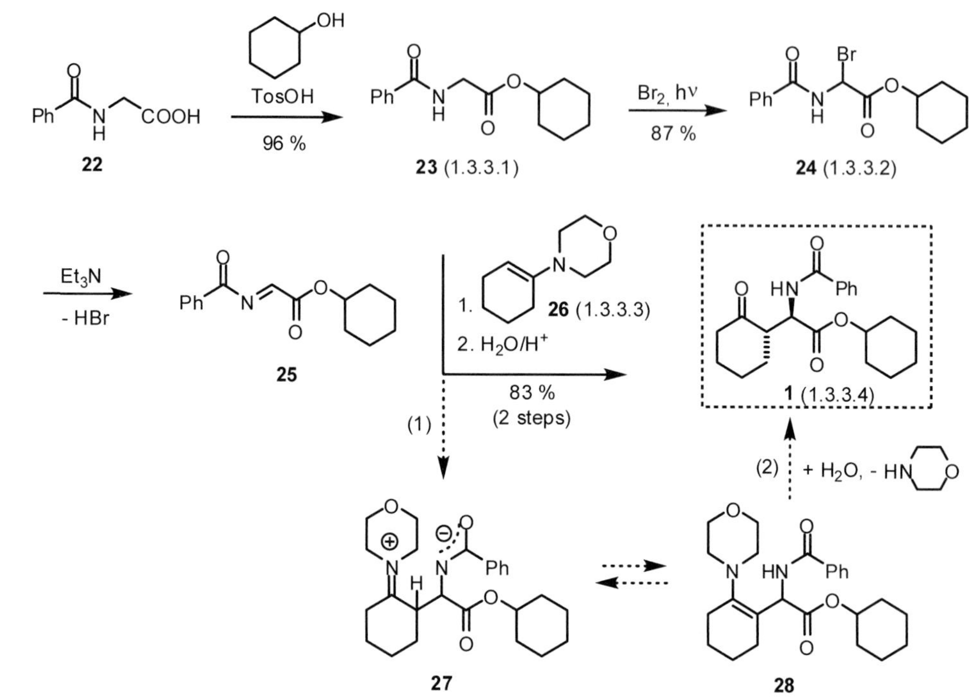

22 → TosOH 96 % → **23** (1.3.3.1) → Br₂, hv 87 % → **24** (1.3.3.2)

Et₃N
– HBr
→ **25**

1. **26** (1.3.3.3)
2. H₂O/H⁺

83 %
(2 steps)
(1)

1 (1.3.3.4)

(2) + H₂O, – HN(morpholine)

27   **28**

The target molecule **1** is obtained in a three-step sequence in an overall yield of 67% (based on hippuric acid (**22**)).

(c) Experimental procedures for the synthesis of 1

1.3.3.1 * Cyclohexyl 2-benzoylaminoacetate [3]

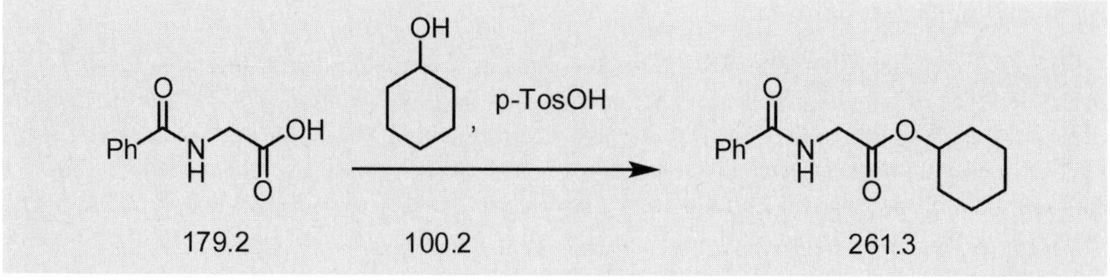

179.2   100.2   261.3

Hippuric acid (35.8 g, 0.20 mol) and cyclohexanol (20.0 g, 0.20 mol) are heated under reflux with *p*-toluenesulfonic acid (1.0 g) in toluene (200 mL) under azeotropic removal of H_2O in a Dean–Stark trap, until the theoretical amount of H_2O is formed.

After cooling to 35–40 °C and diluting with additional EtOAc (200 mL), the organic layer is washed twice with H_2O and dried over $MgSO_4$. The solvent is removed under reduced pressure and the crude product is recrystallized from EtOAc/*n*-hexane (1:1) to give a colorless solid; 50.3 g (96%); mp 102–103 °C; TLC (SiO_2, EtOAc/*n*-hexane, 1:2): $R_f = 0.57$.

UV: λ_{max} (nm) = 224, 194.

IR (KBr): $\tilde{v}$ (cm^{-1}) = 3326, 2955, 2939, 2854, 1748, 1650, 1550, 1494, 1450, 1401, 1380, 1360, 1312, 1251, 1201, 1081, 1013, 949, 733, 692.

^{1}H NMR (300 MHz, CDCl$_3$): δ (ppm) = 7.78–7.38 (m, 5 H, arom-H), 6.70 (s, 1 H, NH), 4.84 (m, 1 H, hex-H$_1$), 4.20 (d, J = 3.3 Hz, 2 H, α-H$_2$), 1.58–1.20 (m, 10 H, hex-H$_2$).

^{13}C NMR (CDCl$_3$): δ (ppm) = 169.5, 167.3, 133.8, 131.7, 128.6, 127.0, 74.3, 42.1, 31.5, 25.2, 23.6.

EI HRMS: m/z = 261.1364 (calcd. 261.1365).

1.3.3.2   **   Cyclohexyl 2-benzoylamino-2-bromoacetate [9]

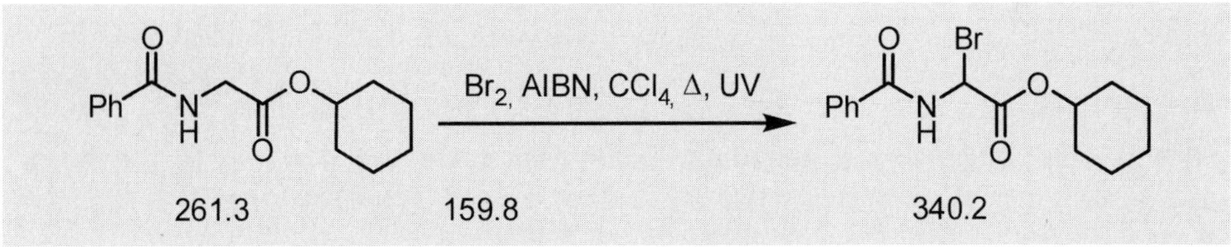

| | Br$_2$, AIBN, CCl$_4$, Δ, UV | |
| 261.3 | 159.8 | 340.2 |

A solution of bromine (3.51 g, 22.0 mmol) in anhydrous carbon tetrachloride (30 mL) is added dropwise over 2 h under UV-irradiation (500 W) to a refluxing solution of the acetate **1.3.3.1** (5.22 g, 20.0 mmol) and azobisisobutyronitrile (50 mg) in carbon tetrachloride (40 mL) to give a light-brown solution. After completion of the addition of bromine, irradiation and refluxing are continued for 3 h.

The solvent is removed under reduced pressure and the product is crystallized from EtOAc/petroleum ether (50–80 °C) (1:1). The water-sensitive product is kept under argon at 4 °C. The product is obtained as a colorless solid; 5.92 g (87%); mp 107–109 °C; TLC (SiO$_2$, EtOAc/n-hexane, 1:2): R_f = 0.27.

UV: λ_{max} (nm) = 231.5, 194.5.

IR (KBr): $\tilde{v}$ (cm^{-1}) = 3298, 3038, 2940, 2861, 1733, 1660, 1602, 1581, 1519, 1490, 1453, 1379, 1358, 1340, 1285, 1240, 1194, 1133, 1009, 934, 719, 691, 530.

^{1}H NMR (300 MHz, CDCl$_3$): δ (ppm) = 7.40–7.80 (m, 5 H, arom-H, 1 H, NH), 6.60 (d, J = 9.9 Hz, 1 H, CH), 4.90 (m, 1 H, c-hex-H$_1$), 1.20–1.90 (m, 10 H, c-hex-H$_2$).

^{13}C NMR (CDCl$_3$, 75.5 Hz): δ (ppm) = 166.3, 165.6, 132.8, 132.4, 128.8, 127.4, 75.9, 50.5, 31.1, 30.6, 25.1, 23.3, 23.2.

1.3.3.3 * 1-Morpholine-1-cyclohexene [10]

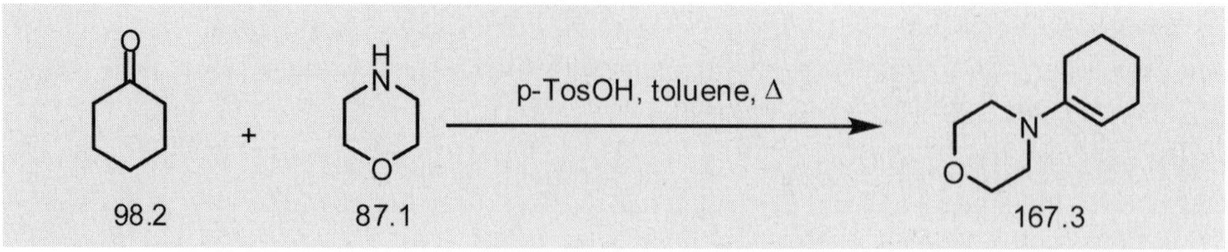
p-TosOH, toluene, Δ
98.2
87.1
167.3

Cyclohexanone (11.8 g, 0.12 mol) and morpholine (12.5 g, 0.14 mol) are heated under reflux with *p*-toluenesulfonic acid (20 mg) in toluene (25 mL) for 10 h with azeotropic removal of H_2O in a Dean–Stark trap.

After cooling to room temperature, the organic layer is washed twice with H_2O until pH 7 is reached and then dried over $MgSO_4$. The solvent is removed and the residue is distilled *in vacuo* to give the enamine as a colorless liquid; 17.5 g (87%), bp_{93} 74–75 °C; TLC (SiO_2, EtOAc/*n*-hexane, 1:2): $R_f = 0.58$.

UV: λ_{max} (nm) = 220.

IR (KBr): $\tilde{\nu}$ (cm^{-1}) = 2926, 2893, 1647, 1450, 1385, 1358, 1264, 1204, 1123, 899, 789.

^{1}H NMR (300 MHz, $CDCl_3$): δ (ppm) = 4.62 (t, J =1.7 Hz, 1 H, 12-H_1), 3.70 (m, 2 H, 2-H_2, 6-H_2), 2.78 (m, 4 H, 3-H_2, 5-H_2), 2.00 (m, 4 H, 9-H_2, 12-H_2), 1.60 (m, 4 H, 10-H_2, 11-H_2).

^{13}C NMR ($CDCl_3$): δ (ppm) = 145.4, 100.4, 66.9, 48.4, 26.8, 24.3, 23.3, 22.7.

EI-HRMS: m/z = 167.1306 (calcd. 167.1310).

1.3.3.4 ** Cyclohexyl 2-benzoylamino-2-(2-oxocyclohexyl) acetate [3]

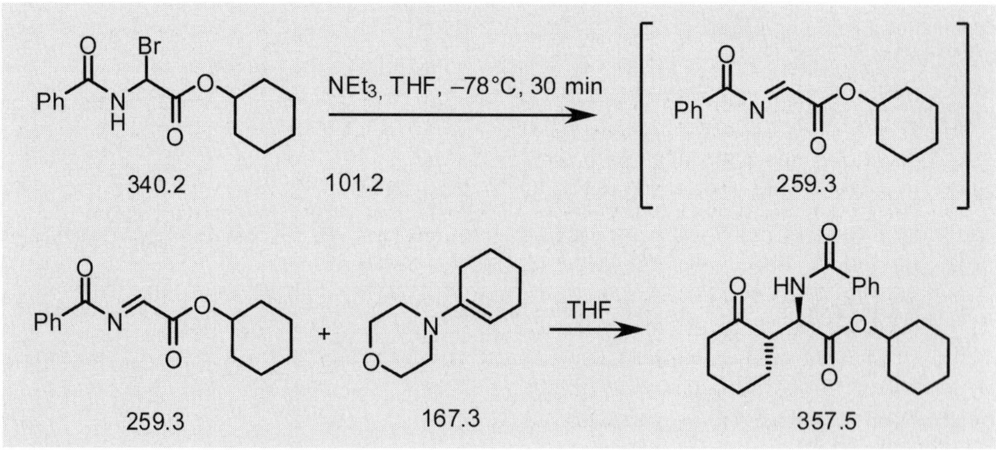
NEt3, THF, –78°C, 30 min
340.2
101.2
259.3
THF
259.3
167.3
357.5

Triethylamine (697 μL, 0.50 g, 5.0 mmol) is added to a solution of the bromoacetate **1.3.3.2** (5.0 mmol) in anhydrous THF (35 mL) under argon at –78 °C. After stirring for 30 min, the solution is cooled to –95 °C and a precooled (–78 °C) solution of the enamine **1.3.3.3** (0.92 g, 5.5 mmol) in THF (10 mL) is carefully added. The temperature is maintained at –95 °C for 6 h and at –78 °C for 6 h thereafter. After warming to room temperature, the mixture is hydrolyzed by the addition of a dilute citric acid solution until the pH reaches 4–5 and stirring is continued for 5 h.

The solvent is then evaporated, the residue is extracted with EtOAc, and the organic layer is washed with H_2O and dried over $MgSO_4$. After removal of the solvent, the product is purified by column chromatography (SiO$_2$, n-hexane/EtOAc, 2:1) and the resulting oil is dissolved in n-hexane and treated in a sonicator for 20 min. Crystallization yields a colorless solid; 1.49 g (83%); mp 106–108 °C; TLC (EtOAc/n-hexane, 2:1): R_f = 0.52; de >98% based on HPLC (RP C18, H_2O/0.1% TFA, CH_3CN/H_2O, 8:2/0.1% TFA; 60 to 90% in 30 min, t_R = 14.6 min).

UV: λ_{max} (nm) = 224.0, 192.5.

IR (KBr): $\tilde{v}$ (cm^{-1}) = 3320, 2936, 2860, 1712, 1654, 1546, 1517, 1488, 1447, 1316, 1281, 1268, 1240, 1208, 1011, 719, 693.

^{1}H NMR (300 MHz, CDCl$_3$): δ (ppm) = 7.78–7.40 (m, arom-H), 7.00 (d, J = 9.65 Hz, 1 H, NH), 4.90 (dd, J = 3.22, 9.59 Hz, 1 H, α-H), 4.80 (td, 1 H, c-hexane), 3.35 (m, 1 H, c-hexanone), 2.38 (m, 4 H, 2 CH$_2$, c-hexanone), 2.28 (m, 2 H, CH$_2$, c-hexanone), 2.10 (m, 2 H, CH$_2$, c-hexanone) 1.90–1.20 (m, 10 H, c-hexane).

EI HRMS: m/z = 358.20131 (calcd. 358.20128).

---

[1]   M. B. Smith, J. March, *March's Advanced Organic Chemistry*, 6th ed., p. 1218, John Wiley & Sons, Inc., New York, **2007**.

[2]   For an example, see: L. F. Tietze, Th. Eicher, *Reaktionen und Synthesen im organisch-chemischen Praktikum und Forschungslaboratorium*, 2nd ed., p. 186, Georg Thieme Verlag, Stuttgart, **1991**.

[3]   a) R. Kober, K. Papadopoulos, W. Miltz, D. Enders, W. Steglich, *Tetrahedron* **1985**, *41*, 1693. b) For the direct organocatalyzed enantioselective α-aminomethylation of ketones, see: I. Ibrahem, J. Casas, A. Cordova, *Angew. Chem.* **2004**, *116*, 6690; *Angew. Chem. Int. Ed.* **2004**, *43*, 6528.

[4]   C. H. Heathcock, in: *Asymmetric Synthesis* (Ed.: J. P. Morrison), Vol. 3, p. 111, Academic Press, New York, **1984**; see also: S. Hauptmann, G. Mann, *Stereochemie*, p. 216, Spektrum Akademischer Verlag, Heidelberg, **1996**.

[5]   S. Masamune, W. Choy, J. S. Petersen, L. R. Sita, *Angew. Chem.* **1985**, *97*, 1; *Angew. Chem. Int. Ed. Engl.* **1985**, *24*, 1.

[6]   R. Kober, W. Hammes, W. Steglich, *Angew. Chem.* **1982**, *94*, 213; *Angew. Chem. Int. Ed. Engl.* **1982**, *21*, 203.

[7]   a) *Houben/Weyl*, Vol. XI/2, p. 309; b) *Organikum*, 21st ed., p. 492, Wiley-VCH, Weinheim, **2001**.

[8]   See ref. [1], p. 553; for a monograph, see: T. A. Hase, *Umpoled Synthons*, John Wiley & Sons, Inc., New York, **1987**.

[9]   R. Kober, W. Steglich, *Liebigs Ann. Chem.* **1983**, 604.

[10] Th. Eicher, L. F. Tietze, *Organisch-Chemisches Grundpraktikum*, 2nd ed., p. 204, Georg Thieme Verlag, Stuttgart, **1995**.

1.3.4   (*S*)-1-Hydroxy-1,3-diphenyl-3-propanone

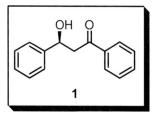

1

Topics:
- Chiral (acyloxy)borane-catalyzed asymmetric Mukaiyama aldol reaction
- Synthesis of chiral β-hydroxy ketones
- Preparation of the CAB ligand from 2,6-dihydroxybenzoic acid and (*S,S*)-(−)-tartaric acid
- Ester and aryl ether formation, cleavage of benzyl esters

(a) General

Aldol reactions are among the most powerful and efficient synthetic methods for the formation of carbon–carbon bonds [1]. In the Mukaiyama aldol reaction (cf. **1.3.3**), silyl enol ethers or silyl ketene acetals are combined with aldehydes in the presence of a Lewis acid (e.g. TiCl₄) to give β-hydroxy ketones (aldols) or β-hydroxy esters, respectively:

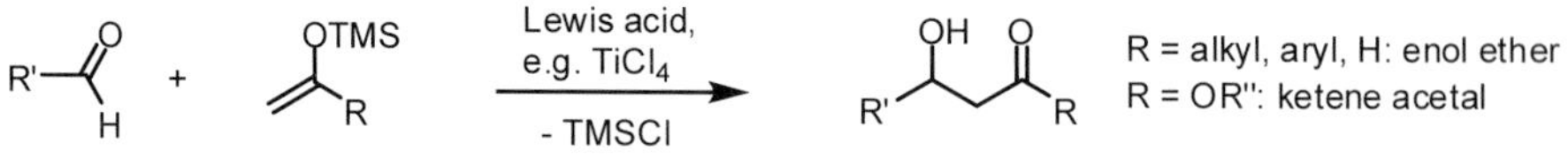

By using a chiral Lewis acid, an asymmetric Mukaiyama aldol reaction can be performed. For this purpose, H. Yamamoto and co-workers developed the chiral (acyloxy)borane (CAB) complexes **2**, which are based on a chiral ligand derived from tartaric acid and aryl boronic acids [2]. They proved to be efficient chiral catalysts for aldol reactions [3] and have also been successfully applied for a variety of other asymmetric transformations such as Diels–Alder reactions [4], hetero-Diels–Alder reactions [5], and allylations [6].

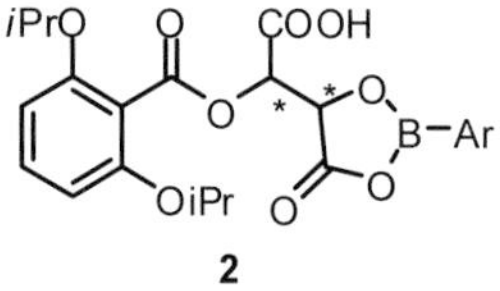

2 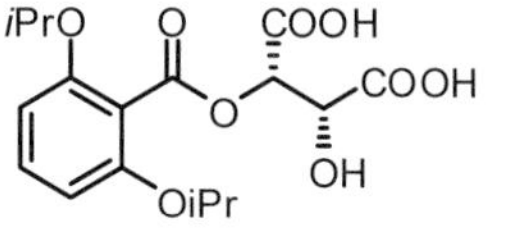   **3**

(b) Synthesis of 1

(**1**) For the synthesis of the catalyst of type **2**, the chiral CAB ligand **3** is prepared in a five-step sequence starting from 2,6-dihydroxybenzoic acid (**4**) and (*S,S*)-(−)-tartaric acid (**6**). Firstly, the *O*-alkylated carboxylic acid **8** is synthesized in a three-step sequence consisting of formation of the methyl ester, its *O*-alkylation to give the bis-isopropyl ether **5**, and saponification of the methyl ester moiety (**5 → 8**) [5c]. Secondly, (*S,S*)-(−)-tartaric acid (**6**) is transformed into the dibenzyl ester **7** by reaction with benzyl alcohol in the presence of a catalytic amount of *p*-toluenesulfonic acid with azeotropic removal of H₂O [7]. Mono-esterification of the 1,2-diol functionality in the dibenzyl ester **7** with the carboxylic acid **8** is accomplished with trifluoroacetic anhydride, probably via the intermediate formation of a mixed anhydride. The final step of the synthesis is cleavage of the dibenzyl ester moieties in **9** by hydrogenation to give the desired CAB ligand **3** [6b].

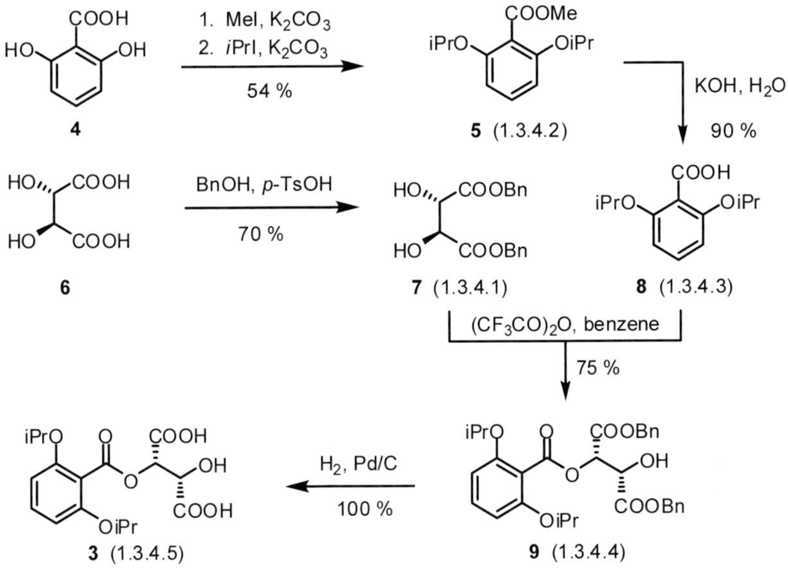

The active CAB species **11** is prepared *in situ* from (2S,3S)-2-O-(2,6-diisopropoxybenzoyl)tartaric acid (**3**) and commercially available 2-phenoxyphenylboronic acid (**10**) in propionitrile [3c]:

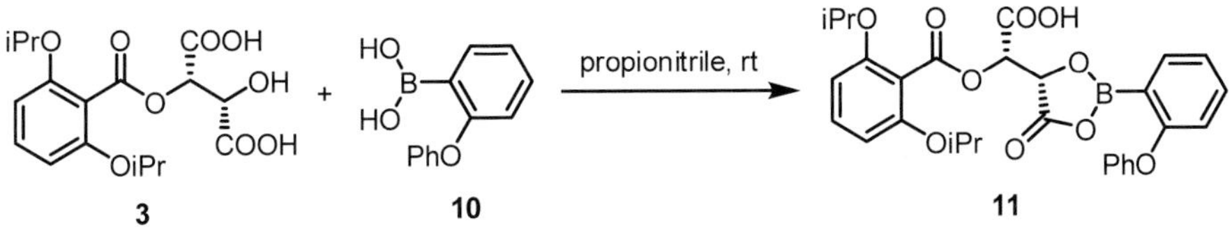

(2) The asymmetric Mukaiyama aldol reaction of benzaldehyde and 1-phenyl-1-(trimethylsilyloxy)ethylene is performed in propionitrile at –78 °C under promotion by 20 mol% of the catalyst and leads to the (S)-enantiomer of 1-hydroxy-1,3-diphenyl-3-propanone (**1**) in 91% chemical yield and 90% *ee*.

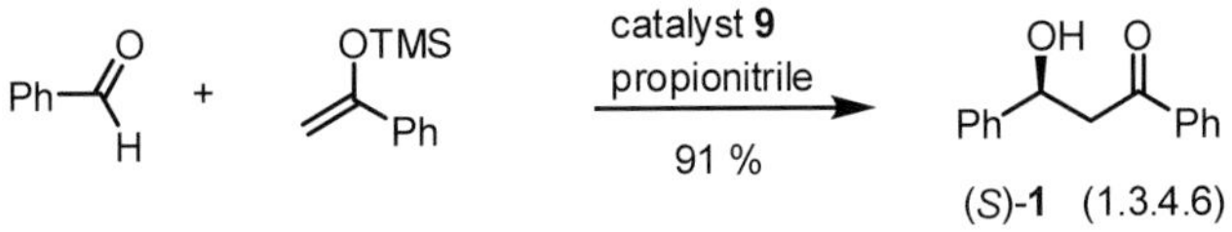

(c) Experimental procedures for the syntheses of 3 and 1

1.3.4.1   *   (S,S)-Dibenzyl tartrate [5]

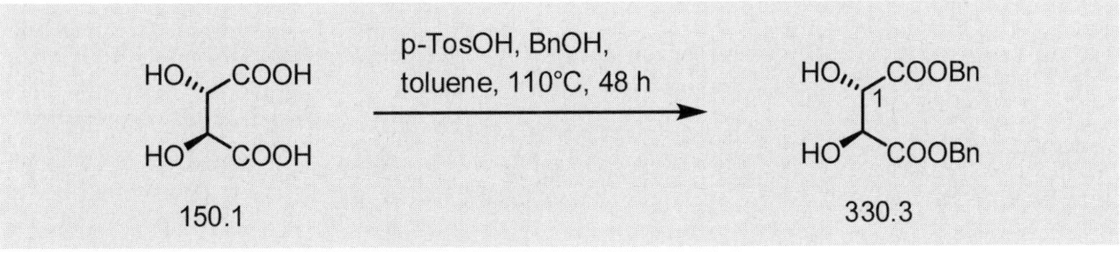

A stirred solution of (*S*,*S*)-(–)-tartaric acid (15.0 g, 100 mmol), benzyl alcohol (20.7 mL, 21.6 g, 200 mmol), and *p*-toluenesulfonic acid monohydrate (476 mg, 2.50 mmol, 2.5 mol%) in toluene (200 mL) in a 500 mL round-bottomed flask equipped with a Dean–Stark trap and an argon bubbler is refluxed for 48 h.

It is then cooled to room temperature, diluted with EtOAc (120 mL), and washed with saturated aqueous NaHCO$_3$ solution (2 × 30 mL) and brine (2 × 30 mL). The organic layer is dried over Na$_2$SO$_4$, filtered, and concentrated *in vacuo*. The residue is dissolved in toluene (80 mL) and the desired product is obtained by precipitation upon addition of isooctane (80 mL). After filtration and drying the residue under high vacuum, the tartrate is obtained as white fibers; 23.2 g (70%), mp 54–55 °C, $[\alpha]_D^{20}$ = +10.0 (*c* = 1.0, CHCl$_3$).

UV (CH$_3$CN): λ_{max} (nm) (lg ε) = 267.0 (2.200), 262.5 (2.439), 251.5 (2.376), 257.0 (2.515), 207.0 (4.207).

IR (KBr): $\bar{\nu}$ (cm^{-1}) = 3464, 3280, 3034, 2946, 1747, 1498, 1455, 1378, 1275, 1218, 1192, 1126, 1093, 1029, 1003, 978, 736, 695, 608, 507, 457.

^{1}H NMR (300 MHz, CDCl$_3$): δ (ppm) = 7.34 (m$_c$, 10 H, 10 × Ph-H), 5.25 (d, *J* = 2.0 Hz, 4 H, 2 × C$\underline{H}_2$Ph), 4.59 (d, *J* = 7.3 Hz, 2 H, 2 × 1-H), 3.17 (d, *J* = 7.3 Hz, 2 H, 2 × OH).

^{13}C NMR (50 MHz, CDCl$_3$): δ (ppm) = 171.3 (2 × –$\underline{C}$OOBn), 134.7 (2 × Ph-$\underline{C}_{quat}$), 128.7 (Ar-$\underline{C}$H), 128.4 (Ar-$\underline{C}$H), 72.05 (2 × C-1), 68.05 (2 × –$\underline{C}$H$_2$Ph).

MS (ESI): *m/z* (%) = 683 (100) [2*M* + Na]$^+$, 353 (22) [*M* + Na]$^+$.

1.3.4.2   **   2,6-Diisopropoxybenzoic acid methyl ester [5]

| 154.1 | 168.2 | 252.3 |

Iodomethane (Caution: reagent might cause cancer!) (17.8 mL, 40.6 g, 286 mmol) is added to a mixture of 2,6-dihydroxybenzoic acid (20.0 g, 130 mmol), anhydrous K$_2$CO$_3$ (19.8 g, 143 mmol), and anhydrous DMF (300 mL) in a 1000 mL round-bottomed flask equipped with a dropping funnel. The mixture is stirred at room temperature for 20 h, then poured into ice-cold HCl (1 M, 300 mL) and extracted with Et$_2$O (3 × 250 mL). The combined organic layers are washed with brine (150 mL), dried over Na$_2$SO$_4$, and concentrated *in vacuo*.

The oily residue (crude 2,6-dihydroxybenzoic acid methyl ester, max. 130 mmol) is dissolved in DMF (300 mL) in a 1000 mL round-bottomed flask equipped with a dropping funnel. First, anhydrous K$_2$CO$_3$ (44.9 g, 325 mmol) is added in one batch, then 2-iodopropane (36.4 mL, 61.9 g, 364 mmol) is added dropwise under continuous stirring at room temperature. Stirring is continued for 2 d, then the mixture is poured into ice-cold HCl (1 M, 300 mL) and extracted with Et$_2$O (2 × 250 mL). The combined organic layers are washed with brine (3 × 150 mL), dried over Na$_2$SO$_4$, and concentrated *in vacuo*. Purification of the residue by column chromatography on silica gel (400 g, *n*-pentane/EtOAc, 20:1) leads to the methyl ester as colorless cuboids; 17.7 g (54% for two steps), mp 57–59 °C, $R_f = 0.30$ (*n*-pentane/EtOAc, 20:1).

UV (CH$_3$CN): λ_{max} (nm) (lg ε) = 280.5 (3.353), 203.0 (4.585).

IR (KBr): $\tilde{\nu}$ (cm^{-1}) = 2981, 1735, 1595, 1467, 1386, 1295, 1255, 1112, 1071, 959, 902, 823, 783, 739, 665.

^{1}H NMR (300 MHz, CDCl$_3$): δ (ppm) = 7.18 (t, J = 8.4 Hz, 1 H, H-4), 6.50 (d, J = 8.4 Hz, 2 H, 2 × H-3), 4.49 (sept, J = 6.2 Hz, 2 H, 2 × OCH(CH$_3$)$_2$), 3.86 (s, 3 H, COOCH$_3$), 1.28 (d, J = 6.2 Hz, 12 H, 2 × OCH(CH$_3$)$_2$).

^{13}C NMR (75.6 MHz, CDCl$_3$): δ (ppm) = 167.3 (COOCH$_3$), 155.9 (C-2), 130.5 (C-4), 116.2 (C-1), 106.5 (2 × C-3), 71.36 (2 × OCH(CH$_3$)$_2$), 52.05 (COOCH$_3$), 22.07 (2 × OCH(CH$_3$)$_2$).

MS (EI, 70 eV): *m/z* (%) = 252 (15) [M]$^+$, 221 (10), 168 (39) [M − 2C$_3$H$_6$]$^+$, 136 (100) [M − 2C$_3$H$_6$ − CH$_3$OH]$^+$, 108 (12), 43 (9) [C$_3$H$_7$]$^+$.

1.3.4.3 ** 2,6-Diisopropoxybenzoic acid [5]

The benzoate **1.3.4.2** (15.6 g, 61.9 mmol) is added to a solution of KOH (28.2 g, 681 mmol) in MeOH (170 mL) and H$_2$O (19 mL). The mixture is heated to 80 °C and stirred for 15 h at this temperature.

After the addition of H$_2$O (200 mL), the MeOH is evaporated under reduced pressure. The aqueous solution is added dropwise to a stirred HCl solution (2 M, 400 mL) at 0 °C to give a white precipitate, which is collected by filtration, washed with ice-cold H$_2$O (3 × 30 mL), and dried *in vacuo*. The benzoic acid is isolated as a colorless amorphous solid; 13.3 g (90%), mp 106–107 °C, $R_f = 0.05$ (*n*-pentane/EtOAc, 10:1).

UV (CH$_3$CN): λ_{max} (nm) (lg ε) = 280.5 (3.343), 204.0 (4.581).

IR (KBr): $\tilde{\nu}$ (cm^{-1}) = 2982, 2934, 2662, 1702, 1597, 1467, 1387, 1340, 1302, 1258, 1173, 1112, 1072, 904, 804, 782, 742, 655, 445.

^{1}H NMR (300 MHz, CDCl$_3$): δ (ppm) = 7.24 (t, J = 8.4 Hz, 1 H, H-4), 6.56 (d, J = 8.4 Hz, 2 H, 2 × H-3), 4.56 (sept, J = 5.9 Hz, 2 H, 2 × OCH(CH$_3$)$_2$), 1.33 (d, J = 5.9 Hz, 12 H, 2 × OCH(CH$_3$)$_2$).

^{13}C NMR (75.6 MHz, CDCl$_3$): δ (ppm) = 168.8 (COOH), 156.7 (C-2), 131.4 (C-4), 114.4 (C-1), 106.9 (2 × C-3), 71.99 (2 × OCH(CH$_3$)$_2$), 22.04 (2 × OCH(CH$_3$)$_2$).

MS (EI, 70 eV): *m/z* (%) = 238 (8) [M]$^+$, 154 (27) [M − 2C$_3$H$_6$]$^+$, 136 (100) [M − OCH(CH$_3$)$_2$ − C$_3$H$_7$]$^+$, 108 (12), 43 (7) [C$_3$H$_7$]$^+$.

1.3.4.4 ** (2S,3S)-2-O-(2,6-Diisopropoxybenzoyl) tartaric acid dibenzyl ester [5]

Trifluoroacetic anhydride (1.96 mL, 2.91 g, 13.9 mmol) is added by means of a syringe over a period of 20 min to a stirred suspension of the acid **1.3.4.3** (3.00 g, 12.6 mmol) and the tartrate **1.3.4.1** (4.16 g, 12.6 mmol) in anhydrous benzene (65 mL) at room temperature. Stirring is continued for 90 min.

The pale-yellow solution is then poured into saturated $NaHCO_3$ solution (100 mL) and the mixture is extracted with Et_2O (3 × 50 mL). The combined organic layers are dried over Na_2SO_4, concentrated *in vacuo*, and the residue is purified by column chromatography on silica gel (210 g, CH_2Cl_2). The tartaric acid dibenzyl ester is obtained as a colorless sticky oil; 5.23 g (75%), $[\alpha]_D^{20} = +33.4$ ($c = 1.0$, $CHCl_3$), $R_f = 0.19$ (CH_2Cl_2).

UV (CH_3CN): λ_{max} (nm) (lg ε) = 282.5 (3.382), 203.0 (4.711).

IR (KBr): $\tilde{v}$ (cm^{-1}) = 3522, 3034, 2979, 2935, 1748, 1595, 1499, 1465, 1385, 1334, 1255, 1114, 1071, 967, 905, 789, 736.

^{1}H NMR (300 MHz, $CDCl_3$): δ (ppm) = 7.38–7.31 (m, 10 H, 10 × Ph-H), 7.25 (t, $J = 8.3$ Hz, 1 H, 4'-H), 6.53 (d, $J = 8.3$ Hz, 2 H, 2 × 3'-H), 5.85 (d, $J = 2.4$ Hz, 1 H, 2-H), 5.33 (d, $J = 12.0$ Hz, 1 H, C$\underline{H}_2$Ph), 5.26 (d, $J = 1.8$ Hz, 2 H, C$\underline{H}_2$Ph), 5.10 (d, $J = 12.0$ Hz, 1 H, C$\underline{H}_2$Ph), 4.82 (dd, $J = 9.0$, 2.4 Hz, 1 H, 3-H), 4.55 (sept, $J = 6.0$ Hz, 2 H, 2 × OC$\underline{H}$(CH$_3$)$_2$), 3.18 (d, $J = 9.0$ Hz, 1 H, OH), 1.30 (d, $J = 6.0$ Hz, 6 H, OCH(C$\underline{H}_3$)$_2$), 1.28 (d, $J = 6.0$ Hz, 6 H, OCH(C$\underline{H}_3$)$_2$).

^{13}C NMR (75.6 MHz, $CDCl_3$): δ (ppm) = 170.2 (COOBn), 166.4 (COOBn), 165.1 (COOAr), 156.4 (2 × C-2'), 135.2 (Ph-C$_{quat}$), 134.7 (Ph-C$_{quat}$), 131.2 (C-4'), 128.6, 128.5, 128.3, 128.2 (Ph-C$\underline{H}$), 114.0 (C-1'), 105.9 (2 × C-3'), 73.01 (C-2), 71.13 (2 × O$\underline{C}$H(CH$_3$)$_2$), 71.00 (C-3), 67.94 ($\underline{C}$H$_2$Ph), 67.32 ($\underline{C}$H$_2$Ph), 21.87 (OCH($\underline{C}$H$_3$)$_2$), 21.81 (OCH($\underline{C}$H$_3$)$_2$).

MS (ESI): m/z (%) = 1124 (100) $[2M + Na]^+$, 573 (25) $[M + Na]^+$.

1.3.4.5 ** (2S,3S)-2-O-(2,6-Diisopropoxybenzoyl)tartaric acid [5]

10% Pd on charcoal (240 mg) is added to a solution of the dibenzyl ester **1.3.4.4** (3.00 g, 5.45 mmol) in EtOAc (25 mL) under an argon atmosphere. The balloon filled with argon is then replaced by a balloon filled with hydrogen and the reaction mixture is stirred at room temperature for 14 h.

The mixture is then filtered through a Celite pad and the filtrate is concentrated to afford the monoacylated tartaric acid in quantitative yield, which is dried *in vacuo* to become a colorless crystalline solid; 2.02 g (100%), mp 76–78 °C, $[\alpha]_D^{20} = +27.8$ (c = 1.0, EtOH).

UV (CH$_3$CN): λ_{max} (nm) (lg ε) = 282.0 (3.376), 202.5 (4.539).

IR (KBr): $\tilde{v}$ (cm^{-1}) = 3495, 2982, 1743, 1596, 1467, 1387, 1254, 1112, 903, 733, 662.

^{1}H NMR (300 MHz, CDCl$_3$): δ (ppm) = 7.25 (t, J = 8.4 Hz, 1 H, 4'-H), 6.53 (d, J = 8.4 Hz, 2 H, 2 × 3'-H), 5.84 (d, J = 1.5 Hz, 1 H, 2-H), 4.87 (d, J = 1.5 Hz, 1 H, 3-H), 4.54 (sept, J = 6.1 Hz, 2 H, 2 × OCH(CH$_3$)$_2$), 1.30 (d, J = 6.1 Hz, 6 H, OCH(CH$_3$)$_2$), 1.25 (d, J = 6.1 Hz, 6 H, OCH(CH$_3$)$_2$).

^{13}C NMR (75.6 MHz, CDCl$_3$): δ (ppm) = 173.1 (COOH), 170.0 (COOH), 164.4 (COOAr), 156.2 (2 × C-2'), 131.8 (C-4'), 114.0 (C-1'), 113.6 (2 × C-3'), 72.63 (C-2), 72.08 (2 × OCH(CH$_3$)$_2$), 70.49 (C-3), 21.88 (OCH(CH$_3$)$_2$), 21.75 (OCH(CH$_3$)$_2$).

MS (ESI): m/z (%) = 763 (87) [2M + Na]$^+$, 393 (100) [M + Na]$^+$.

1.3.4.6   **   (S)-1-Hydroxy-1,3-diphenyl-3-propanone [4]

The monoacylated tartaric acid **1.3.4.5** (74.1 mg, 0.20 mmol) and 2-phenoxyphenylboronic acid (42.8 mg, 0.20 mmol) are dissolved in anhydrous propionitrile (1.0 mL) and stirred at room temperature for 30 min. The reaction mixture is then cooled to –78 °C and benzaldehyde (101 μL, 106 mg, 1.00 mmol) is added by means of a syringe, followed by 1-phenyl-1-(trimethylsiloxy)ethylene (349 μL, 327 mg, 1.70 mmol). The reaction mixture is stirred at –78 °C for 4 h, then an HCl solution (0.25 M, 4.0 mL) is added and the mixture is allowed to warm to room temperature.

The mixture is poured into Et$_2$O (40 mL) and H$_2$O (20 mL), the phases are separated, and the aqueous layer is extracted with Et$_2$O (2 × 20 mL). The combined organic layers are washed with H$_2$O (20 mL) and saturated NaHCO$_3$ solution (20 mL), dried over Na$_2$SO$_4$, and concentrated *in vacuo*. The aldol adduct is obtained as a light-yellow sticky oil after purification by column chromatography on deactivated silica gel (30 g + 0.3 mL NEt$_3$, *n*-pentane/Et$_2$O, 5:1); 206 mg (91%), ee = 90%, $[\alpha]_D^{20}$ = −67.0 (c = 1.0, CHCl$_3$), R_f = 0.14 (*n*-pentane/Et$_2$O, 5:1).

The *ee* value of the aldol **1.3.4.6** is determined by HPLC analysis of the corresponding (+)-MTPA ester, which is obtained by small-scale reaction of **1.3.4.6** with (+)-α-methoxy-α-trifluoromethylphenyl acetyl chloride and anhydrous pyridine in CCl_4 according to Mosher's method [8]. (S)-1-Hydroxy-1,3-diphenyl-3-propanone (25.0 mg, 0.11 mmol) and (+)-α-methoxy-α-trifluoromethylphenyl acetyl chloride (20.7 μL, 27.8 mg, 0.11 mmol) are mixed with CCl_4 (0.1 mL) (Caution: resorption through the skin!) and pyridine (0.1 mL). The reaction mixture is stirred at room temperature for 12 h, poured into Et_2O (10 mL) and H_2O (10 mL), and, after extraction, the phases are separated. The organic layer is dried over Na_2SO_4, concentrated, and the residue obtained is dissolved in EtOAc for HPLC analysis.

UV (CH_3CN): λ_{max} (nm) (lg ε) = 279.0 (3.178), 241.0 (4.103).

IR (KBr): $\tilde{\nu}$ (cm^{-1}) = 3469, 3057, 1670, 1597, 1447, 1393, 1215, 1055, 1020, 916, 872, 747.

^{1}H NMR (300 MHz, [D_6]benzene): δ (ppm) = 7.61 (dd, J = 8.1, 1.5 Hz, 2 H, 2 × 5-H), 7.35 (dd, J = 7.5, 1.8 Hz, 2 H, 2 × 2'-H), 7.20 (t, J = 7.5 Hz, 2 H, 2 × 3'-H), 7.13–7.05 (m, 2 H, 7-H, 4'-H), 6.96 (t, J = 8.1 Hz, 2 H, 2 × 6-H), 5.23 (dd, J = 9.3, 2.9 Hz, 1 H, 1-H), 3.53 (s_{br}, 1 H, OH), 2.94 (dd, J = 17.7, 9.3 Hz, 1 H, 2-H_b), 2.80 (dd, J = 17.7, 2.9 Hz, 1 H, 2-H_a).

^{13}C NMR (75.6 MHz, [D_6]benzene): δ (ppm) = 199.7 (C-3), 144.1 (C-1'), 137.0 (C-4), 133.2 (C-7), 128.5 (2 × C-5, 2 × C-6, C-4'), 127.5 (2 × C-3'), 126.1 (2 × C-2'), 70.02 (C-1), 47.97 (C-2).

MS (EI, 70 eV): m/z (%) = 226 (48) $[M]^+$, 208 (58) $[M - H_2O]^+$, 186 (47), 131 (11) $[M - H_2O - C_6H_5]^+$, 120 (48), 105 (100) $[C_6H_5CHCH_3]^+$, 77 (96) $[C_6H_5]^+$, 51 (33).

HPLC: Chiralcel OD (Daicel); 250 × 4.6 mm ID
       eluent: n-hexane/EtOAc, 40:1; isocratic
       retention time: t_{R1} = 12.8 min (S)-isomer; t_{R2} = 15.1 min (R)-isomer.

---

[1] a) B. Alcaide, P. Almendros, *Eur. J. Org. Chem.* **2002**, 1595; b) C. Palomo, M. Oiarbide, J. M. García, *Chem. Eur. J.* **2002**, *8*, 36.

[2] K. Ishihara, H. Yamamoto, *Eur. J. Org. Chem.* **1999**, 527.

[3] a) K. Ishihara, T. Maruyama, M. Mouri, Q. Gao, K. Furuta, H. Yamamoto, *Bull. Chem. Soc. Jpn.* **1993**, *66*, 3483; b) K. Furuta, T. Maruyama, H. Yamamoto, *J. Am. Chem. Soc.* **1991**, *113*, 1041; c) K. Furuta, T. Maruyama, H. Yamamoto, *Synlett* **1991**, 439.

[4] a) K. Furuta, A. Kanematsu, H. Yamamoto, S. Takaoka, *Tetrahedron Lett.* **1989**, *30*, 7231; b) K. Furuta, Y. Miwa, K. Iwanaga, H. Yamamoto, *J. Am. Chem. Soc.* **1988**, *110*, 6254.

[5] a) Q. Gao, K. Ishihara, T. Maruyama, M. Mouri, H. Yamamoto, *Tetrahedron* **1994**, *50*, 979; b) K. Ishihara, Q. Gao, H. Yamamoto, *J. Org. Chem.* **1993**, *58*, 6917; c) S. Kiyooka, Y. Kaneko, K. Kume, *Tetrahedron Lett.* **1992**, *33*, 4927.

[6] a) K. Ishihara, M. Mouri, Q. Gao, T. Maruyama, K. Furuta, H. Yamamoto, *J. Am. Chem. Soc.* **1993**, *115*, 11490; b) K. Furuta, M. Mouri, H. Yamamoto, *Synlett* **1991**, 561.

[7] a) J. E. Bishop, J. F. O'Connell, H. Rapoport, *J. Org. Chem.* **1991**, *56*, 5079–5091; b) R. Wagner, J. W. Jefferson, W. Tilley, K. Lovey, *Synthesis* **1990**, *9*, 785.

[8] a) J. A. Dale, H. S. Mosher, *J. Am. Chem. Soc.* **1973**, *95*, 512; b) J. A. Dale, D. L. Dull, H. S. Mosher, *J. Org. Chem.* **1969**, *34*, 2543.

1.3.5   [(1*S*,2*R*,6*R*)-2-Hydroxy-4-oxo-2,6-diphenyl]cyclohexanecarboxylic acid ethyl ester

Topics:
- Organocatalysis
- Enantioselective synthesis of an oxocyclohexane carboxylic ester (Michael addition, intramolecular aldol reaction)
- Synthesis of Jørgensen's catalyst from *L*-phenylalanine (amide formation, amines by reduction of amides, imidazolidine formation)

(a) General

Enantioselective catalysis is an important topic in organic synthesis. In the past, this was mainly accomplished by the use of transition metal catalysts containing metals such as Pd [1], Ru, or Rh in the presence of chiral ligands. However, on the basis of the enantioselective synthesis of hydrindens by Wiechert, Eder, Sauer, Hajos, and Parrish [2] over 30 years ago, using *L*-proline as chiral catalyst (cf. **1.3.1**), it has been demonstrated that enantioselective catalysis can also be effected by a wide range of small, metal-free chiral molecules, such as amino acids and their derivatives.

In the meantime, extensive studies have been carried out with regard to the use of such organocatalysts in various reactions [3], such as the aldol [4], Michael [5], Mannich [6], and Diels–Alder reactions [7], as well as hydrogenations [8], with high stereoselectivity. The concept of organocatalysis (or aminocatalysis) is mainly based on electronic similarities between a Lewis-acid-activated carbonyl group and an iminium ion. Thus, an iminium ion is more reactive than a carbonyl moiety due to a lower energy of its LUMO, which is manifested in an increase of its electrophilicity and its α-C–H acidity. In this way, organocatalysis exploits both the higher reactivity of iminium ions and their easy deprotonation to give enamines, which can either react with electrophiles or be used in pericyclic processes.

Among the various asymmetric C–C bond-forming reactions that may be exploited for the formation of chiral building blocks, enantioselective domino reactions [9] are of particular importance as multiple stereogenic centers can be formed during a single transformation. Jørgensen and co-workers recently published a highly diastereo- and enantioselective domino-Michael-aldol reaction of acyclic β-keto esters and α,β-unsaturated ketones in the presence of a chiral organocatalyst easily accessible in a few steps from *L* -phenylalanine [10].

An example of this organocatalyzed domino process, yielding cyclohexanone-4-carboxylates with several stereogenic centers, is presented in section (b), together with the synthesis of the required organocatalyst.

(b) Synthesis of 1

(1) Jørgensen's catalyst **6** is prepared in a four-step sequence, starting from *L*-phenylalanine (**2**), which is transformed into the methyl ester hydrochloride **3** by reaction with SOCl$_2$ in MeOH. Aminolysis of **3** leads to the corresponding methyl amide **4**, which is reduced with lithium aluminum hydride. The 1,2-diamine **5** thus obtained is subjected to cyclocondensation with glyoxylic acid monohydrate to give the desired organocatalyst imidazolidine-2-carboxylic acid **6**.

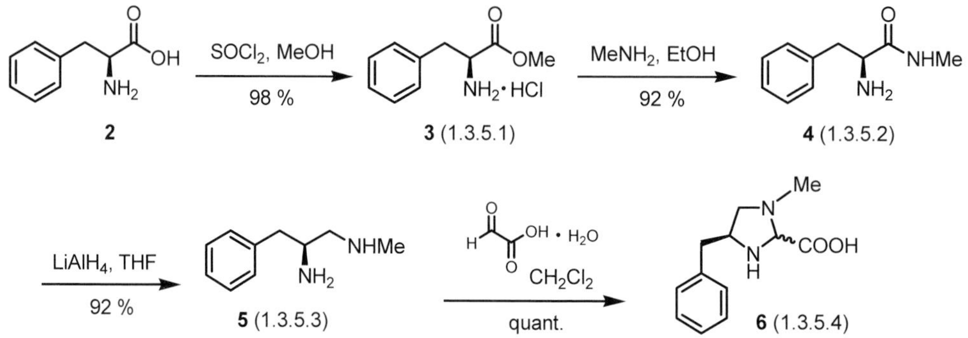

(2) The synthesis of the oxocyclohexanecarboxylic acid ethyl ester **1** in a single process is achieved by a highly diastereo- and enantioselective domino-Michael-aldol reaction of ethyl benzoylacetate (**7**) and benzylidene acetone (**8**) in the presence of the organocatalyst **6**.

Initially, the β-keto ester (**7**) and the enone (**8**) undergo an intermolecular Michael reaction to form the adduct **9**, which subsequently undergoes an intramolecular aldol reaction to give the target molecule **1** with three defined stereogenic centers in a chemical yield of 72% and 88% *ee*.

The catalyst plays a threefold role in this domino process: 1) it activates the Michael acceptor by the formation of an iminium ion (**10**), 2) it generates the active Michael donor by deprotonation of the β-keto ester (**7**), and 3) it acts as a base in the intramolecular aldol reaction.

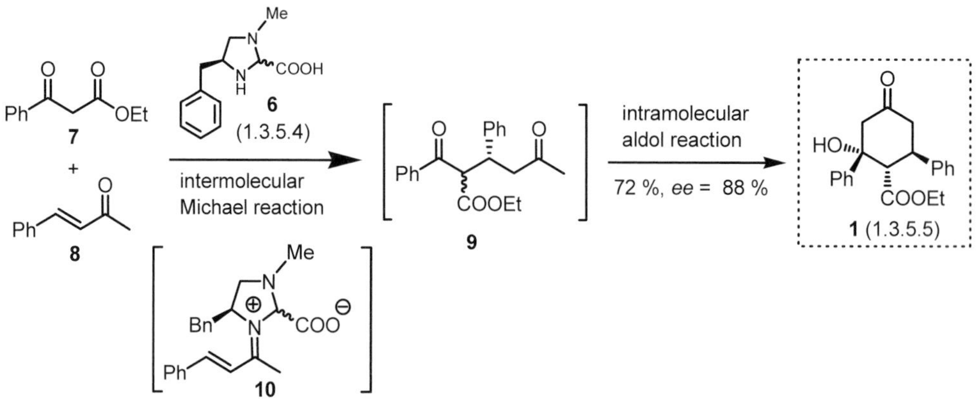

(c) Experimental procedures for the synthesis of 1

1.3.5.1   **   (2S)-2-Amino-3-phenylpropionic acid methyl ester hydrochloride [11]

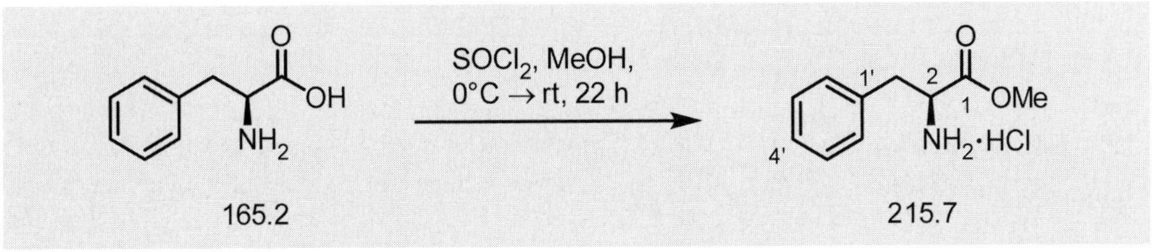

165.2   215.7

Thionyl chloride (18.8 g, 158 mmol, 11.5 mL) is slowly added to a stirred suspension of L-phenylalanine (20.1 g, 122 mmol) in MeOH (120 mL) under an argon atmosphere at 0 °Cand stirring is continued at room temperature for 22 h.

After removal of the solvent under reduced pressure, H_2O (30 mL) is added and evaporated under reduced pressure. This process is repeated three times. After drying *in vacuo*, the hydrochloride is obtained as a colorless solid; 25.7 g (98%), mp 160–161 °C.

> **UV** (CH_3CN): λ_{max} (nm) (lg ε) = 263.5 (2.197), 257.0 (2.298), 252.0 (2.214), 192.5 (4.435).
>
> **IR** (KBr): $\bar{\nu}$ (cm^{-1}) = 2845, 1747, 1584, 1496, 1242, 1146, 1084, 935, 741, 702.
>
> **^{1}H NMR** (300 MHz, D_2O): δ (ppm) = 7.88–7.33 (m, 5 H, Ph-H), 4.70 (s_{br}, NH_2), 4.50 (t, J = 5.9 Hz, 1 H, 2-H), 3.91 (s, 3 H, $COOCH_3$), 3.42 (dd, J = 14.6, 5.9 Hz, 1 H, 3-H), 3.32 (dd, J = 14.3, 7.5 Hz, 1 H, 3-H).
>
> **^{13}C NMR** (75 MHz, D_2O): δ (ppm) = 171.0 (C-1), 134.7 (C-4'), 130.4 (C-2', C-6'), 130.2 (C-3', C-5'), 129.1 (C-1'), 55.08 (C-2), 54.55 ($COOCH_3$), 36.56 (C-3).
>
> **MS** (EI, 70 eV): *m/z* (%) = 179 (2) $[M - HCl]^+$.

1.3.5.2   *   (2S)-2-Amino-3-phenylpropionic acid methyl amide [7c]

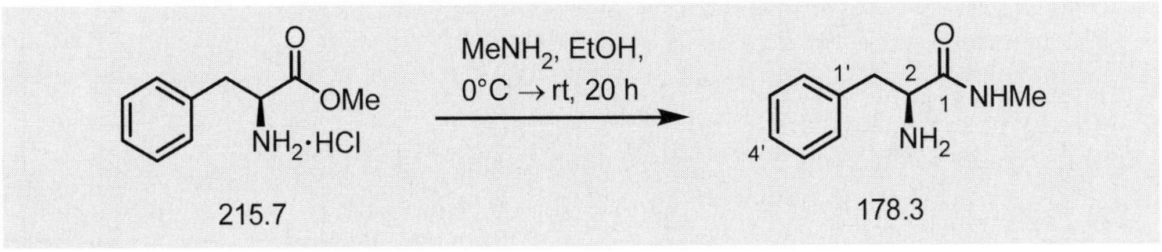

215.7   178.3

A solution of the hydrochloride **1.3.5.1** (25.6 g, 119 mmol) in EtOH (200 mL) is added to a stirred solution of methylamine (8 M, 59.4 mL, 475 mmol) in EtOH under an argon atmosphere at 0 °C and stirring is continued for 20 h at room temperature.

The solvent is then removed *in vacuo*, the residue is suspended in Et_2O (30 mL), and the solvent is again evaporated to remove the excess methylamine. This procedure is repeated twice to give the hydrochloride of the desired product as a white solid. The amide is obtained from the hydrochloride by treating the residue with saturated NaHCO$_3$ solution (100 mL) and extracting with CHCl$_3$ (4 × 100 mL). Washing of the organic layers with brine, drying over Na$_2$SO$_4$, and removal of the solvent *in vacuo* affords the amide as colorless crystals; 19.6 g (92%), mp 55–56 °C, $[\alpha]_D^{20} = -100.5$ (c = 1.0, CHCl$_3$), R_f = 0.39 (EtOAc/MeOH, 1:1).

UV (CH$_3$CN): λ_{max} (nm) (lg ε) = 268 (2.096), 264 (2.190), 258.0 (2.307), 253.0 (2.237), 248.0 (2.130), 192.5 (4.515).

IR (KBr): $\tilde{\nu}$ (cm^{-1}) = 3372, 2939, 1646, 1527, 1399, 1109, 927, 857, 747, 701, 482.

^{1}H NMR (300 MHz, CDCl$_3$): δ (ppm) = 7.35–7.17 (m, 5 H, Ph-H), 3.60 (dd, J = 9.4, 3.8 Hz, 1 H, 3-H$_b$), 3.28 (dd, 1 H, J = 13.8, 4.0 Hz, 3-H$_a$), 2.81 (d, J = 4.9 Hz, 3 H, CH$_3$), 2.67 (dd, J = 13.8, 9.6 Hz, 1 H, 2-H), 1.33 (s$_{br}$, 2 H, NH$_2$).

^{13}C NMR (50 MHz, CDCl$_3$): δ (ppm) = 174.7 (C-1), 137.9 (C-1'), 129.2 (C-2',C-6'), 128.6 (C-3', C-5'), 126.7 (C-4'), 56.38 (C-2), 40.93 (C-3), 25.72 (NH$\underline{C}$H$_3$).

MS (DCI, NH$_3$, 200 eV): m/z (%) = 179 (100) $[M+H]^+$, 196 (45) $[M+NH_4]^+$.

1.3.5.3   **   (2S)-1N-Methyl-3-phenylpropane-1,2-diamine

A solution of the amide **1.3.5.2** (3.79 g, 21.3 mmol) in THF (80 mL) is added dropwise to a stirred suspension of LiAlH$_4$ (2.96 g, 78.0 mmol) in THF (60 mL) under an argon atmosphere and stirring is continued at reflux for 20 h.

After cooling to 0 °C, saturated Na$_2$SO$_4$ solution is added dropwise and the mixture is stirred for 30 min. The white solid is then filtered off and washed with EtOAc. The filtrate is washed with brine, dried over Na$_2$SO$_4$, and the solvent is removed *in vacuo*. Finally, the residue is purified by distillation to yield the diamine as a colorless oil; 3.40 g (97%), bp$_{0.4}$ 120–121 °C, n_D^{20} = 1.528, $[\alpha]_D^{20}$ = –6.0 (c = 1.0, CHCl$_3$), R_f = 0.33 (CHCl$_3$/MeOH, 1:1 + 10% NEt$_3$).

UV (CH$_3$CN): λ_{max} (nm) (lg ε) = 268.0 (2.138), 261.0 (2.268), 192.0 (4.491).

IR (KBr): $\tilde{\nu}$ (cm^{-1}) = 3372, 2939, 1646, 1527, 1399, 1109, 928, 747, 701.

^{1}H NMR (300 MHz, CDCl$_3$): δ (ppm) = 7.30–7.13 (m, 5 H, Ph-H), 2.75 (dd, J = 15.5, 5.0 Hz, 1 H, 3-H$_a$), 2.62 (dd, 1 H, J = 11.4, 3.8 Hz, 3-H$_b$), 2.51–2.41 (m, 2 H, 1-H), 2.40 (s, 3 H, NHC$\underline{H}$$_3$), 1.23 (s$_{br}$, 3 H, NH$_2$, N$\underline{H}CH_3$).

^{13}C NMR (50 MHz, CDCl$_3$): δ (ppm) = 139.1 (C-1'), 129.1 (C-2', C-6'), 128.3 (C-3', C-5'), 126.1 (C-4'), 58.29 (C-1), 52.18 (C-2), 42.72 (C-3), 36.51 (NH$\underline{C}$H$_3$).

MS (DCI, NH$_3$, 200 eV): m/z (%) = 165 (100) $[M+H]^+$.

1.3.5.4   *   (4*R*,2*R/S*)-4-Benzyl-1-methylimidazolidine-2-carboxylic acid
(Jørgensen's catalyst) [5b]

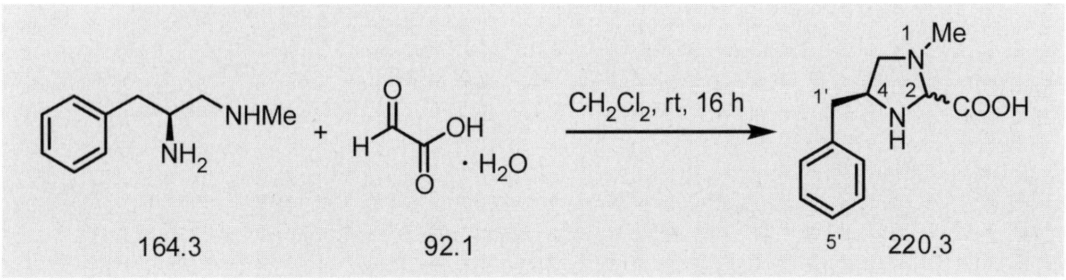

164.3   92.1   220.3

The diamine **1.3.5.3** (2.96 g, 18.05 mmol) is suspended in CH_2Cl_2 (180 mL) under an argon atmosphere. Glyoxylic acid monohydrate (1.66 g, 18.05 mmol) is added and the resulting suspension is stirred at room temperature for 16 h.

Evaporation of the solvent under reduced pressure affords the carboxylic acid in quantitative yield as a colorless solid as a 2:1 mixture of diastereomers; mp 122–123 °C, $[\alpha]_D^{20} = +10.3$ ($c = 1.0$, MeOH), $R_f = 0.47$ (CHCl$_3$/MeOH, 1:1 [+ 10% NEt$_3$]).

UV (CH$_3$CN): λ_{max} (nm) (lg ε) = 267.0 (2.104), 258.0 (2.359), 252.0 (2.330), 248.0 (2.270), 205.0 (3.949).

IR (KBr): $\tilde{v}$ (cm^{-1}) = 3483, 2951, 2786, 1664, 1629, 1573, 1435, 1301, 1205, 1176, 1025, 943, 781, 755, 704, 607.

¹H NMR (300 MHz, CDCl$_3$): major diastereomer: δ (ppm) = 8.10–7.40 (2 × s$_{br}$, 2 H, COOH, NH), 7.32–7.20 (m, 5 H, Ph-H), 4.19 (s, 1 H, 2-H), 3.74 (quintet, $J = 6.8$ Hz, 1 H, 4-H), 3.48–3.41 (m, 1 H, 5-H$_b$), 3.21 (dd, $J = 13.4$, 5.8 Hz, 1 H, 5-H$_a$), 2.93–2.52 (m, 2 H, 1'-H), 2.89 (s, 3 H, N–CH$_3$).

minor diastereomer: δ (ppm) = 8.10–7.40 (2 × s$_{br}$, 2 H, COOH, NH), 7.32–7.20 (m, 5 H, Ph-H), 4.12 (s, 1 H, 2-H), 4.01 (quintet, $J = 6.7$ Hz, 1 H, 4-H), 3.71–3.64 (m, 1 H, 5-H$_b$), 3.01 (dd, $J = 13.4$, 6.3 Hz, 1 H, 5-H$_a$), 2.93–2.52 (m, 2 H, 1'-H), 2.84 (s, 3 H, N–CH$_3$).

¹³C NMR (50 MHz, CDCl$_3$): major diastereomer: δ (ppm) = 168.9 (COOH), 137.4 (C-2'), 128.8 (2 × C-3'), 128.7 (2 × C-4'), 126.8 (C-5'), 84.94 (C-2), 58.44 (C-4), 58.12 (C-5), 40.43 (N–CH$_3$), 38.25 (C-1').

minor diastereomer: δ (ppm) = 169.4 (COOH), 137.3 (C-2'), 129.1 (2 × C-3'), 128.6 (2 × C-4'), 126.7 (C-5'), 81.91 (C-2), 58.92 (C-5), 57.31 (C-4), 39.81 (C-1'), 39.24 (N–CH$_3$).

MS (ESI): m/z (%) = 243 (40) [M + Na]$^+$.

1.3.5.5   **   (1*S*,2*R*,6*R*)-2-Hydroxy-4-oxo-2,6-diphenyl cyclohexane carboxylic acid ethyl ester
[10]

To a stirred solution of benzylidene acetone (77.1 mg, 527 µmol) in CH₃CN (1 mL) are added ethyl benzoylacetate (203 mg, 1.06 mmol) and Jørgensen's catalyst **1.3.5.4** (11.6 mg, 52.7 µmol, 10 mol%) and the resulting solution is stirred for 72 h at room temperature.

The reaction mixture is then diluted with Et₂O (2 mL). After filtration and washing the filter cake with Et₂O (2 mL), the solvent is evaporated *in vacuo* to afford the ethyl ester as a colorless solid; 127 mg (72%), *ee* = 88%, $[\alpha]_D^{20} = -7.6$ (*c* = 1.0, CHCl₃).

UV (CH₃CN): λ_{max} (nm) (lg ε) = 256.5 (0.074), 251.0 (0.022), 201.0 (1.340).

IR (KBr): $\tilde{v}$ (cm⁻¹) = 3348, 1713, 1374, 1225, 1145, 1029, 749, 698.

¹H NMR (300 MHz, CDCl₃): δ (ppm) = 7.55 (d, *J* = 7.5 Hz, 2 H, Ph-H), 7.39–7.22 (m, 7 H, Ph-H), 4.45 (d, *J* = 2.5 Hz, 1 H, OH), 3.86–3.74 (m, 1 H, 5-H), 3.61–3.49 (m, 3 H, 1-H, OCH₂), 2.79–2.70 (m, 4 H, 3-H, 5-H), 0.53 (t, *J* = 7.2 Hz, 3 H, CH₃).

¹³C NMR (50 MHz, CDCl₃): δ (ppm) = 206.0 (C-4), 174.2 (-$\underline{C}$OOR), 144.1 (Ph-C_quat), 140.2 (Ph-C_quat), 128.4 (2 × Ph-C), 127.6 (2 × Ph-C), 127.6 (2 × Ph-C), 127.5 (2 × Ph-C), 124.6 (2 × Ph-C), 60.62 (C-2), 56.61 (C-1), 53.97 (C-3), 47.37 (C-5), 43.28 (C-6), 13.23 (CH₃).

MS (EI, 70 eV): *m/z* (%) = 338 (12) [*M*⁺].

---

[1]   E. Negishi, *Handbook of Organopalladium Chemistry for Organic Synthesis*, Wiley-VCH, Weinheim, **2002**.

[2]   a) Z. G. Hajos, D. R. Parrish, *J. Org. Chem.* **1974**, *39*, 1615; b) U. Eder, G. Sauer, R. Wiechert, *Angew. Chem.* **1971**, *83*, 492; *Angew. Chem. Int. Ed. Engl.* **1971**, *10*, 496.

[3]   a) A. Berkessel, H. Gröger, *Asymmetric Organocatalysis*, Wiley-VCH, Weinheim, **2005**; b) P. I. Dalko, L. Moisan, *Angew. Chem.* **2004**, *116*, 5248; *Angew. Chem. Int. Ed.* **2004**, *43*, 5138; c) B. List, *Tetrahedron* **2002**, *58*, 5573; d) E. R. Jarvo, S. J. Miller, *Tetrahedron* **2002**, *58*, 2481; e) P. I. Dalko, L. Moisan, *Angew. Chem.* **2001**, *113*, 3840; *Angew. Chem. Int. Ed.* **2001**, *40*, 3726; f) H. Gröger, J. Wilken, *Angew. Chem.* **2001**, *113*, 545; *Angew. Chem. Int. Ed.* **2001**, *40*, 529.

[4]   a) N. S. Chowdari, D. B. Ramachary, A. Cordova, C. F. Barbas III, *Tetrahedron Lett.* **2002**, *43*, 9591; b) M. Nakadai, S. Saito, H. Yamamoto, *Tetrahedron* **2002**, *58*, 8167; c) A. B. Northrup, D. W. C. MacMillan, *J. Am. Chem. Soc.* **2002**, *124*, 6798; d) A. Bøgevig, K. Juhl, N. Kumaragurubaran, K. A. Jørgensen, *Chem. Commun.* **2002**, 620; e) A. Cordova, W. Notz, C. F. Barbas III, *J. Org. Chem.* **2002**, *67*, 301; f) K. Sakthivel, W. Notz, T. Bui, C. F. Barbas III, *J. Am. Chem. Soc.* **2001**, *123*, 5260; g) B. List, R. A. Lerner, C. F. Barbas III, *J. Am. Chem. Soc.* **2000**, *122*, 2395.

[5]   a) S. B. Tsogoeva, S. Wei, *Chem. Commun.* **2006**, 1451; b) N. Halland, R. G. Hazell, K. A. Jørgensen, *J. Org. Chem.* **2002**, *67*, 8331; c) N. A. Paras, D. W. C. MacMillan, *J. Am. Chem. Soc.* **2002**, *124*, 7894; d) D. Enders, A. Seki, *Synlett* **2002**, 26; e) J. M. Betancort, K. Sakthivel, R. Thayumanavan, C. F. Barbas III, *Tetrahedron Lett.* **2001**, *42*, 4441; f) J. M. Betancort, C. F. Barbas III, *Org. Lett.* **2001**, *3*, 3737; g) B. List, C. Castello, *Synlett* **2001**, *11*, 1687.

[6]   a) A. Cordova, W. Notz, G. Zhong, J. M. Betancort, C. F. Barbas III, *J. Am. Chem. Soc.* **2002**, *124*, 1842; b) A. Cordova, S. Watanabe, F. Tanaka, W. Notz, C. F. Barbas III, *Org. Lett.* **2002**, *4*, 15; c) W. Notz, K. Sakthivel, T. Bui, C. F. Barbas III, *Tetrahedron Lett.* **2001**, *42*, 199; d) B. List, *J. Am. Chem. Soc.* **2000**, *122*, 9336.

[7]   a) K. Juhl, K. A. Jørgensen, *Angew. Chem.* **2003**, *115*, 1536; *Angew. Chem. Int. Ed.* **2003**, *42*, 1498; b) A. B. Northrup, D. W. C. MacMillan, *J. Am. Chem. Soc* **2002**, *124*, 2458; c) K. A. Ahrendt, C. J. Borths, D. W. C. MacMillan, *J. Am. Chem. Soc.* **2000**, *122*, 4243.

[8]   a) S. Hoffmann, A. M. Seayad, B. List, *Angew. Chem.* **2005**, *117*, 7590; *Angew. Chem. Int. Ed.* **2005**, *44*, 7424; b) M. Rueping, E. Sugiono, C. Azap, T. Theissmann, M. Bolte, *Org. Lett.* **2005**, *7*, 3781; c) H. Adolfsson, *Angew. Chem.* **2005**, *117*, 3404; *Angew. Chem. Int. Ed.* **2005**, *44*, 3340; d) M. Rueping, C. Azap, E. Sugiono, T. Theissmann, *Synlett* **2005**, 2367.

[9]   a) D. Enders, M. R. M. Hüttl, C. Grondal, G. Raabe, *Nature* **2006**, *441*, 861; b) L. F. Tietze, G. Brasche, K. Gericke, *Domino Reactions in Organic Synthesis*, Wiley-VCH, Weinheim, **2006**; c) D. A. Evans, J. Wu, *J. Am. Chem. Soc.* **2003**, *125*, 10162; d) D. B. Ramachary, N. S. Chowdari, C. F. Barbas III, *Angew. Chem.* **2003**, *115*, 4365; *Angew. Chem. Int. Ed.* **2003**, *42*, 4233; e) K. A. Jørgensen, *Angew. Chem.* **2000**, *112*, 3702; *Angew. Chem. Int. Ed.* **2000**, *39*, 3558, and references therein; f) S. Kobayashi, K. A. Jørgensen (Eds.), *Cycloaddition Reactions in Organic Synthesis*, Wiley-VCH, Weinheim, **2000**.

[10] a) N. Halland, P. S. Aburel, K. A. Jørgensen, *Angew. Chem.* **2004**, *116*, 1292; *Angew. Chem. Int. Ed.* **2004**, *43*, 1272. b) A comparable organocatalyzed three-component cyclization of aldehydes, α,β-unsaturated aldehydes, and nitroalkenes gives rise to highly functionalized cyclohexene derivatives, establishing four stereocenters in a one-pot reaction; see ref. [9a].

[11] M. R. Paleo, M. I. Calaza, and F. J. Sardina, *J. Org. Chem.* **1997**, *62*, 6862.

1.4   Electrophilic and nucleophilic acylation

1.4.1   (–)-Ethyl (1*R*)-1-methyl-2-oxocyclopentane-1-carboxylate

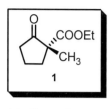

Topics:
- Dieckmann cyclization
- Stereoselective enzymatic reduction of a β-keto ester to a β-hydroxy ester
- Stereoselective α-alkylation of a β-hydroxy ester (Frater–Seebach alkylation)
- Oxidation of a secondary hydroxy group to a keto group

(a) General

The chiral β-keto ester **1** is the starting material for a synthesis of the pheromone frontalin [1]. In general, cyclopentane derivatives are valuable building blocks in the total synthesis of natural products, since many of them, e.g. steroids and iridoids, contain a five-membered ring system.

Retrosynthesis of **1** according to **A** and **B** immediately leads to cyclopentanone or diethyl adipate and two routes **I/II** for the synthesis of **1**:

Route **I** represents a 2-methylation of cyclopentanone-2-carboxylate (**2**), which is easily accessible either from diethyl adipate or from α-acylation of cyclopentanone with dialkyl carbonate. Route **II** requires a 2-acylation of 2-methylcyclopentanone (**3**) with dialkyl carbonate; however, the disadvantage arises that Claisen condensations of **3** are reported to take place preferentially at the less hindered C-5 [2]. Therefore, route **I** is favorable and as the central problem of the synthesis of **1** there remains the enantioselective formation of its stereogenic center.

(b) Synthesis of 1

The starting material of choice for the synthesis of **1** is cyclopentanone carboxylate *rac*-**2**, which is readily prepared from diethyl adipate (**4**) by Dieckmann cyclization in the presence of NaOEt [3]. Since direct enantioselective methylation of *rac*-**2** at the 2-position by application of the SAMP methodology (cf. **1.2.1**) proceeds only with modest stereoselection [4], an indirect approach to **1** is applied [5].

In the first step, the well-established [6] enzymatic reduction of the racemic β-keto ester **2** with Baker's yeast in fermenting aqueous glucose solution is performed, which produces the 2-cyclopentanol-1-carboxylate **5** with (1*R*,2*S*)-configuration in 99% *ee* as a single diastereomer. In the second step, the chiral β-hydroxy ester **5** is deprotonated with two equivalents of LDA and then reacted with methyl iodide in the presence of DMPU. In this process, known as Frater–Seebach alkylation [7], exclusive α-C-alkylation of the β-hydroxy ester is observed to give the product **6** with high stereoselectivity (>98% *de*).

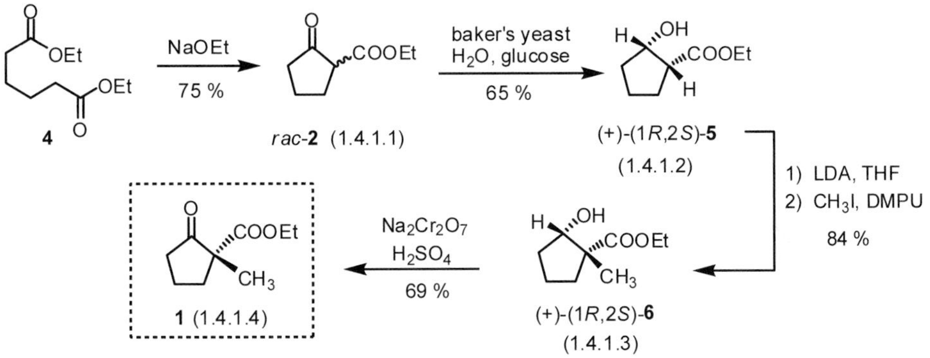
OEt
OEt
4
NaOEt
75 %
COOEt
rac-**2** (1.4.1.1)
baker's yeast
H_2O, glucose
65 %
OH
COOEt
H
(+)-(1*R*,2*S*)-**5**
(1.4.1.2)
1) LDA, THF
2) CH_3I, DMPU
84 %
COOEt
CH_3
1 (1.4.1.4)
$Na_2Cr_2O_7$
H_2SO_4
69 %
(+)-(1*R*,2*S*)-**6**
(1.4.1.3)

Intermediates in the Frater–Seebach alkylation are dianions (here: **7**), which exist as rigid Li-chelated structures. These are thought to be responsible for the stereodifferentiation in the alkylation **5** → **6**, as also described for analogous reactions of open-chain systems (e.g. **8**, "acyclic stereoselection") [8].

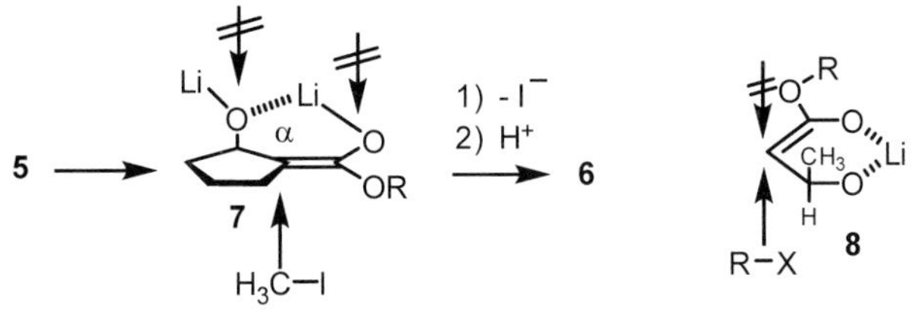
5
Li
Li
α
OR
7
H_3C—I
1) - I^-
2) H^+
6
R
CH_3
Li
H
R–X
8

In the last step, the hydroxy ester **6** is oxidized using $Na_2Cr_2O_7/H_2SO_4$ to provide the chiral β-keto ester **1**.

Thus, the target molecule is obtained in practically enantiopure form (>98% *ee*) in a four-step procedure with an overall yield of 28% (based on **4**).

(c) Experimental procedures for the synthesis of 1

1.4.1.1 * Ethyl 2-oxo cyclopentane-1-carboxylate [3]

202.2   156.2

Sodium ethoxide is prepared by reacting metallic sodium (11.5 g, 0.50 mol) with anhydrous EtOH (150 mL) and distilling off the excess EtOH *in vacuo*. Anhydrous toluene (100 mL) and diethyl adipate (101 g, 0.50 mol) are added, and the resulting suspension is heated under reflux with stirring for 8 h.

The mixture is then cooled to room temperature, HCl (2 M) is added until a clear two-phase system is obtained (ca. 250 mL), and the phases are separated. The organic phase is washed with saturated $NaHCO_3$ and brine (each 100 mL), and dried (Na_2SO_4). The solution is distilled *in vacuo* (20 mbar) and the fraction obtained in the range 100–140 °C is again distilled *in vacuo* (2 mbar, Vigreux column) to give a colorless oil, 58.2 g (75%), bp$_2$ 88–89 °C, n_D^{20} = 1.4519.

> **IR** (film): $\tilde{v}$ (cm^{-1}) = 1740 (ring C=O), 1715 (ester C=O).
>
> **^{1}H NMR** (CDCl$_3$): δ (ppm) = 4.19 (q, *J* = 7.5 Hz, 2 H, OCH$_2$), 3.15 (m, 1 H, CH), 2.6–1.6 (m, 6 H, (CH$_2$)$_3$), 1.29 (t, *J* = 7.5 Hz, 3 H, CH$_3$).

1.4.1.2 * (+)-Ethyl (1*R*,2*S*)-2-hydroxycyclopentane-1-carboxylate [5]

156.1   158.2

In a 3 L Erlenmeyer flask with a stirring or shaking apparatus, baker's yeast (225 g; *Pleser, Darmstadt*) is suspended in H_2O (tap, 1.5 L) and saccharose (225 g) is added. After 0.5 h, ethyl 2-oxocyclopentane-1-carboxylate **1.4.1.1** (22.5 g, 143 mmol) and Triton® X 114 (450 mg, *Fluka*) are added and the mixture is stirred for 48 h at room temperature.

Hyflow Super Cel® (80 g, *Fluka*) is added in portions with stirring and then the mixture is filtered through a G2-frit, saturated with NaCl, and extracted with Et_2O (4 × 300 mL). The ethereal extracts are dried over MgSO$_4$. The extracts of four such experiments are combined, the solvent is removed under normal pressure, and the residue is purified by distillation. The product is obtained as a colorless oil; 62.6 g (65%), bp$_{10}$ 95–96 °C. $[\alpha]_D^{20}$ = +15.1 (*c* = 2.25, CHCl$_3$), lit. [9]: +14.7 (*c* = 2.08, CHCl$_3$).

IR (film): $\tilde{v}$ (cm^{-1}) = 3660, 3450, 2985, 1765.

^{1}H NMR (CDCl$_3$): δ (ppm) = 4.39 (m$_c$, 1 H, C<u>H</u>OH), 4.13 (q, J = 7.1 Hz, 2 H, CH$_2$O), 3.13 (s$_{br}$, 1 H, OH), 2.63 (m$_c$, 1 H, CHCO), 2.02–1.8 (m, 3 H), 1.74 (m$_c$, 2 H), 1.60 (m$_c$, 1 H), 1.27 (t, J = 7.1 Hz, 3 H, OCH$_3$), 1.18 (s, 3 H, CH$_3$).

^{13}C NMR (CDCl$_3$): δ (ppm) = 174.80, 73.81, 60.60, 49.69, 34.12, 26.40, 22.10, 14.24.

1.4.1.3   **   (+)-Ethyl (1*R*,2*S*)-2-hydroxy-1-methylcyclopentane-1-carboxylate [5]

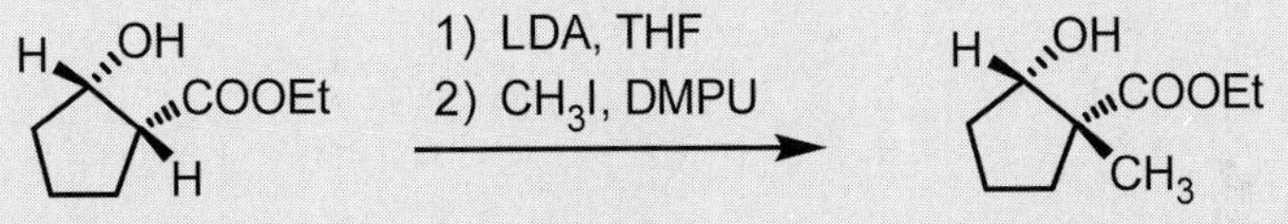

A solution of LDA is prepared by adding methyllithium (375 mL, 0.60 mol, 1.6 M in Et$_2$O) to *N,N*-diisopropylamine (60.7 g, 0.60 mol) in anhydrous THF (225 mL) at –78 °C and then keeping the mixture at 0 °C for 1 h. A solution of the carboxylate **1.4.1.2** (40.1 g, 0.25 mol) in anhydrous THF (60 mL) is then added in one portion to this LDA solution at –50 °C. The temperature rises to –10 °C and stirring is continued for 0.5 h at this temperature. Iodomethane (49.7 g, 0.35 mol; Caution: carcinogenic) in DMPU (125 mL) is then added, whereupon the temperature rises to 40 °C. Stirring is continued for 20 h at room temperature.

The mixture is then poured into saturated NH$_4$Cl solution (1000 mL), extracted with Et$_2$O (4 × 200 mL), and the combined organic layers are washed with brine, dried (MgSO$_4$), and concentrated. The crude product is purified by distillation; 36.1 g (84%), colorless oil, bp$_{10}$ 99–100 °C; $[\alpha]_D^{20}$ = +28.4 (*c* = 1.61, CHCl$_3$).

IR (film): $\tilde{v}$ (cm^{-1}) = 3455 (OH), 1730, 1720, 1705 (C=O).

^{1}H NMR (CDCl$_3$): δ (ppm) = 4.18 (q, J = 7.1 Hz, 2 H), 4.00–3.96 (m, 1 H), 3.19 (m, 1 H), 2.27–2.18 (m, 1 H), 2.02–1.98 (m, 1 H), 1.86–1.82 (m, 1 H), 1.73–1.64 (m, 2 H), 1.59–1.53 (m, 1 H), 1.27 (t, J = 7.1 Hz, 3 H), 1.18 (s, 3 H, CH$_3$).

^{13}C NMR (CDCl$_3$): δ (ppm) = 177.2, 80.0, 60.6, 54.1, 33.3, 32.0, 22.4, 20.5, 14.2.

Note: Methyllithium can be replaced by *n*-butyllithium (1.6 or 2.5 M in *n*-hexane).

1.4.1.4   *   (–)-Ethyl (1*R*)-1-methyl-2-oxocyclopentane-1-carboxylate [5]

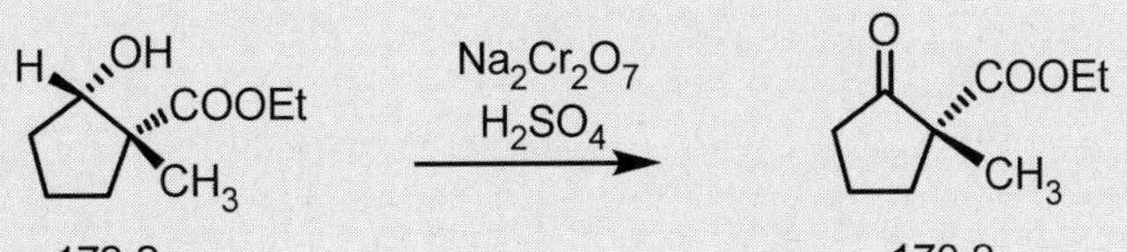

A chromic acid solution prepared from $Na_2Cr_2O_7 \cdot 2\,H_2O$ (89.4 g, 0.30 mol) and concentrated H_2SO_4 (75 g) in H_2O (200 mL) is added dropwise to a solution of (+)-(1R,2S)-**1.4.1.3** (34.4 g, 0.20 mol) in Et_2O (200 mL)at 0–5 °C and stirring is continued for 20 h at room temperature.

H_2O (220 mL) is then added and the mixture is extracted with Et_2O (4 × 200 mL). The combined organic layers are washed with saturated $NaHCO_3$ and NaCl solutions, dried ($MgSO_4$), and concentrated. The crude product is purified by distillation; 23.5 g (69%), colorless oil, bp_{10} 96 °C; $[\alpha]_D^{20} = -13.3$ ($c = 1.09$, $CHCl_3$).

IR (film): $\tilde{\nu}$ (cm^{-1}) = 1750 (C=O), 1735 (C=O).

^{1}H NMR ($CDCl_3$): δ (ppm) = 4.15 (q, $J = 7.1$ Hz, 2 H), 2.53–2.39 (m, 2 H), 2.35–2.27 (m, 1 H), 2.11–2.02 (m, 1 H), 1.96–1.82 (m, 2 H), 1.30 (s, 3 H, CH_3), 1.24 (t, $J = 7.1$ Hz, 3 H).

^{13}C NMR ($CDCl_3$): δ (ppm) = 215.7, 172.4, 61.3, 55.9, 37.6, 36.3, 19.6, 19.4, 14.1.

[1] T. Hosokawa, Y. Makabe, T. Shinohara, S.-I. Murahashi, *Chem. Lett.* **1985**, 1529.

[2] a) T. Hamada, A. Chieffi, J. Ahman, S. L. Buchwald, *J. Am. Chem. Soc.* **2002**, *124*, 1261; b) analogous behavior is shown by 2-methylcyclohexanone: K. W. Baldry, M. J. T. Robinson, *Tetrahedron* **1977**, *33*, 1663.

[3] R. Mayer, M. Kubasch, *J. prakt. Chem.* **1959**, *9*, 43.

[4] D. Enders, A. Zamponi, T. Schäfer, C. Nübling, H. Eichenauer, A. S. Demir, G. Raabe, *Chem. Ber.* **1994**, *127*, 1707.

[5] Th. Eicher, F. Servet, A. Speicher, *Synthesis* **1996**, 863.

[6] a) G. Frater, *Helv. Chim. Acta* **1980**, *63*, 1383. Ru-catalyzed asymmetric reduction of β-keto esters: b) K. Junge, B. Hagemann, S. Enthaler, G. Oehme, M. Michalik, A. Monsees, T. Riermeier, U. Dingerdissen, M. Beller, *Angew. Chem.* **2004**, *116*, 5176; *Angew. Chem. Int. Ed.* **2004**, *43*, 5066; c) M. T. Reetz, X. Li, *Adv. Synth. Catal.* **2006**, *348*, 1157.

[7] a) G. Frater, *Helv. Chim. Acta* **1979**, *62*, 2828; b) D. Seebach, D. Wasmuth, *Helv. Chim. Acta* **1980**, *63*, 197.

[8] G. Frater, U. Müller, W. Günther, *Tetrahedron* **1984**, *40*, 1269.

[9] D. Seebach, D. Roggo, D. T. Maetzke, *Helv. Chim. Acta* **1987**, *70*, 1605.

1.4.2   Ethyl (*S*)- and (*R*)-2-hydroxy-4-phenylbutanoate

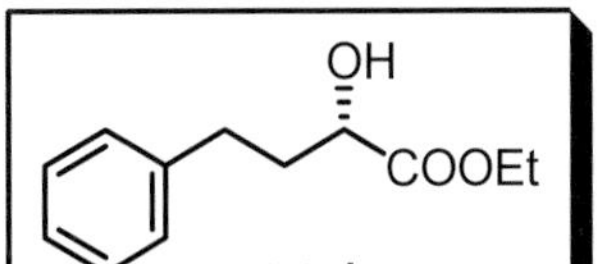

(S)-1

Topics:
- ECP synthesis of both enantiomers of an α-hydroxy-alkanoic ester
- Friedel–Crafts acylation of an arene (succinoylation)
- Catalytic reduction Ph–C=O → Ph–CH$_2$
- Mesylation of an OH function
- Inversion of configuration of a secondary alcohol

(a) General

Syntheses of enantiopure compounds can be performed following two principal strategies:

(1) The construction of the target molecule is conducted in a stereoselective fashion with respect to the required operations, either in an enantioselective way using chiral catalysts or in a diastereoselective way using chiral auxiliaries, which are subsequently removed (*asymmetric synthesis*) [1].

(2) The stereogenic elements required, e.g. one (or several) stereogenic center(s), are introduced by using natural products or other readily available enantiopure compounds as chiral starting materials, which are transformed to the target molecule by stereocontrolled reactions (*ex chiral pool synthesis*, *ECP synthesis*) [2].

ECP syntheses have been performed using a multitude of stereo-defined natural products, e.g. hydroxy- and amino acids, terpenoids, and carbohydrates. For an appropriate choice of a suitable candidate from the chiral pool, the functionalities as well as the number of stereogenic elements and their absolute configuration in the natural product and in the target molecule should correspond. This is demonstrated by the synthesis [3] of (*S*)-1 containing one stereogenic center.

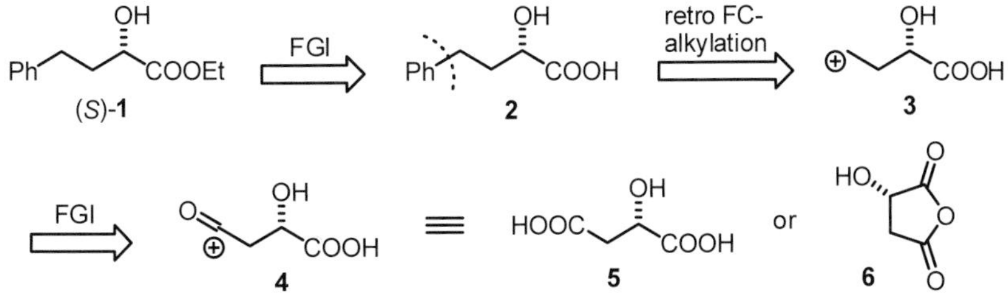

FGI and arene bond disconnection at the benzylic carbon (**1** → **2**) leads to a C$_4$-synthon **3**, which is represented by the acylium ion **4** derived from (*S*)-malic acid (**5**). This chiral (*S*)-configured hydroxy-C$_4$-dicarboxylic acid is a readily available natural product (i.e. from the "chiral pool") that contains the complete carbon side-chain of the target molecule together with the "correct" terminal functionalization and stereochemistry. Therefore, (*S*)-malic acid (**5**), or preferentially its anhydride **6**, are excellent substrates for an ECP synthesis of (*S*)-1 [4, 5].

(b) Synthesis of 1

(1): Reaction of (*S*)-malic acid (**5**) with acetyl chloride gives *O*-acetyl malic anhydride (**7**) by acetylation of the hydroxy group in **5** and elimination of H_2O. Chemoselective Friedel–Crafts acylation (succinoylation) of benzene with the unsymmetrical anhydride **7** in the presence of $AlCl_3$ at the more sterically accessible C=O group affords (*S*)-2-hydroxy-4-oxo-4-phenylbutanoic acid (**8**) in good yield [6]. The acetate in **7** is also cleaved in this process to give a free hydroxy group.

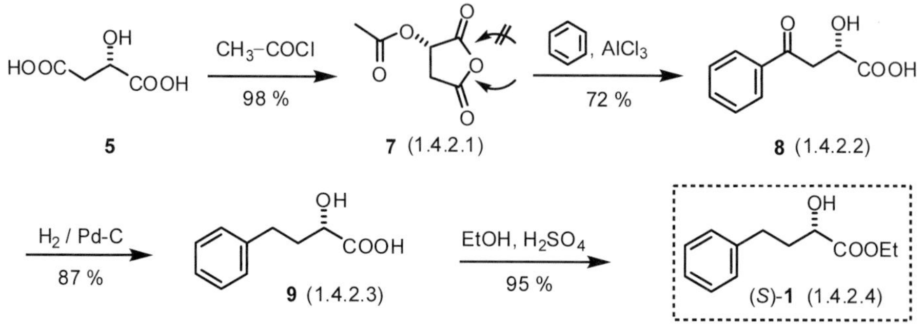

On hydrogenation of the α-oxo acid **8**, the benzylic carbonyl group is readily reduced to a CH_2 unit to yield the α-hydroxy acid **9** in almost quantitative yield and 99% *ee*. Esterification of the acid **9** according to the Fischer method (EtOH/H_2SO_4) yields the ethyl ester (*S*)-**1** in practically enantiopure form (*ee* = 99%).

(2): The simplest way to obtain (*R*)-**1** would be direct inversion of the 2-OH group in the (*S*)-2-hydroxy ester prepared in (1). However, the Mitsunobu reaction (cf. **3.3.4**) as the method of choice gives unsatisfactory results. Thus, even under modified conditions with EtOOC–N=N–COOEt/Ph_3P/$ClCH_2COOH$, followed by hydrolysis with K_2CO_3/H_2O, the acid *ent*-**9** is obtained only in moderate yield [7].

Therefore, the OH group inversion of the stereogenic center in (*S*)-**1** is performed by converting the hydroxy group into a good leaving group by mesylation with CH_3SO_2Cl in pyridine. The mesylate **10**, which is obtained quantitatively, is subjected to an S_N2 displacement (Walden inversion) with potassium propionate in ethanol. The diester **11** is selectively cleaved by alcoholysis with K_2CO_3 in EtOH (due to equilibrium formation of ethanolate) to give the enantiomeric ethyl ester (*R*)-**1** in almost enantiopure form (*ee* = 97%).

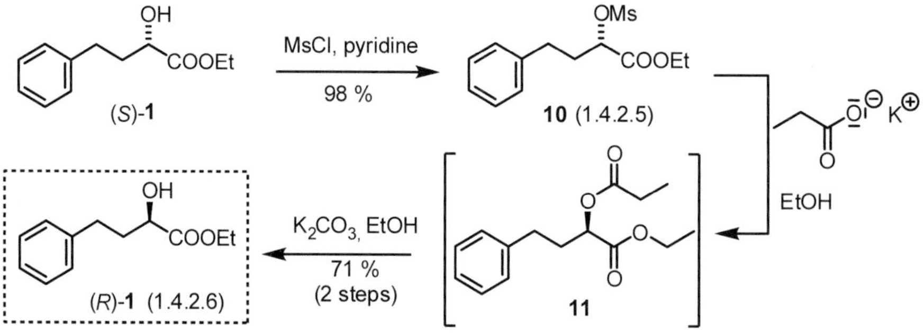

Thus, the target molecule (S)-1 is obtained in a four-step sequence from (S)-malic acid in an overall yield of 63%; the target molecule (R)-1 is obtained from (S)-1 in a three-step sequence in an overall yield of 83% (or from **5** in a seven-step sequence in an overall yield of 52%).

(c) Experimental procedures for the synthesis of 1

1.4.2.1 * (S)-α-Acetoxybutanedioic anhydride [8]

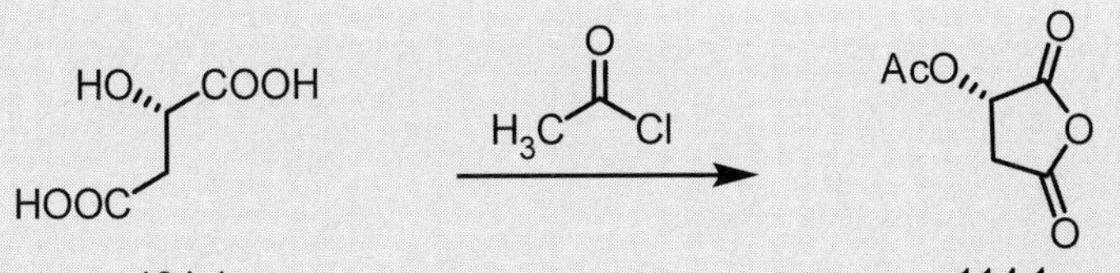

A solution of (S)-malic acid (10.0 g, 0.075 mol) in acetyl chloride (350 mL) is heated under reflux with stirring for 5 h.

The solvent is then removed under reduced pressure to yield (S)-α-acetoxybutanedioic anhydride as a light-yellow solid; 11.6 g (98%); mp 50–52 °C; $[\alpha]_D^{20} = -23.1$ ($c = 5.0$, CHCl$_3$); TLC (SiO$_2$, CHCl$_3$/MeOH/H$_2$O/AcOH = 50:50:3:0.3): $R_f = 0.88$.

> **UV** (MeOH): λ_{max} (nm) = 209.
>
> **IR** (KBr): $\tilde{\nu}$ (cm^{-1}) = 3012, 2962, 1806, 1743, 1405, 1375, 1293, 1216, 1099, 1032, 966, 917, 722, 663, 572.
>
> **^{1}H NMR** (300 MHz, CDCl$_3$): δ (ppm) = 5.51 (dd, $J = 9.6$, 6.3 Hz, 1 H, CH), 3.36 (dd, $J = 19.0$, 9.4 Hz, 1 H, CH$_2$), 3.01 (dd, $J = 19.0$, 6.3 Hz, 1 H, CH$_2$), 2.18 (s, 3 H, CH$_3$).
>
> **^{13}C NMR** (75.5 MHz, CDCl$_3$): δ (ppm) = 170.5, 169.9, 169.5, 137.1, 68.4, 37.7.

1.4.2.2 * (S)-2-Hydroxy-4-oxo-4-phenylbutanoic acid [8]

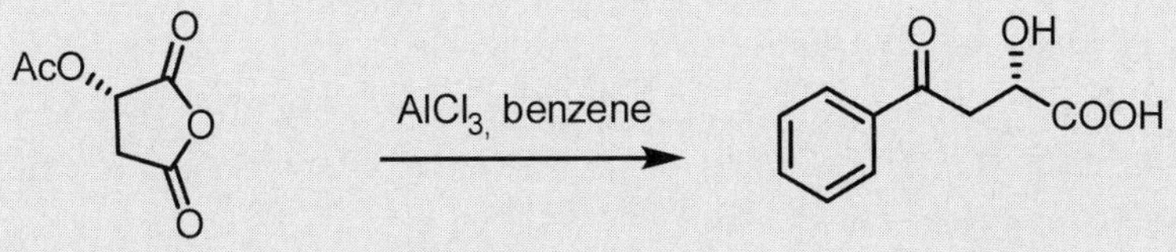

Anhydrous AlCl$_3$ (30 g, 0.15 mol) is added in one portion to a solution of (S)-anhydride **1.4.2.1** (9.5 g, 0.05 mol) in anhydrous benzene (100 mL) at 0 °C. The mixture is heated under reflux with vigorous stirring for 4 h.

It is then poured onto a mixture of crushed ice (100 g) and HCl (1 N, 100 mL). The mixture is stirred for 2 h and extracted with EtOAc (3 × 100 mL). The combined organic phases are washed with brine, dried over Na_2SO_4, filtered, and concentrated under reduced pressure. The crude product is crystallized from EtOAc by the addition of petroleum ether to yield the (S)-2-hydroxy acid as a colorless powder; 8.4 g (72%); mp 136–138 °C; $[\alpha]_D^{20} = -8.75$ (c = 4.0, EtOH); TLC (SiO_2; $CHCl_3$/MeOH/H_2O/AcOH = 70:30:3:0.3): R_f = 0.58.

UV (CH_3OH): λ_{max} (nm) = 278, 241, 201.

IR (KBr): $\tilde{v}$ (cm^{-1}) = 3476, 3083, 3061, 2928, 1734, 1677, 1595, 1451, 1364, 1222, 1194, 1105, 811, 761, 689, 580.

^{1}H NMR (300 MHz, [D_6]DMSO): δ (ppm) = 12.0 (s_{br}, 1 H, COOH), 7.95 (d, J = 7.2 Hz, 2 H, Ar-CH), 7.64 (dd, J = 7.5, 7.2 Hz, 1 H, Ar-CH), 7.53 (t, J = 7.5 Hz, 2 H, Ar-CH), 5.50 (s_{br}, 1 H, OH), 4.50 (d, J = 6.0 Hz, 1 H, CH), 3.32 (d, J = 6.0 Hz, 2 H, CH_2).

^{13}C NMR (75.5 MHz, [D_6]DMSO): δ (ppm) = 197.4, 174.8, 136.7, 128.5, 127.9, 66.6, 42.6.

1.4.2.3   **   (S)-2-Hydroxy-4-phenylbutanoic acid [3]

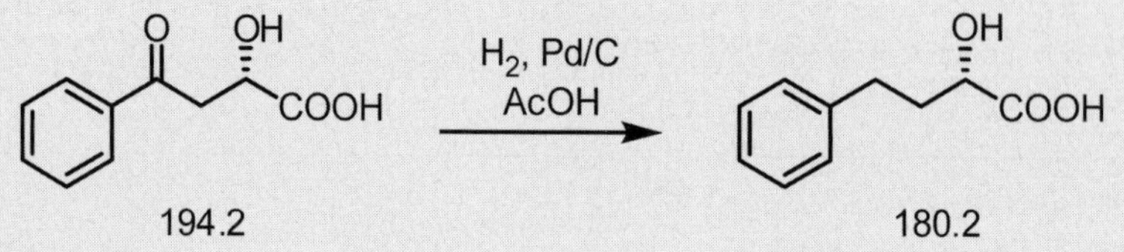

A solution of the hydroxy acid **1.4.2.2** (5.4 g, 0.028 mol) in AcOH (80 mL) is hydrogenated over 10% palladium on carbon (0.7 g) at room temperature under 1 atmosphere pressure of hydrogen.

After complete conversion, as indicated by TLC, the solution is filtered and concentrated under reduced pressure. The crude product is recrystallized from toluene to give (S)-2-hydroxy-4-phenylbutanoic acid as a colorless powder; 4.40 g (87%); mp 65–67 °C; $[\alpha]_D^{20} = +13.4$ (c = 2.5, EtOH); TLC (SiO_2; $CHCl_3$/MeOH/H_2O/AcOH = 70:30:3:0.3): R_f = 0.65.

UV (CH_3OH): λ_{max} (nm) = 267, 258, 242, 207.

IR (KBr): $\tilde{v}$ (cm^{-1}) = 3461, 3027, 2957, 2926, 2861, 2589, 1733, 1497, 1454, 1290, 1270, 1242, 1175, 1097, 1077, 866, 767, 742, 696.

^{1}H NMR (300 MHz, [D_6]DMSO): δ (ppm) = 7.27 (m, 2 H, Ar-CH), 7.17 (m, 3 H, Ar-CH), 3.93 (dd, J = 8.1, 4.5 Hz, 1 H, CH), 2.67 (t, J = 7.8 Hz, 2 H, CH_2), 1.86–1.99 (m, 1 H, CH_2), 1.74–1.85 (m, 1 H, CH_2).

^{13}C NMR (75.5 MHz, [D_6]DMSO): δ (ppm) = 175.5, 141.5, 128.3 (2 CH-Ar), 126.0, 69.0, 35.7, 30.6.

1.4.2.4   *   Ethyl (*S*)-2-hydroxy-4-phenylbutanoate [3]

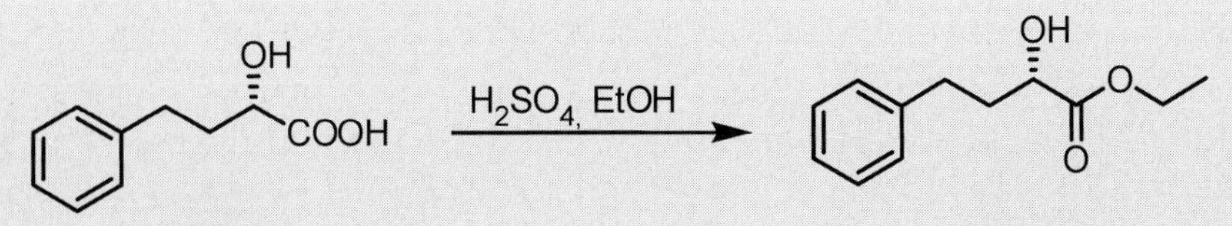

180.2   208.3

Concentrated H_2SO_4 (2 mL) is added to a solution of the hydroxy acid **1.4.2.3** (3.06 g, 0.017 mol) in anhydrous EtOH (200 mL). The mixture is heated under reflux with stirring for 2 h.

The solvent is evaporated under reduced pressure and a mixture of H_2O (50 mL) and EtOAc (200 mL) is added. After shaking, the organic phase is separated, washed with saturated $NaHCO_3$ and brine, and dried over anhydrous Na_2SO_4. Evaporation of the solvent under reduced pressure provides ethyl (*S*)-2-hydroxy-4-phenylbutanoate as a light-yellow oil; 3.36 g (95%); $[\alpha]_D^{20} = +19.8$ ($c = 2.5$, $CHCl_3$); TLC (SiO_2; $CHCl_3$, 0.1% AcOH): $R_f = 0.67$.

UV (CH_3OH): λ_{max} (nm) = 267, 247, 205.

^{1}H NMR (300 MHz, $CDCl_3$): δ (ppm) = 7.33–7.15 (m, 5 H, Ar-CH), 4.23–4.16 (m, 2 H, C$\underline{H}_2$-CH$_3$, 1 H, CH), 2.94 (s$_{br}$, 1 H, OH), 2.84–2.69 (m, 2 H, Ph-C$\underline{H}_2$-CH$_2$), 2.17–2.06 (m, 2 H, Ph-CH$_2$-C$\underline{H}_2$), 2.01–1.88 (m, 2 H, Ph-CH$_2$-C$\underline{H}_2$), 1.27 (t, J = 7.1 Hz, 3 H, CH$_3$).

^{13}C NMR (75.5 MHz, $CDCl_3$): δ (ppm) = 175.1, 141.1, 128.5, 128.3, 125.9, 69.6, 61.6, 35.9, 30.9, 14.1.

1.4.2.5   *   Ethyl (*S*)-2-methanesulfonyloxy-4-phenylbutanoate [3]

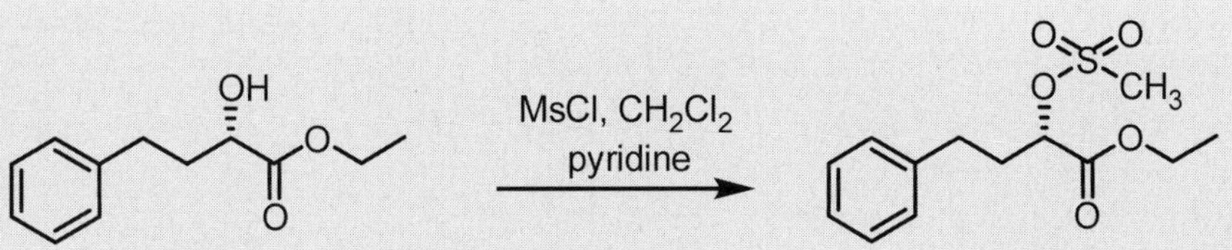

208.3   286.4

Methanesulfonyl chloride (3 mL) is added dropwise to a solution of the hydroxy ester **1.4.2.4** (3.13 g, 0.015 mol) in anhydrous CH_2Cl_2 (10 mL) and anhydrous pyridine (10 mL) at 0 °C. The resulting mixture is stirred at room temperature overnight.

The reaction mixture is then diluted with EtOAc (200 mL) and the resulting solution is washed with H_2O, HCl (2 M), saturated $NaHCO_3$ solution, and brine, and then dried over Na_2SO_4. The solvent is removed under reduced pressure to give the mesylated hydroxy ester as a light-yellow oil, 4.21 g (98%), TLC (SiO_2; $CHCl_3$/MeOH/H_2O/AcOH = 50:50:3:0.3): $R_f = 0.88$.

UV (CH$_3$OH): λ_{max} (nm) = 267, 242, 202.

IR (film): $\tilde{v}$ (cm^{-1}) = 3062, 3029, 2983, 2939, 2869, 1751, 1497, 1455, 1362, 1300, 1252, 1211, 1174, 1039, 964, 864, 844, 820, 747, 701.

^{1}H NMR (300 MHz, [D$_6$]DMSO): δ (ppm) = 7.30 (m, 2 H, Ar-CH), 7.18–7.24 (m, 3 H, Ar-CH), 5.08 (dd, J = 7.5, 5.1 Hz, 1 H, OH), 4.17 (q, J = 7.2 Hz, 2 H, CH$_2$CH$_3$), 4.00 (m$_{br}$, 1 H, CH), 3.27 (s, 3 H, SCH$_3$), 2.71 (t, J = 7.5 Hz, 2 H, CH$_2$), 2.07–2.17 (m, 2 H, CH$_2$), 1.21 (t, J = 7.2 Hz, 3 H, CH$_3$).

^{13}C NMR (75.5 MHz, [D$_6$]DMSO): δ (ppm) = 168.5, 140.1, 128.3, 128.1, 126.1, 76.8, 61.34, 37.9, 33.2, 30.0, 13.8.

1.4.2.6   **   Ethyl (R)-2-hydroxy-4-phenylbutanoate (ent-1) [3]

286.4   208.4

Sodium propionate (4.57 g, 0.016 mol) is added to a solution of the mesylated hydroxy ester **1.4.2.5** (3.92 g, 0.013 mol) in EtOH (130 mL). The resulting mixture is heated under reflux with stirring for 48 h.

The reaction mixture is then cooled to room temperature and filtered; the filtrate is concentrated under reduced pressure. The residue is dissolved in EtOAc (100 mL), and the resulting solution is washed twice with brine and dried over anhydrous Na$_2$SO$_4$. The solvent is removed *in vacuo*. The residue (crude diester **11**) is dissolved in EtOH (250 mL), K$_2$CO$_3$ (5.88 g, 0.043 mol) is added, and the resulting mixture is stirred at room temperature overnight.

It is then filtered, and the filtrate is neutralized with HCl (6 M) and concentrated under reduced pressure. The product is taken up in EtOAc (150 mL) and H$_2$O (50 mL) and the biphasic mixture is shaken. The organic layer is separated, washed with brine, and dried over anhydrous Na$_2$SO$_4$. Evaporation of the solvent gives ethyl (R)-2-hydroxy-4-phenylbutanoate as an oil; 1.92 g (71%), TLC (SiO$_2$; CHCl$_3$, 0.1% AcOH): R_f = 0.67, $[\alpha]_D^{20}$ = –18.8 (c = 2.4 in CHCl$_3$).

UV (CH$_3$OH): λ_{max} (nm) = 267, 258, 242, 205.

IR (film): $\tilde{v}$ (cm^{-1}) = 3429, 3063, 3028, 2980, 2961, 2930, 2865, 1732, 1497, 1454, 1370, 1299, 1247, 1178, 1077, 864, 747, 701.

^{1}H NMR (300 MHz, [D$_6$]DMSO): δ (ppm) = 7.28 (m, 2 H, Ar-CH), 7.14–7.20 (m, 3 H, Ar-CH), 5.41 (s$_{br}$, 1 H, OH), 4.08 (q, J = 7.4 Hz, 2 H, CH$_2$CH$_3$), 4.00 (m$_{br}$, 1 H, CH), 2.66 (t, J = 8.1 Hz, 2 H, CH$_2$), 1.77–1.86 (m, 1 H, CH$_2$), 1.18 (t, J = 7.4 Hz, 3 H, CH$_3$).

^{13}C NMR (75.5 MHz, [D$_6$]DMSO): δ (ppm) = 174.0, 141.5, 128.3, 128.2, 126.8, 68.9, 59.7, 35.7, 30.6, 13.7.

[1]   For monographs, see: a) G. Procter, *Asymmetric Synthesis*, Oxford University Press, Oxford, **1999**; b) R. S. Atkinson, *Stereoselective Synthesis*, John Wiley & Sons, Inc., New York, **1997**; c) M. Christmann, S. Bräse (Eds), *Asymmetric Synthesis–The Essentials*, Wiley-VCH, Weinheim, **2007**.

[2]   a) For a monograph, see: S. Hanessian, *The Chiron Approach*, Pergamon Press, New York, **1983**; b) for chiral pool reagents ("chirons"), see: b) J.-H. Fuhrhop, G. Li, *Organic Synthesis*, 3rd ed., Wiley-VCH, Weinheim, **2003**; c) for a compilation of commercially available enantiopure products, see: M. Breuer, K. Ditrich, T. Habicher, B. Hauer, M. Keßeler, R. Stürmer, T. Zelinsky, *Angew. Chem.* **2004**, *116*, 806; *Angew. Chem. Int. Ed.* **2004**, *43*, 2752.

[3]   W.-Q. Lin, Z. He, Y. Jing, X. Cui, H. Liu, A.-Q. Mi, *Tetrahedron Asymmetry* **2001**, *12*, 1583.

[4]   Conventional retrosynthesis of the α-hydroxy acid **2** leads to hydrocinnammic aldehyde by retro-cyanohydrin transformation. For the problem of asymmetric cyanohydrin synthesis, see: J.-M. Brunel, I. P. Holmes, *Angew. Chem.* **2004**, *116*, 2810; *Angew. Chem. Int. Ed.* **2004**, *43*, 788.

[5]   Other retroanalytical approaches may lead to C_5- and C_6-carbohydrates as chiral starting materials for the synthesis of ethyl esters **1**. However, although readily available, carbohydrates are not considered as substrates for the synthesis of **1**, since transformations sacrificing several stereogenic centers in favor of one are regarded as ineffective with respect to atomic and stereochemical economy.

[6]   Compound **9** has also been prepared by chemo- and enantioselective enzymatic reduction of 4-phenyl-2,4-dioxobutanoic acid: G. Casy, *Tetrahedron Lett.* **1992**, *33*, 8159.

[7]   O. Mitsunobu, *Synthesis* **1981**, 1.

[8]   S. Henrot, M. Larcheveque, Y. Petit, *Synth. Commun.* **1986**, *16*, 183.

1.4.3   Naproxen

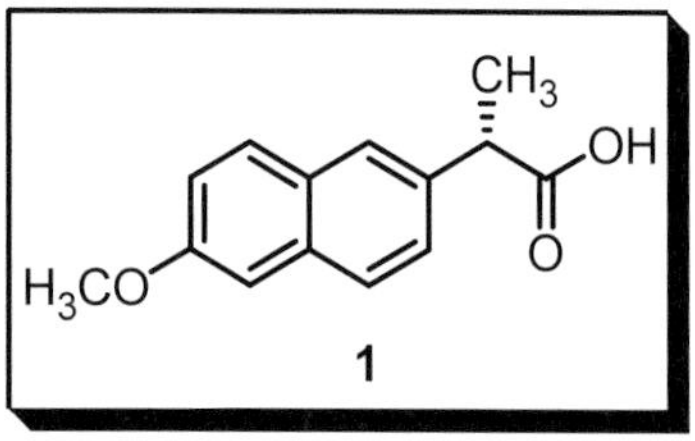

1

Topics:
- Enantioselective synthesis of a drug
- Friedel–Crafts acylation
- Chemoselective α-halogenation of an alkyl aryl ketone
- Acetalization using an orthoester
- Lewis acid-induced rearrangement of an α-halogeno acetal
- Kinetic resolution by enantioselective enzymatic ester hydrolysis
- Resolution by formation of diastereomeric salts

(a) General

Naproxen (**1**, (*S*)-(+)-2-(6'-methoxy-2-naphthyl)propionic acid) is a prominent member of the drugs derived from aryl- and hetaryl-substituted acetic and propionic acids, which exhibit anti-inflammatory, analgetic, and antirheumatic properties [1]. Other important examples are indomethacin (**2**), diclofenac (**3**), ibuprofen (**4**) and tiaprofenic acid (**5**).

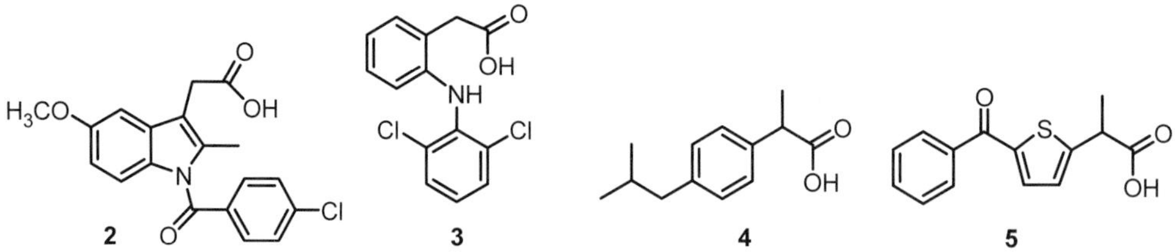

| **2** | **3** | **4** | **5** |

derivatives of acetic acid   derivatives of α-substituted propionic acid

These nonsteroidal anti-inflammatory compounds act as effective inhibitors of prostaglandin biosynthesis [2].

The α-aryl and α-hetaryl propionic acids (e.g. **4** and **5**) are used therapeutically as racemic mixtures. An exception is naproxen, which is marketed and applied as the (*S*)-(+)-enantiomer.

The retrosynthesis of **1** according to pathway **A** leads to 2-methoxynaphthalene via **6** and **8**. Similarly, according to pathway **B**, **1** can be traced back to **7**, which again would be accessible from 2-methoxynaphthalene via **8**. Selective formation of the stereogenic center in **1** could be achieved either by a facially selective alkylation of **6** using the Evans procedure or by enantioselective hydrogenation of **7**. Compounds **6** and **7** should be easily accessible from **8** by classical routes, such as Willgerodt–Kindler or Tl(III)-induced redox transformations (→ **6**) or a cyanohydrin reaction/ CN → COOH hydrolysis/H$_2$O elimination sequence (→ **7**).

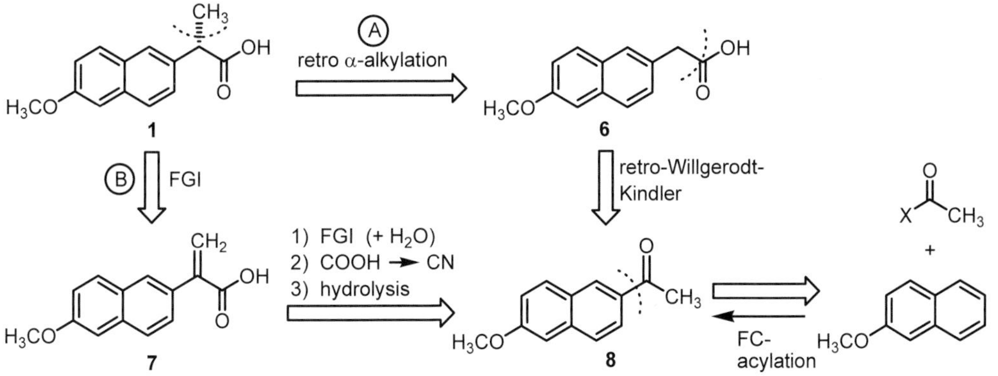

In fact, the first industrial synthesis of naproxen (Syntex) [3] used a Willgerodt–Kindler reaction of 2-acetyl-6-methoxynaphthalene (**8**) to give the morpholide **9**; subsequent hydrolysis led to the arylacetic acid **6**. Its methyl ester **10** was alkylated with CH₃I/NaH (**10 → 11**); saponification of the alkylated ester **11** provided the racemic acid (*rac*-**1**), which was resolved using cinchonidine or other chiral bases to give (*S*)-naproxen (**1**).

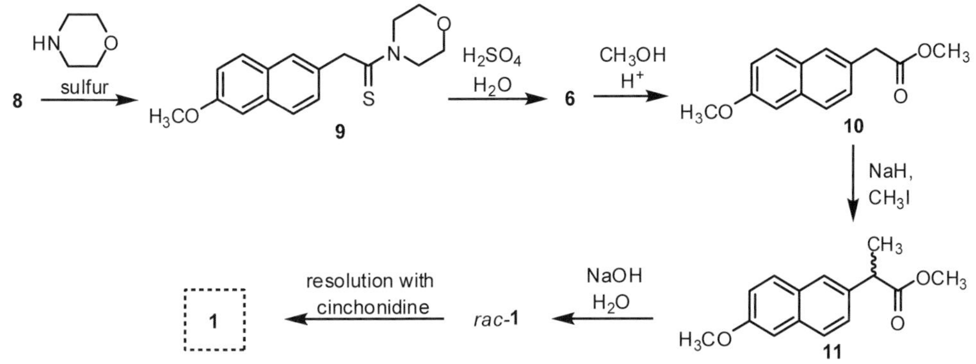

According to the retrosynthetic analysis, (β-naphthyl)acrylic acid **7** is a suitable substrate for the enantioselective formation of **1**, which is readily formed from **7** by homogeneous catalytic hydrogenation in the presence of a chiral Ru complex (Ru(II)-(*S*)-BINAP) [4]. The acrylic acid **7** may be obtained from 2-acetyl-6-methoxynaphthalene (**8**) via the already mentioned cyanohydrin route or more elegantly by an electrocarboxylation. Thus, 2e⁻ cathodic reduction of the carbonyl group in **8** in the presence of CO_2 gives the hydroxy acid **12**, subsequent acid-catalyzed dehydration of which leads to **7** [5].

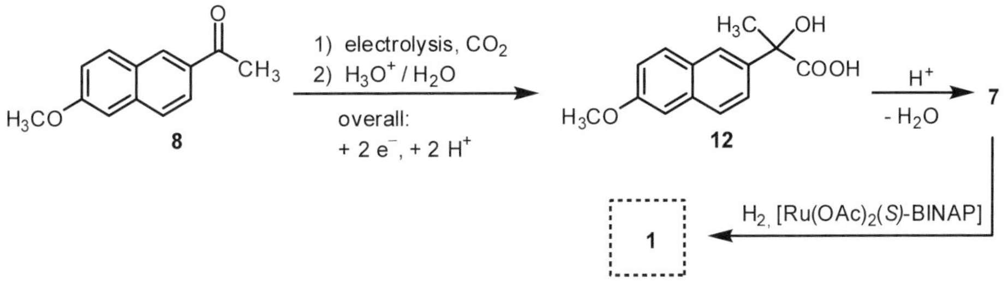

In a strategy that differs from the above retroanalytical considerations, the α-arylpropionate side-chain in **1** can also be established by Lewis acid-induced 1,2-migration of the aryl moiety in aryl ethyl ketals **13** bearing a leaving group at C-2. The reaction leads to esters **14** [6]:

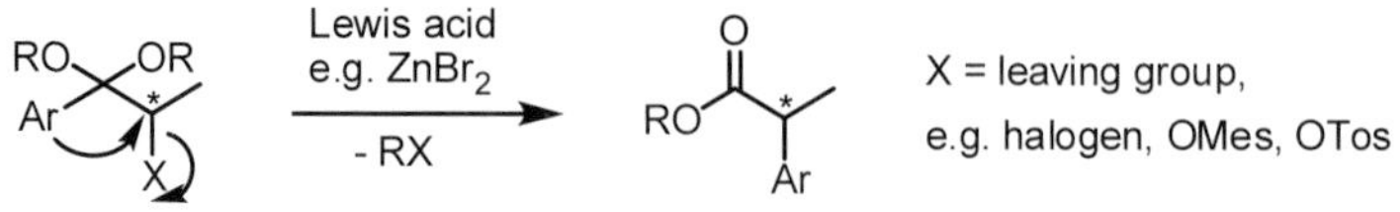

If the leaving group X in **13** assumes a stereodefined position, the 1,2-aryl shift proceeds with preservation of the stereochemical information. This was accomplished in an ECP synthesis of naproxen (**1**) starting with the acid chloride **15** of (S)-O-mesyl lactic acid [3]:

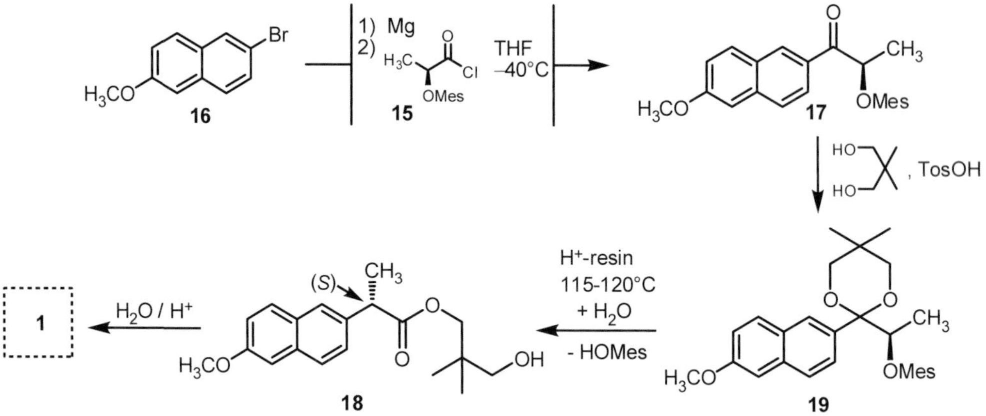

In the first step, the Grignard compound from 2-bromo-6-methoxynaphthalene (**16**) is acylated with the acid chloride **15** to give the naphthyl ketone **17**, the C=O group of which is protected as a 1,3-dioxane with formation of **19**. On treatment with an acidic ion-exchange resin, the O-mesyl acetal moiety in **19** rearranges to give the ester **18**, acid hydrolysis of which provides (S)-naproxen (**1**) in enantiomerically pure form and in 75% overall yield. Remarkably, in the rearrangement step **19 → 18**, migration of the β-naphthyl residue occurs stereoselectively with complete inversion of configuration at the propionate side-chain [7].

For reasons of practicability in the laboratory, a synthesis is presented here that is based on the Lewis acid-promoted rearrangement of an α-bromo ketal to a racemic naproxen ester. This racemic ester is transformed to the chiral target molecule: (1) by enantioselective enzymatic hydrolysis, and (2) by saponification to give the racemic acid and its resolution by cinchonidine salt formation.

(b) Synthesis of 1

The synthesis starts with Friedel–Crafts acylation of 2-methoxynaphthalene (**20**) with propionyl chloride in the presence of AlCl₃. The reaction conditions (nitrobenzene as solvent, 4 d at 0 °C) favor thermodynamic product control in the S_EAr process and direct substitution at the desired β-(2)-position (**20 → 21**) [8]:

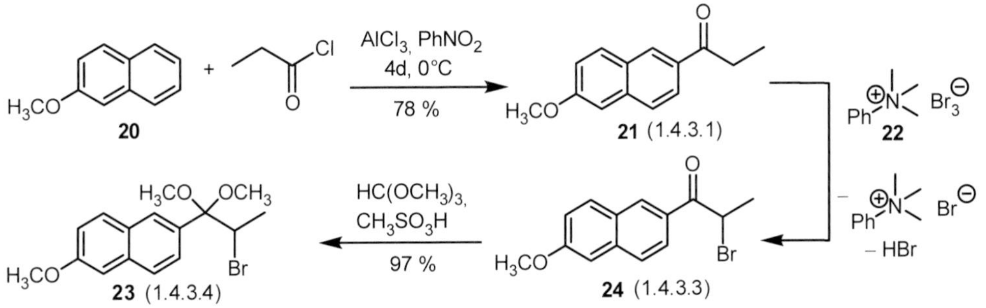

Next, 6-methoxy-2-propionylnaphthalene (21) has to be brominated chemoselectively at the aliphatic side-chain. Since treatment with elemental bromine would lead to additional bromination at the 5-position of the aromatic nucleus [9], trimethylphenylammonium perbromide (22) is used as a specific reagent that leads exclusively to the α-bromo ketone 24 [10].

The perbromide 22 is prepared in two steps from N,N-dimethylaniline via the methanesulfate 25 [11]:

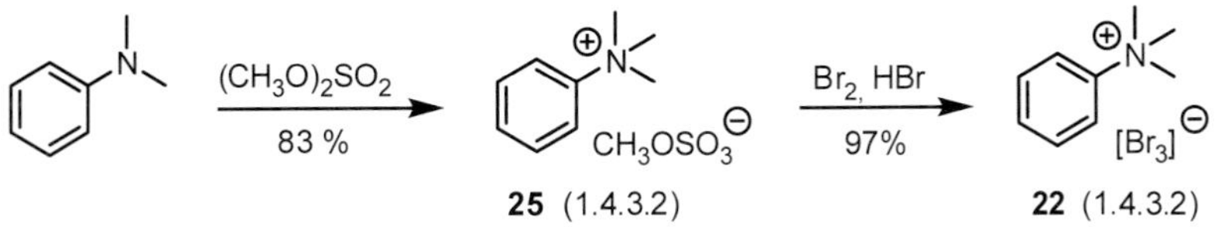

Bromo ketone 24 is transformed to the dimethyl ketal 23 by reaction with trimethyl orthoformate in the presence of CH₃SO₃H. On heating with anhydrous ZnBr₂ in toluene, the α-bromo acetal 23 rearranges to the methyl ester 27 of racemic naproxen. For the Lewis acid-induced 1,2-aryl shift, an arenium ion 26 [12] can be postulated as intermediate, which rearomatizes upon dealkylation with a bromide ion to give CH₃Br and the methyl ester 27 [10, 13].

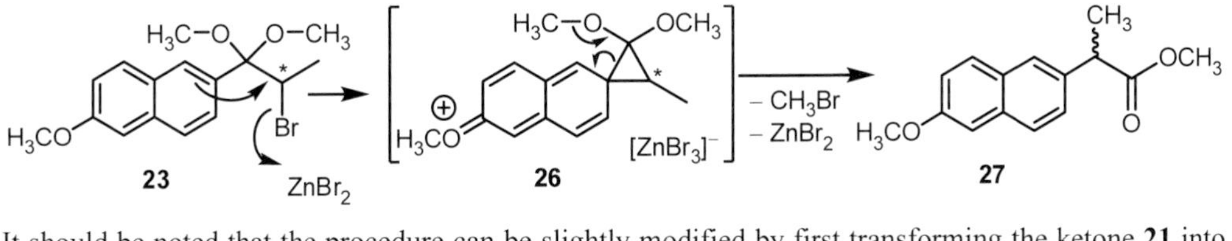

It should be noted that the procedure can be slightly modified by first transforming the ketone 21 into its acetal, which is then brominated using bromine. If one uses a chiral alcohol such as dimethyl tartrate for the acetalization, the subsequent bromination and rearrangement proceed with high induced stereoselectivity [14].

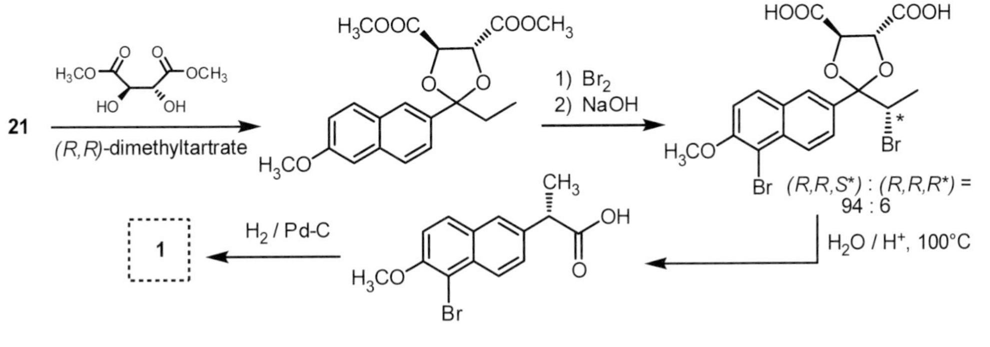

In the final step of the described synthesis of **1**, the racemic methyl ester **27** is subjected to kinetic resolution by enzymatic hydrolytic ester cleavage, using the lipase from *Candida rugosa*. This transformation is conducted up to 40% conversion and gives (*S*)-naproxen (**1**) with 96% *ee* and the (*R*)-ester with 63% *ee* [15]:

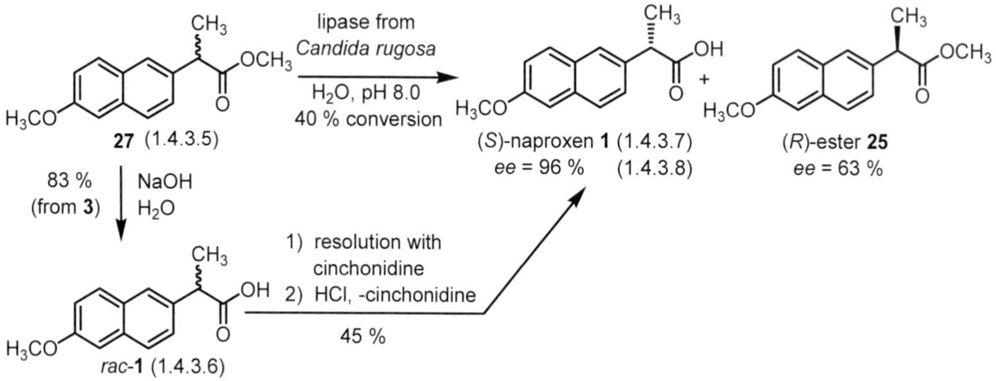

27 (1.4.3.5)

83 %
(from **3**)   NaOH
              H₂O

(*S*)-naproxen **1** (1.4.3.7)
ee = 96 %   (1.4.3.8)

(*R*)-ester **25**
ee = 63 %

1) resolution with cinchonidine
2) HCl, -cinchonidine

45 %

rac-**1** (1.4.3.6)

Alternatively, the racemic methyl ester **27** may be saponified with aqueous NaOH to give racemic naproxen (*rac*-**1**). This is resolved by the formation of diastereomeric salts with the chiral base cinchonidine, which are separated by fractional crystallization. The cinchonidine salt of the (+)-enantiomer is isolated, purified, and cleaved with aqueous HCl to give optically pure (*S*)-(+)-naproxen in 45% yield with $[\alpha]_D^{20} = +68$ (*c* = 0.84, CH₂Cl₂); for the isolation of the (*R*)-(−)-enantiomer, see ref. [16].

(c) Experimental procedures for the synthesis of 1

1.4.3.1   * 6-Methoxy-2-propionylnaphthalene [8]

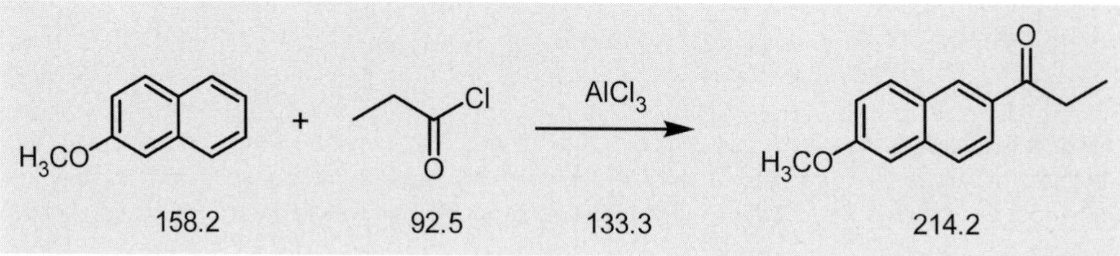

158.2   92.5   133.3   214.2

Anhydrous aluminum trichloride (112 g, 0.84 mol) is dissolved in anhydrous nitrobenzene (1300 mL) and the solution is cooled to 0 to −2 °C (note 1). With vigorous stirring, a solution of 2-methoxynaphthalene (106 g, 0.67 mol) in anhydrous nitrobenzene (340 mL) is added dropwise over 2 h. After stirring for 1 h at 0 °C, propionyl chloride (71.6 g, 0.77 mol, note 2) is added at such a rate that the internal temperature is kept at −3 °C. When the addition is complete, the dark reaction mixture is stirred for 96 h at 0 °C.

It is then poured onto a mixture of crushed ice (ca. 2 kg) and concentrated HCl (225 mL). CH₂Cl₂ is added to provide a clean phase separation, and the aqueous phase is extracted with CH₂Cl₂ (500 mL). The CH₂Cl₂ is distilled off from the combined organic phases and the remaining solution is steam distilled (to remove the nitrobenzene). The solid residue is dissolved in CH₂Cl₂, the solution is dried over Na_2SO_4 and filtered, and the solvent is distilled off. The brownish residue is distilled *in vacuo* (bp$_{0.2}$ 154–156 °C) and the distillation product is recrystallized from MeOH. The acylation product is obtained as colorless needles; 112 g (78%), mp 111–112 °C, TLC (SiO$_2$; benzene): R_f = 0.55.

IR (film): $\tilde{v}$ (cm^{-1}) = 1680 (C=O), 1625, 1600.

^{1}H NMR (CDCl$_3$): δ (ppm) = 7.05 (m, 6 H, aryl-H), 3.92 (s, 3 H, OCH$_3$), 3.05 (q, *J* = 8.0 Hz, 2 H, COCH$_2$), 1.26 (t, *J* = 8.0 Hz, 3 H, CH$_3$).

Notes: (1) It is recommended that the reaction is performed under N$_2$.
(2) Propionyl chloride has to be distilled before use, bp$_{760}$ 78–79 °C.

1.4.3.2   *   Trimethylphenylammonium perbromide [11]

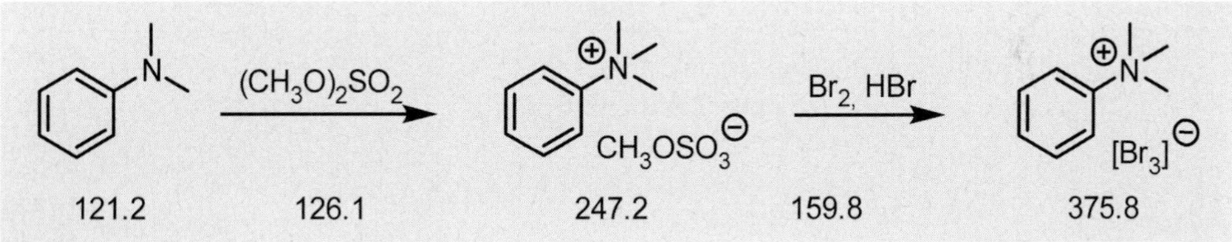

(1) Dimethyl sulfate (63.0 g, 0.50 mol, ~48.0 mL) is added dropwise to a vigorously stirred solution of *N,N*-dimethylaniline (63.0 mL, 0.50 mol) in benzene (120 mL) at 5 °C. During the addition, the temperature of the reaction mixture rises to ca. 75 °C. After stirring for 1 h, the mixture is cooled to 3 °C; the crystalline salt is collected by suction filtration, washed with benzene, and air-dried (hood!). Trimethylphenylammonium methanesulfate is obtained as colorless crystals, 103 g (83%), mp 108–110 °C.

---

^{1}H NMR (CDCl$_3$): δ (ppm) = 8.1–7.9 (m, 2 H, *o*-phenyl-H), 7.8–7.5 (m, 3 H, (*m*+*p*)-phenyl-H), 3.80 (s, 9 H, $^+$N(CH$_3$)$_3$), 3.68 (s, 3 H, CH$_3$OSO$_3^-$).

---

(2) The methanesulfate from (1) (80 g, 0.32 mol) is dissolved in 24% aqueous HBr (320 mL, 1.41 mol), and bromine (74.9 g, 0.47 mol, ~24.0 mL) is added dropwise with intense stirring over 30 min. The precipitated perbromide is collected by suction filtration and recrystallized from acetic acid; yellow needles, 117 g (97%), mp 113–115 °C.

IR (KBr): $\tilde{v}$ (cm^{-1}) = 1600, 1490, 1460, 960.

1.4.3.3   *   2-Bromo-1-(6-methoxy-2-naphthyl)propan-1-one [9]

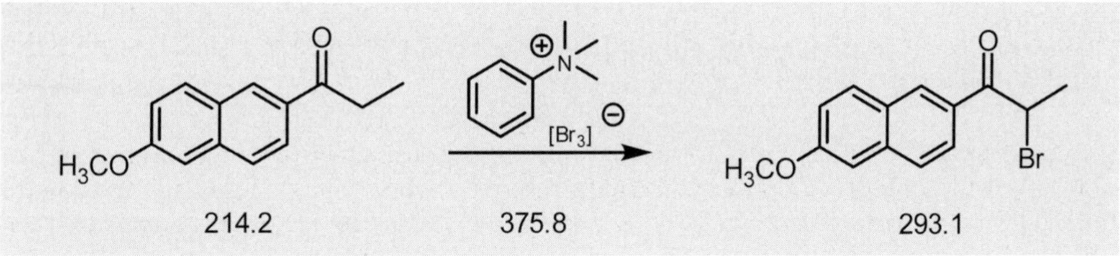

The perbromide **1.4.3.2** (75.2 g, 0.20 mol) is added in one portion to a stirred solution of 6-methoxy-2-propionylnaphthalene (**1.4.3.1**) (42.8 g, 0.20 mol) in tetrahydrofuran (420 mL). A clear orange-red solution results, from which a colorless salt precipitates (note 1) after some minutes; the supernatant solution becomes colorless. Stirring is continued for 30 min at room temperature.

The reaction mixture is diluted with H_2O (1200 mL) and extracted with petroleum ether (2 × 150 mL), and the combined organic extracts are dried (Na_2SO_4). After filtration, the solvent is removed *in vacuo* and the oily residue is triturated with EtOH (400 mL). The bromo ketone crystallizes in fine colorless needles (note 2); it is collected by suction filtration, washed with pre-cooled EtOH, and dried *in vacuo*; 56.0 g (96%), mp 78–79 °C, TLC (SiO_2; benzene): $R_f = 0.65$.

IR (KBr): $\tilde{v}$ (cm^{-1}) = 1685 (C=O), 1620, 1600.

^{1}H NMR (CDCl$_3$): δ (ppm) = 8.5–7.1 (m, 6 H, aryl-H), 5.43 (q, J = 6.0 Hz, 1 H, CH), 3.94 (s, 3 H, OCH$_3$), 1.94 (d, J = 6.0 Hz, 3 H, CH$_3$).

Notes: (1) Trimethylphenylammonium bromide, mp 210–212 °C.
        (2) Crystallization is complete after 12 h in a refrigerator.

1.4.3.4 * 2-Bromo-1-(6-methoxy-2-naphthyl)propan-1-one dimethyl acetal [10]

A suspension of the bromo ketone **1.4.3.3** (41.4 g, 0.14 mol), trimethyl orthoformate (43.5 g, 0.41 mol), and methanesulfonic acid (2.72 g) in anhydrous MeOH (150 mL) is heated to 45 °C for 24 h; a clear solution results.

The reaction mixture is then poured into 2% Na_2CO_3 solution (1000 mL) and is extracted with Et_2O (3 × 250 mL). The combined extracts are dried over Na_2CO_3, filtered, and the solvent is distilled off *in vacuo*. The resulting almost colorless oil is dissolved in MeOH (300 mL) and the solution is cooled (refrigerator, 12 h). The crystallized dimethyl acetal is collected by suction filtration, washed with MeOH at –10 °C, and dried *in vacuo*; fine colorless needles, 46.0 g (97%), mp 87–88 °C, TLC (SiO_2; benzene): $R_f = 0.75$.

IR (KBr): $\tilde{v}$ (cm^{-1}) = 2990, 2970, 2940, 2830 (CH), 1630, 1610.

^{1}H NMR (CDCl$_3$): δ (ppm) = 8.5–7.1 (m, 6 H, aryl-H), 4.54 (q, J = 6.0 Hz, 1 H, CH), 3.49, 3.43, 3.25 (s, 3 H, OCH$_3$), 1.57 (d, J = 6.0 Hz, 3 H, CH$_3$).

1.4.3.5   *   (*R,S*)-Methyl 2-(6-methoxy-2-naphthyl)propionate [13]

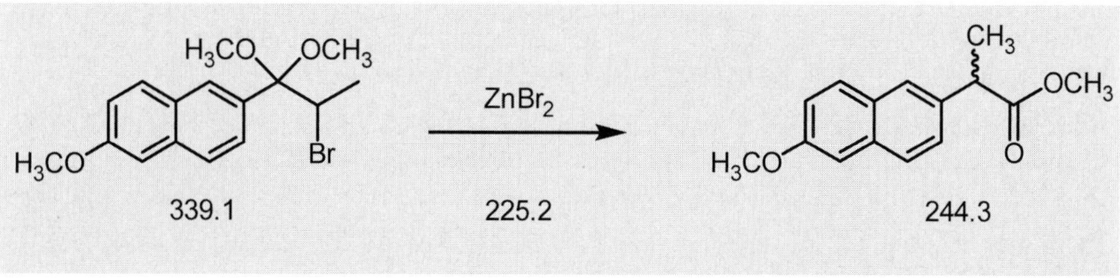

A suspension of the dimethyl acetal **1.4.3.4** (33.9 g, 0.10 mol) and anhydrous zinc bromide (2.25 g, 10.0 mmol) in anhydrous toluene (100 mL) is heated to reflux with stirring under an N_2 atmosphere for 1 h.

After cooling to room temperature, the reaction mixture is poured into H_2O (1000 mL) and extracted with Et_2O (3 × 300 mL). The combined extracts are dried (Na_2SO_4), filtered, and the solvent is removed *in vacuo*.

For enzymatic resolution, a pure sample of the racemic methyl ester **1.4.3.5** is obtained by column chromatography (SiO_2; *n*-hexane/EtOAc, 85:15), mp 89–90 °C.

IR (KBr): $\tilde{v}$ (cm^{-1}) = 3005, 2974, 2932, 1731 ($O=COCH_3$), 1602.

^{1}H NMR (500 MHz, $CDCl_3$): δ (ppm) = 7.70 (d, *J* = 8.5 Hz, 2 H, Ar-H), 7.66 (d, *J* = 1.6 Hz, 1 H, Ar-H), 7.40 (dd, *J* = 8.5, 1.9 Hz, 1 H, Ar-H), 7.14 (dd, *J* = 8.8, 2.5 Hz, 1 H, Ar-H), 7.11 (d, *J* = 2.2 Hz, 1 H, Ar-H), 3.90 (s, 3 H, $PhOCH_3$), 3.85 (q, *J* = 7.3 Hz, 1 H, PhCH), 3.66 (s, 3 H, CO_2CH_3), 1.57 (3 H, d, *J* = 7.3, PhCHC$\underline{H}_3$).

^{13}C NMR (125 MHz, $CDCl_3$): δ (ppm) = 175.13, 157.68, 135.70, 133.73, 129.28, 128.96, 127.18, 126.20, 125.95, 119.00, 105.65, 55.31, 52.02, 45.37, 18.60.

1.4.3.6   *   (*R,S*)-2-(6-Methoxy-2-naphthyl)propionic acid (*rac*-naproxen) [13]

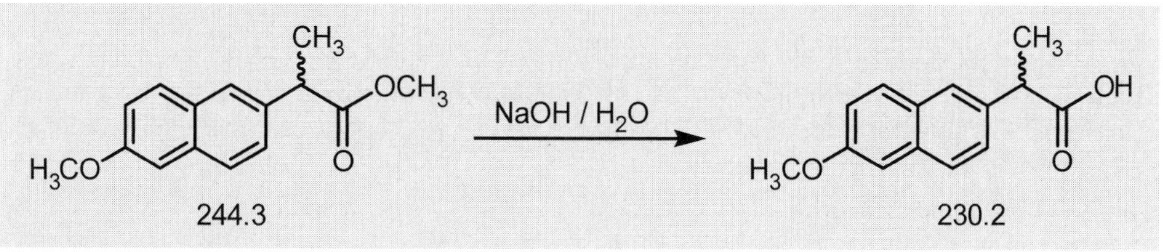

The crude methyl ester **1.4.3.5** is dissolved in MeOH (500 mL), 30% aqueous NaOH (150 mL) is added, and the mixture is heated to reflux for 4 h.

The MeOH is distilled off *in vacuo*, the residue is dissolved in H_2O (ca. 1200 mL), and the alkaline solution is extracted with Et_2O (2 × 400 mL). The organic phase is discarded. The aqueous phase is acidified with concentrated HCl and the precipitated acid is extracted with Et_2O (2 × 400 mL). The combined extracts are dried (Na_2SO_4), filtered, and the solvent is removed. The remaining solid is recrystallized from acetic acid, and a second crop is obtained by (careful) dilution of the mother liquor with H_2O. The racemic acid is obtained as fine colorless needles; 19.0 g (83%), mp 152–153 °C, TLC (SiO_2; Et_2O): $R_f = 0.80$.

IR (KBr): $\tilde{\nu}$ (cm^{-1}) = 3200–2800 (OH), 1710 (C=O), 1605.

^{1}H NMR (500 MHz, $CDCl_3$): δ (ppm) = 10.6 (s_{br}, 1 H, COOH), 7.75–7.65 (m, 3 H, Ar-H), 7.40 (dd, $J = 8.5$, 1.8 Hz, 1 H, Ar-H), 7.13 (d, $J = 8.8$ Hz, 1 H, Ar-H), 7.10 (d, $J = 2.4$ Hz, 1 H, Ar-H), 3.90 (s, 3 H, OCH_3), 3.86 (q, $J = 7.0$ Hz, 1 H, PhCH), 1.58 (d, $J = 7.0$ Hz, 3 H, CH_3).

^{13}C NMR (125 MHz, $CDCl_3$): δ (ppm) = 180.52, 157.70, 134.89, 133.81, 129.30, 128.89, 127.23, 126.19, 126.14, 119.04, 105.59, 55.30, 45.24, 18.14.

1.4.3.7   **   **(S)-2-(6-Methoxy-2-naphthyl)propionic acid [(S)-naproxen]**
           by kinetic enzymatic resolution [15]

Finely powdered racemic naproxen methyl ester **1.4.3.5** (150 mg, 0.65 mmol), mercaptoethanol (1 drop), and polyvinyl alcohol (5 mg) are added to crude *candida rugosa* lipase (EC 3.1.1.3, Type VII, Sigma L-1754, 50 mg, 600 μg of protein) in a 0.2 M phosphate buffer solution at pH 8.0 (1 mL). The suspension is stirred at 30 °C for 120 h. Both the progress of the conversion and the enantiomeric purity can be monitored simultaneously by HPLC analysis (chiral HPLC Lichro Cart 250-4 (S,S)-Whelk-01, 5 μm; hexane/isopropanol/acetic acid, 90:9.5:0.5, 1.2 mL min^{-1}, 254 nm).

The pH of the reaction mixture is adjusted to 2–3 with concentrated HCl and the mixture is extracted with Et_2O (5 × 10 mL). The combined ethereal extracts are extracted with saturated Na_2CO_3 solution (5 × 10 mL) and the combined aqueous layers are re-extracted with Et_2O (3 × 10 mL). The combined ethereal extracts are washed with brine, dried ($MgSO_4$), and the solvent is evaporated under reduced pressure to give the unreacted naproxen methyl ester as a white solid.

The Na_2CO_3 extracts are acidified with HCl (6 N), saturated with NaCl, and extracted with Et_2O (5 × 10 mL). The ethereal phase is washed with brine, dried over $MgSO_4$, and the Et_2O is evaporated under reduced pressure to give (S)-naproxen as a white solid; 53 mg (35% isolated yield); $[\alpha]_D^{20} = +65$ ($c = 1.00$, $CHCl_3$); $ee = 96\%$ [15].

**1.4.3.8   **   (*S*)-2-(6-Methoxy-2-naphthyl)propionic acid [(*S*)-naproxen]
by resolution with cinchonidine [16]

Racemic naproxen **1.4.3.6** (11.5 g, 50.0 mmol) is dissolved in a hot mixture of MeOH (200 mL) and acetone (50 mL). A warm solution of cinchonidine (15.0 g, 51.0 mmol) in MeOH (150 mL)/acetone (100 mL) is added and the mixture is allowed to cool and crystallize over 12 h. The precipitate is filtered off and recrystallized twice from MeOH (350 mL)/acetone (150 mL) allowing a 12 h crystallization time; cinchonidine salt of (*S*)-naproxen, mp 178–179 °C.

The salt is suspended in benzene (160 mL)/6.5 N HCl (160 mL) and the stirred mixture is heated at 30–40 °C until two clear phases are formed (ca. 30 min). The benzene layer is separated, dried (Na$_2$SO$_4$), and concentrated, and the residue is recrystallized from acetone/petroleum ether (40–65 °C); yield 2.60 g (45%) of (*S*)-naproxen, colorless needles, mp 156–157 °C, $[\alpha]_D^{20}$ = +68 (*c* = 0.84, CHCl$_3$).

[1]   A. Kleemann, J. Engel, *Pharmaceutical Substances*, 3rd ed., p. 1304, Thieme Verlag, Stuttgart, **1999**.

[2]   a) H. Auterhoff, J. Knabe, H.-D. Höltje, *Lehrbuch der Pharmazeutischen Chemie*, 13th ed., p. 450, Wissenschaftliche Verlagsgesellschaft mbH, Stuttgart, **1994**; b) J.-J. Li, D. S. Johnson, D. R. Sliscovich, B. D. Roth, *Contemporary Drug Synthesis*, p. 11, John Wiley & Sons, Inc., New York, **2004**.

[3]   P. J. Harrington, E. Lodewijk, *Organic Process Res. and Dev.* **1997**, *1*, 72.

[4]   T. Ohta, H. Takaya, M. Kitamura, K. Nagai, R. Noyori, *J. Org. Chem.* **1987**, *52*, 3174.

[5]   a) A. S. C. Chan, T. T. Huang, J. H. Wagenknecht, R. E. Miller, *J. Org. Chem.* **1995**, *60*, 742. b) Alternatively, **7** has been synthesized from 6-methoxy-2-naphthaldehyde, which involved reaction with ethyl diazoacetate in the presence of [FeCp(CO)$_2$THF]BF$_4$ as catalyst (to give the α-hydroxymethylene arylacetic acid) and reduction to =CH$_2$ with BH$_3$·THF: S. J. Mahmood, C. Brennan, M. M. Hossain, *Synthesis* **2002**, 1807.

[6]   C. Giordano, G. Castaldi, F. Uggeri, *Angew. Chem.* **1984**, *96*, 413; *Angew. Chem. Int. Ed. Engl.* **1984**, *23*, 413.

[7]   In a recent synthesis following this principle, mannitol was used as chiral auxiliary in the transformation of **21** to 1-ester by reaction with HC(OCH$_3$)$_3$/ZnCl$_2$: B. Wang, H. Z. Ma, Q. Z. Shi, *Synth. Commun.* **2002**, *32*, 1697.

[8]   R. T. Rapala, B. W. Robert, W. L. Truett, W. S. Johnson, *J. Org. Chem.* **1962**, *27*, 3814.

[9]   A. Marquet, J. Jaques, *Bull. Soc. Chim. Fr.* **1962**, 90.

[10] C. Giordano, G. Castaldi, F. Casagrande, A. Belli, *J. Chem. Soc., Perkin Trans. I* **1982**, 2575.

[11] A. Marquet, J. Jacques, *Bull. Soc. Chim. Fr.* **1961**, 1822.

[12] Compare: R. Brückner, *Reaktionsmechanismen*, 2nd ed., p. 89, Spektrum Akademischer Verlag, Heidelberg, **2003**.

[13] G. Castaldi, A. Belli, F. Uggeri, C. Giordano, *J. Org. Chem.* **1983**, *48*, 4658.

[14] C. Giordano, G. Castaldi, S. Cavicchiolly, M. Villa, *Tetrahedron* **1989**, *45*, 4243. As a consequence of additional aromatic substitution in the bromination step, reductive dehalogenation is required at the end of the synthesis.

[15] a) Q.-M. Gu, C.-S. Chen, C. J. Sih, *Tetrahedron Lett.* **1986**, *27*, 1763; b) S. Koul, R. Parshad, S. C. Taneja, G. N. Quazi, *Tetrahedron: Asymmetry* **2003**, *14*, 2459.

[16] I. T. Harrison, B. Lewis, P. Nelson, W. Rooks, A. Roszkowski, A. Tomolonis, J. H. Fried, *J. Med. Chem.* **1970**, *13*, 203.

1.4.4   3-Benzoylcyclohexanone

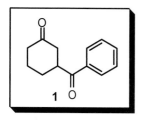

1

Topics:
- (*O*-Trimethylsilyl)cyanohydrin anions as acyl anion equivalents, umpolung of carbonyl groups, nucleophilic acylation according to the Hünig methodology
- Formation of an (*O*-trimethylsilyl)cyanohydrin
- 1,4-Addition of an α-metalated (*O*-trimethylsilyl) cyanohydrin to an enone

(a) General

The concept of "umpolung" [1] has been developed on the basis of reactions in which the polarity of an atom (mainly carbon) in a functional group is changed through chemical transformation. A simple example is provided by the formation of a Grignard compound from a halide R–X through insertion of Mg into the C–X bond (**2 → 3**). Another example is the reaction of a Grignard compound with an elemental halogen (**3 → 2**). In these two processes, an sp^3 C atom changes its polarity from δ^+ to δ^- and *vice versa*:

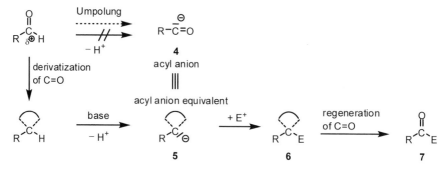

Attractive for synthesis are umpolung reactions at the carbonyl group of aldehydes, in which the polarity of the electrophilic carbon atom of the C=O group is switched to that of an acyl anion **4**:

Direct deprotonation at the aldehyde CH=O group is not possible, since the pK_a of the hydrogen is about 54. However, by derivatization of the C=O group, the acidity of the C–H can be increased to allow the generation of a carbanion **5** as an equivalent of the acyl anion **4**, which is capable of reacting with electrophiles (simplified as E^+, **5 → 6**). Regeneration of the carbonyl group (**6 → 7**) affords a product of type **7** resulting from combination of E^+ with an aldehyde; thus, the process represents a (formal) nucleophilic acylation of an electrophilic system, as illustrated by the following examples.

(1) The acyl anion equivalent **8** formed by addition of cyanide to the carbonyl group of aryl aldehydes is the central intermediate in the combination of two aryl aldehydes to give benzoins **9** (benzoin reaction [2]) or the 1,4-addition of aryl aldehydes to α,β-unsaturated ketones to give 1,4-diones **10** (Stetter reaction [3]), both of which are catalyzed by cyanide:

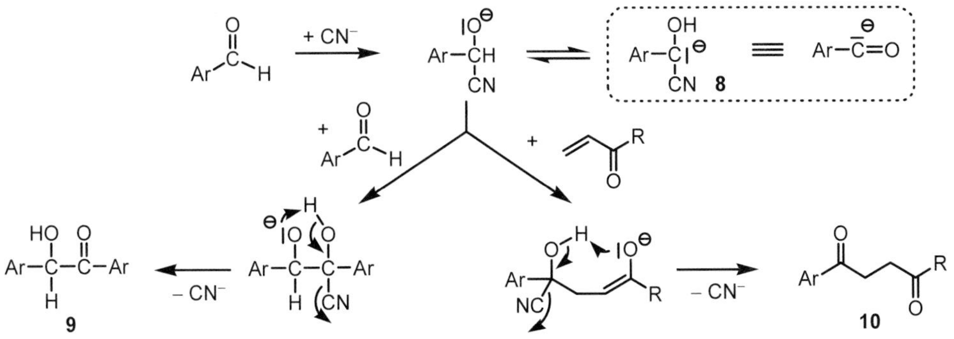

(2) Similar acyl anion equivalents are represented by the α-lithiated *O*-silylcyanohydrins **12**, which result from metalation of *O*-silylcyanohydrins **11** (accessible by addition of trimethylsilyl cyanide to aldehydes) with R–Li (Hünig procedure for nucleophilic acylation [4]). Their reactions with electrophilic systems – e.g. alkylation, addition to aldehydes or ketones, 1,4-addition to enones – lead to **13**, which can easily be transformed into products of type **7** by subsequent desilylation and loss of HCN, as exemplified in section (b):

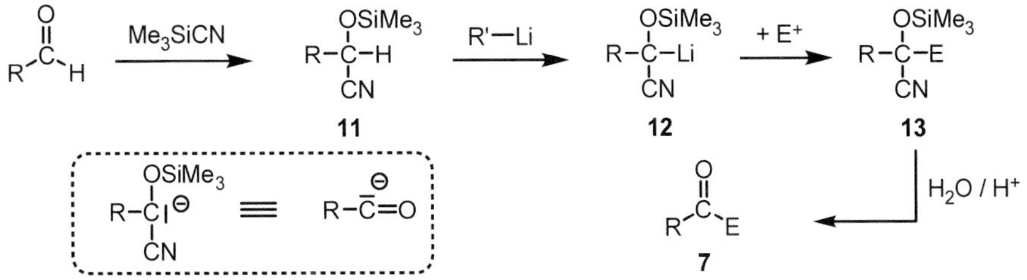

(3) 2-Substituted 1,3-dithianes **14** (cyclic dithioacetals, accessible from aldehydes and propane-1,3-dithiol) can be deprotonated with *n*BuLi to give 2-lithio-1,3-dithianes **15**, which also represent acyl anion equivalents (Corey–Seebach procedure for nucleophilic acylation [5]). As expected, the lithiodithianes **15** are again susceptible to reactions with electrophilic systems, e.g. alkylation, addition to aldehydes or ketones, conjugate addition to enones, ring-opening addition to oxiranes. In the products **16** thus formed, the carbonyl moiety can be regenerated (**16** → **7**) by dethioacetalization, which is preferably carried out by means of an oxidative procedure:

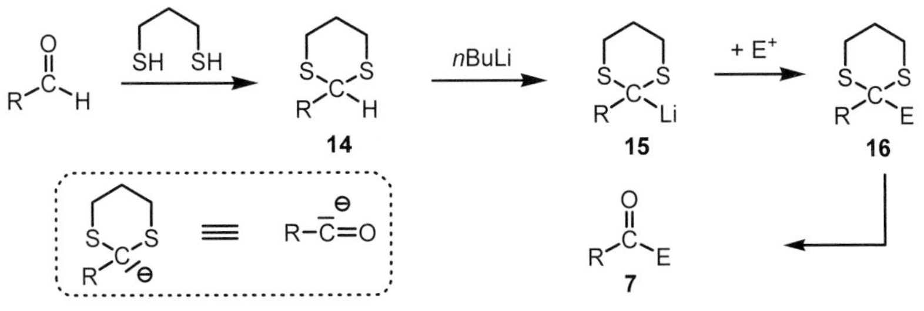

Since 1,3-dithiane chemistry often suffers from the disadvantages of unpleasant odor produced by the sulfur compounds involved and problems in cleaving the thioacetal moiety, the Hünig protocol is often preferred for nucleophilic acylations.

As the result of a simple retrosynthetic analysis, the target molecule **1** should be accessible from cyclohexenone by 1,4-addition of a benzoyl carbanion (**17**) or an equivalent thereof.

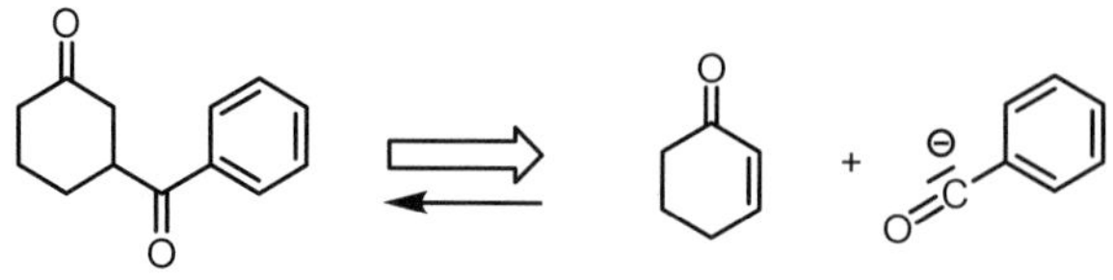

(b) Synthesis of 1

The requisite *O*-trimethylsilylated cyanohydrin **18** is prepared by Lewis acid-catalyzed addition (ZnI$_2$) of trimethylsilyl cyanide to benzaldehyde [6, 7]:

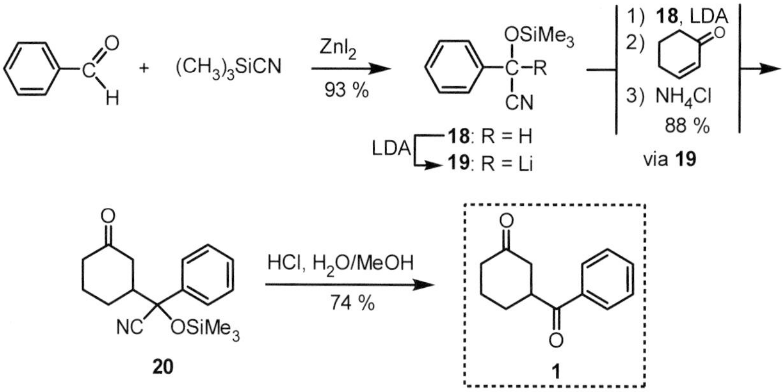

18 is then subjected to metalation with LDA in THF at –78 °C to give the lithiated cyanohydrin **19**, which is reacted *in situ* with cyclohexenone at –78 to –20 °C. 1,4-Addition of the benzoyl anion equivalent **19** to the enone occurs smoothly, leading to the product **20** after work-up with aqueous NH$_4$Cl solution.

On hydrolysis with a strong acid (HCl in H$_2$O/methanol), the cyanohydrin *O*-silyl ether functionality in **20** is cleaved with loss of cyanide to yield the 3-benzoylcyclohexanone (**1**).

Thus, the target molecule **1** is obtained in a three-step sequence in an overall yield of 61% (based on benzaldehyde).

(c) Experimental procedures for the synthesis of 1

1.4.4.1   * Phenyl(trimethylsilyloxy)acetonitrile [8]**

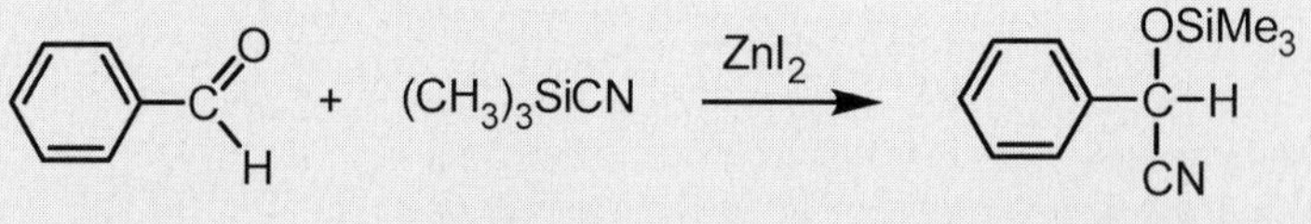

Under nitrogen and with exclusion of moisture, benzaldehyde (7.64 g, 72.0 mmol, note 1) is added dropwise over a period of 20 min to trimethylsilyl cyanide (7.94 g, 80.0 mmol, note 2) and a few mg of anhydrous zinc iodide (note 3) with stirring. The solution is then heated to 80–100 °C for 2 h; the progress of the reaction may be followed by IR.

The product is isolated by fractionating distillation *in vacuo* and is obtained as a colorless oil; 13.7 g (93%), bp$_1$ 62–63 °C, n_D^{20} = 1.4840 (note 4).

IR (film): $\tilde{v}$ (cm^{-1}) = 3070, 3040 (C–H, arom.), 1260, 875, 850, 750 (Si-C).

^{1}H NMR (CDCl$_3$): δ (ppm) = 7.37 (s, 5 H, phenyl-H), 5.45 (s, 1 H, C–H), 0.23 (s, 9 H, Si(CH$_3$)$_3$).

Notes: (1) Benzaldehyde has to be freshly distilled, bp$_{10}$ 62–63 °C.
      (2) Trimethylsilyl cyanide has to be distilled prior to use, bp118–119 °C.
          Caution: (CH$_3$)$_3$SiCN is a toxic compound!
      (3) ZnI$_2$ should be dried *in vacuo* at 100 °C for 5 h before use.
      (4) The cyanohydrin is easily hydrolyzed to give HCN. Caution! Hood!

1.4.4.2   ***   (3-Oxocyclohexyl)phenyl(trimethylsilyloxy)acetonitrile [9]

n-Butyllithium (1.6 M in *n*-hexane, 19.4 mL, 31.0 mmol) is added to a stirred solution of diisopropylamine (3.12 g, 31.0 mmol, note 1) in anhydrous THF (20 mL) at –78 °C under nitrogen and with exclusion of moisture, and the mixture is stirred for 15 min. The silylated cyanohydrin **1.4.4.1** (6.15 g, 30.0 mmol) is added dropwise at the same temperature, which leads to the deposition of a yellow precipitate. Finally, 2-cyclohexen-1-one (2.88 g, 30.0 mmol, note 2) is added dropwise and the temperature of the reaction mixture is slowly increased to –20 °C over a period of 4 h (note 3).

Saturated NH$_4$Cl solution (30 mL) is added and the mixture is stirred for 3 min at room temperature. It is then extracted with Et$_2$O (3×, total of 100 mL), and the combined organic extracts are washed with saturated NH$_4$Cl solution and brine and dried over Na$_2$SO$_4$. The solvents are evaporated and the residue is distilled *in vacuo* in a Kugelrohr apparatus to give the product as a colorless liquid; 8.00 g (88%), bp$_{0.05}$ 140 °C (oven temperature 145 °C), n_D^{20} = 1.5125 (note 4).

IR (film): $\tilde{v}$ (cm^{-1}) = 3080, 3060, 3030 (CH arom.), 2960, 2900, 2870 (CH aliph.), 1720 (C=O), 1260 (Si-C).

^{1}H NMR (CDCl$_3$): δ (ppm) = 7.37 (s, 5 H, phenyl-H), 2.65–1.25 (m, 9 H, cyclohexane-CH + CH$_2$), 0.12 (s, 9 H, (CH$_3$)$_3$Si).

Notes: (1) Diisopropylamine is distilled from CaH_2 before use, bp 84–85 °C.
(2) Cyclohexenone is distilled before use, bp 168–169 °C.
(3) For the preparation of 3-benzoylcyclohexanone, hydrolysis of the reaction mixture is conducted with HCl in H_2O/CH_3OH as described in **1.4.4.3**.
(4) The product is easily hydrolyzed forming HCN. Caution! Hood!

1.4.4.3   ***   3-Benzoylcyclohexanone [9]

OSiMe$_3$	LiN[CH(CH$_3$)$_2$]$_2$	OSiMe$_3$	
205.1	HN[CH(CH$_3$)$_2$]$_2$: 101.2	96.1	202.3

Hydrochloric acid (2 N, 30 mL) and MeOH (15 mL) are added to the reaction mixture (obtained as described in **1.4.4.2**) and stirring is continued for 14 h at room temperature (note).

The mixture is then diluted with water (ca. 100 mL) and extracted with Et_2O (3 × 50 mL). The combined ethereal extracts are washed with NaOH (1 M) and brine, and dried over Na_2SO_4. The solvents are evaporated and the residue is fractionally distilled *in vacuo* to give the product as a colorless oil; 4.48 g (74 %), bp$_{0.01}$ 130–131 °C, n_D^{20} = 1.5574.

IR (film): $\bar{\nu}$ (cm^{-1}) = 3080, 3070, 3030 (CH arom.), 2960, 2880 (CH aliph.), 1710, 1680 (C=O).

^{1}H NMR (CDCl$_3$): δ (ppm) = 8.05–7.15 (m, 5 H, phenyl-H), 4.1–3.5 (m, 1 H, CH–CO), 2.75–1.45 (m, 8 H, cyclohexane-CH$_2$).

Note: HCN is formed during the hydrolysis. Caution! Hood!

[1] T. A. Hase, *Umpoled Synthons*, John Wiley & Sons, Inc., New York, **1987**.

[2] M. B. Smith, J. March, *March's Advanced Organic Chemistry*, 5th ed., p. 1243, John Wiley & Sons, Inc., New York, **2001**.

[3] a) H. Stetter, *Angew. Chem.* **1976**, *88*, 695; *Angew. Chem. Int. Ed. Engl.* **1976**, *15*, 639; b) W. Kreiser, *Nachr. Chem. Tech. Lab.* **1981**, *29*, 172, 445.

[4] S. Hünig, H. Reichelt, *Chem. Ber.* **1986**, *119*, 1772.

[5] a) B. T. Gröbel, D. Seebach, *Synthesis* **1977**, 357; b) P. C. Bulman Page, M. B. van Niel, J. C. Prodger, *Tetrahedron* **1989**, *45*, 7643.

[6] For new efficient catalysts for the cyanotrialkylsilylation of aldehydes and ketones, see: a) B. M. Fetterly, J. G. Verkade, *Tetrahedron Lett.* **2005**, *46*, 8061; b) T. Kano, K. Sasaki, T. Konishi, H. Mii, K Maruoka, *Tetrahedron Lett.* **2006**, *47*, 4615.

[7] For catalytic enantioselective cyanotrialkylsilylation of ketones, see: a) D. E. Fuerst, E. N. Jacobsen, *J. Am. Chem. Soc.* **2005**, *127*, 8964; b) S. S. Kim, J. M. Kwak, *Tetrahedron* **2006**, *62*, 49.

[8] K. Deuchert, U. Hertenstein, S. Hünig, G. Wehner, *Chem. Ber.* **1979**, *112*, 2045.

[9] S. Hünig, G. Wehner, *Chem. Ber.* **1979**, *112*, 2062; *Chem. Ber.* **1980**, *113*, 302, 324.

1.5   Reactions of alkenes via carbenium ions

1.5.1   Piperine

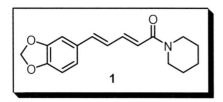

1

Topics:
- Synthesis of a natural product of the arylpolyene type
- Acetal formation by the orthoester method
- Lewis acid-induced C–C bond formation by addition of an acetal to an enol ether
- Acid-catalyzed ROH elimination; acetal hydrolysis
- Knoevenagel condensation/decarboxylation
- Transformation of a carboxylic acid into a carboxylic acid amide

(a) General

Piperine (**1**, 4-(3,4-methylenedioxyphenyl)-1,3-butadiene-1-carboxypiperidide) is a constituent of several pepper species (piperaceae), especially of black pepper (*Piper nigrum* L.) as pungent principle. Like many piper alkaloids, piperine also exhibits antimicrobial properties [1]. Hydrolysis of **1** in a basic medium leads to piperinic acid (**2**) and piperidine, whose name is derived from its natural origin:

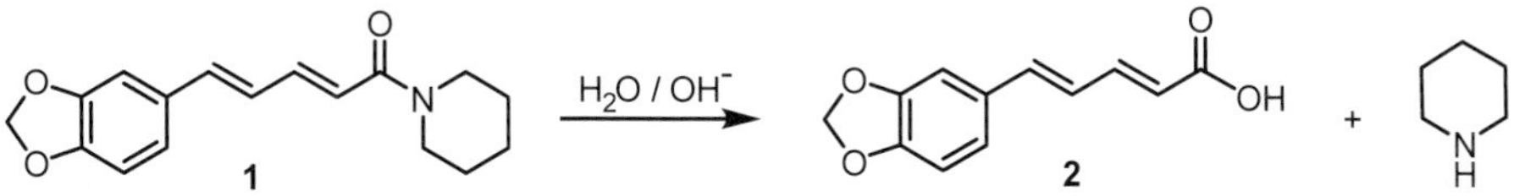

Piperinic acid (**2**) can be regarded as an intermediate for the synthesis of **1**. The retrosynthesis of **2** can be performed in two directions (**A/B**) by disconnections at the double bonds according to a retro-Wittig transformation:

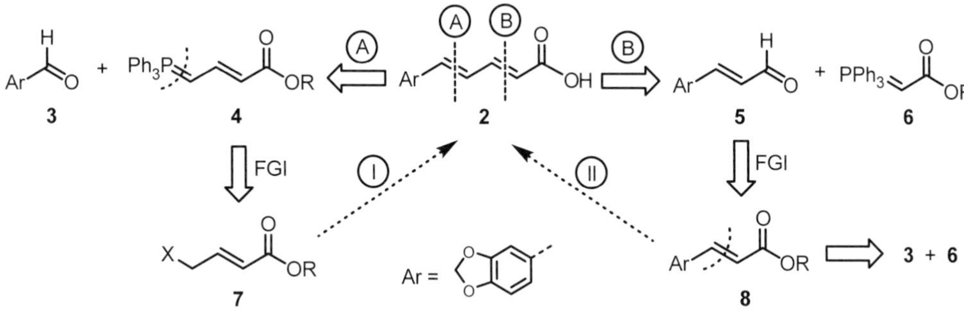

Retrosynthesis according to **A** leads to the aldehyde piperonal (**3**) and a C$_4$-ylide **4**, which is derived from γ-halogeno crotonate **7**, in turn available from crotonate by allylic halogenation (e.g., with NBS).

Retrosynthesis according to **B** leads to the C$_2$-ylide **6** (derived from haloacetate) and 3-arylacrolein **5**, which should be accessible from cinnamate **8** by reduction (e.g., with DIBAL).

Both approaches towards **2** have been described in the literature. However, the carbonyl olefination of **3** and **4** to give **2** (route I) suffers from preparative disadvantages [2]; the same is true for the construction of **2** by two consecutive carbonyl olefinations via **5** (route II) [3]. Therefore, an alternative method is used for the synthesis of **2** [4], which has been effectively applied in the synthesis of polyolefinic systems ([5], cf. **4.2.6**) and which relies on carbenium ion-based C–C bond formation.

(b) Synthesis of 1

In the first part of the synthesis of **1**, piperonal (**3**) is transformed to its diethyl acetal **9** by reaction with triethyl orthoformate in the presence of TosOH. The acetal **9** adds to the C=C double bond of ethyl vinyl ether in the presence of a Lewis acid, e.g. $ZnCl_2$, to give rise to the 3-aryl-1,1,3-triethoxypropane derivative **10**:

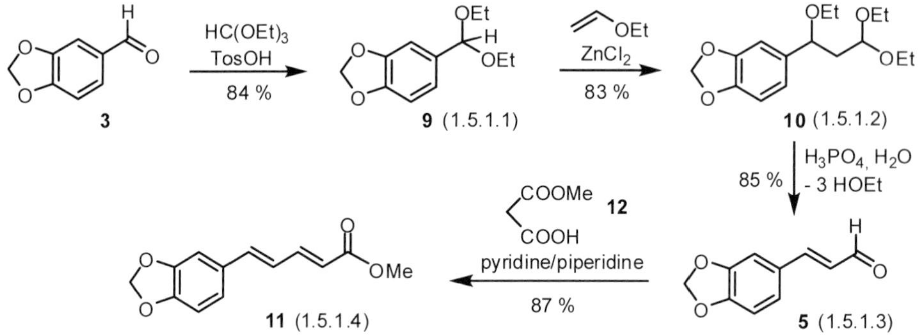

The transformation **9 → 10** can be rationalized by: (1) formation of a carbenium ion from the acetal **9** induced by the Lewis acid, (2) its electrophilic Markownikov-oriented addition to the electron-rich C=C double bond of the vinyl ether, and (3) termination by transfer of OEt to the cationic intermediate **13**:

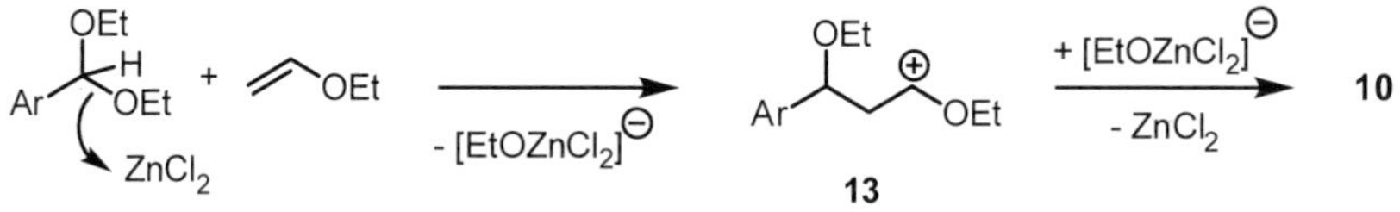

10 is transformed into the α,β-unsaturated aldehyde by acid-catalyzed hydrolysis of the acetal followed by elimination of EtOH. Finally, the C_5-1,3-diene side chain in **1** is completed by Knoevenagel condensation of **5** with monomethyl malonate (**12**) to give the methyl ester **11** as a result of concomitant decarboxylation of the initial condensation product **14** under the reaction conditions [6]:

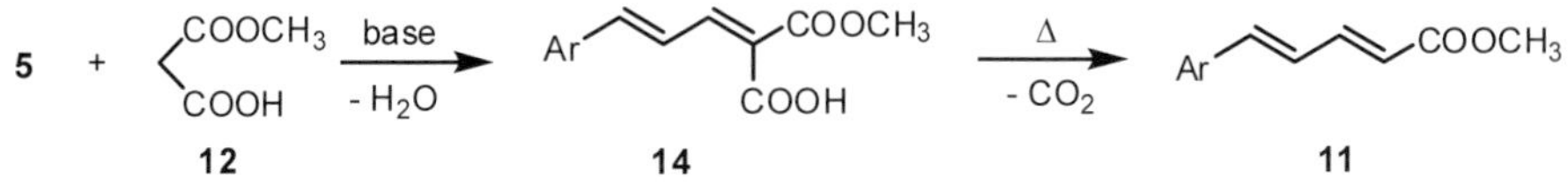

As the last step in the synthesis of **1**, the methyl ester **11** is saponified using KOH in ethanol to give piperinic acid (**2**) and the amide is formed in the conventional manner by reacting **2** with $SOCl_2$ followed by Schotten–Baumann reaction of the intermediate acid chloride with piperidine:

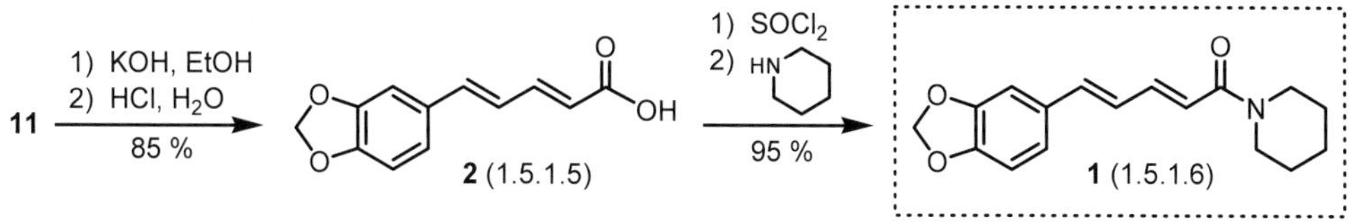

Thus, the target molecule **1** is obtained in a six-step sequence in an overall yield of 42% (based on piperonal).

(c) Experimental procedures for the synthesis of 1

1.5.1.1   * Piperonal diethyl acetal [4]

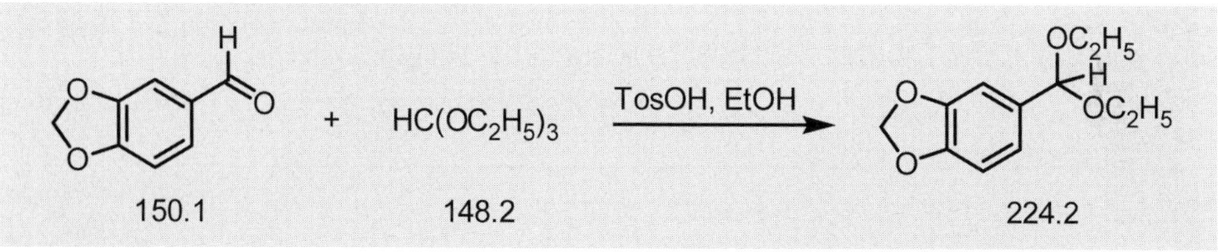

A stirred solution of piperonal (50.0 g, 0.33 mol), triethyl orthoformate (59.3 g, 0.40 mol), and TosOH·H$_2$O (10 mg) in anhydrous EtOH (330 mL) is heated to 80 °C for 1 h with exclusion of moisture.

The excess EtOH is then distilled off, the residue is dissolved in Et$_2$O (200 mL), and the solution is washed several times with H$_2$O (100 mL). The ethereal solution is dried over K$_2$CO$_3$, the solvent is removed, and the residue is fractionated *in vacuo* on a 20 cm Vigreux column. The acetal is obtained as a colorless oil; 62.1 g (84%), bp$_{0.1}$ 84–86 °C.

IR (film): $\tilde{\nu}$ (cm^{-1}) = 2974, 2880, 1504.

^{1}H NMR (CDCl$_3$): δ (ppm) = 6.98 (d, J = 1.3 Hz, 1 H, ArH), 6.93 (dd, J = 7.9, 1.3 Hz, 1 H, ArH), 6.78 (d, J = 7.9 Hz, 1 H, ArH), 5.94 (s, 2 H, OCH$_2$O), 5.39 (s, 1 H, OCHO), 3.60 (dq, J = 9.4, 6.9 Hz, 2 H, OCH$_2$CH$_3$), 3.50 (dq, J = 9.4, 6.9 Hz, 2 H, OCH$_2$CH$_3$), 1.24 (t, J = 6.9 Hz, 6 H, OCH$_2$CH$_3$).

^{13}C NMR (CDCl$_3$): δ (ppm) = 147.66, 147.51, 133.39, 120.25, 107.82, 106.97, 101.42, 101.01, 61.05, 15.19.

1.5.1.2   * 1,1,3-Triethyl-3-(3,4-methylenedioxyphenyl)propane [4]

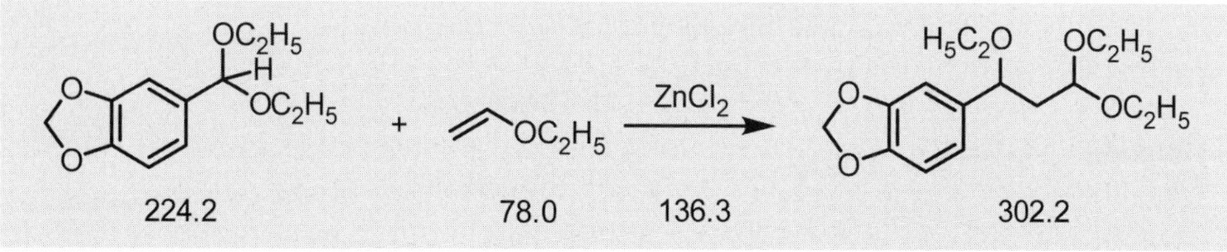

A suspension of anhydrous zinc chloride (0.70 g, note 1) in anhydrous EtOAc (5 mL) is added to the diethyl acetal **1.5.1.1** (56.0 g, 0.25 mol) with stirring and under exclusion of moisture. The mixture is heated to 40 °C and ethyl vinyl ether (19.5 g, 0.25 mol) is added at such a rate that the temperature is maintained between 40 and 45 °C. When the addition is complete, stirring is continued at 40–45 °C for 1 h.

The reaction mixture is then allowed to cool to room temperature and is diluted with Et_2O (130 mL). The ethereal solution is washed with NaOH (2 N, 25 mL) and dried (Na_2SO_4). After removal of the solvent, the residue is fractionated *in vacuo* (20 cm Vigreux column). The product is obtained as a colorless oil; 62.7 g (83%), $bp_{0.1}$ 97–99 °C.

IR (film): $\tilde{v}$ (cm^{-1}) = 2973, 2875, 1503.

^{1}H NMR ($CDCl_3$): δ (ppm) = 6.83 (s_{br}, 1 H, ArH), 6.75 (s_{br}, 2 H, ArH), 5.95 (s, 2 H, OCH_2O), 4.58 (dd, J = 6.9, 4.7 Hz, 1 H, OCHO), 4.28 (dd, J = 8.8, 5.4 Hz, 1 H, Ar–CH–O), 3.22–3.75 (mixture of q, 6 H, OCH_2), 2.08 (ddd, J= 13.9, 6.9, 5.4 Hz, 1 H, C–CH_2–C), 1.84 (ddd, J = 13.6, 8.8, 4.7 Hz, 1 H, C–CH_2–C), 1.21/1.20/1.15 (3 t, J = 6.9 Hz, each 3 H, OCH_2C$\underline{H}_3$).

^{13}C NMR ($CDCl_3$): δ (ppm) = 147.90, 146.93, 136.69, 120.11, 107.97, 106.74, 100.94, 100.39, 78.26, 63.82, 61.34, 42.44, 18.45, 15.43.

Note: Commercially available anhydrous zinc chloride is dried *in vacuo* over P_4O_{10}.

1.5.1.3   *   3-(3,4-Methylenedioxyphenyl)acrolein [4]

A stirred mixture of the triethoxypropane **1.5.1.2** (60.4 g, 0.20 mol), 1,4-dioxane (400 mL), H_2O (140 mL), 90% phosphoric acid (20 mL), and hydroquinone (0.2 g) is heated under reflux for 8 h under an N_2 atmosphere.

After cooling to room temperature, the reaction mixture is poured into ice-cold H_2O (1000 mL). After stirring for 1 h, the precipitated product is collected by suction filtration, and washed with dilute $NaHCO_3$ solution and with H_2O until the washings are neutral. The crude aldehyde is recrystallized from EtOH; 30.0 g (85%), yellow crystals, mp 84–85 °C.

IR (solid): $\tilde{v}$ (cm^{-1}) = 3048, 2992, 2916, 2823, 2729, 2701 (aldehyde H), 1666 (α,β-unsaturated C=O), 1620 (C=C, aryl conjugated), 1597 (C=C, conjugated to CO).

^{1}H NMR ($[D_6]$DMSO): δ (ppm) = 9.59 (d, J = 7.9 Hz, 1 H, CHO), 7.62 (d, J = 15.8 Hz, 1 H, PhC$\underline{H}$=CH), 7.42 (d, J = 1.6 Hz, 1 H, ArH), 7.23 (dd, J = 7.9, 1.6 Hz, 1 H, ArH), 7.00 (d, J = 7.9 Hz, 1 H, ArH), 6.74 (dd, J= 15.8, 7.9 Hz, 1 H, =C$\underline{H}$CHO), 6.09 (s, 2 H, OCH_2O).

^{13}C NMR ($[D_6]$DMSO): δ (ppm) = 194.03, 153.10, 149.97, 148.14, 128.52, 126.74, 125.69, 108.57, 106.83, 101.73.

UV (EtOH): λ_{max} [nm/(log ε)] = 338 (4.29), 297 (4.06), 248 (4.07), 220 (4.06).

1.5.1.4   **   4-(3,4-Methylenedioxyphenyl)-1,3-butadiene-1-carboxylic acid methyl ester [4]

(1) Monomethyl malonate: A solution of KOH (16.8 g, 0.30 mol) in anhydrous MeOH (170 mL) is added dropwise to a stirred solution of dimethyl malonate (40.0 g, 0.30 mol) in anhydrous MeOH (170 mL) at room temperature. Stirring is continued for 24 h and the precipitated potassium salt is collected by suction filtration, washed with Et_2O (50 mL), and dried *in vacuo*.

The salt is dissolved in H_2O (30 mL), and concentrated HCl (58 mL) is added dropwise at 0 °C with stirring. The mixture is then extracted with Et_2O (4 × 50 mL), the combined ethereal extracts are dried (Na_2SO_4), and the solvent is removed. The residue is distilled *in vacuo* and the product is obtained as a colorless oil; 33.6 g (95%), $bp_{0.18}$ 84–85 °C.

---

^{1}H NMR (CDCl$_3$): δ (ppm) = 10.90 (s, 1 H, COOH), 3.78 (s, 3 H, CH$_3$), 3.46 (s, 2 H, CH$_2$).

^{13}C NMR (CDCl$_3$): δ (ppm) = 171.77, 167.12, 52.82, 40.74

---

(2) A mixture of aldehyde **1.5.1.3** (17.6 g, 0.10 mol), monomethyl malonate (17.7 g, 0.10 mol), anhydrous piperidine (1 mL), and anhydrous pyridine (40 mL) is heated at 80 °C for 2 h and at 130 °C for 1 h.

The reaction mixture is then diluted with Et_2O (150 mL), and the ethereal solution is washed several times with H_2O (100 mL). Thereafter, it is washed with HCl (2 N, 100 mL), and then with further H_2O until the washings are neutral. The ethereal solution is dried (MgSO$_4$), the solvent is removed, and the residue is recrystallized from MeOH. The diene ester is obtained as yellow crystals; 20.1 g (87%), mp 142–143 °C.

IR (KBr): $\tilde{v}$ (cm^{-1}) = 2947, 1706 (C=O), 1616 (C=C, aryl conjugated), 1607 (C=C, conjugated to CO), 1505.

^{1}H NMR ([D$_6$]DMSO): δ (ppm) = 7.37 (ddd, J = 15.1, 8.5, 1.9 Hz, 1 H, C$\underline{H}$=CHCOO), 7.22 (d, J = 1.3 Hz, 1 H, ArH), 7.10–7.05 (combined signals, 3 H, ArH, Ar–C$\underline{H}$=C$\underline{H}$), 6.92 (d, J^3 = 7.9 Hz, 1 H, ArH), 6.04 (s, 2 H, OCH$_2$O), 6.00 (d, J = 15.1 Hz, 1 H, =CHCOO), 3.67 (s, 3 H, OCH$_3$).

^{13}C NMR ([D$_6$]DMSO): δ (ppm) = 166.61, 148.18, 147.95, 145.15, 140.51, 130.35, 124.60, 123.18, 119.43, 108.47, 105.67, 101.33, 51.18.

1.5.1.5   *   4-(3,4-Methylenedioxyphenyl)-1,3-butadiene-1-carboxylic acid [4]

A solution of the methyl ester **1.5.1.4** (18.6 g, 80.0 mmol) in 20% KOH in EtOH (100 mL) is heated under reflux for 3 h.

The solvent is then removed *in vacuo* and the residue is dissolved in the minimum amount of hot H_2O (ca. 50 mL). The solution is cooled to 0 °C and acidified by the dropwise addition of concentrated HCl with stirring. The precipitated acid is collected by suction filtration, washed with ice-cold water, dried, and recrystallized from EtOH; 14.5 g (83%), yellow crystals, mp 217–218 °C; TLC (SiO$_2$/Et$_2$O): R_f = 0.60.

IR (KBr): $\tilde{v}$ (cm^{-1}) = 3100–2400 (assoc. OH), 1680 (C=O).

UV (EtOH): λ_{max} [nm/(log ε)] = 343 (4.42), 308 (4.21), 262 (4.07).

¹H NMR ([D$_6$]DMSO): δ = 7.5–7.2 (m, 1 H, vinyl-H-2), 7.2–6.7 (m, 5 H, aryl-H + vinyl-H-3/H-4), 6.01 (s, 2 H, OCH$_2$), 5.93 (d, J = 14 Hz (*trans* coupling), 1 H, vinyl-H-1).

1.5.1.6   *   4-(3,4-Methylenedioxyphenyl)-1,3-butadiene carboxypiperidide (piperine) [2]

(1) The acid **1.5.1.5** (8.72 g, 40.0 mmol) is suspended in anhydrous benzene (180 mL), and then thionyl chloride (10 mL, distilled before use, bp$_{760}$ 78–79 °C) and anhydrous DMF (1.2 mL) are added. The mixture is heated to reflux with stirring for 2 h (N$_2$ atmosphere, hood, evolution of HCl and SO$_2$!). The solvents are removed *in vacuo* and the solid residue (crude acid chloride) is used directly in the next step.

(2) The acid chloride from (1) is dissolved in anhydrous benzene (40 mL) and the solution is cooled to 0 °C. A solution of anhydrous piperidine (14.8 g, 0.17 mol, 16.0 mL) in benzene (40 mL) is then added dropwise with stirring over 20 min; when the addition is complete, stirring is continued for 2 h at room temperature.

H_2O (200 mL) is then added, the aqueous phase is extracted with benzene (3 × 50 mL), and the combined benzene extracts are dried (Na$_2$SO$_4$). The solvent is removed *in vacuo* and the residue (dark oil) is dissolved in hot 4:1 cyclohexane/benzene (80 mL). On cooling to room temperature, the product crystallizes in well-shaped yellowish needles, which are collected by suction filtration, washed with cyclohexane, and dried; 11.0 g (95%), mp 130–132 °C, TLC (SiO$_2$; Et$_2$O): R_f = 0.40.

IR (KBr): $\tilde{\nu}$ (cm^{-1}) = 1640 (C=O amide), 1615, 1590 (C=C).

^{1}H NMR (CDCl$_3$): δ = 7.6–7.3 (m, 1 H, vinyl-H-2), 7.1–6.7 (m, 5 H, aryl-H + vinyl-3-H/4-H), 6.46 (d, J = 14 Hz (*trans* coupling), 1 H, vinyl-H-1), 6.00 (s, 2 H, OCH$_2$), 3.7–3.4 (m, 4 H, NCH$_2$), 1.8–1.4 (m, 6 H, β- and γ-piperidine-CH$_2$).

[1] a) Römpp, Lexikon "*Naturstoffe*" (Eds.: W. Steglich, B. Fugmann, S. Lang-Fugmann), p. 500, Georg Thieme Verlag, Stuttgart, **1997**. b) For the isolation of piperine from black pepper, see: H. Becker, K. P. Adam (Eds.), *Analytik biogener Arzneistoffe*, Pharmazeutische Biologie, Vol. 4, p. 333, Wissenschaftliche Verlagsgesellschaft mbH, Stuttgart, **2000**.

[2] Th. Eicher, H. J. Roth, *Synthese, Gewinnung und Charakterisierung von Arzneistoffen*, p. 303, Georg Thieme Verlag, Stuttgart, **1986**.

[3] R. Pick, Th. Eicher, unpublished results.

[4] F. Dallacker, J. Schubert, *Chem. Ber.* **1975**, *108*, 95.

[5] J. Isler, H. Lindlar, M. Montavon, R. Ruegg, P. Zeller, *Helv. Chim. Acta* **1956**, *39*, 255.

[6] Compare the analogous behavior of malonic acid in the formation of cinnamic acid by decarboxylative Knoevenagel condensation with benzaldehyde: *Organikum*, 21st ed., p. 529, Wiley-VCH, Weinheim, **2001**.

1.5.2   Cicloxilic acid

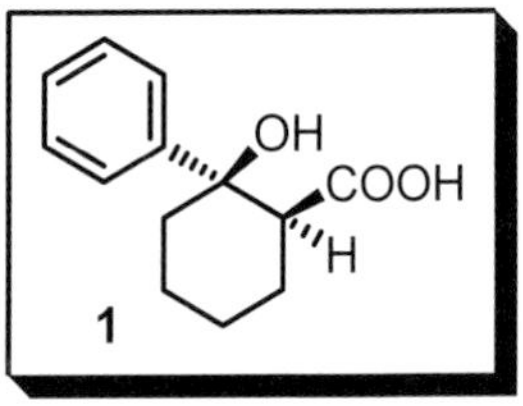

Topics:
- Synthesis of a drug
- Formation of a tertiary alcohol by addition of RMgX to a ketone
- Acid-catalyzed dehydration of a tertiary alcohol
- Stereoselective Prins reaction (acid-catalyzed addition of formaldehyde to an alkene)
- Oxidation of a primary alcohol to a carboxylic acid

(a) General

Cicloxilic acid (**1**, *rac-cis*-2-hydroxy-2-phenylcyclohexane-1-carboxylic acid) is used medicinally as a choleretic and hepatic protectant [1]. Its stereochemistry, with a *cis* relationship of the OH and COOH groups, was established by ^{1}H NMR spectroscopic investigation [2].

A straightforward retrosynthesis of **1** leads to cyclohexanone-2-carboxylic acid (**2**) as starting material, from which **1** might have been considered accessible by a simple Grignard reaction with PhMgBr. However, since this does not work, a somewhat lengthy transformation of the keto ester into **3** was necessary [3], which was then transformed into **1** by Grignard reaction (PhMgBr) followed by oxidation. Further negative aspects of this synthesis are its low yield and its lack of stereoselectivity, giving a mixture of the *cis* and *trans* diastereomers.

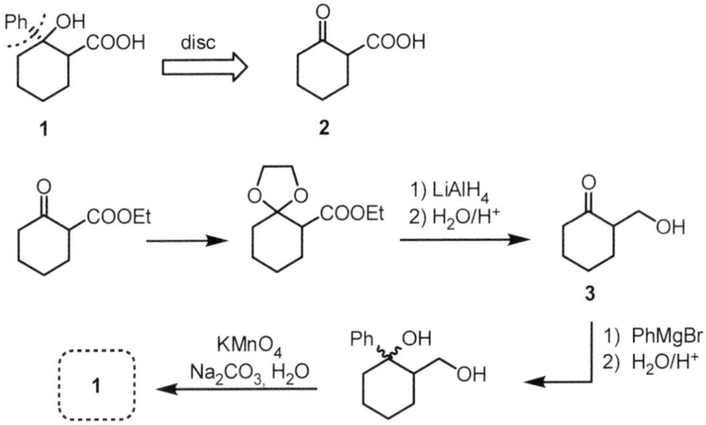

A second, less conventional retrosynthetic analysis leads to 1-phenylcyclohexene and formaldehyde:

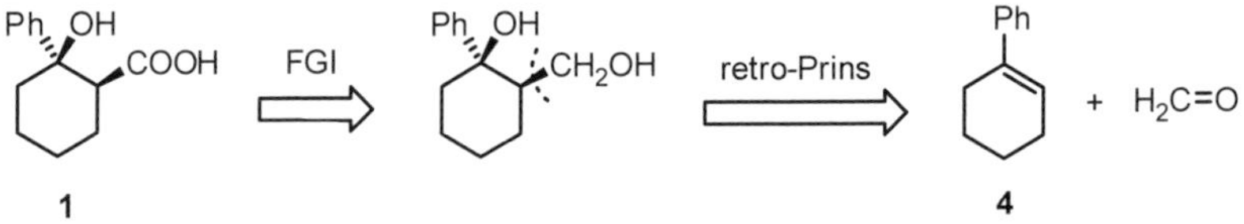

This approach was used in the described synthesis of **1**, with the advantage that it proceeds with high diastereoselectivity.

The acid-catalyzed addition of aldehydes, mainly formaldehyde, to alkenes is known as the Prins reaction. In this process, the carbenium ion derived from addition of the protonated carbonyl source to the alkene C=C bond is the central intermediate; it is intercepted by addition of a nucleophile, preferentially the solvent used (H_2O, formic acid, etc.) to give as products a 1,3-diol and/or its monoester [4, 5].

(b) Synthesis of 1

For the synthesis of **1**, 1-phenylcyclohex-1-ene (**4**) is reacted with formaldehyde in aqueous formic acid (5:95) to give **6** and **7** as the main products, accompanied by a side product **8**, which contains an acetal moiety formed by reaction of **6** with a second molecule of formaldehyde. The formate **7** can easily be transformed into **6** by saponification with NaOH.

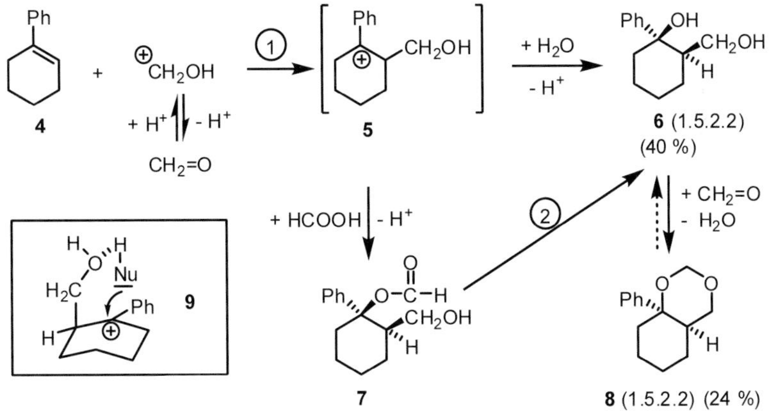

The high stereoselectivity of the Prins reaction, giving the *cis* diastereomers **6** and **7**, can be explained in terms of a pre-orientation through hydrogen bonding between the incoming nucleophile and the hydroxymethyl group in the cation **5** in the transition state [2].

In the final step of the synthesis, the diol **6** is oxidized with $KMnO_4$ in aqueous Na_2CO_3 solution to give cicloxilic acid **1**:

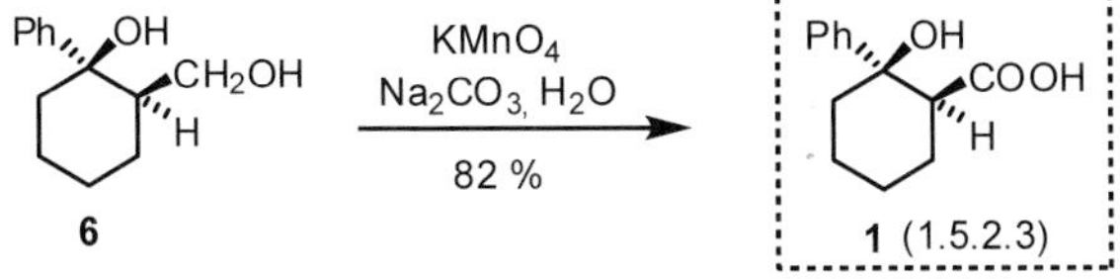

The required substrate, 1-phenylcyclohex-1-ene (**4**), is prepared from cyclohexanone by addition of phenylmagnesium bromide and subsequent acid-catalyzed elimination of H_2O (E1 process) from the formed tertiary benzyl alcohol [6]:

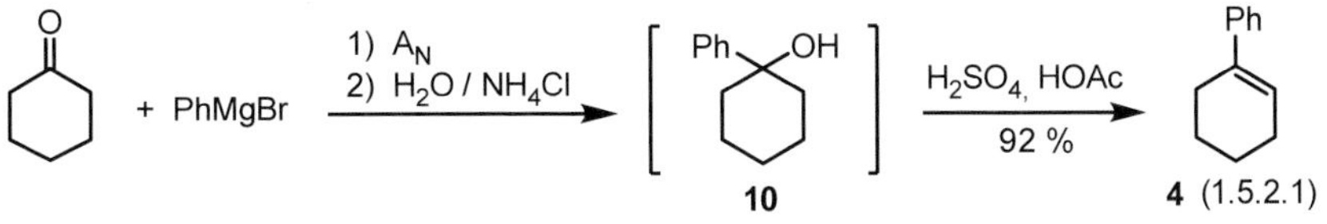

In this way, the target molecule **1** is obtained in a stereoselective three-step sequence in an overall yield of 47% (based on cyclohexanone) [7].

(c) Experimental procedures for the synthesis of 1

1.5.2.1   * 1-Phenylcyclohex-1-ene [6]

PhBr: 157.0
Mg: 24.3   98.1   158.1

The first 20 mL of a solution of bromobenzene (94.5 g, 0.50 mol; note 1) in anhydrous Et_2O (200 mL) and methyl iodide (4–6 drops) are added to magnesium turnings (14.5 g, 0.50 mol) in anhydrous Et_2O (20 mL). When the reaction has started, the rest of the bromobenzene solution is added dropwise with efficient stirring at such a rate that gentle boiling of the reaction mixture is maintained. When the addition is complete, heating under reflux is continued for 2 h.

A solution of cyclohexanone (49.1 g, 0.50 mol; note 2) in anhydrous Et_2O (40 mL) is then added dropwise with efficient stirring to the solution of phenylmagnesium bromide prepared as described above, again at such a rate that gentle boiling of the reaction mixture is maintained. When the addition is complete, heating at reflux is continued for 30 min.

The mixture is then cooled in an ice-bath and an ice-cold saturated solution of NH_4Cl (400 mL) is added dropwise with vigorous stirring. The organic phase is separated and the aqueous phase is extracted with Et_2O (150 mL). The ethereal phases are combined and dried over Na_2SO_4. After filtration, the solvent is removed *in vacuo* and the yellowish residue (note 3) is stirred for 30 s with a mixture of concentrated H_2SO_4 (20 mL) and acetic acid (80 mL) at 50 °C. The mixture is then poured into a two-phase system of H_2O (500 mL) and Et_2O (300 mL) and shaken. The ethereal phase is separated, washed repeatedly with saturated aqueous $NaHCO_3$ solution (4 × 100 mL), and dried over Na_2SO_4. After filtration, the solvent is removed and the residue is distilled *in vacuo*. The product is obtained as a colorless liquid; 72.5 g (92%), $bp_{4.5}$ 90–91 °C, $n_D^{20} = 1.5665$.

IR (film): $\tilde{\nu}$ (cm^{-1}) = 3010, 2910–2835, 1660 (C=C), 1495, 1445.

^{1}H NMR ($CDCl_3$): δ (ppm) = 7.5–7.05 (m, 5 H, phenyl-H), 6.2–5.9 (m, 1 H, vinyl-H), 2.6–2.0 (m, 4 H, CH_2), 2.0–1.4 (m, 4 H, CH_2).

Notes: (1) Bromobenzene is purified by distillation *in vacuo*, bp_{15} 48–49 °C.
          (2) Cyclohexanone has to be distilled before use; bp_{760} 155–156 °C, $n_D^{20} = 1.4500$.
          (3) This residue consists of 1-phenylcyclohexan-1-ol as the crude product and is subjected *in situ* to acid-catalyzed dehydration.

1.5.2.2   **   *cis*-2-Hydroxymethyl-1-phenylcyclohexan-1-ol [4]

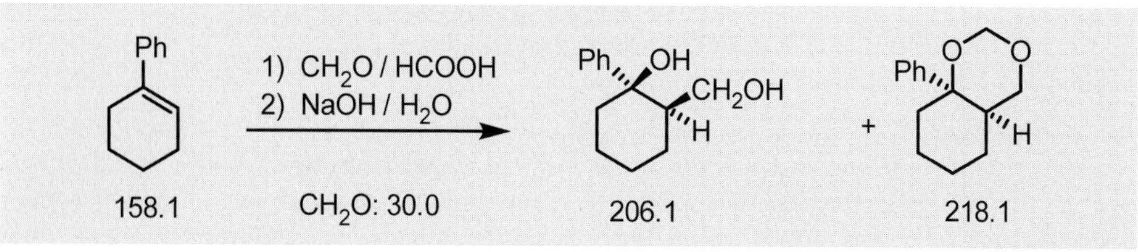

| 158.1 | CH$_2$O: 30.0 | 206.1 | 218.1 |

Phenylcyclohexene (**1.5.2.1**) (66.5 g, 0.42 mol) is suspended in a mixture of formic acid (420 mL) and H$_2$O (15 mL). A 40% formaldehyde solution (44.1 mL, 0.59 mol) is added dropwise over 30 min with stirring. When the addition is complete, the suspension is stirred for 3 h at room temperature.

The solvent is then removed *in vacuo* at an external temperature of 30 °C. A colorless oil is obtained, which is treated with a solution of NaOH (28.0 g) in EtOH (210 mL) with efficient stirring for 12 h at room temperature. The reaction mixture is then diluted with an equal volume of H$_2$O (ca. 250 mL) and extracted with chloroform (2 × 150 mL); the combined extracts are dried (Na$_2$SO$_4$) and filtered, and the solvent is removed *in vacuo*. The oily residue is dissolved in petroleum ether (300 mL) with heating, then the solution is cooled to room temperature and kept in a freezer for 24 h. The crystallized diol is filtered off, retaining also the filtrate (see below), and the purification procedure is repeated. The product is obtained as colorless crystals; 34.5 g (40%), mp 82–83 °C, TLC (SiO$_2$; Et$_2$O): R_f = 0.80.

IR (KBr): $\tilde{v}$ (cm^{-1}) = 3500–3180 (OH), 2930–2840 (CH).

^{1}H NMR (CDCl$_3$): δ (ppm) = 7.6–7.2 (m, 5 H, phenyl-H), 3.73, 2.45 (s, 1 H, OH; exchangeable with D$_2$O), 3.65–3.25 (m, 2 H, OCH$_2$), 2.25–1.25 (m, 9 H, cyclohexyl-H).

The solvent is removed *in vacuo* from the petroleum ether solution of the first crystallization of the diol, the residue is dissolved in the minimum amount of EtOH, and the solution is kept in a freezer for 12 h. The dioxane is obtained as colorless crystals; 22.5 g (24%), mp 62–63 °C, TLC (SiO$_2$; Et$_2$O): R_f ≈ 0.80.

---

^{1}H NMR (CDCl$_3$): δ (ppm) = 7.4–7.25 (m, 5 H, phenyl-H), 4.9–4.75 (m, 2 H, OCH$_2$O), 3.85, 3.53 (d, J = 11.2 Hz, each 1 H, OCH$_2$), 2.5–1.1 (m, 9 H, cyclohexyl-H).

1.5.2.3   *   *cis*-2-Hydroxy-2-phenylcyclohexane carboxylic acid (cicloxilic acid) [4]

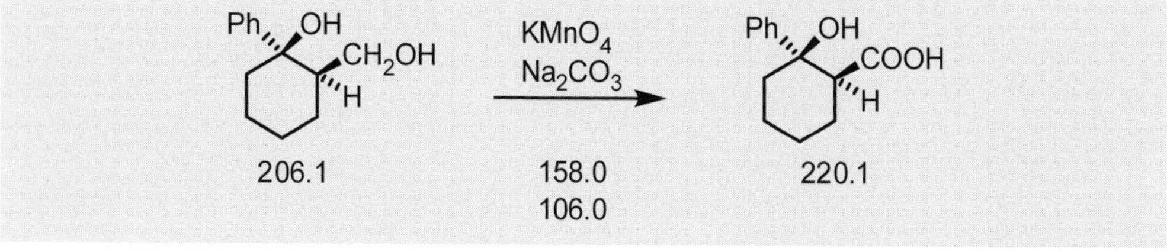

| 206.1 | 158.0
106.0 | 220.1 |

A solution of the diol **1.5.2.2** (29.0 g, 141 mmol) in H_2O (1500 mL) is heated to 85 °C (internal temperature). At this temperature, a mixture of finely powdered $KMnO_4$ (57.5 g, 364 mmol) and anhydrous Na_2CO_3 (29.0 g, 274 mmol) is added in small portions with vigorous stirring. When the addition is complete, stirring at 85 °C is continued for 30 min (note).

The MnO_2 formed is removed by suction filtration and the filter cake is washed with H_2O (3 × 100 mL). Concentrated HCl is added dropwise to the filtrate with stirring until a pH of ca. 1 is reached. The colorless precipitate is collected by suction filtration, washed with a small amount of iced-water, and dried over P_4O_{10} *in vacuo*. Recrystallization from cyclohexane yields 25.5 g (82%) of cicloxilic acid; mp 139–140 °C, TLC (SiO_2; Et_2O): $R_f \approx 0.75$.

IR (KBr): $\tilde{v}$ (cm^{-1}) = 3530 (OH), 3200–2600 (OH), 1680 (C=O).

^{1}H NMR ($CDCl_3$): δ (ppm) = 7.5–7.15 (m, 5 H, phenyl-H), 3.1–2.9 (m, 1 H, CH), 2.15–1.2 (m, 8 H, cyclohexyl-H).

Note: If the reaction mixture still contains an excess of permanganate, MeOH is added until decolorization occurs.

[1] A. Kleemann, J. Engel, *Pharmaceutical Substances*, 3rd ed., p. 434, Thieme Verlag, Stuttgart, **1999**.

[2] L. Turbanti, G. Cerbai, G. Ceccarelli, *Arzneim. Forsch./Drug. Res.* **1978**, *87*, 1249.

[3] US 3 700 775.

[4] DOS 2 607 967 (*Chem. Abstr.* **1977**, *87*, 84688).

[5] a) D. R. Adams, S. P. Bhatnagar, *Synthesis* **1977**, 661; b) J. Thiem, in *Houben-Weyl, Methoden der Organischen Chemie*, Vol. 6/1a, 1980, p. 793. c) For the use of Prins cyclizations in natural product synthesis, see: L. E. Overman, L. D. Pennington, *Chem. Soc. Rev.* **2003**, *32*, 383.

[6] E. W. Garbisch, Jr., *J. Org. Chem.* **1961**, *26*, 4165.

[7] With regard to maximizing the overall yield, it has been considered that the cyclic acetal **8** may be hydrolyzed ($CH_3OH/H_2O/HCl$) to give the diol **6** in practically quantitative yield.

1.5.3   β-Ionone

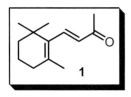

1

Topics: • Synthesis of a terpene-derived C_{13}-dienone
  • α-Alkylation of acetoacetate
  • Ethynylation of a carbonyl compound
  • Acetoacetylation of an alcohol with diketene
  • Carroll reaction of allyl acetoacetates ([3,3]-sigmatropic rearrangement)
  • Cationic cyclization of a 1,5-diene to a cyclohexene derivative

(a) General

Ionones are a group of natural fragrances, which are formed by oxidative degradation of tetraterpenoids (carotins) [1]. α-Ionone **2** is the main component of violet oil, while β- and γ-ionones (**1** and **3**) are found in several essential oils. Structurally related to the ionones are the damascones (e.g., β-damascone **4**, a constituent of rose oil) and irones [e.g., β-irone **5** (2-methyl-β-ionone), a fragrant compound from the oil of iris roots] [2].

β-ionone	(R)-(+)-α-ionone	(R)-(+)-γ-ionone	β-damascone	(R)-(+)-β-irone
1	**2**	**3**	**4**	**5**

β-Ionone is one of the most potent odorous organic compounds (perceptible in concentrations <0.1 ppb); it is an important ingredient of perfumes and is used as a substrate in natural product syntheses, e.g., of damascone [3] and of vitamin A (cf. **4.2.6**).

Three retrosynthetic pathways for β-ionone (**1**) are discussed here.

In **A**, disconnection of the C-6/C-7 bond according to a retro-Heck transformation leads to the cyclohexene **6**, which could formally be obtained by a Diels–Alder reaction. However, this would be an electronically disfavored transformation.

In **B**, the cyclohexene ring is disconnected by a retro-Diels–Alder reaction to give ethylene as dienophile and the 1,3-diene system **8**. However, as discussed before, their [4+2]-cycloaddition is not a favorable process, again due to electronic reasons (no activated dienophile) as well as a lack of regioselectivity (different 1,3-diene moieties exist in **8**).

In **C**, a protonation/deprotonation sequence initiates a ring-opening (**1 → 7 → 9**) to pseudoionone (**10**), which could be obtained by an aldol condensation of citral (**14**) with acetone. Another possible retrosynthesis of **10** includes a retro-Claisen protocol (**10 → 11a → 11b**) leading to dehydrolinalool (**15**) and acetoacetate via dehydrolinalool acetoacetate (**12/13**). **15** may be obtained from methylheptenone **16**, which is accessible from **17** and acetoacetate.

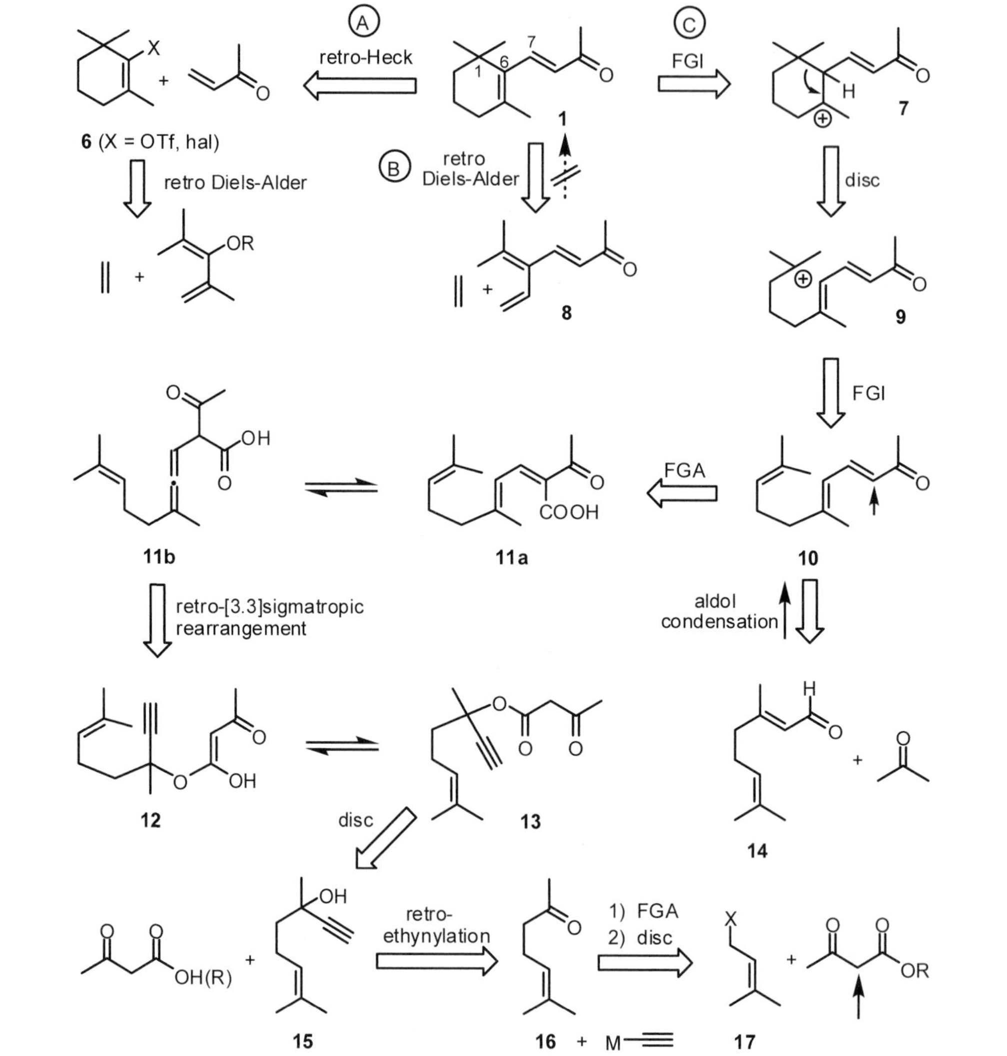

Realizations of the retroanalytical pathways **A–C** for the synthesis of **1** have been reported in the literature.

Thus, a short and efficient approach to **1** utilizes the Heck reaction of the trifluoromethanesulfonate **6** (X = OSO$_2$CF$_3$) of 2,6,6-trimethylcyclohexanone with methyl vinyl ketone [4]:

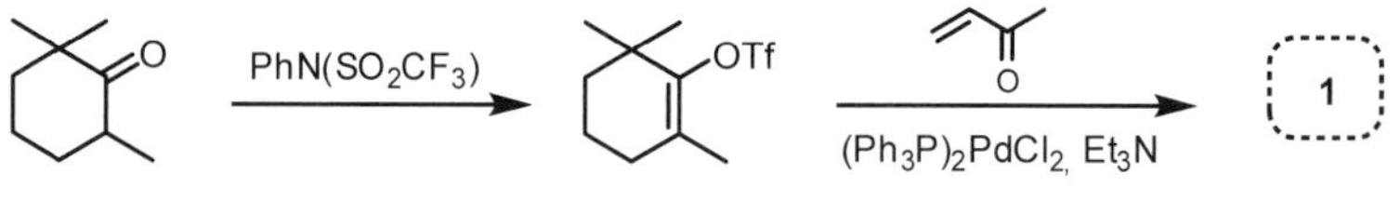

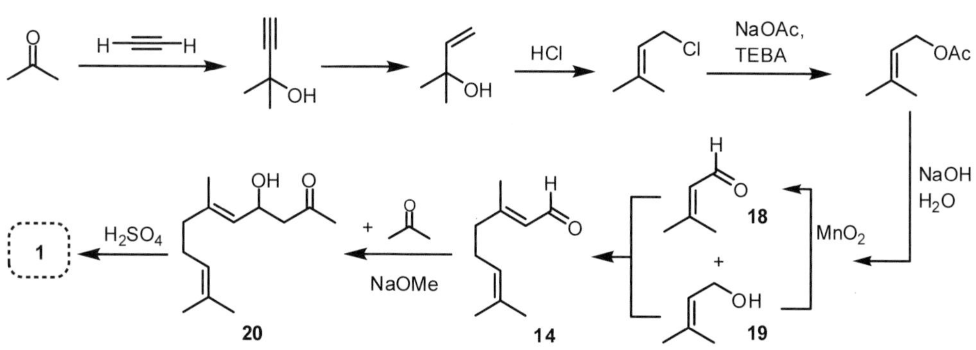

In the second approach towards **1**, acetone is condensed with citral (**14**) [5], which is obtained by a pericyclic domino process of two [3,3]-sigmatropic reactions between the allylic alcohol **19** and the aldehyde **18** via the vinyl allyl ether **21**:

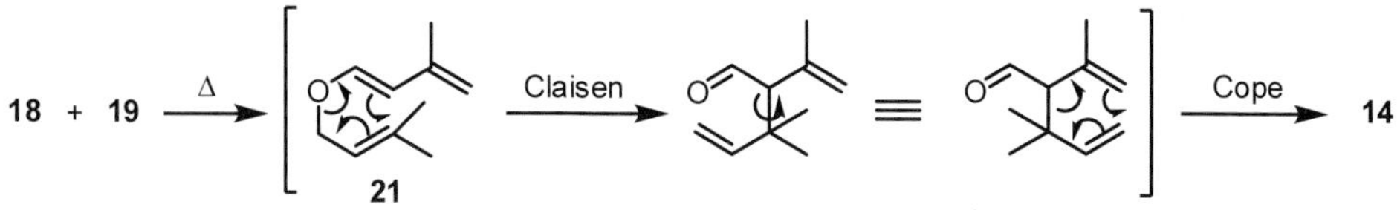

The third approach to the target molecule **1**, according to retrosynthesis **C**, uses elements of the industrial β-ionone synthesis of BASF and is described in detail [6].

(b) Synthesis of 1

First, 6-methylhept-5-ene-2-one (**16**) is prepared from acetoacetate by α-alkylation with prenyl bromide (cf. **4.2.1**), ester hydrolysis, and decarboxylation of the intermediately formed β-keto acid. Ethynylation of methylheptenone **16** with Na acetylide [7] gives the tertiary alcohol dehydrolinalool (**15**), which is esterified with diketene. The propargylic acetoacetate **13** is subjected to a thermal [3,3]-sigmatropic rearrangement in the presence of Al(OiPr)$_3$ with concomitant decarboxylation of the resulting β-allenic acid (**11b**) to give the unsaturated ketone **10** (pseudoionone). The Claisen (oxa-Cope) rearrangement of allylic or propargylic acetoacetates (Carroll reaction) is often used in terpene synthesis (also industrially [8]) as a C$_3$ chain elongation process (here: C$_{10}$ → C$_{13}$).

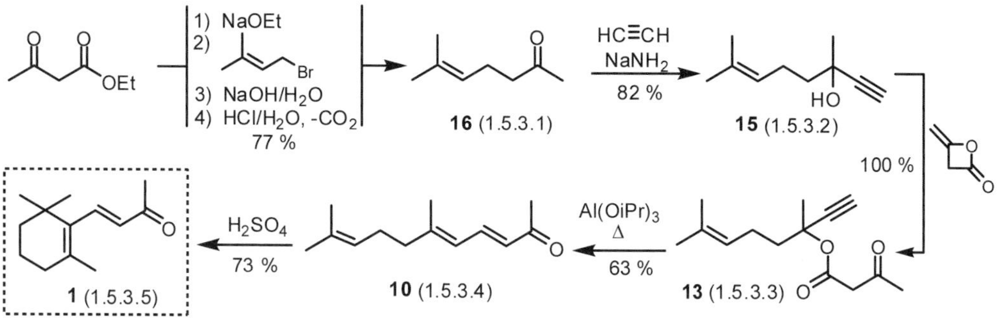

The final step of the synthesis of β-ionone (**1**) is the acid-catalyzed cycloisomerization of pseudoionone (**10**). Mechanistically, a cationic cyclization of the 1,5-diene through carbenium ion formation (by protonation of the terminal C=C double bond) and its addition to an internal olefinic C=C bond resulting in the formation of a cyclohexene can be assumed.

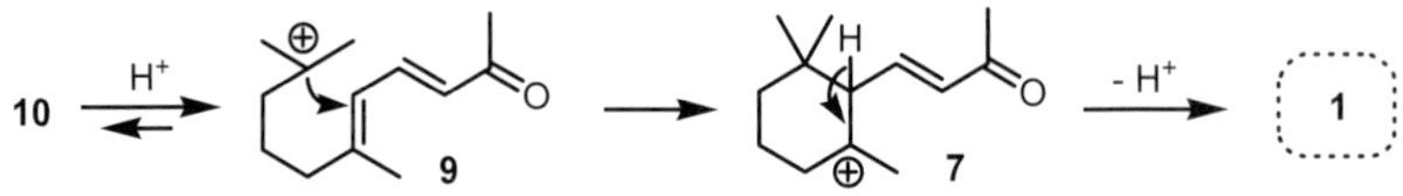

Cyclizations of this type occur with a high degree of stereoselectivity (stereoelectronic control in **9** → **7** as a result of a chair-like transition state) and are involved in the biosynthesis of steroids and other polycycles [9].

Using the described approach, the target molecule **1** is obtained in a five-step sequence with an overall yield of 29% (based on acetoacetate).

(c) Experimental procedures for the synthesis of 1

1.5.3.1 * 6-Methylhept-5-en-2-one [10]

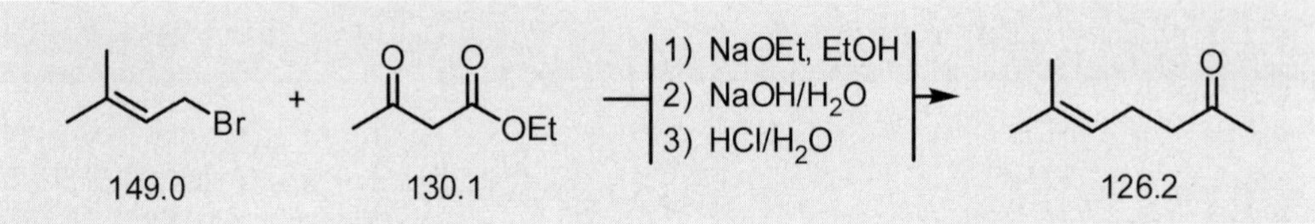

Sodium (12.6 g, 0.55 mol) is added to a stirred solution of ethyl acetoacetate (87.8 g, 0.67 mol) and anhydrous EtOH (150 mL) (formation of H_2). The mixture is cooled to 0 °C and 1-bromo-3-methyl-2-butene (**4.3.1.2**) (74.5 g, 0.50 mol) is added dropwise over 20 min. Stirring is continued at room temperature for 3 h and at 60 °C for 4 h. During this time, a fine crystalline precipitate of sodium bromide forms.

The mixture is then filtered, the filtrate is concentrated *in vacuo*, and the residue is treated with 10% NaOH solution (200 mL). The resulting mixture is stirred at room temperature for 2 h and at 60 °C for 3 h, cooled, and acidified to pH 4 with concentrated HCl. The solution is extracted with Et_2O (3 × 100 mL), and the combined organic phases are washed with saturated $NaHCO_3$ solution (150 mL) and H_2O and dried over $MgSO_4$. The solvent is evaporated *in vacuo* and the residue is fractionally distilled through a Vigreux column to give a colorless oil with a fruit-like odor; 51.7 g (77%), bp_{12} 64–65 °C, n_D^{20} = 1.4404.

IR (film): $\tilde{v}$ (cm^{-1}) = 1720 (C=O), 1360, 1160.

^{1}H NMR (CCl_4): δ (ppm) = 5.00 (m, 1 H, vinyl-H), 2.4–2.1 (m, 4 H, CH_2), 2.04 (s, 3 H, CO–CH_3), 1.63 (m, 6 H, =C(CH_3)$_2$).

1.5.3.2   **   Dehydrolinalool (3,7-dimethyl-1-octyn-6-en-3-ol) [11]

126.2	HC≡CH , NaNH₂	152.2
	39.0	

Finely powdered sodium amide (18.0 g, 0.46 mol, note 1) is added in portions to a stirred solution of methylheptenone **1.5.3.1** (30.0 g, 0.24 mol) in anhydrous Et_2O (150 mL) at –15 °C. After stirring the mixture for 3 h, a rapid stream of acetylene is passed through it for 4 h. The temperature is then held at –20 °C for 15 h. Thereafter, a rapid stream of acetylene is again passed through the mixture at –15 °C for 4 h.

The brown-yellow mixture is poured into well-stirred iced-water (500 mL). The ethereal phase is separated and the aqueous phase is extracted with Et_2O (150 mL). The combined organic phases are dried over $MgSO_4$ and concentrated *in vacuo*. The yellow residue is fractionally distilled to give a colorless oil with an odor similar to that of citral; 29.7 g (82%), bp$_{10}$ 85–88 °C. n_D^{20} = 1.4632 (note 2).

IR (film): $\tilde{v}$ (cm^{-1}) = 3400 (br, OH), 3300 (≡CH), 2970, 2920, 2860 (CH), 1450, 1120.

¹H NMR (CDCl₃): δ (ppm) = 5.15 (t, J = 6.5 Hz, 1 H, vinyl-H), 2.52 [s, 1 H, OH (exchangeable with D₂O)], 2.49 (s, 1 H, ≡CH), 2.35–1.9 (m, 2 H, allyl-CH₂), 1.66 [s, 6 H, =C(CH₃)₂], 1.64 (t, J = 7 Hz, 2 H, CH₂), 1.50 (s, 3 H, CH₃).

Notes: (1) NaNH₂ is obtained by filtering a suspension in toluene (sintered glass filter) and washing twice with Et_2O.
(2) If the product still contains methylheptenone (determined by GC), it is shaken for 15 h with sodium bisulfite solution and redistilled.

1.5.3.3   **   Dehydrolinalool acetoacetate [12]

152.2	84.1	236.3

Sodium methoxide (0.20 g; freshly prepared and dried at 100 °C/0.1 mbar) is added to a solution of dehydrolinalool **(1.5.3.2)** (26.6 g, 175 mmol) in anhydrous toluene (40 mL). Diketene (16.4 g, 195 mmol) is added dropwise to the stirred solution over a period of 2 h, keeping the temperature under 30 °C with occasional cooling if necessary. Stirring is continued at 30 °C for 5 h and at room temperature for 15 h.

The light-brown mixture is then washed with H_2SO_4 (1 M, 50 mL), saturated $NaHCO_3$ solution (50 mL), and H_2O (2 × 50 mL). The toluene solution is dried over Na_2SO_4 and concentrated *in vacuo*. The residual yellow oil is sufficiently pure for further use; yield 41.3 g (100%). Distillation *in vacuo* gives a colorless oil, bp$_{0.005}$ 43–44 °C, n_D^{20} = 1.4652.

IR (film): $\tilde{v}$ (cm^{-1}) = 3290 ($\equiv$CH), 2060 (C$\equiv$C), 1755, 1725 (C=O).

^{1}H NMR (CDCl$_3$): δ (ppm) = 5.25–4.8 (m, vinyl-H + enol-H), 3.34 (s, incl. previous signal 3 H, CO–CH$_2$), 2.56 (s, 1 H, $\equiv$CH), 2.22 (s, 3 H, CO–CH$_3$), 2.1–1.8 (m, 2 H, allyl-CH$_2$), 1.70 [s, 6 H, =C(CH$_3$)$_2$], 1.60 (s, 3 H, CH$_3$), 1.75–1.5 (m, 2 H, CH$_2$).

1.5.3.4 ** Pseudoionone (6,10-dimethylundeca-3,5,9-trien-2-one) [12]

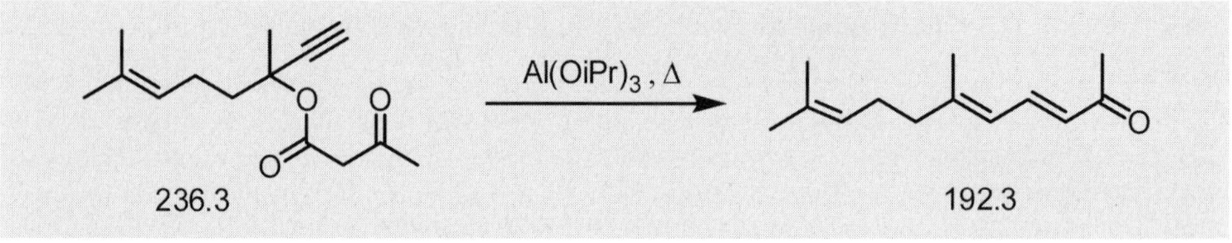

236.3   192.3

A stirred mixture of the crude dehydrolinalool acetoacetate (**1.5.3.3**) (41.3 g, 175 mmol), decalin (50 mL), glacial acetic acid (0.5 mL), and aluminum isopropoxide (40 mg) is heated to 175–190 °C for 2 h with evolution of CO$_2$ (bubble trap).

The mixture is then cooled, washed with H$_2$SO$_4$ (1 M, 50 mL), saturated NaHCO$_3$ solution (3 × 50 mL), and H$_2$O (2 × 50 mL), and dried over CaSO$_4$. The decalin is distilled off at 10 mbar (bp$_{10}$ 70–71 °C). The yellow residue is fractionally distilled to give a pale-yellow oil; 21.2 g (63%), bp$_{0.5}$ 92–95 °C, n_D^{20} = 1.5272. The purity of the product is determined by GC.

IR (film): $\tilde{v}$ (cm^{-1}) = 1685, 1665 (C=O), 1630, 1590 (C=C), 1250, 975.

^{1}H NMR (CDCl$_3$): δ (ppm) = 7.41 (dd, J = 11.0, 3.0 Hz, 1 H, 4-CH), 6.2–5.8 (m, 2 H, 3-CH/5-CH), 5.05 (m, 1 H, 9-CH), 2.4–2.05 (m, 4 H, CH$_2$), 2.27 (s, 3 H, CO–CH$_3$), 1.90 (s, 3 H, 6-CH$_3$), 1.67, 1.61 (s, 6 H, =C(CH$_3$)$_2$).

UV (CH$_3$CN): λ_{max} (log ε) = 284 (4.51), 212 nm (4.14).

1.5.3.5 ** β-Ionone [4-(2,6,6-trimethylcyclohex-1-en-1-yl)but-3-en-2-one] [13]

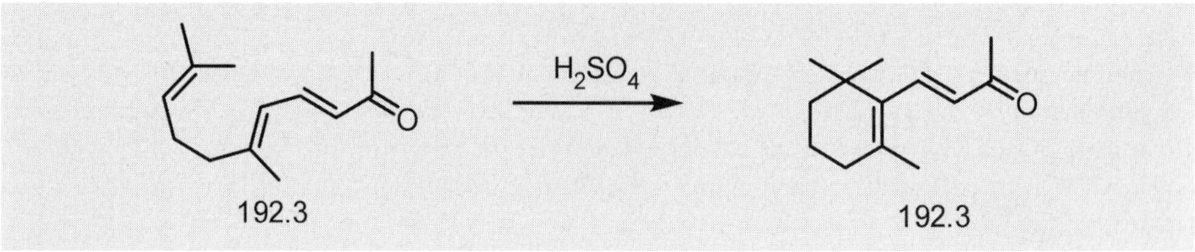

192.3   192.3

Pseudoionone (**1.5.3.4**) (50.0 g, 0.26 mol) is added to a well-stirred mixture of concentrated sulfuric acid (175 g) and glacial acetic acid (75 g) at 5 °C over 40 min, keeping the temperature below 10 °C. Stirring is continued at 10–15 °C for 10 min.

The mixture is then poured into a well-stirred mixture of iced-water (1000 mL) and Et$_2$O (250 mL). The ethereal phase is separated and the aqueous phase is extracted with Et$_2$O (250 mL). The combined organic phases are washed with H$_2$O, 1% Na$_2$CO$_3$ solution, and further H$_2$O. The solvent is evaporated *in vacuo* and the residue is steam distilled. The β-ionone is taken up in Et$_2$O (2 × 250 mL), the ethereal solution is dried over Na$_2$SO$_4$, and then the solvent is evaporated *in vacuo*. The residue is distilled through a 20 cm packed column (Raschig–Ringe) to give a light-yellow oil with a characteristic, pleasant odor; 36.5 g (73%), bp$_{0.7}$ 91–93 °C, n$_D^{20}$ = 1.5198.

IR (film): $\tilde{v}$ (cm^{-1}) = 1700, 1675 (C=O), 1615, 1590 (C=C), 1260.

^{1}H NMR (CDCl$_3$): δ (ppm) = 7.13, 5.99 (d, *J* = 16 Hz, 1 H, 4-CH/3-CH), 2.19 (s, 3 H, CO–CH$_3$), 2.07 (m, 2 H, allyl-CH$_2$), 1.75 (s, 3 H, =C–CH$_3$), 1.8–1.2 (m, 4 H, CH$_2$–CH$_2$), 1.07 (s, 6 H, C(CH$_3$)$_2$).

---

[1]   R. C. Mordi, J. C. Walton, *Tetrahedron* **1993**, *49*, 911.

[2]   Römpp, Lexikon *"Naturstoffe"* (Eds.: W. Steglich, B. Fugmann, S. Lang-Fugmann), 10th ed., p. 334, Georg Thieme Verlag, Stuttgart, **1997**.

[3]   Th. Eicher, S. Hauptmann, *The Chemistry of Heterocycles*, 2nd ed., p. 143, Wiley-VCH, Weinheim, **2003**.

[4]   T. Breining, C. Schmidt, K. Polos, *Synth. Commun.* **1987**, *17*, 85.

[5]   P. V. S. N. Vani, A. S. Chida, R. Srinavasan, M. Chandrasekan, A. K. Singh, *Synth. Commun.* **2001**, *31*, 219.

[6]   W. Reif, H. Grassner, *Chem. Ing. Tech.* **1973**, *45*, 646.

[7]   By mediation with Zn(OTf)$_2$, the addition of terminal acetylenes to aldehydes can be conducted enantioselectively in the presence of chiral amino alcohol ligands: a) N. K. Anaud, E. M. Carreira, *J. Am. Chem. Soc.* **2001**, *123*, 9867; b) D. P. G. Emmerson, W. P. Hems, G. G. Davis, *Org. Lett.* **2006**, 207.

[8]   H. Pommer, A. Nürrenbach, *Pure Appl. Chem.* **1975**, *43*, 527.

[9]   W. S. Johnson, *Angew. Chem.* **1976**, *88*, 33; *Angew. Chem. Int. Ed. Engl.* **1976**, *15*, 9.

[10] In analogy to: *Organikum*, 21st ed., p. 608, Wiley-VCH, Weinheim, **2001**.

[11] W. Kimel, J. D. Surmatis, J. Weber, G. O. Chase, N. W. Sax, A. Ofner, *J. Org. Chem.* **1957**, *22*, 1611.

[12] W. Kimel, N. W. Sax, S. Kaiser, G. G. Eichmann, G. O. Chase, A. Ofner, *J. Org. Chem.* **1958**, *23*, 153.

[13] H. J. V. Krishna, B. N. Joshi, *J. Org. Chem.* **1957**, *22*, 224.

1.6   Transition-metal-mediated reactions

1.6.1   (E)-4-Chlorostilbene

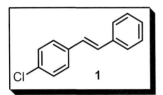
1

Topic: • Pd-catalyzed arylation of an alkene by a Heck reaction

(a) General

The palladium-catalyzed arylation and alkenylation of alkenes is known as the Heck reaction [1]:

$$\text{alkene} \quad + \quad \text{R–X} \quad \xrightarrow[- HX]{\text{"Pd(0)"}} \quad \text{product} \qquad \begin{array}{l} R = \text{aryl, vinyl} \\ X = I, Br, OTf \end{array}$$

This coupling reaction of two sp^2-hybridized carbon centers usually requires the presence of: (a) a mono- or bidentate phosphine as complexing ligand, and (b) a base, often a tertiary amine such as triethylamine or diisopropylethylamine (Hünig base) or an inorganic base such as K_2CO_3, NaOAc, etc.

The generally accepted mechanism for the Heck reaction [2] consists of a catalytic cycle of five consecutive partial steps (1)–(5), as formulated for the reaction of a monosubstituted alkene R^2–CH=CH$_2$ with an aryl or alkenyl halide R^1–X:

$$R^1\text{-X} \quad + \quad \overset{}{\diagup}\!\!\diagdown_{R^2} \quad \xrightarrow[\text{base, } - HX]{nL, \text{ "Pd}^0\text{"}} \quad R^1\diagup\!\!\diagdown\diagup R^2$$

In step (1), an oxidative addition of a 14-electron complex, Pd(0)L$_2$ (**2**), takes place by insertion into the C(sp^2)–X bond of R^1–X to give a tetracoordinated 16-electron Pd(II) complex **3**. **2** is formed *in situ* either by reduction of a Pd(II) source such as Pd(OAc)$_2$ by, for example, a tertiary amine [1e, 2b] or a phosphine or by dissociation of two ligands of a Pd(0)L$_4$ species such as Pd(PPh$_3$)$_4$.

In steps (2) and (3), the alkene coordinates to the Pd(II) species **3** (π-complex **4**) and is inserted into the Pd(II)–R^1 bond. The insertion process (3) is stereoselective and proceeds in a *syn* manner. Since the alkene R^2–CH=CH$_2$ is unsymmetrical, C$_a$ or C$_b$ may be involved in the insertion and two regioisomeric σ-alkyl-Pd(II) species **5** and/or **6** may be formed.

In step (4), Pd-β-hydride elimination — again in a stereoselective *syn* manner, as shown experimentally — leads to the formation of complex **9** and release of the products **7/8** (probably via π-complex formation).

In step (5), the catalytic cycle is completed by regeneration of the catalytic Pd(0)L$_2$ species (**2**) from the Pd(II) hydrido complex **9** by reaction with a base. Steps (4) and (5) are regarded as reductive elimination.

The Suzuki reaction is widely used for the formation of 1,3-butadienes and unsymmetrical biaryls, which are challenging targets in natural product and pharmaceutical chemistry [8–12]:

$$Ar^1-X \ + \ Ar^2-B(OH)_2 \ \xrightarrow{\ Pd^0\ } \ Ar^1-Ar^2$$

X = Br, I, triflate, arylsulfonate

Ar^1 = functionalized aryl (COOR, CN, NO_2 etc.)

Ar^2 = functionalized aryl (ether, acetal etc.)

As Ar^1–X species, mainly bromides, iodides, triflates, and arylsulfonates [13] are used; chlorides can also be coupled in the presence of highly active Pd catalysts such as $HPd(PtBu_3)_3{\cdot}BF_4$ [14]. Moreover, diazonium salts and aromatic carboxylic acids can also be employed [15]. Phosphine-free and palladacycle-based modifications of the Suzuki reaction have also been developed [16]. Generally, in Ar^1–X, functional groups such as NO_2, CN, COOR, etc., are tolerated, whereas the boronic acids Ar^2–$B(OH)_2$ may contain ether or acetal functions.

In section (b), a Suzuki reaction is used for the preparation of **1**, which serves as a substrate for the synthesis of a simple alkaloid (cf. **4.1.3**) containing an unsymmetrical functionalized biaryl unit [17].

(b) Synthesis of 1

First, the substrate for the Suzuki reaction, (2-bromophenyl)acetonitrile (**10**) is prepared by way of a conventional two-step procedure [18] by photobromination of 2-bromotoluene (**8**) to give **9** and S_N displacement of the benzylic bromide by cyanide:

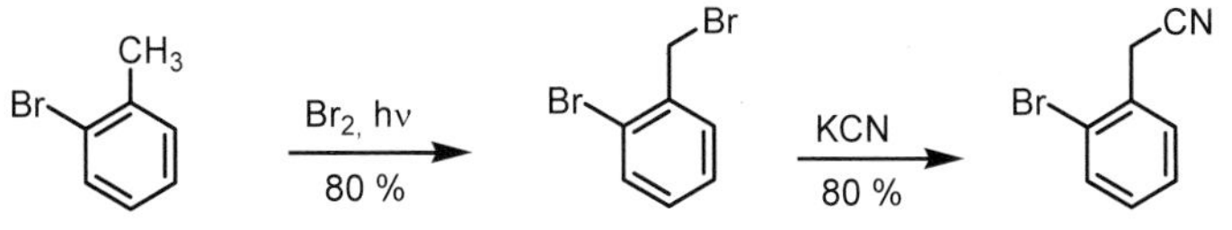

8   9   10 (1.6.2.1)

The second building block is (3,4-dimethoxyphenyl)boronic acid (**12**), which is obtained [19] from 4-bromoveratrole (**11**) by halogen–metal exchange with *t*BuLi (**11** → **13**; with the usually employed *n*BuLi partial *o*-lithiation is observed). Reaction of the formed lithio compound **13** with tri-*n*-butyl borate and subsequent acid hydrolysis of the thus obtained boronate **15** leads to **12**. The formation of **15** probably proceeds via the ate-complex **14** and cleavage thereof with loss of *n*BuOLi:

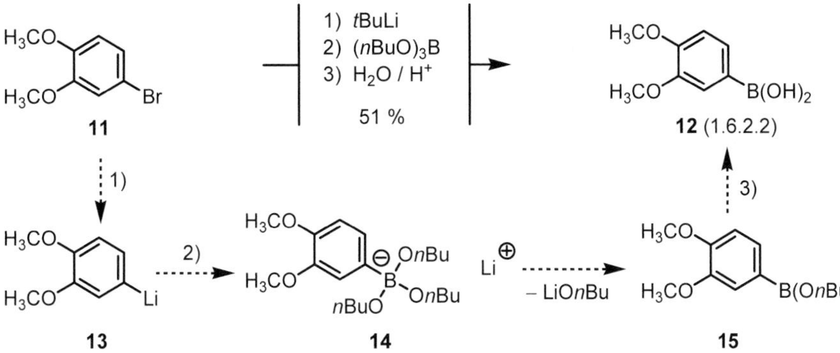

13   14   15

The building blocks **10** and **12** are combined in a Suzuki–Miyaura cross-coupling reaction using Pd(PPh$_3$)$_4$ as catalyst, K$_2$CO$_3$ as base, and CH$_2$Cl$_2$ as solvent. After standard work-up, the biaryl system **1** is isolated in almost quantitative yield.

(c) Experimental procedures for the synthesis of 1

1.6.2.1   **   (2-Bromophenyl)acetonitrile [18]

2-Bromotoluene (8.55 g, 50.0 mmol) is dissolved in CCl$_4$ (250 mL) and the solution is stirred and heated to reflux under irradiation (daylight lamp 500 W). Bromine (8.19 g, 51.3 mmol) is slowly added from a dropping funnel at such a rate that the refluxing CCl$_4$ remains almost colorless. After completion of the reaction, the irradiation is stopped and the solution is cooled to room temperature. The mixture is then rapidly washed with iced water (150 mL), ice-cold saturated NaHCO$_3$ solution (150 mL), and further iced water (150 mL). The organic layer is dried (MgSO$_4$) and concentrated. The residue is distilled *in vacuo*, bp$_{16}$ 130–131 °C. The product is obtained as a colorless liquid, yield 10.0 g (80 %).

2-Bromobenzyl bromide (10.0 g, 40 mmol), sodium cyanide (2.45 g, 50.0 mmol, Caution!), and triethylene glycol (20 mL) are carefully heated to 100 °C with vigorous stirring. The mixture is stirred at this temperature for a further 30 min, then poured into water and extracted with CHCl$_3$ (4 × 20 mL). The isocyanide (formed as a side-product) is removed from the combined organic layers by shaking with 5% H$_2$SO$_4$ (15 mL) for 5 min, and the organic layer is separated and washed sequentially with dilute NaHCO$_3$ solution (30 mL) and water. The organic layer is dried (CaCl$_2$) and concentrated. The residue is purified by distillation; bp$_{17}$ 146–147 °C, colorless liquid; yield 6.27 g (80%).

IR (KBr): $\tilde{v}$ (cm^{-1}) = 3060, 2260, 1565, 1475, 1020, 740.

^{1}H NMR (200 MHz, CDCl$_3$): δ (ppm) = 7.57 (dd, J = 7.8, 1.7 Hz, 1 H, Ar-H), 7.49 (dd, J = 7.8, 1.7 Hz, 1 H, Ar-H), 7.33 (dt, J = 7.8, 1.7 Hz, 1 H, Ar-H), 7.18 (dt, J = 7.8, 1.7 Hz, 1 H, Ar-H), 3.80 (s, 2 H, CH$_2$).

^{13}C NMR (50 MHz, CDCl$_3$): δ (ppm) = 132.8, 129.7, 129.6, 129.5, 127.9, 123.3, 116.7, 24.60.

MS (EI, 70 eV): m/z (%) = 197 (34) [M + H]$^+$, 195 (35) [M – H]$^+$, 171 (8), 169 (9), 116 (100), 89 (36).

1.6.2.2   **   3,4-Dimethoxyphenylboronic acid [19]

217.1	(nBuO)$_3$B: 230.2	182.0

tert-Butyllithium (1.5 M in CH$_2$Cl$_2$, 13.5 mL, 20.2 mmol) is slowly added to a stirred solution of 4-bromoveratrole (4.00 g, 18.4 mmol) in THF (50 mL) at –78 °C over 4 h (the temperature must not exceed –70 °C), followed by trimethyl borate (2.87 g, 27.6 mmol). The mixture is then allowed to warm to room temperature overnight.

HCl (2 M, 25 mL) is added and the aqueous phase is extracted with Et$_2$O (2 × 50 mL). The combined organic layers are extracted with NaOH (2 M, 2 × 50 mL) and then the combined aqueous extracts are acidified with concentrated HCl to pH 1. The aqueous layer is extracted with Et$_2$O (3 × 50 mL), and the combined organic layers are dried (MgSO$_4$) and concentrated to give the boronic acid; yield 1.72 g (51 %), colorless solid, mp 238–240 °C.

^{1}H NMR (500 MHz, CDCl$_3$): δ (ppm) = 7.85 (dd, J = 8.2, 1.3 Hz, 1 H, Ar-H), 7.68 (d, J = 1.3 Hz, 1 H, Ar-H), 7.01 (d, J = 8.2 Hz, 1 H, Ar-H), 4.01 (s, 3 H, OCH$_3$), 3.96 (s, 3 H, OCH$_3$).

^{13}C NMR (125 MHz, CDCl$_3$): δ (ppm) = 153.00, 148.62, 129.92, 117.47, 110.79, 55.91, 55.85.

1.6.2.3   **   2-Cyanomethyl-3',4'-dimethoxybiphenyl [17]

196.0	182.0	253.3

A solution 3,4-dimethoxyphenylboronic acid **1.6.2.2** (2.00 g, 11.0 mmol) in EtOH (60 mL) is added to a mixture of (2-bromophenyl)acetonitrile **1.6.2.1** (1.96 g, 10.0 mmol), toluene (60 mL), Pd(PPh$_3$)$_4$ (348 mg, 0.30 mmol), K$_2$CO$_3$ (4.15 g, 30.0 mmol), and water (40 mL). The mixture is degassed and heated to reflux for 24 h under an inert gas atmosphere.

After cooling to room temperature, water (50 mL) is added and the mixture is extracted with Et$_2$O (3 × 80 mL). The combined organic layers are dried (MgSO$_4$) and concentrated. The residue is purified by column chromatography (SiO$_2$; CH$_2$Cl$_2$) to give 2.43 g (96%) of a yellow oil.

IR (KBr): $\tilde{\nu}$ (cm^{-1}) = 2245 (C≡N).

^{1}H NMR (500 MHz, CDCl$_3$): δ (ppm) = 7.56–7.46 (m, 1 H, Ar-H), 7.44–7.26 (comb. m, 3 H, Ar-H), 6.95 (d, J = 8.5 Hz, 1 H, Ar-H), 6.85 (s$_{br}$, 2 H, Ar-H), 3.93 (s, 3 H, OCH$_3$), 3.90 (s, 3 H, OCH$_3$), 3.63 (s, 2 H, CH$_2$–CN).

^{13}C NMR (125 MHz, CDCl$_3$): δ (ppm) = 149.10, 148.74, 141.87, 132.60, 130.56, 129.11, 128.23, 128.06, 121.21, 118.47, 112.34, 111.36, 56.03, 56.01, 22.14.

[1]   a) M. Beller, C. Bolm (Eds.), *Transition Metals for Organic Synthesis*, two volumes, Wiley-VCH, Weinheim, **2004**;
      b) A. de Meijere, F. Diederich (Eds.), *Metal-Catalyzed Cross-Coupling Reactions*, two volumes, 2nd ed., Wiley-VCH,
      Weinheim, **2004**. c) For an overview of recent advances in the development of Pd catalysts for the Suzuki cross-
      coupling reaction, see: F. Bellina, A. Carpita, R. Rossi, *Synthesis* **2004**, 2415.

[2]   Transition-metal-free Suzuki-type coupling reactions are also known, see: N. E. Leadbeater, M. Marco, *Angew. Chem.*
      2003, *115*, 1445; *Angew. Chem. Int. Ed.* **2003**, *42*, 1407.

[3]   K. C. Nicolaou, E. J. Sorensen, *Classics in Total Synthesis*, p. 586, VCH, Weinheim, **1996**.

[4]   a) N. Krause, *Metallorganische Chemie*, p. 215, Spektrum Akademischer Verlag, Heidelberg, **1996**; b) see also: A. E.
      Jensen, W. Dohle, I. Sapountzis, D. M. Lindsay, V. A. Vu, P. Knochel, *Synthesis* **2002**, 565.

[5]   See ref. [1] and P. Knochel, W. Dohle, N. Gommermann, F. F. Kneisel, F. Kopp, T. Korn, I. Sapountzis, V. A. Vu,
      Angew. Chem. **2003**, *115*, 4438; *Angew. Chem. Int. Ed.* **2003**, *42*, 4302.

[6]   For a review on the Stille reaction, see: V. Farina, V. Krishnamurthy, W. J. Scott, *Organic Reactions*, Vol. 50, p. 1,
      John Wiley & Sons, New York, **1997**.

[7]   C. Dai, G. C. Fu, *J. Am. Chem. Soc.* **2001**, *123*, 2719.

[8]   Vancomycin antibiotics: a) K. C. Nicolaou, C. N. C. Boddy, S. Bräse, N. Winssinger, *Angew. Chem. Int. Ed.* **1999**,
      38, 2096; b) K. C. Nicolaou, S. A. Snyder, *Classics in Total Synthesis II*, p. 239, Wiley-VCH, Weinheim, 2003.

[9]   a) T. Eicher, S. Fey, W. Puhl, E. Büchel, A. Speicher, *Eur. J. Org. Chem.* **1998**, 877; b) A. Speicher, J. Kolz, R. P.
      Sambanje, *Synthesis* **2002**, 2503.

[10] Losartan and other sartans: a) W. H. Birkenhagen, P. W. Leeuw, *Hypertens.* **1999**, *17*, 873; b) see also: A. Kleemann,
      J. Engel, *Pharmaceutical Substances*, 3rd ed., p. 1116 and p. 1024, Thieme, Stuttgart, **1999**.

[11] G. Bringmann, C. Günther, M. Ochse, O. Schupp, S. Tasler, *Biaryls in Nature*, in: *Progress in the Chemistry of
      Organic Natural Products* (Eds.: W. Herz, H. Falk, G. W. Kirby, R. E. Moore), Vol. 82, p. 1, Springer, Wien, New
      York, **2001**.

[12] a) For the Ullmann coupling reaction via organocopper reagents as a tool for biphenyl synthesis, see ref. [4a], p. 195;
      b) for a convenient large-scale synthesis of biaryls by Cu(I)/Pd(II)-catalyzed decarboxylative coupling of
      arylcarboxylic acid salts with bromoarenes, see: L. J. Gooßen, G. Deng, L. M. Levy, *Science* 2006, *313*, 662.

[13] Aryl halides: Development of efficient catalysts: a) review: A. Zapf, M. Beller, *Chem. Commun.* **2005**, 431;
      b) polymer-incarcered Pd: K. Okamoto, R. Akiyama, S. Kobayashi, *Org. Lett.* **2004**, *6*, 1987; c) recyclable and
      reusable Pd(II) acetate/DABCO/PEG-400 system: J.-H. Li, W.-J. Liu, Y.-X. Xie, *J. Org. Chem.* **2005**, *70*, 5409;
      d) ligand-free with Pd(OAc)$_2$/Na$_2$CO$_3$ in aqueous phase: L. Liu, Y. Zhang, B. Xin, *J. Org. Chem.* **2006**, *71*, 3994; e)
      aryl triflates/sulfonates: H. N. Nguyen, X. Huang, S. L. Buchwald, *J. Am. Chem. Soc.* **2003**, *125*, 11818.

[14] a) S. D. Walker, T. E. Barder, J. R. Martinelli, S. L. Buchwald, *Angew. Chem.* **2004**, *116*, 1907; *Angew. Chem. Int.
      Ed.* **2004**, *43*, 1871; b) I. Özdemir, Y. Gök, N. Gürbüz, E. Çetinkaya, B. Çetinkaya, *Synth. Commun.* **2004**, *34*, 4135;
      c) R. B. Bedford, S. L. Hazelwood, M. E. Limmert, *Chem. Commun.* **2002**, *22*, 2610; d) K. W. Anderson, S. L.
      Buchwald, *Angew. Chem.* **2005**, *117*, 6329; *Angew. Chem. Int. Ed.* **2005**, *44*, 6173; e) K. L. Billingsley, K. W.
      Anderson. S. L. Buchwald, *Angew. Chem.* **2006**, *118*, 3564; *Angew. Chem. Int. Ed.* **2006**, *45*, 3484.

[15] M. B. Andruss, C. Song, *Org. Lett.* **2001**, *3*, 3761.

[16] a) T. Mino, Y. Shirae, M. Sakamoto, T. Fujita, *Synlett* **2003**, 882; b) D. Zim, A. S. Gruber, G. Ebeling, J. Dupont, A.
      L. Monteiro, *Org. Lett.* **2000**, *2*, 2881; c) for fluorous biphasic catalysis, see. C. C. Tzschucke, C. Markert, H. Glatz,
      W. Bannwarth, *Angew. Chem.* **2002**, *114*, 4678; *Angew. Chem. Int. Ed.* **2002**, *41*, 4500; for ligandless biaryl coupling,
      see: d) X. Tao, Y. Zhao, D. Shen, *Synlett* **2004**, 359; e) W. Solodenko, H. Weu, S. Leue, F. Stuhlmann, U. Kunz, A.
      Kirschning, *Eur. J. Org. Chem.* **2004**, 3601.

[17] P. Sahakitpichan, S. Ruchirawat, *Tetrahedron Lett.* **2003**, *44*, 5239.

[18] *Organikum*, 21st ed., p. 204/257, Wiley-VCH, Weinheim, **2001**.

[19] G. M. Keserü, G. Mezey-Vándor, M. Nógrádi, B. Vermes, M. Kajtár-Peredy, *Tetrahedron* **1992**, *48*, 913.

1.6.3 (2-Phenylethynyl)aniline

NH2
1

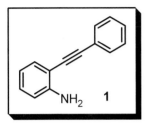

Topic:

- Sonogashira cross-coupling reaction of terminal acetylenes with aryl and vinyl halides

(a) General

The Sonogashira cross-coupling reaction [1] allows C–C bond formation between C(sp) and C(sp^2) centers by a Pd(0)-catalyzed reaction of haloarenes, halohetarenes, and haloalkenes with alkynes to give the coupled products **2**. The reaction is co-catalyzed by Cu(I) iodide and requires the presence of a base, preferentially diethylamine:

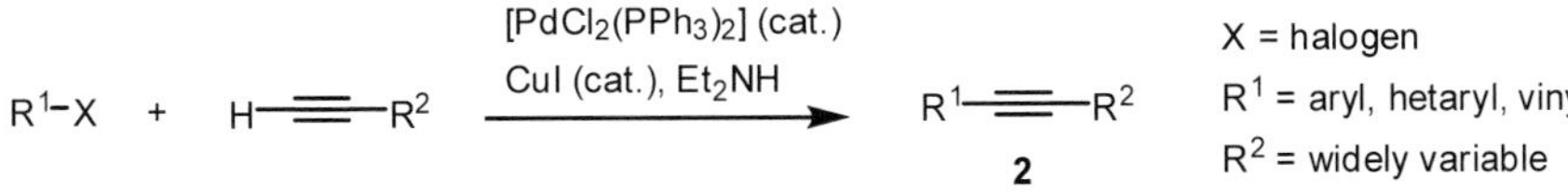

R1-X
+
H—≡—R2
[PdCl2(PPh3)2] (cat.)
CuI (cat.), Et2NH
R1—≡—R2
2
X = halogen
R1 = aryl, hetaryl, vinyl
R2 = widely variable

The Sonogashira coupling is related to the Stephens–Castro reaction [2], the coupling of iodoarenes with copper(I) aryl acetylides:

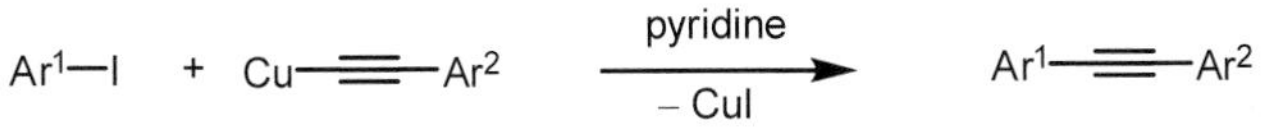

Ar1—I
+
Cu—≡—Ar2
pyridine
– CuI
Ar1—≡—Ar2

The mechanism of the Sonogashira cross-coupling reaction [2] resembles that of the Heck and Suzuki–Miyaura reactions by providing C–C bond formation in the coordination sphere of Pd complexes (L = ligand, e.g. Ph_3P).

L2PdIIX2
3
CuI, 2 Et2NH
2 H—≡—R2
2 (Et2N+H2)X2
L2PdII(—≡—R2)2
4
R2—≡—≡—R2
R1—≡—R2
2
L2Pd0
5
R1-X
③
①
7
6
②
CuI, Et2NH
(Et2N+H2)X
H—≡—R2

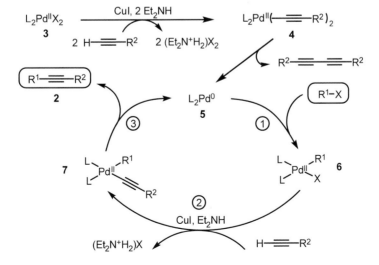

The catalytic cycle is initiated by a Pd(0)L$_2$ species **5**, which is generated, for example, in a preceding sequence from a Pd(II) complex **3** by base-induced exchange of the ligands X with acetylide and subsequent disproportionation of the acetylide Pd(II) complex **4** leading to the formation of Pd(0)L$_2$ (**5**) and, as a side product, a diyne. This can be avoided by using a Pd(0) complex such as Pd(PPh$_3$)$_4$ as catalyst. The Pd(0) species **5** undergoes an oxidative addition to R^1–X (step (1)), which is followed by base-induced substitution of X by acetylide in the Pd(II) complex **6** catalyzed by Cu(I) iodide (step (2)). Finally, a reductive elimination process of the formed Pd(II) complex **7** – most likely from a *syn*-arrangement – (step (3)) leads to the disubstituted acetylene **2** and Pd(0)L$_2$, which again enters into the catalytic cycle.

Usually, iodo compounds are used in the Sonogashira reaction, since these are more reactive than the bromo and chloro compounds [3]. This allows the chemoselective reaction of haloarenes bearing different halogens as substituents. Moreover, the consecutive introduction of two different acetylenic moieties in dihaloalkenes is also possible, which is used in syntheses of analogues of enediyne antibiotics [2]:

As an alternative to the Sonogashira reaction, alkynylation of haloalkenes **9** can be efficiently accomplished by Pd(0)-catalyzed reaction with *in situ* generated alkynyl zinc bromides **8**, which are easily accessible from acetylides and ZnBr$_2$ [4]:

The great synthetic value of the Sonogashira reaction, above all, stems from the fact that further transformations of the alkyne moiety may be performed. This is documented, for example, in a series of syntheses of heterocyclic systems [5].

Thus, in section (b), the preparation of substrate **1** for a Pd-catalyzed indole synthesis (cf. **3.2.4**) by means of a Sonogashira cross-coupling [6] is described.

(b) Synthesis of 1

The synthesis of **1** starts from 2-iodoaniline (**10**), which is commercially available but can also be easily prepared by *ortho*-lithiation of aniline and subsequent quenching with iodine [7]. The Sonogashira cross-coupling reaction of **10** with phenylacetylene in the presence of [PdCl$_2$(PPh$_3$)$_2$] as catalyst, Cu(I) iodide as co-catalyst, and triethylamine as base provides (2-phenylethynyl)aniline (**1**) in almost quantitative yield:

10

1 (1.6.3.1)

(c) Experimental procedure for the synthesis of 1

1.6.3.1   **   2-(Phenylethynyl)aniline [6]

219.0

102.1

701.9
CuI: 190.5

193.2

A mixture of 2-iodoaniline (2.19 g, 10.0 mmol), phenylacetylene (1.12 g, 11.0 mmol), CuI (190 mg, 1.00 mmol), [PdCl$_2$(PPh$_3$)$_2$] (210 mg, 0.30 mmol), and Et$_3$N (20 mL) is stirred for 1 h at room temperature.

The reaction mixture is then diluted with H$_2$O (20 mL) and extracted with CHCl$_3$ (3 × 20 mL). The combined extracts are dried (MgSO$_4$) and concentrated under reduced pressure. The crude product is purified by column chromatography (SiO$_2$; n-hexane/EtOAc, 3:1) and recrystallized from n-hexane/acetone (5:1); 1.84 g (95%), pale-yellow prisms; mp 86–87 °C.

IR (KBr): $\tilde{v}$ (cm^{-1}) = 3500, 2250, 1620.

^{1}H NMR (CDCl$_3$, 500 MHz): δ (ppm) = 7.56–7.51 (m, 2 H), 7.39–7.32 (m, 4 H), 7.14 (dt, J = 7.9, 1.4 Hz, 1 H), 6.79 (t, J = 7.7 Hz, 2 H), 4.39–4.20 (s$_{br}$, 2 H).

[1] a) M. Beller, C. Bolm (Eds.), *Transition Metals for Organic Synthesis*, two volumes, Wiley-VCH, Weinheim, **2004**; b) A. de Meijere, F. Diederich (Eds.), *Metal-Catalyzed Cross-Coupling Reactions*, two volumes, 2nd ed., Wiley-VCH, Weinheim, **2004**. c) Instead of Pd complexes, nanosized Pd(0) (P. Li, L. Wang, H. Li, *Tetrahedron* **2005**, *61*, 8633), ultrafine Ni(0) (L. Wang, P. Li, Y. Zhang, *Chem. Commun.* **2004**, 514), or Pd immobilized on organic–inorganic hybrid materials (P.-M. Li, L. Wang, *Adv. Synth. Catal.* **2006**, 681) has also been applied in the Sonogashira process.

[2] K. C. Nicolaou, E. J. Sorensen, *Classics in Total Synthesis*, p. 582, VCH, Weinheim, **1996**.

[3] a) N. Krause, *Metallorganische Chemie*, p. 216, Spektrum Akademischer Verlag, Heidelberg, **1996**. b) Efficient catalysts have also been developed for the Sonogashira coupling of chloroarenes: A. Köllhofer, T. Pullmann, H. Plenio, *Angew. Chem.* **2003**, *115*, 1086; *Angew. Chem. Int. Ed.* **2001**, *40*, 1056.

[4] a) E.-i. Negishi, M. Qian, F. Zeng, L. Anastasia, D. Babinski, *Org. Lett.* **2003**, *5*, 1597; b) L. Anastasia, E.-I. Negishi, *Org. Lett.* **2001**, *3*, 3111.

[5] See: Th. Eicher, S. Hauptmann, *The Chemistry of Heterocycles*, 2nd ed., p. 280, Wiley-VCH, Weinheim, **2003**.

[6] A. Yasuhara, Y. Kanamori, M. Kaneko, A. Numata, Y. Kondo, T. Sakamoto, *J. Chem. Soc., Perkin Trans. I* **1999**, 529.

[7] V. Snieckus, *Chem. Rev.* **1990**, *90*, 879.

1.6.4   3,3-Dimethylcyclohexanone

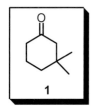

1

Topics: • Domino process: eliminative conjugate addition of a cuprate to a 3-mesyloxy enone/conjugate addition of a cuprate to an enone

• Formation of an *O*-mesylate of an enolizable 1,3-diketone

(a) General

Among the many versatile transition metal-based reagents used in contemporary organic synthesis [1], organocuprates of the general composition $[R^1–Cu–R^2]M$ (M = Li, MgX) play an important role [2].

In general, organocuprates are prepared by transmetalation from lithium organyls or Grignard compounds in donor solvents:

$$Cu(I)X \quad + \quad \underset{(R–MgX)}{R–Li} \quad \xrightarrow[\substack{- LiX \\ (- MgX_2)}]{} \quad R–Cu \quad \xrightarrow{+ R–Li} \quad R_2CuLi$$

According to their structure and reactivity, three types of organocuprates are differentiated, namely (1) homocuprates, (2) heterocuprates, and (3) cyanocuprates.

(1) Homocuprates, e.g. R_2CuLi, contain two identical organic residues R, of which in general only *one* participates in chemical reactions. Homocuprates possess dimeric structures (**2**) with a tetracoordinated (tetrahedral) configuration at Cu associated with solvent molecules:

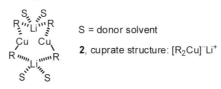

S = donor solvent

2, cuprate structure: $[R_2Cu]^- Li^+$

(2) Heterocuprates ("mixed" cuprates), e.g. [R–Cu–Y]Li, possess an organic residue R, which is transferable in chemical reactions, and a "dummy" ligand Y, which is not transferable; ligands Y can be RO, RS, amide, phosphane, acetylide, 2-thienyl, among others.

(3) Cyanocuprates are subdivided into "low-order cyanocuprates" (**3**) and "high-order cyanocuprates" (**4**), the detailed structures of which, however, are not elucidated [3]. Organocuprates of type **4** are thermally more stable than dialkyl cuprates R_2CuM and also transfer only one organic residue in their reactions, e.g. **5**:

$$R–Li \quad \xrightarrow{+ \ CuCN} \quad \underset{\textbf{3}}{[R–Cu–CN]Li} \quad \xrightarrow{+ \ R–Li} \quad \underset{\textbf{4}}{\substack{[R_2Cu(CN)]Li_2 \\ (\ or \ R_2CuLi \bullet LiCN\ )}} \qquad \underset{\textbf{5}}{\left[\substack{R \\ \diagdown \\ Cu–CN \\ \diagup \\ 2\text{-thienyl}} \right] Li_2}$$

Like other organometallics, organocuprates can participate in substitution and addition reactions. Thus, they undergo Wurtz-type substitution with alkyl, alkenyl, and aryl halides or sulfonates, respectively:

$$R'–X \quad \xrightarrow{R_2CuLi} \quad R'–R \qquad X = Br, I, sulfonate$$

In particular, organocuprates are the organometallics of choice to effect conjugate (1,4-) addition to α,β-unsaturated carbonyl systems (**6 → 10**) as well as to other acceptor-substituted alkenes $H_2C=CH–A$ (A = ester, amide, CN, NO_2, etc.):

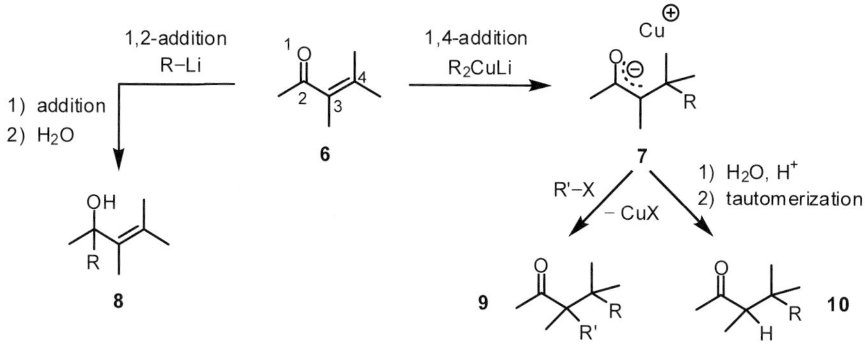

The chemoselectivity in conjugate addition of cuprates to enones is attributed (on the basis of the HSAB principle [4]) to the "soft" character of Cu(I) compared to organometallics with "harder" metal centers (like Li and/or MgX), which favor 1,2-addition (e.g. **6 → 8**). Interestingly, HSAB-differentiated electrophilic reagents give different products with the initially formed enolates **7** of 1,4-addition to enones. Thus, the "soft" R–X attacks at carbon to yield α,β-dialkylated ketones **9** (tandem Michael addition/alkylation; enolate trapping, *trans*-stereochemistry on reaction with cyclic enones), while the "hard" $Me_3Si–Cl$ attacks at oxygen to give the silyl enol ether of **10**.

1,4-Addition to enones is observed on reaction with all types of organocuprates (1)–(3), as shown by the following examples [5]:

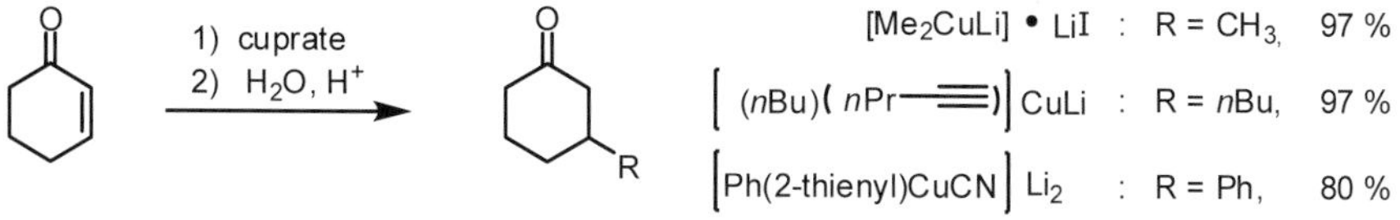

In section (b), the potential of cuprate reagents in conjugate addition reactions is demonstrated by a straightforward synthesis of the target molecule **1** from a simple 1,3-diketone [6].

(b) Synthesis of 1

The starting material for the cuprate reaction is 3-mesyloxy-2-cyclohexenone (**12**), which is prepared from cyclohexane-1,3-dione (**11**) by *O*-mesylation with mesyl chloride in the presence of a base. When the enone **12** is reacted with two equivalents of lithium dimethylcuprate in Et_2O (–78 °C to room temperature), the dimethylcyclohexanone **1** is obtained after acid hydrolysis in 75% yield.

This transfer of two methyl groups to the 3-position of the Michael acceptor **12** is likely to occur as a two-fold domino process. The first cuprate moiety undergoes 1,4-addition (→ **13**) accompanied by elimination of mesylate and intermediate formation of the cycloenone **14**, which is susceptible to 1,4-addition of a second cuprate moiety giving rise to the target molecule **1** after hydrolytic work-up.

The primary cuprate reaction sequence of the mesyloxy enone **12 → 13 → 14** is an example of a so-called eliminative cuprate addition, which generally occurs with enones bearing a leaving group (e.g. halide or sulfonate) in the 3-position.

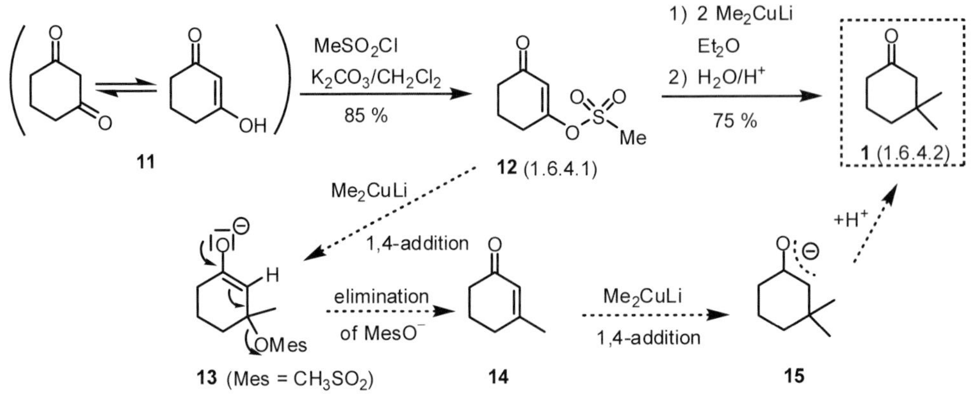

It should be noted that the methylcyclohexenone **14** cannot be obtained in substantial yield using only *one* equivalent of dimethyl cuprate [2b]. In contrast, the 3-chloroenone **17**, derived from dimedone (cf. **3.2.8**), can be either transformed to isophorone (**16**) or to 3,3,5,5-tetramethylcyclohexanone (**18**) [7]:

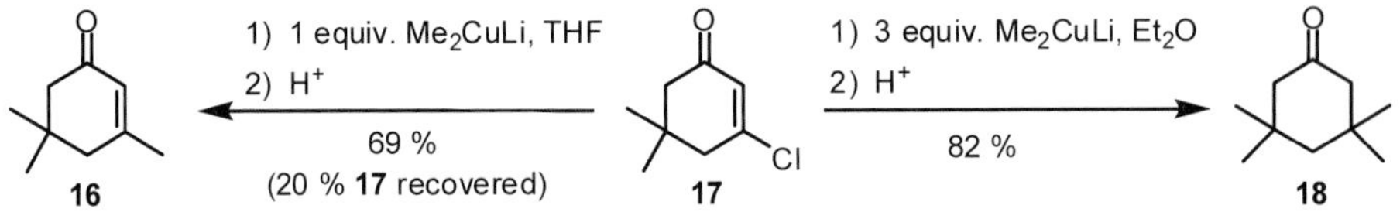

(c) Experimental procedures for the synthesis of 1

1.6.4.1 * 3-Mesyloxy-2-cyclohexenone [6]

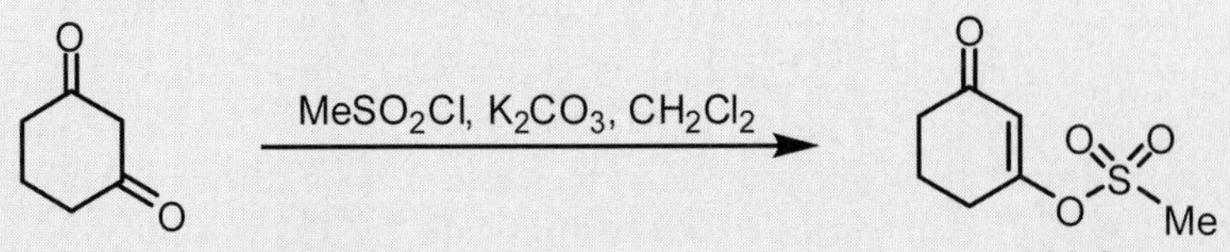

Methanesulfonyl chloride (0.69 mL, 8.92 mmol) is added to a stirred solution of 1,3-cyclohexanedione (1.00 g, 8.92 mmol) in CH_2Cl_2 (45 mL) at room temperature and then anhydrous potassium carbonate (3.70 g, 26.8 mmol) is added.

After stirring for 2 h, the reaction mixture is diluted with CH_2Cl_2 (150 mL) and washed consecutively with H_2O (40 mL) and brine (40 mL). The organic layer is dried over $MgSO_4$, filtered, and concentrated to afford 1.45 g (85%) of the mesylate as a yellow solid (note).

IR (film): $\tilde{\nu}$ (cm^{-1}) = 1680, 1630.

^{1}H NMR (CDCl$_3$, 400 MHz): δ (ppm) = 6.02 (t, J = 1.2 Hz, 1 H, C=C–H), 3.21 (s, 3 H, CH$_3$), 2.60 (td, J = 1.2, 6.3 Hz, 2 H, allyl-CH$_2$), 2.39 (t, J = 7.6/6.3 Hz, 2 H, CO–CH$_2$), 2.06 (quintet, J = 7.6/6.3 Hz, 2 H, CH$_2$).

^{13}C NMR (100 MHz, CDCl$_3$): δ (ppm) = 198.36 (=C–O), 167.62 (C=O), 116.12 (α-C), 38.90, 36.31, 28.48, 20.72.

Note: This material decomposes on standing neat at room temperature for several hours; it may be stored as a solution in CH_2Cl_2 at 5 °C for several days with no evidence of decomposition; in general, freshly prepared mesylate is used for the subsequent reaction.

1.6.4.2   **   3,3-Dimethylcyclohexenone [6]

| 190.2 | CuI: 190.45 | 126.20 |

A solution of methyllithium in Et_2O (1.6 M, 32.9 mL, 52.5 mmol, 10 equiv.) is added dropwise to a stirred suspension of copper(I) iodide (5.00 g, 26.3 mmol, 5 equiv.; note 1) in anhydrous Et_2O (150 mL) under a nitrogen atmosphere at –78 °C. The resulting mixture is allowed to warm to –30 °C for 30 min (a pale-yellow solution results) and is then cooled to –78 °C once more. The mesylate **1.6.4.1** (1.00 g, 5.25 mmol) in anhydrous Et_2O (50 mL) is added over 15 min at < –70 °C. The mixture is allowed to slowly warm to room temperature.

It is diluted with HCl (10%, 250 mL) and then sufficient NH_3 solution (2.5 M) is added to make the aqueous layer just basic. The mixture is filtered and extracted with Et_2O (3 × 250 mL). The combined ethereal extracts are washed consecutively with H_2O (150 mL) and brine (150 mL) and dried over $MgSO_4$. Evaporation of the solvent affords 500 mg (75%) of the ketone as a pale-yellow oil (note 2).

IR (film): $\tilde{v}$ (cm^{-1}) = 1715.

^{1}H NMR ($CDCl_3$, 400 MHz): δ (ppm) = 2.22 (t, J = 6.9 Hz, 2 H), 2.10 (s, 2 H, isolated CH_2), 1.83 (quintet, J = 6.3 Hz, 2 H, CH_2).1.54 (t, J = 6.3 Hz, 2 H), 0.93 (s, 3 H, CH_3).

^{13}C NMR (100 MHz, $CDCl_3$): δ (ppm) = 212.01 (C=O), 54.87, 40.69, 37.86, 35.96, 28.44, 22.40.

Notes: (1) CuI is pre-dried over P_4O_{10} and heated twice in the reaction flask *in vacuo*, flushing with argon.

(2) Distillation of the product does not alter the spectra.

[1]   M. Beller, C. Bolm, *Transition metals for organic synthesis*, Vols. 1/2, 2nd ed., Wiley-VCH, Weinheim, **2004**.

[2]   a) N. Krause, *Modern organocopper chemistry*, Wiley-VCH, Weinheim, **2002**; b) N. Krause, *Metallorganische Chemie*, Spektrum Akademischer Verlag, Heidelberg, **1996**; c) for an overview of recent advances in organocopper chemistry, see: V. Caprio, *Lett. Org. Chem.* **2006**, *3*, 339.

[3]   For compounds **4**, the term "cuprate" is obviously misleading; it should be restricted to the structural type [R^1–Cu–R^2]⁻M⁺.

[4]   M. B. Smith, J. March, *March's Advanced Organic Chemistry*, 5th ed., p. 338, John Wiley & Sons, Inc., New York, **2001**.

[5]   Ref. [2b], p. 178.

[6]   C. J. Kowalski, K. W. Fields, *J. Org. Chem.* **1981**, *46*, 197.

[7]   a) T. Kauffmann, R. Sälker, K.-U. Voß, *Chem. Ber.* **1993**, *126*, 1447; b) E. Piers, K. F. Cheng, I. Nagakura, *Can. J. Chem.* **1982**, *60*, 1256.

1.7   Pericyclic reactions

1.7.1   Tranylcypromine

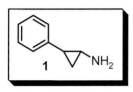

1

Topics:
- Synthesis of an aminocyclopropane-based drug
- [1+2]-Cycloaddition of carbethoxycarbene to styrene, cyclopropanation of an alkene
- Ester saponification
- Separation of stereoisomers by fractional crystallization
- Thermal Curtius degradation

(a) General

Tranylcypromine (**1**) is the racemic mixture of *trans*-(*E*)-2-phenylcyclopropyl-1-amine and is used pharmaceutically as a psychoanaleptic and antidepressant [1]. Tranylcypromine acts as an inhibitor of monoamine oxidase A, thus retarding the metabolic degradation of serotonin, noradrenaline, adrenaline, and other amines [2].

$$Ph \overset{\text{\tiny,,}}{\nabla} NH_2 \qquad H_2N \overset{\text{\tiny,,}}{\nabla} Ph$$

(+)-(*1S,2R*)   **1**   (–)-(*1R,2S*)

Both enantiomers of **1** are known and their structure–activity relationships have been investigated; the (*1S,2R*)-compound shows a ten times higher inhibitory activity than its enantiomer.

For the retrosynthesis of cyclopropanes, the most appropriate bond disconnections generally result from retro-[1+2]-cycloadditions. Thus, the target molecule **1** offers two retroanalytical pathways (**A/B**).

In **A**, after FGI (NH$_2$ → NO$_2$), the resulting phenylnitrocyclopropane **2** should result from [1+2]-cycloaddition of a methylene (CH$_2$) source to (*E*)-2-phenyl-1-nitroethene (**3**), which is easily accessible by nitroaldol condensation (Henry reaction) of benzaldehyde and nitromethane.

In **B**, after FGI of the primary amine function to an acyl azide **4** (retro-Curtius rearrangement) and further on to (*E*)-2-phenylcyclopropane carboxylate **5**, retro-[1+2]-cycloaddition would lead to a carbalkoxycarbene (easily accessible from diazoacetate **7**) and styrene (**6**).

(A) FGI (1)
$$Ph \overset{\text{\tiny,,}}{\nabla} NH_2 \quad \overset{\text{(A)}}{\underset{\text{(1)}}{\rightleftharpoons}} \quad Ph \overset{\text{\tiny,,}}{\nabla} NO_2$$

1   **2**

$$\text{retro} \atop [2+1]CA \Rightarrow \quad Ph \diagup\!\!\!\diagdown NO_2 \quad \textbf{3} \atop + \atop "CH_2"$$

(2) ↕ (B) retro-Curtius

$$Ph \overset{\text{\tiny,,}}{\nabla} CON_3 \quad \overset{FGI}{\Rightarrow} \quad Ph \overset{\text{\tiny,,}}{\nabla} COOR$$

4   **5**

$$\text{retro} \atop [2+1]CA \Rightarrow \quad Ph \diagup\!\!\!\diagdown \quad \textbf{6} \atop + \atop N_2{=}CH{-}COOR \ \ \textbf{7}$$

In fact, phenylnitrocyclopropane **2** can be prepared from phenylnitroethene **3** by cyclopropanation with trimethyloxosulfonium iodide/NaH according to the Corey–Chaykovsky method [3]. Unfortunately, its reduction to tranylcypromine is not described in the literature, thus eliminating this short and straightforward possibility (**I**) for the synthesis for **1**.

However, approach **II** based on retrosynthesis **B** is documented [4] and is described in detail in section (b).

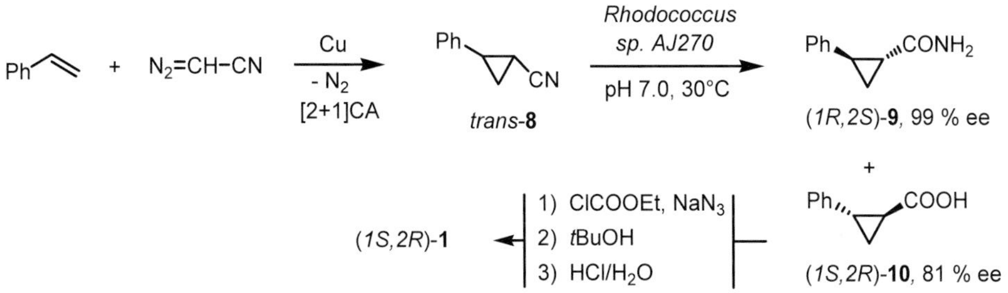

It should be noted that enantiopure (1*R*,2*S*)-**1** can be obtained by way of a chemoenzymatic transformation of (*E*)-2-phenylcyclopropanecarbonitrile (**8**) [5]. *Rhodococcus sp. AJ 270*, a versatile nitrile hydratase/amidase, catalyzes the enantioselective hydrolysis of **8** to afford the corresponding amide **9** and the acid **10** in high enantiomeric excess. The acid **10** is transformed to (+)-(1*S*,2*R*)-tranylcypromine by a modified Curtius rearrangement.

(b) Synthesis of 1

The synthesis of **1** [4] starts with cyclopropanation of styrene with carbethoxycarbene generated by thermolysis of ethyl diazoacetate (**12**). The diazoacetate **12** is prepared from glycine ethyl ester hydrochloride (**11**) by nitrosation with HNO_2 [6]:

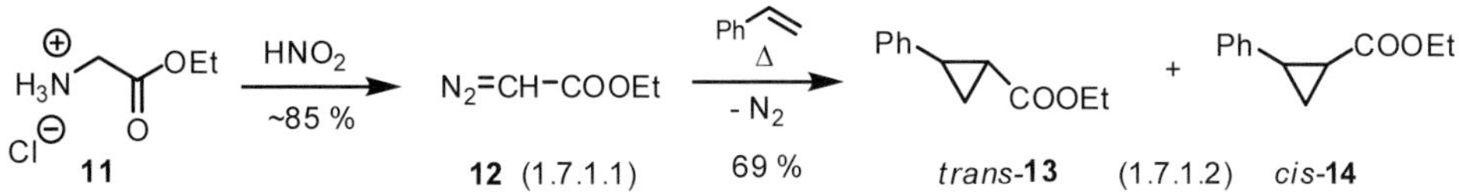

The cyclopropanation leads to a mixture of the diastereomers of ethyl 1-phenylcyclopropane-1-carboxylate (**13** and **14**) as a racemic mixture with a *trans*/*cis* ratio of ca. 2:1. The *cis*/*trans* mixture **13/14** is saponified using aqueous NaOH and the resulting isomeric acids **15/16** are separated by fractional crystallization from H_2O to obtain the pure *trans*-acid **15** required for the further synthesis.

The *trans*-1-phenylcyclopropane-2-carboxylic acid (**15**) is transformed to **1** by Curtius degradation of the corresponding acyl azide. This is achieved by conversion of **15** to the acid chloride (**19**), reaction with NaN_3 to give the azide (**18**), and thermolysis of the azide to afford the isocyanate (**17**) via 1,2-*N*-sextet rearrangement; finally, the isocyanate is transformed in acidic medium (via the corresponding carbamic acid and decarboxylation thereof) to the primary amine **1**:

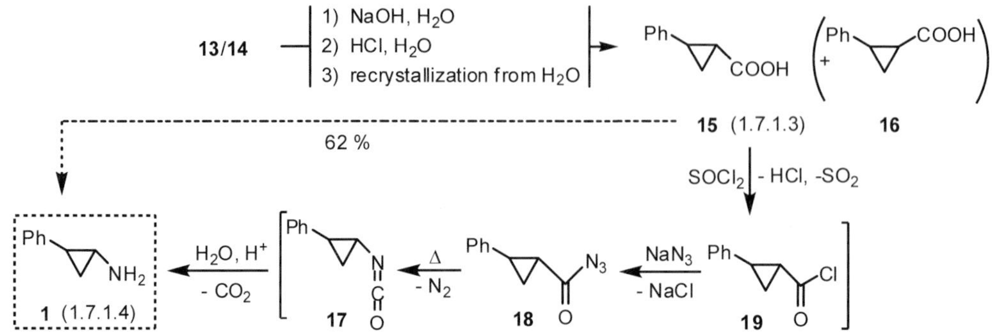

13/14 — 1) NaOH, H_2O / 2) HCl, H_2O / 3) recrystallization from H_2O

Ph COOH   **15** (1.7.1.3)   (Ph COOH + Ph COOH)   **16**

62 %

$SOCl_2$ | - HCl, $-SO_2$

Ph NH₂ **1** (1.7.1.4)   ← H₂O, H⁺ / $- CO_2$ ← [Ph N=C=O **17**] ← Δ / $- N_2$ ← Ph N₃ **18** ← NaN₃ / - NaCl ← Ph Cl **19**

The four-step transformation **15 → 1** is executed in a one-pot procedure with spectroscopic detection of the intermediates **17**–**19**. Tranylcypromine is thus obtained in four separate steps with an overall yield of 19% (based on **11**).

Another possibility is the enantio- and diastereoselective cyclopropanation of styrene with diazoacetates in the presence chiral Co(II) chelate complexes:

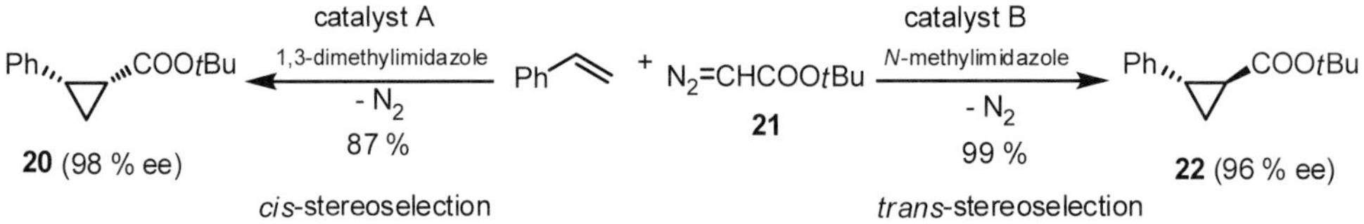

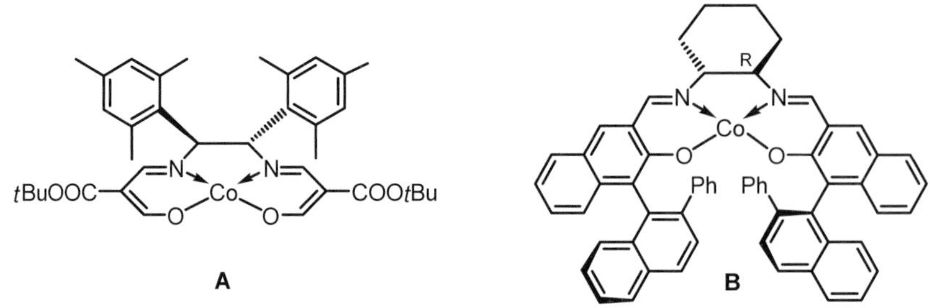

catalyst A   catalyst B

Ph COO*t*Bu   1,3-dimethylimidazole / $- N_2$ / 87 % ← Ph + N₂=CHCOO*t*Bu **21** → N-methylimidazole / $- N_2$ / 99 %   Ph COO*t*Bu

20 (98 % ee)   **22** (96 % ee)

cis-stereoselection   *trans*-stereoselection

*t*BuOOC ... Co ... COO*t*Bu

A

B

Thus, reaction of styrene with *tert*-butyl diazoacetate (**21**) using the β-ketoimidato-Co(II) complex **A** leads to the *cis*-disubstituted cyclopropane **20** [7], whereas using the salen-type Co(II) complex **B** it leads to the *trans*-disubstituted cyclopropane **22**, both with >96% *ee* [8].

(c) Experimental procedures for the synthesis of 1

1.7.1.1 ** Ethyl diazoacetate [6]

$$\underset{139.6}{\overset{\oplus}{H_3N}\overset{}{\underset{Cl^{\ominus}}{\bigwedge}}\underset{O}{\overset{}{\bigvee}}OC_2H_5} \quad \xrightarrow{HNO_2} \quad \underset{114.1}{N_2{=}CH{-}COOC_2H_5}$$

A solution of sodium nitrite (32.8 g, 0.48 mol) in H_2O (100 mL) at –5 °C is added to a well-stirred mixture of glycine ethyl ester hydrochloride (56.0 g, 0.40 mol) in H_2O (100 mL) and CH_2Cl_2 (240 mL) also at –5 °C. The mixture is cooled to –9 °C and 5% sulfuric acid (cold, 38.0 g) is added dropwise over ca. 3 min, keeping the reaction temperature below 1 °C; after the addition, stirring is continued for 15 min.

The mixture is transferred to a cold separatory funnel, the phases are separated, and the aqueous phase is extracted with CH_2Cl_2 (30 mL). The organic phases are combined and shaken with 5% $NaHCO_3$ solution (400 mL in total) until no more acid is present. The CH_2Cl_2 solution is dried over $CaCl_2$ (10 g), filtered, and used directly for the next step without further purification; ca. 280 mL containing 36.0–40.0 g (79–88%) of ethyl diazoacetate.

1.7.1.2 ** Ethyl 2-phenylcyclopropane-1-carboxylate [4]

$$\underset{104.2}{\overset{}{\bigcirc}{\diagdown}} \quad + \quad \underset{114.1}{N_2{=}CH{-}COOC_2H_5} \quad \xrightarrow{\Delta} \quad \underset{}{\overset{Ph}{\diagdown}\underset{COOC_2H_5}{\triangle}} \quad + \quad \underset{190.1}{\overset{Ph}{\diagdown}\underset{}{\triangle}{\diagup}COOC_2H_5}$$

In a three-necked flask fitted with a dropping funnel, a stirrer, a thermometer, and a distillation unit (note 1), styrene (17.7 g, 0.17 mol; note 2) and hydroquinone (0.4 g) are heated to 125 °C. Then, the solution of ethyl diazoacetate prepared in **1.7.1.1**, in which additional styrene (34.4 g, 0.33 mol) and hydroquinone (0.4 g) are dissolved, is added dropwise with vigorous stirring at such a rate that the internal temperature stays at 125–135 °C (external temperature ca. 160 °C). The addition is complete after ca. 3 h.

The CH_2Cl_2 solution (faintly yellow) which distilled off is concentrated *in vacuo* at room temperature to a volume of ca. 40 mL and added dropwise to the reaction mixture (same conditions as above).

For work-up, the CH_2Cl_2 is removed at normal pressure and the excess styrene is distilled off at 15 mbar (18.7 g, 0.18 mol). Distillation is continued at 1 mbar to yield 41.8 g (69% based on reacted styrene) of a colorless liquid, bp_1 106–108 °C, which consists of a mixture of the *cis* and *trans* products (as indicated by ^{1}H NMR).

IR (film): $\tilde{v}$ (cm^{-1}) = 1725 (C=O), 760, 700.

^{1}H NMR (CDCl3): δ (ppm) = 7.4–7.0 (m, 5 H, phenyl-H), 4.16, 3.86 (q, J = 7 Hz, together 2 H, OCH2), 2.7–2.4 (m, 1 H, Ph-CH), 2.2–1.4 (m, 3 H, CH2 and CH-COOEt), 1.26, 0.96 (t, J = 7 Hz, 3 H, CH3; relative intensity 1:2, ratio of the *cis/trans* stereoisomers).

Notes: (1) A reaction flask of volume at least 500 mL should be used, since (especially at the beginning) the production of N$_2$ during the thermolysis of the diazoacetate causes strong foaming.

(2) Styrene has to be distilled over hydroquinone before use; bp$_{12}$ 33–34 °C.

1.7.1.3 ** *trans*-2-Phenylcyclopropane-1-carboxylic acid [4]

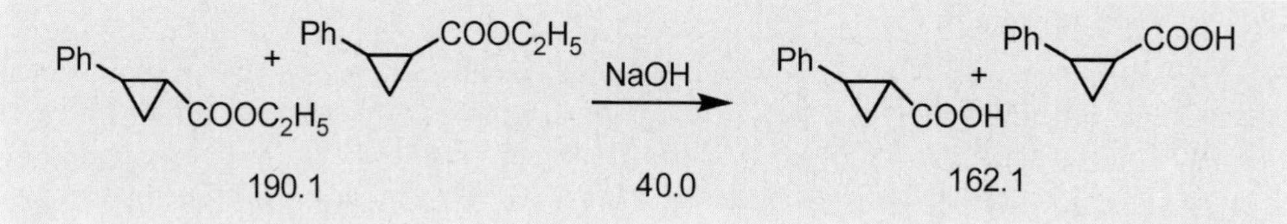

A solution of the *cis/trans* ester mixture from **1.7.1.2** (38.0 g, 0.20 mol) and NaOH (11.8 g, 0.30 mol) in EtOH/H$_2$O (110 mL/15 mL) is heated under reflux for 9 h.

The reddish mixture is then concentrated *in vacuo*, the residue is dissolved in H$_2$O (50 mL), and the solution is cooled in an ice-bath. With efficient stirring, concentrated HCl is added; the precipitated acid is collected by suction filtration, washed with H$_2$O, and recrystallized from H$_2$O (ca. 2.5 L) with the addition of charcoal. After a second recrystallization from H$_2$O, the pure *trans*-acid is obtained; 11.4 g (35%), colorless crystals, mp 92.5–93.5 °C. Concentration of the combined mother liquors from both recrystallizations to a volume of ca. 500 mL yields a second crop; 5.20 g (16%), mp 91–92 °C; total yield of the *trans*-acid 51%, TLC (SiO$_2$; Et$_2$O): $R_f \approx 0.75$.

> **IR** (KBr): $\tilde{v}$ (cm^{-1}) = 3500–2300 (OH, assoc.), 1695 (C=O), 1245.
>
> **^{1}H NMR** (CDCl$_3$): δ (ppm) = 7.4–6.9 (m, 5 H, phenyl-H), 2.7–2.4 (m, 1 H, Ph–CH), 2.9–1.2 (m, 3 H, CH + CH$_2$); COOH is not observed.

1.7.1.4 ** (±)-*trans*-2-Phenylcyclopropyl-1-amine (tranylcypromine) [4]

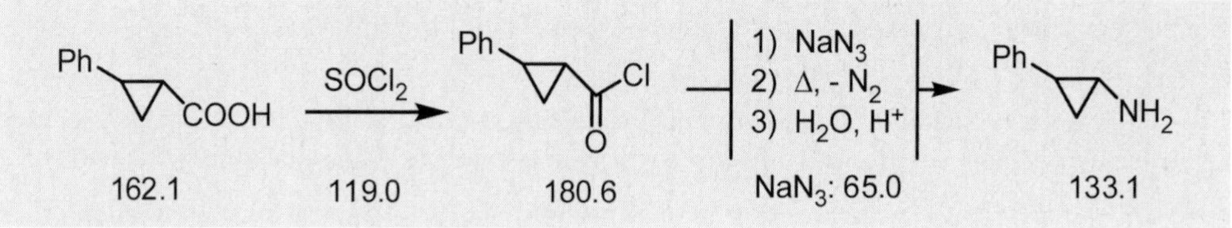

(1) *trans*-2-Phenylcyclopropane-1-carboxylic chloride:

A solution of the *trans*-acid **1.7.1.3** (13.0 g, 0.08 mol) and thionyl chloride (20.3 g, 0.17 mol) in anhydrous benzene (45 mL) is heated under reflux for 5 h (Caution! Hood! Evolution of HCl and SO$_2$!). The reaction mixture is then concentrated *in vacuo*, benzene (30 mL) is added, and distillation *in vacuo* is repeated to remove the excess SOCl$_2$. The yellowish crude acid chloride [IR (film): $\tilde{v}$ = 1780 cm^{-1} (C=O), 13.5 g, (93%)] is used in the next step without further purification.

(2) A solution of the acid chloride from (1) in toluene (70 mL) is added dropwise over 1 h to a well-stirred suspension of sodium azide (21.0 g, 0.32 mol, Caution! Hood! Shield!) in anhydrous toluene at 70 °C (external temperature). The temperature is then slowly increased, whereupon N_2 evolution occurs at 70–80 °C. When the addition is complete, the reaction mixture is heated to reflux until the evolution of N_2 ceases (ca. 4 h).

After cooling to room temperature, the inorganic salts are filtered off by suction and washed with toluene; the solvent is removed from the filtrate *in vacuo*, leaving the isocyanate as the residue [IR (film): $\nu = 2280$ cm^{-1} (N=C=O), 11.9 g]. After cooling to 10 °C, concentrated HCl (135 mL) is added dropwise over 45 min and the solution is heated to reflux for 2 h. The mixture is then cooled to room temperature and ice-cold H_2O (50 mL) is added. After extraction with Et_2O (100 mL), the acidic aqueous phase is concentrated to dryness *in vacuo*. The residue is suspended in Et_2O (100 mL) and, with cooling in an ice bath, a 50% aqueous KOH solution (50 mL) is added. The liberated amine is taken up in Et_2O, the aqueous (alkaline) phase is washed twice with Et_2O (50 mL), and the ethereal extracts are combined and dried ($MgSO_4$). After removal of the solvent, the oily, faintly yellow residue is fractionated *in vacuo* (microdistillation apparatus). Tranylcypromine is obtained as a colorless liquid; 6.60 g (62% based on *trans*-acid **1.7.1.3**), bp$_{0.4}$ 43–45 °C.

IR (film): $\tilde{\nu}$ (cm^{-1}) = 3370, 3300 (NH$_2$), 1605, 1500, 1460, 745, 700.

^{1}H NMR (CDCl$_3$): δ (ppm) = 7.4–6.9 (m, 5 H, phenyl-H), 2.7–2.4 (m, 1 H, Ph–CH), 2.1–1.7 (m, 1 H, CH–N), 1.55 (s, 2 H, NH$_2$), 1.2–0.7 (m, 2 H, cyclopropane CH$_2$).

Derivatives: (1)   Tranylcypromine hydrochloride: The hydrochloride is obtained by passing anhydrous HCl into a solution of the amine in anhydrous Et_2O; colorless crystals, mp 155–157 °C (from MeOH by addition of EtOAc/Et_2O, 1:1).

(2)   *N*-Benzoyl-2-phenyl-1-cyclopropylamine: Tranylcypromine benzoate is obtained by Schotten–Baumann acylation of the amine (1,4-dioxane as solvent, 1 h at +20 °C) with an equimolar amount of benzoyl chloride; colorless needles, mp 120–121 °C (from MeOH).

[1]   A. Kleemann, J. Engel, *Pharmaceutical Substances*, 3rd ed., p. 1923, Thieme Verlag, Stuttgart, **1999**.

[2]   H. Auterhoff, J. Knabe, H.-D. Höltje, *Lehrbuch der Pharmazeutischen Chemie*, 13th ed., p. 389, Wissenschaftliche Verlagsgesellschaft mbH, Stuttgart, **1994**.

[3]   a) J. Asunksis, H. Shechter, *J. Org. Chem.* **1968**, *33*, 1164. b) Cyclopropane α-aminocarboxylates have been obtained by reduction of the corresponding nitrocarboxylates resulting from cycloaddition of nitrodiazoacetates to alkenes: R. P. Wurz, A. B. Charette, *J. Org. Chem.* **2004**, *69*, 1262.

[4]   A. Burger, W. L. Yost, *J. Am. Chem. Soc.* **1948**, *70*, 2198.

[5]   M.-X. Wang, G.-Q. Feng, *New J. Chem.* **2002**, *26*, 1575.

[6]   M. S. Newman, G. F. Ottmann, C. F. Grundmann, *Org. Synth. Coll. Vol. IV*, **1963**, 424.

[7]   T. Niimi, T. Ushida, R. Irie, T. Katsuki, *Tetrahedron Lett.* **2000**, *41*, 3677.

[8]   a) T. Ikeno, M. Sato, T. Yamada, *Chem. Lett.* **1999**, 1445. For other catalyst types and their stereoselectivities, see: b) T. Yamada, T. Ikeno, H. Sekino, M. Sato, *Chem. Lett.* **1999**, 719; c) F. Loeffler, M. Hagen, U. Luening, *Synlett* **1999**, 1826.

1.7.2   11,11-Difluoro-1,6-methano[10]annulene

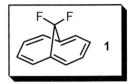

Topics:
- Vogel synthesis of a 1,6-methano-bridged [10]annulene
- Birch reduction of naphthalene
- Chemoselective cyclopropanation ([1+2]-cycloaddition) of a polyolefinic system
- Base-induced dehydrohalogenation
- Norcaradiene–cycloheptatriene rearrangement

(a) General

[10]Annulene is a problematic member of the family of aromatic [*n*]annulenes [1]. Containing ten π-electrons, it formally fulfils the criteria of the Hückel rule for aromatic compounds with $n = 2$. However, in all of its possible double-bond stereoisomers, either high sp^2-bond-angle deformations (as in **2**) or severe steric interactions of hydrogens (as in **4**) exist, which prohibit an approximately planar 10π perimeter geometry required for "aromatic" stabilization.

1,6-methano-[10]annulene

   2   **3**   **4**   **5**

In fact, the diastereomeric all-*cis*- and mono-*trans*-[10]annulenes (**2** and **3**) have been prepared and proved to be unstable, non-planar, and therefore non-aromatic polyenes. As conceived and realized by Vogel, removal of the hydrogen interference in the di-*trans*-form **4** by replacement of the inner hydrogens by a methylene group led to the 1,6-bridged [10]annulene **5**, the spectroscopic data of which correspond to an aromatic 10π-system [2].

The ^{1}H NMR spectrum of **5** shows it to be a diatropic hydrocarbon (ring protons giving rise to an AA'BB' multiplet at δ = 7.27 and 6.95 ppm; CH$_2$ positioned above the π-plane and giving a signal at δ = –0.52 ppm). According to X-ray structural analysis, the sp^2-C perimeter lacks overall planarity, but the average sp^2C–sp^2C distance is of the order seen in benzenoid compounds (137.3–141.9 pm vs. 139.8 pm in benzene) and indicates significant delocalization of the π-system. Chemically, **5** is stable towards oxygen and thermally stable up to 220 °C; it undergoes S_EAr reactions (e.g., bromination, nitration, acylation), as expected for a benzenoid aromatic.

Vogel's synthesis [2] of 1,6-methano[10]annulene (**5**) starts with isotetralin **6**, which is readily available by Birch reduction of naphthalene. Reaction with dichlorocarbene, generated from CHCl$_3$ and *tert*-BuOK, takes place chemoselectively at the internal double bond of **6** to give **7** by cyclopropanation. Dehalogenation of **7** by treatment with Na/liq. NH$_3$ leads to the propellane **8**, which is converted to **5** either (a) by addition of bromine to the double bonds followed by dehydrohalogenation with KOH (via **9**) or (b) by dehydrogenation with DDQ. In both transformations (a) and (b), the intermediate norcaradiene derivative **10** undergoes thermal electrocyclic ring-opening to afford the [10]annulene system.

1.7.3   Dimethyl heptalene-1,2-dicarboxylate

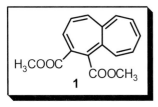

1

Topics:
- Hafner–Ziegler azulene synthesis
- S_NAr of activated haloarenes
- Zincke reaction (pentamethine cyanines from nucleophilic ring-opening of *N*-acceptor-substituted pyridinium ions)
- Formation of a vinylogous 6-aminofulvene and its electrocyclization to azulene
- Dipolar [2+2]-cycloaddition to azulene, ring expansion azulene → heptalene

(a)   General

Heptalene (**4**) belongs to the first members of the group of zero-bridged annulenes (cf. **1.7.2**) [1], the π-electron systems of which formally contain a pentafulvene unit [as in pentalene (**2**) and azulene (**3**)] or a heptafulvene unit [as in heptalene (**4**)] [1]:

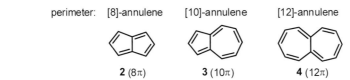

perimeter: [8]-annulene   [10]-annulene   [12]-annulene

2 (8π)   **3** (10π)   **4** (12π)

Heptalene (**4**) itself is a very reactive, oxygen-sensitive, non-planar cyclopolyolefin, temperature-dependent ^{1}H NMR spectroscopic analysis of which is indicative of dynamic interconversions between different conformers. The non-planar structure of the heptalene skeleton has been confirmed by X-ray analysis of stable derivatives such as the dicarboxylate **1**, the two rings of which preferentially adopt a boat-like structure in the crystalline state [2].

Most syntheses of **4** start from 1,4,5,8-tetrahydronaphthalene (**5**), the educt of Vogel's classical methano[10]annulene synthesis (cf. **1.7.2**):

Isotetralin **5** can be epoxidized chemoselectively at the central double bond to give **6**, and subsequent dibromocarbene addition under phase-transfer conditions provides the *anti*-bis-adduct **7**, which is readily dehalogenated and deoxygenated by Li in *tert*-BuOH/THF (→ **10**). NBS bromination of **10** furnishes a mixture of tetrabromides (**9**), reduction of which with zinc gives 3,8-dihydroheptalene (**8**). Dehydrogenation, first by hydride abstraction (Ph₃C⁺ BF₄⁻) and then by deprotonation (Et₃N) (**8** → **4**), completes the synthesis of **4** [3].

(b) Synthesis of 1

For the synthesis of the stable heptalene derivative **1**, a straightforward approach has been reported [4], which is based on the ring expansion of azulene (**3**) by a [2+2]-cycloaddition with dimethyl acetylene dicarboxylate.

Thus, first the synthesis of azulene (**3**) is described [5]. The method of choice is the one-pot, multi-step procedure submitted by Hafner [6] starting from 1-(2,4-dinitrophenyl)pyridinium chloride (**11**), which is prepared *in situ* by S_NAr reaction of 2,4-dinitrochlorobenzene with pyridine.

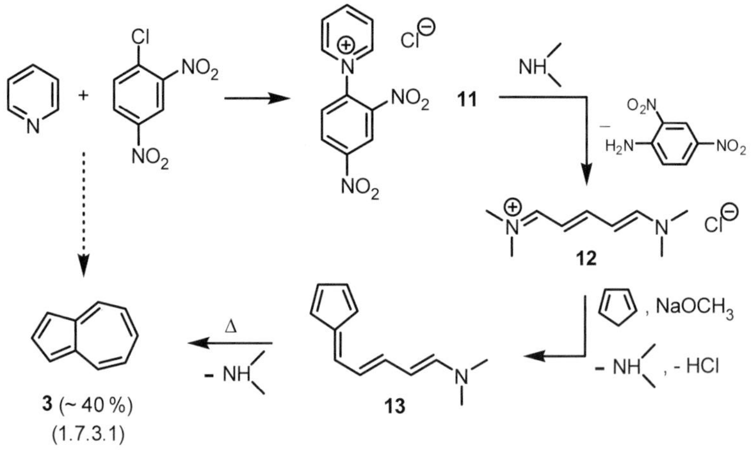

On reaction with dimethylamine, the salt **11** undergoes ring-opening of the pyridine nucleus with amine exchange resulting in the formation of the (symmetrical) pentamethine cyanine **12** as the key intermediate. Condensation of **12** with cyclopentadiene in the presence of NaOCH₃ leads to the bisvinylogous 6-aminofulvene **13**, which, on heating to 125 °C, cyclizes to afford azulene (**3**) with concomitant elimination of HN(CH₃)₂.

The ring-opening process **11** → **12** is an example of the Zincke reaction, which is generally observed when *N*-acceptor-substituted pyridinium salts **14** interact with *O*- or *N*-nucleophiles through initial attack at C-2 (→ **15**) and opening of the N–C-2 bond to furnish 1-azatrienals **16** [7]:

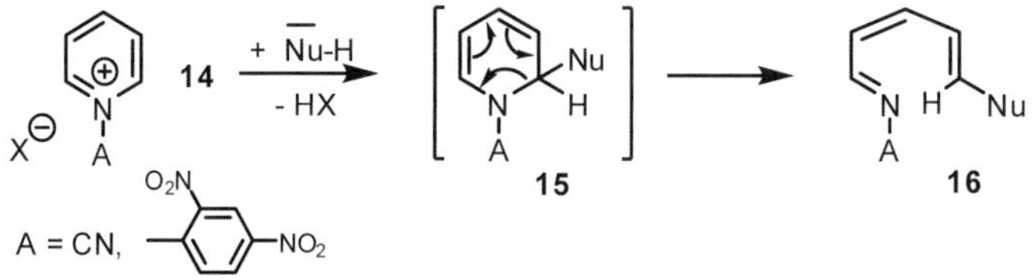

The cyclization of the bisvinylogous 6-aminofulvene **13** to azulene is mechanistically interpreted as an electrocyclic 10π process leading to **17**, which subsequently undergoes loss of the amine moiety in a (thermal) β-elimination (**17 → 3**) [5]:

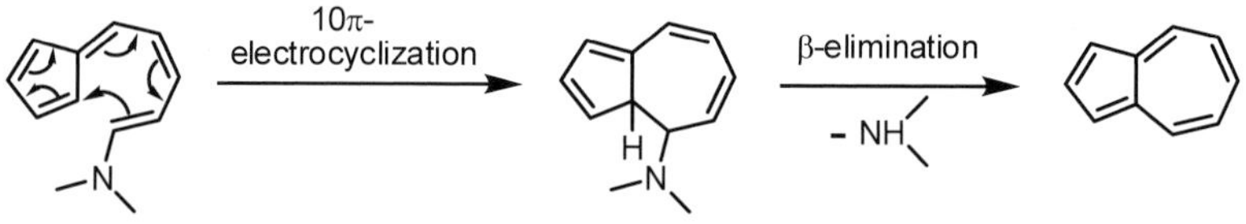

The ring expansion of azulene with dimethyl acetylene dicarboxylate proceeds as a thermal reaction in boiling tetralin. After purification by chromatography (separation from unreacted azulene), the heptalene diester **1** is obtained as air-stable brown-red crystals.

The formation of **1** from azulene (**3**) can be understood as a two-step dipolar [2+2]-cycloaddition [8], since the electron distribution (b) in azulene (**3**) leads to exclusive attack of electrophiles at the five-membered ring with formation of a stabilized tropylium ion in the dipolar intermediate **18**:

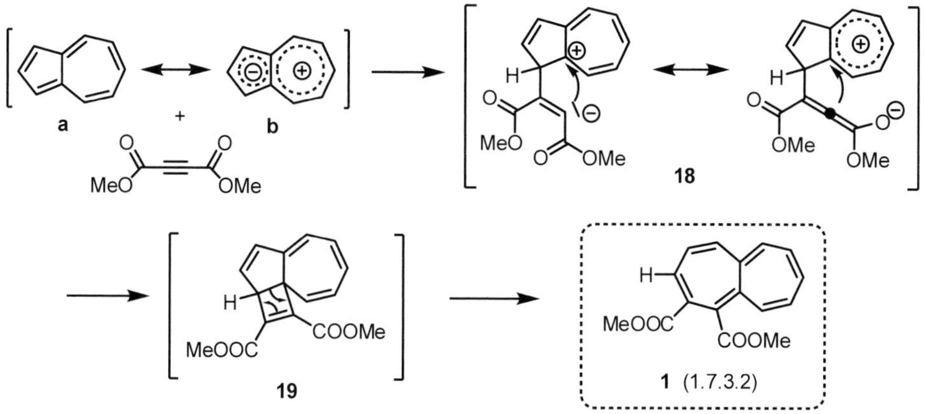

Finally, the formed [2+2]-cycloadduct **19** expands to the heptalene system by 4π-cycloreversion of the cyclobutene subunit to a 1,3-diene.

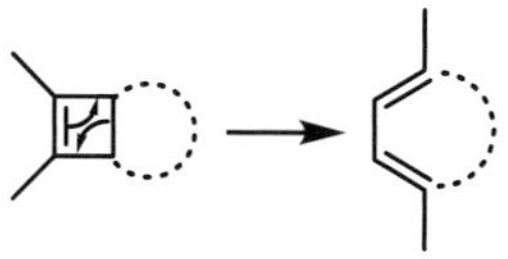

(c) Experimental procedures for the synthesis of 1

1.7.3.1 ** Azulene [6]

2,4-Dinitrochlorobenzene (40.5 g, 0.20 mol, Caution! Caustic!) and anhydrous pyridine (240 mL) are stirred and heated at 85–90 °C (steam bath) for 4 h. N-(2,4-Dinitrophenyl)pyridinium chloride begins to precipitate as a yellow-brown solid after ca. 30 min. The mixture is cooled to 0 °C in an ice-salt bath, and then a solution of dimethylamine (20.0 g, 0.44 mol, see note) in anhydrous pyridine (60 mL) is added dropwise over 30 min. The temperature of the reaction mixture rises to 4 °C. After the addition, the red-brown solution is slowly warmed to room temperature and stirred for 12 h. The drying tube is then replaced with a gas inlet tube and the system is flushed with nitrogen.

Cyclopentadiene (prepared by distillation of the dimer [9]; 14.0 g, 0.21 mol) is added under a nitrogen atmosphere. Sodium methoxide [2.5 M; sodium (4.60 g) in anhydrous MeOH (80 mL)] is added dropwise with stirring over a period of 30 min. The solution warms to 26 °C and is left at room temperature for 15 h. The dropping funnel is then replaced by a distillation head and the reaction mixture is carefully heated (Hood! $(H_3C)_2NH$ is evolved!). Pyridine and MeOH are distilled off until the temperature reaches 105 °C (ca. 150 mL of distillate). The distillation head is removed, anhydrous pyridine (200 mL) is added, and the mixture is heated for 4 d at 125 °C under nitrogen.

The mixture is then cooled to 60 °C and pyridine is distilled off under reduced pressure. The blue-black crystalline residue is extracted with n-hexane (400 mL) in a Soxhlet extractor for 4 h. Traces of pyridine are removed from the blue n-hexane solution by washing it with 10% HCl (3 × 30 mL) and H_2O (30 mL). The organic phase is dried over Na_2SO_4 and the volume is reduced by half by distillation using a 50 cm Vigreux column. The concentrated solution is filtered through a column (30 × 4 cm, 200 g of basic Al_2O_3, activity grade II, n-hexane as eluent). The solvent is evaporated from the eluate to give dark-blue leaflets, 9.10 g (36%, mp 97–98 °C). Further purification can be achieved by sublimation at 90 °C and 10 mbar (mp 99–100 °C).

IR (KBr): $\tilde{v}$ (cm^{-1}) = 1580, 1450, 1395, 1210, 960, 760.

^{1}H NMR (CCl$_4$): δ (ppm) = 8.4–6.8 (m).

UV (n-hexane): λ_{max} (log ε) = 580 (2.46), 352 (2.87), 339 (3.60), 326 (3.48), 315 (3.26), 294 (3.53), 279 (4.66), 274 (4.70), 269 (4.63), 238 (4.24), 222 nm (4.06). The longest wavelength band is indicated as the envelope of an eight-peak absorption band.

Note: The solution of dimethylamine in pyridine is prepared by adding dimethylamine gas (dried over KOH) to anhydrous pyridine under water-free conditions.

1.7.3.2   **   Dimethyl heptalene-1,2-dicarboxylate [6]

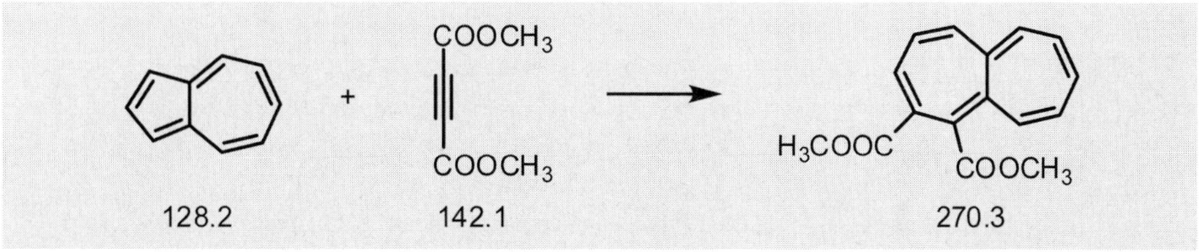

| 128.2 | 142.1 | 270.3 |

Azulene **1.7.3.1** (1.28 g, 10.0 mmol) and dimethyl acetylene dicarboxylate (2.13 g, 15.0 mmol) are heated under reflux in freshly distilled tetralin (20 mL) for 20 min. The solution is cooled, diluted with *n*-hexane (150 mL), and chromatographed on alumina (basic, activity grade IV, 100 g) with *n*-hexane as eluent (fraction 1). Absorbed material is eluted with CH_2Cl_2 until no more product appears in the eluate (TLC) (fraction 2).

The fractions are treated as follows:

Fraction 1: The blue eluate is concentrated *in vacuo* and the residue is chromatographed on alumina (basic, activity grade I, 200 g) eluting with *n*-hexane. Tetralin elutes first, followed by a blue solution containing unreacted azulene. The azulene crystallizes on evaporating the solvent to give a recovered yield of 0.78 g (61%).

Fraction 2: The solvent is evaporated and the residue is chromatographed on an alumina column (basic Al_2O_3, activity grade IV, 500 g) eluting with *n*-hexane/Et_2O (5:3). Purple (1), dark blue (2), blue-green (3), yellow-brown (4), violet (5), and blue (6) fractions are obtained. The product is isolated from fraction 4 by evaporating the solvents *in vacuo* and recrystallizing the residue from *n*-hexane/Et_2O. The yield is 0.26 g (9.6%; 25% based on recovered azulene), mp 112–113 °C.

IR (KBr): $\tilde{\nu}$ (cm^{-1}) = 1720, 1570, 1440, 1260, 1230.

^{1}H NMR ([D_6]actetone): δ (ppm) = 7.27 (d, J = 7 Hz, 1 H, 3-H), 6.7–5.7 (m, 7 H, vinyl-H), 3.71, 3.64 (s, 3 H, OCH_3).

UV (*n*-hexane): λ_{max} (log ε) = 337 (3.63), 266 (4.29) 204 nm (4.36).

---

[1]   For a review, see: H. Hopf, *Classics in Hydrocarbon Chemistry*, p. 285, Wiley-VCH, Weinheim, **2000**.

[2]   K. Hafner, G. L. Knaup, H. J. Lindner, *Bull. Chem. Soc. Jpn.* **1988**, *61*, 155.

[3]   a) E. Vogel, D. Kerimis, N. T. Allison, R. Zellerhoff, J. Wassen, *Angew. Chem.* **1979**, *91*, 579; *Angew. Chem. Int. Ed. Engl.* **1979**, *18*, 545. b) For an alternative synthesis of **4** starting from **5**, see: L. A. Paquette, A. R. Browne, E. Chamot, *Angew. Chem.* **1979**, *91*, 581; *Angew. Chem. Int. Ed. Engl.* **1979**, *18*, 546.

[4]   a) K. Hafner, H. Diehl, H. U. Süß, *Angew. Chem.* **1976**, *88*, 121; *Angew. Chem. Int. Ed. Engl.* **1976**, *15*, 104; b) for comparison, see: X. Jin, A. Linden, H.-J. Hansen, *Helv. Chim. Acta* **2005**, *88*, 873.

[5]   a) For a review on azulene syntheses, see ref. [1], p. 281, and K.-P. Zeller, in Houben-Weyl, *Methoden der Organischen Chemie*, Vol. V/2c, p. 127, Thieme Verlag, Stuttgart, **1985**. b) Recently, a highly flexible route to azulenes was reported, starting with cycloaddition of dichloroketene to cycloheptatriene and ring expansion with diazomethane: S. Carret, A. Blanc, Y. Coquerel, M. Berthod, A. E. Greene, J.-P. Déprés, *Angew. Chem.* **2005**, *117*, 5260; *Angew. Chem. Int. Ed.* **2005**, *44*, 5130.

[6]   K. Hafner, K.-P. Meinhardt, *Org. Synth.* **1984**, *62*, 134; original contribution by K. Hafner, 1981.

[7]   See: Th. Eicher, S. Hauptmann, *The Chemistry of Heterocycles*, 2nd ed., p. 280, Wiley-VCH, Weinheim, **2003**.

[8]   R. Gompper, *Angew. Chem.* **1969**, *81*, 348; *Angew. Chem. Int. Ed. Engl.* **1969**, *8*, 312.

[9]   G. Wilkinson, *Org. Synth. Coll. Vol. IV* **1963**, 475.

1.7.4　Dimethyl 1,8-bishomocubane-4,6-dicarboxylate

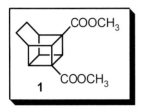

1

Topics:
- Halogenation of a phenol (S_EAr)
- Dehydrogenation of a hydroquinone to a 1,4-benzoquinone
- Use of 1,4-quinones as electron-deficient dienophiles in a Diels–Alder reaction
- Photochemical intramolecular [2+2]-cycloaddition with cyclobutane formation
- Favorskii rearrangement
- Methyl esters from carboxylic acids and diazomethane

(a) General

1,8-Bishomocubane (**2**) is structurally derived from cubane (**3**) by replacement of one cube-edge C–C bond by an ethano (CH_2–CH_2) bridge. For this basket-shaped cage hydrocarbon, the trivial name "basketane" is used.

2　　　　**3**

The synthesis of **2** is based on an elegant approach to cubane [1], developed by Pettit [2], as outlined in the following scheme:

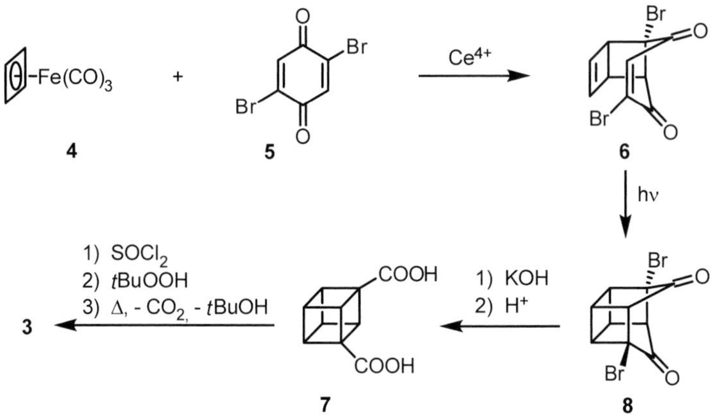

Cyclobutadiene (liberated from its iron complex **4** by oxidation with Ce(IV)) reacts with the dibromoquinone **5** to give the [2+2]-cycloadduct **6**. The two double bonds in **6** are in a *syn*-arrangement, making them suitably predisposed for an intramolecular [2+2]-cycloaddition, which occurs readily upon irradiation and provides the dibromodiketone **8**. This is converted to the cubane-1,3-diacid by ring contraction through twofold Favorskii rearrangement (→ **7**) and further to cubane (**3**) by decarboxylation of its bis-*tert*-butyl perester.

Consequently, the synthesis of **2** includes the essential features of the foregoing strategy, namely: (1) an intramolecular [2+2]-photocyclization of an appropriate precursor containing the handle of the basketane system, and (2) Favorskii ring contraction to transform an α-bromocyclopentanone to a cyclobutane dicarboxylic acid.

(b) Synthesis of 1

The synthesis of **1** starts with the Diels–Alder reaction of cyclohexa-1,3-diene (**9**) with 2,5-dibromo-1,4-benzoquinone (**5**) utilizing the well-established ability of 1,4-quinones to act as electron-deficient dienophiles in [4+2]-cycloadditions [4] (**9** + **5** → **10**):

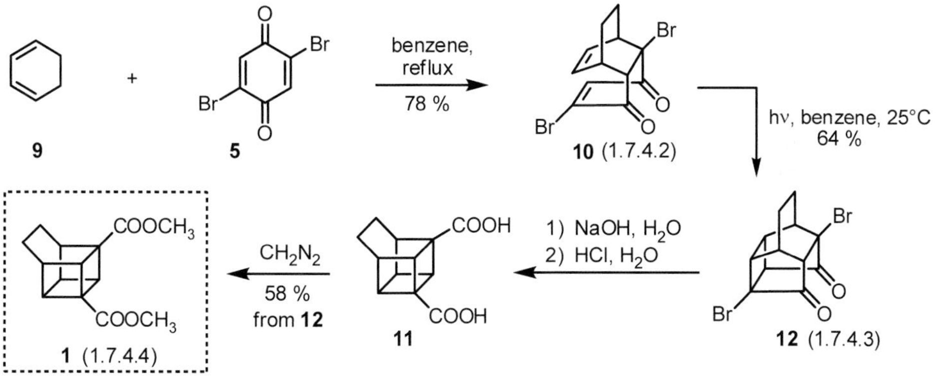

On irradiation of the Diels–Alder adduct **10** in benzene at 25 °C, a photochemically allowed intramolecular [2+2]-cycloaddition of the two *syn*-oriented C=C double bonds occurs with formation of the dibromodione **12**. Treatment of **12** with NaOH leads to the bishomocubane 1,3-dicarboxylic acid **11** by ring contraction via a cyclopropanone [5].

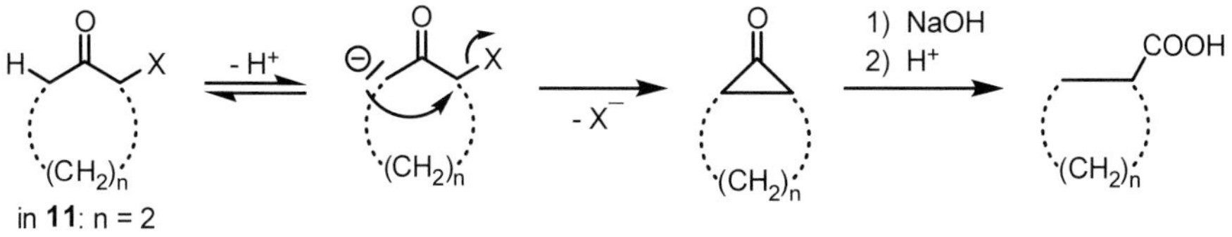

Finally, the diacid **11** is esterified with diazomethane to give the dimethyl ester **1**. The sequence **12** → **11** → **1** is performed as a one-pot procedure.

The requisite dibromo-1,4-benzoquinone **5** is prepared by bromination of hydroquinone, which as an activated arene undergoes symmetrical disubstitution to give **13**. This is oxidized using $FeCl_3$ to afford the dibromoquinone **5**.

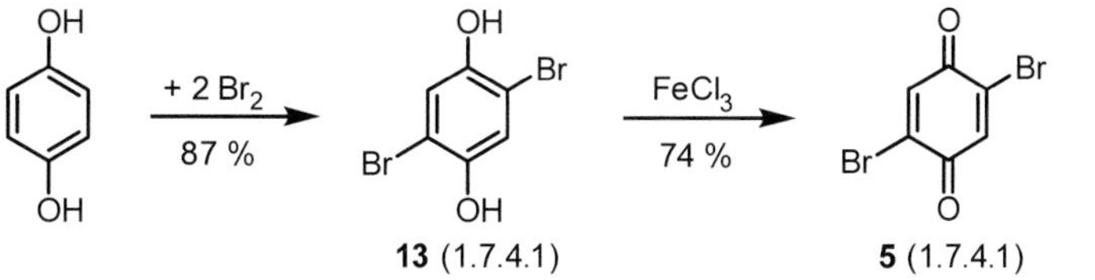

Thus, the basketane diester **1** is obtained in a five-step sequence with an overall yield of 19% (based on hydroquinone).

It should be noted that the hydrocarbon **2** may be obtained from the diacid **11** by a modified Hunsdiecker reaction to give the dibromide **14** followed by reductive debromination with nBu_3SnH (**14 → 2**) [3]. This two-step sequence corresponds to an overall decarboxylation of **11**:

$$
\begin{array}{ccccc}
& \text{HgO, Br}_2 & & nBu_3SnH & \\
& \xrightarrow{\text{CH}_2\text{Br}_2} & & \xrightarrow{\text{AIBN}} & \\
\mathbf{11} & & \mathbf{14} & & \mathbf{2}
\end{array}
$$

(c) Experimental procedures for the synthesis of 1

1.7.4.1 * 2,5-Dibromo-1,4-benzoquinone [6]

| 110.1 | 159.8 | 267.9 | 270.3 | 265.9 |

(1) A solution of bromine (64.0 g, 0.40 mol, ~20.5 mL) in glacial acetic acid (20 mL) is added dropwise to a stirred suspension of hydroquinone (22.0 g, 0.20 mol) in glacial acetic acid (200 mL) at room temperature. The temperature rises to about 30 °C, with the initial formation of a clear solution, followed, after 5–10 min, by the deposition of a colorless precipitate. Stirring is continued for 1 h.

The mixture is then filtered and the solid is washed with a small amount of glacial acetic acid. The mother liquor is concentrated *in vacuo* to around half of its original volume and is allowed to stand for 12 h. The formed crystals are collected and the procedure is repeated to give a third crop. The total yield of crude product is 46.4 g (87%), mp 180–187 °C. Recrystallization from glacial acetic acid gives crystals with mp 188–189 °C. However, the crude product can be used without purification for the next step.

(2) A solution of $FeCl_3 \cdot 6 H_2O$ (65.4 g, 242 mmol) in H_2O (140 mL) is added dropwise to a stirred, refluxing solution of 2,5-dibromohydroquinone prepared in step (1) (27.4 g, 102 mmol) in H_2O (800 mL) over a period of 15 min. The desired *p*-quinone immediately crystallizes from the mixture.

It is collected by filtration after cooling to room temperature, washed with H_2O, and recrystallized from EtOH (800 mL) to give yellow needles; 20.0 g (74%), mp 188–190 °C.

IR (KBr): $\tilde{\nu}$ (cm^{-1}) = 1770, 1760 (C=O).

^{1}H NMR (CDCl$_3$): δ (ppm) = 7.12 (s, C=C–H).

1.7.4.2   * 2,5-Dibromotricyclo[6.2.2.0²·⁷]dodeca-4,9-dien-3,6-dione [3]

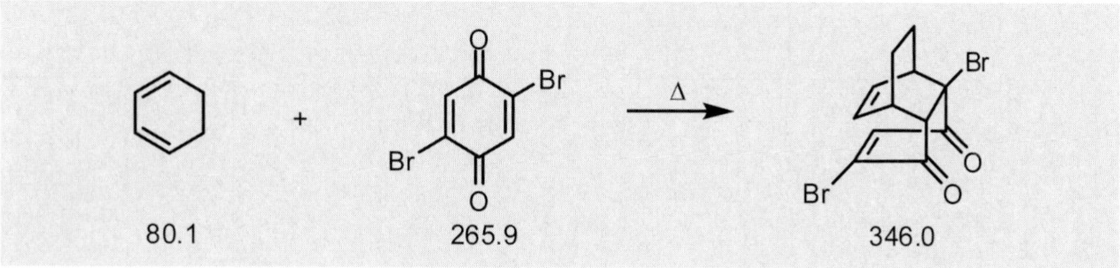

A solution of 2,5-dibromo-*p*-benzoquinone **1.7.4.1** (10.0 g, 37.5 mmol) and cyclohexa-1,3-diene (6.40 g, 80.0 mmol) in anhydrous benzene (20 mL, Caution!) is heated under reflux for 3 h.

The solvent and excess cyclohexa-1,3-diene are then distilled off to leave a thick oil, which crystallizes on scratching. The solid is treated with hot petroleum ether (40–60 °C, 2 × 100 mL). The filtrates are combined and cooled to –10 °C. The product crystallizes as colorless crystals. Concentration of the mother liquor gives a small second crop. The total yield is 10.3 g (78%), mp 116–118 °C.

IR (KBr): $\tilde{v}$ (cm⁻¹) = 1690, 1670 (C=O), 1600 (C=C).

¹H NMR (CDCl₃): δ (ppm) = 7.39 (s, 1 H, vinyl-H), 6.3–6.2 (m, 2 H, vinyl-H), 3.7–3.1 (m, 3 H, bridgehead-H and CO–CH), 2.6–1.2 (m, 4 H, CH₂–CH₂).

1.7.4.3   ** 1,6-Dibromopentacyclo[6.4.0³·⁶.0⁴·¹².0⁵·⁹]dodeca-2,7-dione [3]

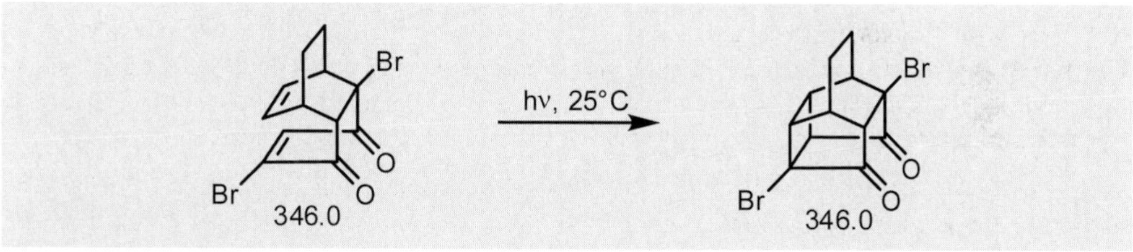

Apparatus: Photolysis apparatus with quartz filter and high-pressure mercury vapor lamp (Philips HPK-125 W or Hanau TQ-150 W).

Dibromodione **1.7.4.2** (10.0 g, 29.9 mmol) in anhydrous benzene (260 mL, Caution!) (note) is flushed with nitrogen for ca. 15 min and irradiated at room temperature for 5 h, during which partial crystallization of the product occurs.

The crystals are filtered off, and the mother liquor is concentrated under reduced pressure to a volume of ca. 30 mL, which leads to the deposition of a second crop of yellowish crystals; total yield 6.40 g (64%), mp 206–208 °C, pure by TLC (silica gel; CH₂Cl₂).

IR (KBr): $\tilde{v}$ (cm⁻¹) = 1780 (C=O).

¹H NMR (CDCl₃): δ (ppm) = 3.41 (s$_{br}$, 3 H), 3.15–3.0 (m, 1 H, CH adjacent to CO and CBr; assignment unclear), 2.5–1.6 (m, 6 H, CH + CH₂).

Note: Irradiation in toluene (under otherwise identical conditions) gives a lower yield (41%).

1.7.4.4 ** Dimethyl 1,8-bishomocubane-4,6-dicarboxylate [3]

346.0   248.3

A stirred mixture of product **1.7.4.3** (6.30 g, 18.2 mmol) and 25% aqueous sodium hydroxide (65 mL) is heated under reflux for 2 h. The cooled solution is acidified with concentrated hydrochloric acid, keeping the temperature below 5 °C. The colorless precipitate is collected by filtration, washed with H_2O, and dried *in vacuo*. The yield is 5.4 g.

The solid is added in small portions to an ethereal solution of diazomethane (prepared from 6.20 g, ~60.0 mmol of nitroso methyl urea [7] at 0 °C). Complete dissolution occurs with nitrogen evolution. The mixture is stirred for 5 min and excess diazomethane is destroyed by slow addition of 2 M acetic acid until N_2 evolution stops.

The organic layer is separated, washed with H_2O, saturated aqueous $NaHCO_3$, and brine (30 mL each), dried over $MgSO_4$, and concentrated *in vacuo*. The dark residue is chromatographed on silica gel (0.06–0.02 mm, 150 g) eluting with *n*-hexane/Et_2O (1:1). The first fraction contains the product, which is obtained as a colorless oil after evaporation of the solvents; it crystallizes from *n*-pentane on cooling to –15 °C. The yield is 2.60 g (58%), mp 54–56 °C (pure by TLC).

IR (KBr): $\tilde{v}$ (cm^{-1}) = 1725 (C=O).

^{1}H NMR (CDCl$_3$): δ (ppm) = 3.73, 3.70 (s, 3 H, COOCH$_3$), 3.5–2.85 (m, 6 H, cyclobutane CH), 1.54 (s$_{br}$, 4 H, CH$_2$–CH$_2$).

---

[1] a) For a review on cubane synthesis, see: H. Hopf, *Classics in Hydrocarbon Chemistry*, p. 53, Wiley-VCH, Weinheim, **2000**. b) The first cubane synthesis was reported by P. E. Eaton, T. W. Cole, *J. Am. Chem. Soc.* **1964**, *86*, 962 and 3157; c) subsequent improvements enable the preparation of **3** on a multigram laboratory scale: M. Bliese, J. Tsanaktsidis, *Aust. J. Chem.* **1997**, *50*, 189.

[2] J. C. Barborek, L. Watts, R. Pettit, *J. Am. Chem. Soc.* **1966**, *88*, 1328.

[3] P. G. Gassmann, R. Yamaguchi, *J. Org. Chem.* **1978**, *43*, 4654.

[4] a) *Houben-Weyl*, Vol. VII/2b, p. 1765; b) *Houben-Weyl*, Vol. VII/3c, p. 23. Diels–Alder reactions of 1,4-benzoquinones have been conducted enantioselectively by catalysis with chiral oxazaborolidinium cations: c) D. H. Ryu, G. Zhou, E. J. Corey, *J. Am. Chem. Soc.* **2004**, *126*, 4800; d) Q.-Y. Hu, G. Zhou, E. J. Corey, *J. Am. Chem. Soc.* **2004**, *126*, 13713.

[5] M. B. Smith, J. March, *March's Advanced Organic Chemistry*, 5th ed., p. 1404, John Wiley & Sons, Inc., New York, **2001**.

[6] J. F. Bagli, Ph. L'Ecuyer, *Can. J. Chem.* **1961**, *39*, 1037.

[7] *Organikum*, 21st ed., p. 647, Wiley-VCH, Weinheim, **2001**.

1.7.5   α-Terpineol

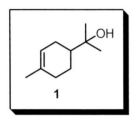

1

Topics: • Synthesis of a cyclic monoterpene alcohol:
(1) in racemic form, (2) in enantiopure form

• Lewis acid-catalyzed regioselective Diels–Alder reaction

• Diastereoselective Diels–Alder reaction by use of a chiral auxiliary (Evans auxiliary)

• Tertiary alcohols from esters and Grignard compounds

• Synthesis and application of an (*S*)-phenylalanine-based Evans auxiliary

(a) General

α-Terpineol (**1**, *p*-menthenol) belongs to the class of cyclic monoterpenes, which are biogenetically derived from two isoprene units via the mevalonate pathway. α-Terpineol is widespread in Nature; its (+)- and (–)-enantiomers have been found in pine oil (etheric oil from *pinus palustris* (Pinaceae)). Due to its odor, which is reminiscent of that of lavender, it is used in perfumery as a fragrance. α-Terpineol is obtained industrially from α-pinene [1].

Retrosynthetic analysis of **1** according to pathway **A** leads to isoprene (**3**) and acrylate (**4a**) or methyl vinyl ketone (**4b**) via the cyclohexene **2**.

(*rac*)-**1**   (A) ⟹   **2a/2b**   retro [4+2] CA ⟹ (I)   **3**   **4a/4b**

(II) retro [4+2] CA   (B)

2a/4a : R = OR' with R' = alkyl
2b/4b : R = CH₃

3   **5**

The formation of **2** from **3** and **4** by a Diels–Alder reaction is a favored process due to favorable electronic interaction (electron-rich diene and electron-poor dienophile). The tertiary alcohol function in **1** can be introduced by reaction of cycloadduct **2a** with two moles of CH₃MgX (see the synthesis in section (b)) or by treating **2b** with one mole of CH₃MgX.

Retrosynthesis according to pathway **B** by direct retro-Diels–Alder disconnection of **1** leads to isoprene (**3**) and the allylic alcohol **5** as dienophile. However, their combination in a thermal [4+2]-cycloaddition (II) is a less favorable process according to frontier orbital interaction considerations since the HOMO (diene)/LUMO (dienophile) energy difference is significantly higher in the case of **5** than it is with **4a** [2].

The [4+2]-cycloaddition of isoprene (**3**) to an acrylic ester **4a** has served as a model reaction in investigating the regioselectivity and stereoselectivity of Diels–Alder reactions.

6/7a/8a: R = Me
7b/8b: R = H

(1) When isoprene (**3**) and methyl acrylate (**6**) are reacted thermally at 80 °C, a 70:30 mixture of the regioisomeric cycloadducts **7a** and **8a** is obtained in 80% yield. The regioselectivity is significantly improved by the addition of a Lewis acid; thus, in the presence of $AlCl_3$, a 95:5 mixture of **7a** and **8a** (77% yield, see section (b)) results [3, 4]. Separation is possible by fractional crystallization of the regioisomeric acids **7b** and **8b** after saponification; the ester **7a** may then be obtained free of its regioisomer by re-esterification of the purified acid **7b** with CH_2N_2.

(2) Asymmetric Diels–Alder reactions have been performed (a) by use of a dienophile connected to a chiral auxiliary, and (b) by use of a chiral Lewis acid as catalyst [5].

Concerning (a), an example is presented in section (b), in which an acrylic acid attached to an Evans auxiliary (cf. **1.2.2**) is utilized for [4+2]-cycloaddition to isoprene [6]. After removal of the auxiliary, an enantiopure ester of type **7a** is produced, which allows the preparation of (R)-**1**.

Concerning (b), a catalytic asymmetric version of the Diels–Alder reaction of isoprene with acrylate has been developed [7] that involves the use of the trifluoroethyl ester **9** and a chiral proline-derived cationic oxazaborolidine derivative **11** (as its triflimide), which gives the cycloadduct **10** in excellent yield (99%) and with high enantioselectivity (ee = 98%). Ester **10** could also be used as a substrate for the synthesis of (R)-**1**.

Catalysts of type **11** have been shown to have a broad spectrum of application in enantioselective [4+2]-cycloadditions [7, 8]. Chiral (acyloxy)boranes [9] (CAB, **12**, cf. **1.3.4**) and chiral bisoxazoline ligands [10] (BOX, **13**) have also proved to be successful catalysts for asymmetric Diels–Alder reactions [11], since they form rigid metal–substrate complexes and provide excellent stereoselectivities. Likewise, for hetero-Diels–Alder reactions with inverse electron demand, efficient catalysts based on Cr, Zr, or Sc complexes are known [12]. Furthermore, enantioselective Diels–Alder reactions [13] using α,β-unsaturated aldehydes as dienophiles and chiral amines as organo-catalysts can be performed by employing, for example, imidazolidinone **14**. The formation of an iminium ion by reaction of the carbonyl moiety of the dienophile with the amine functionality of the organo-catalyst lowers the LUMO energy of the dienophile leading to an acceleration of the Diels–Alder reaction.

(b) Synthesis of 1

(1) Synthesis of (*rac*)-α-terpineol [3, 14]

The Diels–Alder reaction of isoprene with methyl acrylate (**6**) is performed in benzene solution at room temperature in the presence of $AlCl_3$ as Lewis acid. The cycloadduct **7a** (containing a 5% impurity of the regioisomeric ester **8a**; for separation, see section (a)) is reacted with two moles of methylmagnesium iodide. This classical transformation of an ester to a tertiary carbinol containing two equal α-substituents leads to the racemic α-terpineol (*rac*-**1**) after the usual work-up with NH_4Cl solution.

3   **6**   **7a** (1.7.5.1)   *rac*-**1** (1.7.5.2)

(2) Synthesis of (*R*)-(+)-α-terpineol [6]

As chiral auxiliary, an oxazolidin-2-one **17** of the Evans type (cf. **1.2.2**) is used. It is readily prepared from (*S*)-phenylalanine (**15**) by reduction with $LiAlH_4$ and cyclocondensation of the resulting (*S*)-phenylalaninol (**16**) with diethyl carbonate.

15   **16** (1.7.5.3)   **17** (1.7.5.4)

To introduce an acrylic moiety, the chiral auxiliary **17** is equilibrium-deprotonated using LiCl in THF and acylated with acryloyl anhydride prepared *in situ* from acryloyl chloride and acrylic acid in the presence of triethylamine [15] to give the chiral acrylic amide **18**.

Diels–Alder reaction of **17** with isoprene (**3**) proceeds readily at –100 °C in the presence of diethylaluminum chloride in CH_2Cl_2/toluene as solvent to give (after hydrolytic work-up) the cycloadduct **21** with a diastereoselectivity of *de* = 90% via an *s-cis-endo* transition state **19**. The efficient stereodiscrimination can be explained by (1) the Lewis acid Et_2AlCl, which provides for rigid chelation of the dienophile moiety by coordinative interaction with both C=O groups of the *N*-acyloxazolidin-2-one system **20**, and by (2) π-stacking, which causes a stereoelectronic stabilization of the transition state [6]. Thus, improved diastereoselectivity is observed with the benzyl-substituted auxiliary **17** as compared to that seen with auxiliaries with aliphatic residues, e.g. the valinol-derived analogue.

The diastereomeric purity of the cycloadduct **21** can be raised up to > 98% *de* by recrystallization. **21** is cleaved by treatment with CH_3OMgBr (prepared *in situ* by reaction of CH_3MgBr with CH_3OH) to give the chiral methyl ester (*R*)-**22** with >99% *ee*. The chiral auxiliary **17** can be recovered from the reaction mixture, allowing its regenerative use. Finally, the methyl ester (*R*)-**22** is transformed into (*R*)-(+)-α-terpineol ((*R*)-**1**) by reaction with two moles of MeMgI; the chiral monoterpene alcohol (*R*)-**1** is obtained in almost enantiopure form with >98% *ee*.

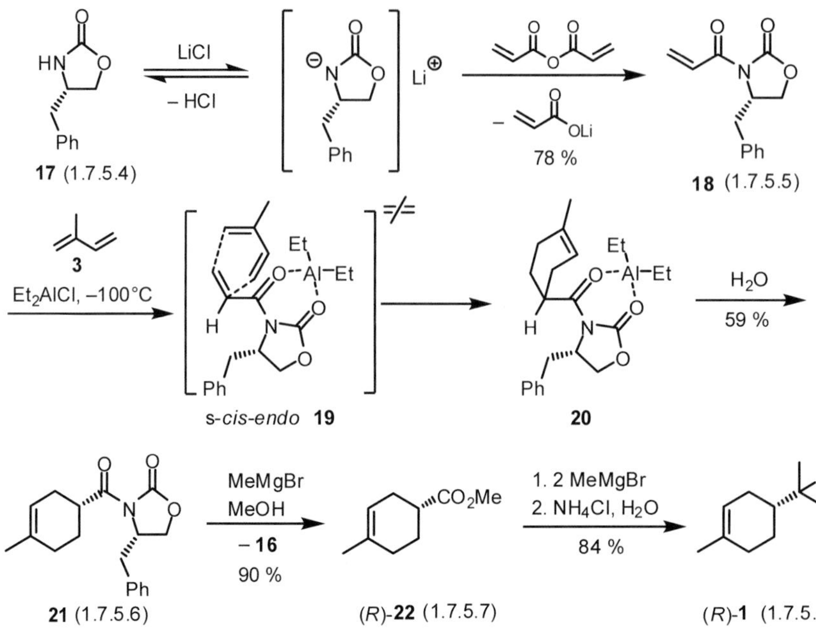

17 (1.7.5.4)

LiCl

– HCl

78 %

18 (1.7.5.5)

3

Et$_2$AlCl, –100 °C

s-cis-endo **19**

20

H$_2$O

59 %

21 (1.7.5.6)

MeMgBr
MeOH

– **16**

90 %

(R)-**22** (1.7.5.7)

1. 2 MeMgBr
2. NH$_4$Cl, H$_2$O

84 %

(R)-**1** (1.7.5.8)

(c) Experimental procedures for the synthesis of 1

1.7.5.1   **   4-Methylcyclohex-3-ene-1-carboxylic acid methyl ester [14]

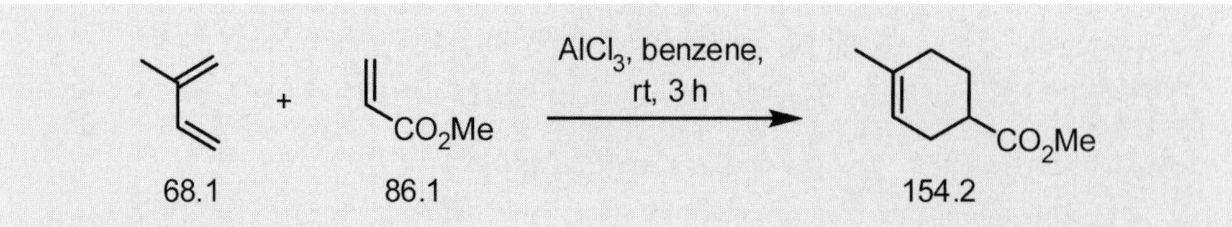

AlCl$_3$, benzene,
rt, 3 h

68.1   86.1   154.2

A solution of methyl acrylate (26.1 g, 303 mmol) (note 1) in anhydrous benzene (30 mL) (Caution: carcinogenic!) is added dropwise to a well-stirred suspension of anhydrous AlCl$_3$ (4.30 g, 32.0 mmol) in anhydrous benzene (250 mL) over 15 min. The temperature rises to 25 °C and the AlCl$_3$ dissolves. A solution of isoprene (20.9 g, 307 mmol) in benzene (50 mL) is then added dropwise over a period of 60 min. The temperature is held at 15–20 °C with occasional cooling during this period. Stirring is continued at room temperature for 3 h.

The solution is poured into HCl (2 M) saturated with NaCl (500 mL), the phases are separated, and the benzene phase is washed with H$_2$O (250 mL) and dried over Na$_2$SO$_4$. The solvent is evaporated under slightly reduced pressure at 70 °C and the residue is fractionally distilled in vacuo. The product is obtained as a colorless oil, 35.7 g (77%), bp$_{17}$ 80–82 °C, n$_D^{20}$ = 1.4630 (note 2).

IR (NaCl): $\tilde{\nu}$ (cm^{-1}) = 1745, 1440, 1175.

^{1}H NMR (200 MHz, CDCl$_3$): δ (ppm) = 5.28 (s$_{br}$, 1 H, vinyl-H), 3.60 (s, 3 H, CO$_2$CH$_3$), 2.70–1.65 (m, 7 H, CH + 3 × CH$_2$), 1.63 (s, 3 H, CH$_3$).

Notes: (1) Methyl acrylate should be freshly distilled over hydroquinone before use, bp_{760} 80–81 °C.

(2) The product is a 95:5 mixture with the regioisomeric 3-methylcyclohex-3-ene-1-carboxylic methyl ester [4] (see section (a)). This mixture is used in the next step.

1.7.5.2   *   *rac*-α-**Terpineol** [3]

An iodine crystal is added to magnesium turnings (7.20 g, 300 mmol) covered with anhydrous Et_2O (30 mL), followed by a few mL of a solution of methyl iodide (42.6 g, 300 mmol) in anhydrous Et_2O (50 mL). The formation of the Grignard reagent starts immediately and the remaining methyl iodide solution is added dropwise at such a rate that the Et_2O refluxes gently (ca. 1 h). After the addition is complete, the solution is heated under reflux for 30 min. Then, a solution of the ester **1.7.5.1** (20.0 g, 130 mmol) in Et_2O (50 mL) is added dropwise with stirring over 40 min. The mixture refluxes vigorously during the addition and a grey precipitate forms. Heating under reflux is continued for 2 h.

The mixture is then cooled in an ice bath and a pre-cooled solution of NH_4Cl (60 g) in H_2O (300 mL) is added. The organic layer is separated and the aqueous layer is extracted with Et_2O (2 × 50 mL). The combined organic layers are dried over Na_2SO_4 and concentrated. The residue is fractionally distilled to give the product as a colorless oil with a turpentine-like odor; 16.4 g (82%), bp_{15} 94–95 °C, n_D^{20} = 1.4790.

IR (NaCl): $\tilde{v}$ (cm^{-1}) = 3600–3200, 2980, 2940, 2850, 1450, 1390, 1375.

^{1}H NMR (200 MHz, $CDCl_3$): δ (ppm) = 5.30 (s_{br}, 1 H, vinyl-H), 2.41 (s, 1 H, OH), 2.20–1.50 (m, 7 H, CH + 3 × CH_2), 1.60 (s, 3 H, CH_3), 1.13 (s, 6 H, $C(CH_3)_2$).

1.7.5.3   **   **(2*S*)-2-Amino-3-phenylpropan-1-ol** [15]

L-Phenylalanine (54.5 g, 330 mmol) is slowly added to a stirred suspension of $LiAlH_4$ (25.0 g, 660 mmol) in anhydrous tetrahydrofuran (400 mL) over 30 min at 0 °C under an N_2 atmosphere. The mixture is warmed to room temperature and then heated under reflux with stirring for 15 h.

The solution is then cooled in an ice bath and H_2O (135 mL) is carefully added dropwise. After filtration, the filtrate is concentrated under reduced pressure and the residue is recrystallized from

THF/H$_2$O (4:1, 200 mL) to yield the phenylalaninol as light-yellow needles; 49.2 g (99%), mp 91–92 °C, $[\alpha]_D^{20}$ = –17.4 (c = 1.0, CHCl$_3$), R_f = 0.14 (EtOAc/MeOH, 10:1).

UV (CH$_3$CN): λ_{max} (nm) (lg ε) = 268.0 (2.229), 258.5 (2.363), 254.0 (2.304), 192.5 (4.473).

IR (KBr): $\tilde{\nu}$ (cm^{-1}) = 3356, 2876, 1577, 1492, 1338, 1122, 1065, 754, 698, 621, 592.

^{1}H NMR (300 MHz, CDCl$_3$): δ (ppm) = 7.32–7.17 (m, 5 H, Ph-H), 3.58 (dd, J = 10.6, 4.2 Hz, 1 H, 1-H$_b$), 3.35 (dd, J = 10.6, 7.3 Hz, 1 H, 1-H$_a$), 3.05 (m$_c$, 1 H, 2-H), 2.75 (dd, J = 13.5, 5.4 Hz, 1 H, 3-H$_b$), 2.48 (s$_{br}$, 3 H, NH$_2$, OH), 2.46 (dd, J = 13.5, 8.8 Hz, 1 H, 3-H$_a$).

^{13}C NMR (50 MHz, CDCl$_3$): δ (ppm) = 138.6, 129.1, 128.5, 126.3 (6 × Ph-C), 65.86 (C-1), 54.10 (C-2), 40.50 (C-3).

MS (DCI, 200 eV): m/z (%) = 152 (100) [M + H]$^+$, 169 (37) [M + NH$_4$]$^+$.

1.7.5.4   **   (4S)-4-Benzyloxazolidin-2-one [15]

A dry, three-necked, round-bottomed flask equipped with a thermometer, a Vigreux column, and a magnetic stirring bar is charged with the amino alcohol **1.7.5.3** (15.1 g, 100 mmol), K$_2$CO$_3$ (1.38 g, 10.0 mmol), and diethyl carbonate (29.5 g, 250 mmol). The mixture is carefully heated to 135–140 °C, and EtOH is allowed to distil as it is formed. After 2 h, 15 mL of distillate has been collected.

The reaction mixture is then diluted with CH$_2$Cl$_2$ (250 mL) and filtered. The solution is washed with saturated NaHCO$_3$ solution (100 mL), dried over MgSO$_4$, and concentrated *in vacuo*. Recrystallization from EtOAc/n-pentane gives the oxazolidinone as colorless needles; 13.3 g (75%), mp 88–89 °C, $[\alpha]_D^{20}$ = –62.5 (c = 1.0, CHCl$_3$), R_f = 0.47 (EtOAc).

UV (CH$_3$CN): λ_{max} (nm) (lg ε) = 263.5 (2.191), 258.0 (2.302), 252.5 (2.218), 206.0 (3.939).

IR (KBr): $\tilde{\nu}$ (cm^{-1}) = 1751, 1404, 1244, 1096, 1021, 942, 757, 708.

^{1}H NMR (300 MHz, CDCl$_3$): δ (ppm) = 7.37–7.17 (m, 5 H, Ph-H), 6.01 (s$_{br}$, 1 H, NH), 4.43 (m$_c$, 1 H, 5-H$_b$), 4.17–4.05 (m, 2 H, 4-H, 5-H$_a$), 2.88 (m$_{cr}$, 2 H, 1'-H$_2$).

^{13}C NMR (50 MHz, CDCl$_3$): δ (ppm) = 159.5 (C-2), 135.9, 129.0, 128.9 (6 × Ph-C), 69.52 (C-5), 53.72 (C-4), 41.33 (C-1').

MS (EI, 70 eV): m/z (%) = 177 (7) [M]$^+$, 86 (86) [M – C$_7$H$_7$]$^+$.

1.7.5.5 **   (4*S*)-3-(1-Acryloyl)-4-benzyl-oxazolidin-2-one [6]

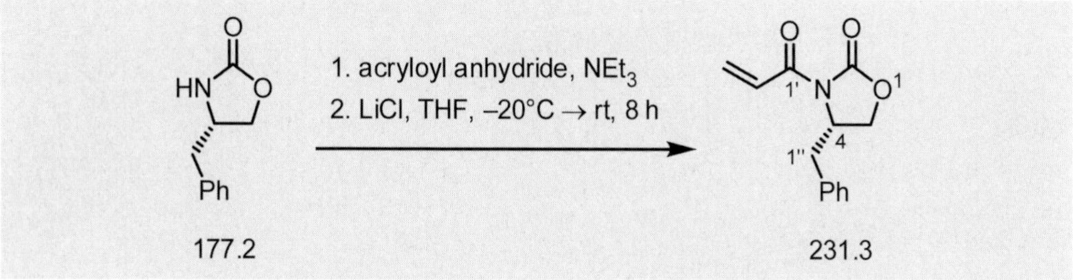

Triethylamine (13.4 g, 132 mmol, 18.4 mL) and acryloyl chloride (5.98 g, 66.0 mmol, 5.34 mL) are added to a stirred solution of acrylic acid (5.13 g, 71.2 mmol, 4.88 mL) in anhydrous tetrahydrofuran (300 mL) at –20 °C and stirring is continued at this temperature for 2 h. LiCl (2.58 g, 61.0 mmol) is added, followed by the oxazolidinone **1.7.5.4** (9.00 g, 50.8 mmol). The mixture is allowed to warm to room temperature and then stirred for 8 h.

The reaction is quenched by the addition of HCl (0.2 M, 70 mL), and the THF is removed *in vacuo*. After addition of EtOAc (100 mL), the mixture is washed with half-saturated NaHCO$_3$ solution (80 mL) and brine (80 mL), dried over Na$_2$SO$_4$, and concentrated *in vacuo*. Column chromatography of the residue on silica gel (*n*-pentane/EtOAc, 4:1) yields the acryloyloxazolidinone as colorless crystals; 9.21 g (78%), mp 74–75 °C, R_f = 0.42 (*n*-pentane/EtOAc, 4:1), $[\alpha]_D^{20}$ = +79.0 (*c* = 1.0, CHCl$_3$).

UV (CH$_3$CN): λ_{max} (nm) (lg ε) = 207.5 (4.316), 191.5 (4.658).

IR (KBr): $\tilde{\nu}$ (cm^{-1}) = 1784, 1682, 1389, 1352, 1313, 1245, 1216, 989, 696.

^{1}H NMR (300 MHz, CDCl$_3$): δ (ppm) = 7.47 (dd, *J* = 17.0, 10.5 Hz, 1 H, 2'-H), 7.37–7.21 (m, 5 H, 5 × Ph-H), 6.58 (dd, *J* = 17.0, 1.7 Hz, 1 H, 3'-H$_b$), 5.92 (dd, *J* = 10.5, 1.7 Hz, 1 H, 3'-H$_a$), 4.70 (m$_c$, 1 H, 4-H), 4.17 (m$_c$, 2 H, 5-H$_2$), 3.33 (dd, *J* = 13.3, 3.3 Hz, 1 H, 1''-H$_b$), 2.78 (dd, *J* = 13.3, 9.4 Hz, 1 H, 1''-H$_a$).

^{13}C NMR (50 MHz, CDCl$_3$): δ (ppm) = 164.8 (C-1'), 153.3 (C-2), 135.2 (C-2''), 131.9 (C-3'), 129.4 (C-2'), 128.9, 127.3 (5 × Ph-C'), 66.20 (C-5), 55.22 (C-4), 37.70 (C-1'').

MS (EI, 70 eV): *m/z* (%) = 231 (27) [*M*]$^+$, 140 (18) [*M* – CH$_2$Ph]$^+$, 55 (100) [*M* – C$_{10}$H$_{10}$NO$_2$]$^+$.

1.7.5.6   * (4*S*,1"*R*)-4-Benzyl-3-(4-methyl-cyclohex-3-enecarbonyl)oxazolidin-2-one [6]**

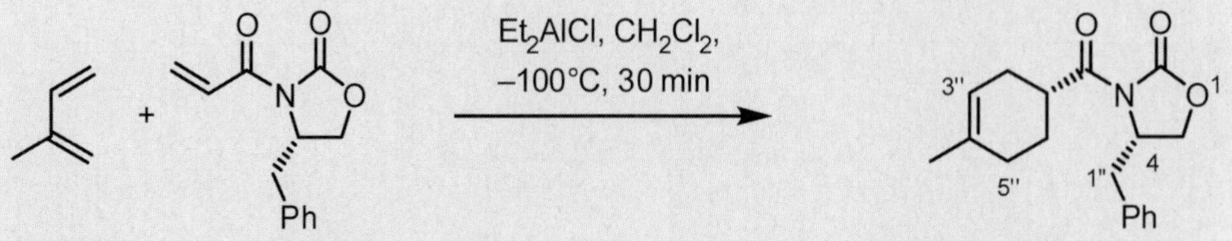

A solution of the acryloyloxazolidinone **1.7.5.5** (6.89 g, 29.8 mmol) and isoprene (70 mL) in anhydrous dichloromethane (70 mL) is cooled to –100 °C. Diethylaluminum chloride (41.7 mL, 1 M in hexanes, 41.7 mmol), cooled to –78 °C, is added via a coolable dropping funnel over a period of 10 min, whereupon the mixture turns yellow. The mixture is stirred at –100 °C for 30 min.

It is then poured into HCl (1 M, 600 mL). After the addition of CH$_2$Cl$_2$ (100 mL), the layers are separated and the aqueous phase is extracted with CH$_2$Cl$_2$ (2 × 200 mL). The combined organic layers are dried over MgSO$_4$, and the solvent is removed under reduced pressure. Purification of the residue by column chromatography on silica gel (*n*-pentane/EtOAc, 4:1) gives the product as white needles; 5.30 g (59%), mp 87–88 °C, $[\alpha]_D^{20}$ = +92.8 (*c* = 1.0, CHCl$_3$), R_f = 0.39 (*n*-pentane/EtOAc, 4:1).

UV (CH$_3$CN): λ_{max} (nm) (lg ε) = 263.5 (2.239), 257.5 (2.392), 252.0 (2.392), 247.0 (2.367).

IR (KBr): $\tilde{v}$ (cm^{-1}) = 3026, 2963, 2835, 1700, 1387, 1238, 1219, 1202.

^{1}H NMR (300 MHz, CDCl$_3$): δ (ppm) = 7.36–7.15 (m, 5 H, 5 × Ph-H), 5.40 (m, 1 H, 3"-H), 4.66 (m, 1 H, 4-H), 4.23–4.10 (m, 2 H, 5-H$_2$), 3.72–3.59 (m, 1 H, 1"-H), 3.24 (dd, *J* = 13.4, 3.2 Hz, 1 H, 1'-H$_b$), 2.75 (dd, *J* = 13.4, 9.5 Hz, 1 H, 1'-H$_a$), 2.35–1.68 (m, 6 H, 2"-H$_2$, 5"-H$_2$, 6"-H$_2$), 1.65 (s, 3 H, 4"-CH$_3$).

^{13}C NMR (75.5 MHz, CDCl$_3$): δ (ppm) = 176.5 (1"-(CO)N), 153.0 (C-2), 135.3, 133.7, 129.4, 128.9, 127.3 (5 × Ph-C, C-4"), 119.0 (C-3"), 66.00 (C-5), 55.24 (C-4), 38.41 (C-1"), 37.88 (Ph-CH$_2$), 29.42 (C-5"), 27.71 (C-2"), 25.68 (C-6"), 23.38 (4"-CH$_3$).

MS (ESI, 70 eV): *m/z* (%) = 622 (6) [2*M* + Na]$^+$, 354 (100) [*M* – H + 2Na]$^+$, 322 (25) [*M* + Na]$^+$.

1.7.5.7   ** (*R*)-4-Methyl-cyclohex-3-enecarboxylic acid methyl ester

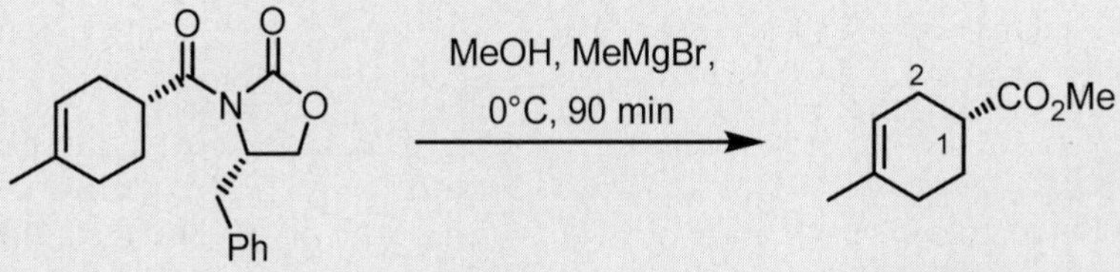

Methylmagnesium bromide (3 M in Et$_2$O, 2.3 mL, 4.68 mmol) is added dropwise to anhydrous methanol (20 mL) at 0 °C and the solution is stirred at this temperature for 5 min. A solution of the Diels–Alder adduct **1.7.5.6** (0.70 g, 2.34 mmol) in MeOH (20 mL) is then added dropwise and the mixture is stirred for 90 min.

The reaction is quenched by the addition of aqueous pH 7 phosphate buffer solution (20 mL) and stirring is continued for a further 30 min at room temperature. The mixture is diluted with half-saturated NH$_4$Cl (40 mL) and NaCl solution (40 mL), and CH$_2$Cl$_2$ (40 mL) is added. The layers are separated and the aqueous layer is extracted with CH$_2$Cl$_2$ (3 × 40 mL). The combined organic layers are dried over MgSO$_4$ and the solvent is removed at room temperature under reduced pressure. The crude product is purified by column chromatography on silica gel (*n*-pentane/Et$_2$O, 14:1) to give the methyl ester as a colorless liquid; 324 mg (90%), n_D^{20} = 1.4624, $[\alpha]_D^{20}$ = +52.2 (*c* = 2.1, CH$_2$Cl$_2$), R_f = 0.46 (*n*-pentane/EtOAc, 20:1).

UV (CH$_3$CN): λ_{max} (nm) (lg ε) = 267.0 (1.934), 263.0 (2.159), 251.5 (2.155), 257.0 (2.269), 191.5 (4.576).

IR (KBr): $\tilde{\nu}$ (cm^{-1}) = 2961, 2928, 1734, 1455, 1442, 1163, 697.

^{1}H NMR (300 MHz, CDCl$_3$): δ (ppm) = 5.41–5.34 (2 × m, 1 H, 3-H), 3.68 (s, 3 H, OCH$_3$), 2.56–2.43 (m, 1 H, 1-H), 1.62–1.78, 1.94–2.05, 2.17–2.26 (m, 9 H, 2-H$_2$, 3-H$_2$, 4-CH$_3$, 5-H$_2$).

^{13}C NMR (75.5 MHz, CDCl$_3$): δ (ppm) = 23.43 (4-CH$_3$), 25.43 (C-6), 27.62 (C-2), 29.24 (C-5), 39.03 (C-1), 51.56 (O-CH$_3$), 119.15 (C-3), 133.69 (C-4), 176.47 (C=O).

MS (EI, 70 eV): *m/z* (%) = 154 (32) [*M*]$^+$, 95 (46) [*M* – CH$_3$ – CO$_2$]$^+$, 94 (100) [*M* – CH$_3$ – CO$_2$ – H]$^+$.

1.7.5.8 ** (*R*)-(+)-α-Terpineol

A solution of the methyl ester **1.7.5.7** (209 mg, 1.36 mmol) in anhydrous Et$_2$O (10 mL) is added dropwise to a solution of methylmagnesium iodide (3 M in Et$_2$O, 1.62 mL, 4.86 mmol) in Et$_2$O (15 mL) at room temperature. The mixture is stirred for 4.5 h at this temperature (TLC control).

The mixture is then poured into saturated NH$_4$Cl solution (30 mL). The layers are separated, and the aqueous layer is extracted with Et$_2$O (5 × 20 mL). The combined organic phases are washed with H$_2$O (20 mL) and brine (20 mL) and dried over MgSO$_4$. After removal of the solvent under reduced pressure, the residue is purified by column chromatography on silica gel (*n*-pentane/Et$_2$O, 7:3) to give (*R*)-(+)-α-terpineol as a colorless oil, which crystallizes upon refrigeration; 177 mg (84%), *ee* = 90%, mp 25–26 °C, $[\alpha]_D^{20}$ = +91.1 (*c* = 1.0, CHCl$_3$), R_f = 0.27 (*n*-pentane/Et$_2$O, 7:3).

IR (KBr): $\tilde{\nu}$ (cm^{-1}) = 3600–3100, 2970, 2924, 2889, 2836, 1438, 1377, 1366, 1158, 1133.

^{1}H NMR (300 MHz, CDCl$_3$): δ (ppm) = 5.42–5.36 (m, 1 H, 3-H), 1.67–1.64 (m, 3 H, 4-CH$_3$), 1.50 (m, 1 H, 1-H), 2.13–1.72, 1.33–1.20 (2 × m, 7 H, 2-H$_2$, 5-H$_2$, 6-H$_2$, OH), 1.91, 1.17 (2 × s, 2 × 3 H, C(OH)(C$\underline{H}_3$)$_2$).

^{13}C NMR (50.3 MHz, CDCl$_3$): δ (ppm) = 133.99 (C-4), 120.51 (C-3), 72.71 (COH), 44.95 (C-1), 30.96 (C-5), 26.85 (C-2), 27.42*, 26.22* (COH(C$\underline{H}_3$)$_2$), 23.93 (C-6), 23.33 (4-CH$_3$).

MS (EI, 70 eV): m/z (%) = 154 (14) [M]$^+$, 136 (69) [M – CH$_3$]$^+$, 121 (55) [M – 2CH$_3$]$^+$.

GC:　　column: WCOT fused silica CP-Chiralsil-DEX CB (25 m × 0.25 mm)
　　　　carrier: H$_2$; temperature: 100 °C

　　　　retention time: t_{R1} = 9.14 min (minor enantiomer); t_{R2} = 9.34 min (major enantiomer)

[1] Römpp Lexikon *"Naturstoffe"* (Eds.: W. Steglich, B. Fugmann, S. Lang-Fugmann), p. 393, Georg Thieme Verlag, Stuttgart, **1997**.

[2] For the principles and rules of Diels–Alder reactions with "normal" and "inverse" electron demand, see: M. B. Smith, J. March, *March's Advanced Organic Chemistry*, 5th ed., p. 1062, John Wiley & Sons, Inc., New York, **2001**.

[3] I. Inukai, M. Kasai, *J. Org. Chem.* **1965**, *30*, 3567.

[4] I. Inukai, T. Kojima, *J. Org. Chem.* **1966**, *31*, 1121.

[5] For a comprehensive overview, see: G. Procter, *Stereoselectivity in Organic Synthesis*, p. 101, Oxford University Press, Oxford, **1998**.

[6] D. A. Evans, K. T. Chapman, J. Bisaha, *J. Am. Chem. Soc.* **1988**, *110*, 1238.

[7] D. H. Ryu, E. J. Corey, *J. Am. Chem. Soc.* **2003**, *125*, 6388.

[8] D. H. Ryu, T. W. Lee, E. J. Corey, *J. Am. Chem. Soc.* **2002**, *124*, 9992.

[9] a) Q. Gao, K. Ishihara, T. Maruyama, M. Mouri, H. Yamamoto, *Tetrahedron* **1994**, *50*, 979; b) K. Ishihara, Q. Gao, H. Yamamoto, *J. Org. Chem.* **1993**, *58*, 6917; c) S. Kiyooka, Y. Kaneko, K. Kume, *Tetrahedron Lett.* **1992**, *33*, 4927; d) K. Furuta, A. Kanematsu, H. Yamamoto, S. Takaoka, *Tetrahedron Lett.* **1989**, *30*, 7231; e) K. Furuta, Y. Miwa, K. Iwanaga, H. Yamamoto, *J. Am. Chem. Soc.* **1988**, *110*, 6254.

[10] a) E. J. Corey, K. Ishihara, *Tetrahedron Lett.* **1992**, *33*, 6807; b) E. J. Corey, N. Imai, H. Y. Zhang, *J. Am. Chem. Soc.* **1991**, *113*, 728.

[11] For reviews, see: a) E. J. Corey, *Angew. Chem.* **2002**, *114*, 1724; *Angew. Chem. Int. Ed.* **2002**, *41*, 1650; b) Y. Hayashi, *Cycloaddit. React. Org. Synth.* **2002**, 5.

[12] a) Y. Yamashita, S. Saito, H. Ishitani, S. Kobayashi, *J. Am. Chem. Soc.* **2003**, *125*, 3793; b) K. Gademann, D. E. Chavez, E. N. Jacobsen, *Angew. Chem.* **2002**, *114*, 3185; *Angew. Chem. Int. Ed.* **2002**, *41*, 3059.

[13] a) K. Juhl, K. A. Jørgensen, *Angew. Chem.* **2003**, *115*, 1536; *Angew. Chem. Int. Ed.* **2003**, *42*, 1498; b) A. B. Northrup, D. W. C. MacMillan, *J. Am. Chem. Soc.* **2002**, *124*, 2458; c) K. A. Ahrendt, C. J. Borths, D. W. C. MacMillan, *J. Am. Chem. Soc.* **2000**, *122*, 4243.

[14] K. Adler, W. Vogt, *Liebigs Ann. Chem.* **1949**, *564*, 109.

[15] C. Schneider, *Ph.D. Thesis*, University of Göttingen, **1992**.

1.7.6   Bicyclo[2.2.2]octene derivative

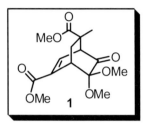

Topics:
- Oxidation of an *o*-methoxy-substituted phenol with a hypervalent iodine compound
- Diels–Alder reaction of a cyclohexa-1,3-dienone with methacrylate

(a) General

In general, hypervalent organoiodine compounds [1–3] are derived from aryl iodides **2** (oxidation level of I: +1) by oxidation at the iodine atom. Thus, **2** can be oxidized to iodoxyarenes **3** and diacylated to give (diacyloxy)iodoarenes **4** (oxidation level of I: +3):

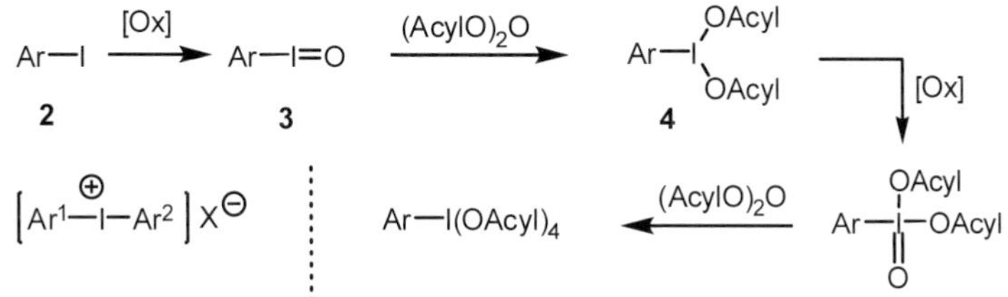

Hypervalent iodine(III) compounds like **4** can be further oxidized to iodine(V) compounds of types **6/7** represented, for example, by Dess–Martin periodinane (**8**, DMP) and *ortho*-iodoxybenzoic acid (**9**, IBX), which are important reagents for the oxidation of primary and secondary alcohols to give aldehydes and ketones, respectively (cf. **2.3.2**).

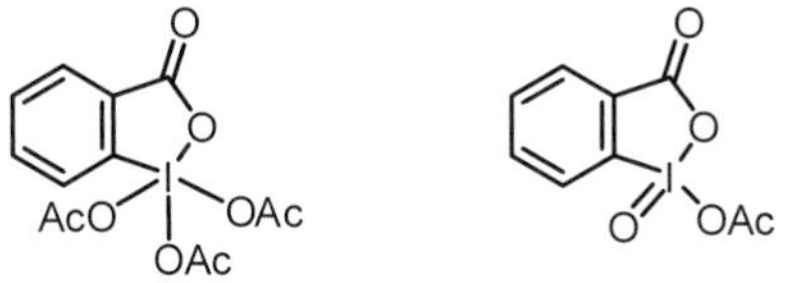

Diaryliodonium compounds **5** are another type of hypervalent iodine compounds used in synthetic chemistry [4].

The trivalent (diacyloxy)iodoarenes **4** (most frequently Ar = Ph, acyl = acetyl or trifluoroacetyl) are often applied for oxidative transformations of organic substrates, leading to the formation of C–C bonds or C–heteroatom bonds of various types [1].

In particular, the oxidation of phenols with **4** leads to a variety of synthetically useful products [2]. *ortho*-Substituted phenols and *o*- or *p*-hydroquinones afford the corresponding benzoquinones, whereas *p*-substituted phenols **10** in the presence of an external or internal nucleophile lead to the corresponding 4,4-disubstituted cyclohexa-2,5-dienones (or spirodienones) **11**:

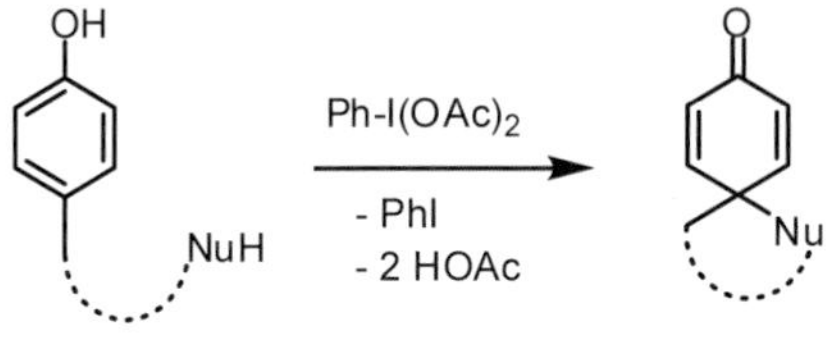

Nu = RO, halogenide anions, electron-rich arenes

Intramolecular phenol oxidations have been widely exploited for the construction of a spirodienone fragment in polycyclic systems [1], especially for the oxidative coupling of two phenolic arene units [5], as illustrated by the following example (**12 → 13**) [6]:

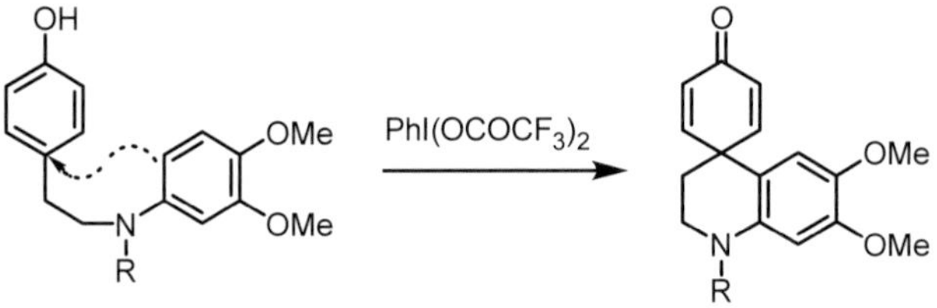

12   (R=H, CH$_3$, acyl, COOR)   **13**

An analogous type of reaction is observed when phenols **14** with a methoxy substituent in an *ortho*-position to the OH function are oxidized with PhI(OAc)$_2$ in methanol as solvent [7, 8]:

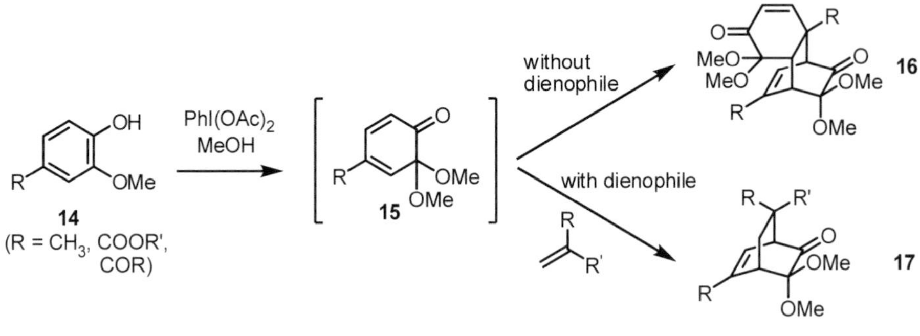

The initial products are the acetal-masked *o*-quinonoid systems **15**, which are of limited stability, but – as potential cyclohexa-1,3-dienes – can be readily trapped by electron-deficient dienophiles in a Diels–Alder reaction to give cycloadducts **17** of the bicyclo[2.2.2]octenone type. In the absence of dienophiles, compounds **15** dimerize to polycycles **16**. As shown in section (b), this remarkable phenol oxidation mediated by a hypervalent iodine source can be conducted as an efficient one-pot synthesis of highly functionalized bicyclo[2.2.2]octane derivatives [8] that are otherwise difficult to obtain.

(b) Synthesis of 1

Methyl vanillate (**18**) is oxidized with (diacetoxy)iodobenzene in the presence of an excess of methyl methacrylate (**20**) in methanol at room temperature. The bicyclo[2.2.2]octen-2-one **1** is obtained (54% overall yield) in a clean, regio- and stereoselective [4+2]-cycloaddition of the dienophile to cyclohexa-2,4-dienone **19** formed *in situ* by oxidation of the electron-rich phenolic substrate **18**.

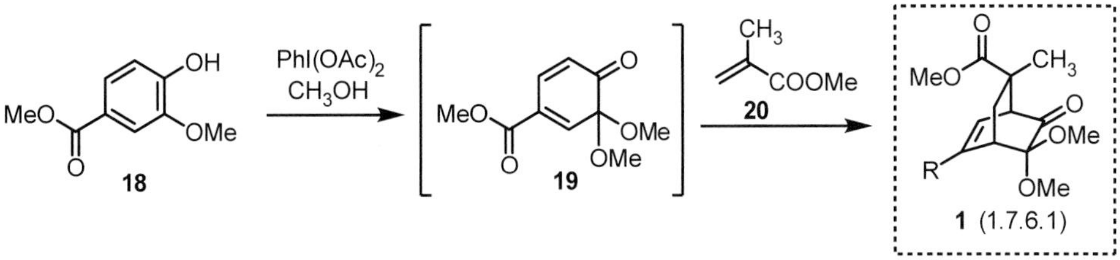

The observed regio- and stereoselectivity of the Diels–Alder reaction **19** + **20** → **1** can be explained in terms of frontier molecular orbital theory [8].

For the oxidative transformation **18** → **19**, two mechanistic alternatives are reasonable [2]. In mechanism A, the phenol **18** is attached to the iodine(III) of PhI(OAc)$_2$ by ligand exchange with extrusion of HOAc to give an intermediate **21**. This undergoes redox disproportionation in an addition/elimination process by attack of CH$_3$OH, resulting in the formation of the cyclohexa-2,4-dienone **19**, iodobenzene, and HOAc. In mechanism B, the phenol **18** is oxidized in a two-electron/one-proton transfer — presumably via a phenoxy radical — to give the (resonance-stabilized) carboxenium ion **22**, which is trapped by addition of CH$_3$OH and loss of a second proton to afford the product **19**. Concomitantly, PhI(OAc)$_2$ is reduced to iodobenzene with formation of two molecules of acetate.

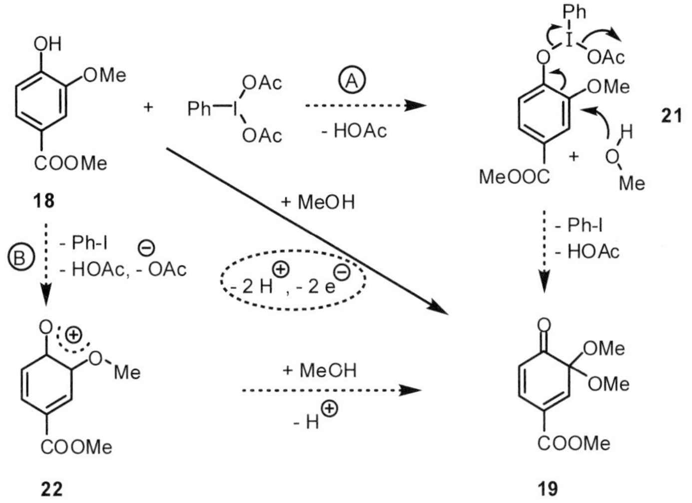

(c) Experimental procedure for the synthesis of 1

1.7.6.1   **   (1*R,4*S**,7*S**)-3,3-Dimethoxy-5,7-bis(methoxycarbonyl)-7-
           methylbicyclo[2.2.2]oct-5-en-2-one [8]**

| 182.2 | 322.1 | 212.2 | 100.1 | 312.3 |

A solution of methyl vanillate (1.00 g, 5.49 mmol) in anhydrous MeOH (70 mL) is added over a period of 8 h by means of a syringe pump to a solution of (diacetoxy)iodobenzene (2.12 g, 6.58 mmol) and methyl methacrylate (14.5 ml, 13.7 g, 137 mmol) in anhydrous MeOH (30 mL) at room temperature under nitrogen atmosphere. Stirring is continued for 2 h.

The solvent, excess dienophile, and other volatile products are evaporated under reduced pressure. Purification by flash chromatography (EtOAc/*n*-hexane, 9:1) gives the product as a colorless liquid; 918 mg (54%); R_f = 0.57 (EtOAc/*n*-hexane, 9:1).

IR (film): $\tilde{\nu}$ (cm^{-1}) = 2975, 1727.

^{1}H NMR (200 MHz, CDCl$_3$): δ (ppm) = 7.09 (dd, J = 6.5, 1.7 Hz, 1 H, C(6)H), 3.65–3.71 (m, 1 H, C(4)H), 3.74 (s, 3 H, C(5)OCH$_3$ or C(7)OCH$_3$), 3.64 (s, 3 H, C(5)OCH$_3$ or C(7)OCH$_3$), 3.48 (d, J = 6.5 Hz, 1 H, C(1)H), 3.34 (s, 3 H, C(3)OCH$_3$), 3.26 (s, 3 H, C(3)OCH$_3$), 2.19 (dd, J = 18.1, 3.1 Hz, 1 H, C(8)H), 1.93 (dd, J = 18.1, 2.2 Hz, 1 H, C(8)H), 1.31 (s, 3 H, C(7)CH$_3$).

^{13}C NMR (75 MHz, CDCl$_3$): δ (ppm) = 201.0 (C-2), 175.7 (C=O), 164.3 (C=O), 137.4 (C-5), 137.4 (C-5), 93.4 (C-3), 57.3 (C-8), 52.4 (C-4), 51.9 (OCH$_3$), 50.0 (H-1), 49.8 (C3-OCH$_3$), 46.7 (C3-OCH$_3$), 38.4 (C-8), 25.4 (C-7).

MS (DCI, NH$_3$, 200 eV): 643 [2M+NH$_4$]$^+$, 330 [2M+NH$_4$]$^+$.

[1] Hypervalent iodine compounds in synthesis: T. Wirth, *Angew. Chem.* **2005**, *117*, 3722; *Angew. Chem. Int. Ed.* **2005**, *44*, 3656.

[2] V. V. Zhdankin, P. J. Stang, *Chem. Rev.* **2002**, *102*, 2523.

[3] For a review on the preparation of hypervalent organoiodine compounds, see: A. Varvoglis, *Top. Curr. Chem.* **2002**, *224*, 69.

[4] For instructive examples on the synthetic utility of diaryliodonium compounds, see ref. [2] and L. F. Tietze, Th. Eicher, *Reaktionen und Synthesen im organisch-chemischen Praktikum und Forschungslaboratorium*, 2nd ed., p. 315, Georg Thieme Verlag, Stuttgart, **1991**.

[5] Phenol oxidation in alkaloid biogenesis: M. Hesse, *Alkaloide*, p. 265, Wiley-VCH, Weinheim, **2000**; see also: J. Fuhrhop, G. Penzlin, *Organic Synthesis*, 2nd ed., p. 293, VCH, Weinheim, **1994**.

[6] a) Compound **13** is a central intermediate in a synthesis of the *Amaryllidacea* alkaloid maritidine: Y. Kita, T. Takada, M. Gyoten, H. Tohma, M. H. Zenk, J. Eichhorn, *J. Org. Chem.* **1996**, *61*, 5857. b) A spirodienone of type **11** is a key intermediate in a synthesis of the anti-Alzheimer agent galanthamine: D. Krikorian, V. Tarpanov, S. Parushev, P. Mechkarova, *Synth. Commun.* **2000**, *30*, 2833.

[7] C.-S. Chu, T.-H. Lee, C.-C. Liao, *Synlett* **1994**, 635.

[8] C.-C. Liao, C.-S. Chu, T.-H. Lee, P. D. Rao, S. Ko, L.-D. Song, H.-C. Shiao, *J. Org. Chem.* **1999**, *64*, 4102.

1.8 Radical reactions

1.8.1 Ethyl 4,6,6,6-tetrachloro-3,3-dimethylhexanoate

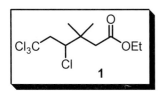

Topics:
- Claisen orthoester reaction, [3,3]-sigmatropic rearrangement
- Telomerization by radical addition of CCl_4 to an alkene

(a) General

The target molecule **1** is the key intermediate in a synthesis of the (dichlorovinyl)cyclopropane carboxylic acid **2**. Esterification of **2** with (3-phenoxy)benzyl alcohol leads to **3**, which is an important insecticide (Permethrin, cf. **4.2.1**). **3** was developed as an analogue of esters of chrysanthemic acid, a group of natural insecticides mainly isolated from the flowers of an aromatic plant of the genus Tannacetum (formerly Chrysanthemum or Pyrethrum).

For the retrosynthesis of **1**, two considerations must be taken into account: (1) the left-hand part of **1** may result from addition of CCl_4 to a C=C double bond, and (2) in the resulting unsaturated ester **4**, the C=C and C=O functionalities are in a 1,5-arrangement and are therefore susceptible to a [3,3]-sigmatropic transformation according to a retro-oxa-Cope rearrangement (**A**) [1]:

Thus, the retrosynthesis of **4** leads to the allylic alcohol **6** and triethyl orthoacetate via **7** and **5** as a simple approach to the γ,δ-unsaturated ester **4** on the basis of a [3,3]-sigmatropic rearrangement. The synthesis of **1** along these lines is described in detail.

(b) Synthesis of 1

In the first step, the ester **4** is prepared by a Claisen orthoester reaction of 3-methyl-2-buten-1-ol (**6**) with triethyl orthoacetate in the presence of phenol [2]:

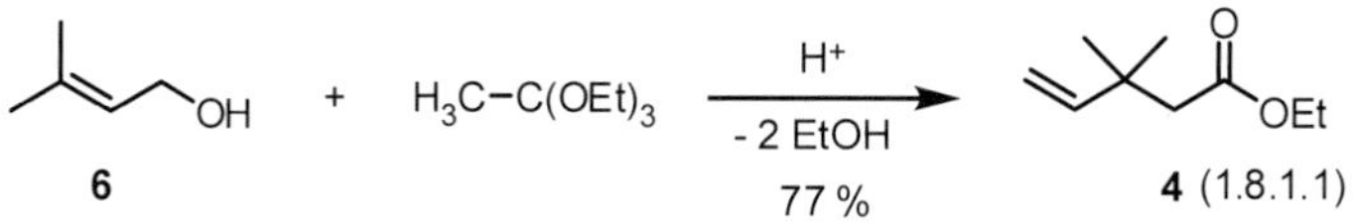

First, one molecule of EtOH in the orthoacetate is exchanged by the allylic alcohol **6** (→ **7**), and then elimination of a second EtOH transforms the orthoester **7** to the ketene acetal **5** [3]; both reactions require H⁺-catalysis. The allyl vinyl ether functionality in **5** is capable of a [3,3]-sigmatropic rearrangement (oxa-Cope reaction; a related rearrangement is the Carroll reaction in **1.5.3**), leading directly to the γ,δ-unsaturated ester **4**.

It should be noted that, as a consequence of a highly ordered chair-like transition state, the oxa-Cope process (e.g., **5** → **4**) can be conducted with high stereoselectivity and transfer of stereogenic information from the substrates to the product. This is exemplified by an instructive example [4] describing the formation of the unsaturated ester **12** with (S)-E-stereochemistry from stereodifferent precursors, namely the (R)-Z-alcohol **8** and the (S)-E-alcohol **9**, by reaction with orthoacetate/propionic acid via the intermediates **10/11** [5]:

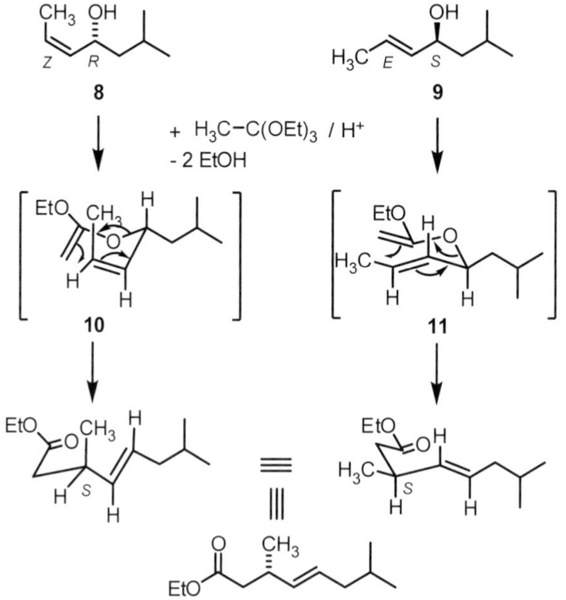

12 (100 % E, ~95 % (S),
~98 % transfer of chirality)

In the second step, CCl$_4$ is added to the unsaturated ester **4** in the presence of dibenzoyl peroxide (DBPO) to yield **1**:

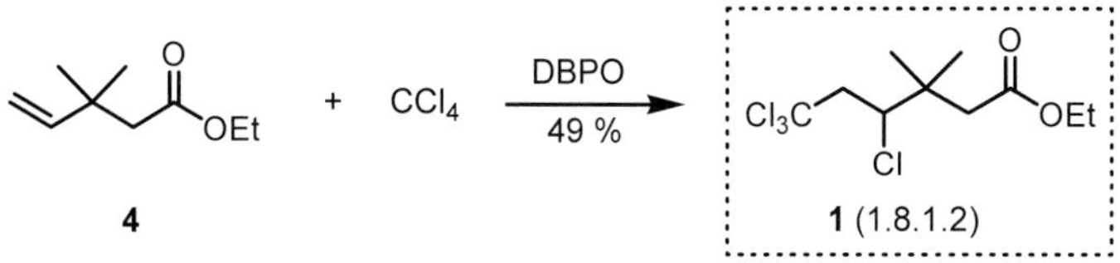

The reaction proceeds by a radical chain process initiated by DBPO:

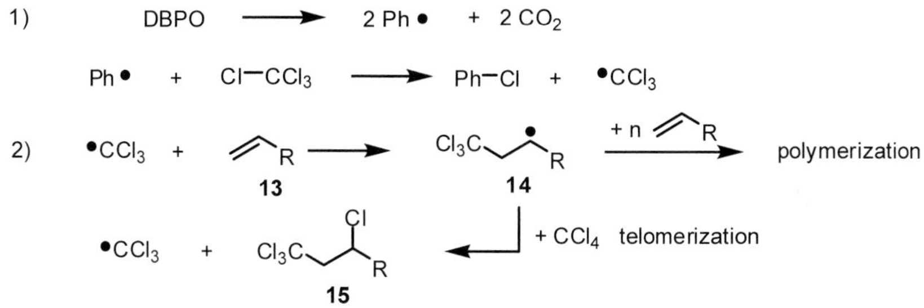

First, DBPO is cleaved thermally to give a phenyl radical, which generates a •CCl$_3$ radical from CCl$_4$. In the chain propagation reaction (2), the •CCl$_3$ radical adds to the terminal carbon atom of the olefinic substrate (**13**) to generate the secondary radical **14**. This radical may either lead to a polymerization of the olefinic substrate or abstract a chlorine atom from CCl$_4$, thus perpetuating chain propagation with formation of the addition product **15** (telomerization [6]). The competition between telomerization and polymerization is controlled by steric factors in the radical intermediate and the olefinic substrate. Increased steric hindrance favors telomerization.

Thus, the target molecule **1** is obtained in a two-step sequence with an overall yield of 38%.

(c) Experimental procedures for the synthesis of 1

1.8.1.1   **   Ethyl 3,3-dimethyl-4-pentenoate [2]

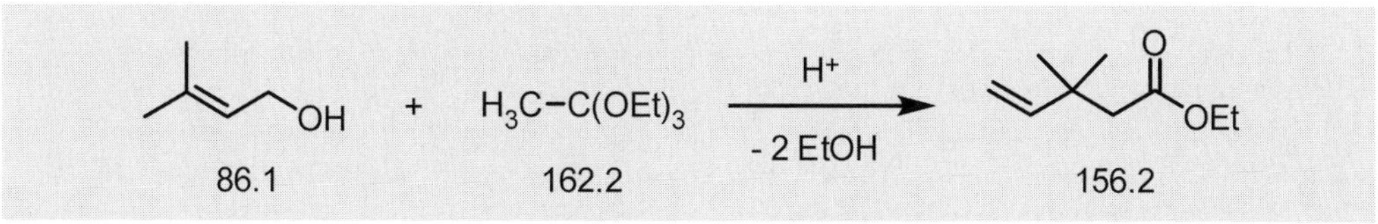

A mixture of 3-methyl-2-buten-1-ol (bp 140 °C; 43.1 g, 0.50 mol), ethyl orthoacetate (distilled, bp 144–146 °C; 97.3 g, 0.60 mol), and phenol (Caution! Irritant! 7.00 g, 74.4 mmol) is heated to 135–140 °C for 10 h with continuous removal of the EtOH formed.

The mixture is then cooled, diluted with Et$_2$O (200 mL), washed sequentially with HCl (1 N, 2 × 100 mL) to hydrolyze the excess of orthoacetate, saturated aqueous NaHCO$_3$ solution, and brine, dried over MgSO$_4$, and concentrated. The residue is distilled over a short column. The yield is 60.4 g (77%), bp$_{11}$ 57–60 °C.

> **IR** (film): $\tilde{v}$ (cm^{-1}) = 3090 (CH, olef.), 1740 (C=O), 1640 (C=C), 1370, 1240.
>
> **^{1}H NMR** (CDCl$_3$): δ (ppm) = 5.90 (dd, J = 18.5, 10.0 Hz, 1 H, C=CH–), 5.15–4.7 (m, 2 H, CH$_2$=C), 4.07 (q, J = 7.0 Hz, 2 H, OCH$_2$), 2.25 (s, 2 H, CH$_2$), 1.20 (t, J = 7 Hz, 3 H, CH$_3$), 1.13 (s, 6 H, CH$_3$).

1.8.1.2   **   Ethyl 4,6,6,6-tetrachloro-3,3-dimethylhexanoate [2]

Ethyl 3,3-dimethyl-4-pentenoate **1.8.1.1** (23.4 g, 150 mmol) and dibenzoyl peroxide (Caution! Explosive! 25% H$_2$O, 2.40 g) in tetrachloromethane (200 mL, Caution!) are heated under reflux for 8 h using a Dean–Stark trap. An additional 2.40 g of moist dibenzoyl peroxide is added and refluxing with removal of H$_2$O is continued for 8 h.

The solution is then cooled, washed twice with ice-cold NaOH (1 N, to remove benzoic acid) and three times with brine. The organic phase is dried over Na$_2$SO$_4$, concentrated, and the residue is distilled *in vacuo* over a short column. The yield is 22.8 g (49%), bp$_{0.2}$ 132–138 °C.

> **IR** (film): $\tilde{v}$ (cm^{-1}) = 2980 (CH), 1730 (C=O), 1465, 1370, 720, 690 (CCl).
>
> **^{1}H NMR** (CDCl$_3$): δ (ppm) = 4.43 (dd, J = 8.0, 3.3 Hz, 1 H, CH–Cl), 4.12 (q, J = 7.0 Hz, 2 H, O–CH$_2$), 3.19 (d, J = 3.3 Hz, 1 H, Cl$_3$C–CH), 3.13 (d, J = 8 Hz, 1 H, Cl$_3$C–CH), 2.66 (d, J = 15.0 Hz, 1 H, CH–CO$_2$Et), 2.26 (d, J = 15.0 Hz, 1 H, CH–CO$_2$Et), 1.24 (t, J = 7.0 Hz, 3 H, O–C–CH$_3$), 1.20 (s, 3 H, CH$_3$), 1.13 (s, 3 H, CH$_3$).

[1]   Other retrosynthetic pathways, e.g. **B** leading to isobutanal and oxirane as educts of a possible synthesis of **1**, are less favorable (criterion of simplicity!).

[2]   Sagami Chemical Research Center, Tokyo, DOS 2 539 895; G. Künast, Bayer AG, private communication, 1981.

[3]   a) M. B. Smith, J. March, *March's Advanced Organic Chemistry*, 5th ed., p. 1451, John Wiley & Sons, Inc., New York, **2001**; b) for a review on Claisen rearrangement, see: A. M. Martin Castro, *Chem. Rev.* **2004**, *104*, 3037.

[4]   K.-K. Chan, N. Cohen, J. P. DeNoble, A. C. Specian, Jr., G. Saucy, *J. Org. Chem.* **1976**, *41*, 3497.

[5]   The chair-like transition states of the pericyclic transformations [(*R*)-*Z*-**10** → (*S*)-*E*-**12** ← (*S*)-*E*-**11**] should be favored by virtue of having the smallest number of nonbonding interactions, i.e. pseudoaxial substituents [4].

[6]   See ref. [3a], p. 977.

1.8.2   3-Bromophenanthrene

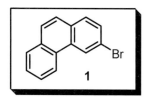

1

Topics:
- Meerwein arylation (radical addition of arenes to activated alkenes)
- 1,2-Elimination
- Photoisomerization of *trans*-stilbenes to *cis*-stilbenes, electrocyclization of *cis*-stilbenes to dihydrophenanthrenes and their dehydrogenation to phenanthrenes

(a) General

For the synthesis of phenanthrenes, three methods are of preparative importance [1].

(1) In the Pschorr phenanthrene synthesis [2], an *o*-amino-*cis*-stilbene carboxylic acid **2** is diazotized, and the resulting diazonium salt is reductively dediazoniated with Cu and cyclized to give a phenanthrene-9-carboxylic acid **3**, which is thermally decarboxylated to a phenanthrene **4**. For the cyclization step, a radical mechanism is likely, in analogy to the Gomberg–Bachmann arylation [2]. The Pschorr method can not be used for the synthesis of **1** [3].

$$
\begin{array}{cccc}
\textbf{2} & \xrightarrow[\substack{-\,N_2 \\ -\,H^+}]{\substack{1)\ HNO_2 \\ 2)\ Cu}} & \textbf{3} & \xrightarrow{-\,CO_2}\ \textbf{4}
\end{array}
$$

2   **3**   **4**

$$
\textbf{5} \xrightarrow[H_2SO_4]{K_2S_2O_8} \textbf{6} \xrightarrow[\substack{+\,HI \\ -\,I_2}]{\Delta}
$$

5   **6**

(2) In an oxidative cyclization [4], α-aryl-*o*-iodocinnamic acids **5** are converted to phenanthrenes by reaction with $K_2S_2O_8$. First, cyclic iodonium salts (iodepinium salts) **6** are formed, which on thermolysis lead to phenanthrene carboxylic acids **3**. These compounds can be decarboxylated to give phenanthrenes **4** as described above. The intermediacy of an iodoso arene species as the initial oxidation product of **5** and its S_EAr_i cyclization to **6** is established, but the mechanism of the cyclization of **6** is not known.

(3) In a photochemical domino process [5], *trans*-stilbenes **7** are photoisomerized to give the corresponding *cis*-stilbenes **8**, which undergo 6π-electrocyclization to dihydrophenanthrenes **9** followed by *in situ* dehydrogenation to phenanthrenes **4**:

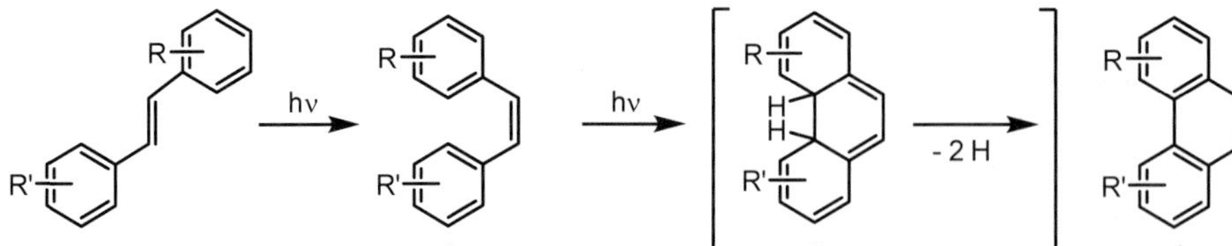

7   8   9   4

The target molecule **1** has been synthesized by application of methods (2) and (3). In section (b), a synthesis based on the photochemical cyclization route [6] is described.

(b) Synthesis of 1

trans-4-Bromostilbene (**12**), the required starting material, is prepared in a two-step sequence from 4-bromoaniline (**10**) by diazotization with HNO_2 and reaction of the diazonium salt with styrene in the presence of Cu(II) chloride in aqueous acetone to afford the bibenzyl derivative **11**:

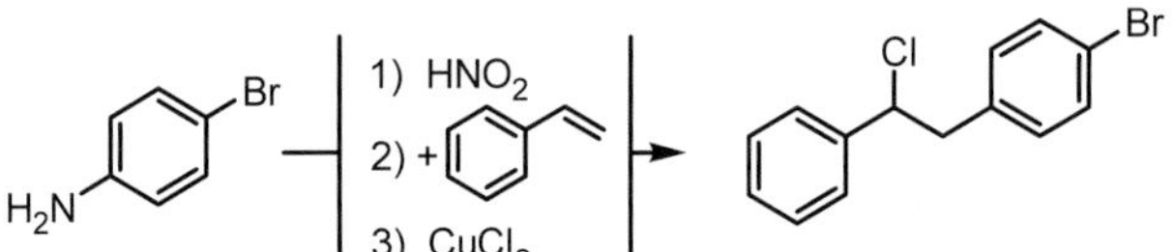

10   11 (1.8.2.1)

1) HNO_2
2) +
3) $CuCl_2$
57 %

The addition of aryl diazonium salts to activated alkenes (besides styrene, acrylonitrile and acrylates are often used) proceeds with loss of N_2 and is catalyzed by Cu(I) (Meerwein arylation [7]). The Meerwein arylation follows a radical chain mechanism related to the Cu(I)-induced Sandmeyer reaction:

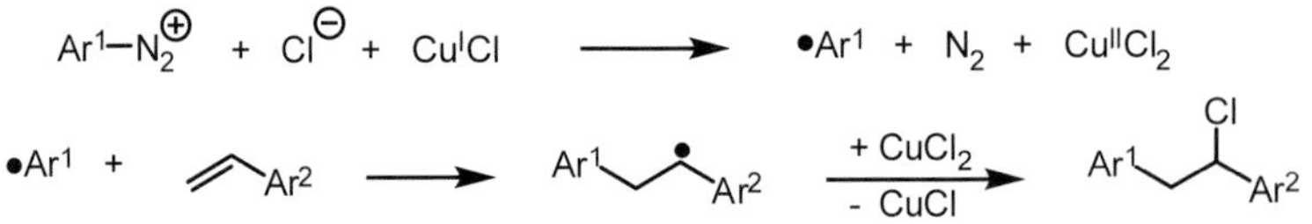

$Ar^1-N_2^{\oplus}$ + $Cl^{\ominus}$ + Cu^ICl ⟶ •Ar^1 + N_2 + $Cu^{II}Cl_2$

•Ar^1 + ⟶ Ar^1 •Ar^2 $\xrightarrow[- CuCl]{+ CuCl_2}$ Ar^1 Ar^2

Dehydrochlorination of the bibenzyl **11** with NaOEt/HOEt leads to *trans*-4-bromostilbene (**12**). This is subjected to photolysis in cyclohexane solution in the presence of iodine, resulting in isomerization to the corresponding *cis*-stilbene, electrocyclic ring-closure to a dihydrophenanthrene, and subsequent dehydrogenation to give the desired 3-bromophenanthrene (**1**).

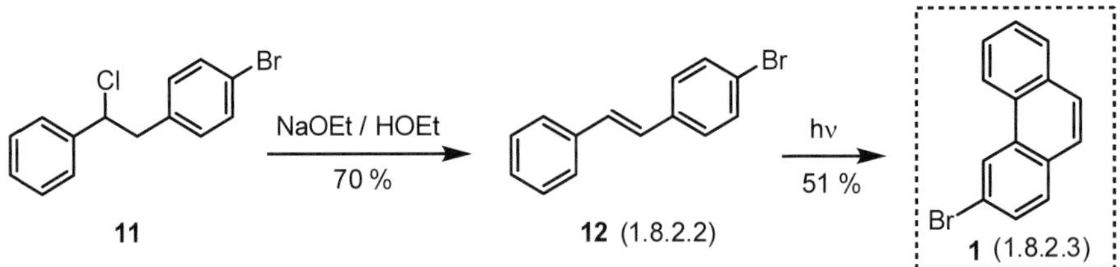

11   12 (1.8.2.2)   1 (1.8.2.3)

NaOEt / HOEt
70 %

hν
51 %

Using this procedure, the target molecule **1** is obtained in a three-step sequence in an overall yield of 20% (based on 4-bromoaniline).

(c) Experimental procedure for the synthesis of 1

1.8.2.1   *   2-(4-Bromophenyl)-1-chloro-1-phenylethane [8]

| 172.0 | 1) HNO₂ 2) + styrene: 104.1 3) CuCl₂x 2 H₂O | 295.6 |

Sodium nitrite (7.00 g, 0.10 mol) in water (35 mL) is added dropwise to a stirred solution of *p*-bromoaniline (17.2 g, 0.10 mol) in hydrochloric acid (5 M, 60 mL) with cooling in an ice-salt bath so as to maintain the reaction temperature below 5 °C (note 1). The solution is brought to pH 4–5 by the addition of solid $NaHCO_3$ (14.3 g) in portions. The solution is then added dropwise over 10 min to a solution of styrene (10.4 g, 0.10 mol; note 2) and $CuCl_2 \cdot 2H_2O$ (4.00 g, 25.0 mmol) in acetone (100 mL). Hydrogen evolution starts slowly, becomes vigorous after ca. 1 h, and ends after 15 h.

Et_2O (100 mL) is added, the dark organic phase is separated, and the aqueous phase is extracted with Et_2O (2 × 50 mL). The combined ethereal phases are dried over $MgSO_4$ and concentrated. The residual brown oil is crystallized by the addition of the minimum volume of petroleum ether (50–70 °C): 16.9 g (57%), mp 81–82 °C. Recrystallization from EtOH gives light-brown needles, mp 87–88 °C.

IR (KBr): $\tilde{\nu}$ (cm^{-1}) = 1585, 800 (*p*-disubst. benzene), 765, 690 (monosubst. benzene).

¹H NMR (CDCl₃): δ (ppm) = 7.30, 6.88 (d, *J* = 8.0 Hz, 2 H, 4-bromophenyl-H), 7.25 (s, 5 H, arom. H), 4.88 (t, *J* = 6.0 Hz, 1 H, Cl–CH), 3.24 (d, *J* = 6.0 Hz, 2 H, CH₂).

UV (CH₂Cl₂): λ_{max} (log ε) = 316 (3.57), 330 nm (sh).

Notes: (1) Completeness of the reaction is tested with starch-iodide paper; the presence of HNO₂ results in a blue color.

(2) Styrene is distilled from hydroquinone, bp₁₂ 33–34 °C, and is stored with hydroquinone.

1.8.2.2   *   *trans*-4-Bromostilbene [8]

| 295.6 | NaOEt / HOEt - HCl | 259.1 |

A sodium ethoxide solution is prepared by dissolving sodium (1.15 g, 0.05 mol) in anhydrous EtOH (50 mL) and then compound **1.8.2.1** (5.90 g, 20.0 mmol) is added with stirring. The suspension is warmed on a steam bath until dissolution is complete. After ca. 2 min, a fine precipitate of sodium chloride begins to form. After ca. 6 min, a voluminous precipitate of the product forms. The mixture is heated under reflux with vigorous stirring for 1 h.

H$_2$O (5 mL) is added to the hot solution, the mixture is cooled with stirring in an ice bath, and the precipitate is collected by filtration and washed with EtOH (10 mL). The crude product (ca. 5.5 g) is recrystallized from isopropanol (decolorizing with activated charcoal) to give colorless needles; 3.60 g (70%), mp 137–138 °C, TLC: single spot (silica gel; CH$_2$Cl$_2$).

IR (KBr): $\tilde{\nu}$ (cm^{-1}) = 1580, 820 (*p*-disubst. benzene), 750, 700, 690 (monosubst. benzene).

^{1}H NMR (CDCl$_3$): δ (ppm) = 7.5–7.05 (m, 9 H, arom. H), 7.05–6.8 (m, 2 H, vinyl-H).

UV (CH$_2$Cl$_2$): λ_{max} (log ε) = 314 (4.52), 427 nm (sh).

1.8.2.3   **   3-Bromophenanthrene [6]

Apparatus: Photolysis apparatus with quartz filter and high-pressure mercury vapor lamp (Philips HPK-125 W or Hanau TQ-150 W).

A solution of *trans*-4-bromostilbene **1.8.2.2** (2.60 g, 10.0 mmol) and iodine (0.13 g, 1.00 mmol) in anhydrous cyclohexane (1000 mL) is irradiated for 16 h while air is passed through it (note).

The solvent is evaporated *in vacuo*, the red residue is dissolved in cyclohexane (50 mL), and the solution is filtered through neutral Al$_2$O$_3$ (25 g, activity grade I). The colorless filtrate is concentrated and the residue (1.35 g, mp 76–78 °C) is recrystallized from EtOH to give colorless needles; 1.30 g (51%), mp 83–84 °C, TLC: single spot (silica gel; CH$_2$Cl$_2$).

IR (KBr): $\tilde{\nu}$ (cm^{-1}) = 1580, 840, 820, 730.

^{1}H NMR (CDCl$_3$): δ (ppm) = 8.8–8.2 (m, 2 H, 2-H/4-H), 7.9–7.2 (m, 7 H, arom. H).

UV (CH$_2$Cl$_2$): λ_{max} (log ε) = 298 (4.17), 286 (4.05), 277 (4.18), 268 (sh), 254 nm (4.95).

Note: Passing oxygen instead of air through the reaction mixture and prolongation of the irradiation time do not improve the yield.

[1]   Another method often mentioned in textbooks, the Haworth synthesis, seems to be of limited scope and is only applicable to phenanthrene itself and some alkyl-substituted derivatives: Beyer-Walter, *Lehrbuch der Organischen Chemie*, 23rd ed., p. 686, S. Hirzel Verlag, Stuttgart, **1998**.

[2]   a) R. A. Abramovitch, *Adv. Free-Radical Chem.* **1966**, *2*, 87; b) R. I. Duclos, Jr., J. S. Tung, H. Rappoport, *J. Org. Chem.* **1984**, *49*, 5243.

[3]   W. E. Bachmann, C. H. Boatner, *J. Am. Chem. Soc.* **1936**, *58*, 2194.

[4]   A. N. Nesmeyanov, T. P. Tolstaya, L. N. Vanchicova, A. V. Petrakov, *Izvestiya Akademii Nauk SSSR, Seriya Kimicheskaya* **1980**, 2530.

[5]   a) F. B. Mallory, C. W. Mallory, *Org. React.* **1984**, *30*, 1; b) W. H. Learhoven, *Org. Photochem.* **1989**, *10*, 163.

[6]   C. S. Wood, F. B. Mallory, *J. Org. Chem.* **1964**, *29*, 3373.

[7]   C. S. Rondestved, Jr., *Org. React.* **1976**, *24*, 225.

[8]   K. G. Tashchuk, A. V. Dombrovski, *Zh. Org. Khim.* **1965**, *1*, 1995; *Chem. Abstr.* **1966**, *64*, 9617

2   Oxidation and reduction

2.1   Epoxidation of C=C bonds

Epoxides (oxiranes) can be used as substrates for a broad spectrum of ring-opening transformations leading to 1,2-disubstituted functionalized alkanes. Therefore, the enantioselective epoxidation of alkenes is of considerable interest in organic synthesis since it allows the generation of stereodefined sp³-carbon centers [1]. Two methods are of general applicability, namely the Sharpless–Katsuki epoxidation [2] (**2.1.1**) and the Jacobsen epoxidation [3a] (**2.1.2**).

2.1.1   Sharpless–Katsuki epoxidation

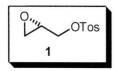

1

Topics: • Asymmetric epoxidation of an allylic alcohol according to Sharpless–Katsuki

• Tosylation of an alcoholic OH group

(a) General

In the Sharpless–Katsuki epoxidation, allyl alcohols **2** are stereoselectively transformed to oxiranes **3** by reaction with a stoichiometric amount of a hydroperoxide (commonly *tert*-butyl or cumene hydroperoxide) as oxidant, in the presence of catalytic amounts of Ti(O*i*Pr)₄ and a dialkyl tartrate, usually DET or DIPT, as a chiral ligand:

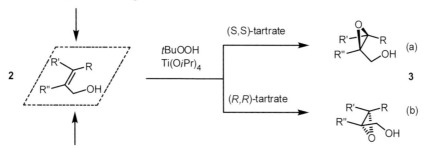

In general, both enantiomers (**a** and **b**) of the oxiranes **3** can be obtained with *ee* values >90% by using either (*R,R*)- or (*S,S*)-dialkyl tartrates. It should be pointed out that:

• allylic alcohol moieties can be epoxidized chemoselectively in the presence of other olefinic double bonds;

• the Sharpless–Katsuki protocol can also be employed to homoallylic alcohols, albeit with lower selectivity;

• racemic mixtures of chiral allylic alcohols may undergo kinetic resolution when subjected to Sharpless–Katsuki epoxidation.

Reactions and Syntheses in the Organic Chemistry Laboratory. L. F. Tietze, Th. Eicher, U. Diederichsen, A. Speicher
Copyright © 2007 WILEY-VCH Verlag GmbH & Co. KGaA, Weinheim
ISBN: 978-3-527-31223-8

As a mechanism of the Sharpless–Katsuki epoxidation, the intermediate formation of a chiral Ti-tartrate complex **4** is proposed [4], which can undergo a substrate–catalyst interaction through coordinative binding to the allylic hydroxyl group, thus allowing a transfer of chiral information to the substrate.

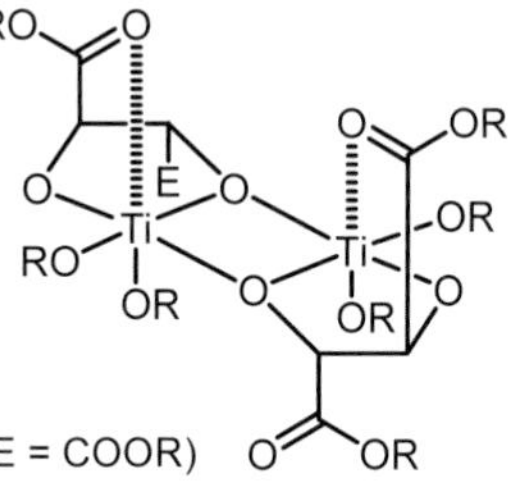

In section (b), the Sharpless–Katsuki epoxidation of allyl alcohol is described [5], which is the key step in a synthesis of (*S*)-propranolol (cf. **3.1.1**).

(b) Synthesis of 1

Allyl alcohol (**5**) is reacted with cumene hydroperoxide in the presence of catalytic amounts of Ti(O*i*Pr)$_4$ and *D*-(–)-diisopropyl tartrate at a temperature below 0 °C in dichloromethane as solvent. The excess hydroperoxide is reduced with trimethyl phosphite. The formed (*S*)-glycidol (**6**) is tosylated *in situ* at the OH function in the presence of triethylamine. Thus, the epoxytosylate **1** is obtained in a one-pot procedure from allyl alcohol.

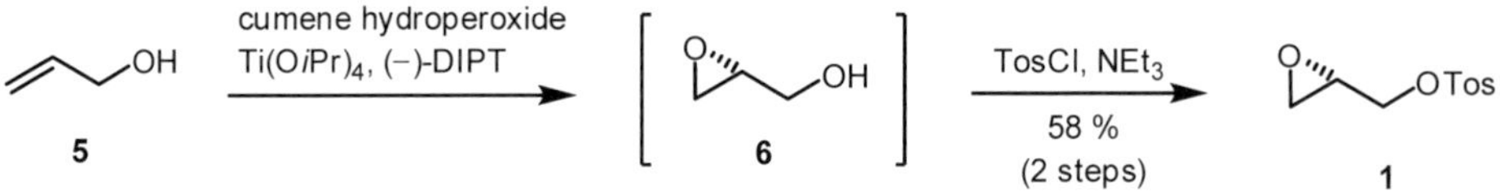

(c) Experimental procedure for the synthesis of 1

2.1.1.1   **   (*S*)-(*O*-Toluenesulfonyl)glycidol [5]

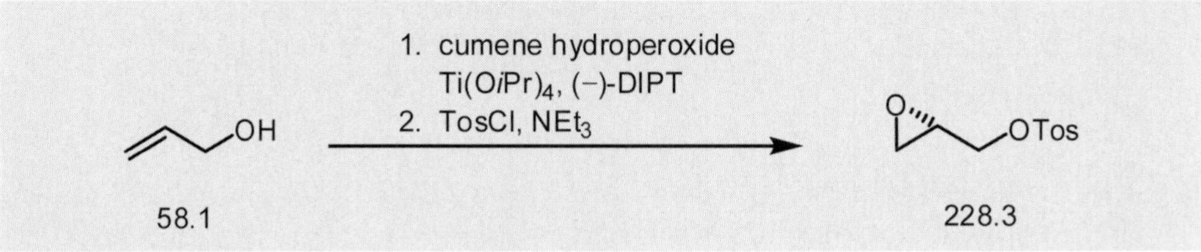

D-(–)-Diisopropyl tartrate (1.21 g, 5.17 mmol) in CH$_2$Cl$_2$ (1.3 mL) is added under nitrogen to activated 3 Å molecular sieves (3 g) in CH$_2$Cl$_2$ (164 mL). Allyl alcohol (5.86 mL, 5.01 g, 86.2 mmol) is then added and the mixture is cooled to –5 °C. Ti(O*i*Pr)$_4$ (1.29 mL, 1.23 g, 4.30 mmol) is then added and the resulting mixture is stirred for 30 min. Precooled (ice bath) cumene hydroperoxide (80%, 30.2 mL, 172 mmol) is added over a period of 45 min, maintaining the internal temperature below –2 °C. Finally, the reaction mixture is stirred vigorously at –5 to 0 °C for 6 h.

After cooling to –20 °C, the excess hydroperoxide is reduced by slow addition of P(OMe)$_3$ (12.7 g, 86.4 mmol) while maintaining the temperature below –10 °C (monitoring by TLC, EtOAc/ n-hexane, 2:3). After 1 h, triethylamine (15.1 mL, 11.0 g, 109 mmol) and thereafter a solution of p-toluenesulfonyl chloride (17.3 g, 90.4 mol) in CH$_2$Cl$_2$ (80 mL) are added. The flask is kept overnight at –20 °C.

The reaction mixture is gradually warmed to room temperature and then filtered through a pad of Celite, which is washed with CH$_2$Cl$_2$. The yellow filtrate is washed sequentially with 10% tartaric acid (2 × 250 mL) and brine (2 × 250 mL), dried over MgSO$_4$, and concentrated under high vacuum to remove cumene, 2-phenyl-2-propanol, P(OMe)$_3$, and PO(OMe)$_3$. The residue is filtered through a short pad of silica gel (25 g) with CH$_2$Cl$_2$ as eluent under nitrogen pressure. Concentration of the filtrate and purification of the residue by flash chromatography gives the product as a light-yellow oil; 11.2 g (58%), R_f = 0.56 (n-pentane/EtOAc); $[\alpha]_D^{20}$ = +17.9 (c = 1.5, CHCl$_3$); ee = 98% (optical purity).

---

^{1}H NMR (300 MHz, CDCl$_3$): δ (ppm) = 7.77 (d, J = 8.4 Hz, 2 H, 3'-H), 7.32 (d, J = 7.8 Hz, 2 H, 2'-H), 4.23 (dd, J = 3.6, 11.3 Hz, 1 H, 3-H), 3.90 (dd, J = 6.3, 11.4 Hz, 1 H, 3-H), 3.12–3.18 (m, 1 H, 2-H), 2.78 (t, J = 4.8 Hz, 1 H, 1-H), 2.56 (dd, J = 2.4, 4.4 Hz, 1 H, 1-H), 2.42 (s, 3 H, CH$_3$).

^{13}C NMR (300 MHz, CDCl$_3$): δ (ppm) = 145.1 (C1'), 132.6 (C4'), 129.9 (C2'), 127.9 (C3'), 70.4 (C3), 48.8 (C2), 44.5 (C1), 21.6 (CH$_3$).

2.1.2   Jacobsen epoxidation

1

Topics: • Enantioselective epoxidation of a trisubstituted alkene using the Jacobsen catalyst and sodium hypochlorite as oxidizing agent

• Synthesis of the Jacobsen Mn-salen-catalyst (Duff formylation, resolution of 1,2-diaminocyclohexane, Schiff-base formation and complexation with Mn(III))

(a) General

The Jacobsen epoxidation allows the transformation of a wide range of unfunctionalized prochiral alkenes into chiral epoxides by reaction with an oxidizing agent, preferentially NaOCl, in the presence of catalytic amounts of the chiral Mn(III)-salen complex **2** (Jacobsen catalyst) [3b]:

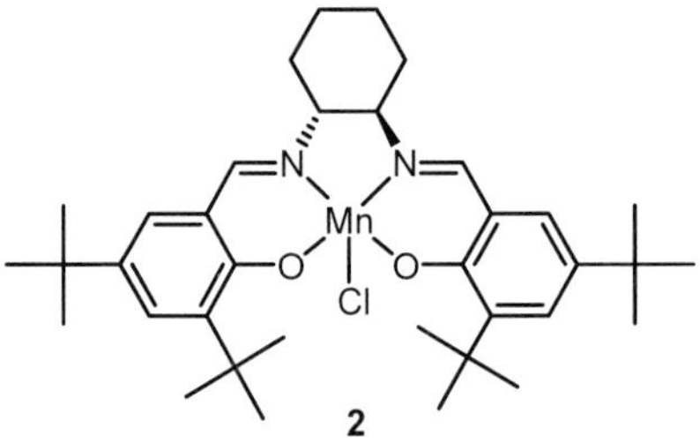

2

Jacobsen Mn-salen complex

Most suitable are disubstituted (*Z*)-alkenes (giving up to 99% *ee*) as well as trisubstituted alkenes, whereas monosubstituted alkenes are poor substrates for the Jacobsen epoxidation.

Interestingly, the Sharpless dihydroxylation (cf. **2.2**) shows a complementary substrate spectrum, since (*E*)-disubstituted alkenes give the best *ee* values (up to 99%) in product formation, although monosubstituted alkenes still give *ee* values of 70–80%.

The Jacobsen catalyst consists of an Mn(III) complex with the chiral salen ligand (salen = bis-salicylaldimine of (*R*,*R*)-diaminocyclohexane) and was introduced in the early 1990s. The manganese core is coordinated by the ligand in a square-planar geometry and is stabilized by a chlorine atom in the axial position [6]. This geometry is most probably responsible for the stereochemical outcome of the asymmetric alkene epoxidation [7]. The oxidant, normally bleach (aqueous NaOCl), is used in stoichiometric amounts [8]. The Jacobsen epoxidation is applied industrially on a tonne scale [9].

New developments include the use of ionic liquids and functionalized solid phases such as zeolites or organo-modified silicates, which allow the chiral catalyst to be immobilized, recovered, and recycled [10].

(b) Synthesis of 1

(1) The Jacobsen Mn(III)-salen complex (2) is prepared in a four-step sequence starting from 2,4-di-*tert*-butylphenol (3) [9]. 3,5-Di-*tert*-butylsalicylaldehyde (4) is obtained by Duff formylation [11] of the phenol 3 by reaction with hexamethylenetetramine in trifluoroacetic acid:

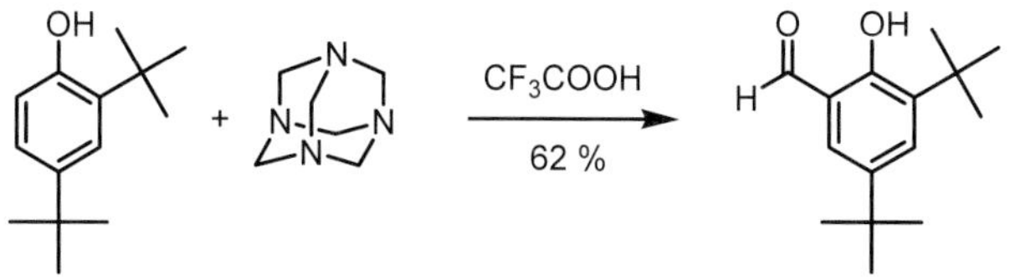

3   **4 (2.1.2.1)**

Resolution of racemic 1,2-diaminocyclohexane (5) is carried out by crystallization of its monoammonium salt 7 with (*R,R*)-(+)-tartaric acid (6), which is thus obtained in high diastereomeric purity.

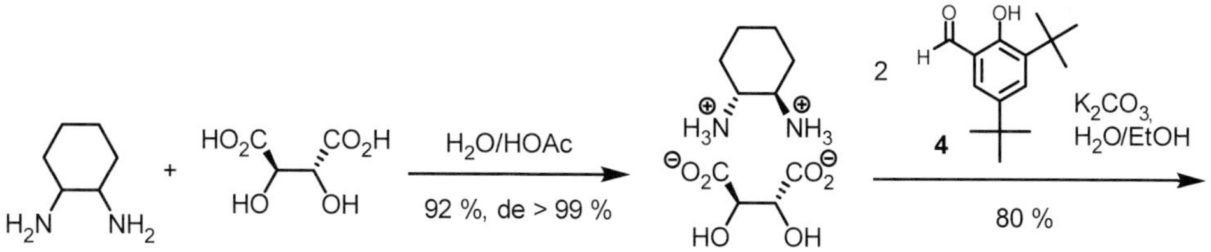

5   **6**   **7 (2.1.2.2)**

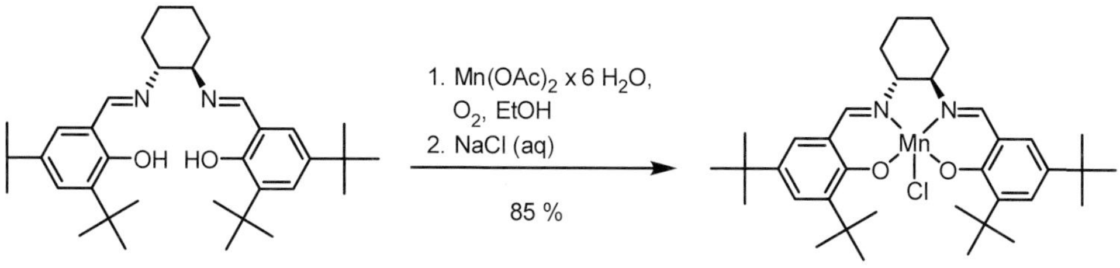

1. Mn(OAc)$_2$ x 6 H$_2$O, O$_2$, EtOH
2. NaCl (aq)

85 %

8 (2.1.2.3)   **2 (2.1.2.4)**

To simplify the procedure, the monotartrate salt 7 is used directly for condensation with the salicylaldehyde 4, thus avoiding an isolation of the water-soluble and air-sensitive free amine. Finally, the Schiff base 8 is transformed to the Mn(III)-salen complex 2 using Mn(II) acetate and oxidation with air; 2 is obtained as a brown air- and water-stable powder.

(2) Triphenylethylene (9) is subjected to enantioselective epoxidation by reaction with aqueous NaOCl solution as oxidant in the presence of catalytic amounts of the Jacobsen Mn(III)-salen complex (2) and 4-phenylpyridine *N*-oxide as co-oxidant in a two-phase system of CH$_2$Cl$_2$/H$_2$O at 0 °C to afford (*R*)-2,2,3-triphenyloxirane (1) in 87% chemical yield with a stereoselectivity of *ee* = 88%.

For comparison, a racemic mixture of 1 may be prepared in 80% yield by epoxidation of 9 with *m*CPBA in CH$_2$Cl$_2$.

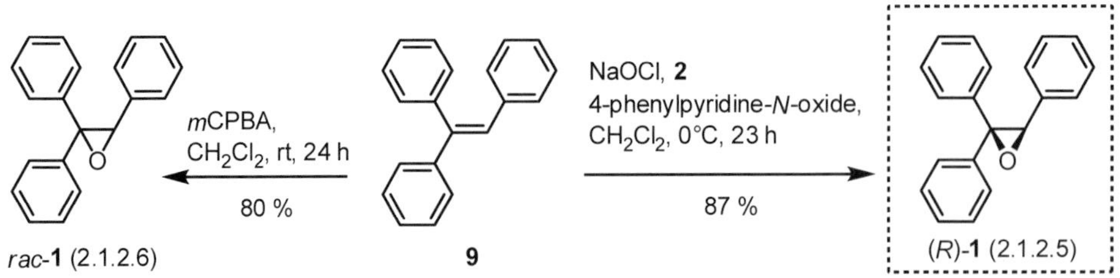

rac-**1** (2.1.2.6)   **9**   (R)-**1** (2.1.2.5)

As the mechanism of the Jacobsen epoxidation, a two-step catalytic cycle is proposed [12]. Oxygen is transferred from the stoichiometric oxidant (NaOCl) to the Mn(III)-salen complex (**2**), generating an intermediate Mn(V)-oxo species (**10** or **11**) [13]. Whether a neutral (**10**) or a cationic (**11**) Mn(V)-oxo species is formed depends on the oxidant and the solvent. In the second step, the oxygen is transferred to the olefinic double bond to give the epoxide and the remaining Mn(III) species **2** or **12** can be reoxidized.

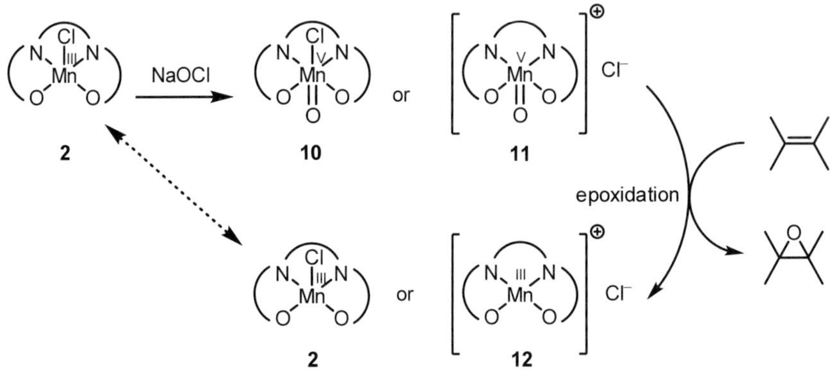

Theoretical studies have shown that the reaction probably proceeds along a radical pathway [14]. In the first step, the alkene approaches the manganese catalyst from the side of the oxo ligand and forms a weakly bonded catalyst–substrate adduct **13**, which reacts via the transition state **14** to give the radical intermediate **15**. The second oxygen–carbon bond is formed via transition state **16** to give a weakly bonded conjugate of the catalyst **17**, which can easily liberate the desired epoxide **18**.

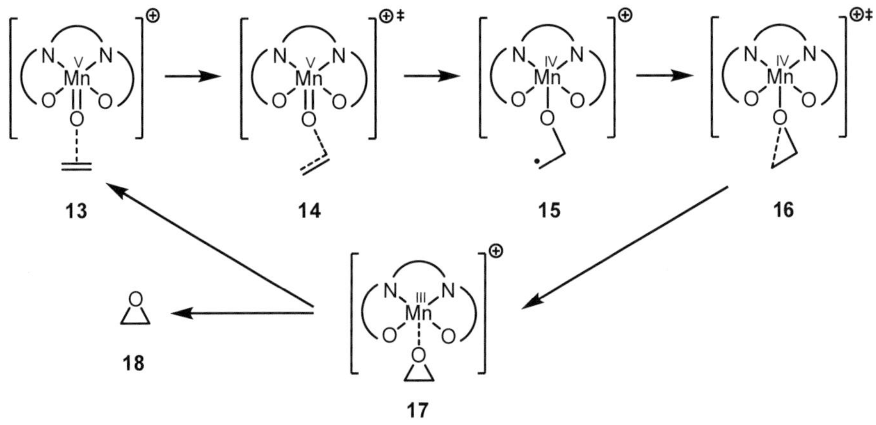

(c) Experimental procedures for the synthesis of 1 and 2

2.1.2.1 ** 3,5-Di-*tert*-butyl-2-hydroxybenzaldehyde [9]

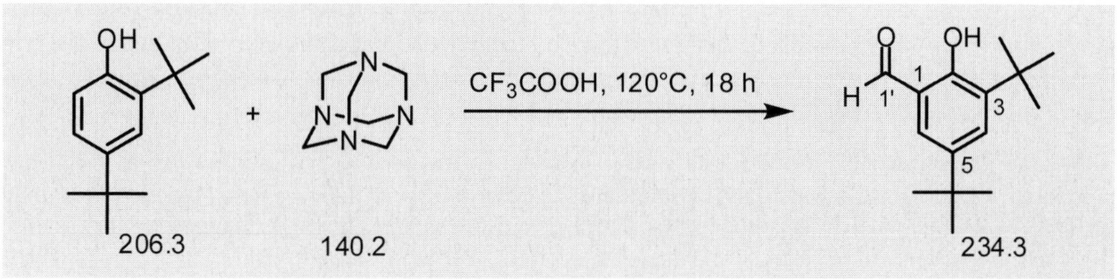

OH
CF3COOH, 120°C, 18 h
H
1'
1
3
5
206.3
140.2
234.3

A dry, three-necked, round-bottomed flask equipped with a reflux condenser is charged with trifluoroacetic acid (200 mL), hexamethylenetetramine (25.1 g, 179 mmol), and 2,4-di-*tert*-butylphenol (28.7 g, 139 mmol) at 0 °C under an argon atmosphere. The mixture is heated to reflux at 120 °C over a period of 45 min and maintained at this temperature for 18 h.

After cooling to 0 °C, H_2O (1200 mL), Na_2CO_3 (140 g, 1.32 mol), and, subsequently, HCl (6 M, 200 mL) are added. The product is extracted with EtOAc (3 × 500 mL), the organic fractions are dried over $MgSO_4$, and the solvent is evaporated. The yellow residue is heated with MeOH (30 mL) and the suspension obtained is subjected to vacuum filtration. The filter cake is washed with MeOH (3 × 50 mL), and the combined filtrates are concentrated under reduced pressure. The solid obtained is recrystallized from EtOAc/*n*-pentane to yield the desired aldehyde as yellow needles. The mother liquor is concentrated to afford a further crop of the aldehyde, giving a total yield of 20.1 g (62%).

UV (CH_3CN): λ_{max} (nm) (lg ε) = 343.5 (3.469), 263.5 (3.998), 219.5 (4.180).

IR (KBr): $\tilde{\nu}$ (cm^{-1}) = 2959, 1650, 1439, 1322, 1206, 894, 829, 737, 534.

^{1}H NMR (300 MHz, $CDCl_3$): δ (ppm) = 11.65 (s, 1 H, OH), 9.87 (s, 1 H, 1'-H), 7.60 (d, 1 H, *J* = 2.6 Hz, 6-H), 7.35 (d, 1 H, *J* = 2.6 Hz, 4-H), 1.43 (s, 9 H, 3-*t*Bu), 1.33 (s, 9 H, 5-*t*Bu).

^{13}C NMR (75 MHz, $CDCl_3$): δ (ppm) = 197.4 (C-1'), 159.1 (C-2), 141.6 (C-5), 137.6 (C-3), 131.9 (C-4), 127.0 (C-6), 119.9 (C-1), 34.99 (3-$\underline{C}(CH_3)_3$), 34.22 (5-$\underline{C}(CH_3)_3$), 31.29 (3-$C(\underline{C}H_3)_3$), 29.23 (5-$C(\underline{C}H_3)_3$).

MS (EI, 70 eV): *m*/*z* (%) = 234.1 (13) $[M]^+$, 219.1 (100) $[M-CH_3]^+$, 57 (62) $[C(CH_3)_3]^+$.

2.1.2.2 ** Resolution of *rac-cis/trans*-1,2-diaminocyclohexane (*R*,*R*)-diaminocyclohexane mono-(+)-tartrate salt [9]

H2N
NH2
HO2C
CO2H
HO
OH
H2O/HOAc, 70°C → 0°C
H3N
NH3
O2C
CO2
HO
OH
116.2
150.1
264.3

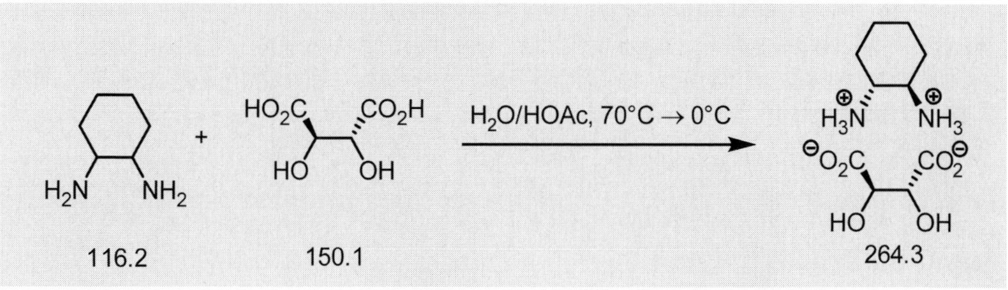

A three-necked, round-bottomed flask equipped with a reflux condenser, an internal thermometer, and an addition funnel is charged with (R,R)-(+)-tartaric acid (75.0 g, 500 mmol) and H_2O (200 mL). The mixture is stirred until the acid has completely dissolved. Then, *rac-cis/trans*-1,2-diaminocyclohexane (120 mL, 970 mmol) is added at such a rate that the reaction temperature reaches 60–65 °C. Glacial acetic acid (50 mL, 875 mmol) is then slowly added to the resulting solution at such a rate that the temperature reaches 65–70 °C. The mixture is cooled to room temperature over a period of 2 h, then cooled to 0 °C for a further 2 h.

The colorless precipitate formed is collected by vacuum filtration and the filter cake is washed with precooled water (50 mL) and MeOH (6 × 50 mL). After drying in high vacuum, the tartrate salt is obtained as a white crystalline solid; 74.7 g (92%), ds > 99:1, $[\alpha]_D^{20}$ = +12.7 (c = 4.0, H_2O).

2.1.2.3　**　(R,R)-N,N'-Bis(3,5-di-*tert*-butylsalicylidene)-1,2-cyclohexanediimine [9]

264.3　　　　　　　　　234.3　　　　　　　　　　　　　　546.8

A three-necked, round-bottomed flask equipped with a reflux condenser and an addition funnel is charged with the tartrate salt **2.1.2.2** (8.33 g, 31.5 mmol), Na_2CO_3 (8.72 g, 63.1 mmol), and H_2O (40 mL). The mixture is stirred until complete dissolution is achieved. EtOH (168 mL) is then added, the solution is heated to reflux, and a solution of the aldehyde **2.1.2.1** (15.0 g, 64.0 mmol) in EtOH (70 mL) is steadily added dropwise over 30 min. After heating to reflux for 2 h, H_2O (42 mL) is added. The mixture is cooled to room temperature over a period of 30 min, and then kept at 0–5 °C overnight.

The precipitate formed is collected by vacuum filtration and the filter cake is washed with EtOH (40 mL). The solid is redissolved in CH_2Cl_2 (140 mL) and the solution is washed with H_2O (3 × 80 mL) and brine (40 mL). The organic layer is dried over $MgSO_4$ and the solvent is removed *in vacuo* to yield the diimine as a yellow powder; 13.8 g (80%), $[\alpha]_D^{20}$ = –283.0 (c = 1.0, CH_2Cl_2).

UV (CH_3CN): λ_{max} (nm) (lg ε) = 328.5 (3.864), 259.5 (4.296), 218.5 (4.663), 194.0 (4.654).

IR (KBr): $\tilde{v}$ (cm^{-1}) = 2961, 2864, 1630, 1439, 1361, 1271, 1203, 1174, 1085, 1038, 879, 773.

^{1}H NMR (300 MHz, $CDCl_3$): δ (ppm) = 13.74 (s, 2 H, OH), 8.32 (s, 2 H, 2 × 1'-H), 7.32 (d, J = 2.4 Hz, 2 H, 2 × 6"-H), 7.00 (d, J = 2.4 Hz, 2 H, 2 × 4"-H), 3.38–3.28 (m, 2 H, 1-H, 2-H), 2.00–1.60 (m, 8 H, 3-H_2, 4-H_2, 5-H_2, 6-H_2), 1.43 (s, 18 H, 2 × 3"-*t*Bu), 1.25 (s, 18 H, 2 × 5"-*t*Bu).

^{13}C NMR (75 MHz, $CDCl_3$): δ (ppm) = 165.8 (C-1'), 158.0 (C-2"), 139.8 (C-5"), 136.3 (C-3"), 126.7 (C-4"), 126.0 (C-6"), 117.8 (C-1"), 72.40 (C-1, C-2), 34.93 (2 × 5"-C(CH$_3$)$_3$), 34.01 (2 × 3"-C(CH$_3$)$_3$), 33.26 (C-3, C-6), 31.39 (2 × 5"-C(CH$_3$)$_3$), 29.40 (2 × 3"-C(CH$_3$)$_3$), 24.34 (C-4, C-5).

MS (EI, 70.0 eV): m/z (%) = 546.3 (63) $[M]^+$, 313.2 (100) $[M - N - CH - C_{14}H_{20}O]^+$.

2.1.2.4 ** [(*R,R*)-*N,N'*-bis(3,5-di-*tert*-butyl-salicylidene)-1,2-cyclohexanediimine] manganese(III) chloride [9]

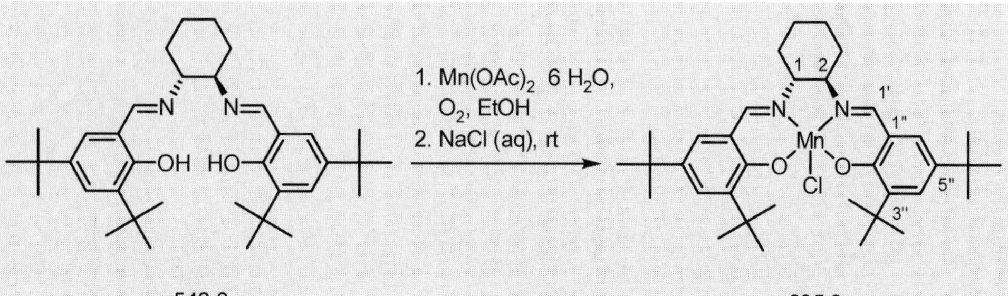

A three-necked, round-bottomed flask equipped with a reflux condenser and an addition funnel is charged with a solution of manganese(II) acetate hexahydrate (4.03 g, 14.4 mmol) in EtOH (36 mL). After heating to reflux, a solution of the Schiff base from **2.1.2.3** (3.00 g, 5.46 mmol) in toluene (20 mL) is slowly but steadily added over 35 min. The mixture is heated to reflux for 2 h, then heating is discontinued and air is bubbled through the solution for 3 h. Brine (6 mL) is then added, and the resulting mixture is stirred at room temperature overnight.

Toluene (20 mL) is added and the solution is washed with H_2O (3 × 50 mL) and brine (50 mL), and dried over $MgSO_4$. The solvent is evaporated *in vacuo* and the solid is redissolved in CH_2Cl_2 (18 mL) and *n*-heptane (18 mL). The solution is concentrated to half of its original volume and left at room temperature overnight. The precipitate formed is collected by vacuum filtration and washed with *n*-heptane (50 mL) to yield the manganese(III) complex as a brown solid; 2.95 g (85%), $[\alpha]_D^{20} = -608.3$ (*c* = 0.012, EtOH).

UV (CH_3CN): λ_{max} (nm) (lg ε) = 421.0 (3.636), 316.5 (4.105), 239.5 (4.578), 197.0 (4.656).

IR (KBr): $\tilde{\nu}$ (cm^{-1}) = 2952, 2866, 1613, 1535, 1433, 1311, 1252, 1175, 1030, 929, 917, 837, 748, 569, 543.

MS (ESI): *m/z* (%) = 679.0 (100) $[M + HCO_2]^+$, 633.3 (17) $[M - H]^+$, 599.5 (100) $[M - Cl]^+$.

2.1.2.5 *** (*R*)-2,2,3-Triphenyloxirane [15]

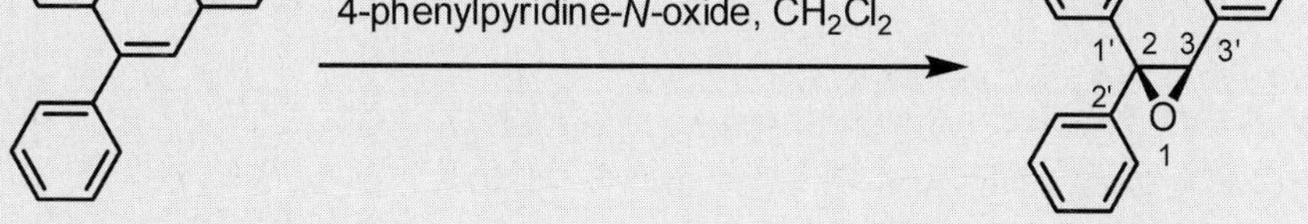

Preparation of the bleach solution:

Sodium hypochlorite solution (10%, 32.8 mL, 55.0 mmol) and aqueous disodium hydrogen phosphate (Na_2HPO_4) solution (0.05 M, 67.2 mL) are mixed and brought to pH 11.3 under pH-meter control by adding 1 M NaOH solution.

A single-necked, round-bottomed flask is charged with triphenylethylene (500 mg, 1.95 mmol) and CH_2Cl_2 (2.5 mL) under an argon atmosphere. 4-Phenylpyridine N-oxide (66.8 mg, 0.39 mmol) and Jacobsen catalyst **2.1.2.4** (61.9 mg, 97.5 µmol) are added and the mixture is cooled to 0 °C, whereupon precooled bleach solution (0.55 M, 5.32 mL, 2.93 mmol) is added and stirring is continued at 0 °C for 23 h. CH_2Cl_2 (10 mL) and H_2O (10 mL) are then added, the organic layer is washed with H_2O (10 mL), and the aqueous layer is extracted with CH_2Cl_2 (3 × 10 mL). The combined organic layers are dried over Na_2SO_4 and the solvent is removed *in vacuo*.

The crude product is purified by column chromatography on silica gel (*n*-pentane/EtOAc, 100:2) to afford the oxirane as a white solid; 464 mg (87%), *ee* = 88%, $[\alpha]_D^{20}$ = +61.2 (*c* = 1.0, $CHCl_3$).

UV (CH_3CN): λ_{max} (nm) (lg ε) = 225.5 (4.273).

IR (KBr): $\tilde{\nu}$ (cm^{-1}) = 3026, 2972, 1887, 1601, 1491, 1447, 1336, 1074, 1029, 903, 865, 822.

^{1}H NMR (200 MHz, $CDCl_3$): δ (ppm) = 7.40–7.00 (m, 15 H, 2 × 2-Ph, 3-Ph), 4.34 (s, 1 H, 3-H).

^{13}C NMR (75 MHz, $CDCl_3$): δ (ppm) = 140.9 (C-3'), 135.7, 135.4 (C-1', C-2'), 129.2, 128.3, 127.8, 127.7, 127.6, 127.5, 126.7, 126.3 (15 × Ph-CH), 68.63 (C-2), 68.05 (C-3).

MS (EI, 70.0 eV): *m/z* (%) = 272.2 (39) $[M]^+$, 165.1 (100) $[M - PhCH_2O]^+$, 105.1 (62) $[C_8H_9]^+$, 77.0 (31) $[Ph]^+$.

HPLC: Chiralpak IA$^®$ (Chiral Technologies Europe); 250 × 4.6 mm
eluent: *n*-hexane/*i*-propanol, 98:2
gradient: isocratic
retention time: t_{R1} = 7.21 min; t_{R2} = 8.06 min.

2.1.2.6   *   (*rac*)-2,2,3-Triphenyloxirane [16]

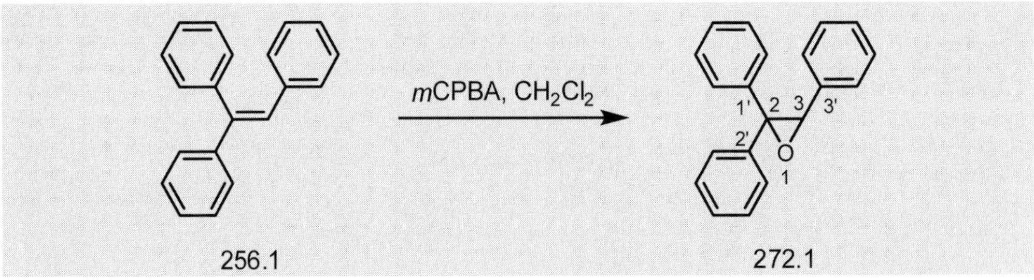

256.1   272.1

A single-necked, round-bottomed flask is charged with triphenylethylene (500 mg, 1.95 mmol) and CH_2Cl_2 (5 mL) under an argon atmosphere at 0 °C. *meta*-Chloroperoxybenzoic acid (70%, 722 mg, 2.93 mmol) is added and the resulting solution is stirred for 24 h at room temperature.

Saturated aqueous $NaHCO_3$ solution (10 mL) is then added and the mixture is stirred for 30 min, whereupon CH_2Cl_2 (20 mL) is added. The solution is transferred to a separatory funnel and, after separation of the layers, the organic layer is washed with saturated NaHCO$_3$ solution (2 × 10 mL) and dried over Na$_2$SO$_4$. The solvent is removed *in vacuo* and the residue is purified by column chromatography on silica gel (*n*-pentane/Et$_2$O, 100:1) to afford the oxirane as a white solid; 13.8 g (80%).

^{1}H NMR (200 MHz, CDCl$_3$): δ (ppm) = 7.40–7.00 (m, 15 H, 2 × 2-Ph, 3-Ph), 4.34 (s, 1 H, 3-H).

[1] a) M. Corsi, *Synlett* **2002**, *12*, 2127; b) E. N. Jacobsen, W. Zhang, A. R. Muci, J. R. Ecker, L. Deng, *J. Am. Chem. Soc.* **1991**, *113*, 7063; c) W. Zhang, J. L. Loebach, S. R. Wilson, E. N. Jacobsen, *J. Am. Chem. Soc.* **1990**, *112*, 2801.

[2] a) A. Pfenninger, *Synthesis* **1986**, 89; b) E. J. Corey, *J. Org. Chem.* **1990**, *55*, 1693; c) P. Besse, H. Veschambre, *Tetrahedron* **1994**, *50*, 8885.

[3] a) W. Zhang, E. N. Jacobsen, *J. Org. Chem.* **1991**, *56*, 2296; b) asymmetric epoxidation of unfunctionalized alkenes may be achieved using hydrogen peroxide as oxidant in the presence of a Ti-salen complex: Y. Sawada, K. Matsumoto, S. Kondo, H. Watanabe, T. Ozawa, K. Suzuki, B. Saito, T. Katsuki, *Angew. Chem.* **2006**, *118*, 3558; *Angew. Chem. Int. Ed.* **2006**, *45*, 3478.

[4] N. Krause, *Metallorganische Chemie*, p. 298, Spektrum Akademischer Verlag, Heidelberg, **1996**.

[5] J. M. Klunder, S. Y. Koo, K. B. Sharpless, *J. Org. Chem.* **1986**, *51*, 3710.

[6] a) T. Katsuki, *Coord. Chem. Rev.* **1995**, *140*, 189; b) R. Irie, K. Noda, Y. Ito, T. Katsuki, *Tetrahedron Lett.* **1991**, *32*, 1055; c) K. Srinivasan, P. Michaud, J. K. Kochi, *J. Am. Chem. Soc.* **1986**, *108*, 2309; d) K. Srinivasan, S. Perrier, J. K. Kochi, *J. Mol. Catal.* **1986**, *36*, 297.

[7] A. Scheurer, H. Maid, F. Hampel, R. W. Saalfrank, L. Toupet, P. Mosset, R. Puchta, N. J. R. van Eikema Hommes, *Eur. J. Org. Chem.* **2005**, *12*, 2566.

[8] a) B. D. Brandes, E. N. Jacobsen, *Tetrahedron Lett.* **1995**, *36*, 5123; b) S. Chang, J. M. Galvin, E. N. Jacobsen, *J. Am. Chem. Soc.* **1994**, *116*, 6937; c) B. D. Brandes, E. N. Jacobsen, *J. Org. Chem.* **1994**, *59*, 4378; d) S. Chang, N. H. Lee, E. N. Jacobsen, *J. Org. Chem.* **1993**, *58*, 6939; e) N. H. Lee, E. N. Jacobsen, *Tetrahedron Lett.* **1991**, *32*, 6533; f) N. H. Lee, A. R. Muci, E. N. Jacobsen, *Tetrahedron Lett.* **1991**, *32*, 5055.

[9] J. F. Larrow, E. N. Jacobsen, Y. Gao, Y. Hong, X. Nie, C. M. Zepp, *J. Org. Chem.* **1994**, *59*, 1939.

[10] a) D.-W. Park, S.-D. Choi, S.-J. Choi, C.-Y. Lee, G.-J. Kim, *Catal. Lett.* **2002**, *78*, 145; b) G. Gbery, A. Zsigmond, K. J. Balkus, Jr., *Catal. Lett.* **2001**, *74*, 77; c) C. E. Song, E. J. Roh, *Chem. Commun.* **2000**, 837.

[11] M. B. Smith, J. March, *March's Advanced Organic Chemistry*, 5th ed., p. 717, John Wiley & Sons, Inc., New York, **2001**.

[12] C. T. Dalton, K. M. Ryan, V. M. Wall, C. Bousquet, D. G. Gilheany, *Top. Catal.* **1998**, *5*, 75.

[13] K. Srinivasan, P. Michaud, J. K. Kochi, *J. Am. Chem. Soc.* **1986**, *108*, 2309.

[14] a) L. Cavallo, H. Jacobsen, *Eur. J. Inorg. Chem.* **2003**, *5*, 892; b) W. Adam, K. J. Roschmann, C. R. Saha-Möller, D. Seebach, *J. Am. Chem. Soc.* **2002**, *124*, 5068.

[15] T. Flessner, S. Doye, *J. Prakt. Chem.* **1999**, *341*, 436;

[16] According to: L. F. Tietze, Th. Eicher, *Reaktionen und Synthesen im organisch-chemischen Praktikum und Forschungslaboratorium*, 2nd ed., p. 476, Georg Thieme Verlag, Stuttgart, **1991**.

2.2   Dihydroxylation of C=C bonds

The dihydroxylation of olefinic C=C bonds to give *cis*-1,2-diols **1** is a fundamental oxidative process in alkene chemistry [1]. 1,2-Diols can also be prepared using oxiranes as precursors (cf. **2.1**) by acid-catalyzed nucleophilic ring-opening with water. In this case, the *trans*-1,2-diols **2** (X = OH) are formed [2]. Formation of the oxirane and ring-opening can be combined in one process.

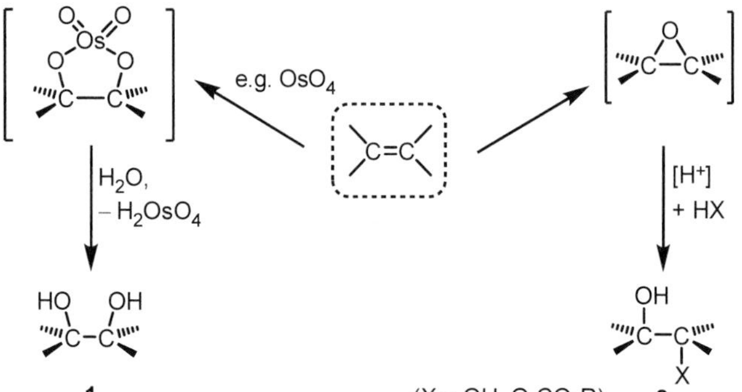

(X = OH, O-CO-R)   **2**

The *cis*-dihydroxylation can be performed with permanganate or osmium tetroxide as oxidants, with cyclic manganate or osmate esters as intermediates. While permanganate shows a tendency for further oxidative transformations [1], the reaction with OsO_4 is a reliable method for the synthesis of *cis*-1,2-diols **1** and is widely used in preparative chemistry [3].

Other methods of dihydroxylation, such as the Prevost method (I_2, Ag benzoate; overall *anti*-addition) and the Woodward method (I_2, Ag acetate, presence of H_2O; overall *syn*-addition) [4], although proceeding stereoselectively, are only of limited preparative importance.

For the enantioselective *cis*-dihydroxylation of alkenes, Sharpless and co-workers [5, 6] have developed an efficient method based on the use of OsO_4, which is presented in Section **2.2.1**.

2.2.1   Sharpless dihydroxylation

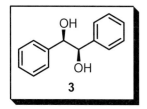

3

Topic: •   Asymmetric *cis*-dihydroxylation of an alkene according to Sharpless

(a) General

The Sharpless dihydroxylation gives excellent selectivities of up to 99.8% *ee* for almost all types of alkenes other than those with (*Z*)-1,2-disubstituted double bonds. In the Sharpless procedure, so-called AD mixes (AD = asymmetric dihydroxylation) are used as reagents; they contain $OsO_2(OH)_4$ as a catalytic non-volatile Os(VIII) source, which is coordinated either by dihydroquinine 1,4-phthalazine-diyl diether (DHQ)$_2$PHAL (**4**) to give AD-mix-α or by dihydroquinidine 1,4-phthalazinediyl diether (DHQD)$_2$PHAL (**5**) to give AD-mix-β, K_2CO_3, and a stoichiometric amount of $K_3[Fe(CN)_6]$ as a reoxidant. The ligand/osmium molar ratio is 2.5:1 and generally only 0.4 mol% of osmium is needed for the catalysis. Ligands based on PHAL [5] show broad applicability towards mono-, 1,1-di-, (*E*)-1,2-di-, tri-, and tetrasubstituted [7] alkenes. For enantioselective Sharpless *cis*-dihydroxylations of (*Z*)-1,2-disubstituted alkenes, special indole ligands belonging to the so-called IND class are used [8]. Furthermore, different ligands [9] have been developed to meet substrate-specific requirements.

(DHQ)$_2$PHAL

4

(DHQD)$_2$PHAL

5

In the PHAL reagents, the osmium atom is coordinated by the ligands **4** or **5** and the resulting complex provides efficient specific recognition of the prochiral alkene faces due to its steric demand.

An empirical rule helps to understand and predict the stereochemical outcome of the Sharpless dihydroxylation. As shown in the model situation **6**, the southeastern and northwestern quadrants are blocked by the ligand, so the smallest and second smallest substituents ought to be in these positions in order to give a favorable transition state for the dihydroxylation. The (DHQD)$_2$PHAL system (AD-mix-β) induces attack from the upper β-face, whereas the (DHQ)$_2$PHAL system (AD-mix-α) induces attack from the lower α-face.

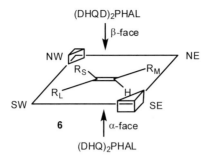

(b) Synthesis of 3

Asymmetric *cis*-dihydroxylation is carried out with (*E*)-stilbene as olefinic substrate using AD-mix-β as oxidation reagent in the presence of methanesulfonamide and *t*BuOH/H₂O as solvent [10]. (*R*,*R*)-(+)-1,2-Diphenyl-1,2-ethanediol (**3**) is obtained in almost quantitative chemical yield and in practically enantiopure form (*ee* > 99%).

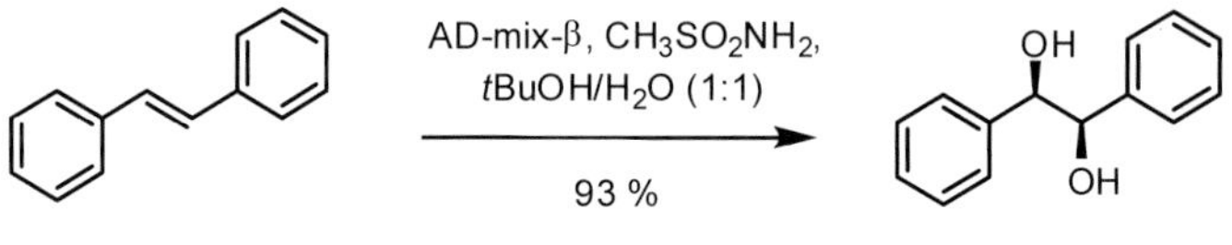

$$\text{AD-mix-}\beta,\ CH_3SO_2NH_2,$$
$$t\text{BuOH/H}_2\text{O (1:1)}$$

93 %

3 (2.2.1.1)

To account for the observed stereoselectivity, the reagent AD-mix-β apparently induces exclusive attack from the β-face of the alkene (cf. (a)).

For the dihydroxylation of a non-terminal alkene such as (*E*)-stilbene, the addition of methanesulfonamide effects an acceleration of the osmate(VI) ester hydrolysis and thus shortens the reaction time.

(c) Experimental procedure for the synthesis of 3

2.2.1.1 ** (*R*,*R*)-(+)-1,2-Diphenyl-1,2-ethanediol [7]

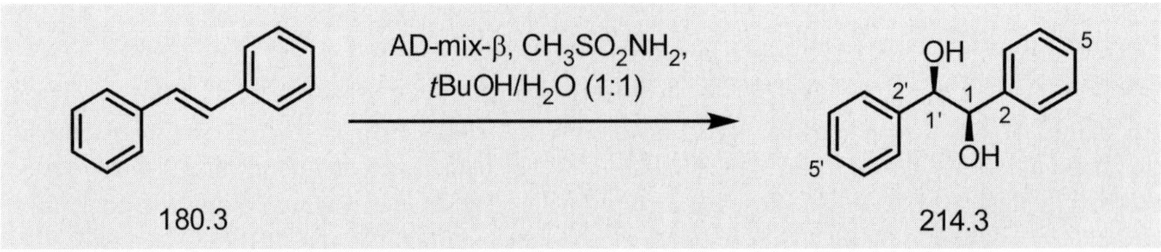

$$\text{AD-mix-}\beta,\ CH_3SO_2NH_2,$$
$$t\text{BuOH/H}_2\text{O (1:1)}$$

180.3   214.3

A 25 mL round-bottomed flask equipped with a magnetic stirrer is charged with *t*BuOH (5 mL), H₂O (5 mL), and AD-mix-β (2.00 g) and the mixture is stirred for 15 min. Methanesulfonamide (158 mg, 1.66 mmol) is then added, stirring is continued for a further 15 min, and then (*E*)-stilbene (250 mg, 1.38 mmol) is added. After vigorous stirring of the slurry at 20 °C for 24 h, sodium sulfite (2.00 g) is added and stirring is continued for a further 60 min.

Thereafter, the mixture is treated with H_2O (30 mL) and extracted with CH_2Cl_2 (3 × 50 mL). The combined organic layers are washed with H_2O (30 mL) and brine (30 mL), dried over $MgSO_4$, and the solvent is evaporated under reduced pressure. The crude product is purified by column chromatography on silica gel (petroleum ether/EtOAc, 3:1) to afford the diol as white crystalline needles. The enantiomeric excess is determined by HPLC on a chiral stationary phase; 279 mg (93%), $ee > 99\%$, mp 124–125 °C, $[\alpha]_D^{20} = +93$ ($c = 0.87$, EtOH), $R_f = 0.5$ (petroleum ether/EtOAc, 2:1).

UV (CH_3CN): λ_{max} (nm) (lg ε) = 263.5 (6.651), 252.0 (6.631), 257.5 (6.641), 194.5 (5.519), 192.5 (6.514).

IR (KBr): $\tilde{v}$ (cm^{-1}) = 3499, 3395, 3063, 2895, 1493, 1452, 1385, 1335, 1252, 1198, 1044, 1012.

^{1}H NMR (300 MHz, $CDCl_3$): δ (ppm) = 7.28–7.20 (m, 6 H, 2 × 4-H, 2 × 4'-H, 5-H, 5'-H), 7.18–7.09 (m, 4 H, 2 × 3-H, 2 × 3'-H), 4.70 (s, 2 H, 1-H, 1'-H), 2.85 (s_{br}, 2 H, 2 × OH).

^{13}C NMR (75 MHz, $CDCl_3$): δ (ppm) = 139.8 (C-2, C-2'), 128.1 (C-4, C-4'), 127.9 (C-5, C-5'), 126.9 (C-3, C-3'), 97.08 (C-1, C-1').

MS (DCI): m/z (%) = 446.3 $[2M + NH_4]^+$, 249.2 $[M + NH_3 + NH_4]^+$, 232.1 $[M + NH_4]^+$, 214.1 $[M - H_2O + NH_4]^+$.

HPLC: Chiralpak IA (Chiral Technologies Europe); 250 × 4.6 mm i.d.
          eluent: *n*-hexane/*i*-propanol, 90:10
          gradient: 0.8 mL min^{-1}.
          retention time: t_R = 15.32 min.

---

[1] *Organikum*, 21st ed., p. 302, Wiley-VCH, Weinheim, **2001**.

[2] When epoxidation is carried out with H_2O_2 in formic acid or acetic acid, monoesters of *trans*-1,2-diols (**2**, X = OCOR) result, which can be saponified to 1,2-diols (**2**, X = OH); compare: L. F. Tietze, Th. Eicher, *Reaktionen und Methoden im organisch-chemischen Praktikum und Forschungslaboratorium*, 2nd ed., p. 62, Georg Thieme Verlag, Stuttgart, **1991**.

[3] a) T. Wirth, *Angew. Chem.* **2000**, *112*, 342; *Angew. Chem. Int. Ed.* **2000**, *39*, 334; b) A. Gypser, D. Michel, D. S. Nirschl, K. B. Sharpless, *J. Org. Chem.* **1998**, *63*, 7322; c) J. K. Cha, N.-S. Kim, *Chem. Rev.* **1995**, *95*, 1761.

[4] M. B. Smith, J. March, *March's Advanced Organic Chemistry*, 5th ed., p. 1050, John Wiley & Sons, Inc., New York, **2001**.

[5] For reviews of the AD, see: a) Q.-H. Fan, Y.-M. Li, A. S. C. Chan, *Chem. Rev.* **2002**, *102*, 3385; b) H. C. Kolb, M. S. VanNieuwenhze, K. B. Sharpless, *Chem. Rev.* **1994**, *94*, 2483.

[6] For the nobel lecture of K. B. Sharpless, see: K. B. Sharpless, *Angew. Chem.* **2002**, *114*, 2126; *Angew. Chem. Int. Ed.* **2002**, *41*, 2024.

[7] K. Morikawa, J. Park, P. G. Andersson, T. Hashiyama, K. B. Sharpless, *J. Am. Chem. Soc.* **1993**, *115*, 8463.

[8] L. Wang, K. B. Sharpless, *J. Am. Chem. Soc.* **1992**, *114*, 7568.

[9] K. B. Sharpless, W. Amberg, M. Beller, H. Chen, J. Hartung, Y. Kawanami, D. Lübben, E. Manoury, Y. Ogino, T. Shibata, T. Ukita, *J. Org. Chem.* **1991**, *56*, 4585.

[10] Z.-M. Wang, K. B. Sharpless, *J. Org. Chem.* **1994**, *59*, 8302.

2.3   Oxidation of alcohols to carbonyl compounds

$$\text{CH-OH} \longrightarrow \text{C=O}$$

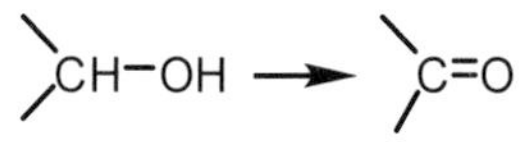

Numerous methods are available for the conversion of primary alcohols to aldehydes and of secondary alcohols to ketones [1]. The chromium(VI) reagents (e.g., chromic acid, dichromate, pyridinium chlorochromate), preferentially used in earlier times, are now set to be replaced almost entirely by less toxic reagents, selected examples of which are indicated in the following scheme:

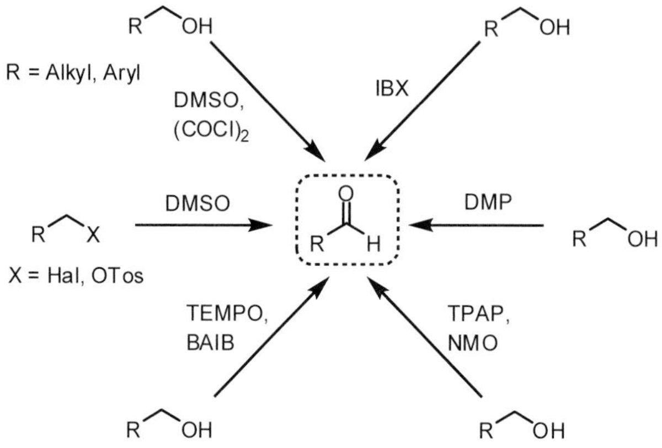

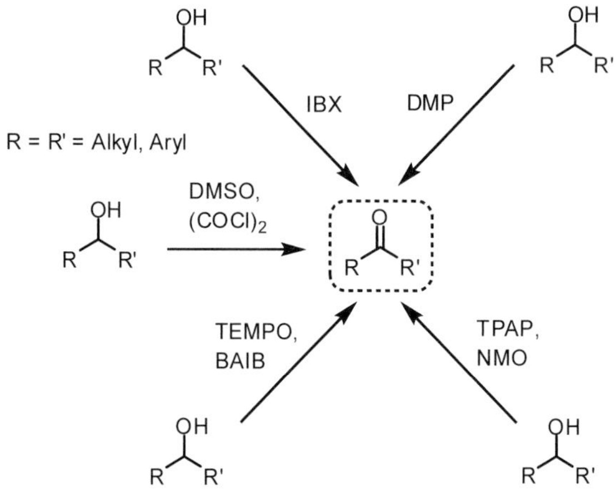

A very common and reliable procedure for the oxidation of primary and secondary alcohols is the Swern oxidation using oxalyl chloride and dimethyl sulfoxide (DMSO) [2]. A disadvantage of this method is that it should not be employed for the transformation of large quantities. DMSO can also be used for the conversion of alkyl halides and tosylates to carbonyl compounds. This procedure is known as Kornblum oxidation ([3], cf. **1.1.1**); it is, however, of limited importance.

Environmental demands concerning chemical processes have encouraged chemists to search for new clean, high-yielding, and selective oxidation methods. Of particular interest are methods using hypervalent iodine compounds (cf. **1.7.6**) as mild and selective oxidizing agents of low toxicity [4]. The Dess–Martin periodinane (DMP) and its direct precursor *o*-iodoxybenzoic acid (IBX) fall into this category, which are employed in organic solvents such as DMSO, CH_2Cl_2, or acetone [5]. However, in spite of their utility, iodine(V) reagents are potentially explosive. Therefore, the use of readily available and relatively stable iodine(III) reagents such as (bisacetoxyiodo)benzene (BAIB) has attracted increased attention [6]. This reagent is used in combination with catalytic amounts of 2,2,6,6-tetramethyl-1-piperdinyloxyl (TEMPO) for regeneration [7]. Sodium hypochlorite can also be used as oxidant instead of BAIB, which is relatively expensive.

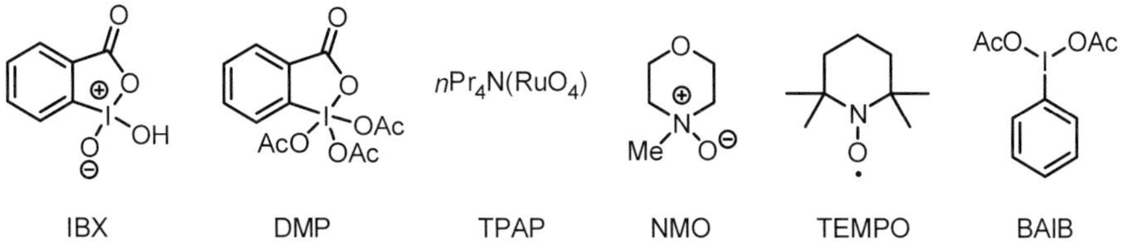

| IBX | DMP | TPAP | NMO | TEMPO | BAIB |

One of the most versatile oxidants is tetrapropylammonium perruthenate (TPAP), which allows a metal-mediated oxidation of alcohols under very mild conditions without any obnoxious or explosive reagents [8]. TPAP is used in catalytic amounts with *N*-methyl-morpholine *N*-oxide (NMO) as a co-oxidant to regenerate the active ruthenium species.

The aforementioned oxidation methods and the preparation of the required reagents are the subjects of the following sections. As a standard reference process, the conversion of the primary alcohol *n*-octanol to the aldehyde *n*-octanal (**1**) is used [9].

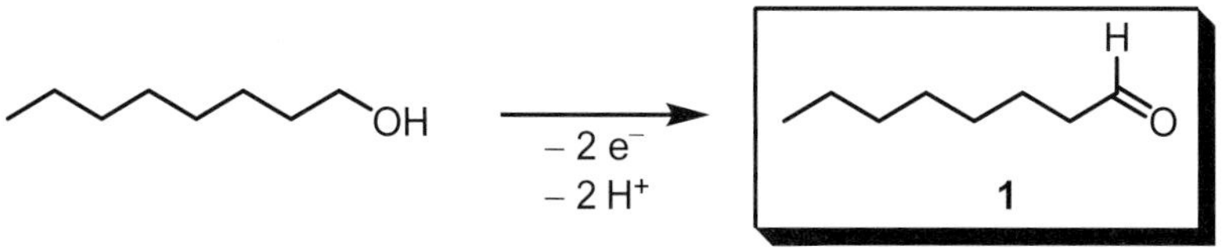

2.3.1   Swern oxidation

(a) General

For the Swern oxidation process, the following mechanism is established:

DMSO (**2**) is activated by *S*-acylation with oxalyl chloride (**3**) followed by loss of CO and CO_2 to give chlorodimethylsulfonium chloride (**4**) as intermediate, which reacts with the alcohol (here: RCH_2OH) to give an alkoxysulfonium ion **6**. Deprotonation of **6** by base (here: triethylamine) generates the ylide **5**, which undergoes proton transfer to the ylide carbon and cleavage at the O–S bond to form the products, the carbonyl compound (here: R–CH=O) and dimethyl sulfide:

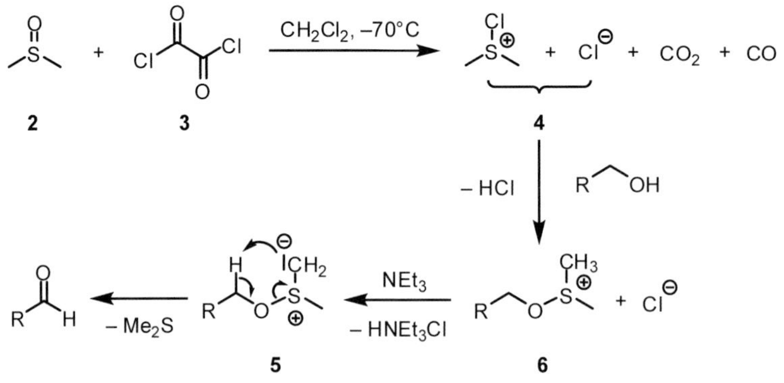

2   **3**   **4**

$- HCl$ R$\diagdown$OH

NEt$_3$

$- Me_2S$ $- HNEt_3Cl$

5   **6**

(b) Swern oxidation of *n*-octanol (experimental procedure)

2.3.1.1   **   *n*-Octanal I [2]

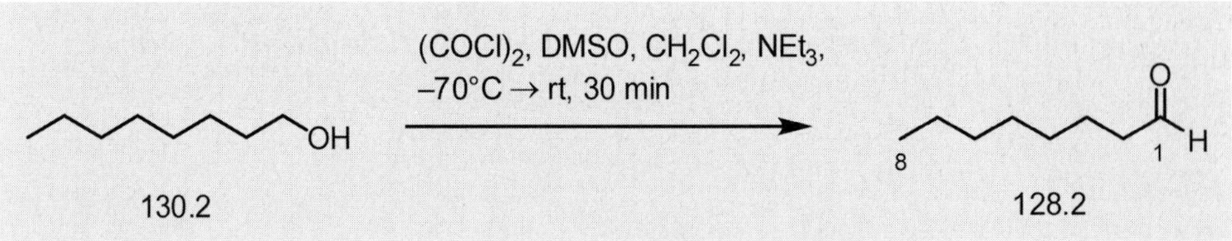

(COCl)$_2$, DMSO, CH$_2$Cl$_2$, NEt$_3$,
$-70°C \rightarrow$ rt, 30 min

130.2   128.2

A solution of oxalyl chloride (15.2 g, 10.3 mL, 120 mmol) in CH$_2$Cl$_2$ (150 mL) is prepared under an argon atmosphere in a dry 500 mL three-necked round-bottomed flask, fitted with a mechanical stirrer, a dropping funnel, and an internal thermometer. The solution is cooled to -70 °C, a solution of DMSO (20.3 g, 18.5 mL, 260 mmol) in CH$_2$Cl$_2$ (30 mL) is added dropwise, and the reaction mixture is stirred for 30 min at -70 °C. A solution of *n*-octanol (13.0 g, 15.8 mL, 100 mmol) in CH$_2$Cl$_2$ (40 mL) is then added dropwise and the resulting mixture is stirred for a further 30 min at -70 °C before NEt$_3$ (50.6 g, 69.5 mL, 500 mmol) is cautiously added. The solution is allowed to warm to room temperature, whereupon H$_2$O (100 mL) is added and stirring is continued for 10 min.

The aqueous layer is extracted with CH$_2$Cl$_2$ (2 × 100 mL), the combined organic layers are dried over Na$_2$SO$_4$, and the solvent is evaporated. The residue is fractionated over a Vigreux column (20 cm) under reduced pressure to afford the aldehyde as a colorless volatile liquid; 12.5 g (98%), bp$_{12}$ 62–63 °C.

IR (KBr): $\tilde{v}$ (cm^{-1}) = 2928, 2858, 1712, 1465, 1414, 1285, 1232, 1109, 938, 725.

^{1}H NMR (300 MHz, CDCl$_3$): δ (ppm) = 9.75 (t, J = 1.8 Hz, 1 H, 8-H), 2.41 (td, J = 7.4, 1.8 Hz, 2 H, 2-H$_2$), 1.61 (m$_c$, 2 H, 3-H$_2$), 1.29 (m$_c$, 8 H, H$_{alkyl}$), 0.87 (t, J = 6.9 Hz, 3 H, 8-H$_3$).

^{13}C NMR (75 MHz, CDCl$_3$): δ (ppm) = 202.9 (C-1), 43.90 (C-2), 31.60 (C-3), 29.11 (C-4), 29.00 (C-5), 22.56 (C-6), 22.07 (C-7), 14.02 (C-8).

MS (DCI, 200 eV): *m/z* (%) = 179 (9) $[M + NH_3 + NH_4]^+$, 162 (100) $[M + NH_4]^+$.

2.3.2 Dess–Martin oxidation

(a) General

In this method, the oxidizing agent is the Dess–Martin periodinane (**1**), a hypervalent organo-iodine(V) compound, which is prepared in a three-step sequence from anthranilic acid (**2**).

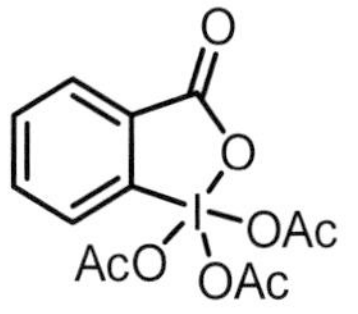
O
O
I
OAc
AcO
OAc

1

(b) Synthesis of Dess–Martin periodinane (1)

Anthranilic acid (**2**) is transformed into *o*-iodobenzoic acid (**3**) by diazotization and subsequent displacement of the diazonium group by iodide (S_NAr process). Reaction of **3** with potassium bromate in H_2SO_4 leads to oxidation at the iodine atom followed by cyclization involving the COOH group to give 1-hydroxy-1,2-benziodoxol-3(1*H*)-one 1-oxide (IBX) (**4**). Peracetylation of **4** with acetic anhydride in acetic acid then affords the desired oxidizing agent 1,1,1-triacetoxy-1,1-dihydro-1,2-benziodoxol-3(1*H*)-one (DMP) (**1**):

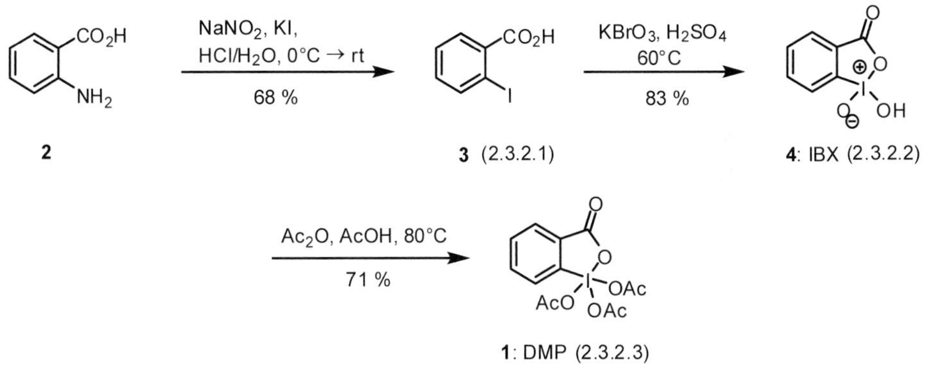
CO2H
NH2
NaNO2, KI,
HCl/H2O, 0°C → rt
68 %
CO2H
I
KBrO3, H2SO4
60°C
83 %
OH
2
3 (2.3.2.1)
4: IBX (2.3.2.2)
Ac2O, AcOH, 80°C
71 %
OAc
AcO
OAc
1: DMP (2.3.2.3)

The Dess–Martin oxidation proceeds according to the following mechanistic pathway. Initially, the alcohol (here: RCH_2OH) replaces one of the acetate residues at the iodine atom in the DMP reagent (**1**). Subsequently, the intermediate **5** (still containing iodine(V)) undergoes a disproportionation process, in which HOAc is lost and the carbonyl compound (here: R–CH=O) as well as the (less reactive) iodine(III) species **6** are formed.

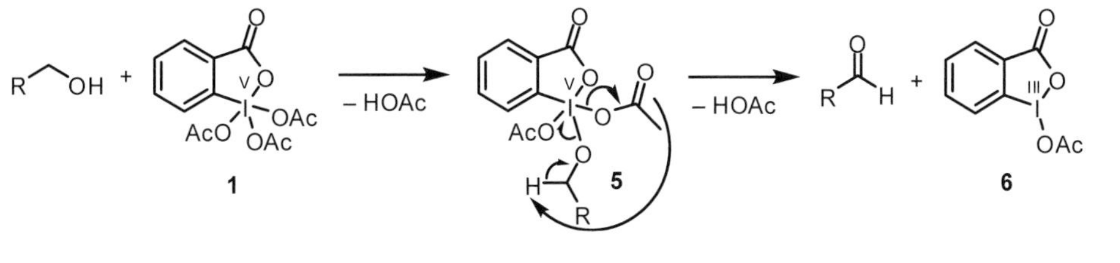
R
OH
+
OAc
AcO
OAc
1
– HOAc
AcO
H
R
5
– HOAc
R
H
+
OAc
6

(c) Experimental procedures for the synthesis of 1

2.3.2.1 ** *o*-Iodobenzoic acid [1]

NaNO$_2$, KI, HCl/H$_2$O,
0 °C → rt, 30 min

137.1   248.0

Under an argon atmosphere, anthranilic acid (34.2 g, 249 mmol) is dissolved in a mixture of H$_2$O (250 mL) and concentrated HCl (62.5 mL). The solution is cooled to 0–5 °C and a solution of sodium nitrite (17.7 g, 257 mmol) in H$_2$O (50 mL) is added dropwise, not letting the temperature exceed 5 °C. The resulting mixture is stirred for 5 min, and then a solution of potassium iodide (42.7 g, 257 mmol) in H$_2$O (65 mL) is added. Stirring is continued for 5 min without cooling and the solution is warmed to 40–50 °C, causing rapid gas evolution and the formation of a brown precipitate.

After 15 min at 40–50 °C, the temperature is increased to 70–80 °C for 10 min and then the solution is cooled with an ice bath. Sodium thiosulfate is added to destroy excess iodine, and the precipitate is collected by filtration and washed with iced water (3 × 200 mL). The solid is dissolved in hot EtOH (175 mL) and treated three times with charcoal. The charcoal is filtered off and the final filtrate is diluted with hot water (80 mL) and heated to reflux. Cold water (100 mL) is added and the solution is left in a refrigerator to give the iodobenzoic acid as yellow-orange needles; 41.2 g (68%), mp 159–160 °C, R_f = 0.39 (*n*-pentane/EtOAc, 4:1, + 1% acetic acid).

UV (CH$_3$CN): λ_{max} (nm) (lg ε) = 285.0 (3.130), 205.0 (4.341).

IR (KBr): $\tilde{v}$ (cm^{-1}) = 2875, 1681, 1581, 1466, 1402, 1295, 1266, 1109, 1014, 896, 739, 678.

^{1}H NMR (300 MHz, CD$_3$OD): δ (ppm) = 8.00 (dd, *J* = 8.0, 1.0 Hz, 1 H, 3-H), 7.79 (dd, *J* = 7.8, 1.7 Hz, 1 H, 6-H), 7.45 (dt, *J* = 7.7, 1.0 Hz, 1 H, 5-H), 7.19 (dt, *J* = 7.8, 1.7 Hz, 1 H, 4-H).

^{13}C NMR (75 MHz, CD$_3$OD): δ (ppm) = 170.1 (CO$_2$H), 142.3 (C-3), 137.8 (C-1), 133.5 (C-4), 131.6 (C-6), 129.1 (C-5), 94.14 (C-2).

MS (EI, 70 eV): *m/z* (%) = 248 (100) [*M*]$^+$, 231 (47) [*M* – OH]$^+$, 203 (10) [*M* – CO$_2$H]$^+$, 121 (2) [*M* – I]$^+$, 76 (10) [C$_6$H$_4$]$^+$.

C$_7$H$_5$IO$_2$ (248.02): calcd: C 33.90, H 2.03; found: C 34.01, H 1.83.

2.3.2.2   **   1-Hydroxy-1,2-benziodoxol-3(1*H*)-one 1-oxide (IBX) [5f]

$$\text{CO}_2\text{H} \quad \xrightarrow[\text{60°C, 4 h}]{\text{KBrO}_3, \text{H}_2\text{SO}_4,}$$

248.0   280.0

Potassium bromate (30.0 g, 180 mmol) is added in small portions to a mechanically stirred solution of **2.3.2.1** (35.0 g, 141 mmol) in H_2SO_4 (300 mL, 0.73 M) (Caution: the temperature of the solution should not exceed 50 °C during the addition!). The resulting mixture is stirred for an additional 20 min at room temperature, then cautiously heated to 60 °C over a period of 1 h and stirred for 3 h at this temperature.

The reaction mixture is then cooled to 0 °C, and the solid compounds are filtered off (Caution: explosive!), washed with cold water (350 mL), EtOH (25 mL), and Et_2O (25 mL), and dried under reduced pressure to afford the oxide as a light-yellow powder; 33.0 g (84%).

¹H NMR (300 MHz, [D₆]DMSO): δ (ppm) = 8.00 (m, 4 H, Ar-H), 4.40 (s_br, 1 H, OH).

Note: Due to its sensitivity, this compound should be immediately subjected to the next step of the reaction sequence.

2.3.2.3   ***   1,1,1-Triacetoxy-1,1-dihydro-1,2-benziodoxol-3(1*H*)-one (DMP) [5f]

$$\xrightarrow[\text{80°C, 2.5 h}]{\text{Ac}_2\text{O, AcOH,}}$$

280.0   424.1

2.3.2.2 (13.0 g, 46.4 mmol) (Caution: explosive!) is added to a mixture of acetic anhydride (16.6 g, 162 mmol) and acetic acid (13.8 g, 230 mmol) maintained at room temperature. The resulting mixture is heated to 80 °C over 1 h and then stirred at this temperature for a further 1.5 h before being slowly cooled to 0 °C. The resulting colorless crystals are washed with Et_2O (6 × 10 mL). The residual solvent is completely removed under reduced pressure to provide the Dess–Martin periodinane as a white crystalline solid; 14.0 g (71%), mp 133–134 °C.

¹H NMR (300 MHz, CDCl₃): δ (ppm) = 8.31, 8.29 (2 × d, *J* = 8.5 Hz, 2 H, 3-H, 6-H), 8.07, 7.80 (2 × dd, *J* = 8.5, 7.3 Hz, 2 H, 4-H, 5-H), 2.33 (s, 3 H, CH₃), 2.01 (s, 6 H, 2 × CH₃).

¹³C NMR (75 MHz, CDCl₃): δ (ppm) = 175.7, 174.0 (3 × COCH₃), 166.1 (C-3), 142.4 (Ar-C), 135.8, 133.8, 131.8, 126.5, 126.0, 20.43 (COCH₃), 20.29 (2 × COCH₃).

(d) Dess–Martin periodinane oxidation of *n*-octanal (experimental procedure)

2.3.2.4 * *n*-Octanal II [5]

Under an argon atmosphere, DMP **2.3.2.3** (489 mg, 1.15 mmol) is added to a solution of *n*-octanol (100 mg, 770 µmol) in CH_2Cl_2 (6.0 mL). The mixture is stirred for 2 h at room temperature, then silica gel is added and the solvent is removed under reduced pressure (max. 40 °C at 700 mbar). Flash chromatography on silica gel (*n*-pentane/Et_2O, 10:1) provides the aldehyde as a colorless volatile liquid; 89.2 mg (90%), R_f = 0.54 (*n*-pentane/Et_2O, 10:1).

> **IR** (KBr): $\tilde{v}$ (cm^{-1}) = 2928, 2858, 1712, 1465, 1414, 1285, 1232, 1109, 938, 725.
>
> **^{1}H NMR** (300 MHz, $CDCl_3$): δ (ppm) = 9.75 (t, *J* = 1.8 Hz, 1 H, 8-H), 2.41 (td, *J* = 7.4, 1.8 Hz, 2 H, 2-H_2), 1.61 (m_c, 2 H, 3-H_2), 1.29 (m_c, 8 H, H_{alkyl}), 0.87 (t, *J* = 6.9 Hz, 3 H, 8-H_3).
>
> **^{13}C NMR** (75 MHz, $CDCl_3$): δ (ppm) = 202.9 (C-1), 43.90 (C-2), 31.60 (C-3), 29.11 (C-4), 29.00 (C-5), 22.56 (C-6), 22.07 (C-7), 14.02 (C-8).
>
> **MS** (DCI, 200 eV): *m/z* (%) = 179 (9) $[M + NH_3 + NH_4]^+$, 162 (100) $[M + NH_4]^+$.

2.3.3 Perruthenate oxidation

(a) General

The catalytic oxidation of primary and secondary alcohols with tetra-*n*-propylammonium perruthenate (TPAP, **1**) requires the use of a stoichiometric amount of *N*-methylmorpholine *N*-oxide (NMO) (**4**) as a co-oxidant. In the oxidation process, the perruthenate anion reacts with the alcohol (here: RCH_2OH) to give the Ru(VII) intermediate **2**, which undergoes disproportionation to an Ru(V) species **5** and the expected aldehyde. Finally, the Ru(V) species is re-oxidized by NMO (which is concomitantly reduced to the tertiary amine **3**), thus regenerating the Ru(VII) anion in **1**.

The reaction is autocatalytic, which means that the rate is initially slow but accelerates strongly as the concentration of the carbonyl product increases and then slows down once more at the end of the reaction. It is suggested that colloidal RuO_2 is responsible for the autocatalysis, which coordinates to $[RuO_4]^-$ to generate an activated $[RuO_4 \cdot n\,RuO_2]^-$ complex. Perruthenate oxidation is very sensitive to water, which must be coercively removed using molecular sieves. The reason for this effect is the binding of water to the RuO_2 particles, which reduces the number of available coordination sites at

[RuO$_4$]$^-$ and thus inhibits the autocatalytic process. Water can also promote the formation of aldehyde hydrates, which could cause an undesired overoxidation to give carboxylic acids.

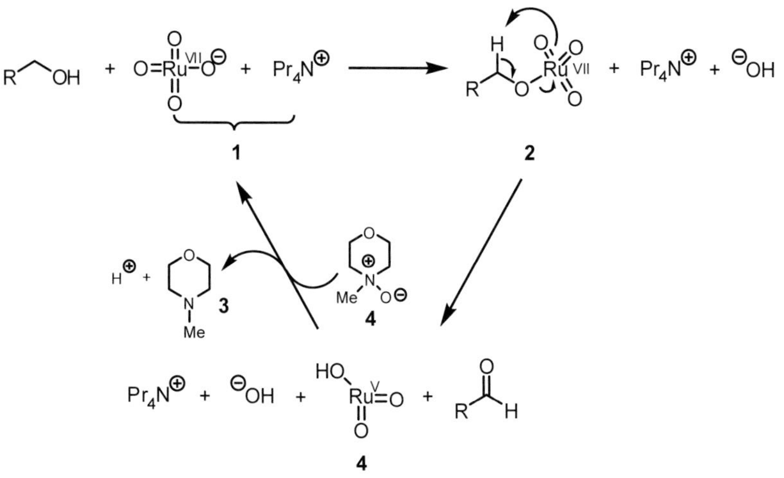

(b) Perruthenate oxidation of *n*-octanol (experimental procedure)

2.3.3.1   *   *n*-Octanal III [8]

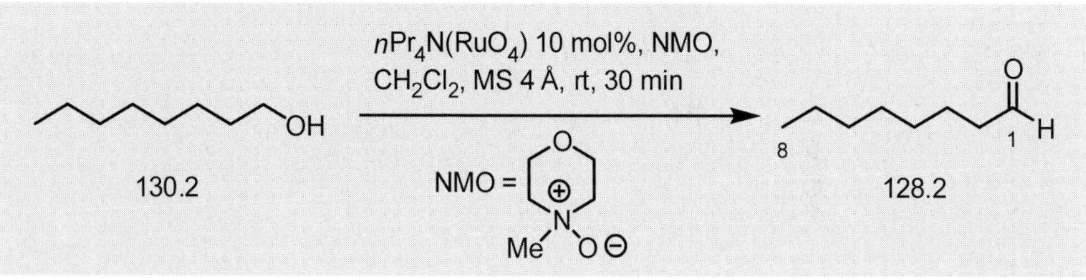

Tetra-*n*-propylammonium perruthenate (TPAP) (28.1 mg, 0.08 mmol, 10 mol%) and *N*-methyl-morpholine *N*-oxide (NMO) (271 mg, 2.31 mmol) are added to a mixture of *n*-octanol (100 mg, 0.77 mmol) and 4 Å molecular sieves (100 mg) in CH$_2$Cl$_2$ (4.0 mL). The mixture is shaken for 30 min at room temperature.

It is then concentrated under reduced pressure (max. 40 °C at 700 mbar) and adsorbed onto silica gel. Column chromatography on silica gel provides the aldehyde as a colorless volatile liquid; 98 mg (99%), R_f = 0.54 (*n*-pentane/Et$_2$O, 10:1).

IR (KBr): $\tilde{\nu}$ (cm^{-1}) = 2928, 2858, 1712, 1465, 1414, 1285, 1232, 1109, 938, 725.

^{1}H NMR (300 MHz, CDCl$_3$): δ (ppm) = 9.75 (t, *J* = 1.8 Hz, 1 H, 8-H), 2.41 (td, *J* = 7.4, 1.8 Hz, 2 H, 2-H$_2$), 1.61 (m$_c$, 2 H, 3-H$_2$), 1.29 (m$_c$, 8 H, H$_{alkyl}$), 0.87 (t, *J* = 6.9 Hz, 3 H, 8-H$_3$).

^{13}C NMR (75 MHz, CDCl$_3$): δ (ppm) = 202.9 (C-1), 43.90 (C-2), 31.60 (C-3), 29.11 (C-4), 29.00 (C-5), 22.56 (C-6), 22.07 (C-7), 14.02 (C-8).

MS (DCI, 200 eV): *m/z* (%) = 179 (9) [*M* + NH$_3$ + NH$_4$]$^+$, 162 (100) [*M* + NH$_4$]$^+$.

2.3.4　TEMPO oxidation

(a) General

TEMPO oxidation is a highly chemoselective oxidation protocol, which makes use of (bisacetoxyiodo)benzene (BAIB, **2**) as a stoichiometric oxidant in combination with tetramethyl-piperidine-nitroxyl (TEMPO, **1**), which is applied in catalytic amounts.

In the reaction, the TEMPO nitroxyl radical **1** is first oxidized by BAIB (**2**) to the N-oxoammonium ion **3**, addition of the alcohol (here: RCH_2OH) to which with loss of HOAc leads to the intermediate **4**. The betaine **4** undergoes disproportionation to the carbonyl compound (here: R-CH=O) and a hydroxylamine derivative **7**, which is re-oxidized to the TEMPO radical **1** by the acetoxy-iodo radical **5** formed in the primary reaction step (**1** + **2**). Iodobenzene (**6**) and HOAc are also obtained in this step. Overall, in the oxidation process RCH_2OH → R–CH=O, the catalyst TEMPO is regenerated and BAIB is consumed, with the formation of iodobenzene (**6**) and acetic acid.

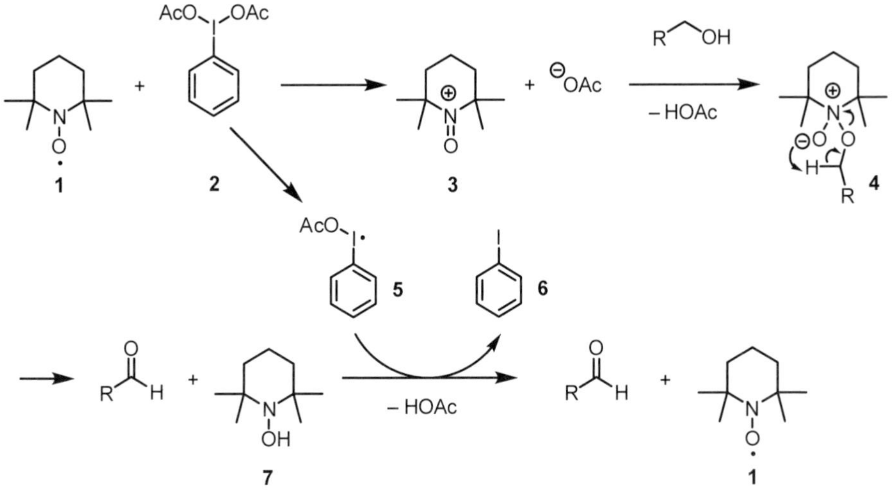

(b)　TEMPO oxidation of n-octanol (experimental procedure)

2.3.4.1　*　n-Octanal IV [7]

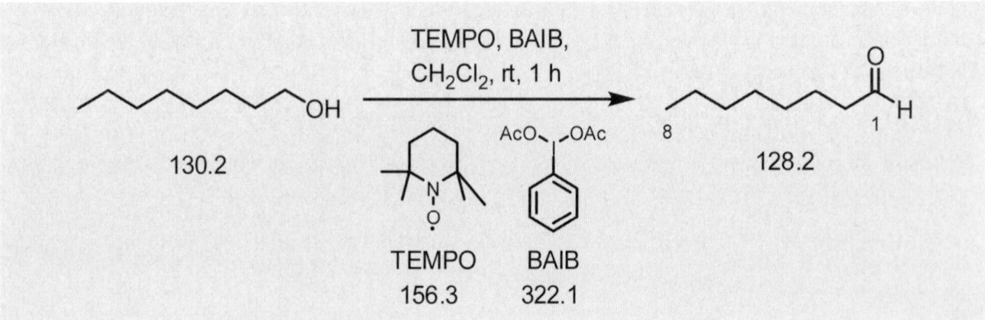

(Bisacetoxyiodo)benzene (BAIB) (272.1 mg, 0.85 mmol) is added to a stirred solution of *n*-octanol (121.4 µL, 100 mg, 0.77 mmol) and 2,2,6,6-tetramethyl-1-piperidinyloxyl (TEMPO) (12.0 mg, 0.07 mmol) in CH_2Cl_2 (0.75 mL) at 20 °C and stirring is continued for 1 h.

The mixture is then diluted with CH_2Cl_2 (0.75 mL) and washed with saturated sodium thiosulfate solution (1 mL), which is re-extracted with CH_2Cl_2 (4 × 1 mL). The combined organic layers are washed with saturated $NaHCO_3$ solution (1 mL) and brine (1 mL), dried over Na_2SO_4, and concentrated under reduced pressure. Flash column chromatography of the residue on silica gel with *n*-pentane/Et_2O (10:1) as eluent yields the aldehyde as a colorless liquid; 88 mg (89%), $R_f = 0.54$ (*n*-pentane/Et_2O, 10:1).

IR (KBr): $\bar{\nu}$ (cm^{-1}) = 2928, 2858, 1712, 1465, 1414, 1285, 1232, 1109, 938, 725.

^{1}H NMR (300 MHz, $CDCl_3$): δ (ppm) = 9.75 (t, $J = 1.8$ Hz, 1 H, 8-H), 2.41 (td, $J = 7.4$, 1.8 Hz, 2 H, 2-H_2), 1.61 (m_c, 2 H, 3-H_2), 1.29 (m_c, 8 H, H_{alkyl}), 0.87 (t, $J = 6.9$ Hz, 3 H, 8-H_3).

^{13}C NMR (75 MHz, $CDCl_3$): δ (ppm) = 202.9 (C-1), 43.90 (C-2), 31.60 (C-3), 29.11 (C-4), 29.00 (C-5), 22.56 (C-6), 22.07 (C-7), 14.02 (C-8).

MS (DCI, 200 eV): m/z (%) = 179 (9) $[M + NH_3 + NH_4]^+$, 162 (100) $[M + NH_4]^+$.

---

[1]   J.-E. Bäckvall, *Modern Oxidation Methods*, Wiley-VCH, Weinheim, **2003**.

[2]   a) M. Marx, T. T. Tidwell, *J. Org. Chem.* **1984**, *49*, 788; b) A. J. Mancuso, D. Swern, *Synthesis* **1981**, 165.

[3]   N. Kornblum, J. W. Powers, G. J. Anderson, W. J. Jones, H. O. Larson, O. Levand, W. M. Weaver, *J. Am. Chem. Soc.* **1957**, *79*, 6562.

[4]   a) V. V. Zhdankin, P. J. Stang, *Chem. Rev.* **2002**, *102*, 2523; b) T. Wirth, U. H. Hirt, *Synthesis* **1999**, 1271; c) A. Varvoglis, *Hypervalent Iodine in Organic Chemistry*, Academic Press, London, **1997**; d) A. Varvoglis, *Tetrahedron* **1997**, *53*, 379; e) P. J. Stang, V. V. Zhdankin, *Chem. Rev.* **1996**, *96*, 1123.

[5]   a) K. Surendra, N. S. Krishnaveni, M. A. Reddy, Y. V. D. Nageswar, K. Rama Rao, *J. Org. Chem.* **2003**, *68*, 2058; b) A. P. Thottumakra, T. K. Vinod, *Tetrahedron Lett.* **2002**, *43*, 569; c) M. Frigerio, M. Santagostino, S. Sputore, G. Palmisano, *J. Org. Chem.* **1995**, *60*, 7272; d) E. J. Corey, A. Palani, *Tetrahedron Lett.* **1995**, *36*, 3485; e) S. D. Meyer, S. L. Schreiber, *J. Org. Chem.* **1994**, *59*, 7549; f) D. B. Dess, J. C. Martin, *J. Am. Chem. Soc.* **1991**, *113*, 7277; g) D. B. Dess, J. C. Martin, *J. Org. Chem.* **1983**, *48*, 4155; h) A. Speicher, V. Bomm, T. Eicher, *J. prakt. Chem.* **1996**, *338*, 588.

[6]   a) W. Qian, E. Jin, W. Bao, Y. Zhang, *Angew. Chem.* **2005**, *117*, 974; *Angew. Chem. Int. Ed.* **2005**, *44*, 952; b) H. Tohma, S. Takizawa, T. Maegawa, Y. Kita, *Angew. Chem.* **2000**, *112*, 1362; *Angew. Chem. Int. Ed.* **2000**, *39*, 1306; c) T. Yokoo, K. Matsumoto, K. Oshima, K. Utimoto, *Chem. Lett.* **1993**, 571.

[7]   A. de Mico, R. Margarita, L. Parlanti, A. Vescovi, G. Piancatelli, *J. Org. Chem.* **1997**, *62*, 6974.

[8]   S. V. Ley, J. Norman, W. P. Griffith, S. P. Marsden, *Synthesis* **1994**, 639.

[9]   For comparison, see the earlier oxidation methods for *n*-octanol → *n*-octanal in: L. F. Tietze, Th. Eicher, *Reaktionen und Synthesen im organisch-chemischen Praktikum und Forschungslaboratorium*, 2nd ed., p. 96, Georg Thieme Verlag, Stuttgart, **1991**. It should be noted that in application of the oxidation methods **2.3.1–2.3.4** to the above reference process, further oxidation aldehyde → carboxylic acid is suppressed.

[10] L. F. Fieser, K. L. Williamson, *Organic Experiments*, D. C. Heath & Co., Lexington (MA)/Toronto, **1979**, p. 280.

2.4 Enantioselective reduction of ketones

The reduction of aldehydes and ketones with aluminum and boron hydrides to give primary and secondary alcohols, respectively, is a general procedure and has found wide application in organic synthesis. For the enantioselective reduction of prochiral ketones, modified aluminum and boron hydrides associated with a chiral ligand can be employed. In addition, enzymes [1] and hydrogenations in the presence of chiral catalysts [2] have been used.

The most widely applied chiral aluminum and boron compounds are (R)-BINAL-H (**1**), (+)-alpine borane (**2**), diisopinocampheyl chloroborane (**3**), and the so-called CBS reagent (R)-**4**, as well as their enantiomers. The reagent **1**, which was developed by Noyori and co-workers, has special advantages for the reduction of aryl and unsaturated ketones [3]. It is formed *in situ* by adding an equimolar amount of (R)-BINOL to LiAlH$_4$, followed by one equivalent of an alcohol, usually EtOH, although MeOH may also be used. One can assume that the reaction proceeds via transition state TS-**1**.

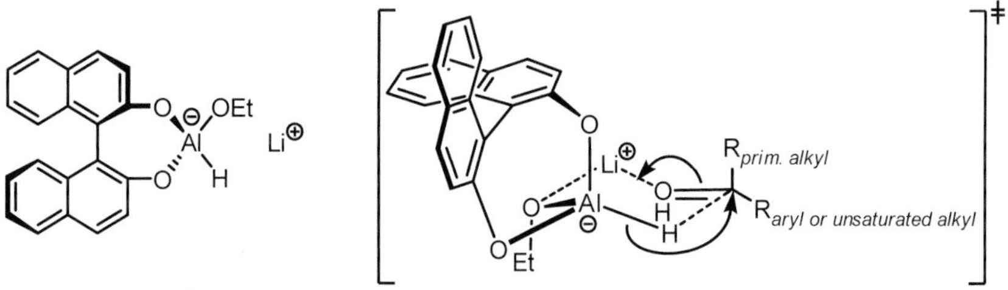

(R)-BINAL–H (**1**)   TS-**1**

The terpene-modified boron compounds **2** and **3** developed by Midland and Brown and co-workers are very useful for the reduction of alkynyl ketones [4]. Reagent **3** usually gives better *ee* values, since it is the stronger Lewis acid and therefore the proposed transition structure TS-**3** is tighter.

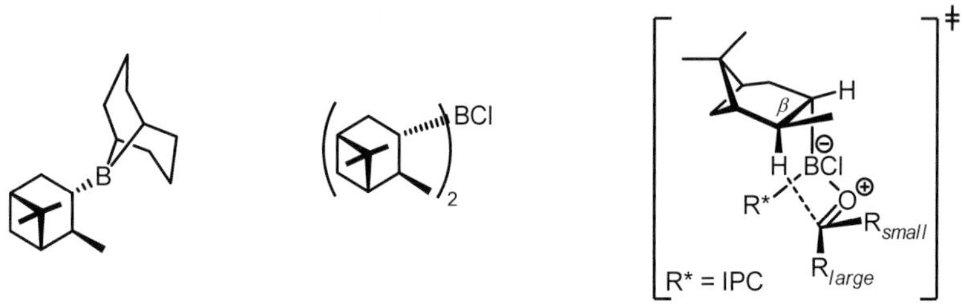

(+)-Alpine Borane (**2**)   (+)-(IPC)$_2$BCl (**3**)   TS-**3**

All three reagents have to be applied in equimolar amounts, which is clearly a disadvantage compared to the use of oxazaborolidine **4**, which is employed in catalytic amounts with a BH$_3$ derivative as stoichiometric reducing agent. The latter reaction is based on the work of Itsuno and co-workers, who succeeded in reducing various ketones by employing mixtures of chiral amino alcohols and BH$_3$·THF [5].

Corey, Bakshi, and Shibata then developed the structurally defined chiral oxazaborolidines (R)-**4** starting from *L*-proline, and used them for the enantioselective reduction of prochiral ketones via transition state TS-**4** [6].

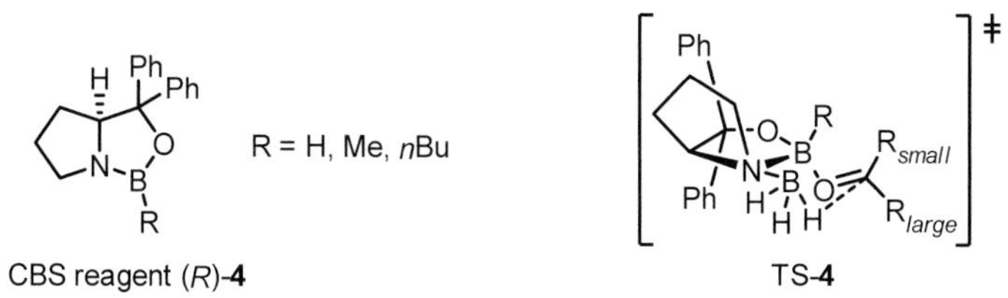

CBS reagent (R)-**4**

R = H, Me, nBu

TS-**4**

Due to its high sensitivity to air and moisture, the initially employed catalyst (R)-**4a** was replaced by the more stable and therefore nowadays commonly applied methyl and butyl derivatives (R)-**4b** and (R)-**4c**, respectively:

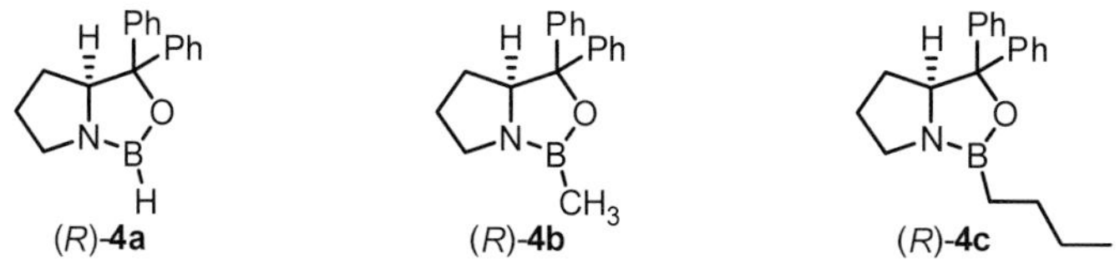

(R)-**4a**   (R)-**4b**   (R)-**4c**

As stoichiometric reducing agents, one uses borane complexes such as $BH_3 \cdot THF$ or $BH_3 \cdot SMe_2$ as well as catecholborane [7]. The latter is usually applied in combination with (R)-**4c** at low temperatures and shows a very low "background reduction rate". In a systematic temperature assay by Stone, the best selectivities were obtained between 30 and 50 °C using catalysts (R)-**4b** and (R)-**4c** and 1,4-thioxane-BH_3 complex [8].

The CBS methodology has been successfully employed for a broad variety of unsymmetrically substituted carbonyl compounds, including alkyl, alkenyl, alkynyl, and aryl ketones. The observed enantioselectivities are generally high if the steric and/or electronic features of the carbonyl substituents are sufficiently different; thus, a large (R_{large}) and a small (R_{small}) substituent in substrate **5** are required [9]. In the important case of the reduction of diaryl ketones with sterically comparable substituents, high enantiomeric excess is obtained for aryl substituents with different electronic properties [10]. The stereochemical outcome of the CBS reduction can be predicted according to the reaction of **5** with (R)-**4b** to give **6** via TS-**4b**.

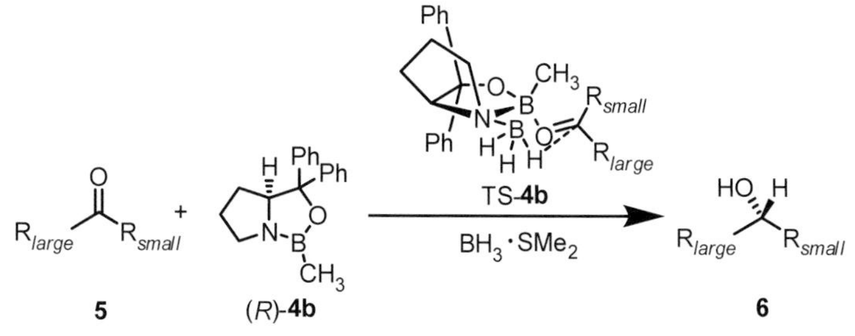

TS-**4b**

$BH_3 \cdot SMe_2$

5   (R)-**4b**   **6**

The proposed intermediates are shown in the following general mechanistic scheme [2]:

The initial step is the coordination of BH_3 to the nitrogen atom at the α face of the oxazaborolidine **4b** to form the cis-fused complex **7**, in which the Lewis basic nitrogen activates BH_3 as a hydride donor while the Lewis acidity of the endocyclic boron atom is strongly increased. The prochiral ketone **5** binds to the endocyclic boron atom through the sterically more accessible non-bonding electron pair a at the oxygen and cis to the vicinal BH_3 group to align the carbonyl group and the coordinated BH_3 in a favorable six-membered chair-like transition state TS-**4b**. The hydride transfer then takes place from above to give the complex **8**. Regeneration of the catalyst may occur by two different pathways:

1) dissociation of **8** into the catalyst **4b** and the borinate **9**, or 2) addition of BH₃ to **8** to form a six-membered BH₃-bridged species **10**, which decomposes to give the catalyst–BH₃ complex **7** and the borinate **9**. Facile disproportionation of **9** to afford the corresponding dialkoxyborane **11** and BH₃ permits the efficient use of the hydrogen atoms of the stoichiometric reductant. Finally, the desired chiral alcohol **6** is obtained from the dialkoxyborane **11** upon work-up with aqueous MeOH.

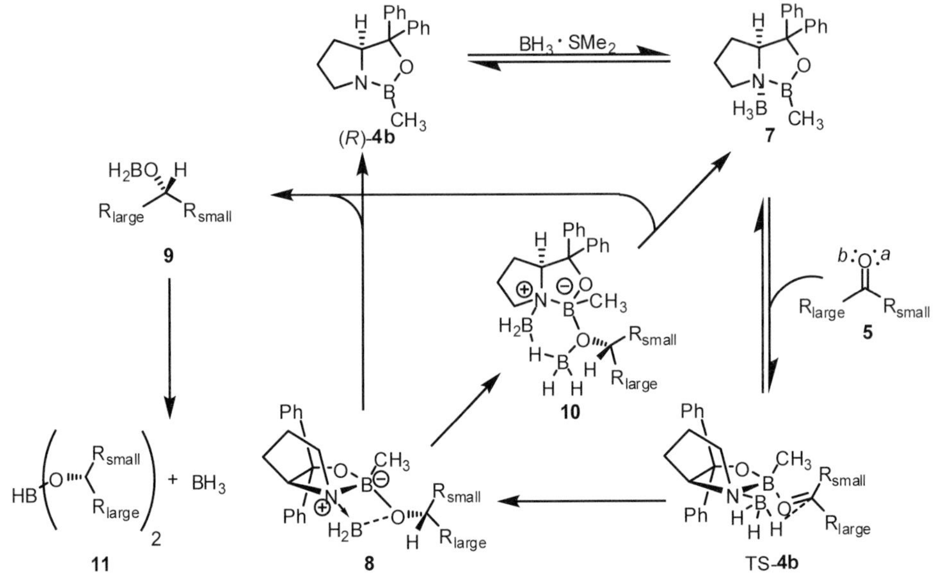

In Sections 2.4.1 and 2.4.2, details are given of the enantioselective reduction of phenyl alkyl ketones **12a/12b** as prochiral substrates to the corresponding benzyl alcohols **13/14** using the chiral complex aluminum hydride/BINAL-H **1** (prepared *in situ*) and the hydride donor combination Me₂S-BH₃/ (*R*)-methyl-CBS **4b**. These two reagents give complementary results with respect to the stereochemistry of product formation: butyrophenone (**12a**) is reduced by **1** to give (*S*)-1-phenylbutanol ((*S*)-**13**), whereas acetophenone (**12b**) is reduced utilizing the CBS reagent **4b** as catalyst to give exclusively (*R*)-1-phenylethanol ((*R*)-**14**):

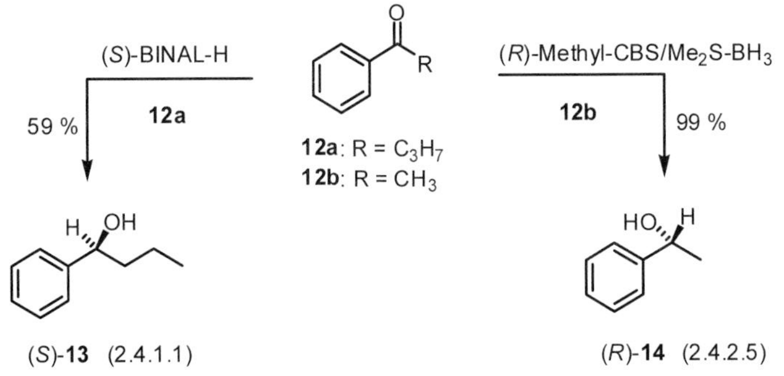

(*S*)-**13** (2.4.1.1)   (*R*)-**14** (2.4.2.5)

2.4.1   BINAL-H reduction of butyrophenone

The following experimental procedure describes the enantioselective reduction of butyrophenone using a stoichiometric amount of the aluminum hydrido complex obtained by reaction of (S)-(+)-BINOL and ethanol with $LiAlH_4$ to give almost enantiopure (S)-(–)-1-phenylbutanol.

2.4.1.1   ***   (S)-(–)-1-Phenylbutanol [3]

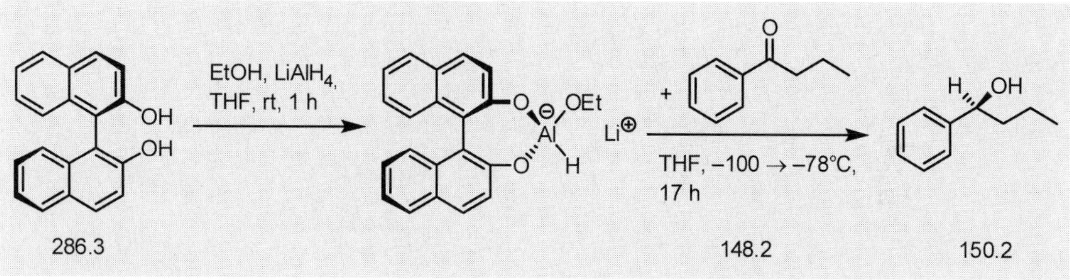

286.3   148.2   150.2

A solution of EtOH (483 mg, 9.50 mmol, 0.56 mL) in THF (5 mL) is added dropwise to a stirred suspension of $LiAlH_4$ (361 mg, 9.50 mmol) in anhydrous THF (20 mL) at 0 °C under nitrogen. (–)-(S)-Binaphthol (2.72 g, 9.50 mmol) in THF (30 mL) is then added and the mixture is stirred for 1 h at room temperature. After cooling to –100 °C, butyrophenone (415 mg, 2.80 mmol, 0.42 mL) in THF (5 mL) is added with stirring, and stirring is continued for 3 h at –100 °C and for 14 h at –78 °C.

At –78 °C, HCl (2 M, 32 mL) and Et_2O (100 mL) are added and the phases are separated. The aqueous phase is extracted with Et_2O (3 × 30 mL) and the combined organic phases are washed with brine, dried over Na_2SO_4, and concentrated. The residue is distilled in a Kugelrohr to give the alcohol as a colorless liquid; 250 mg (59%), bp_{14} 112 °C (oven temperature 120 °C), $n_D^{20} = 1.5138$, $[\alpha]_D^{20} = -43$ (c = 1, benzene). The optical rotation of the pure enantiomer is $[\alpha]_D^{20} = -45$.

Recrystallization of the residue of the distillation from benzene allows recovery of ~80% of the (–)-(S)-binaphthol used.

IR (NaCl): $\tilde{\nu}$ (cm^{-1}) = 3400 (OH), 1030 (C–O).

^{1}H NMR (CDCl$_3$): δ (ppm) = 7.40–7.20 (m, 5 H, arom. H), 4.50 (t, J = 6.0 Hz, 1 H, CH–OH), 3.15 (s, 1 H, OH), 1.09–0.70 (m, 5 H, CH$_2$, CH$_3$).

Note: It is important that the reaction is performed with rigorous exclusion of moisture. The flask and other apparatus must be dried at ca. 130 °C before use. Solutions are best added by means of a syringe (cannula). In several attempts to carry out this reaction, the product was found to be contaminated with butyrophenone.

2.4.2 CBS reduction of acetophenone

(a) Synthesis of the methyl-CBS catalyst (R)-4b

As catalyst for enantioselective reductions, methyl-CBS (R)-4b is prepared from (S)-proline in a linear four-step sequence [11].

In the first step, the NH function of (S)-proline (15) is protected as carbamate 16 using benzyl chloroformate (CbzCl) in alkaline aqueous solution. Formation of the methyl ester 17 is carried out using methanol in the presence of boron trifluoride etherate (BF$_3$·OEt$_2$) as an activator of the carbonyl group. Grignard reaction of the N-Cbz-protected proline methyl ester 17 with phenylmagnesium chloride in THF directly provides the tertiary amino alcohol 18 under cleavage of the carbonyl benzyloxy protecting group. For purification, the crude product is converted into its hydrochloride salt employing HCl gas, and this is recrystallized from MeOH/Et$_2$O. Finally, the pure free amino alcohol is obtained by deprotonation of the amine hydrochloride with NaOH and recrystallization.

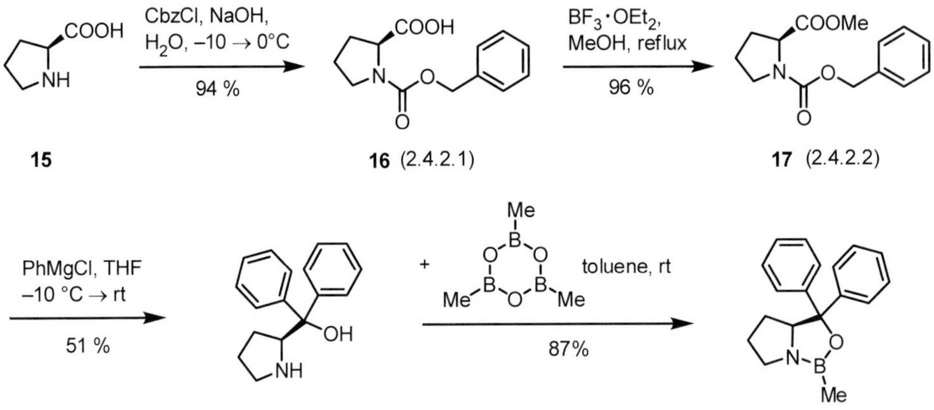

The final step of the oxazaborolidine synthesis is the reaction of the chiral (S)-proline-derived amino alcohol 18 with trimethylboroxine in toluene. The desired methyl-CBS catalyst (R)-4b is obtained after high-vacuum distillation as a white solid that can be stored in closed containers at room temperature and weighed or transferred in air.

(b) Experimental procedures for the synthesis of (R)-4b

2.4.2.1 ** N-(Benzyloxycarbonyl)-(S)-proline [11]

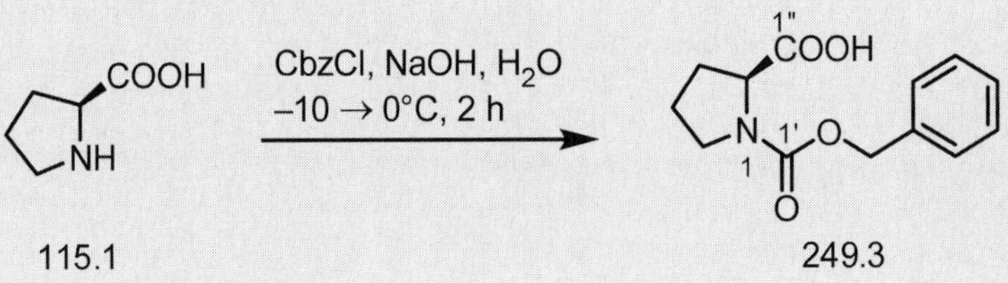

(S)-Proline (23.0 g, 0.20 mol) is added to a magnetically stirred 2 M solution of NaOH in H_2O (100 mL, 0.20 mol) at –10 °C and then benzyl chloroformate (40.9 g, 0.24 mol, 36.4 mL) is added dropwise over 1 h at –5 to 0 °C (internal thermometer). NaOH solution (4 M, 0.28 mol, 70 mL) is then added and stirring is continued for a further 1 h at –5 to 0 °C.

The mixture is then washed with Et_2O (2 × 50 mL). The aqueous solution is acidified to pH 2 with ice-cold 6 M HCl, then saturated with Na_2SO_4 and extracted with EtOAc (3 × 100 mL). The combined extracts are dried twice over anhydrous Na_2SO_4 and concentrated under reduced pressure to give the protected proline carbamate as a colorless oil; 46.9 g (94%), $[\alpha]_D^{20} = -37.0$ ($c = 2.0$, EtOH), $R_f = 0.17$ (CH_2Cl_2).

UV (CH_3CN): λ_{max} (nm) (lg ε) =263 (2.136), 257 (2.241), 252 (2.124), 204 (3.980).

IR (KBr): $\tilde{v}$ (cm^{-1}) = 2958, 1707, 1499, 1428, 1359, 1179.

^{1}H NMR (300 MHz, $CDCl_3$): δ (ppm) = 10.52 (s_{br}, 1 H, COOH), 7.47–7.20 (m, 5 H, Ph-H), 5.24–5.02 (m, 2 H, Bn-H_2), 4.47–4.31 (m, 1 H, 2-H), 3.70–3.37 (m, 2 H, 5-H_2), 2.37–1.78 (m, 4 H, 3-H_2, 4-H_2).

^{13}C NMR (75.5 MHz, $CDCl_3$) (ratio of rotamers: 1:1.5): δ (ppm) = 178.1 + 176.7 (1"-C), 155.6 + 154.4 (1'-C), 136.4 + 136.3 (Ph-C), 128.4 + 128.3 (Ph-C), 128.0 + 127.9 (Ph-C), 127.8 + 127.6 (Ph-C), 127.5 + 127.0 (Ph-C), 67.38 + 67.07 (Bn-C), 59.20 + 58.59 (C-2), 46.85 + 46.53 (C-5), 30.81 + 29.42 (C-3), 24.20 + 23.37 (C-4).

MS (EI, 70 eV): m/z (%) = 249 (12) $[M]^+$, 204 (22) $[M - CO_2H]^+$, 160 (28) $[M - C_7H_6]^+$, 114 (36) $[M - Cbz]^+$, 91 (100) $[C_7H_7]^+$.

2.4.2.2 ** *N*-(Benzyloxycarbonyl)-(*S*)-proline methyl ester [11]

Boron trifluoride etherate (28.4 g, 0.203 mol, 24.6 mL) is added to a magnetically stirred solution of the proline carbamate **2.4.2.1** (33.7 g, 0.135 mol) in anhydrous MeOH (400 mL). The resulting solution is heated to reflux for 1 h.

After evaporation of the solvent, the residue is stirred with iced water (200 mL) and extracted with EtOAc (3 × 100 mL). The combined extracts are successively washed with brine (100 mL), sodium hydrogencarbonate solution (1 M, 100 mL), and further brine (100 mL), and then dried over anhydrous Na_2SO_4. Evaporation of the solvent *in vacuo* yields the Cbz-protected proline methyl ester as a colorless oil; 34.2 g (96%), $[\alpha]_D^{20} = -55.7$ ($c = 1.0$, MeOH), $R_f = 0.32$ (CH_2Cl_2).

UV (CH$_3$CN): λ_{max} (nm) (lg ε) = 267 (2.177), 263 (2.335), 257 (2.442), 252 (2.368), 204 (3.994).

IR (KBr): $\tilde{v}$ (cm^{-1}) = 3479, 3033, 2954, 2882, 1748, 1707, 1416.

^{1}H NMR (300 MHz, CDCl$_3$) (ratio of rotamers: 1:1): δ (ppm) = 7.43–7.23 (m, 5 H, Ph-H), 5.25–5.00 (m, 2 H, Bn-H), 4.45–4.30 (m, 1 H, 2-H), 3.74 + 3.58 (2 s, 3 H, OCH$_3$), 3.70–3.41 (m, 1 H, 3-H$_b$), 2.33–2.11 (m, 1 H, 3-H$_a$), 2.09–1.81 (m, 2 H, 4-H$_2$).

^{13}C NMR (75.5 MHz, CDCl$_3$) (ratio of rotamers: 1:1): δ (ppm) = 173.2 + 173.1 (C-1''), 154.8 + 154.2 (C-1'), 136.6 + 136.5 (Ph-C), 128.4 + 128.3 (Ph-C), 128.3 + 128.1 (Ph-C), 127.9 + 127.8 (Ph-C), 127.8 + 127.7 (Ph-C), 127.4 + 126.9 (Ph-C), 66.92 + 66.85 (Bn-C), 52.15 + 51.96 (C-2), 46.83 + 46.34 (C-5), 30.84 + 29.82 (C-3), 24.23 + 23.45 (C-4).

MS (EI, 70 eV): m/z (%) = 263 (4) [M]$^+$, 204 (10) [M – CO$_2$CH$_3$]$^+$, 160 (12) [M – CH$_3$ – C$_7$H$_6$]$^+$, 108 (100) [M – C$_7$H$_8$O]$^+$, 91 (46) [C$_7$H$_7$]$^+$, 77 (30) [C$_6$H$_5$]$^+$.

2.4.2.3   ***   (S)-2-(Diphenylhydroxymethyl)pyrrolidine [11]

A solution of the Cbz-protected proline methyl ester **2.4.2.2** (26.3 g, 100 mmol) in anhydrous THF (100 mL) is added to a stirred 2 M solution of phenylmagnesium chloride in THF (400 mL, 800 mmol) over 1 h at –10 to 0 °C. The cooling bath is then removed and stirring is continued for 16 h.

The reaction mixture is then poured onto a mixture of ice (300 g), NH$_4$Cl (60 g), and H$_2$O (100 mL) and the resulting mixture is concentrated *in vacuo* to a volume of about 500 mL. After extraction with Et$_2$O (4 × 300 mL), the combined extracts are washed with brine (400 mL), dried over anhydrous K$_2$CO$_3$, and concentrated under reduced pressure to a volume of about 500 mL. Dry HCl gas is then bubbled into the solution until the mixture is acidic. The precipitated amine hydrochloride is collected by filtration, washed with Et$_2$O, and recrystallized from MeOH/Et$_2$O (1:4). The hydrochloride is then suspended in Et$_2$O (300 mL) and treated with NaOH solution (2 M, 60 mL) for 45 min. The resulting solution is extracted with Et$_2$O (3 × 200 mL). The combined extracts are washed with brine (400 mL), dried over anhydrous K$_2$CO$_3$, and concentrated under reduced pressure to give a yellow solid, which is recrystallized from MeOH/H$_2$O to afford the pyrrolidine as colorless crystals; 12.7 g (51%), $[\alpha]_D^{20}$ = –56.0 (c = 3.0, MeOH), R_f = 0.34 (EtOAc).

UV (CH$_3$CN): λ_{max} (nm) (lg ε) = 258 (2.678), 252.5 (2.619), 201.5 (4.365).

IR (KBr): $\tilde{v}$ (cm^{-1}) = 3406, 3329, 2972, 1493, 1446, 1403.

^{1}H NMR (300 MHz, CDCl$_3$): δ (ppm) = 7.65–7.42 (m, 4 H, Ph-H), 7.37–7.05 (m, 6 H, Ph-H), 4.31–4.17 (m, 1 H, 2-H), 3.09–2.84 (m, 2 H, 5-H$_2$), 1.83–1.45 (m, 4 H, 4-H$_2$, 3-H$_2$).

^{13}C NMR (75.5 MHz, CDCl$_3$): δ (ppm) = 148.2, 145.4, 128.2, 127.9, 126.4, 126.3, 125.8, 125.5 (12 × Ph-C), 77.06 (C-1'), 64.45 (C-2), 46.74 (C-5), 26.25 (C-3), 25.48 (C-4).

MS (DCI): m/z (%) = 254 (100) [M + H]$^+$.

2.4.2.4 * (S)-2-Tetrahydro-1-methyl-3,3-diphenyl-1H,3H-pyrrolo[1,2-c][1,3,2] oxazaborolidine [11]**

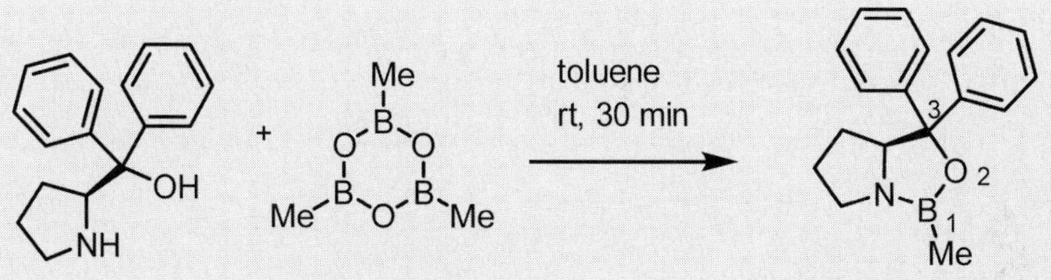

Trimethylboroxine (0.33 g, 2.64 mmol) is added to a solution of the pyrrolidine **2.4.2.3** (1.00 g, 3.95 mmol) in toluene (10 mL). After 2 min, a white precipitate is formed.

After 30 min, further toluene (10 mL) is added, and a portion of toluene (13 mL) is distilled off for azeotropic removal of trimethylboroxine (with air-cooling to avoid crystallization of the trimethylboroxine). Fresh toluene (10 mL) is added and the mixture is distilled to dryness; the process is repeated once more. The obtained yellow oil is distilled (150 °C, 0.05 mbar, sublimation apparatus) to give the oxazaborolidine as a white solid; 0.95 g (87%).

MS (EI, 70 eV): m/z (%) = 277 (72) [M]$^+$, 165 (60) [C$_8$H$_{12}$BNO$_2$]$^+$, 70 (100) [C$_4$H$_8$N]$^+$.

(c) CBS reduction of acetophenone

As a standard procedure, the enantioselective reduction of acetophenone to give almost enantiopure (R)-(+)-1-phenylethanol is described, which involves the use of a catalytic amount of methyl-CBS and a stoichiometric amount of borane dimethyl sulfide complex.

2.4.2.5 ** (R)-(+)-1-Phenylethanol [9a]

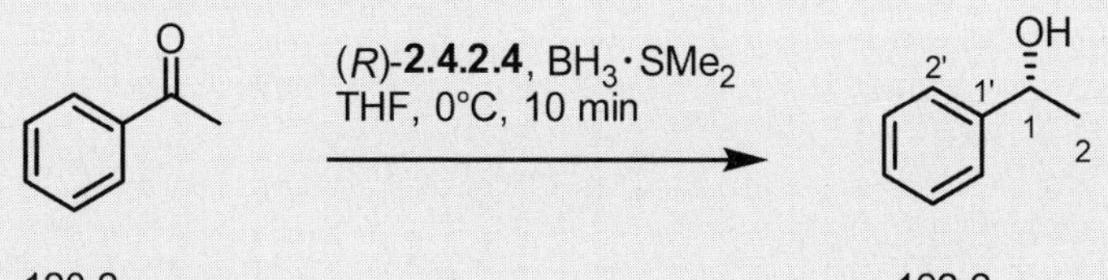

A solution of acetophenone (240.3 mg, 2.00 mmol) in THF (20 mL) is added to a stirred ice-cold solution of the oxazaborolidine **2.4.2.4** (55.4 mg, 0.200 mmol) and borane dimethyl sulfide complex (2 M in THF, 0.600 mL, 1.20 mmol) in THF (20 mL). Stirring is continued at 0 °C for 10 min.

MeOH (2 mL) is then added, the solvent is removed *in vacuo*, and the residue is adsorbed on silica (0.5 g). After column chromatography eluting with CH_2Cl_2, the phenylethanol is obtained as a colorless liquid, 241.8 mg (99%), *ee* = 97.3%, $[\alpha]_D^{20}$ = +41.0 (*c* = 5.0, MeOH), R_f = 0.37 (CH_2Cl_2).

UV (CH_3OH): λ_{max} (nm) (lg ε) = 263 (2.339), 257 (2.441), 252 (2.381), 247 (2.299), 207 (3.970).

IR (KBr): $\tilde{\nu}$ (cm^{-1}) = 3354, 2973, 2927, 1603, 1493, 1452, 1302, 1204, 1078.

^{1}H NMR (300 MHz, $CDCl_3$): δ (ppm) = 7.39–7.22 (m, 5 H, Ph-H), 4.86 (q, 1 H, *J* = 6.3 Hz, 1-H), 2.09 (1 H, OH), 1.48 (d, *J* = 6.3 Hz, 3 H, 2-H).

^{13}C NMR (75.5 MHz, $CDCl_3$): δ (ppm) = 145.8 (C-1'), 128.4 (C-3', C-5'), 127.4 (C-4'), 125.3 (C-2', C-6'), 70.32 (C-1), 25.09 (C-2).

MS (EI, 70 eV): *m/z* (%) = 122 (44) $[M]^+$, 107 (100) $[C_7H_7]^+$, 79 (87) $[C_6H_7]^+$.

GC: column: WCOT fused silica CP-Chirasil-DEX CB (25 m × 0.25 mm)
             carrier: H_2
             temperature: 135 °C
             retention time: t_{R1} = 1.04 min (major enantiomer)
                                 t_{R2} = 1.11 min (minor enantiomer)

---

[1]   W. Stumpfer, B. Kosjek, C. Moitzi, W. Kroutil, K. Faber, *Angew. Chem.* **2002**, *114*, 1056; *Angew. Chem. Int. Ed.* **2002**, *41*, 1014.

[2]   a) R. Noyori, T. Ohkuma, *Pure Appl. Chem.* **1999**, *71*, 1493; b) T. Benincori, E. Cesarotti, O. Piccolo, F. Sannicolo, *J. Org. Chem.* **2000**, *65*, 2043; T. Benincori, O. Piccolo, S. Rizzo, F. Sannicolo, *J. Org. Chem.* **2000**, *65*, 8340.

[3]   R. Noyori, I. Tomino, Y. Tanimoto, *J. Am. Chem. Soc.* **1979**, *101*, 3129.

[4]   a) M. M. Midland, *Chem. Rev.* **1989**, *89*, 1553; b) H. C. Brown, V. Ramachandran, *Acc. Chem. Res.* **1992**, *25*, 16.

[5]   a) A. Hirao, S. Itsuno, S. Nakahama, N. Yamazaki, *J. Chem. Soc., Chem. Commun.* **1981**, 315; b) S. Itsuno, A. Hirao, S. Nakahama, N. Yamazaki, *J. Chem. Soc., Perkin Trans. 1* **1983**, 1673.

[6]   E. J. Corey, C. J. Helal, *Angew. Chem.* **1998**, *110*, 2092; *Angew. Chem. Int. Ed.* **1998**, *37*, 1986.

[7]   a) E. J. Corey, J. O. Link, *Tetrahedron Lett.* **1989**, *30*, 6275; b) E. J. Corey, R. K. Bakshi, *Tetrahedron Lett.* **1990**, *31*, 611.

[8]   B. Stone, *Tetrahedron Asymmetry* **1994**, *5*, 465.

[9]   a) E. J. Corey, R. K. Bakshi, S. Shibata, C.-P. Chen, V. K. Singh, *J. Am. Chem. Soc.* **1987**, *109*, 7925; b) A. M. Salunkhe, E. R. Burkhardt, *Tetrahedron Lett.* **1997**, *38*, 1523; c) E. J. Corey, C. J. Helal, *Tetrahedron Lett.* **1995**, *36*, 9153; d) P. Wipf, S. Lim, *J. Am. Chem. Soc.* **1995**, *117*, 558; e) K. A. Parker, M. W. Ledeboer, *J. Org. Chem.* **1996**, *61*, 3214.

[10] a) E. J. Corey, C. J. Helal, *Tetrahedron Lett.* **1995**, *36*, 9153; b) E. J. Corey, C. J. Helal, *Tetrahedron Lett.* **1996**, *37*, 5675.

[11] a) E. J. Corey, S. Shibata, R. K. Bakshi, *J. Org. Chem.* **1988**, *53*, 2861; b) G. A. Reichard, C. Stengone, S. Paliwal, I. Mergelsberg, S. Majmundar, C. Wang, R. Tiberi, A. T. McPhail, J. J. Piwinski, N.-Y. Shih, *Org. Lett.* **2003**, *5*, 424.

3   Heterocyclic compounds

Introduction

Heterocycles constitute the largest group of organic compounds and are found in a vast number of natural products (e.g., vitamins, alkaloids, antibiotics) as well as in products with biological, medicinal, and technical relevance (e.g., pharmaceuticals, biocides, dyes). In the following chapters, **3.1**–**3.5**, selected syntheses and reactions of a series of aromatic and non-aromatic heterocycles are presented, which are organized according to ring size and structural complexity.

Chapter **3.1** deals with *three- and four-membered heterocycles*.

The selected examples are:

- the synthesis of propranolol (**3.1.1**) as a racemic mixture and as its (*S*)-enantiomer (the latter is obtained by enantioselective epoxidation using the Sharpless–Katsuki reaction, cf. **2.1.1**) employing a regio- and stereoselective ring-opening of an oxirane;

- the synthesis of an oxetane (**3.1.2**) by photochemical [2+2]-cycloaddition of a carbonyl compound to an alkene (Paterno–Büchi reaction);

- the synthesis of a β-lactam (**3.1.3**) by a *cis*-stereoselective thermal [2+2]-cycloaddition of a ketene and an imine (Staudinger reaction).

Chapter **3.2** deals with *five-membered heterocycles* and their benzo derivatives.

The selected examples are:

- the synthesis of 2,4-diphenylfuran (**3.2.1**), in which a 2,4,6-triphenylpyrylium cation undergoes a base-induced ring-opening and recyclization;

- the synthesis of 3,4-dimethylpyrrole (**3.2.2**) utilizing the cyclocondensation of isocyanoacetate and 2-nitro-2-butene according to the Barton–Zard pyrrole synthesis;

- the synthesis of a benzothiophene (**3.2.3**) by a novel sequence consisting of a directed *ortho*-metalation of a benzamide, Ar-Li trapping by $(CH_3S)_2$, and intramolecular acylation of a metalated Ar–S–CH$_3$ species;

- the synthesis of 2-phenylindole (**3.2.4**) by TBAF-catalyzed cyclization of (*o*-phenylethynyl)aniline obtained by a Sonogashira cross-coupling (cf. **1.6.3**);

- the synthesis of melatonin (**3.2.5**) using a domino process including a Japp–Klingemann reaction and a Fischer indole synthesis of an α-(3-phthalimidopropyl)acetoacetate;

- the synthesis of a dihydropyrazole derivative (**3.2.6**) utilizing the thermal conversion of a 1,2,4-oxadiazole in a 1,5-electrocyclization;

Reactions and Syntheses in the Organic Chemistry Laboratory. L. F. Tietze, Th. Eicher, U. Diederichsen, A. Speicher
Copyright © 2007 WILEY-VCH Verlag GmbH & Co. KGaA, Weinheim
ISBN: 978-3-527-31223-8

- the synthesis of camalexin (**3.2.7**) based on the Fürstner indole synthesis involving an intramolecular McMurry reaction of an *ortho*-*N*-diacylaniline;

- the formation of a tetrahydroindazolone from a 1,3-diketone and an aryl hydrazine (**3.2.8**) as an example of a microwave-assisted heterocycle synthesis.

Chapter **3.3** deals with *six-membered heterocycles* and their benzo derivatives.

The selected examples are:

- azine and diazine syntheses with acetoacetate as a building block (**3.3.1**), which include the Hantzsch dihydropyridine synthesis, $SnCl_4$-mediated pyridine synthesis with β-enamino-nitriles, and the Biginelli synthesis of 3,4-dihydropyrimidin-2(1*H*)-ones;

- the synthesis of (*R*)-salsolidine (**3.3.2**) through application of an Ru-catalyzed enantioselective transfer hydrogenation (Noyori hydrogenation) of the corresponding 3,4-dihydroisoquinoline obtained by Bischler–Napieralski synthesis;

- the synthesis of epirizole, a 2-pyrimidinyl-substituted pyrazolone (**3.3.3**), in which the two heterocyclic systems are consecutively formed by application of the Pinner pyrimidine synthesis and the Knorr pyrazolone synthesis;

- the convergent multi-step synthesis of a Ras FT inhibitor (**3.3.4**), which is constructed from a piperazinone part (obtained by the Mitsunobu protocol) and an imidazole part (obtained by application of the Marckwald imidazole synthesis);

- the linear multi-step synthesis of *rac*-dihydrexidine (**3.3.5**), which utilizes a Heck reaction (cf. **1.6.1**) and a Dieckmann cyclization for the construction of a β-tetralone and a photocyclization of a protected enamide to give a partially hydrogenated benzophenanthridine system.

Chapter **3.4** deals with a series of *condensed heterocyclic compounds*, in which different heterocycles are combined by annelation.

The selected examples are:

- the synthesis of a naphthoindolizinequinone (**3.4.1**), the tetracyclic system of which is built from an indolizine derivative (obtained from a pyridinium betaine by way of a 1,3-dipolar cycloaddition with ADE) and a 1,4-quinone (obtained by intramolecular Friedel–Crafts cyclization);

- the synthesis of a pyrrolo[2,3-*d*]pyrimidine (**3.4.2**), in which a domino process based on a Dakin–West reaction and Dimroth rearrangement is applied;

- the synthesis of a 2,6-naphthyridine (**3.4.3**) using a halogenoazine metalation, Sonogashira cross-coupling, and isoquinoline synthesis according to the Larock principle;

- the synthesis of caffeine (**3.4.4**) utilizing the Traube purine synthesis in the Bredereck modification via 4-amino-1,3-dimethyluracil and theophylline;

- the multi-step synthesis of a nedocromil analogon (**3.4.5**), the heterotricyclic structure of which is constructed by linear annelation of a 4-chromone and 4-quinolone unit;

- the synthesis of a polyheterocycle (**3.4.6**) by high-pressure cyclization of the Knoevenagel product of an *o*-homoallyloxy-substituted benzaldehyde with dimethylbarbituric acid.

Chapter **3.5** deals with *further heterocyclic systems* and with *heterocyclic dyes*.

The selected examples are:

- the synthesis of the furofuran samin (**3.5.1**), in which a Wittig reaction and an *exo-trig* radical cyclization of a (β-bromoethyl)propargyl ether moiety represent the key steps;

- the synthesis of a dibenzopyridino[18]crown-6 (**3.5.2**) through macrocyclization by cycloalkylation of two bisphenol units with 2,6-bis(bromomethyl)pyridine under high-dilution conditions;

- the synthesis of indigo (**3.5.3**) from *o*-nitrobenzaldehyde;

- the synthesis of pyrvinium iodide (**3.5.4**), in which an unsymmetrical cyanine system is formed by aldol condensation of a pyrrole-3-carbaldehyde (obtained by Paal–Knorr pyrrole synthesis) with a 2-methylquinolinium system (obtained by Doebner–Miller quinoline synthesis);

- the synthesis of a porphyrin derivative (**3.5.5**) by acid-catalyzed cyclotetramerization of 3,4-dimethylpyrrole (cf. **3.2.2**) after hydroxymethylation with formaldehyde;

- the synthesis of a rotaxane (**3.5.6**) consisting of a crown ether and a blocked linear bis-(4,4'-dipyridylium)ethane unit.

3.1   Three- and four-membered heterocycles

3.1.1   (S)-Propranolol

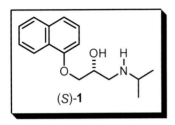

(S)-1

Topics:
- Synthesis of a drug (a) as a racemic mixture and (b) as the (S)-enantiomer
- O-Alkylation of a phenol, O-tosylation of glycidol
- Asymmetric epoxidation of allyl alcohol by the Sharpless–Katsuki reaction
- Stereo- and regioselective oxirane opening by a secondary amine

(a) General

Propranolol (**1**) belongs to the class of aroyloxy- or hetaroyloxy-propanolamines **2**, which are medicinally applied as β-adrenolytics ("β-blocking agents") and which in their activity profile show cardioselectivity, intrinsic sympathomimetic activity, and membrane effects. Propranolol is manufactured as a racemic mixture, although it is known that the (S)-enantiomer is far more effective than the (R)-enantiomer [1–3].

$$\text{2}$$

Ⓧ = aryl, hetaryl
R = H, CH3
⋯ = stereogenic center

Retrosynthesis of propranolol (**2**, X = α-naphthyl, R = H) follows the conventional disconnection pattern for 1,2-diol derivatives. Thus, the synthesis could start with O-alkylation of α-naphthol (**4**) by using glycidol **5** containing a leaving group and may proceed via an oxirane ring-opening of the formed glycidol ether **3** by isopropylamine. Glycidol **5** can be obtained by epoxidation of an allylic compound **6** (**6** → **5**), which can be performed in an enantioselective way thus allowing the stereocontrolled formation of the stereogenic center in **1**.

Ar = α-naphthyl

3

disc.

Ar—OH + X △ ⇌ X ⌇

4   **5**   **6**

The described concept is used for the preparation of *rac*-propranolol as well as of its (S)-enantiomer.

A different enantioselective synthesis of **1** was developed using an asymmetric nitroaldol reaction of aldehyde **7** with nitromethane catalyzed by a chiral La(III)-Li-(R)-BINOL complex [4]. The nitroaldol

8 possessing (S)-configuration at the stereogenic center was transformed into (S)-propranolol by catalytic reduction NO₂ → NH₂ with H₂/PtO₂ and subsequent introduction of the isopropyl moiety by reductive amination with acetone:

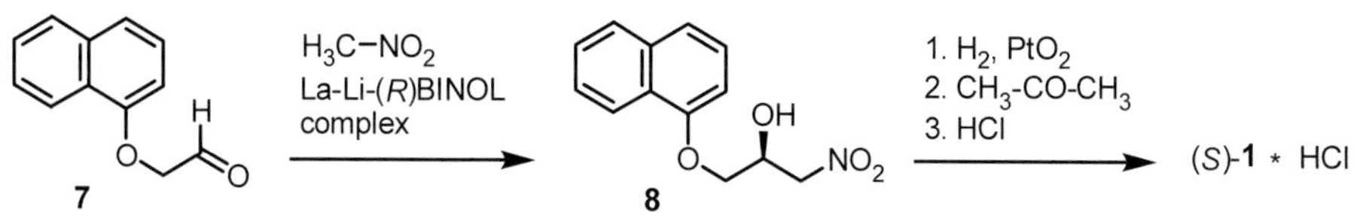

(b) Synthesis of 1

The synthesis of racemic propranolol (rac-**1**) [5, 6] starts with O-alkylation of α-naphthol (**9**) with epichlorohydrin, which proceeds via an S_N reaction of the α-naphtholate anion to give the glycidol ether **10**:

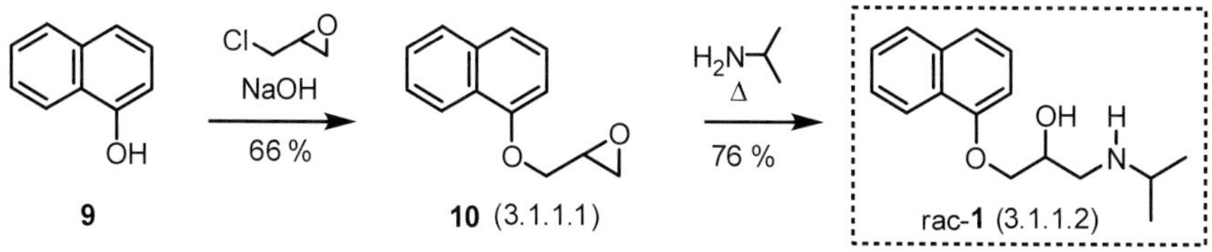

Subsequent reaction of **10** with isopropylamine leads to rac-propranolol (rac-**1**); obviously, opening of the oxirane ring in **10** occurs regioselectively by attack of the N-nucleophile at the sterically less hindered CH_2 site.

The asymmetric synthesis of (S)-**1** [7] starts with the enantioselective epoxidation of allyl alcohol using cumene hydroperoxide in the presence of catalytic amounts of Ti(OiPr)₄ and (−)-diisopropyl tartrate according to the Sharpless–Katsuki protocol (for details, see **2.1.1** and refs. [7, 8]). The initially formed (S)-glycidol (**11**) is sulfonylated in situ with TosCl/ NEt₃ to give the (2S)-glycidyl tosylate (**12**). Treatment of the tosylate **12** with sodium 1-naphthoxide in DMF affords the chiral α-naphthyl epoxy ether **13**, which is subjected in situ to regioselective oxirane ring-opening by reaction with isopropylamine to give (S)-propranolol ((S)-**1**):

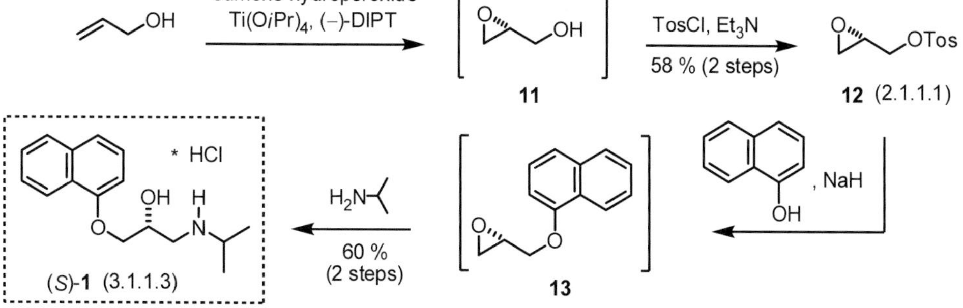

It should be noted that (S)-propranolol can also be obtained from rac-**1** by chemoenzymatic resolution [9]. Enantioselective esterification of rac-**1** with succinic anhydride mediated by lipase (PS-D) leads directly to (S)-**1** and the hemisuccinate **14** of (R)-**1**. In this way, (R)-propranolol may also be prepared, namely by hydrolysis of **14** in MeOH/H₂O in the presence of K₂CO₃:

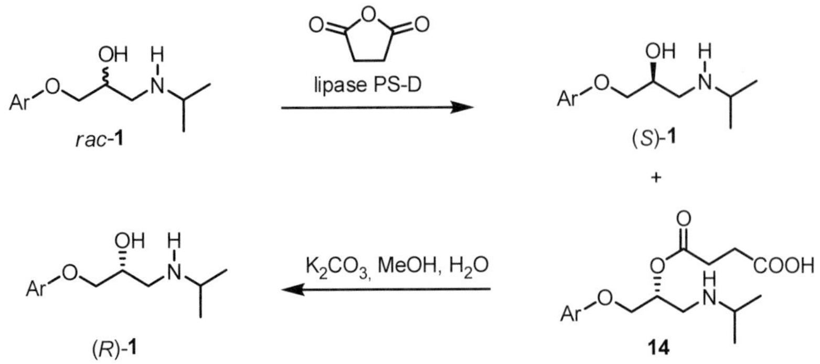

The described procedure allows the synthesis of *rac*-**1** in a two-step sequence with an overall yield of 50% (based on α-naphthol), and of the (*S*)-enantiomer in a three-step sequence (two separate operations) with an overall yield of 32% (based on allylic alcohol).

(c) Experimental procedures for the synthesis of 1

3.1.1.1   *   2,3-Epoxypropyl-1-(1'-naphthyl) ether [5]

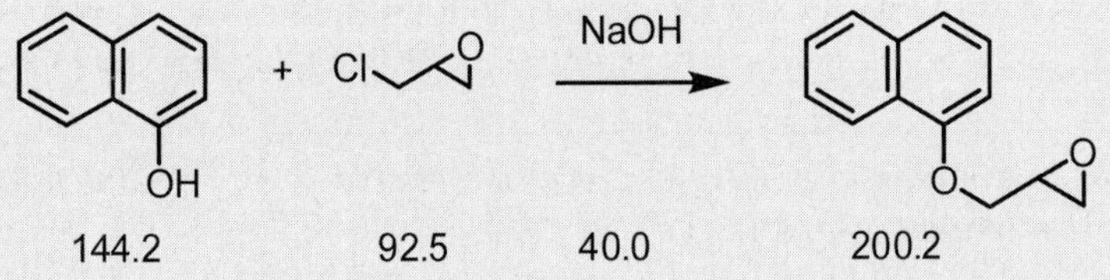

1-Naphthol (7.21 g, 50.0 mmol) is dissolved in a solution of NaOH (2.00 g, 50.0 mmol) in H_2O (10 mL). With vigorous stirring, epichlorohydrin (4.67 g, 50.0 mmol) (note) is added dropwise at such a rate that the internal temperature does not exceed 35 °C. When the addition is complete, stirring is continued for 12 h at 20 °C.

The reaction product (which separates as an oil) is extracted with Et_2O (2 × 20 mL). The ethereal extracts are combined, extracted with 1 N NaOH (2 × 20 mL), and dried (Na_2SO_4). After filtration, the solvent is removed and the residue is fractionated *in vacuo*. The product is obtained as a colorless oil; 6.63 g (66%), bp$_{0.5}$ 124–125 °C, TLC (SiO_2; $CHCl_3$): $R_f = 0.75$.

IR (film): $\tilde{\nu}$ (cm^{-1}) = 3070, 2930 (CH), 1600, 1590, 1520 (C=C), 1410, 1280, 1110, 800, 780.

^{1}H NMR (300 MHz, CDCl$_3$): δ (ppm) = 8.25–8.34 (m, 1 H, 8-H), 7.78–7.82 (m, 1 H, 2-H), 7.31–7.57 (m, 4 H, 4-H, 5-H, 6-H, 7-H), 6.80 (dd, *J* = 9.0, 1.0 Hz, 1 H, 3-H), 4.39 (dd, *J* = 11.0, 3.0 Hz, 1 H, O–CH$_2$), 4.14 (dd, *J* = 11.0, 6.0 Hz, 1 H, O–CH$_2$), 3.43–3.52 (m, 1 H, oxirane-CH), 2.96 (dd, *J* = 5.0, 4.0 Hz, 1 H, oxirane-CH$_2$), 2.84 (dd, *J* = 5.0, 1.0 Hz, 1 H, oxirane-CH$_2$).

Note: Epichlorohydrin has to be distilled before use, bp$_{760}$ 116–117 °C, n$_D^{20}$ = 1.4380.

3.1.1.2   *   1-Isopropylamino-3-(1'-naphthyloxy)-2-propanol (*rac*-propranolol) [6]

200.2   59.1   259.4

A mixture of epoxy ether **3.1.1.1** (5.52 g, 28.0 mmol) and isopropylamine (5.10 g, 86.0 mmol) (note 1) is heated under reflux for 16 h.

The excess amine is then removed *in vacuo*, the residue is poured into HCl (2 N, 75 mL), and the acidic phase is extracted with Et$_2$O (2 × 30 mL) (note 2). The acidic phase is added dropwise with stirring to ice-cold NaOH (2 N, 150 mL). The product precipitates as colorless crystals, which are collected by suction filtration, dried *in vacuo* over P$_4$O$_{10}$, and recrystallized from cyclohexane. *rac*-Propranolol is obtained as colorless needles; 5.45 g (76%), mp 94–96 °C, TLC (SiO$_2$; MeOH): R_f = 0.20.

IR (KBr): $\tilde{\nu}$ (cm^{-1}) = 3280 (NH), 1600, 1590, 1410, 1280, 1110.

^{1}H NMR (300 MHz, CDCl$_3$): δ (ppm) = 8.20–8.22 (m, 1 H, 8-H), 7.78–7.82 (m, 1 H, 2-H), 7.31–7.50 (m, 4 H, 4-H, 5-H, 6-H, 7-H), 6.80 (dd, J = 7.5, 1.0 Hz, 1 H, 3-H), 4.10–4.20 (m, 3 H, CH$_2$–O, CH–O), 2.80–3.02 (m, 3 H, CH$_2$–N, CH–N), 1.80 (s$_{br}$, 2 H, OH, NH), 1.21 (d, J = 6 Hz, 6 H, CH$_3$).

Notes: (1)   Isopropylamine is dried over solid KOH and distilled before use, bp$_{760}$ 33–34 °C.
          (2)   This extraction removes neutral impurities.

3.1.1.3   **   (S)-Propranolol [7]

144.2   228.3   200.2   259.4

1-Naphthol (1.35 g, 9.4 mmol) in DMF (5 mL) is added to a stirred suspension of NaH (60%, 428 mg, 10.7 mmol) in DMF (9 mL) under an argon atmosphere at room temperature to produce a foamy green sludge. After 30 min, (2S)-glycidyl tosylate **2.1.1.1** (2.0 g, 8.9 mmol) is added and stirring is continued for 3.5 h (the reaction is monitored by TLC, EtOAc/n-hexane, 2:3). Isopropylamine (7.58 mL, 89.2 mmol) and H$_2$O (0.76 mL, 42.4 mmol) are added and the mixture is heated to reflux for 3.5 h (the reaction is monitored by TLC, CH$_2$Cl$_2$/n-hexane, 1:1).

After cooling to room temperature, the reaction mixture is diluted with H$_2$O (25 mL) and extracted with Et$_2$O (3 × 25 mL). The combined organic extracts are washed with NaOH (1 N, 50 mL) and brine (50 mL). After extraction with HCl (2 N, 40 mL), NaOH (2 N, 50 mL) is added. The residue is collected by suction filtration and recrystallized from n-hexane/Et$_2$O to give (2S)-propranolol as colorless needles; 1.29 g (56%), mp 72–73 °C, $[\alpha]_D^{20} = -8.5$ ($c = 1.0$, EtOH), $ee = 85\%$, $R_f = 0.22$ (MeOH/EtOAc, 1:1). The enantiomeric excess is determined by HPLC on a chiral stationary phase.

---

^{1}H NMR (300 MHz, CDCl$_3$): δ (ppm) = 8.21–8.27 (m, 1 H, 9'-H), 7.75–7.82 (m, 1 H, 6'-H), 7.40–7.20 (m, 3 H, 4'-H, 7'-H, 8'-H), 7.37 (t, $J = 12.0$ Hz, 1 H, 3'-H), 6.80 (dd, $J = 1.2$, 6.8 Hz, 1 H, 2'-H), 4.08–4.20 (m, 3 H, 2-H, 1-H), 2.98 (dd, $J = 3.6$, 12.0 Hz, 1 H, 3-H), 2.77–2.89 (m, 2 H, 3-H, 2''-H), 1.10 (d, $J = 6.3$ Hz, 6 H, CH$_3$).

^{13}C NMR (300 MHz, CDCl$_3$): δ (ppm) = 154.3 (C1'), 134.4 (C2'), 127.5 (C6'), 126.4 (C7'), 125.8 (C3'), 125.5 (C8'), 125.2 (C10'), 121.8 (C9'), 120.5 (C4'), 104.8 (C2'), 70.7 (C1), 68.5 (C2), 49.6 (C3), 48.9 (C2''), 23.1 (Me), 23.0 (Me).

HPLC: Chiralpak IB (Daicel); 250 × 4.6 mm
        eluent: *n*-hexane/EtOAc, 1:1, isocratic
        flow: 1.0 mL min^{-1}
        detection: UV 270 nm
        retention time: $t_R = 3.54$ min (minor enantiomer), 3.98 min (major enantiomer)

---

[1]   A. Kleemann, J. Engel, *Pharmaceutical Substances*, 3rd ed., p. 1613, Georg Thieme Verlag, Stuttgart, **1999**.

[2]   H.-D. Höltje, H. Auterhoff, J. Knabe, *Lehrbuch der Pharmazeutischen Chemie*, 13th ed., p. 504, Wissenschaftliche Verlagsgesellschaft mbH, Stuttgart, **1994**.

[3]   H. J. Roth, H. Fenner, *Arzneistoffe, Struktur – Bioreaktivität – Wirkungsbezogene Eigenschaften*, p. 422, Georg Thieme Verlag, Stuttgart, **1988**.

[4]   H. Sasai, T. Suzuki, N. Itoh, M. Shibasaki, *Appl. Organomet. Chem.* **1995**, *9*, 421.

[5]   Th. v. Lindemann, *Ber. Dtsch. Chem. Ges.* **1891**, *24*, 2145; see also: *Houben/Weyl*, Vol. 6/3, p. 424, **1965**.

[6]   A. F. Crowther, L. H. Smith, *J. Med. Chem.* **1968**, *11*, 1009; USP 3 337 628 (*Chem. Abstr.* **1969**, *70*, 3607j).

[7]   J. M. Klunder, S. Y. Ko, K. B. Sharpless, *J. Org. Chem.* **1986**, *51*, 3710.

[8]   M. B. Smith, J. March, *March's Advanced Organic Chemistry*, 5th ed., p. 1053, John Wiley & Sons, Inc., New York, **2001**.

[9]   S. V. Damle, P. N. Patil, M. M. Salunkhe, *Synth. Commun.* **1999**, *29*, 3855.

3.1.2   Oxetane derivative

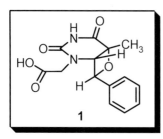

1

Topic: • Formation of an oxetane derivative by photochemical [2+2]-cycloaddition of a C=O group and a C=C double bond

(a) General

Oxetanes are the higher homologues of oxiranes and possess a slightly distorted, non-planar square structure [1]. For the synthesis of oxetanes [2], two methods (1 and 2) are useful.

(1): Alcohols with a leaving group in the γ-position can be cyclized to oxetanes by S_Ni processes upon interaction with bases, viz.:

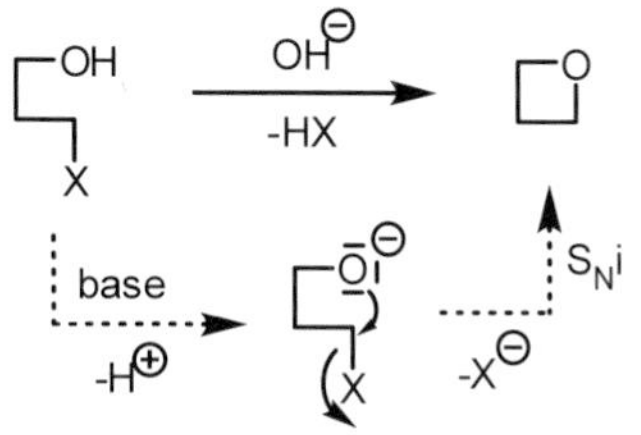

(2): The photochemical [2+2]-cycloaddition of carbonyl compounds (aldehydes/ketones) and alkenes, known as the Paterno–Büchi reaction [3], likewise leads to oxetanes:

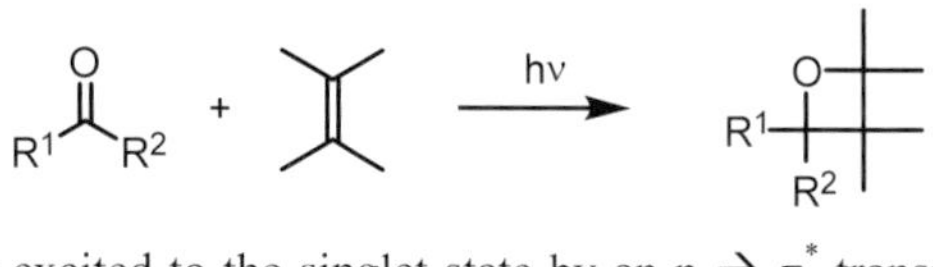

The carbonyl group is first excited to the singlet state by an n → π^* transition, which is followed by intersystem crossing to the lower-energy triplet state; if the addition to the C=C bond takes place from the singlet state, according to the Woodward–Hoffman rules, the cycloaddition should be a concerted, stereospecific process.

In fact, stereospecificity in oxetane formation (e.g. → **2**) is observed with alkenes bearing electron-withdrawing substituents:

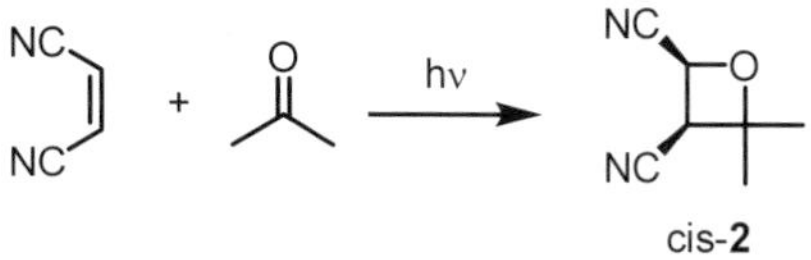

cis-**2**

In contrast, alkenes with donor substituents react in a non-stereoselective manner, as shown for the formation of the *cis/trans* isomeric oxetanes **4**.

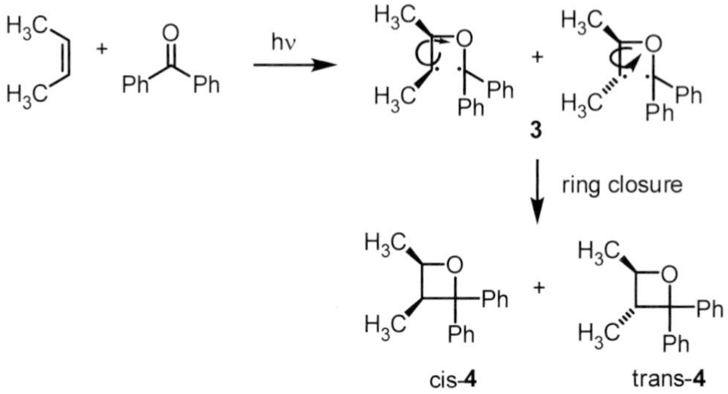

This is interpreted in terms of the intermediacy of diradical species **3** and their capacity to undergo rotation about the C-2/C-3 bond before ring-closure.

Oxetane formation by [2+2]-photocycloaddition is generally possible with acceptor-substituted alkenes, quinones, and electron-deficient heterocycles. A typical example of a Paterno–Büchi reaction of such heterocyclic systems is presented in section (b).

(b) Synthesis of 1

The substrate for the Paterno–Büchi reaction is thymine-1-acetic acid (**7**), which is easily accessible from thymine (**5**) by *N*-alkylation with ethyl bromoacetate to give **6** and subsequent saponification of the ester group:

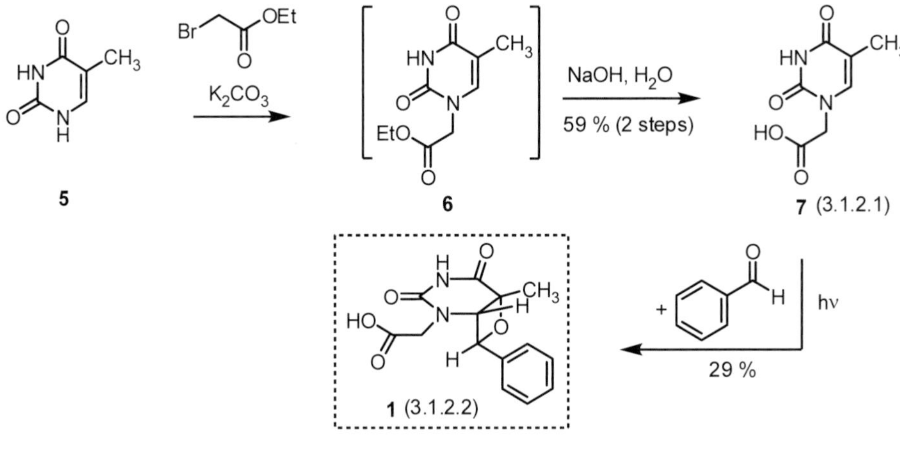

When thymine-1-acetic acid (**7**) is irradiated together with benzaldehyde in H_2O/acetonitrile solution, the bicyclic oxetane derivative **1** is formed as a cycloadduct, in which the phenyl residue at the four-membered ring is oriented exclusively in an *exo*-configuration according to its spectroscopic data [4]. Since the photochemically induced [2+2]-cycloaddition of the aldehyde C=O group to the thymine 5,6-C=C double bond apparently occurs with complete regio- and stereospecificity, it is likely that oxetane formation proceeds in a concerted manner.

(c) Experimental procedures for the synthesis of 1

3.1.2.1   **   Thymine-1-acetic acid [5, 6]

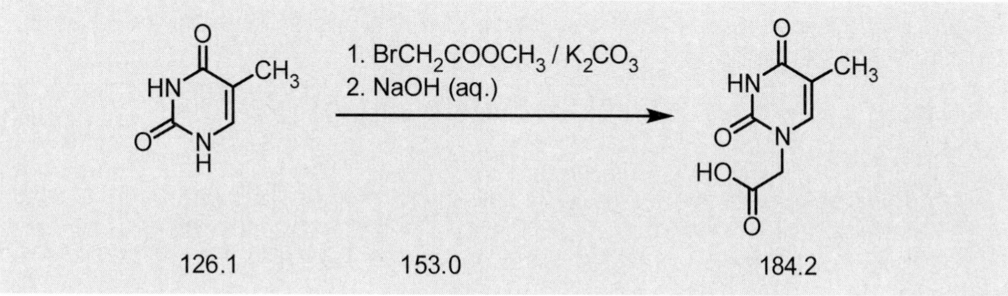

126.1   153.0   184.2

Methyl bromoacetate (11 mL, 119 mmol) is added to a suspension of thymine (15.0 g, 119 mmol) and
K_2CO_3 (16.5 g, 119 mmol) in anhydrous DMF (300 mL) and the mixture is vigorously stirred
overnight under an N_2 atmosphere.

The mixture is then filtered and concentrated to dryness *in vacuo*. The solid residue is treated with ice-
cold H_2O (250 mL) and HCl (2 N, 10 mL) and the resulting mixture is stirred for 30 min at 0 °C. The
precipitate is collected by filtration and washed with ice-cold H_2O (3 × 100 mL). It is then suspended
in H_2O (120 mL) and NaOH (7.2 g) is added. The reaction mixture is heated to 100 °C for 10 min,
cooled to 0 °C, treated with HCl (36%, 12 mL), and stirred for 30 min at 0 °C. The precipitate formed
is collected by filtration, washed with ice-cold H_2O (3 × 100 mL), and dried over P_2O_5 in a desiccator;
19.9 g (59%), mp 260–261 °C.

¹H NMR (200 MHz, [D₆]DMSO): δ (ppm) = 11.33 (s, 1 H, OH), 7.49 (s, 1 H, CH-6), 4.38 (s, 2 H,
CH₂), 1.76 (s, 3 H, CH₃).

¹³C NMR (50 MHz, [D₆]DMSO): δ (ppm) = 169.6 (COOH), 164.3 (C-4), 150.9 (C-2), 141.7 (C-
6H), 108.3 (C-5), 48.4 (CH₂), 11.8 (CH₃).

3.1.2.2   ***   (6-Methyl-3,5-dioxo-8-phenyl-7-oxa-2,4-diazabicyclo[4.2.0]oct-2-yl)-acetic acid [4]

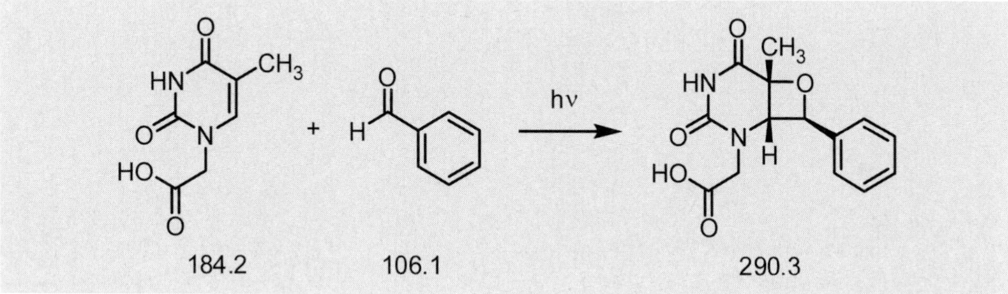

184.2   106.1   290.3

Thymine-1-acetic acid **3.1.2.1** (4.00 g, 21.7 mmol) and benzaldehyde (10 mL, 95 mmol) are suspended
in CH₃CN (240 mL). H_2O (40 mL) is added under vigorous stirring and with heating to 45 °C until a
clear solution is obtained. The solution is degassed under argon and irradiated for 50 h with a 300 W
Hg lamp without the use of filters.

The solvent is then removed *in vacuo* and the residue is dissolved in EtOAc (500 mL). The organic phase is extracted with aqueous $KHCO_3$ solution (0.5 M, 5 × 100 mL). The combined aqueous phases are re-extracted with EtOAc (3 × 150 mL). The aqueous phase is cooled and treated with ice-cold HCl (12 N) until pH 2 is reached. The precipitate formed is collected by filtration, washed with HCl (0.1 N, 100 mL, 0 °C), and dried *in vacuo*. The desired oxetane derivative is obtained as the pure *exo* product; 1.81 g (28%); mp 240–241 °C.

IR (KBr): $\tilde{\nu}$ (cm^{-1}) = 3430, 1717, 1671, 1489, 1406, 1282, 1237, 886, 774.

UV (MeOH/H_2O, 1:1): λ_{max} (nm) = 225.

^{1}H NMR (300 MHz, [D_6]DMSO): δ (ppm) = 12.81 (s, 1 H; OH), 10.84 (s, 1 H; NH), 7.46–7.31 (m, 5 H; phenyl), 5.64 (d, J = 6.6 Hz, 1 H; O–CH), 4.35 (d, J = 6.6 Hz, 1 H; H-6), 4.09 (d, J = 17.7 Hz, 1 H; CH_2), 3.80 (d, J = 17.4 Hz, 1 H; CH_2), 1.66 (s, 3 H; CH_3).

^{13}C NMR (75.5 MHz, [D_6]DMSO): δ (ppm) = 170.3, 170.2, 151.3 (C-2), 139.2 (phenyl), 128.5 (phenyl), 128.3 (phenyl), 126.4 (phenyl), 85.5 (O–CH), 77.1 (C-5), 64.3 (CH-6), 48.2, 22.3.

ESI-HRMS: m/z = 291.09774 $[M + H]^+$.

[1] Th. Eicher, S. Hauptmann, *The Chemistry of Heterocycles*, 2nd ed., p. 38, Wiley-VCH, Weinheim, **2003**.
[2] Ref. [1], p. 39.
[3] M. B. Smith, J. March, *March's Advanced Organic Chemistry*, 5th ed., p. 1249, John Wiley & Sons, Inc., New York, **2001**.
[4] (a) T. Stafforst, U. Diederichsen, *Chem. Commun.* **2005**, *27*, 3430; (b) T. Stafforst, U. Diederichsen, *Eur. J. Org. Chem.* **2007**, 681.
[5] K. L. Dueholm, M. Egholm, C. Behrens, L. Christensen, H. F. Hansen, T. Vulpius, K. H. Petersen, R. H. Berg, P. E. Nielsen, O. Buchardt, *J. Org. Chem.* **1994**, 5767.
[6] J. L. Rabinowitz, S. Gurin, *J. Am. Chem. Soc.* **1953**, *75*, 5758.

3.1.3   Azetidin-2-one derivative

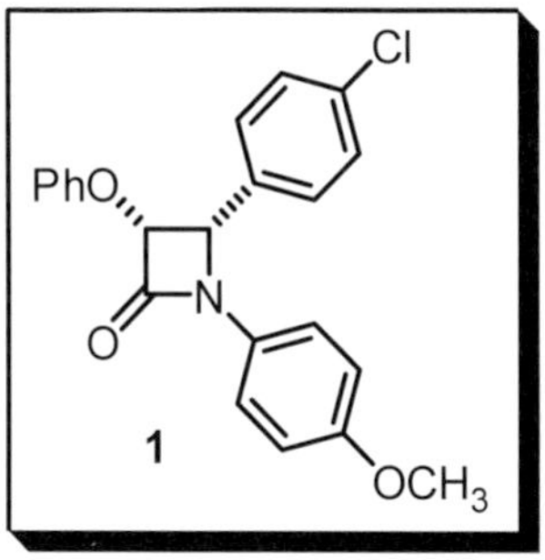

1

Topics:
- Imine formation from Ar–NH$_2$ and Ar–CH=O
- Dehydration of a carboxylic acid R–CH$_2$–COOH to a ketene R–CH=C=O using the Mukaiyama reagent; preparation of the Mukaiyama reagent
- [2+2]-Cycloaddition of a ketene to an imine, *cis*-stereoselective β-lactam formation (Staudinger reaction)

(a) General

The β-lactam moiety (azetidin-2-one) represents the pharmacophoric structural subunit in penicillins, cephalosporins, and related antibiotics [1]. For the synthesis of azetidin-2-ones, the following methods are representative [2]:

(1) Cyclization of β-amino carboxylic acids (**2**) or the corresponding esters (**3**), viz.:

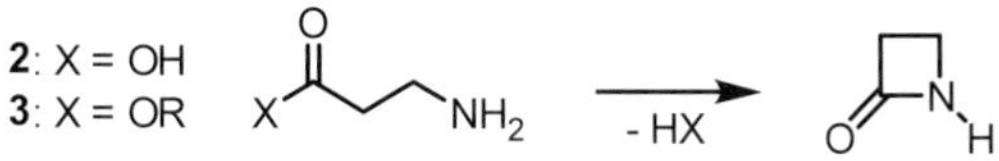

2: X = OH
3: X = OR

Cyclodehydration of **2** can be performed with, e.g., CH$_3$SO$_2$Cl/NaHCO$_3$, while cyclization of **3** may be accomplished by treatment with a strong base, e.g. a Grignard reagent or an amide. If the β-amino ester carries defined stereochemical information, stereodefined β-lactams are obtained [3].

(2) [2+2]-Cycloaddition of ketenes to imines (Staudinger reaction):

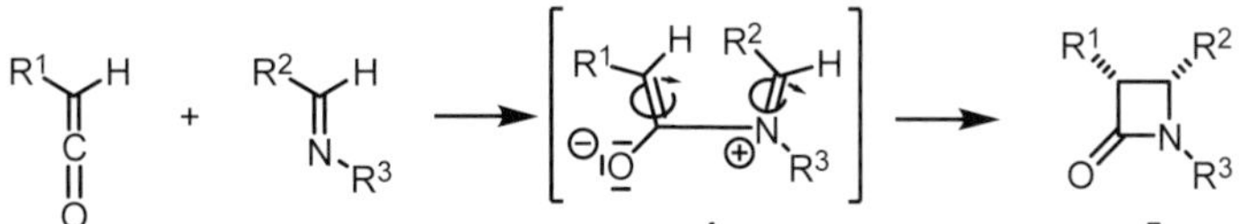

4   **5**

As in similar cycloadditions of ketenes and alkenes, the Staudinger reaction is likely to occur via dipolar intermediates **4** in a thermally allowed two-step [2+2] process [4] to give *cis*-3,4-disubstituted azetidin-2-ones **5** stereoselectively.

(3) [2+2]-Cycloaddition of alkenes to isocyanates, preferentially ClSO$_2$–N=C=O:

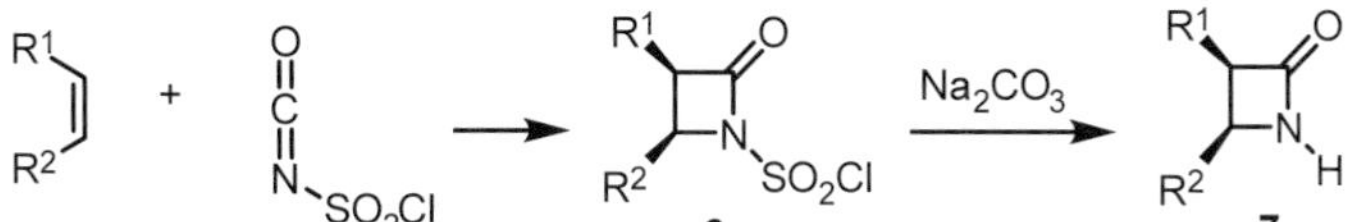

6   **7**

As in (2), this cycloaddition also occurs stereoselectively; thus, the stereochemistry of the alkene is transferred to the NH-azetidin-2-one **7**, which is obtained from the cycloadduct **6** by removal of the chlorosulfonyl group with base.

In (b), an example of the Staudinger method (2) of β-lactam formation is presented, which can also be performed in an asymmetric way [5].

(b) Synthesis of 1

As precursors of the ketenes required in the Staudinger reaction, either acid chlorides in the presence of a base or carboxylic acids R–CH$_2$–COOH in the presence of a dehydrating agent are used. Thus, 2-chloro-1-methylpyridinium salts (e.g. 10) – known as Mukaiyama reagents [6] – have been shown to allow dehydration of carboxylic acids to ketenes under mild conditions in an efficient one-pot procedure [5]. 10, originally developed for the synthesis of esters and amides from carboxylic acids, can be obtained from 2-chloropyridine by N-alkylation with methyl iodide.

Thus, the reaction of phenoxyacetic acid (8) with the imine 9 in the presence of triethylamine and 2-chloro-1-methylpyridinium iodide (10) proceeds smoothly at room temperature in CH$_2$Cl$_2$ and affords the azetidin-2-one 1 in almost quantitative yield with complete cis-stereoselectivity (ratio of cis/trans-diastereomers of 99:1):

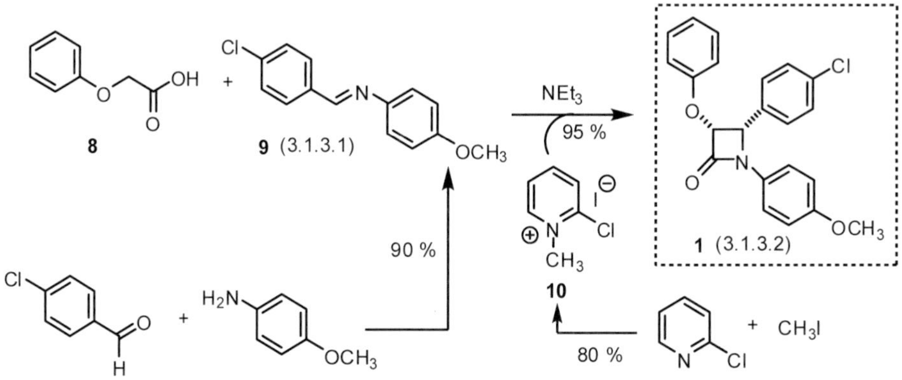

Initially, the pyridinium ion 10 undergoes a nucleophilic displacement (addition/elimination) of the 2-chloro substituent by the carboxylate (formed from 8 with NEt$_3$) to give the 2-acyloxypyridinium ion 12; subsequently, 12 is deprotonated and cleaved to give 1-methyl-2-pyridone (14) and phenoxyketene (13). 13 then undergoes the cis-selective [2+2]-cycloaddition with the imine 9 to give 1.

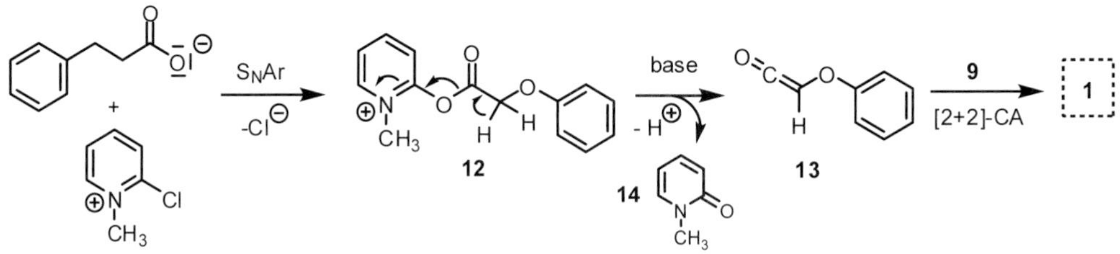

The imine 9 is prepared by condensation of 4-chlorobenzaldehyde with p-anisidine in the presence of MgSO$_4$.

It should be noted that an asymmetric version of the Staudinger reaction with 10 has been developed [5] by utilizing (+)-erythro-2-amino-1,2-diphenylethanol (15) as an auxiliary. Thus, a chiral glycine derivative 17 is prepared from oxazolidinone 16 (obtained from 15 by cyclization with diethyl carbonate) by N-alkylation with ethyl bromoacetate and saponification of the ester group:

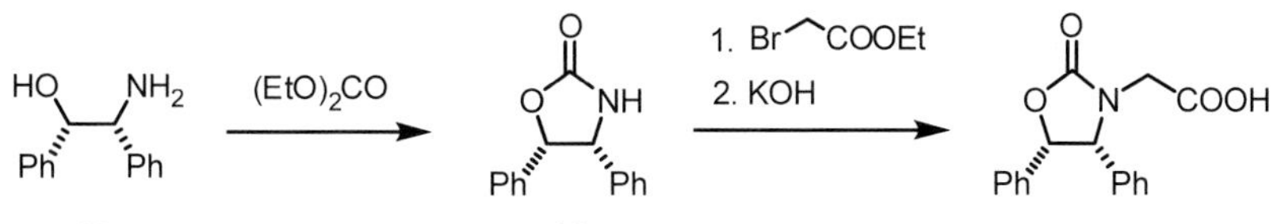

When the chiral carboxylic acid **17** is reacted with the imine **9** in the presence of 2-chloro-1-methylpyridinium *p*-toluenesulfonate **10** (TosO⁻ instead of I⁻) and NEt₃ under the same conditions as described above, the *cis*-β-lactam **18** is obtained in diastereomerically pure form [6, 7]:

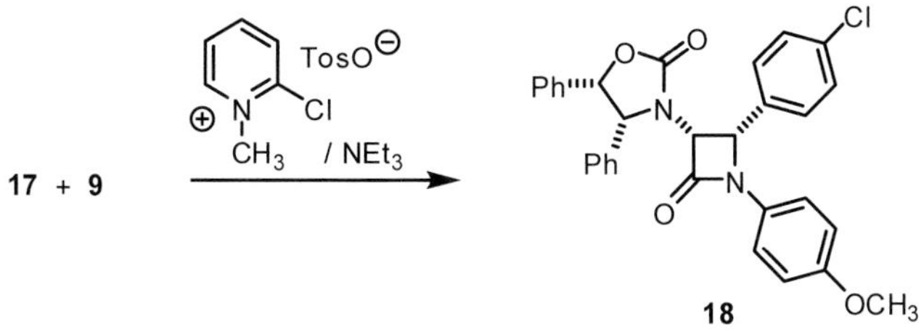

(c) Experimental procedures for the synthesis of 1

3.1.3.1   **   (*p*-Chlorobenzylidene)-(*p*-methoxyphenyl)-imine [5]

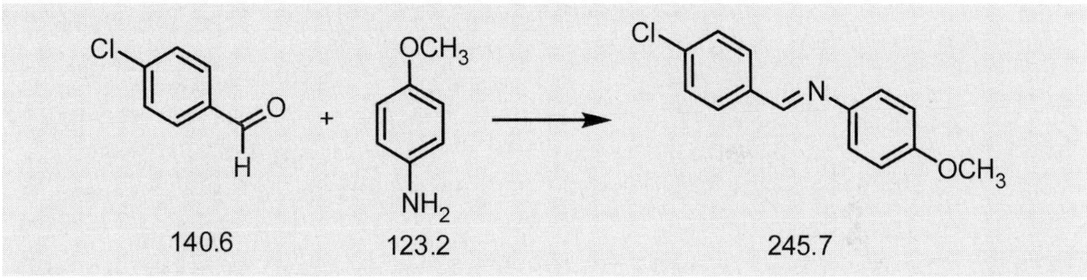

A solution of *p*-anisidine (2.46 g, 20.0 mmol) in anhydrous CH₂Cl₂ (10 mL) is added over a period of 5 min to a stirred solution of *p*-chlorobenzaldehyde (2.81 g, 20.0 mmol) in anhydrous CH₂Cl₂ (20 mL) under argon atmosphere. A large excess of magnesium sulfate (3.00 g) is added and the resulting mixture is stirred for 17 h at room temperature.

The solution is then filtered and the solvent is evaporated to afford the crude imine. Recrystallization from *n*-hexane/CH₂Cl₂ (1:1) gives the product as a colorless solid; 3.98 g (81%), mp 113–115 °C; TLC (CHCl₃/MeOH, 19:1): $R_f = 0.76$.

UV: λ_{max} (nm) = 336, 271, 239, 199.

IR (KBr) $\tilde{\nu}$ (cm⁻¹) = 1620, 1506, 1255, 839.

¹H NMR (CDCl₃, 300 MHz): δ (ppm) = 8.42 (s, 1 H, N=C*H*), 7.85–7.75 (m, 2 H, arom), 7.47–7.35 (m, 2 H, arom), 7.25–7.17 (m, 2 H, arom), 6.94–6.86 (m, 2 H, arom), 3.82 (s, 3 H, OCH₃).

¹³C NMR (CDCl₃, 75.5 MHz): δ (ppm) = 158.4 (C_qCH=N), 156.6 (N=CH), 144.3 (C_q), 136.8 (C_q), 134.8 (C_q), 129.6 (Ph), 128.9 (Ph), 122.2 (Ph), 114.3 (Ph), 55.43 (OCH₃).

MS (EI, 70 eV): *m/z* = 245 [*M* + H⁺].

3.1.3.2 ** *cis*-4-(*p*-Chlorophenyl)-1-(*p*-methoxyphenyl)-3-phenoxy-azetidin-2-one [5]

245.7   152.2   255.6   363.9

Under an argon atmosphere, triethylamine (0.64 mL, 4.59 mmol) is added to a stirred solution of phenoxyacetic acid (198.8 mg, 1.96 mmol) and 2-chloro-1-methylpyridinium iodide (515.2 mg, 2.02 mmol) in anhydrous CH_2Cl_2 (5.0 mL) at room temperature. The imine **3.1.3.1** (562.8 mg, 2.29 mmol) in CH_2Cl_2 (2.0 mL) is then added and stirring is continued for 17 h (after a few minutes, a suspension is formed).

H_2O (5.0 mL) is then added and the mixture is extracted with CH_2Cl_2 (3 × 30 mL). The combined organic layers are washed with brine (30 mL), dried over $MgSO_4$, and filtered, and the solvent is evaporated. Column chromatography (CH_2Cl_2) gives the product as a colorless solid; 702.5 mg (95%), mp 168–170 °C; TLC (CH_2Cl_2): $R_f = 0.37$; $[\alpha]_D^{20} = 0$ ($c = 0.5$, DMSO).

UV (MeOH): λ_{max} (nm) = 262, 201.

IR (KBr) $\tilde{v}$ (cm^{-1}) = 1745, 1598, 1514, 1394, 1240, 1113, 839, 751.

^{1}H NMR (CDCl$_3$, 50 °C, 300 MHz): δ (ppm) = 7.36–7.11 (m, 8 H, arom), 6.97–6.75 (m, 5 H, arom), 5.53 (d, $J = 4.6$ Hz, 1 H, PhOCH), 5.31 (d, $J = 4.6$ Hz, 1 H, *p*-ClC$_6$H$_4$CH), 3.74 (s, 3 H, OCH$_3$).

^{13}C NMR (CDCl$_3$, 75.5 MHz): δ (ppm) = 55.5 (OCH$_3$), 61.5 (*p*-ClC$_6$H$_4$CH), 81.4 (PhOCH), 114.6 (2 C, Ph), 115.8 (2 C, Ph), 118.9 (2 C, Ph), 122.4 (Ph), 128.7 (2 C, Ph), 129.4 (2 C, Ph), 129.5 (2 C, Ph), 130.4 (Ph), 131.6 (C$_q$), 134.7 (C$_q$), 156.8 (C$_q$), 157.0 (C$_q$), 162.3 (C$_q$).

MS (ESI): *m/z* (%) = 781 (100) [2 *M* + Na]$^+$, 402 (88) [*M* + Na]$^+$.

---

[1]   a) Th. Eicher, S. Hauptmann, *The Chemistry of Heterocycles*, 2nd ed., p. 44/159/389, Wiley-VCH, Weinheim, **2003**; b) C. Coates, J. Kabir, E. Turos, *Sci. Synth.* **2005**, *21*, 609.

[2]   M. J. Miller, *Tetrahedron* **2000**, *56*, 5553.

[3]   For instructive examples, see: (a) S. D. Bull, S. G. Davies, P. M. Kelly, M. Gianotti, A. D. Smith, *J. Chem. Soc., Perkin Trans. 1*, **2001**, 3106; (b) A. Córdova, S. Watanabe, F. Tanaka, W. Notz, C. F. Barbas III, *J. Am. Chem. Soc.* **2002**, *124*, 1866.

[4]   M. B. Smith, J. March, *March's Advanced Organic Chemistry*, 5th ed., p. 1250, John Wiley & Sons, Inc., New York, **2001**.

[5]   S. Matsui, Y. Hashimoto, K. Saigo, *Synthesis* **1998**, 1161.

[6]   Alternative methods for the stereoselective formation of *cis*-3,4-disubstituted β-lactams by the Staudinger reaction have been reported: a) S. France, H. Wack, A. M. Hafez, A. E. Taggi, D. R. Witsil, T. Lectka, *Org. Lett.* **2002**, *4*, 1603; b) M. M. Shah, S. France, T. Lectka, *Synlett* **2003**, 1937.

[7]   For the formation of *trans*-3,4-disubstituted β-lactams by the Staudinger reaction, see: a) M. Nahmany, A. Melman, *J. Org. Chem.* **2006**, *71*, 5804; b) Q. Yuan, S.-Z. Jiang, Y.-G. Wang, *Synlett* **2006**, 1113 (auxiliary-assisted enantioselective synthesis).

3.2   Five-membered heterocycles

3.2.1   2,4-Diphenylfuran

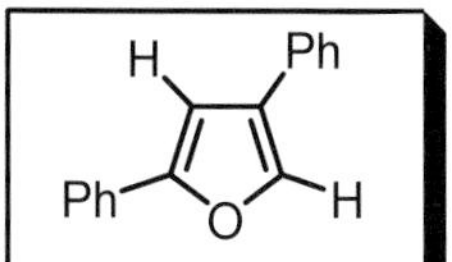

1

Topics: • Transformation of pyrylium salts to furans
  • Ring-opening of pyrylium ions and recyclization
  • Haller–Bauer cleavage of non-enolizable ketones

(a) General

Pyrylium ions such as **2** are positively charged aromatic heterocycles with one oxygen in a six-membered ring system. They can easily undergo addition of a nucleophile at the positions 2, 4, and 6.

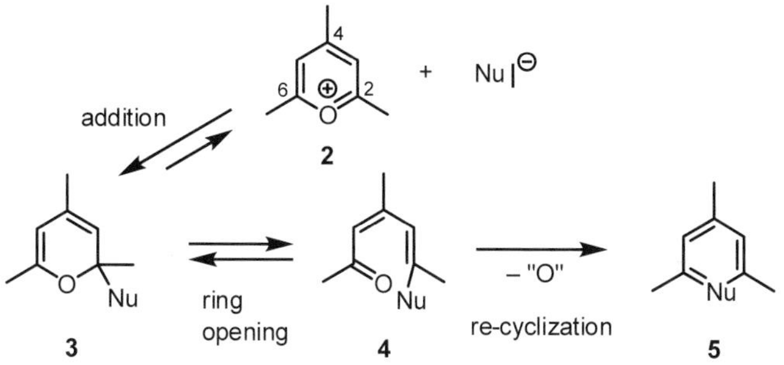

Attack of the nucleophile at the 2-position leads to *2H*-pyrans **3**, which by electrocyclic ring-opening give products of type **4**. If the reacting nucleophile is OH⁻, the pyrylium system is reconstituted in acid media; however, if N-, S-, P- or C-containing nucleophiles are used, new heterocyclic or carbocyclic systems **5** are formed with the incorporation of the attacking atom. As illustrated by the examples **6-8**, the transformation of pyrylium ions is of considerable preparative value [1]:

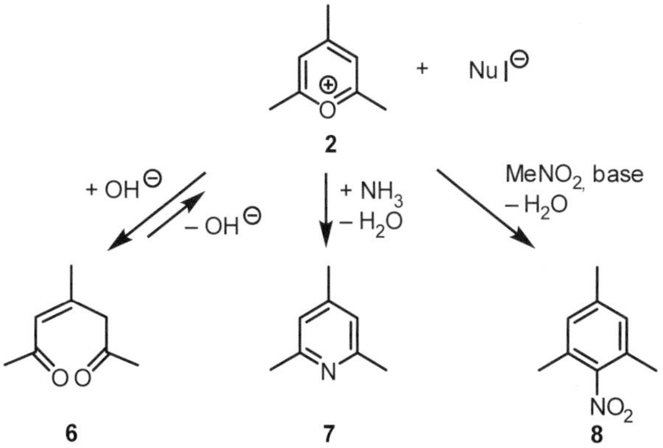

In this way, reaction of the pyrylium ion **9** with iodine in the presence of sodium carbonate in aqueous medium allows a convenient and regioselective synthesis of 2,4-disubstituted furans [2], which are otherwise difficult to obtain [3].

(b) Synthesis of 1

2,4,6-Triphenylpyrylium tetrafluoroborate (**13**) is prepared by means of a modified Dilthey synthesis [2,4]. In this method of pyrylium ion formation, an aryl aldehyde is cyclocondensed with two molecules of an acetophenone in the presence of BF_3-etherate; intermediates are the chalcones **10** (resulting from aldol condensation of the first acetophenone with the aryl aldehyde), the pentane-1,5-diones **11** (resulting from Michael addition of the second acetophenone to **10**), and the 4H-pyrans **12** (resulting from cyclodehydration of **11**); the final step is (formal) hydride abstraction from **12** by the Lewis acid, providing the 2,4,6-triarylpyrylium salt **9**:

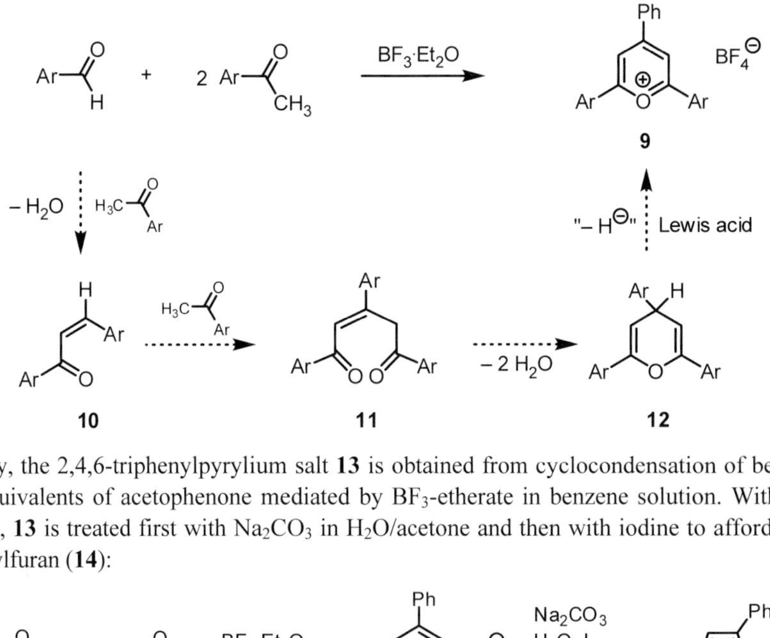

Accordingly, the 2,4,6-triphenylpyrylium salt **13** is obtained from cyclocondensation of benzaldehyde and two equivalents of acetophenone mediated by BF_3-etherate in benzene solution. Without further purification, **13** is treated first with Na_2CO_3 in H_2O/acetone and then with iodine to afford 2-benzoyl-3,5-diphenylfuran (**14**):

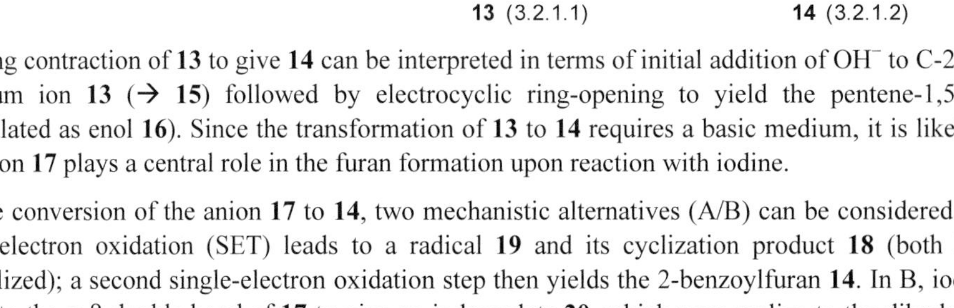

13 (3.2.1.1)   14 (3.2.1.2)

The ring contraction of **13** to give **14** can be interpreted in terms of initial addition of OH^- to C-2 of the pyrylium ion **13** ($\rightarrow$ **15**) followed by electrocyclic ring-opening to yield the pentene-1,5-dione (formulated as enol **16**). Since the transformation of **13** to **14** requires a basic medium, it is likely that the anion **17** plays a central role in the furan formation upon reaction with iodine.

For the conversion of the anion **17** to **14**, two mechanistic alternatives (A/B) can be considered. In A, single-electron oxidation (SET) leads to a radical **19** and its cyclization product **18** (both highly delocalized); a second single-electron oxidation step then yields the 2-benzoylfuran **14**. In B, iodine is added to the α,β-double bond of **17** to give an iodoenolate **20**, which may cyclize to the dihydrofuran

21 by $S_{N}i$ displacement of iodide and aromatize to **14** by base-induced elimination of HI. For both pathways A/B, the overall stoichiometry is that one molecule of I_2 is consumed and two I^- and two H^+ ions are formed [5].

Mechanism for the transformation **13 ➙ 14**:

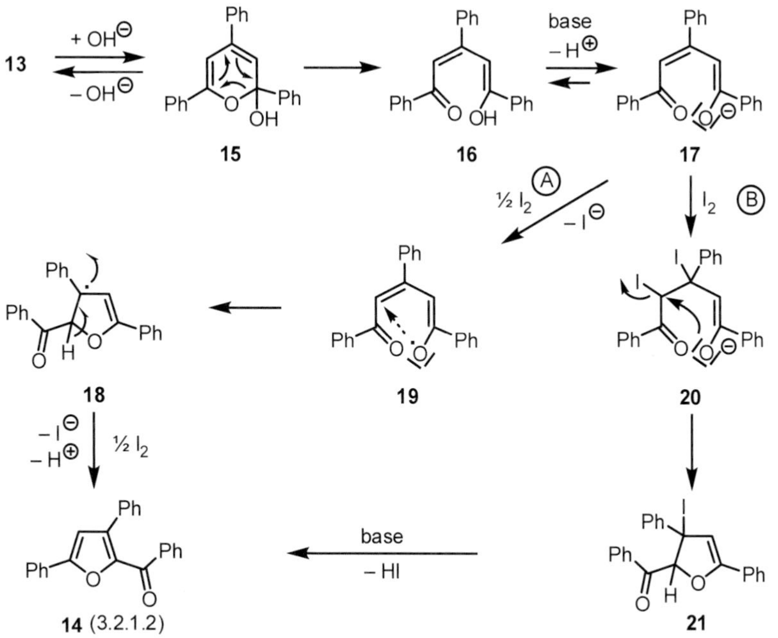

Finally, the 2-benzoyl group in the furan **14** has to be replaced by hydrogen. This transformation is accomplished by application of a Haller–Bauer-type reaction of **14** with KO*t*Bu in DMSO leading to 2,4-diphenylfuran (**1**) upon hydrolysis:

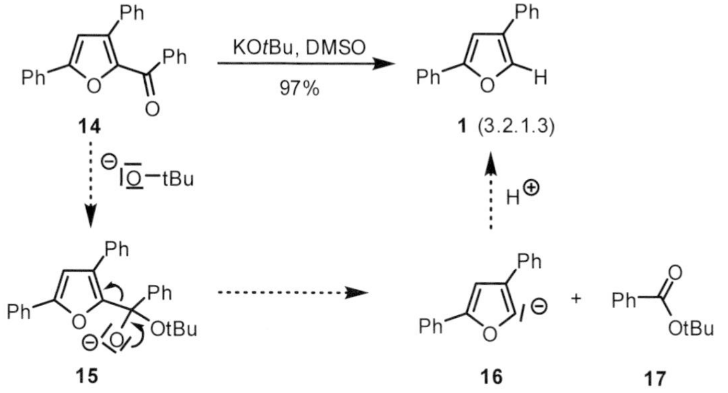

The Haller–Bauer reaction [6] involves a base-promoted acyl cleavage of a non-enolizable ketone; traditionally, the reaction is carried out with sodium amide as the base, thus representing a synthesis of amides from ketones. Cleavage of unsymmetrical ketones such as **18** is directed by the stability of the anion formed:

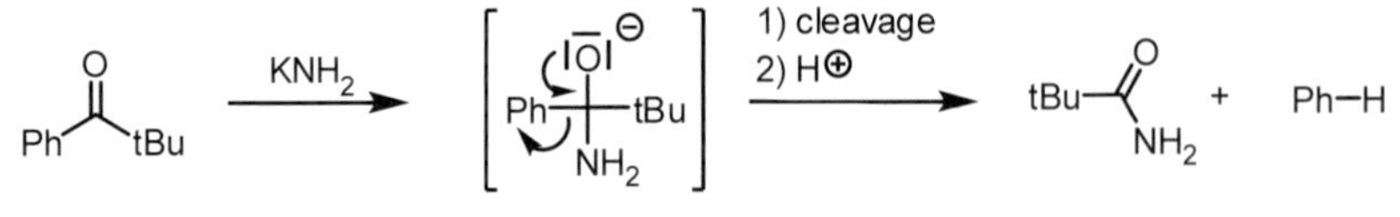

Due to the better stabilization of a phenyl anion compared to a *tert*-butyl anion, the intermediate **19**, which is formed by addition of the amide to ketone **18**, delivers benzene and the amide **20**.

In the transformation of **14** with KO*t*Bu, an α-furyl carbanion **16** or a phenyl anion might result by cleavage of the primary adduct **15**. Since **16** is more stable than the phenyl anion, furan **1** and *tert*-butyl benzoate (**17**) are formed as the only products.

(c) Experimental procedures for the synthesis of 1

3.2.1.1 * 2,4,6-Triphenylpyrylium tetrafluoroborate [3]

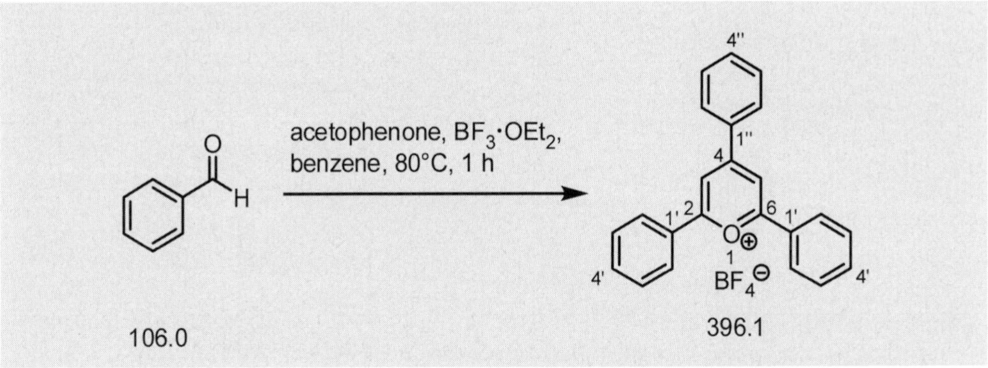

Freshly distilled benzaldehyde (1.00 mL, 9.42 mmol) and acetophenone (2.50 mL, 20.7 mmol) in anhydrous benzene (10 mL) (Caution: carcinogenic!) are added to a stirred solution of boron trifluoride etherate (6.00 mL, 22.6 mmol) under an N_2 atmosphere. The mixture is heated to reflux for 2 h.

After cooling to 20 °C, acetone (10 mL) is added and the dark-red solution is poured into Et_2O (100 mL). The yellow precipitate of the pyrylium salt is collected by filtration, washed with Et_2O, and dried under vacuum for 12 h; 0.972 g (60%), mp 247–248 °C.

UV (CH_3CN): λ_{max} (nm) (lg ε) = 406 (7.219), 353.5 (7.159), 275.5 (7.051).

IR (KBr): $\tilde{v}$ (cm^{-1}) = 3070, 1745, 1624, 1593, 1579, 1527, 1497, 1470, 1273, 1248, 1194, 1167, 1057.

^{1}H NMR (300 MHz, $CDCl_3$): δ (ppm) = 9.20 (s, 2 H, 5-H, 3-H), 8.80–8.70 (m, 4 H, 4 × 2'-H), 8.65–8.55 (m, 2 H, 2 × 2''-H), 8.00–7.75 (m, 9 H, 9 × Ph-H).

^{13}C NMR (75 MHz, $CDCl_3$): δ (ppm) = 172.1 (C-2, C-6), 167.4 (2 × C-1'), 136.1, 133.9 (C-4), 130.9, 130.7, 130.2 (C-1''), 129.7, 116.4 (C-3, C-5).

MS (EI, 70 eV): *m/z* (%) = 396 (10) [*M*]$^+$, 309 (100) [*M* – BF_4]$^+$, 202 (46) [*M* – $C_7H_7BF_4O$]$^+$, 105 (69) [*M* – $C_{16}H_{12}BF_4$]$^+$, 77 (100) [*M* – $C_{17}H_{12}BF_4O$]$^+$, 49 (41) [*M* – $C_{19}H_{16}BF_4O$]$^+$.

3.2.1.2   *   2-Benzoyl-3,5-diphenylfuran [3]

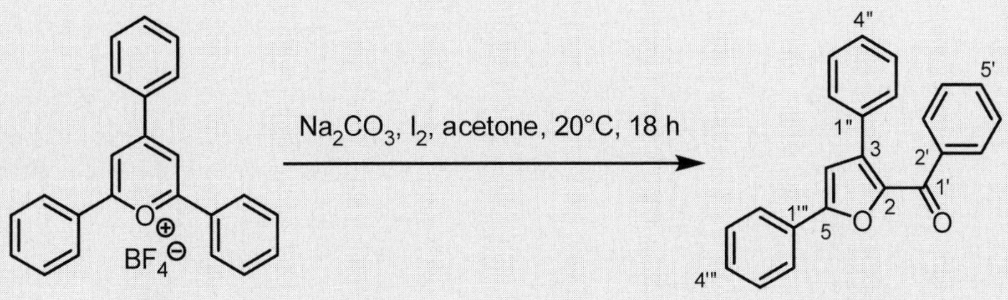

Na$_2$CO$_3$, I$_2$, acetone, 20°C, 18 h

396.1   324.1

A solution of Na$_2$CO$_3$ (500 mg, 4.71 mmol) in H$_2$O (1.6 mL) is added to a stirred suspension of the pyrylium salt **3.2.1.1** (972 mg, 2.46 mmol) in acetone (15 mL) and the mixture is stirred for 2 h at 20 °C. Iodine (1.00 g, 3.93 mmol) is added and stirring is continued for 16 h.

The dark mixture is then poured into a solution of Na$_2$S$_2$O$_3$ (6.21 g) in H$_2$O (75 mL) and the aqueous layer is extracted with CH$_2$Cl$_2$ (3 × 50 mL). The combined organic layers are washed with H$_2$O and brine, dried over MgSO$_4$, and the solvent is evaporated under reduced pressure. The residue is purified by column chromatography on silica gel (petroleum ether/EtOAc, 10:1) to afford the product as a yellow solid; 0.542 g (68%), mp 119–120 °C, R_f = 0.4 (petroleum ether/EtOAc, 10:1).

UV (CH$_3$CN): λ_{max} (nm) (lg ε) = 341.5 (6.954), 265 (6.844).

IR (KBr): $\tilde{v}$ (cm^{-1}) = 3051, 2924, 2854, 1965, 1641, 1599, 1576, 1570, 1523, 1472.

^{1}H NMR (300 MHz, CDCl$_3$): δ (ppm) = 7.93–7.85 (m, 2 H, 2 × 3'-H), 7.78–7.68 (m, 2 H, 2 × 2'''-H), 7.58–7.56 (m, 2 H, 2 × 2''-H), 7.47–7.25 (m, 9 H, 9 × Ar-H), 6.92 (s, 1 H, 4-H).

^{13}C NMR (75 MHz, CDCl$_3$): δ (ppm) = 183.5 (C-1'), 156.0 (C-2), 146.0 (C-5), 138.0 (C-1'''), 137.7 (C-3), 132.2 (C-1''), 132.2 (C-5'), 129.7, 129.3, 129.2, 129.0 (C-4'''), 128.4 (C-4''), 128.2, 128.1, 125.0, 109.8 (C-4).

MS (EI, 70 eV): m/z (%) = 324 (100) [M]$^+$, 247 (6) [M – C$_6$H$_5$]$^+$, 191 (10) [M – C$_8$H$_5$O$_2$]$^+$, 105 (8.5) [M – C$_{16}$H$_{11}$O]$^+$, 77 (17) [M – C$_{17}$H$_{11}$O$_2$]$^+$, 51 (10) [M – C$_{21}$H$_{14}$O$_2$]$^+$

3.2.1.3   *   2,4-Diphenylfuran [3]

$tBuOK$, dioxane, 20°C, 45 min

324.1   220.1

H_2O (34.0 μL (!)) and the benzoyldiphenylfuran **3.2.1.2** (200 mg, 0.62 mmol) are added to a stirred suspension of potassium *tert*-butoxide (800 mg, 7.08 mmol) in 1,4-dioxane (5.00 mL), and stirring is continued for 30 min.

The mixture is then slowly poured into iced water (30 mL) and stirred for 15 min. The product is extracted with CH_2Cl_2 (3 × 10 mL), and the combined organic layers are washed with H_2O and brine. After drying over $MgSO_4$, the solvent is removed under reduced pressure and the residue is purified by column chromatography on silica gel (petroleum ether/EtOAc, 10:1) to afford 225 mg (97%) of 2,4-diphenylfuran as colorless needles; mp 109–110 °C, R_f = 0.60 (SiO_2, petroleum ether/EtOAc, 10:1).

UV (CH_3CN): λ_{max} (nm) (lg ε) = 275.5 (6.696), 242.0 (6.64), 226.5 (6.611), 199.5 (6.556), 197.5 (6.551).

IR (KBr): $\tilde{v}$ (cm^{-1}) = 3441, 3135, 3106, 3036, 1609, 1538, 1490, 1453, 1199.

^{1}H NMR (300 MHz, $CDCl_3$): δ (ppm) = 7.70 (s, 1 H, 5-H), 7.68–7.62 (m, 2 H, 2 × 2'-H), 7.50–7.44 (m, 2 H, 2 × 2"-H), 7.40–7.28 (m, 4 H, 2 × 3'-H, 2 × 3"-H), 7.26–7.20 (m, 2 H, 4'-H, 4"-H), 6.88 (s, 1 H, 4'-H, 4"-H).

^{13}C NMR (75 MHz, $CDCl_3$): δ (ppm) = 154.8 (C-2), 137.9 (C-5), 132.3 (C-1'), 130.6 (C-4), 128.8, 128.7, 128.3 (C-1"), 127.6, 127.1, 123.8, 123.9, 103.9 (C-3).

MS (EI, 70 eV): m/z (%) = 220 (100) $[M]^+$, 192 (15) $[M-CO]^+$, 191 (53) $[M-CHO]^+$, 189 (16) $[M-CH_3O]^+$; 165 (7) $[M-C_{13}H_9]^+$.

---

[1]   a) Th. Eicher, S. Hauptmann, *The Chemistry of Heterocycles*, 2nd ed., p. 222, Wiley-VCH, Weinheim, **2003**; b) T. S. Balaban, A. T. Balaban, *Sci. Synth.* **2003**, *14*, 11; c) Houben-Weyl, *Methoden der organischen Chemie*, 4th ed., Vol. E 7b, p. 755, Georg Thieme Verlag, Stuttgart, **1992**.

[2]   For the chemistry and synthesis of furans, see: a) B. Koenig, *Sci. Synth.* **2002**, *9*, 183; b) Houben-Weyl, *Methoden der organischen Chemie*, 4th ed., Vol. E 6a, p. 1, Georg Thieme Verlag, Stuttgart, **1992**.

[3]   I. Francesconi, A. Patel, P. W. Boykin, *Synthesis* **1999**, 61.

[4]   R. Lombard, J.-P. Stephan, *Bull. Soc. Chim. Fr.* **1958**, 1458.

[5]   When the pyrylium ion **13** is reacted with Na_2S and the product is oxidized with I_2 or air, the thiophene compound analogous to **14** is obtained: C. L. Pedersen, *Acta Chem. Scand. B* **1975**, *29*, 791.

[6]   For a review, see: J. P. Gilday, L. A. Paquette, *Org. Prep. Proced. Int.* **1990**, *22*, 167.

3.2.2   3,4-Dimethylpyrrole

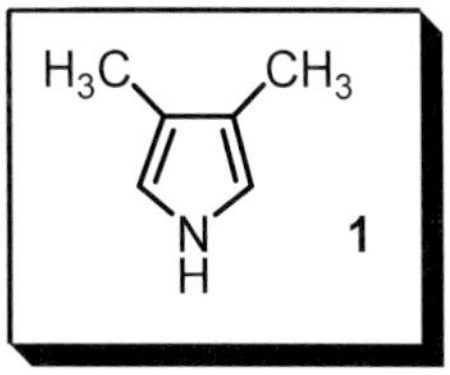

Topics:
- Barton–Zard synthesis of pyrroles
- Nitroaldol addition
- Acetylation of a nitroaldol
- Saponification of a carboxylic ester, decarboxylation of a pyrrole carboxylic acid
- Formation of an isocyanoacetate by dehydration of an N-formyl-α-amino ester

(a) General

Numerous methods are available for the synthesis of pyrroles [1]. While the classical Paal–Knorr synthesis (cf. **3.5.5**) is still the method of choice for the preparation of 2,5-disubstituted pyrroles, 3,4-disubstituted pyrroles such as **1** are conveniently obtained by isocyanide-based methods, among them the Barton–Zard synthesis and the van Leusen synthesis.

In the Barton–Zard synthesis [2], 5-unsubstituted 3,4-disubstituted pyrrole-2-carboxylates **3** are formed from isocyanoacetates and nitroalkenes **2** in the presence of a base (e.g. DBU, TMG, K_2CO_3 [3]):

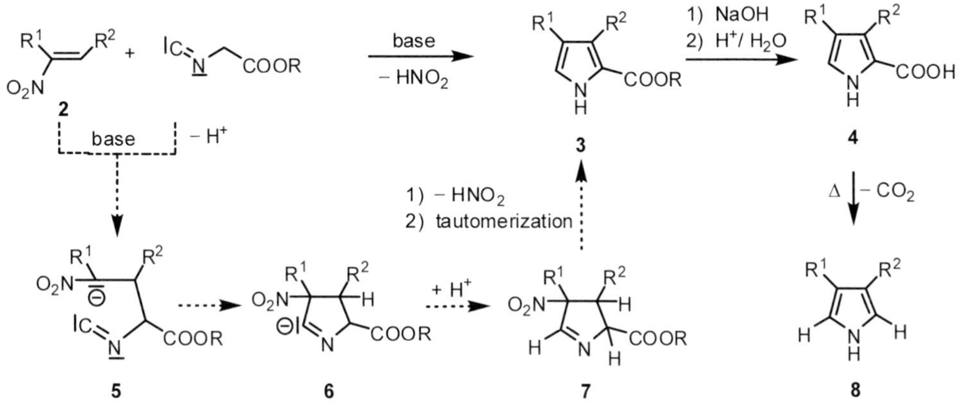

This cyclocondensation is thought to proceed via Michael addition of deprotonated isocyanoacetate to the nitroalkene, intramolecular trapping of the resulting nitronate anion **5** by the isocyanide functionality (→ **6**), reprotonation (**6** → **7**) and base-induced elimination of HNO_2, followed by tautomerization of the resulting 2*H*-pyrroles to 1*H*-pyrroles **3**. The pyrrole-2-carboxylates **3** can be saponified to give the pyrrole-2-carboxylic acids **4**, which undergo thermal decarboxylation to the 3,4-disubstituted pyrroles **8**.

Closely related to the Barton–Zard method is the van Leusen pyrrole synthesis [4], in which (tosylmethyl)isocyanide (TosMIC) is cyclocondensed with α,β-unsaturated ketones in the presence of a base, e.g. NaH, to give 3-acyl-4-substituted pyrroles **9**:

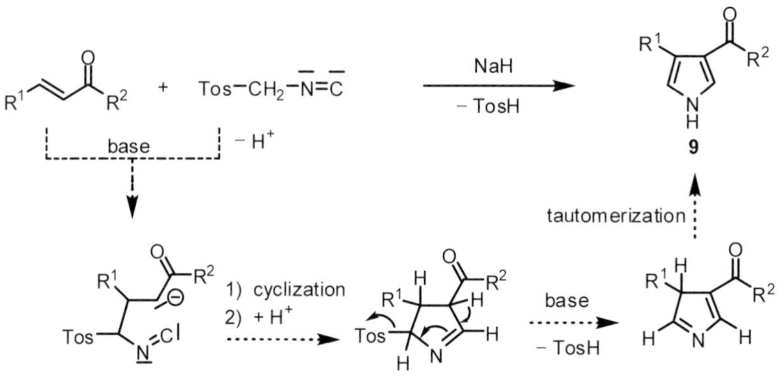

In the mechanism of the van Leusen synthesis, conjugate addition of deprotonated TosMIC to the enone produces an enolate **10**, which is intramolecularly trapped by the isocyanide function with ring-closure and re-protonation (**10 → 11**); deprotonation of **11** at 3-CH gives an enolate, which undergoes 1,4-elimination of sulfinate and formation of a 3*H*-pyrrole **12**, which finally tautomerizes to the 2,5-unsubstituted 1*H*-pyrrole **9**.

3,4-Dimethylpyrrole has been synthesized via several routes [5]. One of them utilizes a Diels–Alder reaction of the sulfoximine **13** with 2,3-dimethylbuta-1,3-diene followed by base-induced ring-contraction of the cycloadduct **14** [6]:

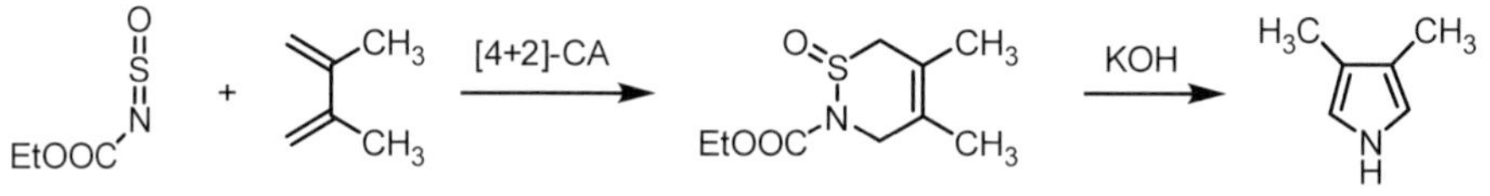

However, due to the preparative disadvantages of this conceptually elegant approach, the Barton–Zard protocol [5] is preferred for the synthesis of **1** and is thus presented in section (b).

(b) Synthesis of 1

First, 2-acetoxy-3-nitrobutane (**16**) is prepared by fluoride-mediated nitroaldol addition (Henry reaction) of nitroethane to acetaldehyde (**→ 15**) and subsequent DMAP-catalyzed acetylation of the nitroaldol **15** with acetic anhydride.

Second, ethyl isocyanoacetate (**20**) is prepared by *N*-formylation of ethyl glycinate hydrochloride (**18**) by aminolysis of ethyl formate (**→ 19**) followed by dehydration of *N*-formyl glycine **19** with POCl₃. The dehydration of primary *N*-formyl amines, preferentially with POCl₃, is a widely used method for isonitrile formation [7].

Then, 2-acetoxy-3-nitrobutane (**16**) and ethyl isocyanoacetate (**20**) are subjected to the Barton–Zard cyclocondensation in the presence of tetramethylguanidine (TMG) as a base to give the pyrrole-2-carboxylate **22**. 2-Nitro-2-butene (**17**) required for the Barton–Zard process is initially produced *in situ* by base-induced elimination of acetic acid from the acetoxybutane **16**:

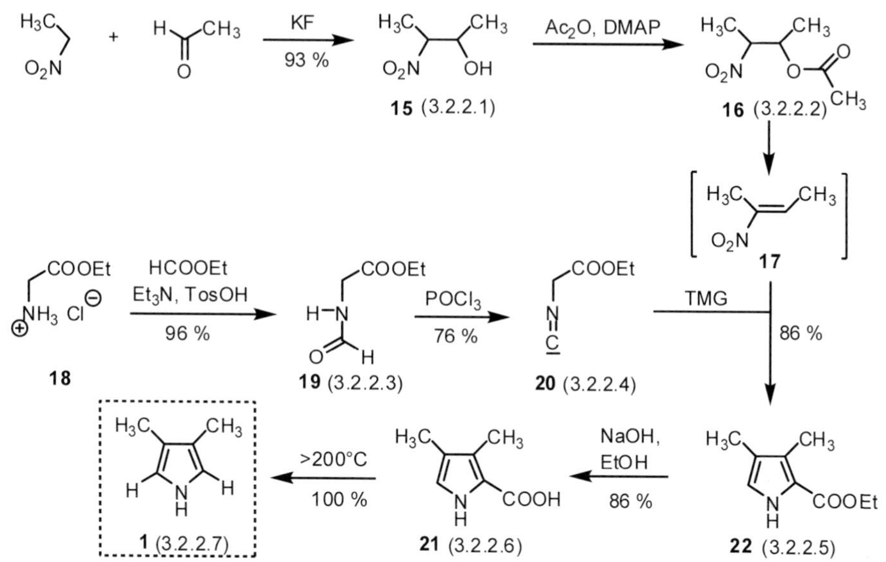

H$_3$C + H–CH$_3$ →(KF, 93 %) H$_3$C–CH$_3$ (O$_2$N, OH) **15** (3.2.2.1) →(Ac$_2$O, DMAP) H$_3$C–CH$_3$ **16** (3.2.2.2)

[H$_3$C═CH$_3$ **17**]

18 →(HCOOEt, Et$_3$N, TosOH, 96 %) **19** (3.2.2.3) →(POCl$_3$, 76 %) **20** (3.2.2.4) →(TMG, 86 %)

1 (3.2.2.7) ←(>200°C, 100 %) **21** (3.2.2.6) ←(NaOH, EtOH, 86 %) **22** (3.2.2.5)

The synthesis of **1** is completed by saponification of the ester **22** using NaOH in ethanol followed by thermal decarboxylation of the thus formed pyrrole-2-carboxylic acid **21** to yield the 3,4-dimethyl-pyrrole (**1**) [10].

(c) Experimental procedures for the synthesis of 1

3.2.2.1   *   2-Nitrobutan-3-ol [7]

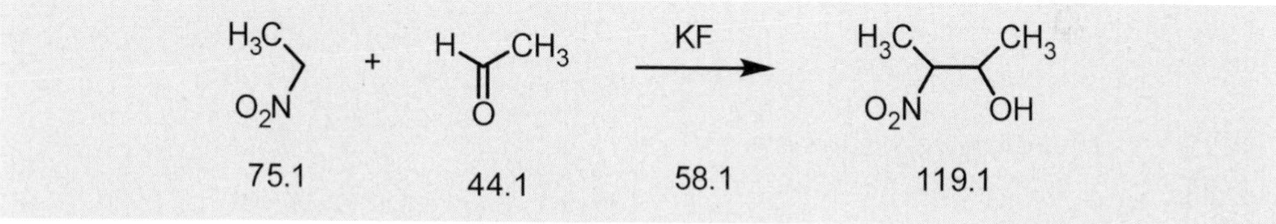

H$_3$C (O$_2$N) + H–CH$_3$ (O) →(KF) H$_3$C–CH$_3$ (O$_2$N, OH)

75.1   44.1   58.1   119.1

Nitroethane (75.1 g, 1.00 mol) is added dropwise over 30 min to a stirred mixture of acetaldehyde (44.1 g, 56.5 mL, 1.00 mol), potassium fluoride (2.95 g, 50.0 mmol), and isopropanol (40 mL) at 0 °C. The temperature is raised to 20 °C to start the weakly exothermic aldol addition reaction, and then the mixture is maintained at 35–40 °C by cooling with an ice/water mixture. After 30 min, stirring is continued for 12 h at room temperature.

After evaporation of the solvent *in vacuo*, the residue is dissolved in CH$_2$Cl$_2$ (100 mL), the solution is filtered through a G-4 fritted disc, and the solvent is removed *in vacuo*. The oily residue (110.6 g, 93%) consists of the almost pure product, which is used in the next step without further purification; TLC (CH$_2$Cl$_2$): $R_f = 0.25$; for NMR data, see ref. [7].

3.2.2.2 ** 3-Acetoxy-2-nitrobutane [7]

| 119.1 | 102.1 | 161.2 |

Acetic anhydride (77.2 g, 0.75 mol) is added dropwise to a stirred solution of the nitro alcohol **3.2.2.1** (59.5 g, 0.50 mol) and DMAP (ca. 2.0 g) in anhydrous CH_2Cl_2 (100 mL). An exothermic reaction occurs, and the temperature should be kept below 40 °C (occasional cooling with an ice bath). When the addition is complete, stirring is continued at 35–40 °C (inner temperature) for 2 h; the solution develops a green color.

The reaction mixture is brought to 30 °C, whereupon methanol (32.0 g, 1 mol) is added dropwise and stirring is continued for 2 h at room temperature. The solvent is evaporated *in vacuo*, the residue is dissolved in CH_2Cl_2 (20 mL), and DMAP is removed by rapid filtration through SiO_2 (eluent: CH_2Cl_2). After evaporation of the solvent, the residue is distilled *in vacuo*, whereby the bath temperature should not exceed 60 °C (shield); 58.0 g (72%), blue (!) oil, $bp_{0.01}$ 50–53 °C; R_f (SiO_2; CH_2Cl_2) = 0.75 (note).

IR (film): $\tilde{v}$ (cm^{-1}) = 3471, 2954, 2860, 1743 (C=O), 1555, 1455 (NO_2) [7].

^{1}H NMR ($CDCl_3$, 400 MHz): δ (ppm) = 1.25, 1.5 (m, 3 H, CH_3), 1.97, 2.01 (s, together 3 H, $OCOCH_3$), 4.6, 5.25 (m, 1 H, CH) (mixture of diastereomers).

Note: In ref. [7], it is reported that an explosion occurred upon attempted vacuum distillation on a small scale. Therefore, in ref. [7], the crude product was used for pyrrole synthesis (→ **3.2.2.5**). In our experience, the nitroacetate can be safely handled under the above-described conditions for distillative purification. The origin of the blue color is unknown.

3.2.2.3 * *N*-Formylglycine ethyl ester [8]

| 139.6 | 74.1 | 101.2 | 131.1 |

Triethylamine (111 g, 1.10 mol) is added dropwise to a refluxing solution of glycine ethyl ester hydrochloride (140 g, 1.00 mol) and *p*-toluenesulfonic acid monohydrate (100 mg) in ethyl formate (500 mL). Heating under reflux is continued for 20 h.

The solution is then cooled to 20 °C, the precipitated triethylamine hydrochloride is removed by filtration, and the solution is concentrated *in vacuo* to a volume of ca. 150 mL. On cooling to –5 °C, more hydrochloride precipitates and is removed by filtration. The filtrate is distilled *in vacuo* to give a colorless oil; 126 g (96%), $bp_{0.1}$ 110–111 °C.

IR (film): $\tilde{\nu}$ (cm^{-1}) = 3300 (NH), 1740 (C=O, ester), 1655 (C=O, amide).

^{1}H NMR (CCl$_4$): δ (ppm) = 8.21 (s, 1 H, CHO), 7.2 (s$_{br}$, 1 H, NH), 4.20 (q, J = 7 Hz, 2 H, OCH$_2$), 4.05 (d, J = 6 Hz, 2 H, NCH$_2$), 1.30 (t, J = 7 Hz, 3 H, CH$_3$).

3.2.2.4   **   Ethyl isocyanoacetate [9]

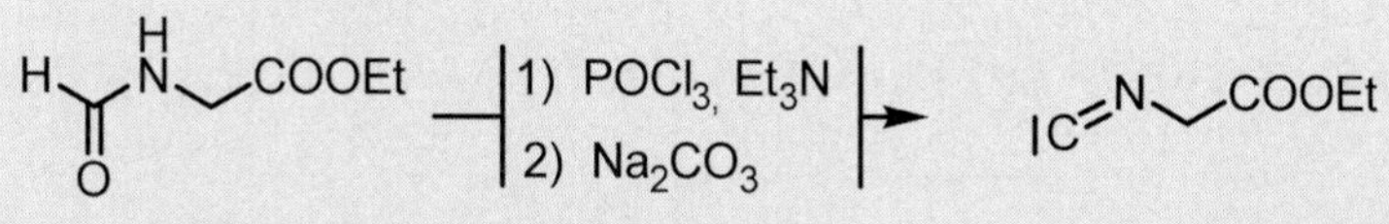

Phosphorus oxychloride (76.5 g, 0.50 mol) is added dropwise to a stirred solution of the N-formyl ester **3.2.2.3** (65.5 g, 0.50 mol) and NEt$_3$ (125 g, 1.24 mol) in CH$_2$Cl$_2$ (500 mL) at 0 °C. Stirring is continued for 1 h.

Anhydrous sodium acetate (100 g) in water (400 mL) is then added, keeping the temperature at 20–25 °C with rapid stirring (Caution: foaming!). Stirring is continued at room temperature for 30 min.

The aqueous phase is diluted with water (to 1000 mL) and extracted with CH$_2$Cl$_2$ (2 × 250 mL). The organic phase is washed with brine, dried over K$_2$CO$_3$, and concentrated. The residue is distilled *in vacuo* to give the product as a colorless oil; 43.1 g (76%), bp$_{12}$ 80–82 °C (note).

IR (film): $\tilde{\nu}$ (cm^{-1}) = 2150 (NC), 1750 (C=O).

^{1}H NMR (CCl$_4$): δ (ppm) = 4.28 (q, J = 7 Hz, 2 H, OCH$_2$), 4.25 (s, 2 H, NCH$_2$), 1.33 (t, J = 7 Hz, 3 H, CH$_3$).

Note: Caution! All reactions must be conducted in a hood and rubber gloves must be worn!

3.2.2.5   **   3,4-Dimethylpyrrole-2-carboxylic acid ethyl ester [7]

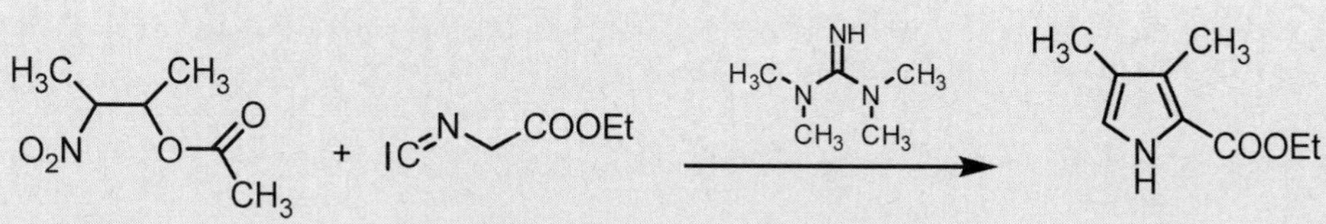

The acetoxynitrobutane **3.2.2.2** (40.0 g, 248 mmol) is added dropwise to a stirred solution of the isocyano ester **3.2.2.4** (22.6 g, 200 mmol) and tetramethylguanidine (60.0 g, 508 mmol) in anhydrous THF (200 mL) and isopropanol (200 mL) at 0 °C. An exothermic reaction occurs, and the inner temperature is kept at 0 °C by cooling (solid CO$_2$/methanol bath). When the addition is complete, stirring is continued for 12 h at room temperature.

Water (1000 mL) is then added to the reaction mixture and the product is extracted with CH_2Cl_2 (2 × 200 mL). The organic phase is dried (Na_2SO_4), the solvent is removed, the residue is dissolved in Et_2O (20 mL), and the solution obtained is rapidly filtered through silica gel (eluent: Et_2O). After removal of the solvent, the solid residue is washed with cold n-hexane; 28.7 g (86%), colorless crystals, mp 82–83 °C; R_f (SiO_2; CH_2Cl_2) = 0.30 (note).

IR (KBr): $\tilde{v}$ (cm^{-1}) = 1743 (C=O) [7].

^{1}H NMR ($CDCl_3$, 400 MHz): δ (ppm) = 6.63 (s, 1 H, pyrrole-5-H), 4.29 (q, J = 14 Hz, 2 H, OCH_2), 2.25, 1.99 (s, 3 H, CH_3), 1.33 (t, J = 14 Hz, 3 H, CH_3).

^{13}C NMR ($CDCl_3$, 400 MHz): δ (ppm) = 161.7 (C=O), 126.5, 120.5, 120.0, 119.2 (pyrrole-C), 59.7 (OCH_2), 14.5, 10.2, 9.85 (CH_3).

Note: The ester **3.2.2.5** can be recrystallized from CH_2Cl_2/n-hexane, mp 92–94 °C [7]. However, the above product does not show any impurities (^{1}H NMR and TLC).

When crude **3.2.2.2** was used for the above pyrrole synthesis (see note under **3.2.2.2**), the yield of **3.2.2.5** dropped to 58%.

3.2.2.6 * 3,4-Dimethylpyrrole-2-carboxylic acid [5]

The pyrrole ester **3.2.2.5** (25.0 g, 150 mmol) is suspended in EtOH (100 mL) and 30% aqueous potassium hydroxide solution (140 mL) is added with stirring. The mixture is heated to reflux for 2 h to give a yellowish solution.

After cooling to 0 °C, the solution is brought to pH 1 by the addition of ice-cold concentrated hydrochloric acid with stirring. A microcrystalline precipitate of the pyrrole carboxylic acid is formed, which cannot be conveniently filtered off by suction; it is therefore taken up in Et_2O (4 × 1 L). The combined extracts are dried (Na_2SO_4) and the solvent is evaporated. The crystalline residue is triturated with Et_2O (50 mL) at –30 °C, collected by suction filtration, and washed with a small amount of cold Et_2O; 18.0 g (86%), colorless crystals, mp >200 °C (dec.); TLC (Et_2O): R_f = 0.30.

---

^{1}H NMR ([D_6]DMSO, 400 MHz): δ (ppm) = 10.97 (s_{br}, 1 H, COOH), 6.64 (s, 1 H, pyrrole-5-H), 2.14, 1.90 (s, 3 H, CH_3).

^{13}C NMR ([D_6]DMSO, 400 MHz): δ (ppm) = 162.76 (C=O), 125.24, 125.92, 119.13, 118.90 (pyrrole-C), 10.37, 10.02 (CH_3).

3.2.2.7 ** 3,4-Dimethylpyrrole [5]

The pyrrole carboxylic acid **3.2.2.6** (15.9 g, 100 mmol) is placed in a 50 mL round-bottomed flask connected to a short-path distillation apparatus and heated to ca. 300 °C *in vacuo* (ca. 20 mbar). Gas evolution (decarboxylation) occurs and the product distils at ca. 160 °C as a colorless oil, which solidifies on standing in a refrigerator to give long needles; 9.50 g (100%); TLC (CH$_2$Cl$_2$): R_f = 0.65.

^{1}H NMR (CDCl$_3$, 400 MHz): δ (ppm) = 1.755 (s$_{br}$, 1 H, NH), 6.55 (s, 2 H, pyrrole-2- and 5-H), 2.095 (s, 6 H, CH$_3$).

^{13}C NMR (CDCl$_3$, 400 MHz): δ (ppm) = 118.10, 115.49 (pyrrole-C), 9.91 (CH$_3$).

[1]   a) D. S. Black, *Sci. Synth.* **2002**, *9*, 44; b) *Houben/Weyl*, Vol. E 6a; c) J. A. Joule, K. Mills, *Heterocyclic Chemistry*, 4th ed., p. 255, Blackwell Science, Oxford, **2000**. For an overview on pyrrole synthesis, see: d) R. J. Sundberg, in *Comprehensive Heterocyclic Chemistry* (Eds.: A. R. Katritzky, C. W. Rees, E. F. V. Scriven), Vol. 2, Pergamon, Oxford, **1996**; e) V. F. Ferreira, M. C. B. V. De Souza, A. C. Cunha, L. O. R. Pereira, M. L. G. Ferreira, *Org. Prep. Proced. Int.* **2001**, 33, 411; f) pyrrole synthesis by MCR: G. Balme, *Angew. Chem.* **2004**, *116*, 6396; *Angew. Chem. Int. Ed.* **2004**, *43*, 6238.

[2]   D. H. R. Barton, J. Kervagoret, S. J. Zard, *Tetrahedron* **1990**, *46*, 7587.

[3]   For an improved procedure for the Barton–Zard synthesis, see: P. Bobal, D. A. Lightner, *J. Heterocycl. Chem.* **2001**, *38*, 527.

[4]   a) O. Possel, A. M. van Leusen, *Heterocycles* **1977**, *7*, 77; b) N. P. Parvi, M. L. Trudell, *J. Org. Chem.* **1997**, *62*, 2649.

[5]   Q. Chen, T. Wang, Y. Zhang, Q. Wang, J. Ma, *Synth. Commun.* **2002**, *32*, 1031.

[6]   M. D'Aurias, D. E. Luca, G. Mauriello, R. Racioppi, G. J. Sleiter, *J. Chem. Soc., Perkin Trans. 1* **1997**, 2260.

[7]   M. B. Smith, J. March, *March's Advanced Organic Chemistry*, 5th ed., p. 1350, John Wiley & Sons, New York, **2001**.

[8]   U. Schöllkopf, D. Hoppe, private communication, 1981.

[9]   a) D. Hoppe, U. Schöllkopf, *Liebigs Ann. Chem.* **1972**, *763*, 1; b) U. Schöllkopf, D. Hoppe, original contribution 1981.

[10] A remarkable improvement of yield in Barton–Zard syntheses by the use of *tert*-butyl methyl ether as solvent has recently been reported: A. Bhattacharya, S. Cherukuri, R. E. Plata, N. Patel Jr., V. Tamez, J. A. Grosso, M. Peddicord, V. A. Palaniswamy, *Tetrahedron Lett.* **2006**, *47*, 5481.

3.2.3   4,6-Dimethoxybenzo[*b*]thiophene

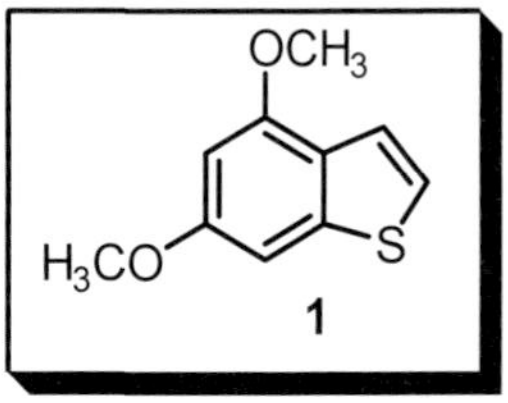

Topics:
- Synthesis of a benzothiophene derivative
- Transformation of a carboxylic acid to an amide
- Directed metalation of an arene
- α-Metalation of an alkyl aryl thioether
- Intramolecular acylation of an organolithium compound by an amide
- Reduction C=O → CH–OH, dehydration of a carbinol

(a) General

Benzo[*b*]thiophene derivatives exhibit various biological activities. The system shows bioisosterism with naphthalene and indole. Thus, compound **2** is an insecticide, like the corresponding naphthalene compound; **3** acts as a plant growth inhibitor like indolyl-3-acetic acid, and **4** has an even stronger effect on the central nervous system than the indole analogue tryptamine [1]:

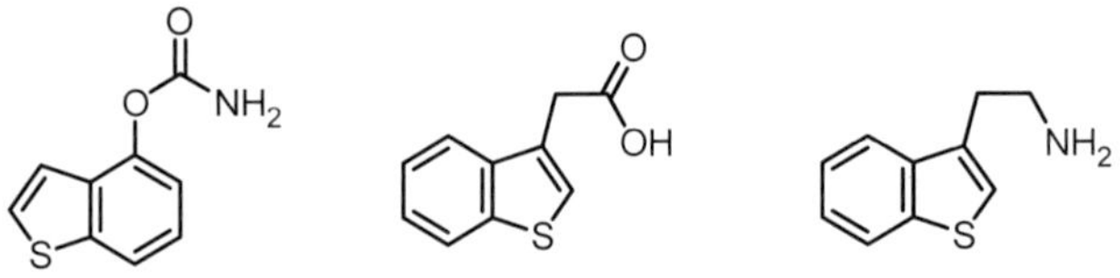

There are numerous methods for the synthesis of benzo[*b*]thiophenes, utilizing aryl thiols as starting materials [2], but their general scope and applicability is limited. One of these methods [1, 2] starts with *S*-alkylation of thiophenolate **5** with α-halogeno carbonyl compounds to give the (α-phenylthio) carbonyl compounds **6**, which undergo an intramolecular hydroxyalkylation (S_EAr) and subsequent dehydration in the presence of $ZnCl_2$ to afford benzo[*b*]thiophenes **7**:

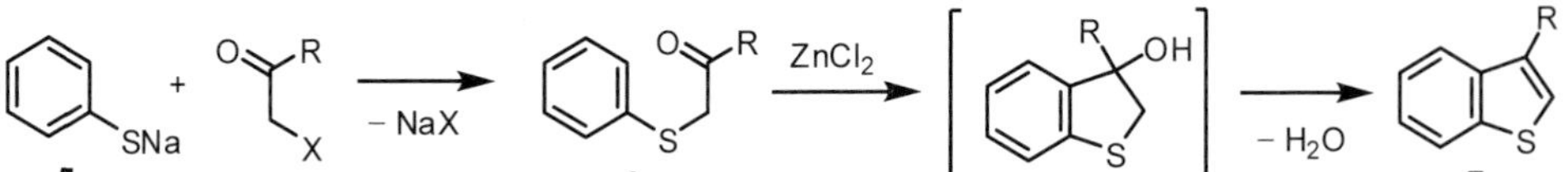

Needless to say, this method is restricted to derivatives of type **7** substituted at the hetero ring, since the presence of substituents at the benzo ring can cause regioselectivity problems in the S_EAr ring closure.

In a recent method [3], the use of the expensive and environmentally offensive thiophenol is avoided; instead, the readily available *N,N*-dialkylbenzamides **8** are used as substrates. A methylsulfanyl group is introduced at a position *ortho* to the amide function by directed metalation [4] and electrophilic reaction of the Ar-*ortho*-Li intermediate with CH_3–S–S–CH_3 (→ **9**). The SCH_3 group in **9** is then subjected to another metalation, which is followed by cyclization through intramolecular acylation of the formed thiomethyl carbanion by the adjacent amide function. The thioindoxyls **10** [5] thus obtained are reduced with $NaBH_4$ to the carbinols **11**, which spontaneously eliminate H_2O to provide benzo[*b*]thiophenes **12**:

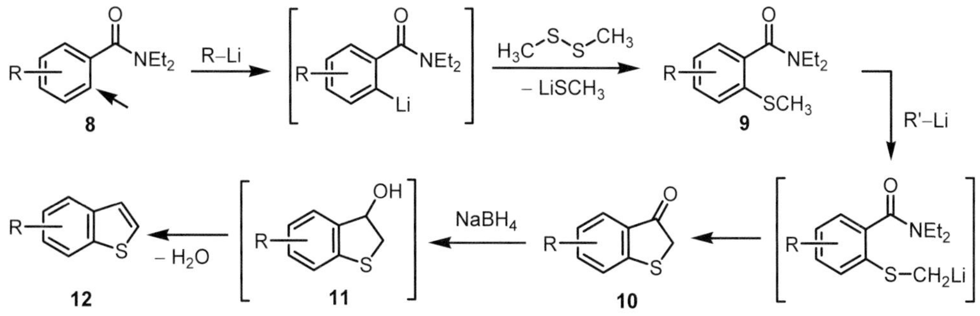

Directed metalation of arenes and hetarenes generally occurs regioselectively at positions *ortho* to functional groups capable of stabilizing the lithio arenes that result from hydrogen-metal exchange through intramolecular complexation. Even methoxy groups may serve as complexing donor systems (e.g. **13** → **14**), but of greater directing power are *N,N*-dialkylamide and oxazoline functions (e.g. **8** → **15** and **17** → **18**) [6], as demonstrated by the regiochemical outcome of arene metalation in the synthesis of the target molecule **1** outlined in (b).

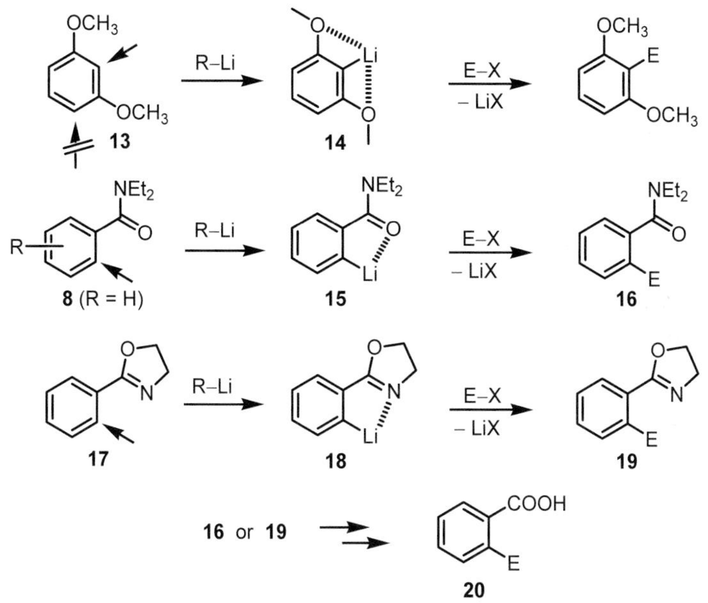

Since substrates **8** and **17** are derived from benzoic acids, their *ortho*-metalation and concomitant electrophilic transformation (**15** → **16**; **18** → **19**) represents a directed regioselective *ortho*-functionalization of these acids (**16/19** → **20**). Possible functionalizations are alkylation, acylation, carboxylation, alkylsulfamation, etc.

It should be noted that acylation of organolithium compounds (as in **9** → **10**) is advantageously performed with Weinreb amides (*N*-methyl-*N*-alkoxyamides) **21** [7], which prevent the addition of a second R–Li moiety to the carboxyl function (as observed in the case of esters and acid chlorides) by chelate-stabilization of the primary addition product **22** [8]:

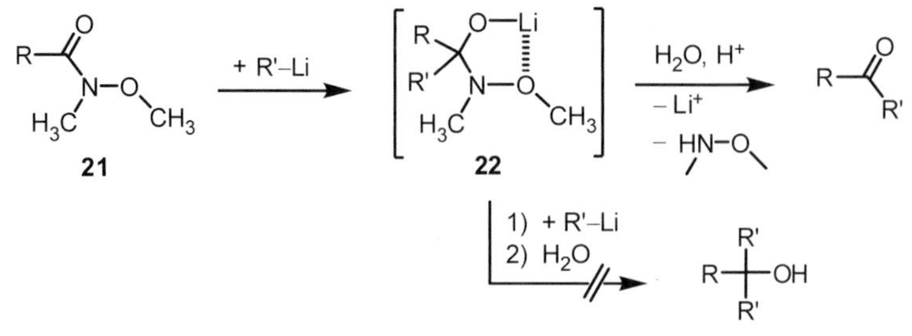

(b) Synthesis of 1

N,N-Diethyl-2,4-dimethoxybenzamide (**24**) is conveniently prepared by aminolysis of the acid chloride from 2,4-dimethoxybenzoic acid (**23**) with HNEt$_2$ [9]:

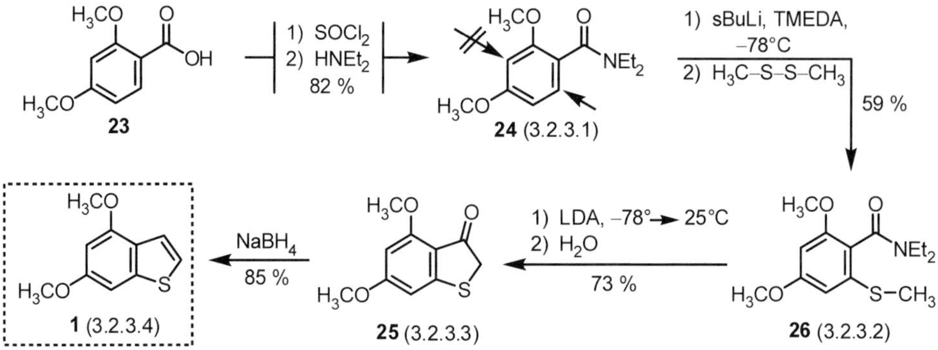

The *N,N*-diethylbenzamide **24** is metalated regioselectively using *sec*-BuLi at –78 °C in the presence of TMEDA and the *ortho*-Li compound intermediately formed is quenched by reaction with dimethyl disulfide to give the *o*-(methylsulfanyl)benzamide **26**. In general, the addition of TMEDA (*N,N,N',N'*-tetramethylethylenediamine) enhances metalation reactions, since the oligomeric organolithium compounds are monomerized (and thus activated) by complex formation (**27**) with this reagent [10]:

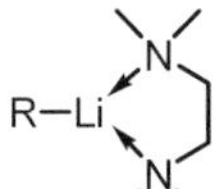

Cyclization of the *o*-(methylsulfanyl)benzamide **26** to the thioindoxyl **25** occurs readily on metalation with LDA at –78 °C and subsequent intramolecular acylation of the initially formed carbanion in the side chain. When **25** is reacted with NaBH$_4$, reduction of the carbonyl group and concomitant H$_2$O elimination takes place to yield the benzo[*b*]thiophene **1**.

Thus, the target molecule **1** is obtained in a four-step sequence with an overall yield of 30% (based on 2,4-dimethoxybenzoic acid (**23**)).

(c) Experimental procedures for the synthesis of 1

3.2.3.1   **   *N,N*-Diethyl-2,4-dimethoxybenzamide [9]

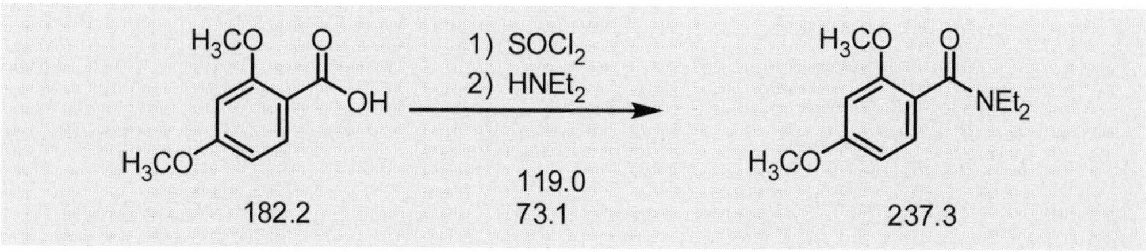

2,4-Dimethoxybenzoic acid (3.20 g, 17.6 mmol) and thionyl chloride (12.7 g, 107 mmol) are dissolved in anhydrous benzene (100 mL) and heated to reflux with stirring for 2 h. The solvent and excess thionyl chloride are removed *in vacuo*.

The crude acid chloride (3.60 g, 17.6 mmol) is dissolved in anhydrous benzene (40 mL; Caution: carcinogenic!) and the reaction mixture is cooled to 0 °C. A solution of diethylamine (3.96 g, 54.1 mmol) in anhydrous benzene (15 mL) is slowly added at 0 °C with stirring. The reaction mixture is then stirred for 2 h at 0 °C and thereafter for 15 h at room temperature.

After removal of the solvent under reduced pressure, the residue is extracted with CH_2Cl_2 (3 × 40 mL). The combined organic layers are washed sequentially with 5% $NaHCO_3$ (100 mL), 5% HCl (100 mL), and water (100 mL), and dried (Na_2SO_4). Removal of the solvent affords the amide as a yellow liquid; 3.49 g (82%).

IR (film): $\tilde{v}$ (cm^{-1}) = 2838, 1606 (C=O), 1427, 1277, 1207, 1157, 1028.

¹H NMR (500 MHz, CDCl₃): δ (ppm) = 7.12 (d, *J* = 8.3 Hz, 1 H), 6.49 (dd, *J* = 8.3, 2.2 Hz, 1 H), 6.46 (d, *J* = 2.2 Hz, 1 H), 3.81 (s, 3 H, OCH₃), 3.79 (s, 3 H, OCH₃), 3.52 (q, *J* = 7.1 Hz, 2 H, CH₂CH₃), 3.16 (q, *J* = 7.1 Hz, 2 H, CH₂CH₃), 1.23 (t, *J* = 7.1 Hz, 3 H, CH₂CH₃), 1.03 (t, *J* = 7.1 Hz, 3 H, CH₂CH₃).

¹³C NMR (500 MHz, CDCl₃): δ (ppm) = 168.8, 161.2, 156.6, 128.4, 119.9, 104.6, 98.7, 55.5, 55.4, 42.9, 38.9, 14.0, 12.9.

3.2.3.2   ***   *N,N*-Diethyl-2-methylsulfanyl-4,6-dimethoxybenzamide [3]

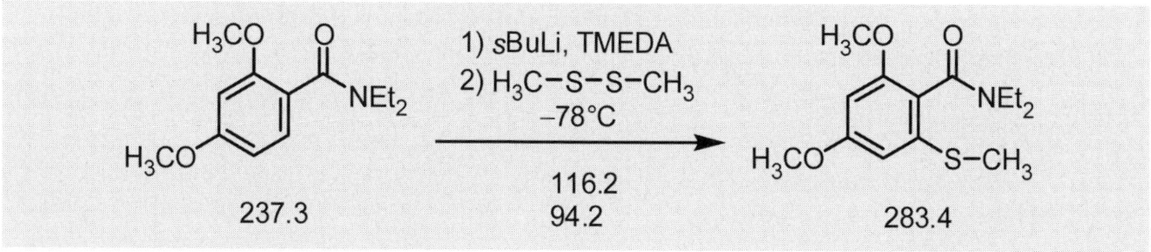

sec-BuLi (9.08 mL, 1.3 M in cyclohexane, 1.5 equiv.) is introduced through a septum by means of a syringe into a stirred solution of TMEDA (1.37 g, 11.8 mmol, 1.5 equiv.) in anhydrous THF (30.0 mL) at –78 °C. Stirring is continued for 20 min and then a solution of the benzamide **3.2.3.1** (1.87 g, 7.88 mmol) in anhydrous THF (15 mL) is added to the reaction mixture at –78 °C (note 1). After stirring for an additional 30 min, dimethyl disulfide (2.08 g, 22.1 mmol) in anhydrous THF (10 mL) is added at –78 °C. Stirring is continued for 15 min, and then the reaction mixture is allowed to warm to room temperature and left under ambient conditions for 15 h.

The solvent is then removed under reduced pressure, saturated NH_4Cl solution (100 mL) is added, and the product is extracted with Et_2O (3 × 40 mL). The combined organic layers are washed with brine (100 mL) and dried (Na_2SO_4). Removal of the solvent affords the thioether as a colorless solid (note 2), mp 83–84 °C; 1.31 g (59%).

IR (solid): $\tilde{v}$ (cm^{-1}) = 2932, 1619 (C=O), 1395, 1277, 1152, 831.

^{1}H NMR (500 MHz, CDCl$_3$): δ (ppm) = 6.43 (d, J = 2.1 Hz, 1 H), 6.29 (d, J = 2.1 Hz, 1 H), 3.82 (s, 3 H, OCH$_3$), 3.77 (s, 3 H, OCH$_3$), 3.77 (dq (superimposed), J = 15.0, 7.1 Hz, 1 H, CH$_2$CH$_3$), 3.40 (dq, J = 15.0, 7.1 Hz, 1 H, CH$_2$CH$_3$), 3.13 (q, J = 7.1 Hz, 2 H, CH$_2$CH$_3$), 2.45 (s, 3 H, SCH$_3$), 1.25 (t, J = 7.1 Hz, 3 H, CH$_2$CH$_3$), 1.04 (t, J = 7.1 Hz, 3 H, CH$_2$CH$_3$).

^{13}C NMR (500 MHz, CDCl$_3$): δ (ppm) = 166.8, 160.8, 156.8, 137.4, 119.4, 104.0, 95.7, 55.7, 55.5, 42.6, 38.7, 16.4, 13.9, 12.6.

Notes: (1) The use of 1.5 equivalents of *sec*-BuLi proved to be essential for the success of this reaction step.
   (2) If necessary, the product can be purified by column chromatography (EtOAc/*n*-hexane, 2:1, R_f = 0.21).

3.2.3.3   ***   2,3-Dihydro-4,6-dimethoxybenzo[*b*]thiophen-3-one [3]

*n*BuLi (1.38 mL, 2.5 M in *n*-hexane, 1.5 equiv.) is added through a septum by means of a syringe to a stirred solution of diisopropylamine (580 mg, 5.73 mmol, 2.5 equiv.) in anhydrous THF (20 mL) at 0 °C. Stirring is continued for 20 min at 0 °C and then the reaction mixture is cooled to –78 °C. A solution of the 2-methylsulfanyl benzamide **3.2.3.2** (0.65 g, 2.29 mmol) in anhydrous THF (5 mL) is added dropwise and stirring is continued for 30 min at –78 °C. The reaction mixture is then allowed to warm to room temperature and is stirred under ambient conditions for 15 h.

Saturated NH_4Cl solution (100 mL) is then added and the mixture is extracted with Et_2O (3 × 40 mL). The combined organic layers are dried (Na_2SO_4), the solvent is removed under reduced pressure, and the crude thioindoxyl is crystallized from EtOH; colorless needles (notes 1 and 2), mp 110–112 °C; 350 mg (73%).

IR (solid): $\tilde{v}$ (cm^{-1}) = 2933, 1568 (C=O), 1277, 1209, 1152, 831.

^{1}H NMR (500 MHz, CDCl$_3$): δ (ppm) = 6.45 (d, J = 1.9 Hz, 1 H), 6.15 (d, J = 1.9 Hz, 1 H), 3.91 (s, 3 H, OCH$_3$), 3.87 (s, 3 H, OCH$_3$), 3.76 (s, 2 H, SCH$_2$).

^{13}C NMR (500 MHz, CDCl$_3$): δ (ppm) = 196.0, 167.4, 161.5, 159.7, 113.6, 99.8, 95.7, 55.9, 39.7.

Notes: (1) The product can be purified by column chromatography (EtOAc/*n*-hexane, 2:1; R_f = 0.53).
(2) The thioindoxyl is air-sensitive and should be kept under argon in a refrigerator.

3.2.3.4   *   4,6-Dimethoxybenzo[*b*]thiophene [3]

| 210.2 | 37.8 | 194.3 |

A solution of NaBH₄ (70 mg, 1.90 mmol) in MeOH/10% aq. NaOH (10:3) (15 mL) is added dropwise to a stirred solution of the thioindoxyl **3.2.3.3** (200 mg, 0.95 mmol) in MeOH/10% aq. NaOH (6:1) (20 mL). The reaction mixture is heated to reflux with stirring for 16 h.

The solvent is then removed under reduced pressure, 10% H_2SO_4 (30 mL) is added to the residue, and the resulting mixture is extracted with Et₂O (3 × 40 mL). The combined organic layers are washed with water (100 mL) and dried (Na₂SO₄). Removal of the solvent affords the benzo[*b*]thiophene as yellow crystals (note), mp 77–79 °C; 155 mg (85%).

¹H NMR (500 MHz, CDCl₃): δ (ppm) = 7.38 (d, *J* = 5.6 Hz, 1 H), 7.15 (d, *J* = 5.5 Hz, 1 H), 6.93 (d, *J* = 1.5 Hz, 1 H), 6.41 (d, *J* = 1.9 Hz, 1 H), 3.92 (s, 3 H, OCH₃), 3.86 (s, 3 H, OCH₃).

¹³C NMR (500 MHz, CDCl₃): δ (ppm) = 158.8, 155.4, 142.1, 125.0, 121.8, 120.2, 96.2, 95.8, 55.7, 55.4

Note: The product can be purified by column chromatography (EtOAc/hexane, 2:1; R_f = 0.53).

[1]   Th. Eicher, S. Hauptmann, *The Chemistry of Heterocycles*, 2ⁿᵈ ed., p. 81, Wiley-VCH, Weinheim, **2003**.

[2]   S. Rajappa, *Comprehensive Heterocyclic Chemistry* (Eds.: A. R. Katritzky, C. W. Rees), Vol. 4, Pergamon, New York, **1984**.

[3]   C. Mukherjee, S. Kamila, A. De, *Tetrahedron* **2003**, *59*, 4767.

[4]   a) N. Krause, *Metallorganische Chemie*, p. 45, Spektrum Akademischer Verlag, Heidelberg, **1996**; b) hydrogen-metal permutation as a tool of planned metal-organic synthesis: M. Schlosser, *Angew. Chem.* **2005**, *117*, 380; *Angew. Chem. Int. Ed.* **2005**, *44*, 376..

[5]   Thioindoxyls are the precursors of the thioindigo dyes, which are formed by their oxidative dimerization [1].

[6]   V. Snieckus, *Chem. Rev.* **1990**, *90*, 879.

[7]   S. Nahm, S. M. Weinreb, *Tetrahedron Lett.* **1981**, *22*, 3815.

[8]   An instructive example utilizing directed *ortho*-metalation and acylation of an organolithium compound with a Weinreb amide as key transformations is the synthesis of the natural product mamanutha quinone: T. Yoon, S. J. Danishefsky, S. de Gala, *Angew. Chem.* **1994**, *106*, 923; *Angew. Chem. Int. Ed. Engl.* **1994**, *33*, 853.

[9]   C. Mukherjee, S. Kamila, A. De, S. S. Mondal *Tetrahedron* **2003**, *59*, 1339.

[10] Ref. [4a], p. 14.

3.2.4   2-Phenylindole

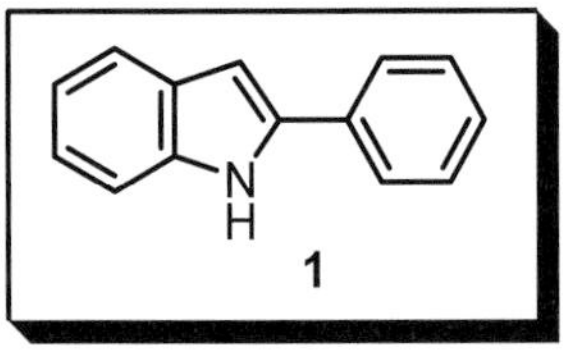

Topics:
- Transition-metal-mediated indole synthesis
- Sonogashira cross-coupling reaction
- Acetylation of a primary amine
- Formation of indoles from (*o*-ethynyl)anilines

(a) General

Indole is one of the most important heterocycles, since a large number of natural products (e.g., the amino acid tryptophan and many alkaloids) and pharmaceuticals are derived from this ring system. Therefore, numerous synthetic approaches to indoles have been developed [1].

For the synthesis of 2-substituted indoles **4** (as in the case of the target molecule **1**), two of the classical methods are relevant, namely the Reissert synthesis (reductive cyclization of *o*-nitrobenzyl ketones **2**) and the Fischer synthesis (acid-catalyzed eliminative cyclization of *N*-arylhydrazones of methyl ketones **3**):

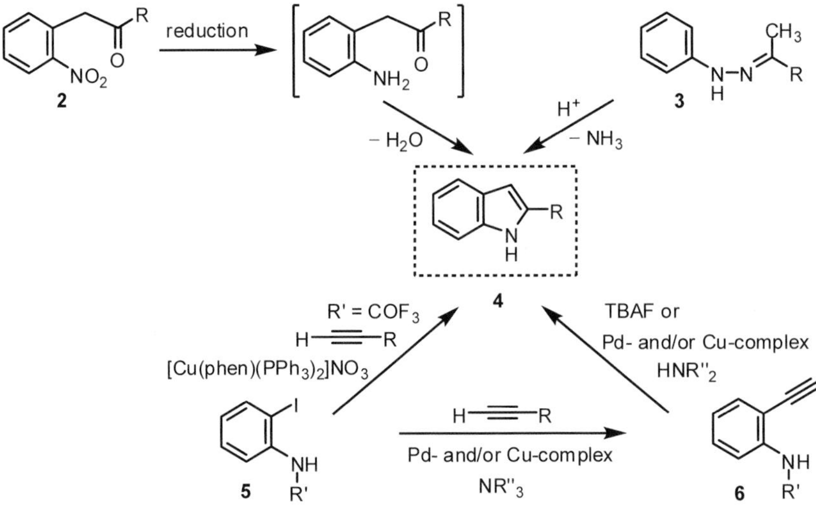

The repertoire of methods for the synthesis of indoles of type **4** is considerably enlarged by a series of transition-metal-mediated reactions [2]. Thus, the *N*-acyl or *N*-sulfonyl derivatives of (*o*-ethynyl)-arylamines **6** cyclize to the corresponding 2-substituted indoles upon interaction with TBAF or Pd and/or Cu complexes. The (*o*-ethynyl)arylamines **6** are conveniently obtained from Pd-mediated Sonogashira cross-coupling reactions (cf. **1.6.3**) of *o*-iodoanilines **5** with terminal acetylenes [3–5]. The potential of this method is demonstrated by a solid-phase version utilizing polymer-bound (*o*-halogeno)anilines [6]. Recently, a one-step procedure (**5 → 4**) starting from *N*-trifluoroacetates **5** (R' = COCF$_3$) was introduced utilizing a Cu complex as catalyst [7].

Another transition-metal-assisted formation of 2-substituted indoles is the Pd-catalyzed annulation of iodoanilines and ketones [8]:

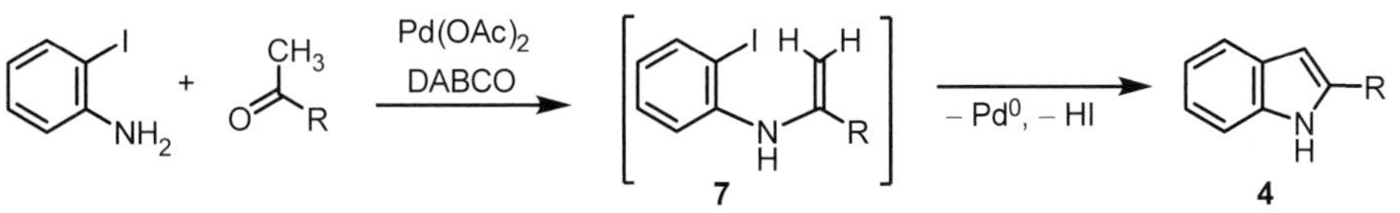

This transformation was shown to proceed via enamine formation ($\rightarrow$**7**) and subsequent intramolecular Heck reaction (cf. **1.6.1**) (**7 $\rightarrow$ 4**).

The target molecule **1** has been prepared from the phenylhydrazone of acetophenone (**3**, R = Ph) by application of the Fischer indole synthesis [9]. Here, as a modern alternative, a Pd-mediated approach [3] is presented (section (b)).

(b) Synthesis of 1

The starting material for the synthesis of **1** is the commercially available *o*-iodoaniline (**8**), which is subjected to a Sonogashira cross-coupling reaction with phenylacetylene in the presence of (Ph₃P)₂PdCl₂, CuI, and triethylamine, providing an almost quantitative yield of (*o*-phenyl-ethynyl)aniline (**9**, cf. **1.6.3**):

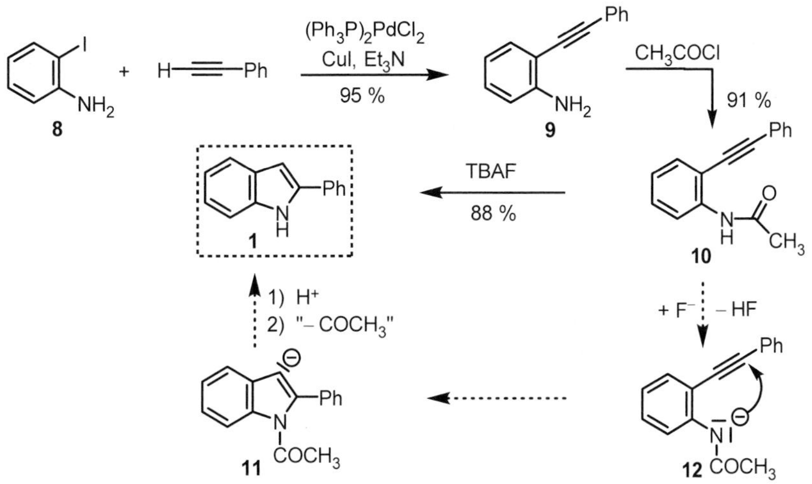

The aniline derivative **9** is acylated using acetyl chloride to give the *N*-acetyl compound **10**, which readily cyclizes to 2-phenylindole (**1**) on treatment with TBAF in refluxing tetrahydrofuran.

In contrast to the Pd-assisted cyclization [1, 7], the mechanism of the TBAF-promoted process (**10 $\rightarrow$ 1**) remains speculative. It was found that anilines **6** (R' = H) do not cyclize under these conditions. Moreover, for the reaction of **10** to give **1**, three equivalents of TBAF have to be used. It can therefore be assumed that deprotonation of **10** to afford the anion **12** initiates the cyclization (**12 $\rightarrow$ 11**), which is completed by reprotonation and fluoride-induced *N*-deacylation of the indole (**11 $\rightarrow$ 1**).

Alternatively, (2-phenylethynyl)aniline (**9**) can be directly cyclized to 2-phenylindole (**1**) upon treatment with the stronger base KO*t*Bu in NMP (79% yield) [10].

Using the first approach, the target molecule **1** is prepared in a three-step sequence with an overall yield of 73% (based on *o*-iodoaniline (**8**)).

(c)　Experimental procedures for the synthesis of 1

3.2.4.1　*　*N*-Acetyl-2-(phenylethynyl)aniline [3]

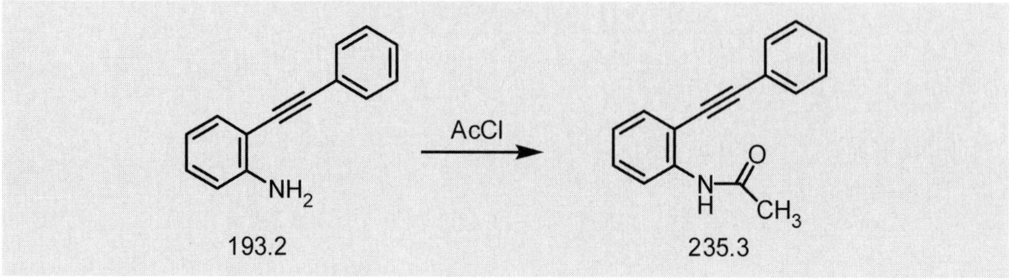

193.2　　　　　　　　　　　　　　235.3

Acetyl chloride (785 mg, 10.0 mmol) is added dropwise to a stirred solution of the aniline **1.6.3.1** (1.75 g, 9.05 mmol) in a pyridine/THF mixture (1:2, 15 mL) and stirring is continued for 24 h at room temperature.

The reaction mixture is then diluted with H_2O (20 mL), extracted with $CHCl_3$ (3 × 20 mL), and the combined extracts are dried ($MgSO_4$). The solvent is evaporated under reduced pressure and the residue is purified by recrystallization from *n*-hexane/acetone (5:1) to give colorless needles; yield 1.83 g (86%); mp 119–121 °C.

> **IR** (KBr): $\tilde{\nu}$ $(cm^{-1}) = 3300, 1660$.
>
> **^{1}H NMR** (CDCl$_3$, 500 MHz): δ (ppm) = 8.41 (d, J = 8.2 Hz, 1 H, ArH), 7.98 (s_{br}, 1 H, NH), 7.56–7.26 (comb. m, 7 H, ArH), 7.07 (t, J = 7.7 Hz, 1 H, ArH), 2.24 (s, 3 H, CH$_3$).
>
> **^{13}C NMR** (CDCl$_3$, 125 MHz): δ (ppm) = 168.13, 138.92, 131.66, 131.52, 129.77, 128.96, 128.62, 123.41, 122.36, 119.32, 111.80, 96.41, 84.30, 24.99.
>
> **MS** (EI, 70 eV): m/z (%) = 235 (31) $[M]^+$, 193 (100), 65 (16).

3.2.4.2　*　2-Phenylindole [3]

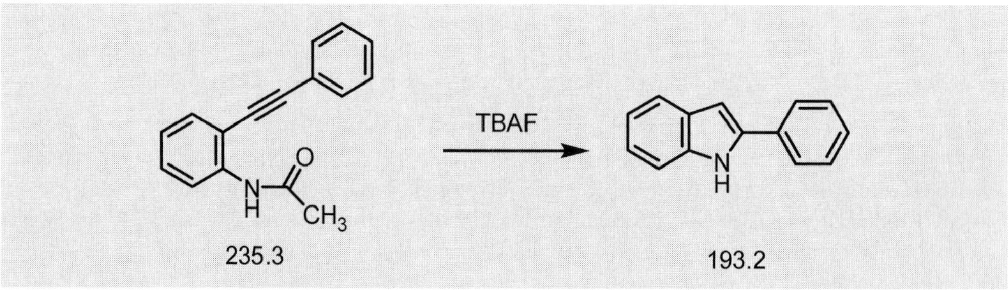

235.3　　　　　　　　　　　　　　　193.2

A mixture of the acetamide **3.2.4.1** (1.65 g, 7.00 mmol) and TBAF (1 M soln. in THF, 14.0 mmol) in THF (35 mL) is heated to reflux for 12 h.

After removal of the THF *in vacuo*, H_2O (50 mL) is added to the residue and the resulting mixture is extracted with EtOAc (3 × 20 mL). The combined EtOAc extracts are dried ($MgSO_4$) and concentrated. The residue is purified by column chromatography (SiO$_2$; CH$_2$Cl$_2$); colorless crystals; yield 1.19 g (88%); mp 185–187 °C.

IR (KBr): $\tilde{\nu}$ (cm^{-1}) = 3445, 1655.

^{1}H NMR (CDCl$_3$, 500 MHz): δ (ppm) = 8.42–8.28 (br s, 1 H, NH), 7.68–7.62 (comb. m, 3 H, ArH), 7.48–7.22 (comb. m, 4 H, ArH), 7.20 (dt, J = 8.2/1.1 Hz, 1 H, ArH), 7.12 (dt, J = 7.1/1.1 Hz, 1 H, ArH), 6.83 (dd, J = 1.1/1.9 Hz, 1 H, ArH).

^{13}C NMR (CDCl$_3$, 125 MHz): δ (ppm) = 137.87, 136.80, 132.37, 129.26, 129.02, 127.71, 125.15, 122.36, 120.66, 120.27, 110.89, 99.98.

MS (EI, 70 eV): m/z (%) = 193 (100) $[M]^+$, 165 (19).

---

[1] a) Th. Eicher, S. Hauptmann, *The Chemistry of Heterocycles*, 2nd ed., p. 102, Wiley-VCH, Weinheim, **2003**; b) for a review on practical methodologies for synthesizing indoles, see: G. R. Humphrey, J. T. Kuethe, *Chem. Rev.* **2006**, *106*, 2875.

[2] a) As a standard reference, see: M. Beller, C. Bolm (Eds.), *Transition Metals for Organic Synthesis*, two volumes, Wiley-VCH, Weinheim, **2004**; b) 1,2-disubstituted indoles are obtained from (*o*-halogenoaryl)alkynes and primary amines by a Ti/Pd-catalyzed domino hydroamination/N-arylation reaction: H. Siebeneicher, I. Bytschkov, S. Doye, *Angew. Chem.* **2003**, *115*, 3151; *Angew. Chem. Int. Ed.* **2003**, *42*, 3042; c) for a Neber route to 2-substituted indoles, see: D. F. Taber, W. Tian, *J. Am. Chem. Soc.* **2006**, *128*, 1058.

[3] A. Yasuhara, Y. Kanamori, M. Kaneko, A. Numata, Y. Kondo, T. Sakamoto, *J. Chem. Soc., Perkin Trans. 1* **1999**, 529.

[4] A. Arcadi, S. Cacchi, F. Marinelli, *Tetrahedron Lett.* **1989**, *30*, 2581.

[5] 2-Iodoanilines (**5**, R' = H) are easily prepared by *ortho*-lithiation of anilines and subsequent iodination: V. Snieckus, *Chem. Rev.* **1990**, *90*, 879.

[6] a) M. D. Collini, J. W. Ellingboe, *Tetrahedron Lett.* **1997**, *38*, 7963; b) H.-Ch. Zhang, H. Ye, A. F. Moretto, K. K. Brumfield, B. E. Maryanoff, *Org. Lett.* **2000**, *2*, 89.

[7] S. Cacchi, G. Fabrizi, L. M. Parisi, *Org. Lett.* **2003**, *5*, 3843.

[8] C. Chen, D. R. Lieberman, R. D. Larsen, T. R. Verhoeven, P. J. Reider, *J. Org. Chem.* **1997**, *62*, 2676.

[9] A. Guy, J. P. Guette, G. Lang, *Synthesis* **1980**, 222.

[10] A. L. Rodriguez, C. Koradin, W. Dohle, P. Knochel, *Angew. Chem.* **2000**, *112*, 2607; *Angew. Chem. Int. Ed.* **2000**, *39*, 2488.

3.2.5 Melatonin

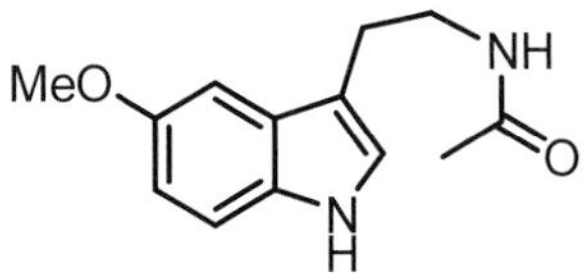

1

Topics:
- Synthesis of an indole-derived natural product
- Japp–Klingemann reaction
- Fischer indole synthesis
- Ester and amide hydrolysis
- Thermal decarboxylation, *N*-acetylation

(a) General

Melatonin (**1**, 5-methoxy-(*N*-acetyl)tryptamine) is a naturally occurring hormone produced by the pineal gland in vertebrates. Its formation and secretion is increased at night, which leads to the onset of sleep. Melatonin has been shown to exhibit medicinally useful activities in treating sleep disorders, in protection against oxidative stress, and as an inhibitor of the onset of Alzheimer's disease [1].

Retrosynthesis of melatonin (**1**) using a retro-Fischer indole synthesis [2] approach leads to the hydrazine **3** and the aldehyde **4** via the hydrazone **2**; **4** is accessible from the dihydropyrrole **5**.

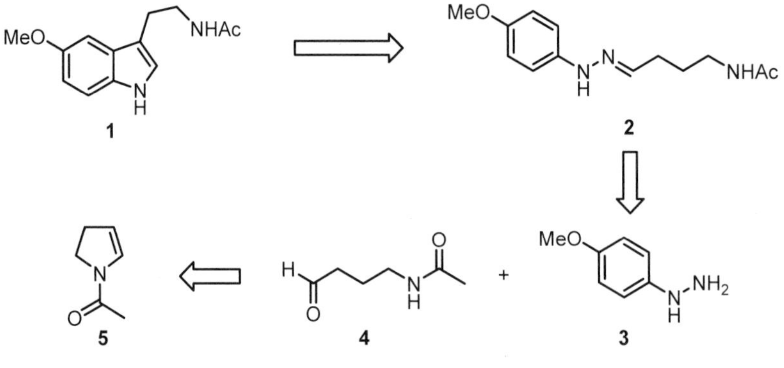

Numerous syntheses of melatonin (**1**) have been performed according to this retrosynthetic analysis [3].

For example, **1** has been formed by reaction of the enamide **5** (as precursor of the aldehyde **4**) with the hydrochloride of **3**. Compound **5** can be prepared by oxidation of pyrrolidine (**6**) with persulfate to give the trimer **7** and subsequent *N*-acetylation of dihydropyrrole **8** formed by thermal cleavage of **7** [4]:

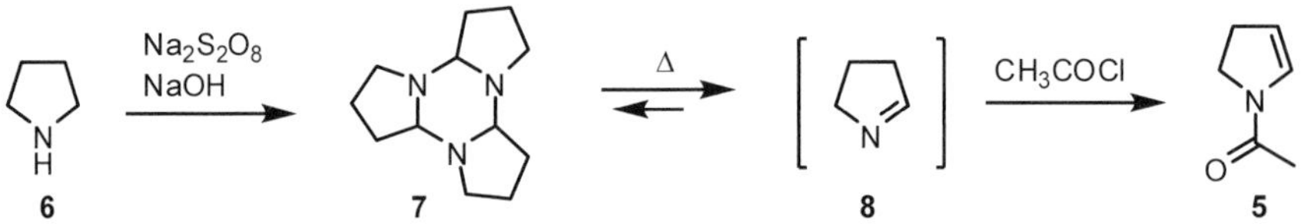

Another synthesis according to a Fischer strategy is presented in section (b).

A different approach (not discussed in the retrosynthesis) utilizes radical-based indole formation [1]. The key intermediate is the acetylene **11**, which is obtained by Mitsunobu reaction (cf. **3.3.4**) of 2-iodo-(*N*-mesyl)-*p*-anisidine (**9**) with the phthalimide **10**. The acetylene **11** undergoes an *exo-trig* radical cyclization mediated by (Me₃Si)₃SiH and AIBN to give a mixture of the indolenine **12** and the indole **13**. The indolenine is converted *in situ* to the indole by treatment with TosOH. Removal of the phthalyl protecting group by hydrazinolysis and the mesyl group by cleavage with KOH liberates the NH-indole moiety and the primary amino function of the tryptamine **14**, which is then acetylated to furnish melatonin (**1**).

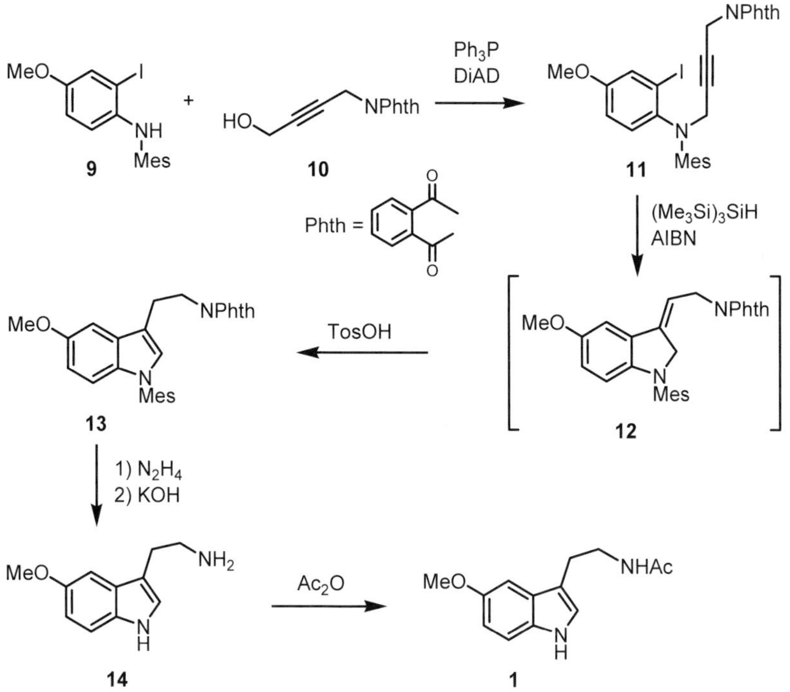

(b) Synthesis of 1

The synthesis of melatonin (**1**) described here is easy to perform, avoids expensive educts, and is scalable to an industrial process [5, 6].

Phthalimide (**15**) is alkylated with 1,3-dibromopropane under microwave irradiation to give the 1-bromopropyl compound **16**, which, after Finkelstein exchange of bromide by iodide, is used for α-alkylation of ethyl acetoacetate to give **18**. Subsequent Japp–Klingemann reaction with the (4-methoxyphenyl)diazonium salt **17** affords the indole-2-carboxylate **21** directly. This domino process [7] includes the formation of the hydrazone **20** via the azo compound **19** followed by a Fischer indole synthesis to give **21**. Subsequent hydrolysis of the phthalimide and the ester moiety (via the unstable carboxylic acid **22**) furnishes 5-methoxytryptamine (**14**), which in the final step is acetylated at the primary amino group to give melatonin (**1**).

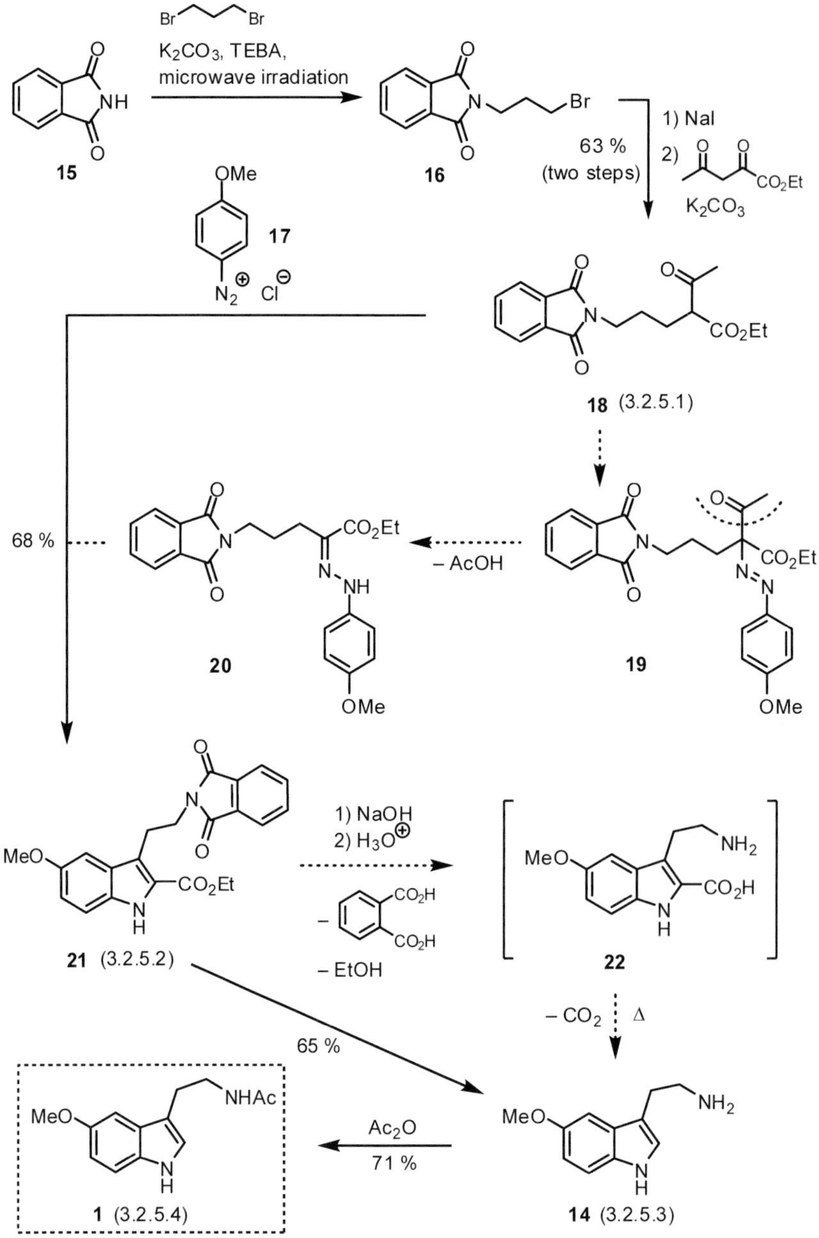

Thus, for the synthesis of the target molecule, a linear sequence consisting of four individual steps is performed, which leads to **1** in an overall yield of 20% (based on **15**).

(c) Experimental procedures for the synthesis of 1

3.2.5.1 ** Ethyl 2-(3-phthalimidopropyl)acetoacetate [6]

185.2   317.3

A suspension of potassium phthalimide (1.27 g, 6.84 mmol), 1,3-dibromopropane (2.77 g, 13.7 mmol), and triethylbenzylamine chloride (TEBA, 154 mg, 0.670 mmol, 10.0 mol%) in CH_3CN (1.60 mL) is heated to 100 °C for 10 min under microwave irradiation.

After cooling to 20 °C, Et_2O (25 mL) is added and the precipitate (KBr and the side-product bisphthalimidopropane) is removed by filtration. The filtrate is concentrated *in vacuo*, the residue (crude *N*-(3-bromopropyl)phthalimide) is dissolved in CH_3CN (2.00 mL), and K_2CO_3 (4.73 g, 34.2 mmol) and ethyl acetoacetate (980 mg, 7.53 mmol) are added. The resulting suspension is heated to reflux for 2 h.

After cooling to room temperature, acetone (25 mL) is added and the mixture is filtered. The filtrate is concentrated *in vacuo* and the residue is recrystallized from EtOAc/ligroin to give the product as colorless discs; 1.37 g (63%), mp 65–66 °C, R_f = 0.45 (*n*-pentane/EtOAc, 1:1).

UV (CH_3CN): λ_{max} (nm) (lg ε) = 292 (3.248), 241 (4.039), 232 (4.163), 219 (4.624).

IR (KBr): $\tilde{v}$ (cm^{-1}) = 3459, 2969, 2934, 1772, 1738, 1713, 1613, 1463, 1438, 1402, 1368, 1368, 1341, 1283, 1243, 1192, 1144, 1124, 1091, 1043, 882, 848, 831, 795, 724, 632, 532.

^{1}H NMR (300 MHz, $CDCl_3$): δ (ppm) = 7.83–7.79 (m, 2 H, 3"-H, 6"-H), 7.70–7.67 (m, 2 H, 4"-H, 5"-H), 4.15 (q, *J* = 7.0 Hz, 2 H, OCH_2CH_3), 3.68 (t, *J* = 7.0 Hz, 2 H, 3'-H_2), 3.47 (t, *J* = 7.0 Hz, 1 H, 3'-H), 2.20 (s, 3 H, 1-H_3), 1.92–1.80 (m, 2 H, 2'-H_2), 1.70–1.55 (m, 2 H, 1'-H_2), 1.24 (t, *J* = 7.0 Hz, 3 H, OCH_2CH_3).

^{13}C NMR (300 MHz, $CDCl_3$): δ (ppm) = 202.6 (C-2), 169.4 (C-4), 168.3 (C-2", C-7"), 133.9 (C-4", C-5"), 132.0 (C-2"a, C-7"a), 123.2 (C-3", C-6"), 61.44 (OCH_2CH_3), 58.94 (C-3), 37.24 (C-3'), 28.94 (C-2'), 26.22 (C-1), 25.05 (C-1'), 14.04 (OCH_2CH_3).

MS (EI, 70 eV): *m/z* (%) = 317 (3) $[M]^+$, 275 (16) $[M - CH_3CO]^+$, 201 (41) $[M - CH_3CO - CO_2Et]^+$, 160 (100) $[Phth–CH_2]^+$, 77 (16) $[Ph]^+$, 43 (60) $[CH_3CO]^+$.

3.2.5.2   **   5-Methoxy-3-(2-phthalimidoethyl)-indole-2-carboxylic acid ethyl ester [6]

1. *p*-anisidine, HCl, NaNO₂
2. NaOAc, EtOH, 0 °C, 5 h

317.3   392.4

A solution of sodium nitrite (104 mg, 1.51 mmol) in H_2O (0.4 mL) is added dropwise to a stirred solution of *p*-anisidine (185 mg, 1.50 mmol) in H_2O (3.4 mL) and concentrated HCl (1.1 mL) at 0 °C and stirring is continued for 30 min at 0 °C (solution A). A solution of the ester **3.2.5.1** (512 mg, 1.62 mmol) in EtOH (2.6 mL) is added dropwise to a stirred suspension of NaOAc (1.38 g, 16.8 mmol) in EtOH (2.6 mL) at 0 °C and stirring is continued for 30 min, whereupon ice (5 g) is added (solution B). Solution A is then added to solution B by transfer cannulation at 0 °C, and the mixture is allowed to warm to room temperature and is stirred for a further 3 h.

The reaction mixture is then basified by slow addition of saturated Na_2CO_3 solution at 0 °C and extracted with CH_2Cl_2 (3 × 25 mL). The combined organic layers are washed with H_2O (25 mL), dried over $MgSO_4$, and the solvent is removed *in vacuo*. The red residue is dissolved in anhydrous EtOH (20 mL), treated with a saturated solution of HCl in anhydrous EtOH [2 mL; prepared from acetyl chloride (1.18 g, 15.0 mmol) and EtOH (692 mg, 15.0 mmol)], and the mixture is heated to reflux for 1 h. After cooling to room temperature, the solvent is removed under reduced pressure and the residue is partitioned between H_2O (10 mL) and CH_2Cl_2 (25 mL). The aqueous layer is basified by the addition of saturated Na_2CO_3 solution and extracted with CH_2Cl_2 (3 × 25 mL). The combined organic layers are washed with brine (10 mL), dried over $MgSO_4$, and concentrated *in vacuo*. Crystallization from EtOH provides the indole ester as a yellow solid; 398 mg (68%), mp 238–239 °C, $R_f = 0.58$ (*n*-pentane/EtOAc, 1:1).

UV (CH₃CN): λ_{max} (nm) (lg ε) = 326 (3.763), 299 (4.293), 240 (4.271), 218 (4.777).

IR (KBr): $\tilde{v}$ (cm⁻¹) = 3322, 2940, 1771, 1719, 1682, 1545, 1467, 1437, 1394, 1355, 1261, 1220, 1016, 808, 716, 653, 530.

¹H NMR (300 MHz, CDCl₃): δ (ppm) = 8.65 (s$_{br}$, 1 H, 1-H), 7.83–7.79 (m, 2 H, 3″-H, 6″-H), 7.67–7.64 (m, 2 H, 4″-H, 5″-H), 7.21 (d, J = 8.9, 1 H, 7-H), 7.06 (d, J = 2.3 Hz, 1 H, 4-H), 6.90 (dd, J = 8.9, 2.3 Hz, 1 H, 6-H), 4.40 (q, J = 7.2 Hz, 2 H, OC$\underline{H}_2$CH₃), 3.77 (s, 3 H, OCH₃), 3.98 (t, J = 7.9 Hz, 2 H, 2′-H₂), 3.42 (t, J = 7.9 Hz, 2 H, 1′-H₂), 1.43 (t, J = 7.2 Hz, 3 H, OCH₂C$\underline{H}_3$).

¹³C NMR (300 MHz, CDCl₃): δ (ppm) = 168.3 (C-2″, C-7″), 162.1 (CO₂Et), 154.5 (C-5), 133.8 (C-4″, C-5″), 132.2 (C-2″a, C-7″a), 131.0 (C-3a), 128.4 (C-7a), 124.5 (C-2), 123.0 (C-3″, C-6″), 119.3 (C-3), 117.4 (C-7), 112.8 (C-6), 110.1 (C-4), 60.97 (O$\underline{C}$H₂CH₃), 55.56 (OCH₃), 38.13 (C-2′), 24.03 (C-1′), 14.40 (OCH₂$\underline{C}$H₃).

MS (EI, 70 eV): m/z (%) = 392 (32) [M]⁺, 232 (40) [M – CH₂Phth]⁺, 186 (100) [M – OCH₃ – (CH₂)₂Phth]⁺, 77 (6) [Ph]⁺.

3.2.5.3   *   5-Methoxytryptamine [6]

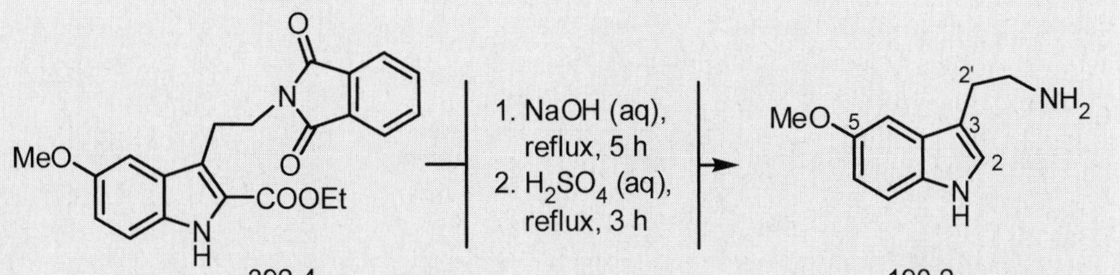

A mixture of the indole ester **3.2.5.2** (1.00 g, 2.55 mmol) and NaOH solution (2 M, 25 mL) is heated to reflux for 5 h to provide a homogeneous solution. H_2SO_4 (20% (v/v), 50 mL) is then added dropwise over 20 min and the reaction mixture is heated to reflux for a further 3 h.

The solution is cooled for 3 h in an ice bath and the precipitated phthalic acid is removed by filtration. The solution is made alkaline by the addition of NaOH (30% (v/v)) and extracted with CH_2Cl_2 (5 × 10 mL). The combined organic layers are washed with H_2O (10 mL) and brine (10 mL), and dried over anhydrous Na_2SO_4. The solvent is removed *in vacuo* to give the amine as light-yellow crystals; 317 mg (65%), mp 121–122 °C.

UV (CH$_3$CN): λ_{max} (nm) (lg ε) = 296.5 (3.677), 278.0 (3.789), 224.5 (4.369), 202.0 (4.418).

IR (KBr): $\tilde{\nu}$ (cm^{-1}) = 3335, 2595, 1586, 1492, 1305, 1218, 1048, 1010, 957, 922, 791, 638.

^{1}H NMR (300 MHz, CDCl$_3$): δ (ppm) = 8.27 (s$_{br}$, 1 H, NH), 7.24 (dd, J = 8.8, 0.6 Hz, 1 H, 7-H), 7.04 (d, J = 2.2 Hz, 1 H, 4-H), 6.99 (d, J = 2.1 Hz, 1 H, 2-H), 6.86 (dd, J = 8.8, 2.4 Hz, 1 H, 6-H), 3.86 (s, 3 H, OMe), 3.03 (t, J = 6.8 Hz, 2 H, 2'-H$_2$), 2.88 (t, J = 6.8 Hz, 2 H, 1'-H$_2$), 1.35 (s$_{br}$, 2 H, NH$_2$).

^{13}C NMR (300 MHz, CDCl$_3$): δ (ppm) = 153.8 (C-5), 131.5 (C-7a), 127.8 (C-3a), 122.9 (C-2), 113.3 (C-3), 112.1 (C-7), 111.8 (C-6), 100.6 (C-4), 55.89 (OCH$_3$), 42.21 (C-2'), 29.42 (C-1').

MS (EI, 70 eV): m/z (%) = 190 (36) $[M]^+$, 160 (100) $[M - CH_3NH_2]^+$, 145 (28) $[M - CH_3NH_2 - CH_3]^+$.

3.2.5.4 * *N*-[2-(5-Methoxy-1*H*-indol-3-yl)ethyl]acetamide (melatonin) [6]

MeO⟨indole⟩—NH₂ → CH₂Cl₂, NEt₃, Ac₂O, 0 °C → rt, 20 min → MeO⟨indole⟩—NHAc

190.2   232.3

NEt₃ (33.2 mg, 32.8 µmol) and acetic anhydride (40.7 mg, 39.9 µmol) are added dropwise to a solution of the amine **3.2.5.4** (50.0 mg, 26.3 µmol) in CH_2Cl_2 (2 mL) at 0 °C. The ice bath is removed and the solution is stirred at room temperature for 20 min and then poured into iced water (5 mL). Melatonin precipitates as a colorless solid and is collected by filtration. The compound is then dried *in vacuo*; 43.4 mg (71%), mp 117–118 °C.

UV (CH_3CN): λ_{max} (nm) (lg ε) = 297.0 (3.592), 275.5 (3.701), 223.5 (4.272), 200.5 (4.420).

IR (KBr): $\tilde{v}$ (cm^{-1}) = 3294, 2934, 1651, 1486, 1217, 1036.

¹H NMR (300 MHz, CDCl₃): δ (ppm) = 8.44 (s$_{br}$, 1 H, NH), 7.24 (d, J = 8.8 Hz, 1 H, 7-H), 7.01 (d, J = 2.5 Hz, 1 H, 4-H), 6.83 (dd, J = 8.8, 2.7 Hz, 1 H, 6-H), 5.71 (s$_{br}$, 1 H, NHAc), 3.83 (s, 3 H, OMe), 3.56 (t, J = 6.8 Hz, 1 H, 2'-H$_b$), 3.54 (t, J = 6.8 Hz, 1 H, 2'-H$_a$), 2.93 (t, J = 6.8 Hz, 2 H, 1'-H₂), 1.90 (s, 3 H, 1''-H₃).

¹³C NMR (300 MHz, CDCl₃): δ (ppm) = 170.2 (C-2''), 154.0 (C-5), 131.6 (C-7a), 127.7 (C-3a), 122.9 (C-2), 112.5 (C-3), 112.3 (C-7), 112.1 (C-6), 100.4 (C-4), 55.9 (OCH₃), 39.8 (C-2'), 25.3 (C-1'), 23.3 (C-1'').

MS (EI, 70 eV): m/z (%) = 232 (30) [M]$^+$, 173 (100) [M – Ac – CH₃]$^+$, 160 (93) [M – CH₃NHAc]$^+$, 145 (15) [M – CH₃NHAc – CH₃]$^+$.

[1] D. W. Thomson, A. G. J. Commeureuc, S. Berlin, J. A. Murphy, *Synth. Commun.* **2003**, *33*, 3631.
[2] a) For discussion of the Fischer indole synthesis and its detailed mechanism, see: Th. Eicher, S. Hauptmann, *The Chemistry of Heterocycles*, 2nd ed., p. 106, Wiley-VCH, Weinheim, **2003**; b) for further information on indole synthesis, see: Houben-Weyl, Methoden der organischen Chemie, 4th ed., Vol. E 6b₁/E 6b₂, Georg Thieme Verlag, Stuttgart, **1994**.
[3] H. M. Hügel, F. Nurlawis, *Heterocycles* **2003**, *60*, 2349.
[4] W. Marais, C. W. Holzapfel, *Synth. Commun.* **1998**, *28*, 3681.
[5] C. Prabhakar, N. V. Kumar, M. R. Reddy, M. R. Sarma, G. Om Reddy, *Organic Process Research & Development* **1999**, *3*, 155.
[6] L. He, J.-L. Li, J.-J. Zhang, P. Su, S.-L. Zheng, *Synth. Commun.* **2003**, *33*, 741; the microwave-assisted indole formation described in this paper proved to be irreproducible.
[7] a) L. F. Tietze, *Chem. Rev.* **1996**, *96*, 115; b) L. F. Tietze, G. Brasche, K. Gericke, *Domino Reactions in Organic Synthesis*, Wiley-VCH, Weinheim, **2006**.

3.2.6   3-(4-Methylbenzoylamino)-1-phenyl-4,5-dihydropyrazole

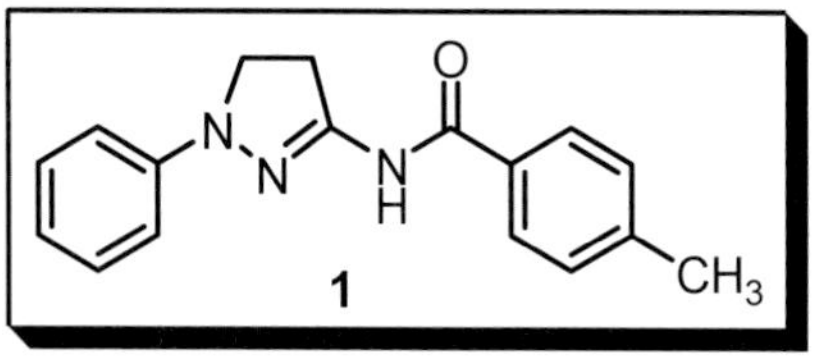

Topics:
- Cyanoethylation of a primary amine
- Formation of an amidoxime
- Formation of a 1,2,4-oxadiazole
- Thermal heterocycle isomerization:
  3-(β-aminoethyl)-1,2,4-oxadiazole
  → (3-arylamino)-4,5-dihydropyrazole

(a) General

The formation of five-membered heterocycles with two or more heteroatoms (of the general formula **3**) is often accomplished by ring closure of 1,5-dipolar acyclic species of the general formula **2**. This process has been interpreted in terms of an — in principle reversible — 6π-electrocyclization [1–3]:

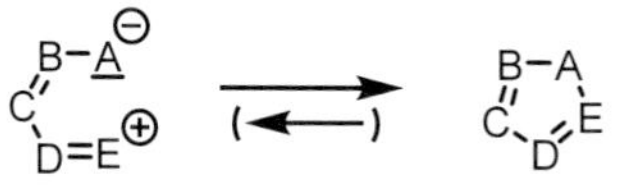

1,5-Dipolar species can be envisaged as reactive intermediates, which are generated (1) by combination of suitable acyclic precursors, or (2) by ring-opening reactions of heterocycles, as illustrated by the following examples.

(1) α-Diazocarbonyl compounds undergo addition to nitriles with elimination of N_2 catalyzed by Lewis acids or transition metal ions (Cu(II), Pd(II), and especially Rh(II)) to give 1,3-oxazoles **5**. The reaction is likely to proceed via intermediary formation of nitrile ylides **4** and their ring closure to **5** in 1,5-dipolar fashion [3]:

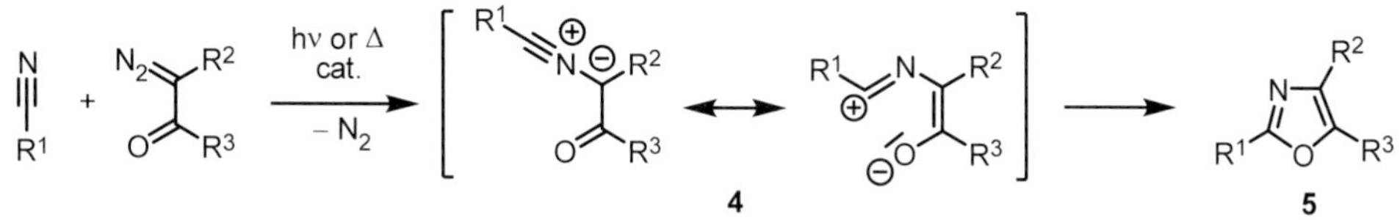

(2) 1,2-Oxazoles (e.g. **6**) are isomerized photochemically to give 1,3-oxazoles (e.g. **7**). In this rearrangement, an (isolable) 3-acylazirine (e.g. **8**) is involved as an intermediate, which is transformed to the 1,3-oxazole system by 1,5-electrocyclization of the nitrile ylide **9** [4, 5]:

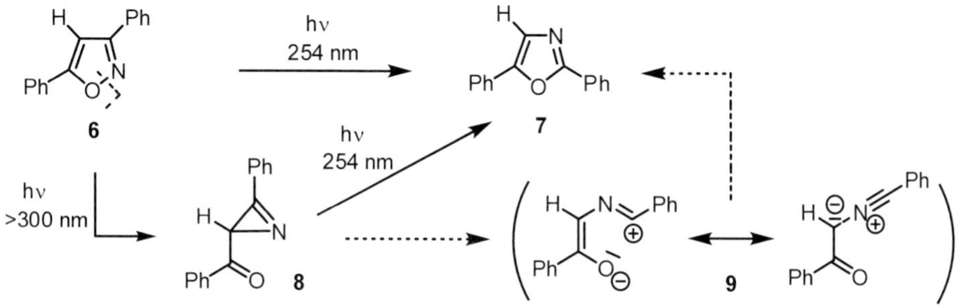

For further examples of 1,5-electrocyclic processes, see ref. [1].

(b) Synthesis of 1

(3-Acylamino)-4,5-dihydropyrazoles like **1** can be prepared by thermal isomerization of 3-(β-aminoethyl)-1,2,4-oxadiazoles [6]. In general, 1,2,4-oxadiazoles **11** are obtained by cyclocondensation of amidoximes **10** with carboxylic esters in the presence of a base:

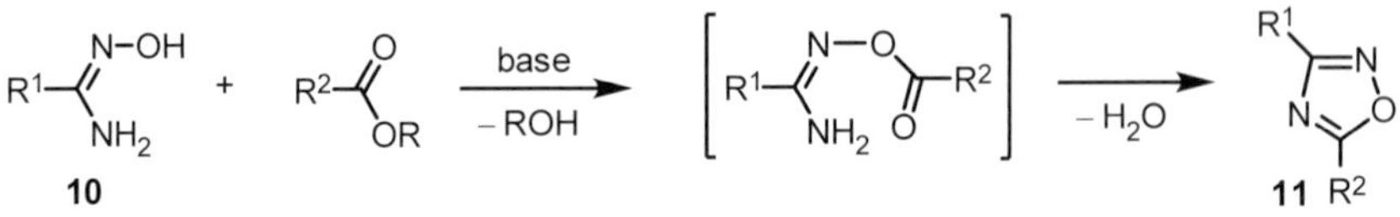

Therefore, in the first part of the synthesis of **1**, the amidoxime **13** is synthesized.

Aniline is mono-alkylated by conjugate addition to acrylonitrile in the presence of Cu(OAc)$_2$ (cyanoethylation [7]) yielding β-anilinopropionitrile (**12**), from which the required amidoxime **13** is obtained by addition of hydroxylamine to the C≡N triple bond:

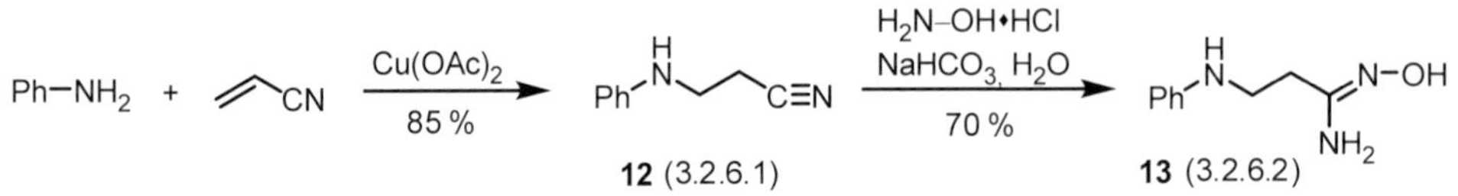

In the second part of the synthesis, the amidoxime **13** is cyclocondensed with ethyl *p*-toluate, which occurs readily on treatment with sodium ethoxide in refluxing EtOH and provides the 1,2,4-oxadiazole **14** in 83% yield. When the 3-(β-aminoethyl)-1,2,4-oxadiazole **14** is heated in *n*-butanol solution, it readily isomerizes to afford the (3-acylamino)-4,5-dihydropyrazole **1**:

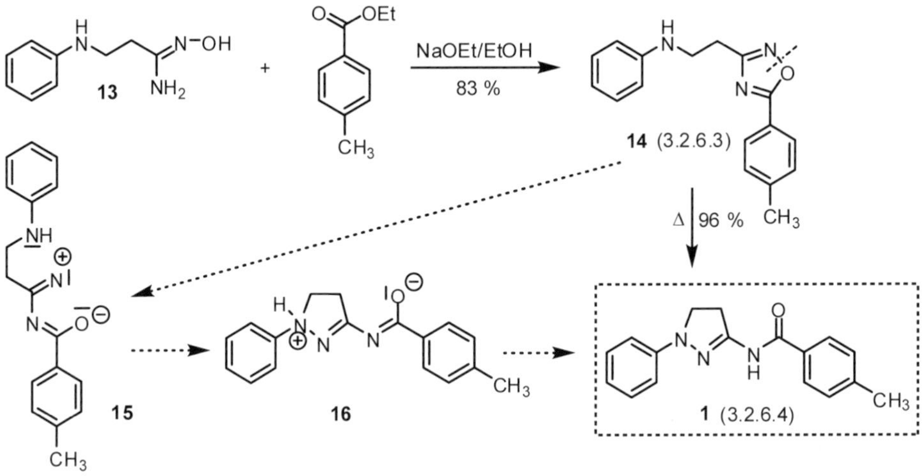

The thermal isomerization **14 → 1** can be rationalized mechanistically in analogy to the examples in section (a). Opening of the 1,2,4-oxadiazole system in **14** at the (relatively weak) O–N bond produces a 1,5-dipole **15**, which is intercepted by intramolecular nucleophilic attack of the aniline nitrogen (probably favored by proximity effects) to yield the 4,5-dihydropyrazole **16** and thereafter the product **1** by prototropy.

In this way, the target molecule is obtained by a four-step sequence in an overall yield of 47% (based on aniline).

(c) Experimental procedures for the synthesis of 1

3.2.6.1   * β-Anilinopropionitrile [8]

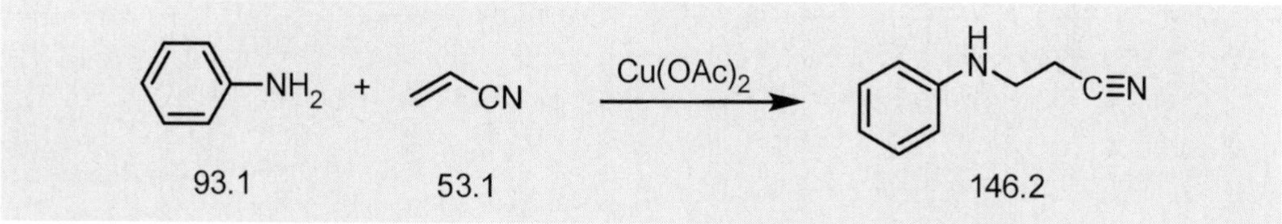

A stirred mixture of aniline (93.1 g, 1.00 mol), acrylonitrile (53.1 g, 1.00 mol) (*note*), and Cu(II) acetate (1.85 g) is heated to reflux (ca. 95 °C). The bath temperature is raised to 110 °C over a period of 30 min and is held there for 1 h.

The mixture is then cooled to ca. 80 °C and unreacted aniline and propionitrile are distilled off under reduced pressure (20 mbar); ca. 29 g of aniline is recovered. The dark residue is fractionally distilled to give a forerun, followed by a yellowish oil, which crystallizes in the receiver. Recrystallization from EtOH gives colorless needles; 84.5 g (85%, based on reacted aniline), bp$_{0.02}$ 115–120 °C, mp 50–51 °C.

> **IR** (KBr): $\tilde{v}$ (cm^{-1}) = 3360 (NH), 2260 (CN).
>
> **^{1}H NMR**: δ (ppm) = 7.25–6.3 (m, 5 H, arom. H), 3.95 (s$_{br}$, 1 H, NH (exchangeable)), 3.36 (q, J = 6 Hz, 2 H, CH$_2$, t after addition of D$_2$O), 2.48 (t, J = 6 Hz, 2 H, CH$_2$).

Note: Aniline (bp$_{20}$ 84–85 °C) and acrylonitrile (bp$_{760}$ 74–75 °C) are distilled before use.

3.2.6.2   * β-Anilinopropionamide oxime [6]

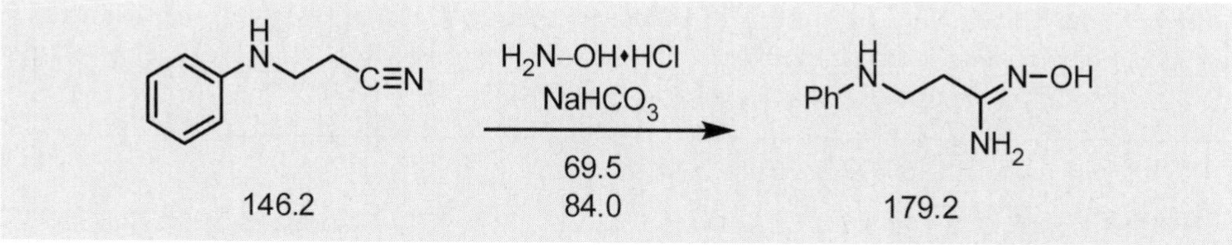

Sodium hydrogencarbonate (16.8 g, 0.20 mol) is added in portions to a solution of hydroxylamine hydrochloride (14.0 g, 0.20 mol) in water (50 mL). A solution of β-anilinopropionitrile **3.2.6.1** (14.6 g, 0.10 mol) in EtOH (100 mL) is then added and the mixture is heated under reflux for 6 h.

The solution is then concentrated *in vacuo* to one-third of its original volume to give a greenish oil, which is taken up in Et$_2$O and dried over Na$_2$SO$_4$. The ethereal solution is concentrated *in vacuo* to give an oily residue (14.6 g), which is pure by TLC and can be crystallized from *n*-hexane/EtOAc (1:1; 80 mL); 12.6 g (70%), mp 84–86 °C. Recrystallization from the same solvent mixture gives pale-reddish needles, mp 90–92 °C (*note*).

IR (KBr): $\tilde{v}$ (cm^{-1}) = 3500 (NH), 3370, 3390 (NH$_2$), 1660 (C=N).

^{1}H NMR: δ (ppm) = 7.3–6.3 (m, 5 H, arom. H), 4.7 (s$_{br}$, 1 H, NH), 3.24, 2.30 (t, J = 6 Hz, 2 H, CH$_3$).

Note: If the crystallization is unsuccessful, the residue is chromatographed on silica gel (200 g) eluting with Et$_2$O. One recrystallization of the product gives colorless needles, mp 91–92 °C.

3.2.6.3   *   3-(β-Anilinoethyl)-5-(*p*-tolyl)-1,2,4-oxadiazole [6]

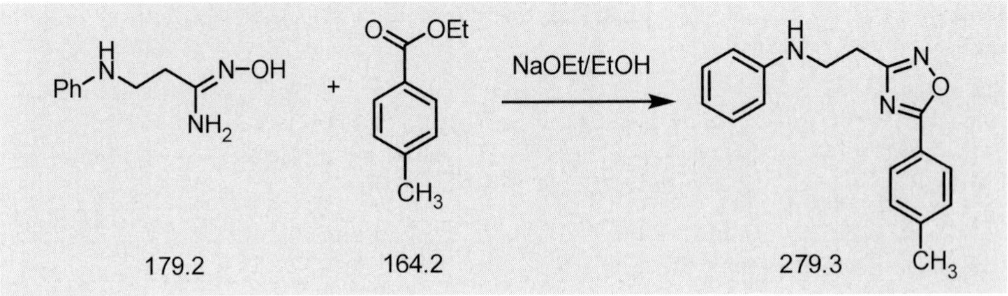

| 179.2 | 164.2 | 279.3 |

A mixture of amidoxime **3.2.6.2** (8.95 g, 50.0 mmol) and ethyl *p*-toluate (16.4 g, 0.10 mol) in anhydrous EtOH (50 mL) is added over 3 min to a stirred solution of sodium ethoxide (1.20 g, 52.0 mmol, of sodium in anhydrous EtOH (50 mL)). The solution becomes yellow and a crystalline precipitate begins to form after ca. 10 min. The mixture is refluxed for 8 h.

It is then cooled and filtered, and the solid is washed with EtOH. The solid is then suspended in H$_2$O (250 mL), stirred for 10 min, collected by filtration, and dried; 8.42 g, mp 96–99 °C. The EtOH mother liquor is concentrated *in vacuo*, the residue is taken up in H$_2$O (100 mL), and the resulting solution is extracted with CH$_2$Cl$_2$ (3 × 100 mL). The combined extracts are dried (MgSO$_4$), concentrated *in vacuo*, and the residue is crystallized from EtOH; 3.20 g, mp 94–98 °C. The total yield is 11.6 g (83%); recrystallization from EtOH gives colorless platelets, mp 101–102 °C.

IR (KBr): $\tilde{v}$ (cm^{-1}) = 3400 (NH), 1630 (C=N).

^{1}H NMR: δ (ppm) = 8.1–6.5 (m, 9 H, arom. H), 4.05 (s$_{br}$, 1 H, NH (exchangeable)), 3.57 (m, J = 6 Hz, 2 H, CH$_2$; t after D$_2$O addition), 3.05 (t, J = 6 Hz, 2 H, CH$_2$), 2.30 (s, 3 H, *p*-tolyl-CH$_3$).

3.2.6.4   *   3-(4-Methylbenzoylamino)-1-phenyl-4,5-dihydropyrazole [6]

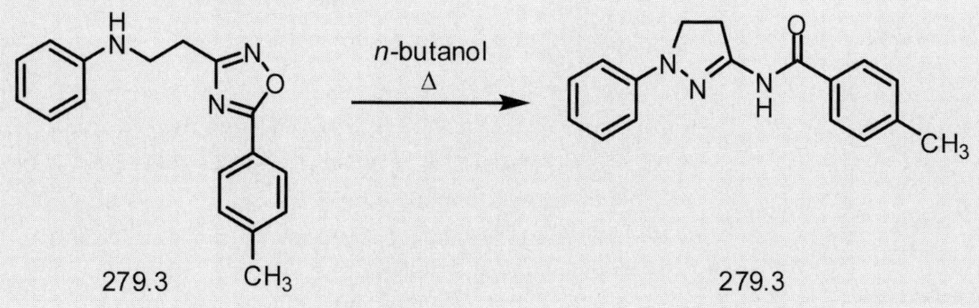

279.3   CH₃   279.3

The oxadiazole **3.2.6.3** (5.60 g, 20.0 mmol) is heated under reflux in anhydrous *n*-butanol (30 mL) for 8 h.

The solution is then cooled to room temperature and the acylaminodihydropyrazole crystallizes as yellow needles in pure form (TLC). The yield is 5.41 g (96%), mp 182–183 °C (*note*).

IR (KBr): $\tilde{v}$ (cm^{-1}) = 3310 (NH), 1665, 1620 (C=O/C=N).

^{1}H NMR: δ (ppm) = 8.6 (s$_{br}$, 1 H, NH), 7.6–6.7 (m, 9 H, arom. H), 3.62 (m$_c$, 4 H, CH$_2$–CH$_2$), 2.35 (s, 3 H, *p*-tolyl-CH$_3$).

Note: Recrystallization from *n*-butanol does not change the melting point.

[1]   R. Huisgen, *Angew. Chem.* **1980**, *92*, 979; *Angew. Chem. Int. Ed. Engl.* **1980**, *19*, 947.
[2]   E. C. Taylor, I. J. Turchi, *Chem. Rev.* **1979**, *79*, 181.
[3]   C. J. Moody, K. J. Doyle, *Progr. Heterocycl. Chem.* **1997**, *9*, 1.
[4]   B. Singh, E. F. Ullman, *J. Am. Chem. Soc.* **1967**, *89*, 6911.
[5]   Th. Eicher, S. Hauptmann, *The Chemistry of Heterocycles*, 2nd ed., p. 144, Wiley-VCH, Weinheim, **2003**.
[6]   D. Korbonits, E. M. Bako, K. Horvath, *J. Chem. Research (S)* **1979**, 64.
[7]   *Houben-Weyl*, Vol. XI/1, p. 272.
[8]   S. A. Heininger, *J. Org. Chem.* **1957**, *22*, 1213.

3.2.7   Camalexin

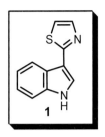

1

Topics:
- Synthesis of a phytoalexin
- Halogen–metal exchange
- Addition of an organolithium compound to an aryl aldehyde
- Oxidation CH–OH → C=O, reduction Ar–NO$_2$ → Ar–NH$_2$
- N-Formylation of a primary amine
- Fürstner indole synthesis: low-valent titanium-induced reductive cyclization of (2-acyl)anilides

(a) General

Camalexin (**1**, 3-(2-thiazolyl)indole) belongs to the class of phytoalexins, which play an important role in the antimicrobial defense mechanism of plants. Camalexin and its 6-methoxy derivative are produced in the leaves of the false flax (*Camelina sativa*) in response to infection by *Alternaria brassicae* and thus display antifungal activity [1].

Among the numerous concepts for the construction of indoles [2], the Fürstner indole synthesis has been successfully applied to the synthesis of **1** and other indole-based natural products [1, 3]:

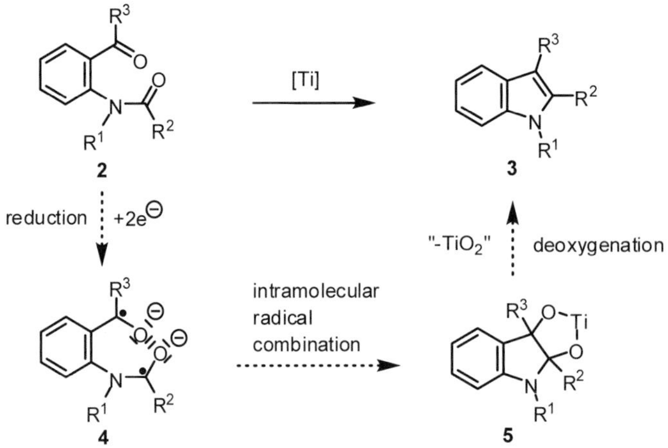

In the Fürstner indole synthesis, (2-acyl)anilides **2** are cyclized by intramolecular reductive coupling of the two carbonyl groups mediated by "low-valent titanium" (abbreviated as [Ti]), thus creating the C-2/C-3 bond of indoles **3**. This process is similar to the classical McMurry reaction [4], the reductive dimerization of aldehydes and ketones forming olefinic C=C double bonds. Its mechanism presumably involves one-electron transfers to the carbonyl groups (**2** → **4**) and intramolecular radical combination of **4** to give a titanium dioxygen species **5**, which finally is deoxygenated to give the indole system **3** [5].

The Fürstner method is highly flexible with respect to the substitution pattern in the heterocyclic indole part and has proved to be compatible with a great number of Lewis acidic and reducible functional groups in the substrate **2**. The low-valent titanium species [Ti] can be generated from TiCl$_3$ and reducing agents such as Zn, Mg, or potassium-graphite laminate C$_8$K [3].

In the synthesis of **1** presented in section (b), the titanium reagent is prepared directly ("instant method") in the presence of the substrate **6** required for the reductive cyclization.

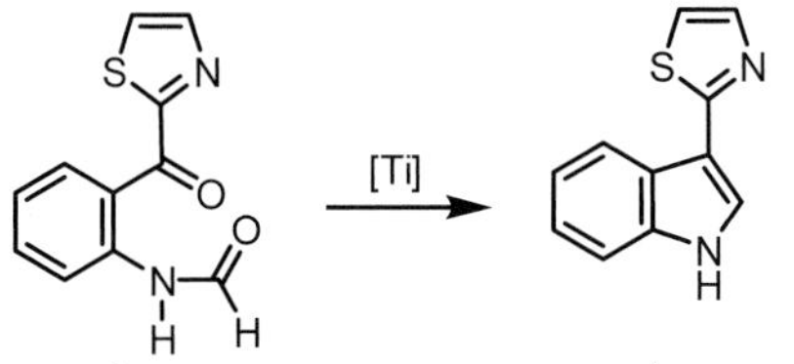

6 **1**

(b) Synthesis of 1

Commercially available 2-bromothiazole (**7**) is subjected to halogen–metal exchange by reaction with *n*-butyllithium. The resulting 2-lithiothiazole (**8**) is subsequently added to the carbonyl group of 2-nitrobenzaldehyde at –78 °C in Et_2O to give (after hydrolytic work-up) the secondary carbinol **9**, which is oxidized to the ketone **10** by pyridinium dichromate:

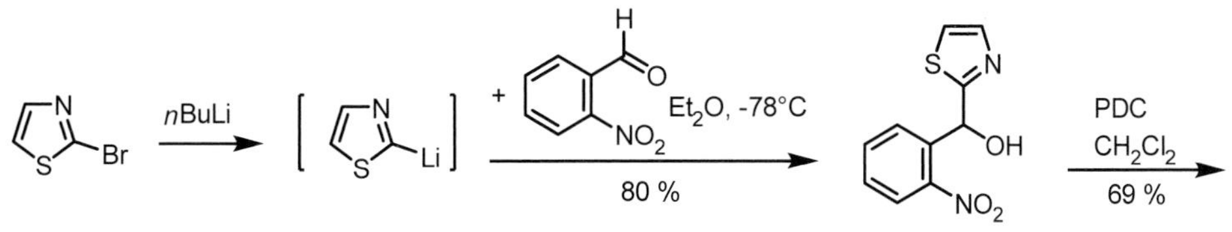

7 **8** **9** (3.2.7.1)

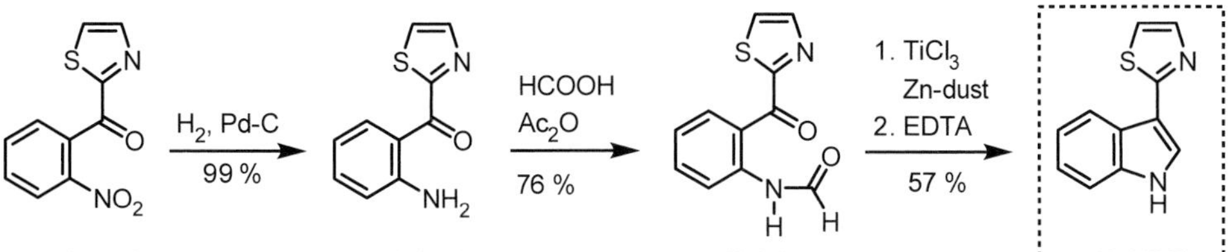

10 (3.2.7.2) **11** (3.2.7.3) **6** (3.2.7.4) **1** (3.2.7.5)

Chemoselective catalytic reduction of the nitro group in **10** by H_2 over Pd on charcoal followed by formylation of the NH_2 group in the amino ketone **11** yields the (2-acyl)-*N*-formylaniline **6**, which is reductively cyclized by treatment with $TiCl_3$/zinc dust in THF. Work-up with EDTA in order to decomplex the Lewis-acidic titanium salts from the basic thiazole nitrogen affords camalexin (**1**).

It should be noted that, alternatively, the (2-nitrophenyl)-(2-thiazolyl)-ketone (**10**) can be obtained from C–Si acylation of (2-trimethylsilyl)thiazole (**12**) by (2-nitro)benzoyl chloride [1]:

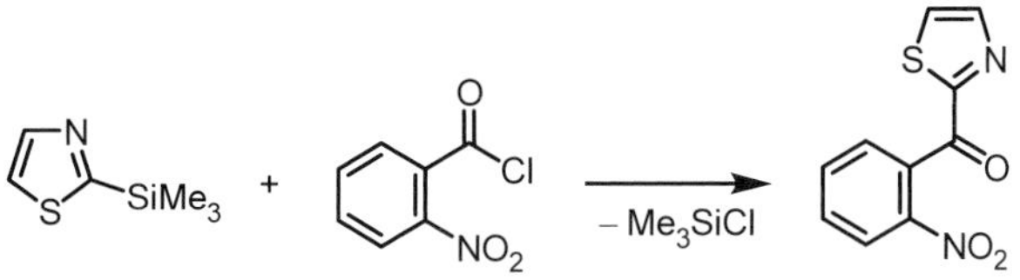

12 **10**

Since **12** ("Dondoni's thiazole") is a noxious compound and is difficult to separate from the product **10**, the preparation of **10** starting with 2-nitrobenzaldehyde (as described above) is preferred.

In this way, the target molecule **1** is prepared in a five-step sequence with an overall yield of 24% (based on **7**).

(c) Experimental procedures for the synthesis of 1

3.2.7.1 ** (2-Nitrophenyl)-(2-thiazolyl)methanol [1]

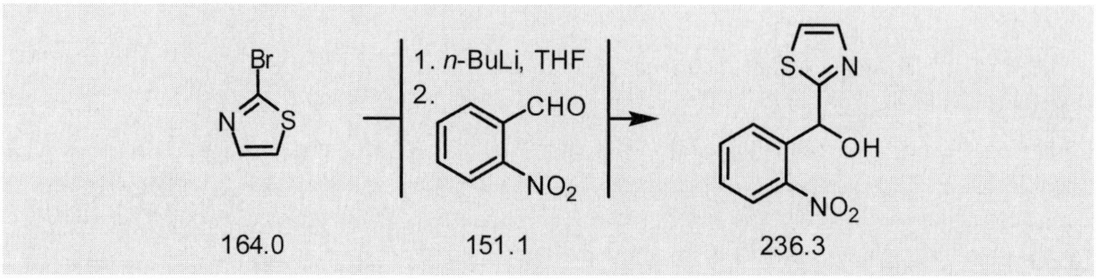

164.0	151.1	236.3

A solution of 2-bromothiazole (5.0 g, 30.5 mmol) in anhydrous Et$_2$O/THF (2:1; 20 mL) is added to a stirred solution of nBuLi (1.6 M in n-hexane, 20 mL, 32.0 mmol) in Et$_2$O (80 mL) at –78 °C under argon over a period of 45 min. After stirring for a further 15 min at –78 °C, a solution of 2-nitrobenzaldehyde (4.50 g, 30.0 mmol) in THF (20 mL) is added dropwise to the reaction mixture over a period of 45 min. Stirring is continued for a further 30 min at –78 °C.

The cold mixture is then carefully added to a 10% NH$_4$Cl solution (100 mL), the aqueous layer is extracted with EtOAc (3 × 30 mL), and the combined organic phases are washed with brine (20 mL), dried over MgSO$_4$, and concentrated *in vacuo*. The residue is recrystallized from toluene to give the product as pale-yellow crystals; 4.62 g (80%), mp 130–131 °C; R_f = 0.17 (n-pentane/EtOAc, 4:1).

> **^{1}H NMR** (200 MHz, CDCl$_3$): δ (ppm) = 8.03 (d, J = 8 Hz, 1 H, 1-H), 7.58 (m, 3 H, 2-H, 3-H, 4-H), 7.48 (m, 1 H, 6-H), 7.32 (d, J = 8 Hz 1 H, 7-H), 6.62 (s, 1 H, 5-H), 4.50 (s$_{br}$, 1 H, 8-H).

3.2.7.2 * (2-Nitrophenyl)-(2-thiazolyl) ketone [1]

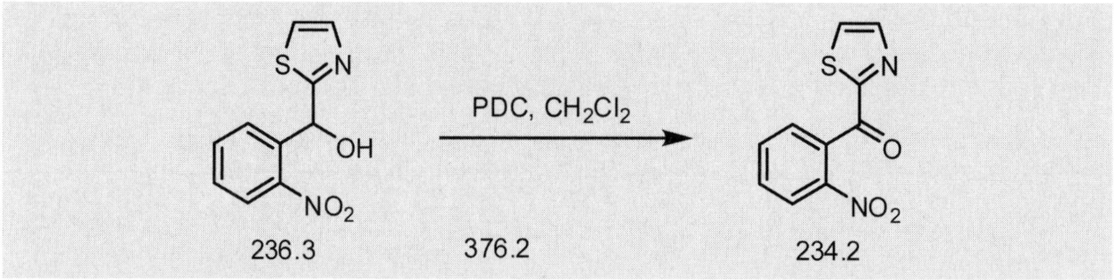

236.3	376.2	234.2

PDC (12.1 g, 32.1 mmol) is added to a solution of the carbinol **3.2.7.1** (3.8 g, 16.1 mmol) in CH$_2$Cl$_2$ (150 mL) under an argon atmosphere. The suspension is stirred for 5 h at room temperature.

It is then filtered through a short pad of Celite, the pad is washed with CH$_2$Cl$_2$ (250 mL), and the combined filtrate and washings are dried over MgSO$_4$. After evaporation of the solvent *in vacuo*, the residue is recrystallized from MeOH to give (2-nitrophenyl)-(2-thiazolyl) ketone as pale-yellow crystals; 2.60 g (69%), mp 119–120 °C; R_f = 0.39 (n-pentane/EtOAc, 4:1).

(c) Experimental procedures for the synthesis of 1

3.2.8.1   **   5,5-Dimethylcyclohexane-1,3-dione (dimedone) [8]

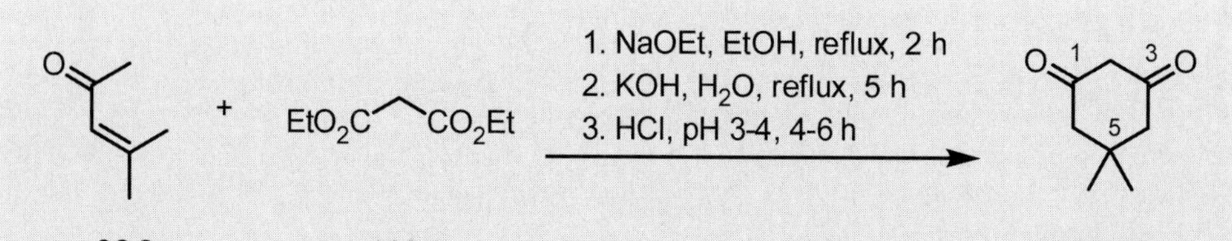

| 98.2 | 160.2 | 140.2 |

Anhydrous EtOH (40 mL) is added dropwise to small pieces of sodium (2.30 g, 100 mmol) in a three-necked round-bottomed flask. The mixture is heated to reflux for 1 h until a homogeneous solution is obtained. Diethyl malonate (17.0 g, 106 mmol) is added dropwise and, after 10 min, mesityl oxide (10.0 g, 102 mmol) is slowly added to the reaction mixture, which is then heated to reflux for 2 h with stirring. KOH (12.5 g, 223 mmol) and H_2O (60 mL) are added, and the solution is stirred at reflux for an additional 2–3 h and then left at room temperature for 24 h.

The reaction mixture is then heated to reflux once more for an additional 2–3 h and the hot solution is adjusted to pH 3–4 by the addition of HCl (4 M). About 50 mL of the solvent is removed under reduced pressure and the remaining solution is again adjusted to pH 3–4 by the addition of HCl (4 M), heated for a short time, and cooled to room temperature. Dimedone crystallizes and is collected by filtration. Subsequently, it is washed with cold H_2O and Et_2O. Recrystallization from hot acetone (10 mL per 1 g of crude product) yields dimedone as a white crystalline solid; 9.10 g (65%), mp 147–148 °C.

> **IR** (KBr): $\tilde{v}$ (cm^{-1}) = 2956, 2532, 1907, 1621, 1519, 1472, 1412, 1348, 1305, 1248, 1227, 1146, 984, 874, 829, 613, 592, 578, 562, 467, 445.
>
> **MS** (EI, 70 eV): m/z (%) = 140 (40) $[M]^+$, 83 (100) $[M-C_3H_5O]^+$, 56 (60) $[C_3H_4O]^+$, 55 (38), 41 (23).

3.2.8.2   **   1-(4-Methoxyphenyl)-4,5,6,7-tetrahydro-6,6-dimethylindazol-4-one [7]

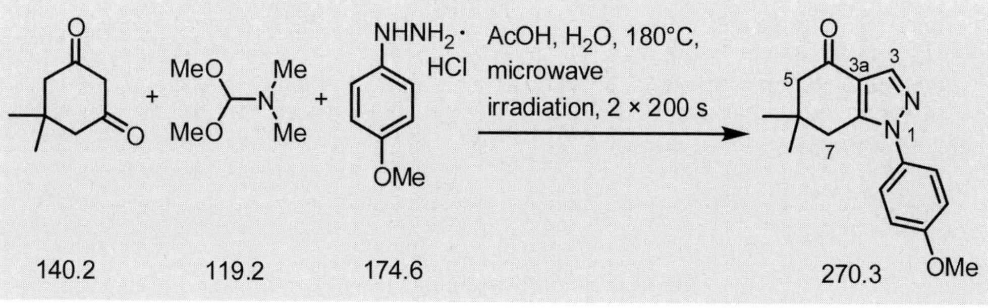

| 140.2 | 119.2 | 174.6 | 270.3 |

A Smith process vial (5 mL) for microwave irradiation is charged with dimedone (175 mg, 1.25 mmol), *N,N*-dimethylformamide dimethyl acetal (199 μL, 1.50 mmol), (4-methoxyphenyl)-hydrazinium chloride (198 mg, 1.25 mmol), demineralized H_2O (3 mL), and acetic acid (186 μL, 3.25 mmol). The vial is sealed with an appropriate cap and submitted to microwave irradiation (2 × 200 s, 180 °C) using a Smith Creator.

The reaction mixture is then diluted with H_2O (40 mL) and the aqueous layer is extracted with CH_2Cl_2 (5 × 10 mL). The combined organic layers are washed with brine (2 × 50 mL), dried over Na_2SO_4, and the solvent is evaporated under reduced pressure. After column chromatography (petroleum ether/ EtOAc, 8:2 to 7:3), the imidazolone is obtained as a yellow solid; 200 mg (59%), mp 117–118 °C, R_f = 0.13 (*n*-pentane/EtOAc, 7:3).

UV (MeOH): λ_{max} (nm) (lg ε) = 355.0 (3.04), 254.5 (4.20), 202.0 (4.40).

IR (KBr): $\tilde{v}$ (cm^{-1}) = 3103, 2961, 2836, 1677, 1594, 1544, 1518, 1489, 1445, 1399, 1368, 1303, 1260, 1222, 1185, 1168, 1113, 1059, 1034, 970, 900, 887, 837, 811, 767, 682, 671, 648, 624, 583, 532.

^{1}H NMR (300 MHz, $CDCl_3$): δ (ppm) = 8.03 (s, 1 H, 3-H), 7.38 (dd, *J* = 9.0, 2.2 Hz, 2 H, 2'-H, 6'-H), 6.99 (dd, *J* = 9.0, 2.2 Hz, 2 H, 3'-H, 5'-H), 3.86 (s, 3 H, OCH_3), 2.75 (s, 2 H, 7 H), 2.40 (s, 2 H, 5-H), 1.10 (s, 6 H, 2 × CH_3).

^{13}C NMR (50 MHz, $CDCl_3$): δ (ppm) = 192.7 (C-4), 159.4 (C-7a), 148.2 (C-4'), 137.9 (C-3), 131.7 (C-1'), 125.2 (C-2', C-6'), 119.2 (C-3a), 114.5 (C-3', C-5'), 55.58 (OCH_3), 52.04 (C-5), 36.72 (C-7), 35.82 (C-6), 28.35 (2 × CH_3).

MS (ESI): *m/z* (%) = 563 (18) $[(2M + Na)^+]$, 371 (16) $[(M + C_6H_6 + Na)^+]$, 325 (100) $[(2M - 2C_7H_9O + H)^+]$, 324 (20) $[(2M - 2C_7H_7ON + Na)^+]$, 271 (11) $[(M + H)^+]$.

---

[1] a) A. Loupy, *Microwaves in Organic Synthesis*, Wiley-VCH, Weinheim, **2002**; b) P. Lidström, J. P. Tierney, *Microwave-Assisted Organic Synthesis*, Blackwell, Oxford, **2004**.

[2] a) D. Adam, *Nature* **2003**, *421*, 571; b) C. O. Kappe, *Angew. Chem.* **2004**, *116*, 6408; *Angew. Chem. Int. Ed.* **2004**, *43*, 6250.

[3] a) D. R. Baghurst, D. M. P. Mingos, *J. Chem. Soc., Chem. Commun.* **1992**, 674; b) R. Saillard, M. Poux, J. Berlan, M. Audhuy-Peaudecerf, *Tetrahedron* **1995**, *51*, 4033.

[4] a) D. Bogdal, M. Lukasiewicz, J. Pielichowski, A. Miciak, Sz. Bednarz, *Tetrahedron* **2003**, *59*, 649; b) M. Lukasiewicz, D. Bogdal, J. Pielichowski, *Adv. Synth. Catal.* **2003**, *345*, 1269; c) X. Zhang, D. O. Hayward, D. M. P. Mingos, *Catal. Lett.* **2003**, *88*, 33; d) X. Zhang, C. S.-M. Lee, D. M. P. Mingos, D. O. Hayward, *Catal. Lett.* **2003**, *88*, 129.

[5] a) M. Larhed, A. Hallberg, *J. Org. Chem.* **1996**, *61*, 9582; b) K. Olofsson, S.-Y. Kim, M. Larhed, D. P. Curran, A. Hallberg, *J. Org. Chem.* **1999**, *64*, 453.

[6] a) L. Perreux, A. Loupy, *Tetrahedron* **2001**, *57*, 9199; b) N. Kuhnert, *Angew. Chem.* **2002**, *114*, 1943; *Angew. Chem. Int. Ed.* **2002**, *41*, 1863; c) C. R. Strauss, *Angew. Chem.* **2002**, *114*, 3741; *Angew. Chem. Int. Ed.* **2002**, *41*, 3589.

[7] V. Molteni, M. M. Hamilton, L. Mao, C. M. Crane, A. P. Termin, D. M. Wilson, *Synthesis* **2002**, 1669.

[8] Adapted from: Th. Eicher, L. F. Tietze, *Organisch-Chemisches Grundpraktikum*, 2nd ed., p. 284, Georg Thieme Verlag, Stuttgart, **1995**.

3.3 Six-membered heterocycles

3.3.1 Azine and diazine syntheses with acetoacetate

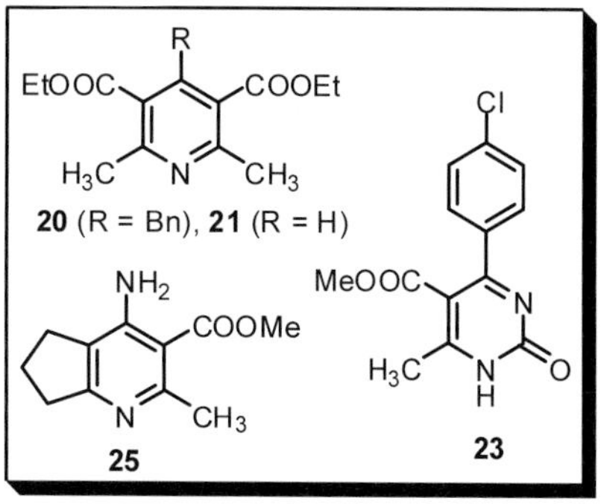

EtOOC
COOEt
R
H3C
N
CH3
20 (R = Bn), 21 (R = H)
NH2
COOMe
N
CH3
25
Cl
MeOOC
N
H3C
N
H
O
23

Topics:
- Hantzsch synthesis of pyridines
- Tin-mediated synthesis of 4-aminopyridines
- Biginelli synthesis of 3,4-dihydropyrimidin-2-ones
- Dehydrogenation of a dihydropyrimidinone to a pyrimidinone

(a) General

Several highly efficient synthetic methods are known for the construction of six-membered heterocycles, in which β-dicarbonyl compounds, preferentially β-keto esters, are incorporated as C_2-building blocks by – in general – one-pot multi-component processes. Three of these methods are illustrated in (1)–(3).

(1): In the Hantzsch synthesis of pyridines, two molecules of a β-dicarbonyl compound (β-keto-ester or β-diketone), an aldehyde, and ammonia are combined in a four-component cyclocondensation to give 1,4-dihydropyridines **1**, which can be oxidized to pyridines **2** [1]:

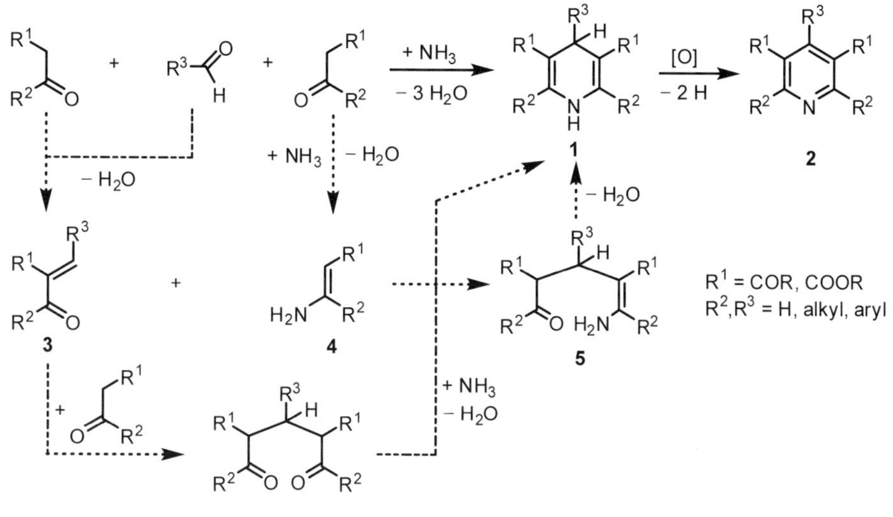

+ NH3
- 3 H2O
[O]
- 2 H
1
2
- H2O
+ NH3
- H2O
3
4
5
+ NH3
- H2O
R1 = COR, COOR
R2,R3 = H, alkyl, aryl

The process can be performed in two different ways. First, one molecule of the β-dicarbonyl compound combines with the aldehyde to form the corresponding Knoevenagel product **3**, which then reacts with the β-enaminone (**4**) preformed by addition of NH_3 (or a primary amine) to another molecule of the β-dicarbonyl compound. This Michael addition yields 5-aminopent-4-enone **5**, cyclization of which provides the 1,4-dihydropyridine **1**.

The procedure allows the preparation of unsymmetrical 1,4-dihydropyridines. Thus, in this modification of the Hantzsch synthesis using preformed β-enaminones, different β-dicarbonyl compounds for the formation of the β-enaminones and for the Knoevenagel condensation can be used [2]. In the second approach, the two molecules of the β-dicarbonyl compound react with the aldehyde in a domino Knoevenagel–Michael addition process to yield the 1,5-dicarbonyl system **6**, which undergoes cyclocondensation with NH_3 leading to **1**. Only symmetrical compounds can be prepared by this route.

It should be noted that 1,4-dihydropyridines such as nifedipine (**7**) and analogues are potent Ca antagonists and coronary dilators and are therefore medicinally important as antihypertensives [3].

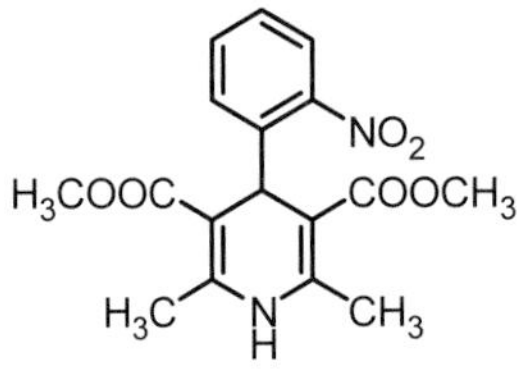

(2): In the Biginelli synthesis of 3,4-dihydropyrimidinones such as **8** [4], a β-keto ester, an aldehyde, and urea undergo an acid- or metal ion-catalyzed three-component cyclocondensation:

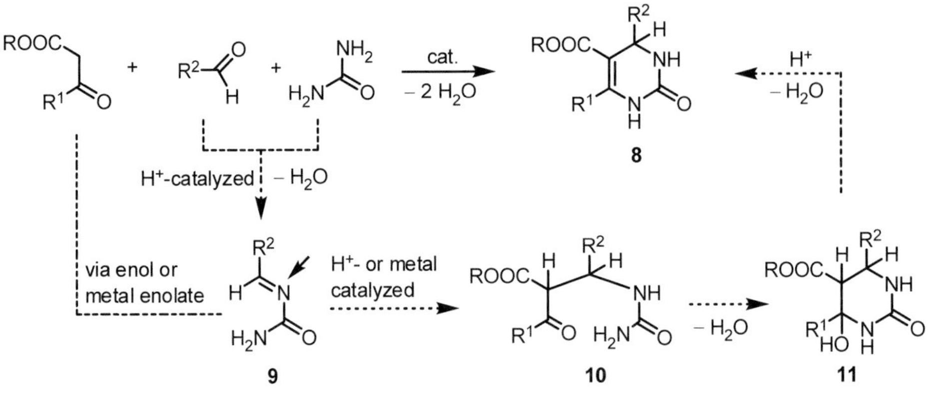

Despite the formal resemblance to the synthesis of 1,4-dihydropyridines described in (1), the mechanism of the Biginelli reaction is quite different [5]. The rate-determining step is the acid-catalyzed formation of an acylimine intermediate **9** from the aldehyde and urea. By *N*-protonation or *N*-coordination with metal ions (Fe(III), Ni(II), etc.), the acylimine **9** can be activated as an iminium ion and intercepted by the β-keto ester (as an enol or metal enolate) to produce an open-chain ureide **10**, which subsequently cyclizes (via the cyclic ureide **11** and its dehydration) to give the 3,4-dihydropyrimidinone **8**.

A solution of methyl acetoacetate (2.32 g, 20.0 mmol), 4-chlorobenzaldehyde (2.81 g, 20.0 mmol), urea (1.80 g, 30.0 mmol), and $FeCl_3 \cdot 6\,H_2O$ (1.35 g, 5.00 mmol) in EtOH (40 mL) containing four drops of concentrated HCl is heated under reflux for 5 h.

After cooling, the reaction mixture is poured onto crushed ice (200 g) and the resulting mixture is stirred for 15 min. The precipitate formed is collected by filtration and washed first with cold H_2O (2 × 50 mL) and then with a mixture of $EtOH/H_2O$ (1:1, 3 × 40 mL). The crude product is dried and recrystallized from EtOH; colorless crystals, 4.85 g (83%), mp 200–201 °C.

IR (KBr): $\tilde{\nu}$ (cm^{-1}) = 3364, 3218, 3093, 2947, 1712, 1687, 1633, 1488.

^{1}H NMR (500 MHz, [D_6]DMSO): δ (ppm) = 9.26 (s, 1 H), 7.78 (s, 1 H), 7.39 (d, J = 8.4 Hz, 2 H), 7.25 (d, J = 8.4 Hz, 2 H), 5.14 (d, J = 3.5 Hz, 1 H), 3.53 (s, 3 H), 2.25 (s, 3 H).

^{13}C NMR (125 MHz, [D_6]DMSO): δ (ppm) = 165.7, 151.9, 149.0, 143.6, 131.8, 128.4, 128.1, 98.6, 53.2, 50.8, 17.8.

3.3.1.5   *   4-(*p*-Chlorophenyl)-5-methoxycarbonyl-6-methylpyrimidin-2(1*H*)-one [11]

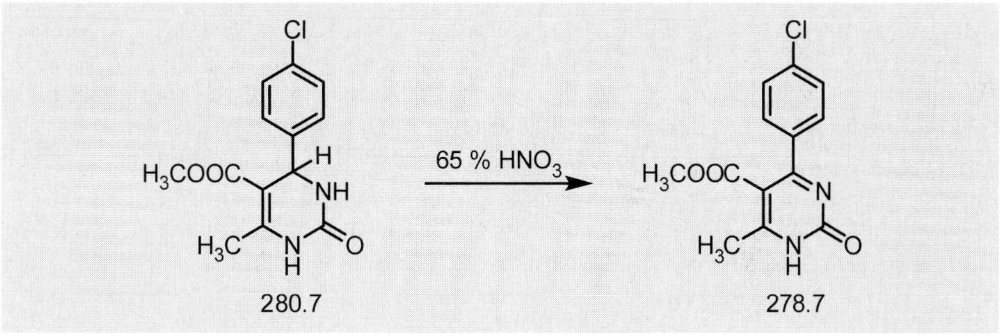

280.7   278.7

Nitric acid (65%) is cooled to 0 °C and the dihydropyrimidinone **3.3.1.4** (2.81 g, 10.0 mmol) is added in portions over 5 min. The mixture is stirred for an additional 2 min at 0 °C resulting in a yellow solution, which is allowed to warm to room temperature over 15 min.

The solution is immediately poured onto crushed ice (50 g) and brought to pH 8 with solid K_2CO_3 (CO_2 evolution!). The resulting mixture is extracted with $CHCl_3$ (4 × 100 mL) and the combined organic layers are washed with H_2O (100 mL), dried ($MgSO_4$), and concentrated. The crude product is recrystallized from EtOH; 2.45 g (88%), yellow-green solid, mp 173–174 °C.

---

^{1}H NMR (500 MHz, [D_6]DMSO): δ (ppm) = 8.31 (s, 1 H, NH), 7.54 (d, J = 8.4 Hz, 2 Ar-H), 7.47 (d, J = 8.4 Hz, 2 H, Ar-H), 3.52 (s, 3 H, OCH_3), 2.40 (s, 3 H, CH_3).

^{13}C NMR (125 MHz, [D_6]DMSO): δ (ppm) = 169.38, 166.29, 162.22, 155.59, 136.74, 135.04, 129.38, 128.36, 108.51, 51.99, 18.58

3.3.1.6   *   2-Aminocyclopent-1-ene carbonitrile [12]

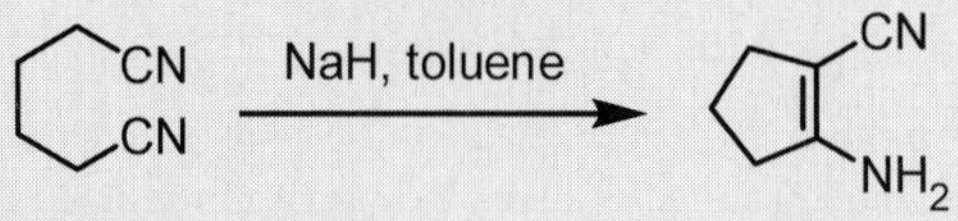

108.1   24.0   108.1

A solution of adipodinitrile (10.8 g, 0.10 mol) in toluene (100 mL) is added dropwise to a slurry of sodium hydride (60% suspension in oil, 4.20 g, 0.105 mol) in anhydrous toluene (30 mL). The mixture is heated to reflux for 15 h.

Thereafter, EtOH (5 mL), H_2O (35 mL), and acetic acid (5 mL) are carefully added. The organic layer is separated and the aqueous phase is extracted with EtOAc (3 × 50 mL). The combined organic layers are washed with H_2O (50 mL), dried ($MgSO_4$), and concentrated *in vacuo*. The crude product is recrystallized by dissolving it in hot toluene (25 mL) and then adding petroleum ether (150 mL); colorless solid, 6.60 g (61%), mp 130–131 °C.

IR (solid): $\tilde{v}$ (cm^{-1}) = 3460, 3425, 3257, 2188, 1647, 1615.

^{1}H NMR (500 MHz, $CDCl_3$): δ (ppm) = 4.53 (s_{br}, 2 H, NH_2), 2.53 (t, J = 7.0 Hz, 2 H), 2.46 (t, J = 7.0 Hz, 2 H), 1.93 (q, J = 7.0 Hz, 2 H).

^{13}C NMR (125 MHz, $CDCl_3$): δ (ppm) = 162.4, 119.0, 74.43, 34.29, 31.26, 22.00.

MS (EI, 70 eV): m/z (%) = 107 (100) $[M – H]^+$, 93 (1), 80 (20), 67 (3), 53 (10).

3.3.1.7   *   4-Amino-3-methoxycarbonyl-2-methyl-5,6-trimethylene pyridine [6]

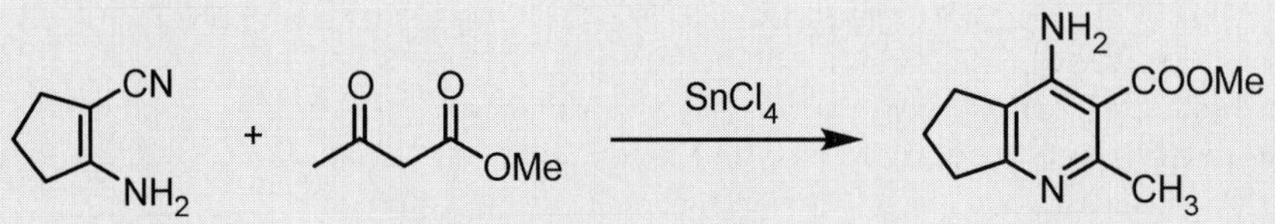

108.1   116.1   260.5   206.2

The aminocyanoalkene 3.3.1.6 (2.16 g, 20.0 mmol) and tin(IV) chloride (10.4 g, 40.0 mmol) are added to a stirred solution of methyl acetoacetate (2.32 g, 20.0 mmol) in anhydrous toluene (50 mL). The reaction mixture is stirred under a nitrogen atmosphere at room temperature for 30 min and then heated to reflux for 3 h.

The mixture is then filtered and the solvent is removed under reduced pressure. The residue is stirred for 30 min with saturated Na_2CO_3 solution (160 mL). The resulting suspension is extracted with EtOAc (3 × 100 mL) and the combined extracts are washed with brine (2 × 100 mL), dried ($MgSO_4$), and concentrated to give pale-yellow crystals; 2.96 g (72%), mp 110–111 °C.

IR (solid): $\tilde{\nu}$ (cm^{-1}) = 3422, 3340–3260, 3220–3160, 2950, 2925, 2847, 1687, 1577, 1432, 1245.

^{1}H NMR (500 MHz, CDCl$_3$): δ (ppm) = 5.73 (s$_{br}$, 2 H, NH$_2$), 3.90 (s, 3 H, OCH$_3$), 2.94 (t, J = 7.7 Hz, 2 H, CH$_2$), 2.69 (t, J = 7.7 Hz, 2 H, CH$_2$), 2.65 (s, 3 H, CH$_3$), 2.14 (q, J = 7.7 Hz, 2 H, CH$_2$).

^{13}C NMR (125 MHz, CDCl$_3$): δ (ppm) = 169.7, 165.7, 160.2, 151.6, 118.4, 106.4, 51.52, 35.05, 27.27, 26.67, 21.98.

[1]   Th. Eicher, S. Hauptmann, *The Chemistry of Heterocycles*, 2nd ed., p. 299, Wiley-VCH, Weinheim, **2003**.

[2]   F. Bossert, H. Meyer, E. Wehinger, *Angew. Chem.* **1981**, *93*, 755; *Angew. Chem. Int. Ed. Engl.* **1981**, *20*, 762.

[3]   S. Goldmann, J. Stoltefuß, *Angew. Chem.* **1991**, *103*, 1587; *Angew. Chem. Int. Ed. Engl.* **1991**, *30*, 1559.

[4]   C. O. Kappe, *Tetrahedron* **1993**, *49*, 6937.

[5]   a) J. Lu, Y. Bai, *Synthesis* **2002**, 466; b) Bi(III) triflate as catalyst: R. Varala, M. Alam, S. R. Adapa, *Synlett* **2002**, 67; c) Cu(II) triflate as a reusable catalyst: *Tetrahedron Lett.* **2003**, *44*, 3305; d) green protocol for the Biginelli reaction (Ag$_3$PW$_{12}$O$_{40}$ as a water-tolerant catalyst): J. Yadav, B. V. S. Reddy, P. Sridhar, J. S. S. Reddy, K. Nagaiah, N. Lingaiah, P. S. Saiprasad, *Eur. J. Org. Chem.* **2004**, 552.

[6]   A. C. Veronese, R. Callegary, C. F. Morelli, *Tetrahedron* **1995**, *51*, 12277.

[7]   Ref. [1], p. 295.

[8]   B. Loev, K. M. Snader, *J. Org. Chem.* **1965**, *30*, 1914.

[9]   E. E. Ayling, *J. Chem. Soc.* **1938**, 1014.

[10] K. H. Lee, K.-Y. Koo, *Bull. Korean Chem. Soc.* **2002**, *23*, 1505.

[11] a) F.-A. Kang, J. Kodah, Q. Guan, X. Li, W. V. Murray, *J. Org. Chem.* **2005**, *70*, 1957; b) A. Puchala, F. Belaj, J. Bergman, C. O. Kappe, *J. Heterocyclic Chem.* **2001**, *38*, 1345.

[12] a) M. Winkler, L. Martinkova, A. C. Knall, S. Krahulec, N. Klempier, *Tetrahedron* **2005**, *61*, 4249; b) J. K. Williams, *J. Org. Chem.* **1963**, *28*, 1054.

[13] a) Q. E. Thompson, *J. Am. Chem. Soc.* **1958**, *80*, 5483; b) L. Rodriguez-Hahn, M. Parra, M. Martinez, *Synth. Commun.* **1984**, *14*, 967.

[14] E. C. Taylor, A. McKillop, *Adv. Org. Chem.* **1970**, *7*, 1.

[15] *Houben-Weyl*, Vol. VII/1, p. 484.

3.3.2 (R)-Salsolidine

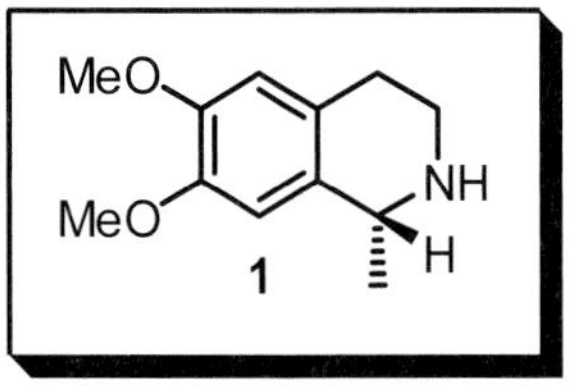

Topics:
- Synthesis of an isoquinoline alkaloid
- Bischler–Napieralski synthesis of a 3,4-dihydroisoquinoline
- Noyori hydrogenation of an imine (here: 3,4-dihydroisoquinoline → 1,2,3,4-tetrahydroisoquinoline) by use of a chiral Ru catalyst

(a) General

Salsolidine (1) belongs to the group of anhalonium alkaloids and is found in *Salsola Richteri* (Chenopodiaceae). In general, anhalonium alkaloids are constituents of the Mexican peyotl cactus; further representatives of these highly toxic isoquinoline-based alkaloids are anhalonine (2) and carnegine (3) [1, 2]:

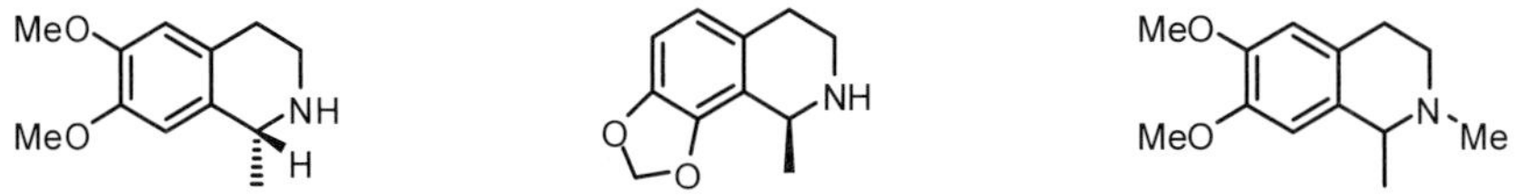

1: (+)-(R)-salsolidine   2: (–)-(S)-anhalonine   3: (±)-carnegine

Retrosynthesis of *rac*-1 can be conducted in two directions (A/B). According to (A), reversal of a Pictet–Spengler synthesis [3,4] (the most common method for 1,2,3,4-tetrahydro-β-carboline formation) leads to the iminium ion 4 and thereafter to the β-arylethylamine 5 and acetaldehyde as substrates (I) [5]. According to (B), FGI (dehydrogenation at the N–C-1 bond) leads to the 3,4-dihydroisoquinoline 6, which may be obtained by cyclization of the amide 7, the *N*-acetyl derivative of 5, by means of a Bischler–Napieralski reaction (II) [6]. Moreover, approach II offers the possibility of synthesizing (R)-1 by asymmetric catalytic hydrogenation of the imine moiety in 6 using the Noyori method [7].

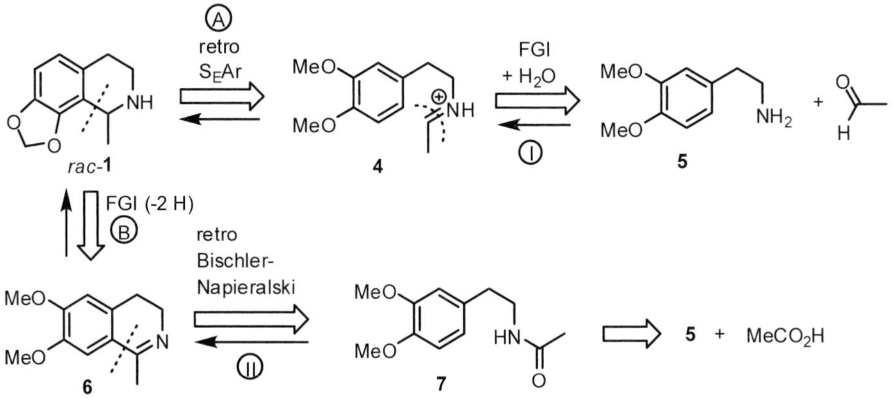

Another enantioselective approach for the synthesis of (S)-salsolidine ((S)-1) has recently been reported [8]. The substrate is 6,7-dimethoxy-3,4-dihydroisoquinoline, which is subjected to an enantioselective Strecker reaction [9] by hydrocyanation in the presence of the Jacobsen catalyst 10 and trifluoroacetic anhydride:

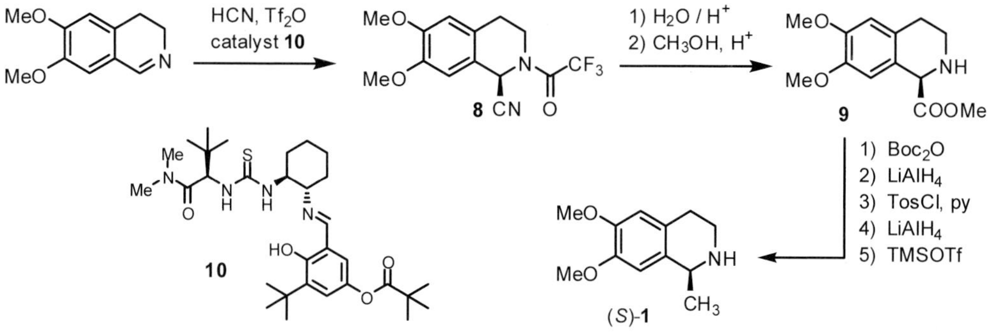

The crucial intermediate **8** is obtained in high chemical yield and enantiomeric excess (86%, 95% *ee*) and can be transformed into (S)-**1** via ester **9** and reduction (COOMe → CH$_3$).

(b) Synthesis of 1

2-(3,4-Dimethoxyphenyl)ethylamine (**5**) is *N*-acylated with acetic anhydride in the presence of triethylamine and a catalytic amount of DMAP to give the amide **7**, which is transformed to the dihydroisoquinoline **6** by cyclization with POCl$_3$ (Bischler–Napieralski reaction) [10, 11]. Instead of POCl$_3$, polyphosphoric acid, H$_2$SO$_4$, CF$_3$CO$_2$H, or CF$_3$SO$_3$H may also be used.

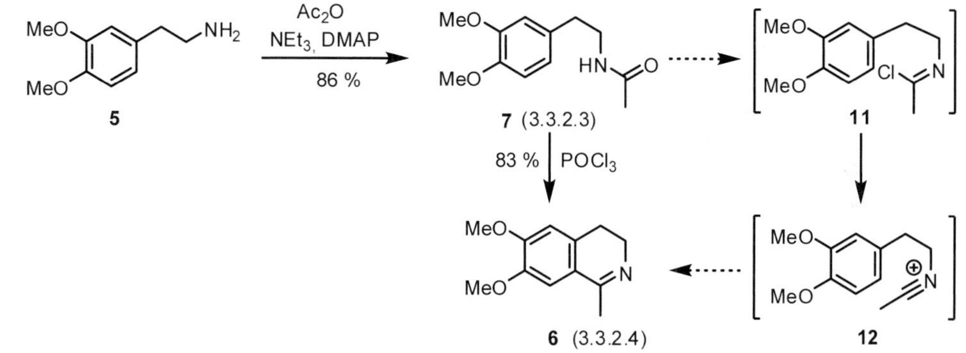

Presumably, in the POCl$_3$-mediated Bischler–Napieralski reaction, chloroimines (such as **11**) and the corresponding nitrilium ions (such as **12**) are intermediates, which cyclize to the 3,4-dihydro-isoquinoline system in an intramolecular S_EAr process. Chloroimine **11** resembles the Vilsmeier reagent [12].

For the synthesis of (R)-salsolidine (**1**), the 1-methyl-3,4-dihydroisoquinoline **6** is subjected to transfer hydrogenation with formic acid/NEt$_3$ in the presence of the chiral Ru catalyst **16** [7], which leads to the chiral 1-methyl-1,2,3,4-tetrahydroisoquinoline **1** possessing (R)-configuration at the stereogenic center C-1 with 95% *ee* and a chemical yield of 81%.

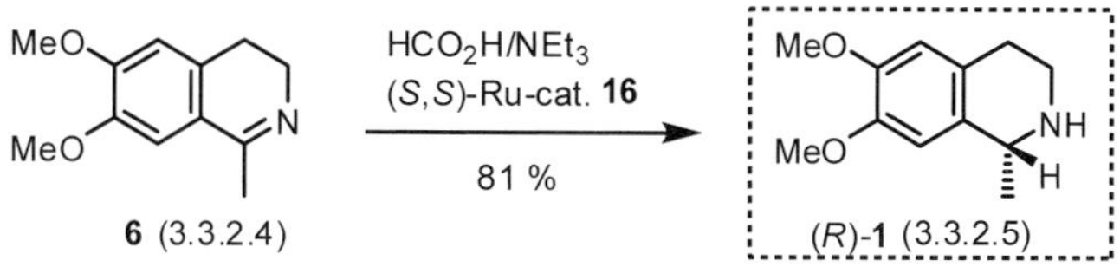

The required chiral ruthenium complex **16** can be obtained in two steps: (1S,2S)-1,2-diphenylethylenediamine (**13**) is monotosylated with p-toluenesulfonyl chloride in the presence of NEt$_3$ to give the sulfonamide **14**. The chiral Ru complex (**16**) is prepared *in situ* by addition of the chiral sulfonamide **14** to the achiral ruthenium complex [RuCl$_2$(η^6-p-cymene)]$_2$ (**15**) [7].

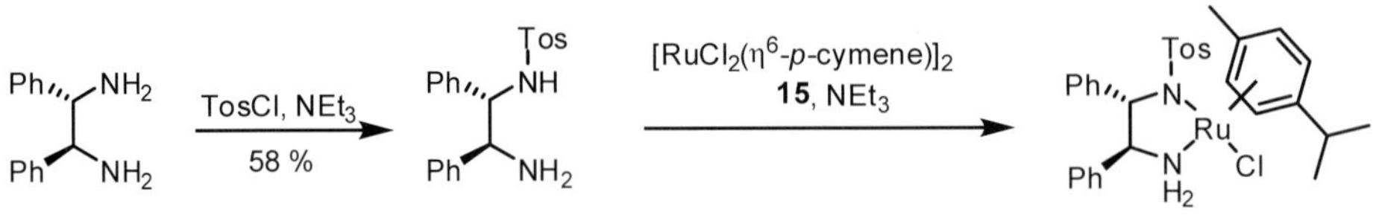

13   **14** (3.3.2.1)   **16** (3.3.2.2)

The direction of asymmetric transfer hydrogenation catalyzed by the chiral ruthenium complex is illustrated schematically in the following figure:

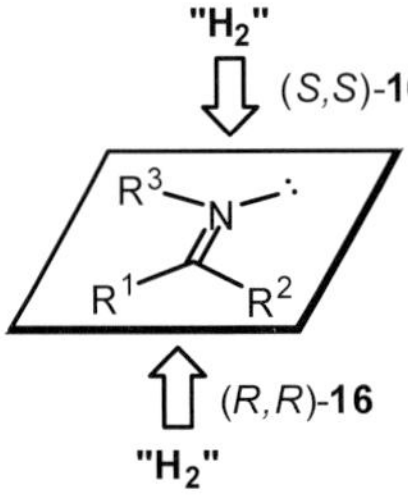

(c) Experimental procedures for the synthesis of 1

3.3.2.1   *   (1S,2S)-N-Tosyl-1,2-diphenylethylenediamine [13]

A solution of p-toluenesulfonyl chloride (450 mg, 2.40 mmol) in THF (5 mL) is added to a solution of (1S,2S)-(–)-1,2-diphenylethylenediamine (500 mg, 2.40 mmol) in THF (20 mL) and NEt$_3$ (1 mL) over a period of 0.5 h at 0 °C and the mixture is stirred for 12 h at room temperature.

The solvent is removed under reduced pressure and the residue is treated with saturated NaHCO$_3$ solution (40 mL) and CH$_2$Cl$_2$ (40 mL). The organic phase is separated, washed with brine, dried over Na$_2$SO$_4$, and then concentrated under reduced pressure. The crude product is purified by chromatography eluting with EtOAc to give the monosulfonamide as a white solid; 790 mg (90%), $[\alpha]_D^{20} = +25.0$ ($c = 0.2$, CHCl$_3$).

> ^{1}H NMR (200 MHz, CDCl$_3$, D$_2$O exchange): δ (ppm) = 7.55–7.00 (m, 10 H, Ph-H), 7.25 (d, $J = 8.0$ Hz, 2 H, Ar-H), 6.88 (d, $J = 8.0$ Hz, 2 H, Ar-H), 4.30 (d, $J = 5.5$ Hz, 1 H, C$\underline{H}$–NHSO$_2$), 4.05 (d, $J = 5.5$ Hz, 1 H, C$\underline{H}$-NH$_2$), 2.25 (s, 3 H, CH$_3$).
>
> ^{13}C NMR (50.3 MHz, CDCl$_3$): δ (ppm) = 142.4, 141.4, 139.3, 137.1, 129.1, 128.3, 128.2, 127.3, 127.2, 126.9, 126.8, 126.5 (Ar-C), 63.3, 60.4 (CH$_2$), 21.4 (CH$_3$).
>
> **MS** (DCI): m/z (%) = 367 [M + H]$^+$, 384 [M + NH$_4$]$^+$.

3.3.2.2   ***   (S,S)-Ruthenium catalyst [13]

366.5   651.2

A mixture of [RuCl$_2$(η^6-*p*-cymene)]$_2$ (202 mg, 330 μmol), the monosulfonamide **3.3.2.1** (290 mg, 792 μmol), and NEt$_3$ (0.18 mL) in CH$_3$CN (3.3 mL) is heated at 80 °C for 1 h. The warm, orange solution is then used immediately for the transfer hydrogenation of the dihydroisoquinoline **3.3.2.4**.

3.3.2.3   *   *N*-Acetyl-2-(3,4-dimethoxyphenyl)ethylamine [10]

181.2   223.3

NEt$_3$ (12.5 mL) is added to a stirred solution of 2-(3,4-dimethoxyphenyl)ethylamine (4.50 g, 24.8 mmol) and 4-dimethylaminopyridine (303 mg, 2.48 mmol) in anhydrous CH$_2$Cl$_2$ (25 mL) at 0 °C. Acetic anhydride (2.50 mL, 26.4 mmol) is then added dropwise and stirring is continued for 24 h at room temperature.

The mixture is then washed with H$_2$O (200 mL), HCl (2 M, 100 mL), saturated NaHCO$_3$ solution (2 × 200 mL), and brine (200 mL), and dried over MgSO$_4$. After filtration, the solvent is evaporated under reduced pressure and the crude product thus obtained is crystallized from EtOAc/*n*-pentane to give the *N*-acetylamine as colorless needles. A further batch of the product can be obtained by evaporation of the solvent and a second crystallization from EtOAc/*n*-pentane; 4.74 g (86%), mp 94–95 °C, R_f = 0.71 (CH$_2$Cl$_2$/MeOH, 7:1).

UV (CH$_3$CN): λ_{max} (nm) (lg ε) = 280.0 (0.164), 230.0 (0.468), 201.5 (2.611).

IR (KBr): $\tilde{\nu}$ (cm^{-1}) = 3254, 1634, 1518, 1263, 1156, 1139, 1020, 815, 767, 611.

^{1}H NMR (300 MHz, CDCl$_3$): δ (ppm) = 6.82–6.71 (m, 3 H, 3 × Ar-H), 5.93 (s$_{br}$, 1 H, NH), 3.86 (s, 3 H, OCH$_3$), 3.85 (s, 3 H, OCH$_3$), 3.48 (dt, J = 7.0, 6.0 Hz, 2 H, 2-H$_2$), 2.76 (t, J = 7.0 Hz, 2 H, 1-H$_2$), 1.90 (s, 3 H, CH$_3$).

^{13}C NMR (50.3 MHz, CDCl$_3$): δ (ppm) = 170.0 (CO), 148.8 (C-3'), 147.4 (C-4'), 131.2 (C-1'), 120.4, 111.7, 111.2 (C-2', C-5', C-6'), 55.71, 55.65 (2 × OCH$_3$), 40.64 (C-2), 35.01 (C-1), 23.08 (CH$_3$).

MS (DCI, NH$_3$, 200 eV): *m/z* (%) = 241 (100) [*M* + 18]$^+$, 447 (11) [2*M* + 1]$^+$, 464 (11) [2*M* + 18]$^+$.

3.3.2.4 ** 6,7-Dimethoxy-1-methyl-3,4-dihydroxyisoquinoline [11]

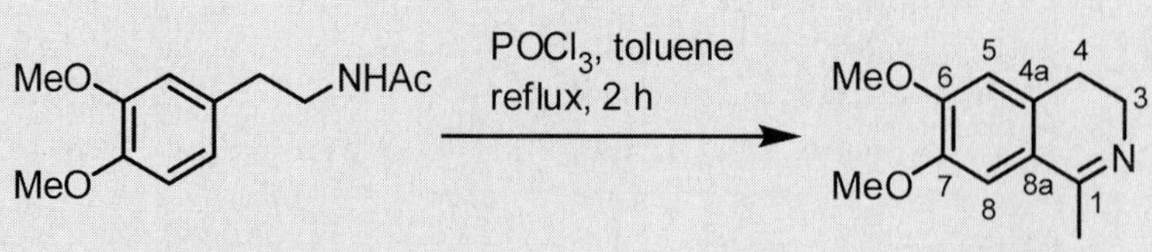

223.3   205.3

Phosphorus oxychloride (6.0 mL) is added dropwise to a stirred solution of the N-acetylamine **3.3.2.3** (6.00 g, 26.9 mmol) in toluene (30 mL) over 15 min. The resulting solution is stirred under reflux for 2 h and then stored at 4 °C for 12 h to give a yellow precipitate.

The precipitate is collected by filtration and washed with cold MeOH and EtOAc. Recrystallization from MeOH/EtOAc gives the dihydroxyisoquinoline as a white powder. Another batch of the product can be obtained by evaporation of the solvent and crystallization from EtOAc/n-pentane; 4.55 g (83%), mp 202–203 °C, R_f = 0.69 (CH_2Cl_2/MeOH, 7:1).

UV (CH_3CN): λ_{max} (nm) (lg ε) = 352.0 (0.309), 301.5 (0.398), 243.5 (0.671), 231.0 (0.551).

IR (KBr): $\tilde{v}$ (cm^{-1}) = 2611, 1656, 1565, 1335, 1278, 1167, 1069.

^{1}H NMR (300 MHz, $CDCl_3$): δ (ppm) = 6.98 (s, 1 H, 8-H), 6.68 (s, 1 H, 5-H), 3.92 (s, 3 H, OCH_3), 3.88 (s, 3 H, OCH_3), 3.63 (td, J = 7.0, 1.5 Hz, 2 H, 3-H_2), 2.64 (t, J = 7.0 Hz, 2 H, 4-H_2), 2.36 (s, 3 H, 1-CH_3).

^{13}C NMR (50.3 MHz, $CDCl_3$): δ (ppm) = 173.5 (C-1), 156.1 (C-6), 148.5 (C-7), 132.7 (C-10), 117.7 (C-), 111.0 (C-5), 110.6 (C-8), 56.38, 56.24 (2 × OCH_3), 40.43 (C-3), 24.89 (C-4), 19.38 (CH_3).

MS (EI, 70 eV): m/z (%) = 205 (100) $[M]^+$, 190 (48) $[M - CH_3]^+$, 174 (9) $[M - OCH_3]^+$.

3.3.2.5 *** (R)-6,7-Dimethoxy-1-methyl-1,2,3,4-tetrahydroisoquinoline ((R)-salsolidine) [7a]

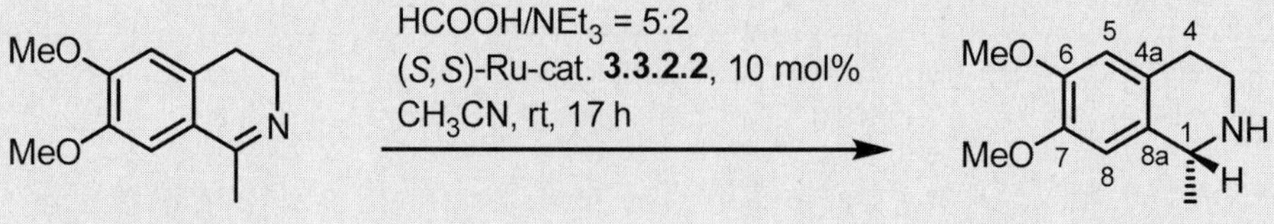

205.3   207.3

A mixture of formic acid and NEt_3 (5:2, 3.3 mL) is added to a stirred solution of the dihydroisoquinoline **3.3.2.4** (1.35 g, 6.60 mmol) and the preformed (S,S)-ruthenium catalyst **3.3.2.2** (430 mg, 0.66 mmol, 10.0 mol%) in CH_3CN (13 mL). Stirring is continued for 17 h at room temperature.

The mixture is then basified to pH 8–9 by the addition of saturated Na_2CO_3 solution and extracted with EtOAc (3 × 20 mL). The combined organic layers are washed with brine (1 × 20 mL), dried over $MgSO_4$, and concentrated under reduced pressure. The residue is purified by flash chromatography on silica gel (EtOAc/MeOH/NEt$_3$, 92:5:3) to give (R)-salsolidine as a brown oil; 1.11 g (81%), $[\alpha]_D^{20}$ = +51.1 (c = 2.70, EtOH), R_f = 0.38 (CH$_2$Cl$_2$/MeOH, 7:1). The measured optical rotation corresponds to an ee value of 95%.

UV (CH$_3$CN): λ_{max} (nm) (lg ε) = 282.5 (0.186), 201.0 (1.759).

IR (KBr): $\tilde{v}$ (cm^{-1}) = 2932, 1610, 1512, 1464, 1372, 1256, 1126, 1030, 857, 790.

^{1}H NMR (300 MHz, CDCl$_3$): δ (ppm) = 6.62, 6.57 (2 × s, 2 H, 5-H, 8-H), 4.04 (q, J = 6.7 Hz, 1 H, 1-H), 3.86 (s, 3 H, OCH$_3$), 3.85 (s, 3 H, OCH$_3$), 3.25 (dt, J = 12.0, 4.5 Hz, 1 H, 3-H$_b$), 2.99 (m$_c$, 1 H, 3-H$_a$), 2.88–2.55 (m, 2 H, 4-H$_2$), 1.66 (s$_{br}$, 1 H, NH), 1.42 (d, J = 6.7 Hz, 3 H, CH$_3$).

^{13}C NMR (75.5 MHz, CDCl$_3$): δ (ppm) = 147.2 (C-7), 147.1 (C-6), 132.4 (C-8a), 126.7 (C-4a), 111.6 (C-5), 108.9 (C-8), 55.86, 55.73 (2 × OCH$_3$), 51.13 (C-1), 41.77 (C-3), 29.48 (C-4), 22.78 (CH$_3$).

MS (EI, 70 eV): m/z (%) = 192 (100) $[M - CH_3]^+$, 207 (10) $[M]^+$.

---

[1]   Römpp Lexikon "Naturstoffe" (Eds.: W. Steglich, B. Fugmann, S. Lang-Fugmann), p. 38, Georg Thieme Verlag, Stuttgart, **1997**.

[2]   M. Hesse, Alkaloide, p. 36, Wiley-VCH, Weinheim, **2000**.

[3]   Th. Eicher, S. Hauptmann, The Chemistry of Heterocycles, 2nd ed., p. 346, Wiley-VCH, Weinheim, **2003**.

[4]   The Pictet–Spengler reaction is verified by Nature in the biogenesis of 1,2,3,4-tetrahydroisoquinoline alkaloids, cf. refs. [2], [3], and C. Gremmen, M. J. Wanner, G.-J. Koomen, Tetrahedron Lett. **2001**, 8885.

[5]   In fact, rac-salsolidine has been obtained by Pictet–Spengler cyclization of the β-arylethylamine **5** with acetaldehyde or acetaldehyde equivalents: H. Singh, K. Singh, Indian J. Chem. Sect. B **1989**, 28, 802, and literature cited therein.

[6]   Ref. [3], p. 338.

[7]   a) N. Uematsu, A. Fuji, S. Hashiguchi, T. Ikariya, R. Noyori, J. Am. Chem. Soc. **1996**, 118, 4916; b) review on asymmetric transfer hydrogenation: S. Gladiali, E. Alberico, Chem. Soc. Rev. **2006**, 35, 218; c) alternative asymmetric synthesis of (R)-salsolidine: D. Taniyama, M. Hasegawa, K. Tomioka, Tetrahedron Lett. **2000**, 41, 5533; d) for an alternative asymmetric synthesis of (S)-salsolidine, see J. Wu, F. Wang, Y. Ma, X. Cui, L. Cun, J. Zhu, J. Deng, B. Yu, Chem. Commun. **2006**, 1766.

[8]   K. Itoh, S. Akashi, B. Saito, T. Katsuki, Synlett **2006**, 1595.

[9]   M. B. Smith, J. March, March's Advanced Organic Chemistry, 5th ed., p. 1240, John Wiley & Sons, Inc., New York, **2001**.

[10] P. J. Garratt, S. Travard, S. Vonhoff, J. Med. Chem. **1996**, 39, 1796.

[11] F. Zhang, G. Dryhurst, J. Med. Chem. **1993**, 36, 11.

[12] Ref. [9], p. 715.

[13] N. Rackelmann, Ph.D. Thesis, University of Göttingen, **2004**.

3.3.3　Epirizole

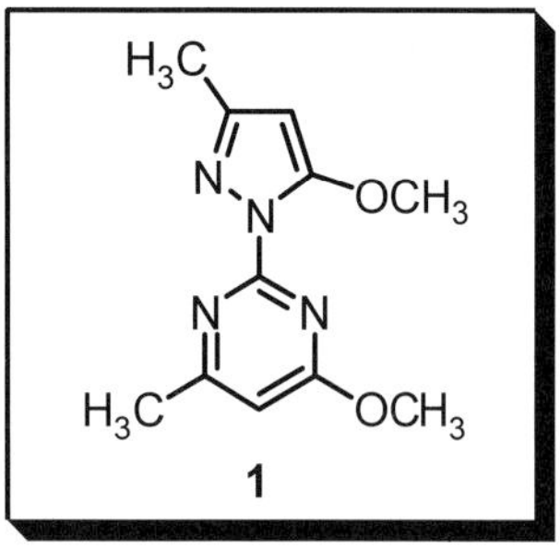

1

Topics:
- Synthesis of a drug
- Pinner synthesis of pyrimidines
- Knorr synthesis of pyrazolones
- *O*-Methylation at the pyrimidinone and pyrazolone system

(a) General

Epirizole (**1**), a (2-pyrimidinyl)-substituted methoxypyrazole, belongs to the large family of pyrazolone derivatives, which are medicinally applied as antipyretics and antirheumatics and which were widely used in the last century. Epirizole exhibits analgesic and anti-inflammatory properties [1]. Other examples of this family are antipyrine (**2**), pyramidone (**3**), and metamizole (**4**):

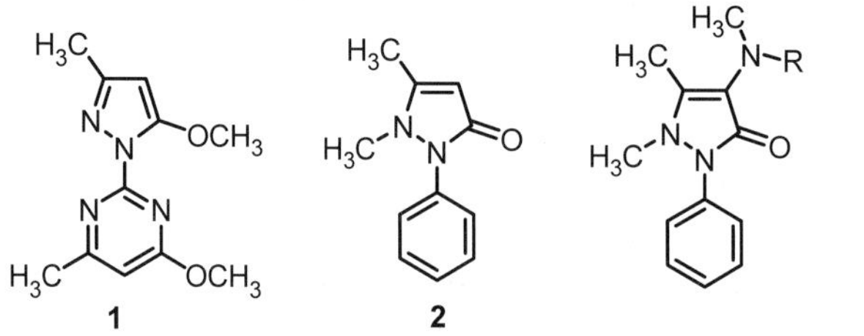

3: R = CH$_3$
4: R = CH$_2$-SO$_3^{\ominus}$ Na$^{\oplus}$

Retrosynthesis of **1** starts with *O*-demethylation (FGI), transforming **1** to the pyrimidinone-pyrazolone **5**. Further disconnections can be performed (1) at the pyrimidine and (2) at the pyrazolone part of **5**, making use of a retro-Pinner and a retro-Knorr approach. The Pinner and the Knorr methods are valuable and widely used procedures for the synthesis of pyrimidines and pyrazolones, respectively.

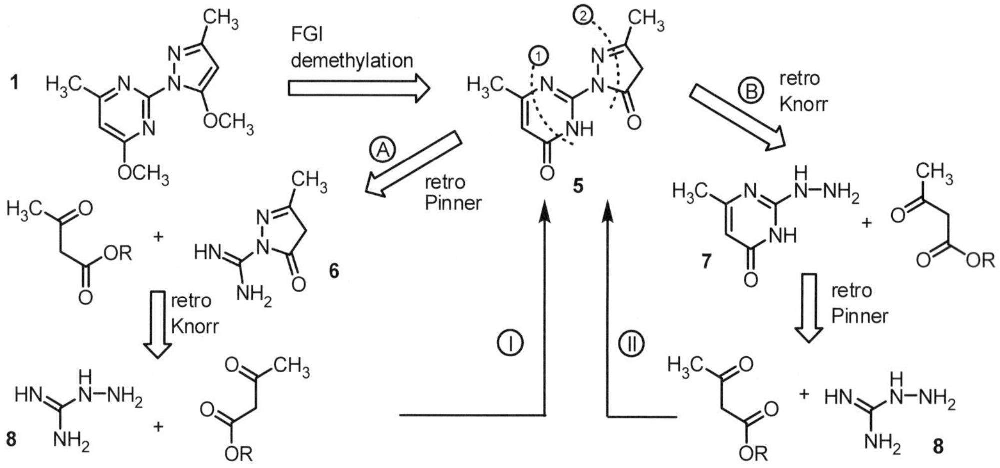

In the Pinner synthesis [2], 1,3-diketones are cyclocondensed with N–C–N building blocks such as ureas, thioureas, amidines, and guanidines to give pyrimidine derivatives of types **9** or **10**. β-Keto esters react analogously; thus, with amidines, pyrimidin-4(3*H*)-ones **11** are formed, which display the 2,4,6-substitution pattern at the pyrimidine part of the key intermediate **5** in the epirizole synthesis.

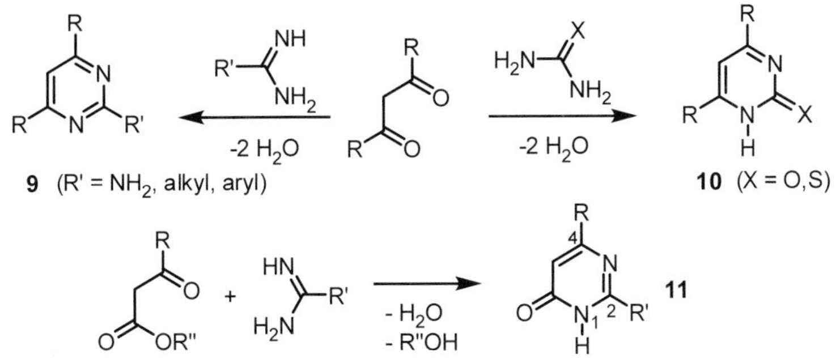

9 (R' = NH$_2$, alkyl, aryl)   **10** (X = O,S)

11

In the Knorr synthesis [3], β-keto esters undergo cyclocondensation with hydrazine or monosubstituted hydrazines to give 2,4-dihydro-3*H*-pyrazol-3-ones **12** (via hydrazones or their enehydrazine tautomers as intermediates). Notably, **12** corresponds structurally to the pyrazolone part of the key compound **5**.

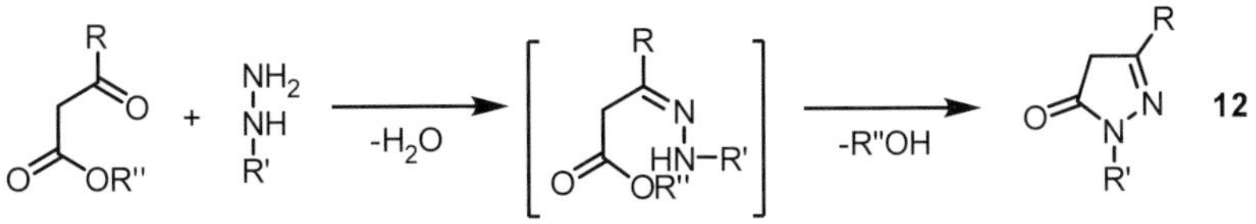

12

As a consequence, for the synthesis of epirizole, two approaches (I/II) using Pinner and Knorr cyclizations must be considered. In approach I, the pyrazolone part of **1** is constructed first and the pyrimidine part follows; in approach II, the pyrimidine part is constructed first, followed by the pyrazolone part; in both approaches, *O*-methylation concludes the synthesis. However, only approach I can be realized, since aminoguanidine (**8**) and acetoacetate — identified as starting materials for both approaches I and II according to the retrosynthesis — lead exclusively to pyrazolone formation (**6**) via participation of the hydrazine functionality of **8** and *not* to pyrimidinone formation (**7**) via the amidine function [4].

Alternatively, epirizole has been synthesized by a strategy different from I/II [1, 4], starting from a suitable pyrimidine building block:

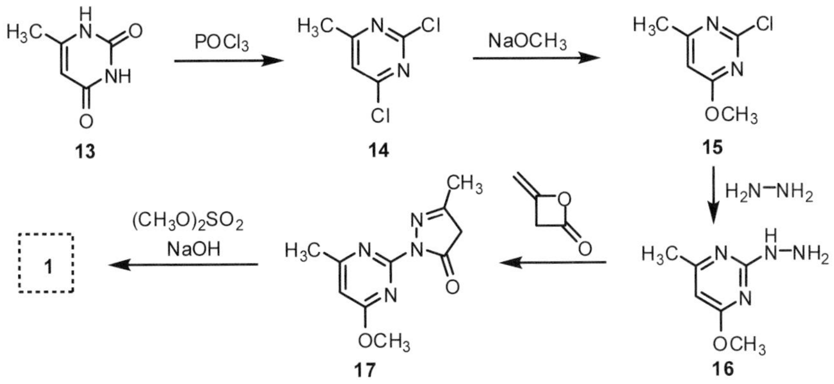

13   **14**   **15**

1   **17**   **16**

Thus, 6-methyluracil (**13**), easily accessible by Pinner condensation of acetoacetate with urea, is transformed into the 2,4-dichloropyrimidine **14** by treatment with $POCl_3$. In an S_NAr reaction with methoxide, the 4-Cl substituent of **14**, being more reactive than 2-Cl, is displaced chemoselectively to give **15** [5]. Subsequently, the 2-Cl is substituted by hydrazine to afford **16**, which undergoes a Knorr cyclocondensation with diketene (as an acetoacetate equivalent) providing the pyrazolone part of intermediate **17**. Finally, O-methylation of **17** with dimethyl sulfate yields epirizole (**1**).

(b) Synthesis of 1

The presented laboratory synthesis of epirizole (**1**) [4] is based on the retrosynthesis A realizing approach I. In the first step, Knorr cyclocondensation is performed with acetoacetamide (**19**) and aminoguanidine (as hydrogencarbonate **18**) to give the 1-guanidino-substituted pyrazolone **6**:

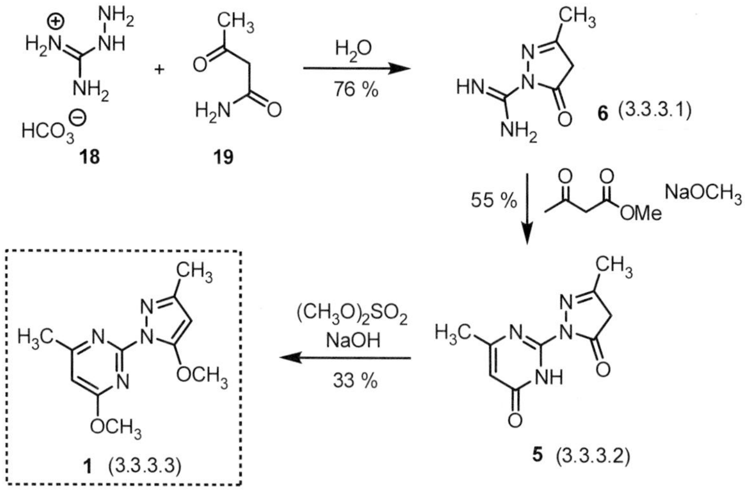

In the second step, methyl acetoacetate is cyclocondensed with the guanidinopyrazolone **6** in the presence of $NaOCH_3$, which, by way of a Pinner synthesis, leads to formation of the pyrimidine part of the key intermediate **5**. In the last step, O-methylation of both the pyrazolone and pyrimidinone subunits of **5** is accomplished by reaction with dimethyl sulfate in alkaline medium furnishing epirizole (**1**).

Thus, the target molecule **1** is obtained in a three-step sequence from low-cost substrates in an overall yield of 52% based on **19**.

(c) Experimental procedures for the synthesis of 1

3.3.3.1 ** 2-Formamidino-5-methyl-2,4-dihydro-3*H*-pyrazol-3-one [4]

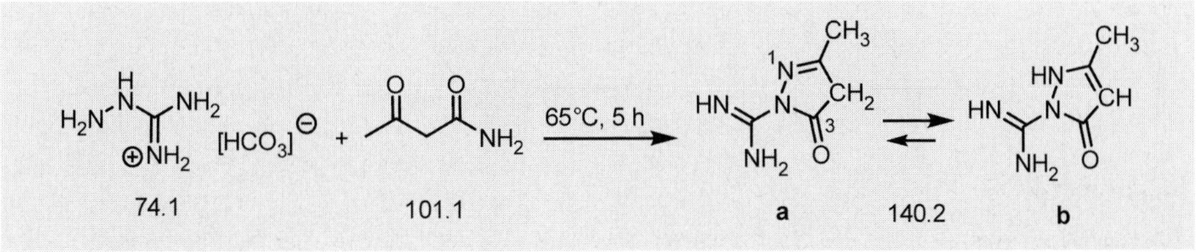

Aminoguanidine hydrogencarbonate (5.00 g, 36.8 mmol) is added to a solution of acetoacetamide (3.08 g, 30.6 mmol) in water (70 mL) and the reaction mixture is heated to 60–70 °C for 5 h.

After cooling to room temperature, the precipitate is filtered off, washed with H_2O, and dried. The product is obtained as a colorless solid; 6.77 g (76%), mp 180–181 °C.

1H NMR (200 MHz, [D6]DMSO): δ (ppm) = 10.0–9.6 (s_{br}, 1 H, NH), 8.60–7.80 (s_{br}, 2 H, NH), 4.45 (s, 1 H, H3), 1.98 (s, 3 H, CH3) [6].

13C NMR (75 MHz, [D6]DMSO): δ (ppm) = 167.5 (CO), 155.9 (NH–C–NH2), 153.0 (C-4, CN), 82.1 (C-3, CH), 14.9 (CH3).

MS (EI, 70 eV): $m/z = 140\ [M]^+$.

3.3.3.2 ** 2-(5-Methyl-2,4-dihydro-3*H*-pyrazol-3-on-2-yl)-6-methylpyrimidin-4(3*H*)-one [4]

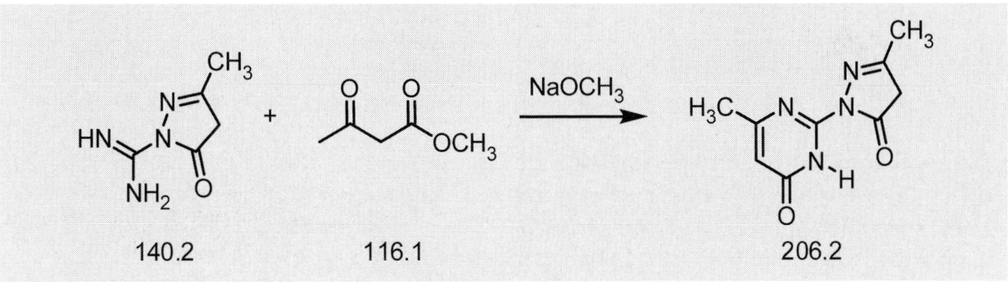

A solution of sodium methoxide is prepared from methanol (12.5 mL) and sodium (0.44 g, 19.0 mmol). Pyrazolone **3.3.3.1** (2.61 g, 19.2 mmol) and methyl acetoacetate (2.17 g, 18.7 mmol) are then added and the solution is heated under reflux for 4 h.

The solvent is then evaporated, the residue is dissolved in water H_2O (40 mL), and the pH is adjusted to 3 by adding 10% aqueous HCl. The precipitate formed is filtered off, washed with water H_2O, and dried. The product is obtained as a colorless solid; 2.16 g (55%), mp 165–166 °C.

1H NMR (300 MHz, [D6]DMSO): δ (ppm) = 6.09 (s, 1 H, H-3'), 5.25 (s, 1 H, H-3), 2.22 (s, 3 H, CH3), 2.19 (s, 3 H, CH3) [6].

13C NMR (75 MHz, [D6]DMSO): δ (ppm) = 107.4 (C-3'), 90.9 (C-3), 12.3 (CH3), 3.0 (CH3).

MS (ESI): $m/z = 435\ [2M + Na]^+$, 229 $[M + Na]^+$, 207 $[M + H]^+$, 205 $[M - H]^+$.

3.3.3.3   ******   **4-Methoxy-2-(5-methoxy-3-methyl-1H-pyrazol-1-yl)-6-methylpyrimidine (epirizole)** [4]

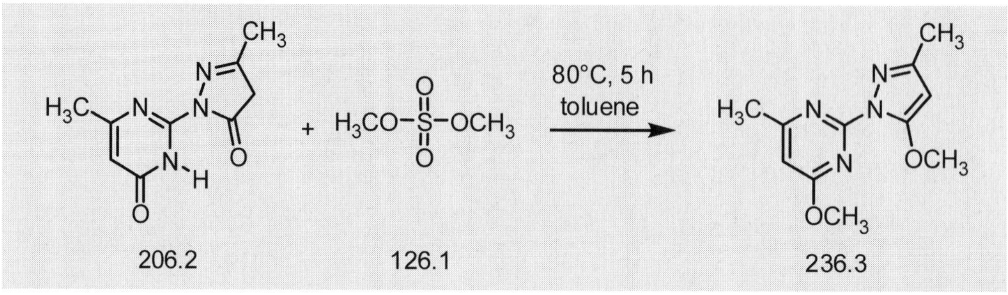

| 206.2 | 126.1 | 236.3 |

The pyrimidinone **3.3.3.2** (500 mg, 2.43 mmol) is dissolved in toluene (25 mL) and heated to 80 °C, and then dimethyl sulfate (1.82 g, 14.4 mmol) is added dropwise over 10 min. The reaction mixture is heated under reflux for 5 h.

The solvent is then evaporated *in vacuo* and the residue is purified by flash chromatography (EtOAc/MeOH, 4:1). Epirizole is obtained as a yellowish solid; 192 mg (33%), mp 217–218 °C.

---

^{1}H NMR (300 MHz, [D$_6$]DMSO): δ (ppm) = 6.85 (1 H, H-3'), 5.41 (1 H, H-3), 3.98 (3 H, CH$_3$), 3.30 (3 H, CH$_3$), 2.45 (3 H, CH$_3$), 2.28 (3 H, CH$_3$).

^{13}C NMR (75 MHz, [D$_6$]DMSO): δ (ppm) = 104.1 (C-3'), 94.1 (C-3), 54.5 (OCH$_3$), 52.7 (OCH$_3$), 22.4 (CH$_3$), 12.4 (CH$_3$).

MS (ESI): m/z = 491 [2M + Na$^+$], 235 [M + H$^+$].

---

[1]   A. Kleemann, J. Engel, *Pharmaceutical Substances*, 3rd ed., p. 709, Georg Thieme Verlag, Stuttgart, **1999**.

[2]   Th. Eicher, S. Hauptmann, *The Chemistry of Heterocycles*, 2nd ed., p. 403, Wiley-VCH, Weinheim, **2003**.

[3]   Ref. [2], p. 188.

[4]   German patent DE 22 37 632 (Daiichi Seiyaku Co., Ltd., Tokyo, 1973).

[5]   For selectivities in S$_N$Ar displacement reactions of halogenopyrimidines, see ref. [2], p. 400.

[6]   According to NMR, pyrazolones **3.3.3.1** and **3.3.3.2** predominantly exist in the tautomeric form **b** in DMSO solution. For the issue of pyrazolone tautomerism and the systematic nomenclature of pyrazolones, see ref. [2], p. 187.

3.3.4   Ras farnesyltransferase inhibitor

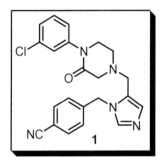

Topics: • Synthesis of a clinically investigated drug of the imidazole-piperazinone type
- Marckwald imidazole synthesis
- Formation of a piperazinone
- Delepine reaction
- Oxidative dethionation of a mercapto heterocycle
- Mitsunobu reaction

(a) General

Mutant ras proteins, the products of ras oncogenes, are involved in a significant proportion of human cancers. The enzyme farnesyl-protein transferase (FPTase) catalyzes farnesylation of the ras protein, thereby activating it. Thus, FPTase inhibitors are currently the subject of intense interest as novel and improved anticancer agents [1]. The imidazole/piperazinone-based compound **1** (as its hydrochloride) has been identified as an FPTase inhibitor showing efficacy in animal models with a high therapeutic index and has been tested in phase I and phase II clinical studies.

A possible retrosynthesis of **1** leads to a disconnection at the non-amide piperazinone nitrogen to give two heterocyclic building blocks, namely the 1-benzyl-5-halogenomethylimidazole **2** and the *N*-arylpiperazinone **3**. The synthetic step would be a simple N-alkylation.

1   2   +   3

The retrosynthesis of the building block **2** affords *p*-cyanobenzylamine, dihydroxyacetone, and thiocyanate. Thus, for the synthesis of the 1,5-disubstituted imidazole, a modified Marckwald approach [2] would be most suitable.

For building block **3**, a series of FGI and disconnections leads to *m*-chloroaniline, chloroacetyl chloride, and ethanolamine as suitable substrates for *N*-aryl piperazinone formation.

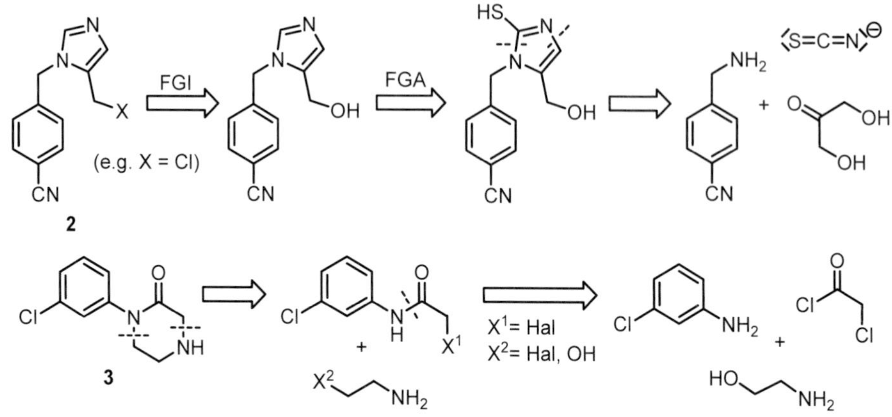

As shown in section (b), the synthesis of **1–3** [1] is accomplished on the basis of these retroanalytical considerations.

(b) Synthesis of 1

(1) Building block **2** is synthesized via the preparation of 4-cyanobenzylamine (**5**) by Delepine reaction [3] of 4-cyanobenzyl bromide (**4**) [4] with hexamethylenetetramine followed by treatment with H_3PO_4:

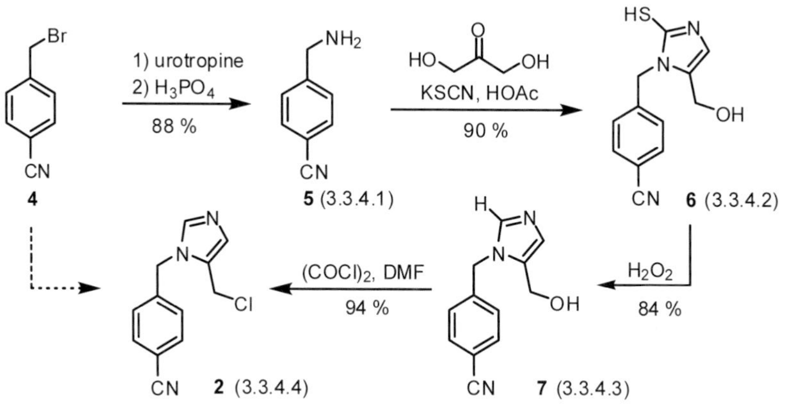

Then, a Marckwald synthesis of the 5-hydroxymethyl-2-mercaptoimidazole **6** is performed by cyclocondensation of the benzylamine **5** (as its H_3PO_4 salt) with 1,3-dihydroxyacetone and potassium thiocyanate in the presence of acetic acid.

A reasonable mechanistic interpretation of the reaction **5 → 6** (as given in ref. [1]) starts with the formation of an imine from **5** and 1,3-dihydroxyacetone, which is in tautomeric equilibrium with an α-amino carbonyl compound as the key intermediate of the Marckwald process. This is followed by addition of thiocyanate with concomitant ring closure to give the imidazole **6**. It can be assumed that a heterocumulene acts as an intermediate.

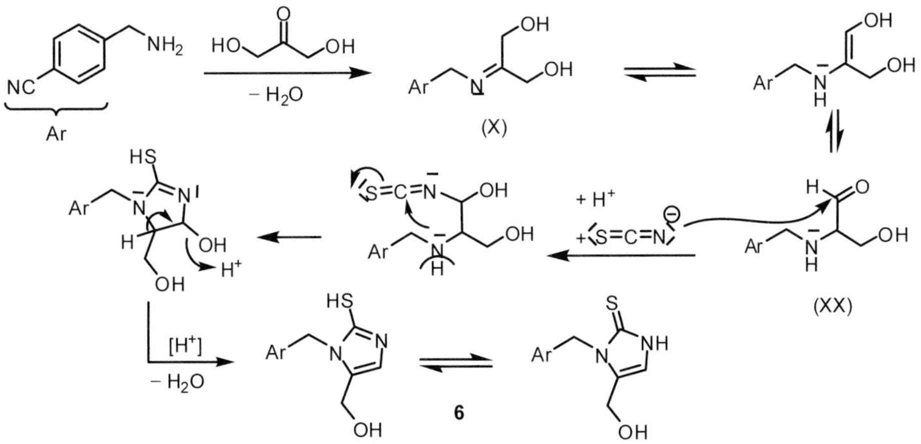

To obtain the desired **2** from **6**, the thiol group in **6** has to be replaced by hydrogen. This is accomplished by oxidation of the thiol to a sulfinic acid moiety, which thermally eliminates SO_2; for this oxidative dethionation (**6 → 7**), hydrogen peroxide in aqueous acetic acid is the reagent of choice:

$$Het-SH \xrightarrow{H_2O_2} Het-SO_2H \xrightarrow[-SO_2]{\Delta} Het-H$$

Finally, the hydroxy group in the side chain in **7** is substituted by a chlorine atom using oxalyl chloride in dimethylformamide. It has been shown [1] that the chlorinating agent is the Vilsmeier reagent **8** derived from $(COCl)_2$ and DMF, which converts **7** into an iminium ion **9** as intermediate. This is subsequently dealkylated by attack of chloride at the heterobenzylic position to give **2**:

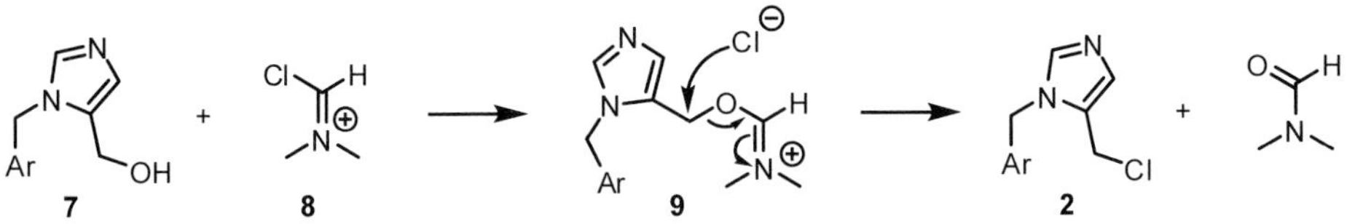

(2) Building block **3** is synthesized in a two-step procedure. First, *m*-chloroaniline is acylated under Schotten–Baumann conditions with chloroacetyl chloride in the two-phase system isopropyl acetate/aqueous $KHCO_3$ solution to give a quantitative yield of the chloroacetamide **10**. Without isolation, this is treated with ethanolamine to give the hydroxy amide **11** in an S_N process. Second, **11** is cyclodehydrated under Mitsunobu conditions [5, 6] with diisopropylazodicarboxylate (DIAD) and tri-*n*-butylphosphine to yield the piperazinone **3** (isolated as its HCl salt):

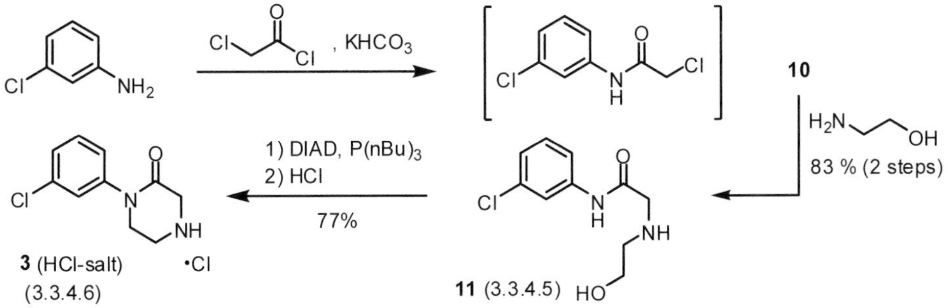

The generally accepted mechanism of the Mitsunobu reaction is as follows. The tertiary phosphine and azodicarboxylate (ADE) initially form a betaine **12**, which is transformed to an alkoxyphosphonium salt **14** by reaction with an alcohol R–OH. This then reacts with a nucleophile H–Nu in a disproportionation process, in which the substitution product R–Nu (**13**), a phosphine oxide, and hydrazinodicarboxylate are formed. Overall, a nucleophilic substitution at the OH-bearing C-atom of the alcohol takes place; if this C-atom is a stereogenic center of defined stereochemistry, the Mitsunobu reaction leads to inversion of configuration.

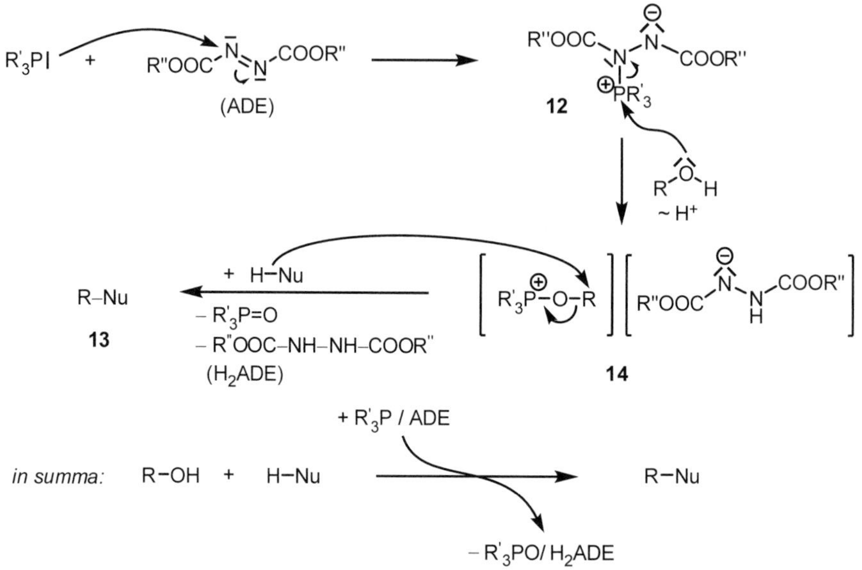

(3) Finally, the piperazinone **3** is combined with the chloromethylimidazole **2** by alkylation at the secondary NH group in acetonitrile in the presence of diisopropylethylamine (Hünig base) to give **1** in 83% yield:

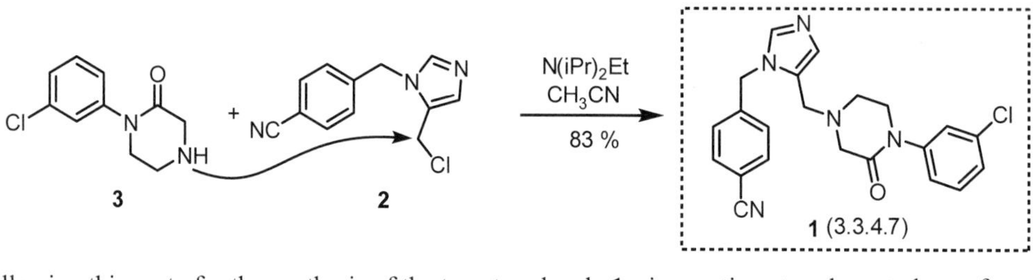

Following this route for the synthesis of the target molecule **1**, six reaction steps have to be performed. Building block **2** is obtained in four steps in an overall yield of 63% (based on **4**) and building block **3** is obtained in two steps in an overall yield of 77% (based on *m*-chloroaniline).

(c) Experimental procedures for the synthesis of 1

3.3.4.1   **   4-Cyanobenzylamine [1]

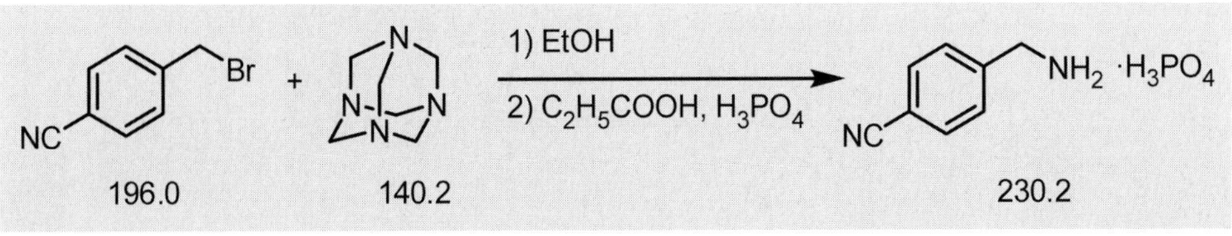

| 196.0 | 140.2 | 230.2 |

A slurry of urotropine (hexamethylenetetramine, HMTA) (3.65 g, 26.0 mmol) in EtOH (25 mL) is added to a stirred slurry of 4-cyanobenzyl bromide (5.00 g, 25.5 mmol) in EtOH (25 mL) maintained at 50 °C over 10 min. EtOH (2 × 10 mL) is then added and the reaction mixture is heated to 70 °C for 1.5 h.

The reaction mixture is then cooled to 55 °C and propionic acid (20.6 mL) is added. Concentrated phosphoric acid (6.5 mL) is gradually added, maintaining the temperature below 65 °C; the mixture is then kept at 70 °C for 30 min. It is allowed to cool to room temperature over 1 h and then stirred for a further 1 h. The reaction slurry is filtered, and the filter cake is washed with EtOH (4 × 15 mL), H_2O (5 × 8 mL), and CH_3CN (2 × 3 mL), and dried. The yield is 4.66 g (79%) of a colorless, crystalline solid.

IR (FT-IR, solid): $\tilde{\nu}$ (cm^{-1}) = 2642 (O=P–O–H), 2359, 2233 (C≡N), 1653, 1496, 1332, 1233, 1109, 960, 922, 889, 844, 584.

3.3.4.2   *   4-(5-Hydroxymethyl-2-mercaptoimidazol-1-ylmethyl)benzonitrile [1]

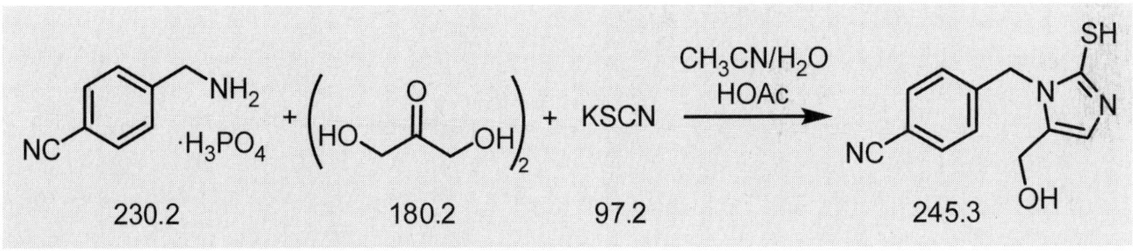

| 230.2 | 180.2 | 97.2 | 245.3 |

A slurry of the ammonium salt **3.3.4.1** (4.07 g, 17.7 mmol), potassium thiocyanate (2.58 g, 26.6 mmol), and dihydroxyacetone dimer (1.75 g, 9.7 mmol) in CH_3CN/H_2O (93:7, 18 mL) and acetic acid (2.0 mL) is stirred for 18 h at 55 °C.

The reaction mixture is then cooled to room temperature and the precipitate formed is filtered off, washed with CH_3CN (17 mL), H_2O (35 mL), and EtOAc (17 mL), and dried *in vacuo* to give 3.31 g (76%) of a light-tan solid.

IR (FT-IR, solid): $\tilde{\nu}$ (cm^{-1}) = 3042 (OH), 2925, 2360, 2222 (C≡N), 1606, 1488 (C=N), 1292, 1024, 815, 631, 521.

^{1}H NMR ([D$_6$]DMSO): δ (ppm) = 12.24 (s, 1 H, SH), 7.79 (d, J = 8.5 Hz, 2 H, Ar-H), 7.36 (d, J = 8.5 Hz, 2 H, Ar-H), 6.90 (s, 1 H, C=C–H), 5.37 (s, 2 H, Ar–CH$_2$), 5.23 (s$_{br}$, 1 H, CH$_2$–O$\underline{H}$), 4.15 (s, 2 H, C$\underline{H}_2$–OH).

^{13}C NMR ([D$_6$]DMSO): δ (ppm) = 162.7, 142.8, 132.3, 130.2, 127.7, 118.7, 113.1, 109.9, 53.1, 46.3.

3.3.4.3   *   4-(5-Hydroxymethylimidazol-1-ylmethyl)benzonitrile [1]

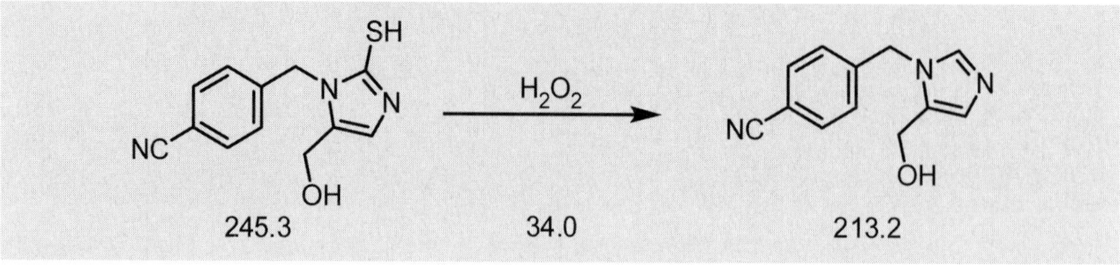

$$245.3 \qquad\qquad 34.0 \qquad\qquad 213.2$$

A 35% aqueous solution of H_2O_2 (3.72 g, 38.3 mmol) is added dropwise to a stirred solution of the mercapto compound **3.3.4.2** (2.85 g, 11.6 mmol) in acetic acid (5.5 mL) and H_2O (2.5 mL), with the temperature being maintained between 30 and 40 °C (cooling with an ice bath). The resulting yellow-orange solution is stirred for 30 min at 40 °C and then cooled to room temperature.

The reaction is quenched by the addition of 10% aqueous sodium sulfite (2 mL) and the mixture is treated with activated charcoal (0.2 g) and stirred for 30 min. The slurry is filtered and the filtrate is basified to pH 9 with 25% aqueous ammonia (ca. 11 mL) at 20 °C. The resulting slurry is stirred for 30 min and filtered. The solid is washed with H_2O (2 × 15 mL) and H_2O/MeOH (2:1, 15 mL) and dried. The yield is 1.75 g (71%) of a brown solid, mp 163–164 °C.

IR (FT-IR, solid): $\tilde{v}$ (cm^{-1}) = 3122 (OH), 3057, 2836, 2745, 2359, 2232 (C≡N), 1698, 1495 (C=N), 1326, 1247, 1105, 1027, 831, 780, 660, 556.

^{1}H NMR ([D$_6$]DMSO): δ (ppm) = 7.81 (s, 1 H, N–CH=N), 7.71 (d, J = 8.5 Hz, 2 H, Ar-H), 7.29 (d, J = 8.5 Hz, 2 H, Ar-H), 6.85 (s, 1 H, C=CH–N), 5.34 (s, 2 H, Ar–CH$_2$), 5.11 (s$_{br}$, 1 H, CH$_2$–OH), 4.29 (s, 2 H, CH$_2$–OH).

^{13}C NMR ([D$_6$]DMSO): δ (ppm) = 143.4, 138.6, 132.5, 131.6, 127.7, 127.6, 118.6, 110.3, 52.7, 47.1.

3.3.4.4   **   4-(5-Chloromethylimidazol-1-ylmethyl)benzonitrile HCl salt [1]

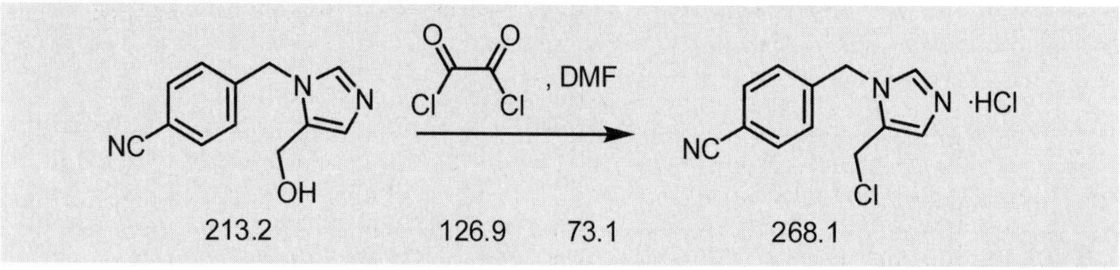

$$213.2 \qquad\qquad 126.9 \qquad 73.1 \qquad\qquad 268.1$$

Oxalyl chloride (1.52 g, 12.0 mmol) is slowly added to a stirred solution of dimethylformamide (1.75 g, 24.0 mmol) in CH$_3$CN (20 mL), maintaining the temperature below 10 °C (cooling with an ice bath). The white slurry containing the "Vilsmeier reagent" is slowly added to a stirred suspension of the hydroxymethyl compound **3.3.4.3** (2.02 g, 9.5 mmol) in CH$_3$CN (15 mL), keeping the temperature below 6 °C (cooling with an ice bath). Finally, further CH$_3$CN (5 mL) is added and the reaction mixture is warmed to 25 °C and stirred for 3 h.

The slurry is then cooled to 0 °C and stirred for 1 h. After filtration, the solid is washed with ice-cold CH₃CN (8 mL) and dried *in vacuo* to give 2.20 g (80%) of a light-tan solid; mp 204–206 °C.

IR (FT-IR, solid): $\tilde{v}$ (cm⁻¹) = 3004, 2814, 2231 (C≡N), 1459 (C=N), 1319, 820, 765, 684, 548.

¹H NMR ([D₆]DMSO): δ (ppm) = 9.41 (d, *J* = 1.2 Hz, 1 H, N–CH=N), 7.88 (d, *J* = 8.5 Hz, 2 H, Ar-H), 7.88 (s, 1 H, C=CH–N), 7.54 (d, *J* = 8.5 Hz, 2 H, Ar-H), 5.68 (s, 2 H, Ar–CH₂), 4.92 (s, 2 H, CH₂–Cl).

¹³C NMR ([D₆]DMSO): δ (ppm) = 139.7, 137.7, 132.8, 130.1, 128.8, 120.8, 118.5, 111.3, 49.0, 33.1.

3.3.4.5   **   *N*-(3-Chlorophenyl)-2-(2-hydroxyethylamino)acetamide [1]

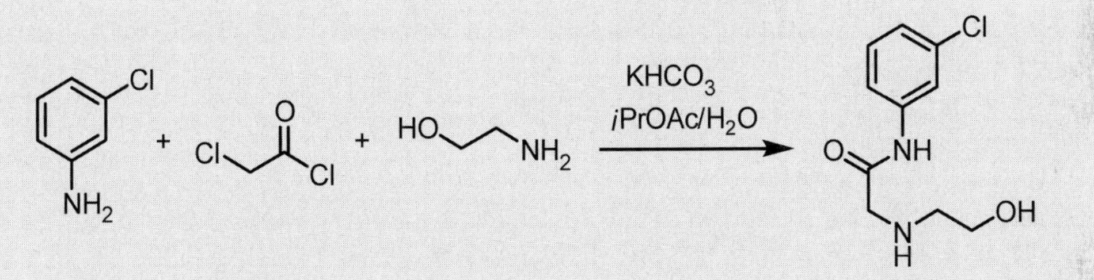

Chloroacetyl chloride (3.58 g, 31.7 mmol) is added dropwise to a stirred biphasic mixture of 3-chloroaniline (3.00 g, 23.5 mmol) in isopropyl acetate (23 mL) and potassium hydrogencarbonate (3.91 g) in H₂O (16 mL) at below 10 °C. The organic phase is separated, treated with ethanolamine (4.7 mL, 31.7 mmol), and the resulting mixture is heated to 60 °C for 1 h.

After the addition of water (7 mL), the organic phase is separated and cooled to 5 °C over 1 h. A crystalline precipitate is formed, which is collected by filtration, washed with isopropyl acetate (2 × 5 mL), and dried *in vacuo*. The yield is 3.69 g (69%) of colorless crystals, mp 99–101 °C.

IR (FT-IR, solid): $\tilde{v}$ (cm⁻¹) = 3310 (OH), 3057, 2930, 2873, 1683 (O=CNHR), 1593, 1542, 1418, 1056, 770, 679.

¹H NMR ([D₆]DMSO): δ (ppm) = 10.1 (s_br, 1 H, CO–NH), 7.84 (dd (t), *J* = 2.1 Hz, 1 H, Ar-H), 7.51 (ddd, *J* = 8.1, 2.1, 0.9 Hz, 1 H, Ar-H), 7.32 (dd (t), *J* = 8.1 Hz, 1 H, Ar-H), 7.10 (ddd, *J* = 8.1, 2.1, 0.9 Hz, 1 H, Ar-H), 4.62 (s_br, 1 H), 3.46 (t, *J* = 5.5 Hz, 2 H, CH₂CH₂OH), 3.29 (s, 2 H, CO–CH₂), 2.60 (t, *J* = 5.5 Hz, 2 H, CH₂CH₂OH).

¹³C NMR ([D₆]DMSO): δ (ppm) = 171.0, 140.2, 133.1, 130.4, 123.0, 118.7, 117.6, 60.4, 52.8,

3.3.4.6   **   1-(3-Chlorophenyl)piperazin-2-one · HCl [1]

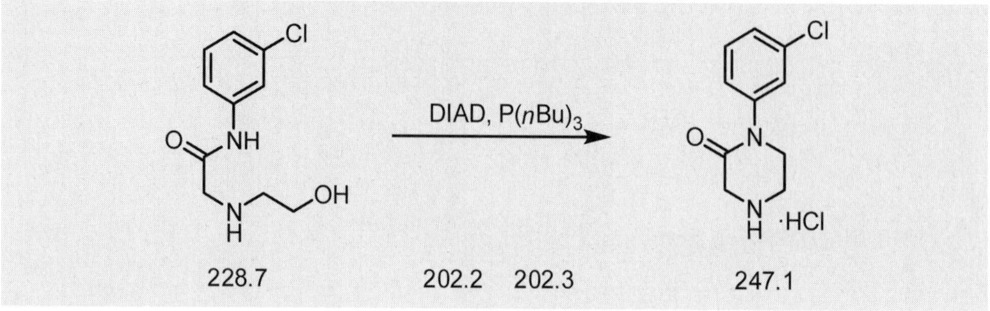

228.7   202.2   202.3   247.1

Diisopropyl azodicarboxylate (DIAD) (7.20 g, 35.6 mmol) is added dropwise to a stirred solution of tri-*n*-butylphosphine (90%) (8.00 g, 35.6 mmol) in EtOAc (15 mL), keeping the temperature below 0 °C (cooling with an ice/salt bath). Stirring is continued for 30 min at 0 °C, and then the resulting yellow solution is added dropwise over 1 h to a stirred slurry of the amide **3.3.4.5** (6.00 g, 26.2 mmol) in EtOAc (35 mL), maintaining the temperature below 5 °C. The solution is warmed to room temperature over 1 h, then to 40 °C, whereupon 3.55 M anhydrous ethanolic HCl (7.3 mL, 26.2 mmol, note) is added over 1 h.

The resulting slurry is cooled to 0 °C within 1 h; the deposited hydrochloride is collected by filtration, washed with ice-cold EtOAc (2 × 10 mL), and dried *in vacuo* to yield 4.40 g (68%) of a colorless crystalline solid, mp 230–232 °C.

IR (FT-IR, solid): $\tilde{\nu}$ (cm^{-1}) = 3052, 2646, 1655 (O=CONR$_2$), 1590, 1494, 1406, 1333, 889, 785, 695, 513.

^{1}H NMR ([D$_6$]DMSO): δ (ppm) = 10.1 (s$_{br}$, 2 H, NH), 7.47 (dd (t), J = 2.1 Hz, 1 H, Ar-H), 7.44 (m$_c$, 1 H, Ar-H), 7.38 (ddd, J = 8.1, 2.1, 0.9 Hz, 1 H, Ar-H), 7.31 (ddd, J = 8.1, 2.1, 0.9 Hz, 1 H, Ar-H), 3.91 (t, J = 5.7 Hz, 2 H, CH$_2$), 3.83 (s, 2 H, CO–CH$_2$), 3.50 (t, J = 5.7 Hz, 2 H, CH$_2$).

^{13}C NMR ([D$_6$]DMSO): δ (ppm) = 162.1, 142.7, 132.9, 130.7, 127.0, 126.0, 46.1, 44.9, 39.9.

Note: The ethanolic HCl is prepared by passing gaseous hydrogen chloride (16.6 g, 455 mmol, lecture bottle) into anhydrous EtOH (128.1 mL) with stirring and ice-cooling.

3.3.4.7   *   4-{5-[4-(3-Chlorophenyl)-3-oxo-piperazin-1-ylmethyl]imidazol-1-ylmethyl} benzonitrile [1]

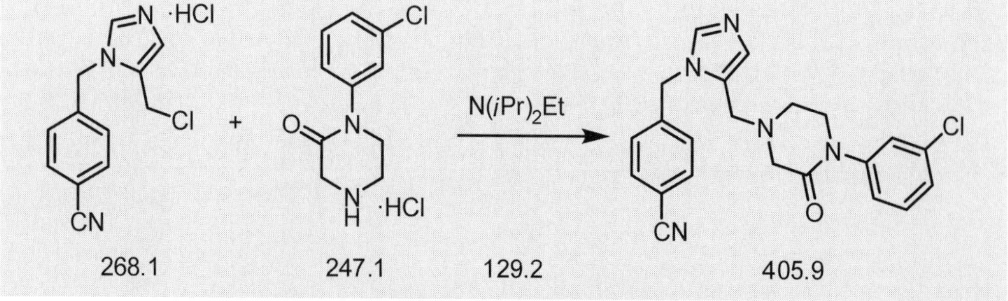

268.1   247.1   129.2   405.9

A mixture of the salts **3.3.4.4** (990 mg, 3.70 mmol) and **3.3.4.6** (890 mg, 3.60 mmol) in CH_3CN (5 mL) and ethyldiisopropylamine (1.85 mL) is stirred for 30 h at 0 °C.

H_2O (15 mL) is then added to give a light-brown precipitate, which is collected by filtration (note). It is washed with CH_3CN/H_2O (1:5; 6 mL) and CH_3CN/H_2O (1:9; 2 × 5 mL) and dried to yield 1.06 g (73%) of the desired product, mp 140–141 °C.

IR (solid): $\tilde{v}$ (cm^{-1}) = 2812, 2230 (C≡N), 1651 (O=CNR$_2$), 1423, 1338, 1320, 1105, 1077, 820, 783, 698, 663, 550.

^{1}H NMR ([D$_6$]DMSO): δ (ppm) = 7.83 (s, 1 H, N–CH=N), 7.79 (d, J = 8.2 Hz, 2 H, Ar-H), 7.40 (dd (t), J = 8.1 Hz, 1 H, Ar-H), 7.35 (dd (t), J =1.9 Hz, 1 H, Ar-H), 7.30 (ddd, J = 8.1, 1.9, 1.0 Hz, 1 H, Ar-H), 7.28 (d, J = 8.2 Hz, 2 H, Ar-H), 7.20 (ddd, J = 8.1, 1.9, 1.0 Hz, 1 H, Ar-H), 6.92 (s, 1 H, C=CH–N), 5.39 (s, 2 H, Ar–CH$_2$), 3.44 (s, 2 H), 3.32 (t, J = 5.4 Hz, 2 H), 3.02 (s, 2 H), 2.60 (t, J = 5.4 Hz, 2 H).

^{13}C NMR ([D$_6$]DMSO): δ (ppm) = 165.7, 144.0, 142.3, 139.5, 132.8, 132.3, 130.3, 129.4, 127.5, 126.8, 126.2, 125.6, 124.0, 118.6, 110.0, 56.7, 49.2, 48.8, 48.1, 47.3.

Note: If the product does not precipitate, it can be extracted with CH_2Cl_2 and purified after removal of the solvent by column chromatography (SiO$_2$; CH_2Cl_2/MeOH, 95:5, R_f = 0.33).

[1]   a) P. E. Maligres, M. S. Waters, S. A. Weissman, J. C. McWilliams, S. Lewis, J. Cowen, R. A. Reamer, R. P. Volante, P. J. Reider, D. Askin, *J. Heterocycl. Chem.* **2003**, *40*, 229. b) For an overview on farnesyltransferase inhibitors, see: I. M. Bell, *J. Med. Chem.* **2004**, *47*, 1869. c) Recently, structurally simple FTPase inhibitors have been shown to arrest the growth of malaria parasites: M. P. Glenn, S.-Y. Chang, O. Hucke, C. L. M. J. Verlinde, K. Rivas, C. Hornéy, K. Yokoyama, F. S. Buckner, P. R. Pendyala, D. Chakrabarti, M. Gelb, W. C. Van Voorhis, S. M. Sebti, A. D. Hamilton, *Angew. Chem.* **2005**, *117*, 4981; *Angew. Chem. Int. Ed.* **2005**, *44*, 4903.

[2]   Th. Eicher, S. Hauptmann, *The Chemistry of Heterocycles*, 2nd ed., p. 172, Wiley-VCH, Weinheim, **2003**.

[3]   The Delepine reaction (cf. *Houben-Weyl*, Vol. XI/1, p. 106) is useful for the preparation of benzylamines from benzyl halides.

[4]   4-Cyanobenzyl bromide (**4**) is available commercially or can be prepared by NBS bromination of 4-cyanotoluene: a) L. Wen, M. Li, J. B. Schlenoff, *J. Am. Chem. Soc.* **1997**, *33*, 7726; b) *Organikum*, 21st ed., p. 205, Wiley-VCH, Weinheim, **2001**.

[5]   D. L. Hughes, *Org. React.* **1992**, *42*, 335.

[6]   Mitsunobu reaction in catalytic fashion: T. Y. S. But, P. H. Toy, *J. Am. Chem. Soc.* **2006**, *128*, 9636.

3.3.5 (±)-Dihydrexidine

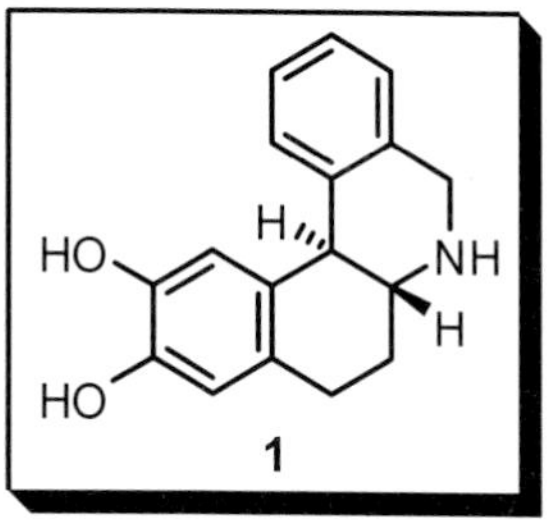

Topics:
- Synthesis of a dopamine agonist
- Iodination (S_EAr)
- Esterification, catalytic hydrogenation
- Heck reaction
- Dieckmann cyclization
- Ester hydrolysis, Krapcho cleavage of a β-keto ester
- Enamine formation with a primary amine and subsequent N-acylation
- Photocyclization of an enamide
- Reduction R–CO–NH$_2$ → R–CH$_2$–NH$_2$ by diborane
- Catalytic debenzylation, demethylation with HBr

(a) General

(±)-Dihydrexidine (**1**, *trans*-10,11-dihydroxy-5,6,6a,7,8,12b-hexahydrobenzo[*a*]phenanthridine) is a highly potent and selective agonist of the dopamine D$_1$ receptor [1, 2]. For its synthesis, 6,7-dimethoxy-2-tetralone (**3**) is a suitable substrate, which can be prepared by several methods [1]. The most widely used procedure is based on a Friedel–Crafts reaction of the acid chloride of (3,4-dimethoxyphenyl)acetic acid (**2**) with ethylene followed by cyclization in the presence of AlCl$_3$ [3]:

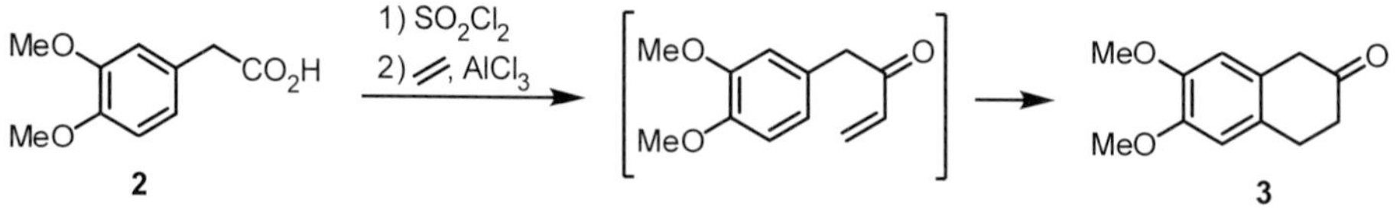

However, this approach suffers from preparative difficulties and a modest overall yield. Here, a new synthesis of the β-tetralone **3** starting from **2** is described [4], which raises the overall yield to 35%.

(b) Synthesis of 1

In the first part of the synthesis, (3,4-dimethoxyphenyl)acetic acid (**2**) is chemoselectively iodinated with iodine monochloride to give the iodo acid **4**, which is esterified by reaction with SOCl$_2$ in MeOH to afford the methyl ester **5** [4]. Heck reaction (cf. **1.6.1**) of **5** with methyl acrylate in the presence of NEt$_3$ and dichlorobis(triphenylphosphine)palladium(II) – a highly stable and low-cost Pd(II) catalyst, which is transformed *in situ* into the required Pd(0) catalyst Pd(PPh$_3$)$_2$ – affords the cinnamate **6** in 86% yield. This is catalytically hydrogenated over Pd/C to yield the propionate **7** in an overall yield of 62% over four steps from **2**. The subsequent Dieckmann cyclization of the diester **7** using KO*t*Bu proceeds chemoselectively due to the higher acidity of the benzylic CH$_2$–CO$_2$R group and leads to the β-keto ester **8**. Removal of the ester moiety in **8** with H$_2$O/DMSO/LiCl according to the Krapcho method [5] leads to the desired tetralone **3**. It can be purified by column chromatography on polyamide or via its bisulfite adduct [4]:

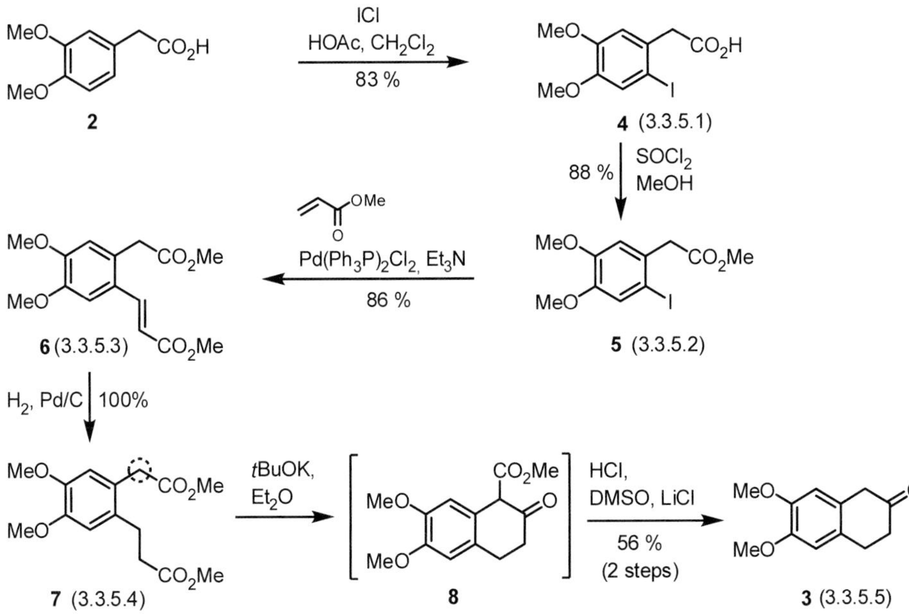

In the second part of the synthesis, the β-tetralone **3** is reacted with benzylamine to give the enamine **9**, which is *N*-acylated with benzoyl chloride. Without isolation, the resulting enamide **10** is irradiated in THF solution to give the tetracyclic lactam **11** as the product of a photocyclization (cf. **1.8.2**):

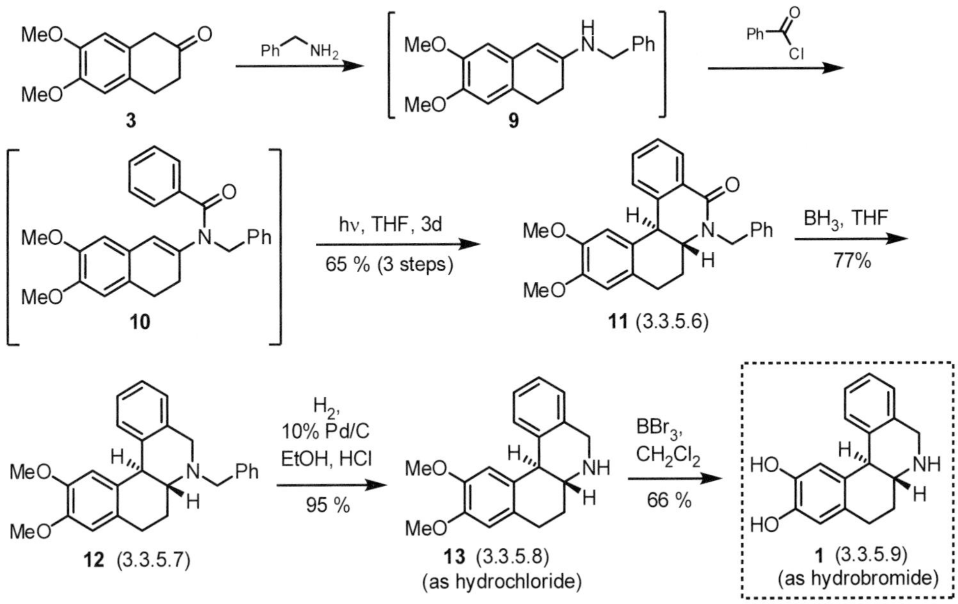

Since the relative stereochemistry of the hydrogen atoms at the B/C ring junction in the cyclization product **11** was shown to be *trans* [1], a plausible interpretation of the formation of **11** might be a photochemically allowed conrotatory 6π-electrocyclization of **10** to give **14** followed by a thermal suprafacial 1,5-sigmatropic hydrogen shift leading to **11** [6].

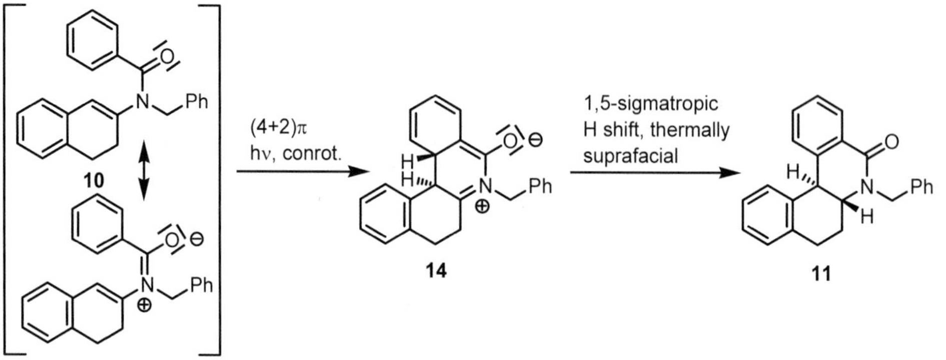

Finally, the lactam **11** is reduced by diborane in THF to give the tertiary amine **12**, which is subjected to *N*-deprotection by catalytic debenzylation with H_2 over Pd/C in EtOH/HCl and to *O*-deprotection by demethylation with BBr_3 in CH_2Cl_2 to afford **1** via **13**. For reasons of crystallizability, the final product *trans*-**1** is isolated as its hydrobromide.

Thus, the target molecule **1** is obtained in five steps from **3** in an overall yield of 31% or in eleven steps from **2** in an overall yield of 11%.

It should be noted that the originally described demethylation of **13** with HBr gives only low yields.

(c) Experimental procedures for the synthesis of 1

3.3.5.1 ∗∗ (2-Iodo-4,5-dimethoxyphenyl)acetic acid [4]

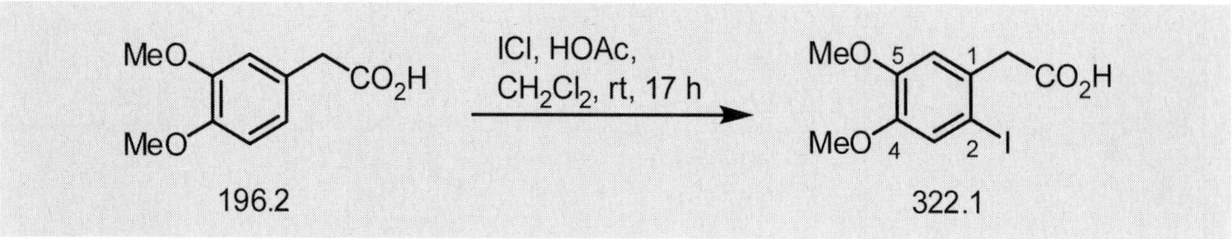

Iodine monochloride (87.8 g, 540 mmol) is added dropwise from a dropping funnel over 3 h to a stirred solution of (3,4-dimethoxyphenyl)acetic acid (100 g, 510 mmol) in anhydrous CH_2Cl_2 (850 mL) and glacial acetic acid (100 mL) at room temperature under an argon atmosphere. Stirring is continued for 17 h at the same temperature.

The reaction is then quenched by the addition of saturated aqueous sodium thiosulfate solution (400 mL). The organic layer is separated, washed with sodium thiosulfate solution (2 × 400 mL) as well as HCl (2 M, 1 × 400 mL), and dried over $MgSO_4$. Any precipitate of the product in the separatory funnel is dissolved by adding some CH_2Cl_2. The solvent is evaporated under reduced pressure, the residue is suspended in Et_2O (500 mL), and the suspension is stirred for 30 min. The iodo acid is obtained as a white solid by filtration; 137 g (83%), mp 165–166 °C, $R_f = 0.44$ (*n*-pentane/ EtOAc, 1:1).

UV (CH$_3$CN): λ_{max} (nm) (lg ε) = 285 (3.462), 239 (4.082), 211 (4.579).

IR (KBr): $\tilde{v}$ (cm^{-1}) = 3546, 3334, 3006, 2938, 2592, 1708, 1507, 1463, 1384, 1325, 1166, 1019, 860.

^{1}H NMR (300 MHz, [D$_6$]acetone): δ (ppm) = 7.33 (s, 1 H, 3-H), 7.08 (s, 1 H, 6-H), 3.85 (s, 3 H, OCH$_3$), 3.83 (s, 3 H, OCH$_3$), 3.77 (s, 2 H, CH$_2$).

^{13}C NMR (50 MHz, [D$_6$]acetone): δ (ppm) = 171.2 (CO$_2$H), 150.0 (C-5), 149.3 (C-4), 131.1 (C-2), 122.0 (C-6), 114.6 (C-3), 88.80 (C-1), 55.79 (5-OCH$_3$), 55.54 (4-OCH$_3$), 45.03 (CH$_2$).

MS (EI, 70 eV): m/z (%) = 322 (100) $[M]^+$, 277 (72) $[M-CO_2H]^+$, 195 (44) $[M-I]^+$, 150 (11) $[M-CO_2H-I]^+$.

3.3.5.2   *   (2-Iodo-4,5-dimethoxyphenyl)acetic acid methyl ester [4]

322.1   336.1

Thionyl chloride (60 mL) is added dropwise from a dropping funnel over 3 h to a stirred solution of the iodo acid **3.3.5.1** (110 g, 342 mmol) in anhydrous MeOH (800 mL) at room temperature under an argon atmosphere. Stirring is continued for 15 h at the same temperature.

Thereafter, the solvent is evaporated under reduced pressure, and the residue is dissolved in CH$_2$Cl$_2$ (400 mL) and treated with saturated NaHCO$_3$ solution. After stirring for 15 min at room temperature, the organic layer is separated and subsequently washed with saturated NaHCO$_3$ solution (2 × 200 mL), H$_2$O (1 × 200 mL), and brine (1 × 200 mL). The organic layer is dried over MgSO$_4$, filtered, and the solvent is evaporated under reduced pressure. Recrystallization of the residue from EtOAc/n-hexane affords the methyl ester as white needles; 101 g (88%), mp 77–78 °C, R_f = 0.27 (n-pentane/Et$_2$O, 2:1).

UV (CH$_3$CN): λ_{max} (nm) (lg ε) = 285 (3.468), 239 (4.090), 211 (4.574).

IR (KBr): $\tilde{v}$ (cm^{-1}) = 3079, 2992, 2934, 1722, 1507, 1437, 1329, 1218, 1165, 1029.

^{1}H NMR (300 MHz, CDCl$_3$): δ (ppm) = 7.21 (s, 1 H, 3-H), 6.79 (s, 1 H, 6-H), 3.83 (s, 6 H, 2 × OCH$_3$), 3.72 (s, 2 H, CH$_2$), 3.70 (s, 3 H, CO$_2$CH$_3$).

^{13}C NMR (50 MHz, CDCl$_3$): δ (ppm) = 171.1 (CO$_2$CH$_3$), 149.2 (C-5), 148.5 (C-4), 129.8 (C-2), 121.4 (C-6), 113.1 (C-3), 88.73 (C-1), 56.03 (5-OCH$_3$), 55.82 (4-OCH$_3$), 52.06 (CO$_2$CH$_3$), 45.43 (CH$_2$).

MS (EI, 70 eV): m/z (%) = 336 (100) $[M]^+$, 277 (92) $[M-CO_2CH_3]^+$, 209 (83) $[M-I]^+$, 150 (10) $[M-CO_2CH_3-I]^+$.

3.3.5.3 ** (*E*)-3-(4,5-Dimethoxy-2-methoxycarbonylmethyl-phenyl)acrylic acid methyl ester [4]

A solution of the methyl ester **3.3.5.2** (80.0 g, 248 mmol), methyl acrylate (86.0 mL, 82.0 g, 953 mmol), and NEt$_3$ (100 mL, 72.6 g, 718 mmol) in anhydrous CH$_3$CN (300 mL) is thoroughly deoxygenated by bubbling an argon stream through it for 45 min, and then dichloro-bis(triphenylphosphine)palladium(II) (1.00 g, 1.42 mmol, 0.6 mol%) is added. The reaction mixture is heated to reflux for 5 h.

The solvent is then evaporated under reduced pressure and the residue is taken up in EtOAc (1000 mL). The solution is washed with H$_2$O (1 × 400 mL), HCl (2 M, 2 × 400 mL), and again with H$_2$O (1 × 400 mL). After drying over MgSO$_4$, the solution is concentrated *in vacuo* to a volume of 300 mL and, after the addition of charcoal (1.0 g), is heated to boiling for decolorization. The mixture is filtered and the solvent is evaporated under reduced pressure. Crystallization of the residue from EtOH (200 mL) affords the acrylic ester as colorless needles; 62.7 g (86%), mp 96–97 °C, R_f = 0.45 (*n*-pentane/EtOAc, 7:3).

UV (CH$_3$CN): λ_{max} (nm) (lg ε) = 327 (4.187), 296 (4.158), 238 (4.066), 219 (4.160).

IR (KBr): $\tilde{\nu}$ (cm^{-1}) = 3082, 2998, 2953, 1730, 1603, 1516, 1428, 1272, 1095, 1001, 860.

^{1}H NMR (300 MHz, CDCl$_3$): δ (ppm) = 7.87 (d, *J* = 15.8 Hz, 1 H, 3-H), 7.07 (s, 1 H, 6'-H), 6.73 (s, 1 H, 3'-H), 6.28 (d, *J* = 15.8 Hz, 1 H, 2-H), 3.88 (s, 6 H, 2 × OCH$_3$), 3.78 (s, 3 H, 2-CO$_2$CH$_3$), 3.72 (s, 2 H, 1''-H$_2$), 3.68 (s, 3 H, 1''-CO$_2$CH$_3$).

^{13}C NMR (75 MHz, CDCl$_3$): δ (ppm) = 171.4 (C-1), 167.3 (C-2''), 150.7 (C-5'), 148.4 (C-4'), 141.3 (C-3'), 127.2 (C-2'), 125.9 (C-1'), 117.4 (C-2), 113.4 (C-3), 108.8 (C-6'), 55.89 (4'-OCH$_3$), 55.84 (5'-OCH$_3$), 52.14 (2-CO$_2$CH$_3$), 51.59 (1''-CO$_2$CH$_3$), 37.93 (C-1'').

MS (EI, 70 eV): *m/z* (%) = 294 (100) [*M*]$^+$, 262 (21) [*M* – OCH$_3$ – H]$^+$, 234 (30) [*M* – CO$_2$CH$_3$ – H]$^+$, 221 (22) [*M* – CHCO$_2$CH$_3$ – H]$^+$, 203 (50) [*M* – CO$_2$CH$_3$ – OCH$_3$ – H]$^+$, 175 (51) [*M* – CO$_2$CH$_3$ – CO$_2$CH$_3$ – H]$^+$, 161 (17) [*M* – CO$_2$CH$_3$ – CO$_2$CH$_3$ – CH$_3$]$^+$, 59 (18) [CO$_2$CH$_3$]$^+$.

3.3.5.4   **   3-(4,5-Dimethoxy-2-methoxycarbonylmethyl-phenyl)propionic acid methyl ester [4]

The acrylic ester **3.3.5.3** (50.0 g, 170 mmol) is dissolved in hot EtOH (1000 mL) and 10% palladium on charcoal (5.00 g) is added. The flask is flushed with hydrogen for 30 min and hydrogenation is carried out at ambient pressure until hydrogen uptake ceases (approx. 27 h).

The suspension is then filtered through a pad of Celite® (approx. 1 cm), which is subsequently washed with Et$_2$O (1000 mL). The solvents are evaporated under reduced pressure to afford the product as a colorless oil in quantitative yield; 50.3 g (100%), R_f = 0.45 (n-pentane/EtOAc, 7:3).

UV (CH$_3$CN): λ_{max} (nm) (lg ε) = 284 (3.527), 233 (3.957), 203 (4.671).

IR (KBr): $\tilde{\nu}$ (cm^{-1}) = 3449, 2998, 2953, 2849, 1736, 1610, 1521, 1437, 1276, 1162, 1099, 1013.

^{1}H NMR (300 MHz, CDCl$_3$): δ (ppm) = 6.71 (s, 1 H, 3'-H), 6.69 (s, 1 H, 6'-H), 3.84 (s, 6 H, 2 × OCH$_3$), 3.68 (s, 3 H, 1"-CO$_2$CH$_3$), 3.66 (s, 3 H, 2-CO$_2$CH$_3$), 3.60 (s, 2 H, 1"-H$_2$), 2.90 (t, J = 7.9 Hz, 2 H, 3-H$_2$), 2.57 (t, J = 7.9 Hz, 2 H, 2-H$_2$).

^{13}C NMR (75 MHz, CDCl$_3$): δ (ppm) = 173.3 (C-1), 172.2 (C-2"), 148.2 (C-5'), 147.4 (C-4'), 131.3 (C-1'), 124.0 (C-2'), 113.5 (C-6'), 112.3 (C-3'), 55.86 (4'-OCH$_3$), 55.82 (5'-OCH$_3$), 52.03 (1"-CO$_2$CH$_3$), 51.61 (2-CO$_2$CH$_3$), 37.87 (C-1"), 35.34 (C-2), 27.68 (C-3).

MS (EI, 70 eV): m/z (%) = 296 (100) [M]$^+$, 264 (56) [M – OCH$_3$ – H]$^+$, 237 (47) [M – CO$_2$CH$_3$]$^+$, 223 (27) [M – CH$_2$CO$_2$CH$_3$]$^+$, 165 (51) [M – CO$_2$CH$_3$ – CH$_2$CO$_2$CH$_3$ + H]$^+$.

3.3.5.5   **   6,7-Dimethoxy-3,4-dihydro-1H-naphthalen-2-one (6,7-dimethoxy-β-tetralone) [4]

A solution of the propionic ester **3.3.5.4** (45.7 g, 154 mmol) in anhydrous Et$_2$O (500 mL) is added dropwise to a well stirred suspension of KOtBu (19.0 g, 170 mmol) in Et$_2$O (1000 mL) over 1.5 h at room temperature under an argon atmosphere. Stirring is continued for 1 h, and then the resulting suspension is filtered and the filter cake is washed with Et$_2$O (500 mL). The potassium enolate thus collected is dried in high vacuum and is obtained in quantitative yield.

A solution of the potassium salt (10.0 g, 33.1 mmol) and anhydrous LiCl (1.68 g, 39.7 mmol) in anhydrous DMSO is deoxygenated by bubbling an argon stream through it for 15 min, and then concentrated HCl (3.30 mL, 40.0 mmol) is rapidly added with stirring. The flask is placed in an oil bath, preheated to 125 °C, and stirring is continued at this temperature for 5 h.

After cooling the reaction mixture, it is diluted with EtOAc (500 mL) and the organic layer is washed with H_2O (3 × 200 mL) and dried over $MgSO_4$, and the solvent is evaporated under reduced pressure. The residue is purified by column chromatography on polyamide (15% EtOAc/n-pentane) and then recrystallized from EtOAc/n-hexane to afford the β-tetralone as a white solid; 3.81 g (56%), mp 85–86 °C; R_f = 0.45 (n-pentane/EtOAc, 7:3).

UV (CH_3CN): λ_{max} (nm) (lg ε) = 285 (3.571), 202 (4.597).

IR (KBr): $\tilde{v}$ (cm^{-1}) = 3014, 2998, 2958, 2851, 1717, 1515, 1462, 1346, 1248, 1113, 880.

^{1}H NMR (300 MHz, $CDCl_3$): δ (ppm) = 6.74 (s, 1 H, 8-H), 6.62 (s, 1 H, 5-H), 3.88 (s, 3 H, OCH_3), 3.86 (s, 3 H, OCH_3), 3.51 (s, 2 H, 1-H_2), 3.00 (t, J = 6.7 Hz, 2 H, 4-H_2), 2.55 (t, J = 6.7 Hz, 2 H, 3-H_2).

^{13}C NMR (75 MHz, $CDCl_3$): δ (ppm) = 210.8 (C-2), 147.9 (C-7), 147.7 (C-6), 128.4 (C-4a), 125.0 (C-8a), 111.3 (C-8), 111.1 (C-5), 56.03 (2 × OCH_3), 44.20 (C-1), 38.58 (C-3), 28.11 (C-4).

MS (EI, 70 eV): m/z (%) = 206 (100) $[M]^+$, 164 (40) $[M – C_2H_2O]^+$, 57 (57), 43 (21), 41 (28).

3.3.5.6 ** *trans*-6-Benzyl-10,11-dimethoxy-5,6,6a,7,8,12b-hexahydrobenzo[*a*]-phenanthridin-5-one [1]

A solution of the tetralone **3.3.5.5** (1.03 g, 5.00 mmol) and benzylamine (0.56 g, 5.23 mmol) in toluene (20 mL) is heated to reflux for 5 h under an argon atmosphere with continuous removal of H_2O using a Dean–Stark trap. The solution is then cooled to about 80 °C and benzoyl chloride (0.77 g, 5.48 mmol) and triethylamine (0.56 g, 5.53 mmol) are added dropwise. The resulting mixture is stirred at room temperature for 2 h.

Toluene is then removed under reduced pressure and CH_2Cl_2 (100 mL) and H_2O (50 mL) are added. The mixture is shaken, the phases are separated, and the aqueous layer is extracted with CH_2Cl_2

(50 mL). The combined organic phases are dried over Na_2SO_4. Column chromatography on silica gel yields the crude enamide as a yellow solid (2.01 g). Small amounts of remaining 6,7-dimethoxy-β-tetralone do not interfere with the next step.

To remove the remaining 6,7-dimethoxy-β-tetralone, sodium borohydride (50 mg) is added to a solution of the crude product in EtOH (50 mL). The mixture is stirred for 30 min at 50 °C, the EtOH is evaporated under reduced pressure, and CH_2Cl_2 (100 mL) and H_2O (30 mL) are added. The organic layer is separated, dried over Na_2SO_4, and the solvent is removed under reduced pressure. The residue is dissolved in THF (60 mL) and irradiated for 3 d with a 300 W mercury high-pressure lamp in a ring reactor.

The product is purified by column chromatography on silica gel and recrystallized from Et_2O/ n-pentane to give the phenanthridin-5-one as colorless needles; 1.29 g (65% over three steps), mp 191–195 °C, R_f = 0.16 (n-pentane/EtOAc 3:1).

UV (CH_3CN): λ_{max} (nm) (lg ε) = 282.0 (3.76), 200.0 (4.88).

IR (KBr): $\tilde{v}$ (cm^{-1}) = 2934, 1655, 1514, 1460, 1403, 1261, 1115, 1023, 742.

^{1}H NMR (300 MHz, $CDCl_3$): δ (ppm) = 8.20 (m, 1 H, Ar-H), 7.53 (m, 1 H, Ar-H), 7.49–7.39 (m, 2 H, 2 × Ar-H), 7.32–7.19 (m, 5 H, 5 × Ar-H), 6.92 (s, 1 H, Ar-H), 6.63 (s, 1 H, Ar-H), 5.34 (d, J = 15.9 Hz, 1 H, 1'-H_b), 4.78 (d, J = 15.9 Hz, 1 H, 1'-H_a), 4.36 (d, J = 11.4 Hz, 1 H, 12b-H), 3.89, 3.87 (2s, 6 H, OCH_3), 3.78 (m_c, 1 H, 6^a-H), 2.67 (m_c, 2 H, 8-H_2), 2.26 (m_c, 1 H, 7-H_b), 1.75 (m_c, 1 H, 7-H_a).

^{13}C NMR (75 MHz, $CDCl_3$): δ (ppm) = 166.2 (C-5), 147.6, 146.8, 141.5, 138.5, 131.2, 129.4, 123.6 (C-4a, C-2', C-8a, C-10, C-11, C-12a, C-12c), 130.9, 129.1, 128.6, 126.8, 126.5, 122.8 (C-1, C-2, C-3, C-4, C-3', C-4', C-5', C-6', C-7'), 112.7, 111.7 (C-9, C-12), 59.99 (C-6a), 56.03 (OCH_3), 55.76 (OCH_3), 45.84 (C-12b), 45.06 (C-1'), 29.09 (C-8), 26.20 (C-7).

MS (ESI): m/z (%) = 821.0 (100) $[2M + Na]^+$, 1219.6 (74) $[3M + Na]^+$, 422.2 (15) $[M + Na]^+$.

3.3.5.7   **   *trans*-6-Benzyl-10,11-dimethoxy-5,6,6a,7,8,12b-hexahydrobenzo[a]-phenanthridine [1]

A solution of borane·THF complex in THF (1 M, 5.50 mL, 5.50 mmol) is slowly added to a stirred ice-cold solution of the phenanthridin-5-one **3.3.5.6** (734 mg, 1.84 mmol) in anhydrous THF (60 mL) and then the reaction mixture is heated to reflux for 16 h.

After cooling to room temperature, H_2O (6 mL) is slowly added and the solvent is evaporated under reduced pressure. The residue is taken up in toluene (30 mL), methanesulfonic acid (0.6 mL) is added, and the resulting mixture is heated to 70 °C with stirring for 1 h. The mixture is then diluted with H_2O (25 mL) and the aqueous layer is separated. The toluene layer is extracted with HCl (6 M, 4 × 30 mL) and the combined aqueous phases are cooled in an ice bath and slowly basified with concentrated NH_4OH (120 mL). The free base is extracted with CH_2Cl_2 (3 × 50 mL) and the combined organic phases are dried over $MgSO_4$. After filtration, the organic solution is concentrated under reduced pressure to afford the N-benzylphenanthridine as a yellow solid; 545 mg (77%), mp (HCl salt) 230–232 °C, R_f = 0.43 (n-pentane/EtOAc, 2:1).

UV (CH_3CN): λ_{max} (nm) (lg ε) = 286.5 (3.614), 194.5 (4.890), 192.5 (4.898).

IR (KBr): $\tilde{v}$ (cm^{-1}) = 3442, 3006, 2944, 2854, 1608, 1514, 1463, 1347, 1257, 1236, 1195, 1125, 1087, 1014, 874, 765, 701.

^{1}H NMR (300 MHz, $CDCl_3$): δ (ppm) = 7.44–7.08 (m, 9 H, 1-H, 2-H, 3-H, 4-H, 3'-H, 4'-H, 5'-H, 6'-H, 7'-H), 6.89 (s, 1 H, 12-H), 6.73 (s, 1 H, 9-H), 4.06 (d, J = 10.6 Hz, 1 H, 12b-H), 3.89 (s, 3 H, OCH_3), 3.95–3.82 (m, 2 H, 5-H_b, 1'-H_b), 3.78 (s, 3 H, OCH_3), 3.52 (d, J = 15.2 Hz, 1 H, 1'-H_a), 3.29 (d, J = 13.3 Hz, 1 H, 5-H_a), 2.86 (m_c, 2 H, 8-H_2), 2.35 (m_c, 1 H, 6a-H), 2.22 (m_c, 1 H, 7-H_b), 2.03–1.86 (m, 1 H, 7-H_a).

^{13}C NMR (75 MHz, $CDCl_3$): δ (ppm) = 147.2, 146.7, 139.5, 137.6, 136.0, 130.5, 129.7 (C-4a, C-8a, C-10, C-11, C-12a, C-12c, C-2'), 129.0 (C-3', C-7'), 128.3 (C-4', C-6'), 127.2, 127.0, 126.6, 126.3, 126.0 (C-1, C-2, C-3, C-4, C-5'), 111.7, 110.9 (C-9, C-12), 65.30 (C-6a), 57.70 (C-5), 56.03 (OCH_3), 55.93 (OCH_3), 53.37 (C-1'), 43.43 (C-12b), 28.11 (C-8), 27.59 (C-7).

MS (ESI): m/z (%) = 386 (100) $[M + H]^+$.

3.3.5.8 ** trans-10,11-Dimethoxy-5,6,6a,7,8,12b-hexahydrobenzo[a]phenanthridine hydrochloride [1]

A suspension of the N-benzylphenanthridine **3.3.5.7** (392 mg, 1.02 mmol) in EtOH (100 mL) is carefully acidified with concentrated HCl (0.40 mL) and then the solvent is removed under reduced pressure. The residue is dissolved in EtOH (80 mL), 10% Pd/C catalyst (100 mg) is added, and the mixture is shaken at room temperature under hydrogen (3.5 bar) for 8 h.

After removal of the catalyst by filtration, the solvent is evaporated under reduced pressure. The residue is recrystallized from CH₃CN/MeOH to afford the hydrochloride as a light-yellow crystalline solid; 320 mg (95%), mp 238–239 °C, $R_f = 0.62$ (CH₂Cl₂/MeOH, 4:1).

UV (MeOH): λ_{max} (nm) (lg ε) = 285.0 (3.585), 204.5 (4.687).

IR (KBr): $\tilde{\nu}$ (cm⁻¹) = 3420, 2937, 2775, 1607, 1515, 1446, 1205, 1128, 1092, 1038, 871, 750.

¹H NMR (300 MHz, [D₆]DMSO): δ (ppm) = 10.0 (s$_{br}$, 2 H, NH₂), 7.45–7.25 (m, 4 H, 1-H, 2-H, 3-H, 4-H), 6.87, 6.84 (2s, 2 H, 9-H, 12-H), 4.36 (s, 2 H, 5-H₂), 4.26 (d, J = 10.9 Hz, 1 H, 12b-H), 3.75 (s, 3 H, OCH₃), 3.68 (s, 3 H, OCH₃), 2.95 (m$_c$, 1 H, 6a-H), 2.88–2.69 (m, 2 H, 8-H₂), 2.28–2.14 (m, 1 H, 7-H$_b$), 2.08–1.90 (m, 1 H, 7-H$_a$).

¹³C NMR (75 MHz, [D₆]DMSO): δ (ppm) = 147.6, 146.7, 137.2, 130.5, 129.5 (C-8a, C-10, C-11, C-12a, C-12c), 127.8, 127.6, 126.8, 125.2 (C-1, C-2, C-3, C-4), 124.9 (C-4a), 112.6, 112.0 (C-9, C-12), 56.55 (C-6a), 55.63 (OCH₃), 55.52 (OCH₃), 43.47 (C-5), 40.60 (C-12b), 26.89 (C-8), 25.20 (C-7).

MS (ESI): m/z (%) = 296 (100) $[M - \text{Cl}]^+$.

3.3.5.9   **   *trans*-10,11-Dihydroxy-5,6,6a,7,8,12b-hexahydrobenzo[*a*]phenanthridine hydrobromide [1]

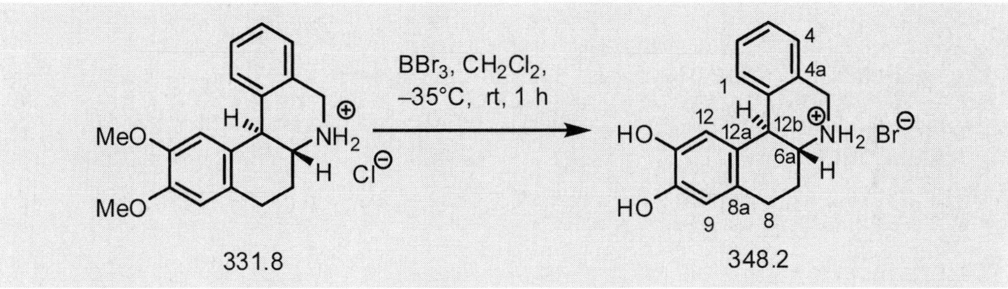

The phenanthridine hydrochloride **3.3.5.8** (83.0 mg, 0.25 mmol) is taken up in saturated aqueous NaHCO₃ solution (10 mL) and the liberated free amine is extracted with CH₂Cl₂ (3 × 10 mL). The combined organic phases are washed with brine, dried over Na₂SO₄, and concentrated *in vacuo*.

Boron tribromide (1 M in CH₂Cl₂, 0.75 mL, 0.75 mmol) is added dropwise to a solution of the residue (73.8 mg, 0.25 mmol) in CH₂Cl₂ (5 mL) at –35 °C. The mixture is stirred at room temperature for 1 h.

Et₂O (4 mL) and MeOH (0.1 mL) are then added, the solvents are removed under reduced pressure, and Et₂O is added to precipitate the crude product. Recrystallization from CH₃CN affords the desired (±)-dihydrexidine hydrobromide as yellow needles; 57 mg (66%), mp 185–186 °C, $R_f = 0.10$ (CH₂Cl₂/MeOH, 10:1).

UV (CH$_3$OH): λ_{max} (nm) (lg ε) = 288.5 (3.55), 202.5 (4.63).

IR (KBr): $\tilde{v}$ (cm^{-1}) = 3224, 2937, 1521, 1276, 750.

^{1}H NMR (300 MHz, [D$_6$]DMSO): δ (ppm) = 9.59 (s$_{br}$, 2 H, NH$_2$), 9.38 (s$_{br}$, 2 H, OH), 7.37 (m, 4 H, Ar-H), 6.74 (s, 1 H, Ar-H), 6.64 (s, 1 H, Ar-H), 4.40 (s, 2 H, 5-H$_2$), 4.19 (d, J = 10.8 Hz, 1 H, 12b-H), 2.99 (m, 1 H, 6a-H), 2.73 (m, 2 H, 8-H$_2$), 2.19 (m, 1 H, 7-H$_b$), 1.94 (m, 1 H, 7-H$_a$).

^{13}C NMR (75 MHz, [D$_6$]DMSO): δ (ppm) = 144.0, 143.1, 136.3, 130.3, 127.7, 124.3 (C-4a, C-8a, C-10, C-11, C-12a, C-12c), 127.7, 127.5, 126.8, 126.2 (C-1, C-2, C-3, C-4), 115.9, 114.6 (C-9, C-12), 56.74 (C-6a), 43.99 (C-5), 40.30 (C-12b), 26.29 (C-8), 25.31 (C-7).

MS (ESI): m/z (%) = 268.1 (100) [M + H]$^+$.

[1]   W. K. Brewster, D. E. Nichols, R. M. Riggs, D. M. Mottola, T. W. Lovenberg, M. H. Lewis, R. B. Mailman, *J. Med. Chem.* **1990**, *33*, 1756.

[2]   T. A. Knoerzer, V. J. Watts, D. E. Nichols, R. B. Mailman, *J. Med. Chem.* **1995**, *38*, 3062.

[3]   a) Alex A. Cordi, J.-M. Lacoste, J.-J. Descombes, C. Courchay, P. M. Vanhoutte, M. Laubie, T. J. Verbeuren, *J. Med. Chem.* **1995**, *38*, 4056; b) compare: L. F. Tietze, Th. Eicher, *Reaktionen und Synthesen im organisch-chemischen Praktikum und Forschungslaboratorium*, 2nd ed., p. 279, Georg Thieme Verlag, Stuttgart, **1991**; c) for a review on approaches to 2-tetralones, see: C. C. Silveira, A. L. Braga, T. S. Kaufman, E. J. Lenardão, *Tetrahedron* **2004**, *38*, 8295.

[4]   A. M. Qandil, D. W. Miller, D. E. Nichols, *Synthesis* **1999**, 2033.

[5]   A. Kouvrakis, H. E. Katerinopoulos, *Synth. Commun.* **1995**, *25*, 3035; Krapcho cleavage leads to (formal) decarboxylation of β-keto esters, malonates, and cyanoacetates.

[6]   M. B. Smith, J. March, *March's Advanced Organic Chemistry*, 5th ed., p. 1426 and p. 1436, John Wiley & Sons, Inc., New York, **2001**.

3.4   Condensed heterocycles

3.4.1   6-Ethoxycarbonylnaphtho[2,3-*a*]indolizine-7,12-quinone

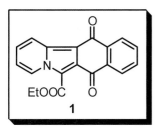

1

Topics:
- Indolizine synthesis
- N–CH–EWG-substituted pyridinium betaines as 1,3-dipoles in 1,3-dipolar cycloaddition
- Chemoselective cleavage of *tert*-butyl esters
- Formation of a cyclic dicarboxylic anhydride
- Regioselective cleavage of an anhydride by a Grignard compound
- Formation of a quinone by intramolecular Friedel–Crafts acylation

(a) General

Indolizine (**2**), one of the three benzopyrroles (**2**–**4**), constitutes the core structure of many naturally occurring alkaloids. The chemistry of indolizines is not as well known as that of indoles (**3**); however, it has attracted much interest in recent years [1, 2]. On the other hand, isoindoles (**4**) are usually less stable compounds due to their *ortho*-quinoid structure, though **4** can be isolated as colorless needles at low temperature.

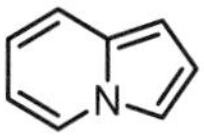

indolizine	indole	isoindole
benzo[a]pyrrole	benzo[b]pyrrole	benzo[c]pyrrole
2	**3**	**4**

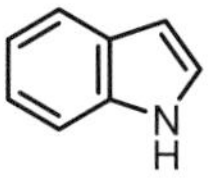

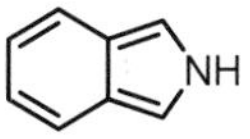

There are three important methods for the synthesis of indolizines:

(1) 2-Methyl-*N*-phenacylpyridinium ions **5**, easily accessible from 2-picolines by alkylation, cyclize upon treatment with a base to give 2-phenylindolizines **7**. Evidently, of the two CH-acidic centers in **5**, the 2-methyl group is deprotonated selectively to provide the enamines **6** as intermediates, which lead to the indolizines **7** by intramolecular aldol condensation [3]:

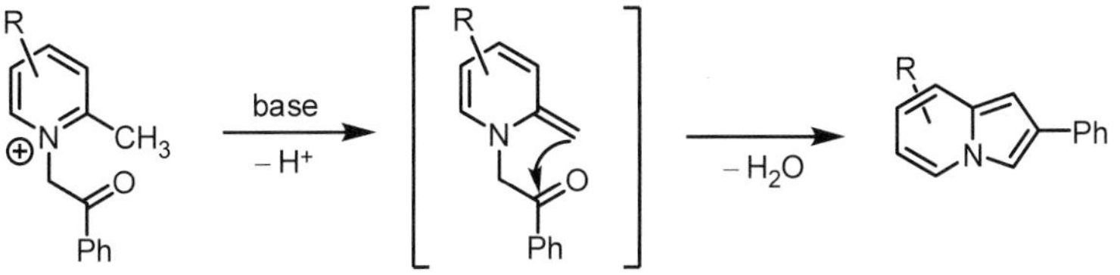

5   **6**   **7**

(2) The Baylis–Hillman reaction (cf. **1.1.6**) of pyridine-2-carbaldehyde **8** with acceptor-substituted alkenes yields the adducts **9a**; the corresponding acetate derivatives **9b** lead to 2-acceptor-substituted indolizines **10** by a thermally induced intramolecular cyclization [4]:

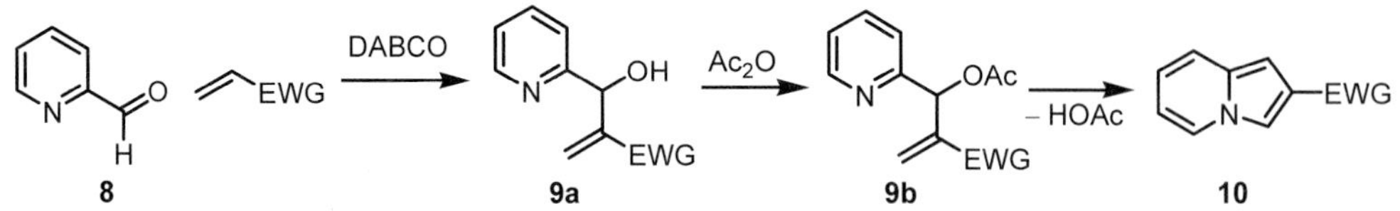

With only a few exceptions, methods (1) and (2) are restricted to the formation of 2-substituted indolizines and are therefore of limited scope and applicability.

(3) EWG-CH-substituted pyridinium-*N*-betaines **12** undergo 1,3-dipolar cycloaddition with acetylene dicarboxylate, propiolate, or maleinate, or with electron-deficient alkenes as dipolarophiles. With alkynes, the cycloadducts (**13/14**) undergo spontaneous dehydrogenation to give indolizines, which are of the 1,2,3-trisubstituted type **16** or (indicating a regioselective cycloaddition) of the 1,3-disubstituted type **17** [5–9]. With olefinic substrates, the presence of an oxidant for additional dehydrogenation of the primary cycloadducts (e.g. **15 → 16**) is required [10]:

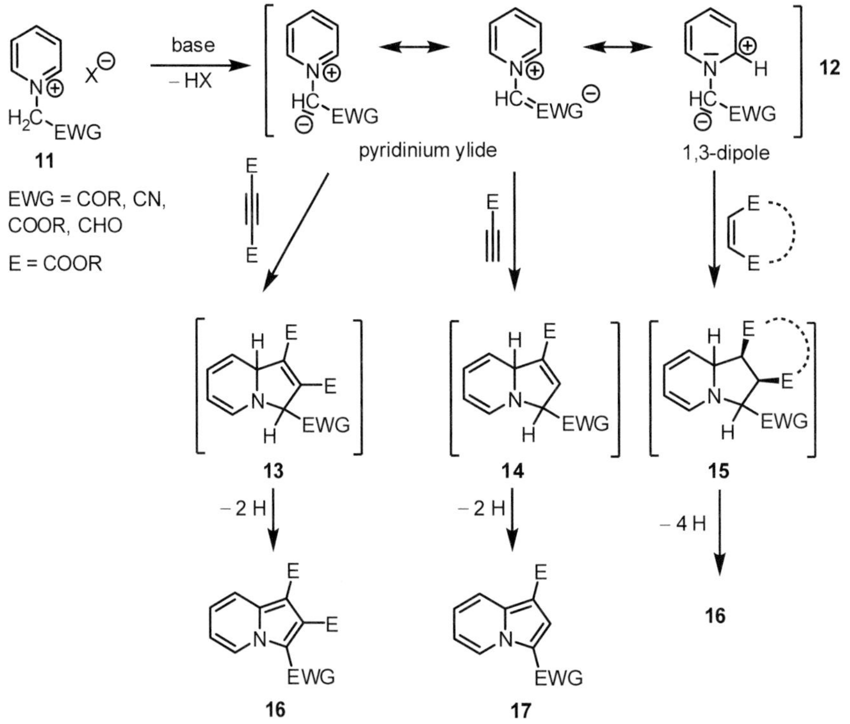

Since the pyridinium ylides **12** result from deprotonation of the corresponding *N*-alkylpyridinium ions **11**, cycloaddition is easily performed with **11**, the dipolarophile, and a base, preferentially in PTC [5,6] or microwave-assisted [9] MCR versions.

Here, method (3) is chosen for the preparation of the indolizine **1**.

The 1,3-dipolar cycloaddition is a very versatile and effective method for the preparation of five-membered heterocycles [11]. In general, 1,3-dipolar cycloadditions proceed as concerted 6π-processes, in which a 1,3-dipole **19** (containing usually one or more nitrogen atoms) reacts with an electron-acceptor-substituted alkyne (a) or alkene (b), i.e.:

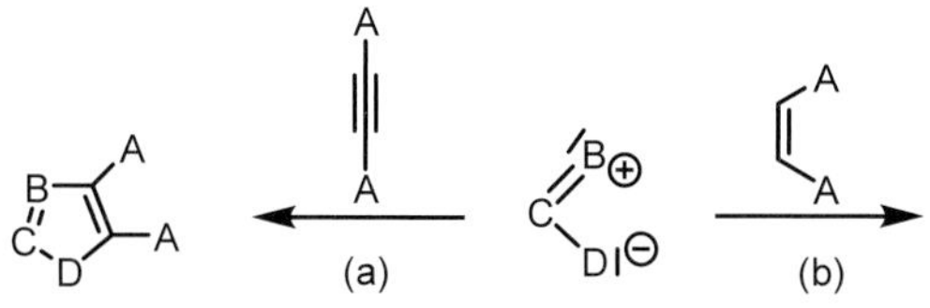

Electron-donor-substituted dipolarophiles can also be used, but generally give lower yields. In case (a), aromatic azoles **18** are the products; in case (b), dihydroazoles **20** result, in which the alkene configuration is transferred to the sp^3 centers in the product, thus indicating that the cycloaddition proceeds in a stereoselective manner.

Classical examples of (a) are the formation of pyrazoles/1,2-oxazoles/1,2,3-triazoles (**21–23**) by 1,3-dipolar cycloaddition of diazoalkanes/nitrile oxides/azides to alkynes:

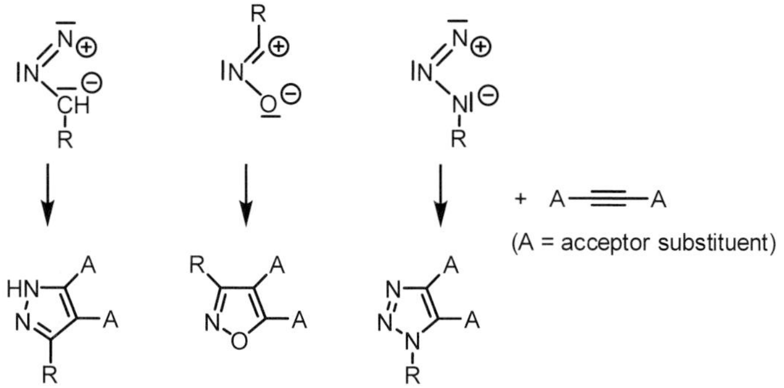

Other types of 1,3-dipoles are represented by nitrones **24**, azomethine ylides **25**, and mesoionic compounds **26**, which also undergo cycloaddition to olefinic and acetylenic dipolarophiles to allow the synthesis of a large number of different heterocyclic systems [12]:

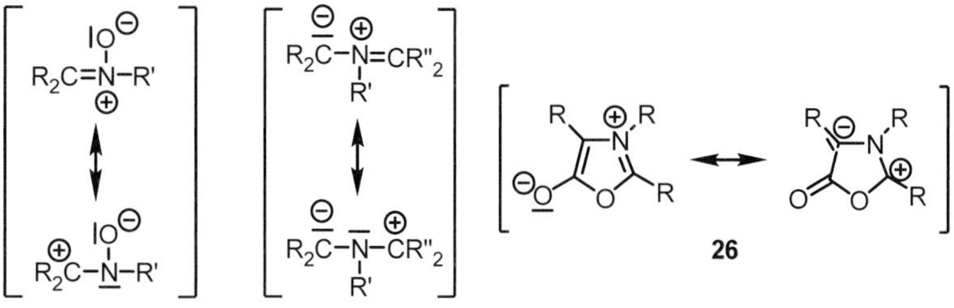

In this context, indolizine formation **12** → **16/17** can be classified as 1,3-dipolar cycloaddition with a dipole of the azomethine ylide type.

(b) Synthesis of 1

For the synthesis of **1** containing a quinone moiety, the indolizine-2,3-dicarboxylic acid **30** was chosen as key intermediate. It was planned to prepare the quinone moiety in analogy to anthraquinone formation from phthalic anhydride via *o*-benzoylbenzoic acid [13].

The synthesis of **1** [14] starts with the reaction of (*N*-ethoxycarbonylmethyl)pyridinium bromide (**27**), readily available from pyridine and ethyl bromoacetate [15], with di-*tert*-butyl acetylenedicarboxylate in the presence of K_2CO_3 as a base. Under these conditions, the indolizine-1,2,3-triester **29** is obtained by a 1,3-dipolar cycloaddition of the alkyne to the initially formed pyridinium ylide **28**:

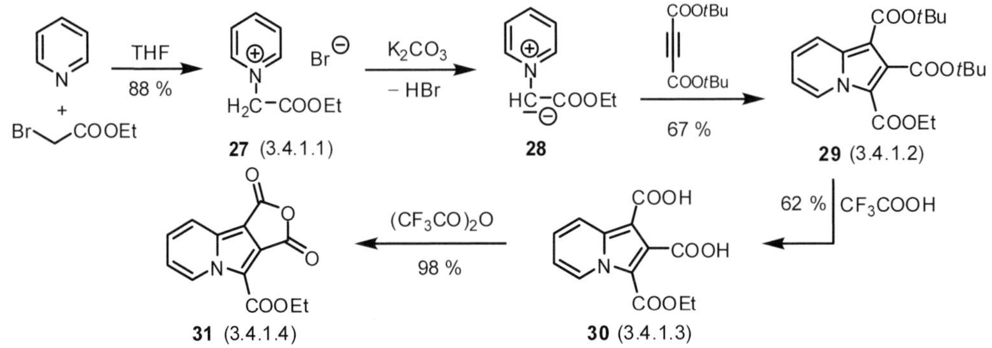

Then, the *tert*-butyl ester groups in **29** are cleaved chemoselectively using CF_3COOH to give the dicarboxylic acid **30**, which is transformed into the anhydride **31** on treatment with trifluoroacetic anhydride.

Reaction of the anhydride **31** with phenylmagnesium bromide in THF at –78 °C leads exclusively to the 2-benzoylindolizine-1-carboxylic acid **32**; thus, the regioisomeric acid **33** is not formed. The reason for this unexpected regioselectivity of anhydride opening (cf. **1.1.4**) is not known. However, one can assume that a complexation of the Grignard reagent with the adjacent less reactive COOEt group favors attack at the carbonyl group at C-2.

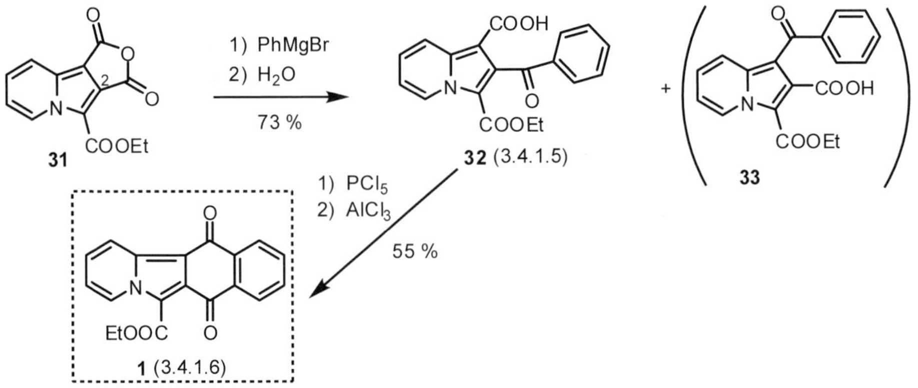

Finally, treatment of the carboxylic acid **32** with PCl_5 followed by $AlCl_3$ affords the quinone **1** by an intramolecular Friedel–Crafts acylation via the intermediately formed acid chloride.

Thus, the target molecule **1** is obtained from the pyridinium salt **27** in a six-step sequence in an overall yield of 16%.

(c) Experimental procedures for the synthesis of 1

3.4.1.1 * *N*-(Ethoxycarbonylmethyl)pyridinium bromide [15]

79.1   167.0   246.1

Ethyl bromoacetate (6.68 g, 45.0 mmol) is added dropwise to a stirred solution of anhydrous pyridine (3.56 g, 45.0 mmol) in anhydrous THF (200 mL) and the mixture is stirred at 25 °C for 12 h. The solvent is distilled off to give the crude product as a beige solid; 8.65 g (88%), mp 125–127 °C.

FT-IR: $\tilde{\nu}$ (cm^{-1}) = 1737 (C=O).

^{1}H NMR (500 MHz, [D$_6$]DMSO): δ (ppm) = 9.09 (d, J = 6.6 Hz, 2 H, Ar-H), 8.72 (t, J = 7.9 Hz, 1 H, Ar-H), 8.25 (t, J = 7.9 Hz, 2 H, Ar-H), 5.71 (s, 2 H, N–CH$_2$), 4.24 (q, J = 7.3 Hz, 2 H, CH$_2$), 1.25 (t, J = 7.3 Hz, 3 H, CH$_3$).

^{13}C NMR (125 MHz, [D$_6$]DMSO): δ (ppm) = 166.4, 146.8, 146.2, 127.8, 62.3, 60.3, 13.9.

3.4.1.2 ** Di-*tert*-butyl 3-ethoxycarbonylindolizine-1,2-dicarboxylate [14]

246.1   226.3   138.2   389.4

Anhydrous K$_2$CO$_3$ (7.28 g, 52.7 mmol) and di-*tert*-butyl acetylenedicarboxylate (7.91 g, 35.0 mmol) are added to a suspension of the pyridinium bromide **3.4.1.1** (8.61 g, 35.0 mmol) in anhydrous THF (340 mL) and the mixture is stirred at room temperature for 4 d.

The solid is then filtered off and the filtrate is concentrated *in vacuo* to give an oily residue (crude product, 11.9 g), which is purified by column chromatography on SiO$_2$ (*n*-hexane/Et$_2$O, 10:1, R_f = 0.36). The indolizine tricarboxylate is obtained as a yellow solid; 9.13 g (67%), mp 124–126 °C.

FT-IR: $\tilde{\nu}$ (cm^{-1}) = 1734 (C=O, COOEt), 1677 (C=O, COO*t*Bu).

^{1}H NMR (500 MHz, CDCl$_3$): δ (ppm) = 9.55 (dt, J = 7.3, 1.0 Hz, 1 H, Ar-H), 8.19 (dt, J = 9.1, 1.0 Hz, 1 H, Ar-H), 7.28 (ddd, J = 9.1, 7.0, 1.2 Hz, 1 H, Ar-H), 6.96 (dt, J = 7.0, 1.2 Hz, 1 H, Ar-H), 4.42 (q, J = 7.0 Hz, 2 H, OCH$_2$), 1.65 (s, 9 H, C(CH$_3$)$_3$), 1.63 (s, 9 H, C(CH$_3$)$_3$), 1.39 (t, J = 7.0 Hz, 3 H, CH$_3$).

^{13}C NMR (125 MHz, [D$_6$]DMSO): δ (ppm) = 164.4, 162.1, 160.6, 137.0, 132.2, 127.9, 125.8, 120.1, 114.7, 111.7, 105.3, 82.4, 81.1, 60.7, 28.6, 28.3, 14.7.

3.4.1.3   *   3-Ethoxycarbonylindolizine-1,2-dicarboxylic acid [14]

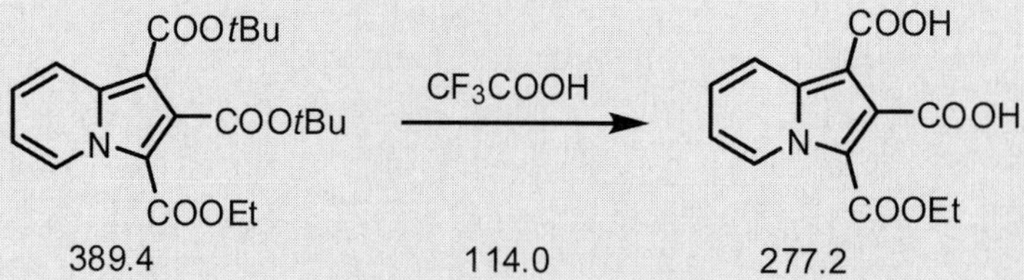

389.4   114.0   277.2

Trifluoroacetic acid (15.7 g, 137 mmol) is added to a solution of the indolizine tricarboxylate **3.4.1.2** (5.35 g, 13.7 mmol) in CH$_2$Cl$_2$ (55 mL) and the mixture is stirred at room temperature overnight.

The orange-brown solid that separates is collected by filtration, washed with *n*-hexane (40 mL), and dried *in vacuo* to yield the monocarboxylate as a yellow solid; 2.35 g (62%), mp 207–208 °C.

FT-IR: $\tilde{v}$ (cm^{-1}) = 1701 (C=O, COOEt), 1658 (C=O, COOH).

^{1}H NMR (500 MHz, [D$_6$]DMSO): δ (ppm) = 12.86 (s, 2 H, OH), 9.41 (dt, *J* = 7.0, 1.0 Hz, 1 H, Ar-H), 8.28 (dt, *J* = 8.8, 1.2 Hz, 1 H, Ar-H), 7.51 (ddd, *J* = 8.8, 6.7, 0.9 Hz, 1 H, Ar-H), 7.23 (td, *J* = 7.0, 1.2 Hz, 1 H, Ar-H), 4.31 (q, *J* = 7.0 Hz, 2 H, OCH$_2$), 1.28 (t, *J* = 7.0 Hz, 3 H, CH$_3$).

^{13}C NMR (125 MHz, [D$_6$]DMSO): δ (ppm) = 166.0, 163.8, 159.7, 137.2, 132.2, 127.6, 127.1, 119.3, 115.7, 110.6, 102.5, 60.5, 13.9.

3.4.1.4   *   3-Ethoxycarbonylindolizine-1,2-dicarboxylic anhydride [14]

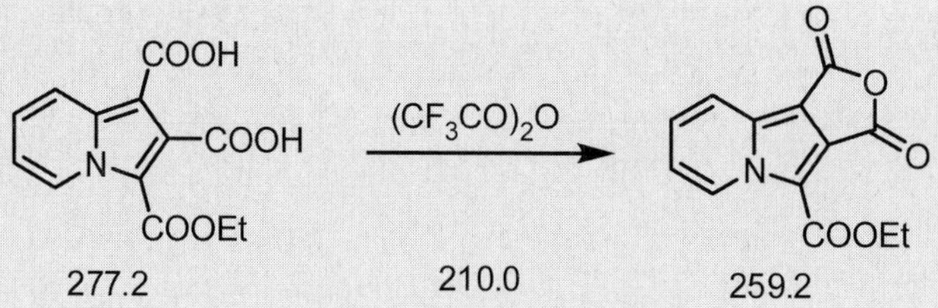

277.2   210.0   259.2

A suspension of the 1,2-dicarboxylic acid **3.4.1.3** (2.08 g, 7.50 mmol) in CH$_2$Cl$_2$ (25 mL) and trifluoroacetic anhydride (4.73 g, 22.5 mmol) is heated to reflux for 2 h.

The solvent is then distilled off and the yellow-green residue is suspended in *n*-hexane/Et$_2$O (1:1). The solid is collected by filtration and dried *in vacuo* to give the anhydride as a yellow solid; 1.91 g (98%), mp 167–168 °C.

FT-IR: $\tilde{v}$ (cm^{-1}) = 1828 (C=O, anhydride), 1764 (C=O, anhydride), 1693 (C=O, ester).

^{1}H NMR (500 MHz, [D$_6$]DMSO): δ (ppm) = 9.70 (d, *J* = 7.3 Hz, 1 H, Ar-H), 7.98 (d, *J* = 9.1 Hz, 1 H, Ar-H), 7.57 (t, *J* = 8.9 Hz, 1 H, Ar-H), 7.26 (td, *J* = 8.0, 1.2 Hz, 1 H, Ar-H), 4.50 (q, *J* = 7.0 Hz, 2 H, OCH$_2$), 1.50 (t, *J* = 7.0 Hz, 3 H, CH$_3$).

^{13}C NMR (125 MHz, [D$_6$]DMSO): δ (ppm) = 159.3, 158.0, 157.4, 132.1, 129.6, 129.2, 128.4, 119.1, 117.5, 110.9, 109.5, 61.8, 14.2.

3.4.1.5   **   2-Benzoyl-3-ethoxycarbonylindolizine-1-carboxylic acid [14]

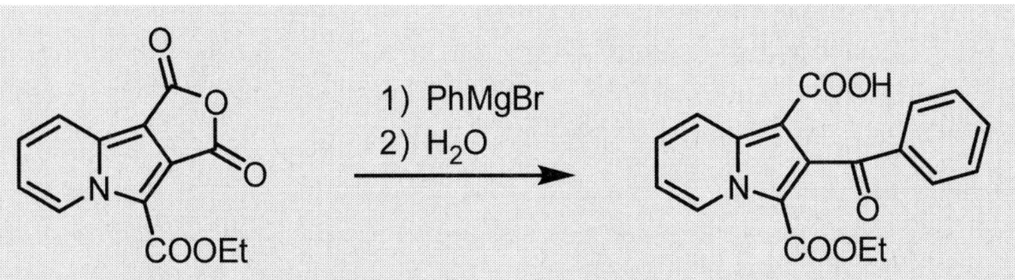

Phenylmagnesium bromide (15 mL, 15.0 mmol, 1.0 M solution in THF, Aldrich, note) is slowly added to the indolizine-1,2-dicarboxylic anhydride **3.4.1.4** (1.95 g, 7.50 mmol) in anhydrous THF (40 mL) at –78 °C and the brown mixture is stirred for 15 min. The mixture is then allowed to warm to 0 °C and stirred for an additional 30 min.

The reaction mixture is subsequently acidified with 10% HCl, the organic layer is separated, and the aqueous layer is extracted with CH_2Cl_2 (3 × 150 mL); insoluble material is filtered off. The combined organic layers are washed with H_2O (2 × 50 mL), dried (Na_2SO_4), and the solvent is evaporated. The brown solid (crude product: 2.02 g) is purified by column chromatography on silica gel ($CHCl_3$/MeOH, 50:1, R_f = 0.17) to give the benzoylindolizine as a yellow solid; 1.85 g (73%), mp 203–204 °C.

> **FT-IR**: $\tilde{\nu}$ (cm^{-1}) = 1654 (C=O, COOH), 1597 (C=O, COOEt).
>
> **^{1}H NMR** (500 MHz, $CDCl_3$): δ (ppm) = 9.61 (d, J = 7.3 Hz, 1 H, Ar-H), 8.40 (d, J = 9.1 Hz, 1 H, Ar-H), 7.85 (d, J = 7.3 Hz, 2 H, Ar-H), 7.55 (t, J = 7.3 Hz, 1 H, Ar-H), 7.45–7.39 (m, 3 H, Ar-H), 7.09 (td, J = 6.9, 1.2 Hz, 1 H, Ar-H), 4.08 (q, J = 7.3 Hz, 2 H, OCH_2), 0.86 (t, J = 7.3 Hz, 3 H, CH_3).
>
> **^{13}C NMR** (125 MHz, $CDCl_3$): δ (ppm) = 192.2, 168.2, 160.2, 139.0, 137.5, 137.4, 133.1, 129.3, 128.3, 128.2, 128.1, 127.3, 120.1, 115.5, 112.9, 60.8, 13.4.

Note: Phenylmagnesium bromide can be prepared by the standard procedure from bromobenzene and Mg turnings in THF.

3.4.1.6   **   6-Ethoxycarbonylnaphtho[2,3-*a*]indolizine-7,12-quinone [14]

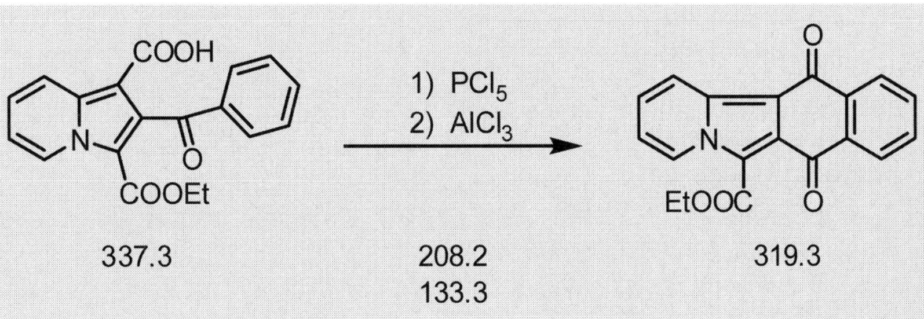

A suspension of the carboxylic acid **3.4.1.5** (0.41 g, 1.23 mmol) and phosphorus pentachloride (1.29 g, 6.19 mmol) in 1,2-dichloroethane (10 mL) is stirred at room temperature overnight. Aluminum(III) chloride (0.83 g, 6.22 mmol) is then added to the red-colored solution and the mixture is heated at 50 °C for 2 h (color change to green). Additional aluminum(III) chloride (0.33 g, 2.47 mmol) is added and the mixture is heated at 50 °C for a further 2 h.

The mixture is then diluted with H_2O (15 mL) and extracted with CH_2Cl_2 (3 × 75 mL). The combined organic layers are washed with water and dried over Na_2SO_4. The solvent is evaporated and the residue (crude product: 0.35 g) is purified by column chromatography (silica gel; *n*-hexane/EtOAc, 5:1, R_f = 0.29) to give the indolizinequinone as an orange solid, 0.21 g (55%), mp 151–152 °C.

FT-IR: $\tilde{\nu}$ (cm^{-1}) = 1691 (C=O; quinone), 1670 (C=O, quinone), 1640 (C=O, ester).

^{1}H NMR (500 MHz, [D$_6$]DMSO): δ (ppm) = 9.29 (dt, ^3J = 7.3, 1.0 Hz, 1 H, Ar-H), 8.63 (dt, J = 8.8, 1.1 Hz, 1 H, Ar-H), 8.22 (m$_c$, 2 H, Ar-H, Ar-H), 7.71 (m$_c$, 2 H, Ar-H), 7.44 (ddd, J = 8.8, 6.9, 1.0 Hz, 1 H, Ar-H), 7.10 (td, J = 6.9, 1.3 Hz, 1 H, Ar-H), 4.56 (q, J = 6.9 Hz, 2 H, OCH$_2$), 1.54 (t, J = 7.0 Hz, 3 H, CH$_3$).

^{13}C NMR (125 MHz, [D$_6$]DMSO): δ (ppm) = 180.2, 179.6, 161.3, 136.4, 135.5, 134.9, 133.4, 132.9, 128.0, 127.4, 127.5, 127.4, 126.1, 121.2, 117.1, 115.0, 112.3, 61.8, 14.2.

---

[1]   M. Shipman, *Science of Synthesis* **2001**, *10*, 745.

[2]   A. R. Katritzky, C. W. Rees, E. F. V. Scriven (Eds.), *Comprehensive Heterocyclic Chemistry II*, Vol. 8, p. 237, Elsevier, Oxford, **1996**.

[3]   a) F. Kröhnke, *Ber. Dtsch. Chem. Ges.* **1933**, *66*, 604; b) F. Kröhnke, *Angew. Chem.* **1953**, *65*, 605; c) A. E. Tschitschibabin, *Ber. Dtsch. Chem. Ges.* **1927**, *60*, 1607.

[4]   a) M. L. Bode, P. T. Kaye, *J. Chem. Soc., Perkin Trans. 1* **1993**, 1809; b) D. Basavaiah, A. J. Rao, *Chem. Commun.* **2003**, 604.

[5]   J. Alvarez-Builla, M. G. Quintanilla, C. Abril, M. T. Gandasegui, *J. Chem. Res. (S)* **1986**, 202.

[6]   M. T. Gandasegui, J. Alvarez-Builla, *J. Chem. Res. (S)* **1986**, 74.

[7]   L. Zhang, F. Liang, L. Sun, Y. Hu, H. Hu, *Synthesis* **2000**, 1733; when **11** (EWG = COOH) is reacted with ADE, additional decarboxylation occurs and 2,3-disubstituted indolizines (EWG = H) are obtained.

[8]   J. M. Minguez, J. J. Vaquero, J. Alvarez-Builla, O. Castano, *J. Org. Chem.* **1999**, *64*, 7788.

[9]   U. Bora, A. Saikia, R. C. Boruah, *Org. Lett.* **2003**, *5*, 435.

[10] X. Wei, Y. Hu, T. Li, H. Hu, *J. Chem. Soc., Perkin Trans. 1* **1993**, 2487.

[11] A. Padwa, W. H. Pearson (Eds.), *Synthetic Applications of 1,3-Dipolar Cycloaddition Chemistry Toward Heterocycles and Natural Products*, John Wiley & Sons, Inc., New York, **2002**.

[12] For details, see textbooks on heterocyclic chemistry, e.g. Th. Eicher, S. Hauptmann, *The Chemistry of Heterocycles*, 2nd ed., p. 129/148/183/204, Wiley-VCH, Weinheim, **2003**.

[13] L. F. Fieser, *Organic Syntheses, Coll. Vol. 1*, p. 517; *Vol. 4*, p. 73

[14] Y. Miki, N. Nakamura, R. Yamakawa, H. Hachiken, K. Matsushita, *Heterocycles* **2000**, *53*, 2143.

[15] a) P. N. Praveen Rao, H. Mini, H. Li, A. G. Habeeb, E. E. Knaus, *J. Med. Chem.* **2003**, *46*, 4872; b) Z. Dega-Szafran, G. Schroeder, M. Szafran, *J. Phys. Org. Chem.* **1999**, *12*, 39.

3.4.2   EGF-R-Pyrrolo[2,3-*d*]pyrimidine

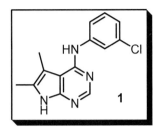

Topics:
- Technical heterocycle synthesis using domino and MCR processes
- Formation of a 2-amino-3-cyanopyrrole by Dakin–West reaction of alanine und subsequent cyclocondensation with malonodinitrile
- Pyrimidine formation by (modified) Remfry–Hull synthesis
- Dimroth rearrangement

(a) General

Derivatives of pyrrolo[2,3-*d*]pyrimidine have been shown to bind to the epidermal growth factor receptor (EGF-R) causing inhibition of tyrosine kinase. By variation of the substitution pattern, the selectivity and the biological profile could be optimized, leading to target molecule **1**, which has attained development status as an antitumor agent [1].

To support further biological profiling as well as to supply the drug for initial clinical trials, the first laboratory synthesis had to be elaborated to a technical large-scale process.

The research synthesis was conducted along the following lines:

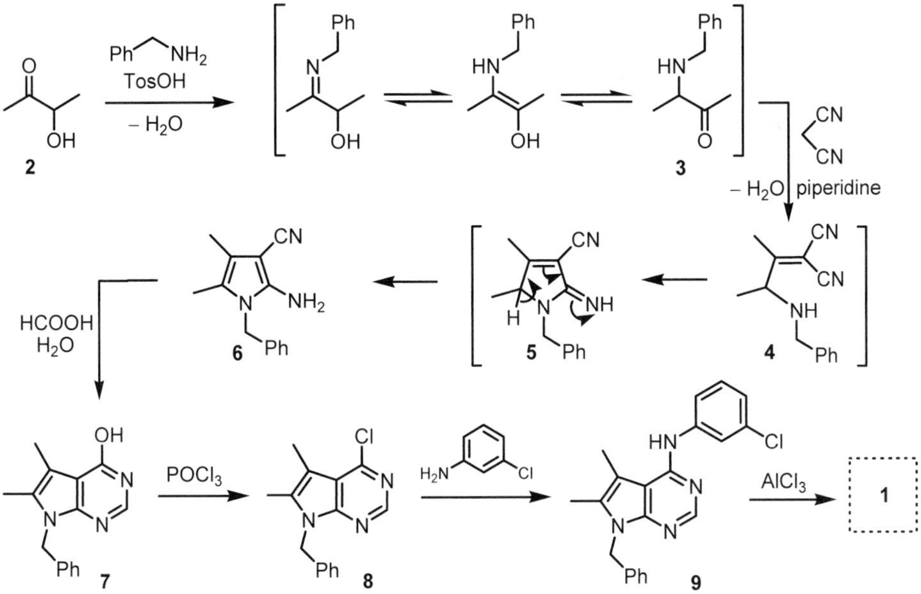

The α-hydroxy ketone **2** (acyloin) is first condensed with benzylamine in the presence of TosOH, then with malonodinitrile in the presence of piperidine to give the *N*-benzyl-protected 2-amino-3-cyanopyrrole **6** (via the α-amino ketone **3** and **4** formed by Knoevenagel condensation of **3** with CH₂(CN)₂, which cyclizes to give **5**). A pyrimidine ring is then annelated to the pyrrole by a modified

Remfry–Hull cyclocondensation [2] with aqueous formic acid (→ **7**). This is followed by replacement of the hydroxy group in **7** by a chloro substituent (**7** → **8**) and an S_NAr displacement of Cl by *m*-chloroaniline (**8** → **9**) to give the pyrrolopyrimidine **9**, *N*-debenzylation of which using AlCl₃ yields **1**.

However, several problems arose in scaling-up the synthesis, necessitating alteration of the following steps:

(1) The synthesis of **1** outlined above employs *N*-benzyl protection of the pyrrole moiety. However, removal of the benzyl group turned out to be technically difficult due to the requirement of a large excess of AlCl₃.

(2) The pyrimidine formation in boiling formic acid raised safety concerns and also resulted in dark coloration of the product, which necessitated a cumbersome purification.

(3) Low solubilities of the hydroxy and chloropyrimidine intermediates made it necessary to work in dilute solutions and to use a large excess of POCl₃.

Therefore, for the technical synthesis of **1**, a slightly different approach has been developed.

(1) It turned out to be favorable to still use a 2-amino-3-cyanopyrrole as key intermediate, but in the *N*-unprotected form **11**.

(2) A simple one-step synthesis of **11** is known [3]; moreover, **11** is sufficiently stable to be handled.

(3) For the formation of the pyrimidine moiety in **1**, formic acid was replaced by ethyl orthoformate as a suitable formic acid derivative.

The improved procedure [1] is presented in section (b).

(b) Synthesis of 1

The key intermediate 2-amino-3-cyano-4,5-dimethylpyrrole (**11**) can be prepared [3] by cyclocondensation of 3-amino-2-butanone (**10**) and malonodinitrile in the presence of a base.

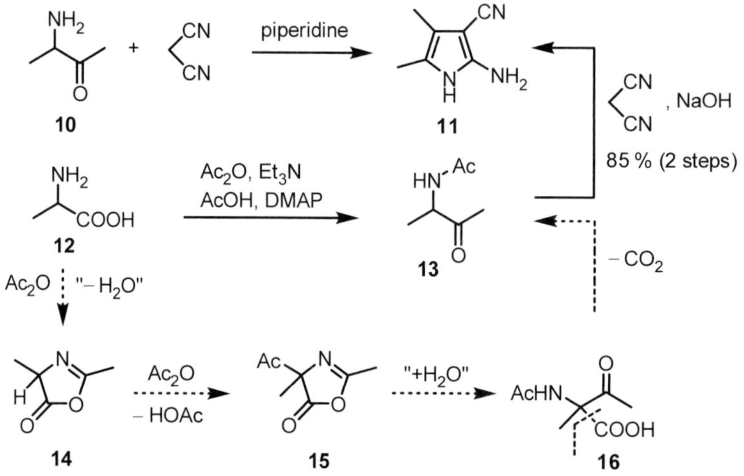

However, amino ketone **10** is unstable, and therefore it is replaced by the *N*-acetyl amino derivative **13**. The preparation of **13** by a Dakin–West reaction [4] of *rac*-alanine (**12**) can be efficiently combined with the reaction with malonodinitrile in a one-pot multi-component process (MCR [5]) to afford the pyrrole **11** in high yield.

In the Dakin–West reaction, α-acylamino ketones (e.g. **13**) are produced from α-amino acids and acid anhydrides. As mechanistic investigations [6] have shown, azlactones (e.g. **14**) are formed initially, which subsequently undergo C-acylation at C-5 (**14 → 15**), ring-opening hydrolysis to a β-keto acid (**15 → 16**), and decarboxylation (**16 → 13**).

For the annelation of the pyrimidine part to the pyrrole **11**, a one-pot three-component cyclocondensation was developed, in which **11**, triethyl orthoformate (as formic acid equivalent), and *m*-chloroaniline react to yield the iminopyrrolopyrimidine **17**:

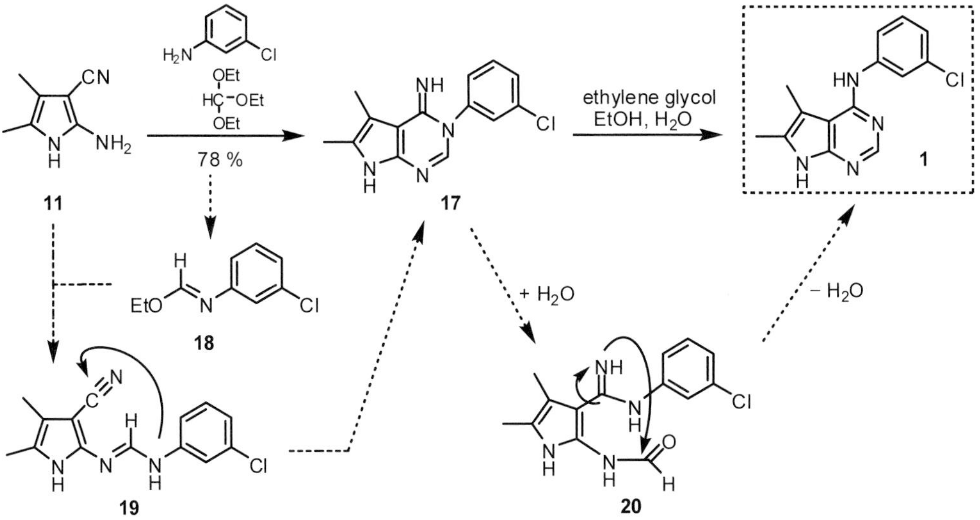

The formation of **17** can be rationalized by a domino process [7], in which an imidate **18** is first formed from the orthoester and *m*-chloroaniline; by reaction with the amino function of pyrrole **11** and prototropy, the imidate **18** is transformed into the amidine **19**, which cyclizes by intramolecular addition of the NH group to the nitrile function [8]. Finally, the formed 4-imino-3-arylpyrimidine **17** isomerizes to **1** containing a 4-aminoaryl moiety on heating it in an ethylene glycol/H$_2$O mixture. The isomerization **17 → 1**, which can be referred to as a Dimroth-type rearrangement, is not truly a rearrangement, but rather a hydrolysis – addition of H$_2$O and ring-opening – to give the amidine **20**, which undergoes recyclization by addition of the imino group to the formyl group followed by elimination of water.

In general, in Dimroth rearrangements [9], isomerizing interconversions of heterocycles take place by ring-opening and recyclization leading to an exchange of an "inner" with an "outer" heteroatom. Another instructive example is the thermal rearrangement of 5-amino-1-phenyl-1,2,3-triazole (**21**) to 5-phenylamino-1,2,3-triazole (**22**):

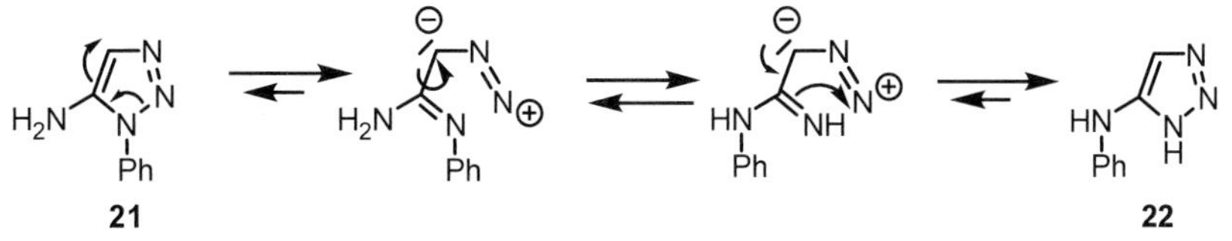

The new procedure led to an improvement in the preparation of **1** from the initial five-step laboratory synthesis to a three-step synthesis using multi-component domino processes to give **1** in 61% overall yield (based on *rac*-alanine (**12**)). Most importantly, the new approach could be employed on a large scale.

(c) Experimental procedures for the synthesis of 1

3.4.2.1   **   2-Amino-3-cyano-4,5-dimethylpyrrole [1]

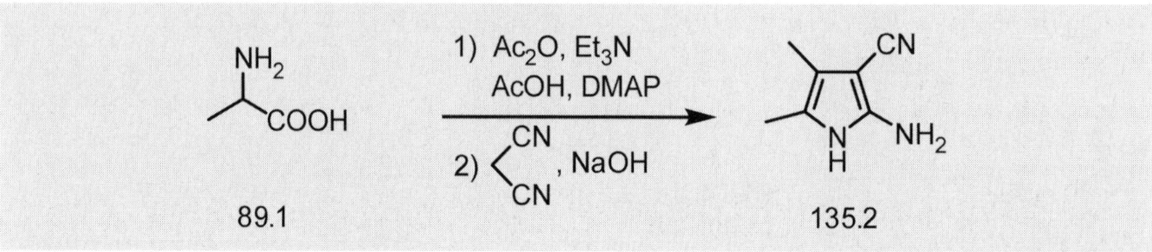

A mixture of acetic anhydride (16.9 g, 165 mmol), acetic acid (2.26 g, 37.6 mmol), triethylamine (19.0 g, 188 mmol), and 4-(dimethylamino)pyridine (0.10 g, 0.75 mmol) is heated to 50 °C. Then *D,L*-alanine (6.79 g, 76.2 mmol) is added in small portions over 4 h with stirring, keeping the reaction temperature between 45 and 55 °C; stirring of the red-colored mixture is continued for 8 h at 50 °C.

Acetic anhydride, acetic acid, and triethylamine are then distilled off (15–20 mbar), gradually increasing the bath temperature up to 100 °C. The residue is cooled to room temperature and diluted with water (40 mL). Malonodinitrile (4.71 g, 71.3 mmol) is added and the mixture is slowly poured into 30% NaOH solution (25 mL), keeping the temperature below 60 °C.

The mixture is cooled to 0 °C and the orange precipitate formed is filtered off, washed with water (45 mL), and dried *in vacuo* to give the pyrrole as a beige solid: 6.65 g (66%), mp 162–164 °C.

> **¹H NMR** ([D$_6$]DMSO): δ (ppm) = 9.78 (s, 1 H, NH), 5.30 (s, 2 H, NH$_2$), 1.90 (s, 3 H, CH$_3$), 1.81 (s, 3 H, CH$_3$).
>
> **¹³C NMR** ([D$_6$]DMSO): δ (ppm) = 146.3, 118.7, 115.0, 111.0, 71.1, 10.1, 9.4.
>
> **FT-IR** (solid): $\tilde{\nu}$ (cm^{-1}) = 3408 (NH), 3279 (NH), 2188 (C≡N), 1636 (N–H), 1581 (NH), 1498, 1440.

3.4.2.2   **   3-(3-Chlorophenyl)-5,6-dimethyl-4*H*-pyrrolo[2,3-*d*]pyrimidine-4-imine [1]

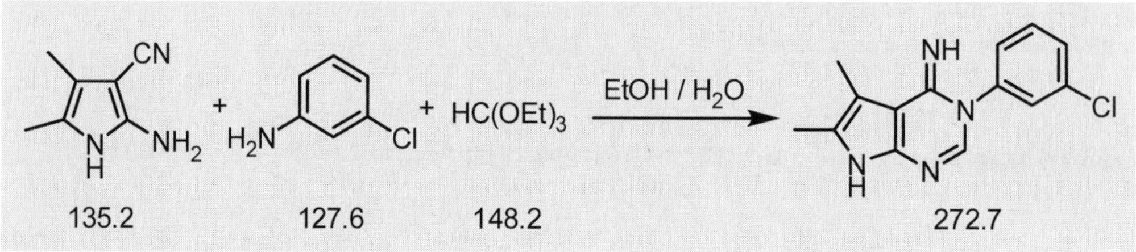

A solution of triethyl orthoformate (2.34 g, 16.5 mmol) and 3-chloroaniline (2.68 g, 21.0 mmol) in anhydrous EtOH (15 mL) is acidified to pH 5–5.5 by adding 2–3 drops of acetic acid. The mixture is then heated to 50 °C and aminocyanopyrrole **3.4.2.1** (2.03 g, 15.0 mmol) is added in small portions over 4 h, keeping the temperature at 45–50 °C. The mixture is then kept at 50 °C for a further 4 h and thereafter is left at room temperature for 8 h.

Water (1.5 mL) is then added, and the mixture is cooled to 0 °C and kept at this temperature for 30 min. The precipitate formed is filtered off, washed with EtOH/H_2O (4:1, 150 mL), and dried *in vacuo* to yield the pyrimidine imine as a yellow solid: 2.32 g (57%), mp 150–152 °C.

---

^{1}H NMR ([D_6]DMSO): δ (ppm) = 10.92 (s, 1 H, NH), 10.21 (s, 1 H, NH), 8.53 (s, 1 H, H-2), 7.33–7.02 (m, 4 H, Ar-H), 2.05 (s, 3 H, CH_3), 1.93 (s, 3 H, CH_3).

^{13}C NMR ([D_6]DMSO): δ (ppm) = 147.8, 145.0, 141.7, 133.7, 130.8, 121.7, 120.2, 118.0, 113.8, 10.3, 9.4.

FT-IR (solid): $\tilde{v}$ (cm^{-1}) = 3259 (NH indole), 2201, 1664, 1594, 1523, 1499, 1328, 1285.

3.4.2.3   *   4-(3-Chlorophenyl)-5,6-dimethyl-7*H*-pyrrolo[2,3-*d*]pyrimidine (EGF-R-pyrrolo[2,3-*d*]pyrimidine) [1]

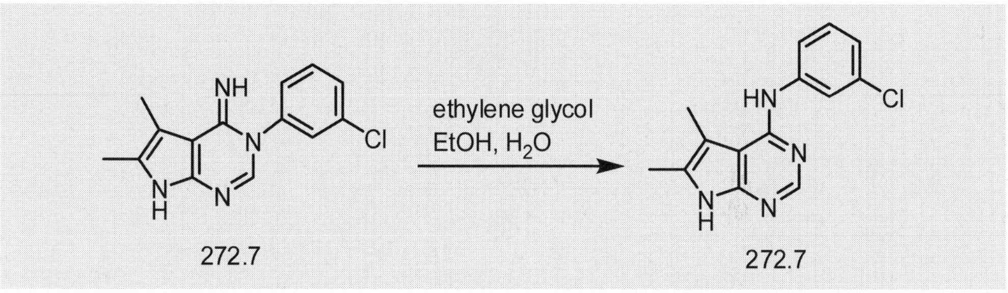

A suspension of the pyrrolopyrimidine imine **3.4.2.2** (1.91 g, 7.00 mmol) in H_2O (5 mL), EtOH (10 mL), and ethylene glycol (10 mL) is heated to 95 °C for 4 h.

After cooling the mixture to room temperature over 60 min, the precipitate formed is collected by filtration, washed with H_2O (40 mL), and dried *in vacuo* at 50 °C to yield the pyrrolopyrimidine as a yellow solid: 1.57 g, (83%), mp 240–242 °C.

---

^{1}H NMR ([D_6]DMSO): δ (ppm) = 11.48 (s, 1 H, NH), 8.18 (s, 1 H, NH), 8.11 (s, 1 H), 7.93 (s, 1 H), 7.67 (d, *J* = 8.2 Hz, 1 H, Ar-H), 7.30 (t, *J* = 8.2 Hz, 1 H, Ar-H), 7.01 (d, *J* = 8.2 Hz, 1 H, Ar-H), 2.39 (s, 3 H, CH_3), 2.26 (s, 3 H, CH_3).

^{13}C NMR ([D_6]DMSO): δ (ppm) = 152.6, 150.7, 149.3, 142.1, 132.7, 129.8, 129.3, 121.2, 119.6, 118.8, 104.8, 103.5, 10.7, 10.3.

FT-IR (solid): $\tilde{v}$ (cm^{-1}) = 3445 (NH), 1607 (NH), 1595 (NH), 1447.

[1]   R. W. Fischer, M. Misun, *Organic Process Research & Development* **2001**, *5*, 581.

[2]   Th. Eicher, S. Hauptmann, *The Chemistry of Heterocycles*, 2nd ed., p. 402, Wiley-VCH, Weinheim, **2003**.

[3]   K. Gewald, *Z. Chem.* **1961**, *1*, 349.

[4]   G. L. Buchanan, *Chem. Soc. Rev.* **1988**, *17*, 91.

[5]   See: J. Zhu, H. Bienayme, *Multicomponent Reactions*, Wiley-VCH, Weinheim, **2004**.

[6]   a) W. Steglich, G. Höfle, A. Prox, *Chem. Ber.* **1972**, *105*, 1718; b) W. Steglich, G. Höfle, *Angew. Chem.* **1969**, *81*, 1001; *Angew. Chem. Int. Ed. Engl.* **1969**, *8*, 981.

[7]   a) L. F. Tietze, *Chem. Rev.* **1996**, *96*, 195; b) L. F. Tietze, G. Brasche, K. Gericke, *Domino Reactions in Organic Synthesis*, Wiley-VCH, Weinheim, **2006**.

[8]   The described mechanism of pyrimidine formation is simplified. In reality, it is more complicated [1] and involves H^+-catalyzed orthoester–imino ester–amidine equilibria, in which the symmetrical amidine Ar–NH–CH=N–Ar is likely to play a central role. The equilibria are driven to the product side by removal of **1** from the solution due to its low solubility and not by thermodynamic factors.

[9]   Ref. [2], p. 202 and 410.

3.4.3   7-Phenyl-1,6-naphthyridine

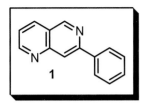

1

Topics:
- Metalation of 2-bromopyridine
- Formylation of a lithiopyridine
- Sonogashira coupling
- Larock isoquinoline synthesis, application to naphthyridine formation

(a) General

Naphthyridines (pyridopyridines) can be viewed as C–C condensation products of two pyridines. Of the six possible naphthyridines, four relate topologically to isoquinoline by replacement of a CH unit in the benzo part by a nitrogen atom:

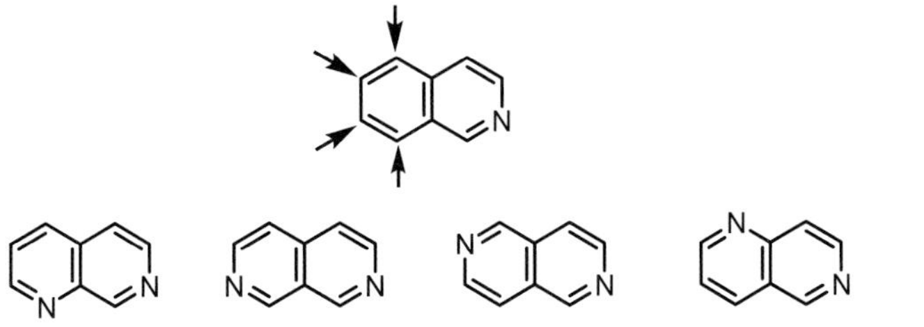

1,7-   2,7-   2,6-   1,6-naphthyridine

In principle, naphthyridines may be synthesized by applying the methods for quinoline or isoquinoline formation but starting with pyridine derivatives. However, since most of these (like the Doebner–Miller and Bischler–Napieralski syntheses, cf. **3.5.5** and **3.3.2**) proceed through S_EAr cyclizations, adaptation to the electron-deficient pyridine system is limited [1].

In a novel method for isoquinoline synthesis [2], various 2-substituted isoquinolines **3** have been synthesized by transition-metal-mediated cyclization of (2-ethynyl)benzaldimines **2**, which are easily accessible from 2-halogenobenzaldehydes [3] by Sonogashira cross-coupling reactions (Larock isoquinoline synthesis):

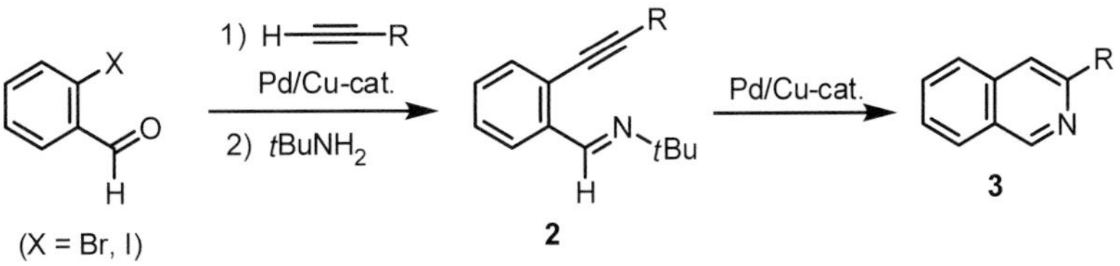

1) H≡≡R
Pd/Cu-cat.
2) *t*BuNH₂

Pd/Cu-cat.

(X = Br, I)   **2**   **3**

This versatile method (and a modified procedure [4] amenable to **4**, R' = H) has been successfully applied for the construction of 1,6-, 1,7-, 2,6-, and 2,7-naphthyridines **5** by cyclization of imines derived from ethynylpyridinecarbaldehydes **4**:

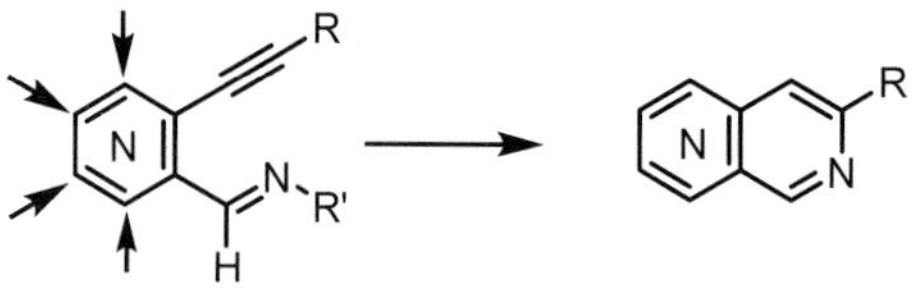

4 (R' = H, *t*Bu)   **5**

Since the 7-substituted 1,6-naphthyridine structure of the target molecule **1** corresponds to **5**, the synthesis of **1** presented in section (b) follows the described principle [2b].

(b) Synthesis of 1

As key intermediate for the synthesis of **1**, 2-bromopyridine-3-carbaldehyde (**7**) is used, which is prepared [4, 5] by metalation of 2-bromopyridine **6** at the 3-position with LDA [6] to give the 3-lithio-2-bromopyridine (**6a**), which is intercepted with DMF. The aldehyde **7** thus formed is transformed to the imine **8** by condensation with *tert*-butylamine:

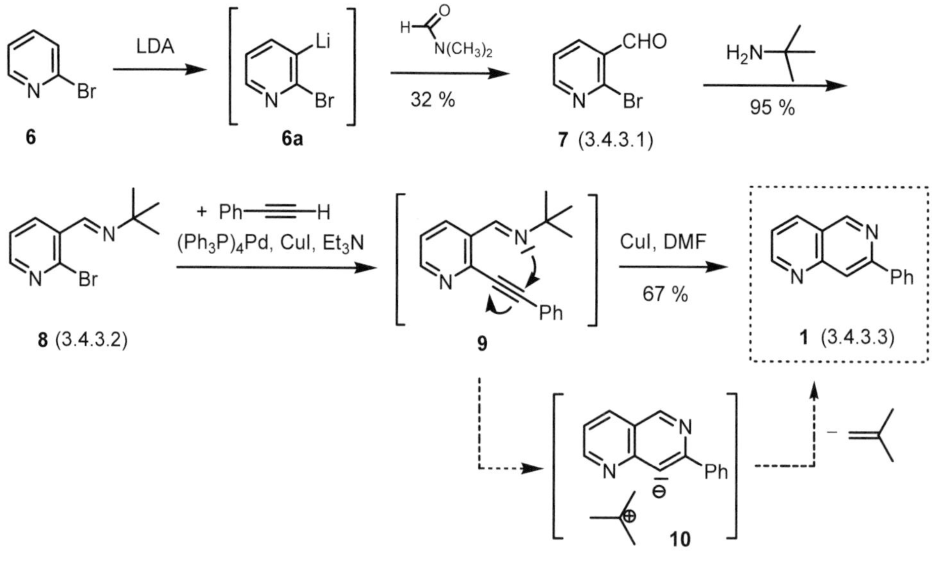

Sonogashira cross-coupling (cf. **1.6.3**) of **8** with phenylacetylene occurs readily by using the palladium catalyst (Ph₃P)₄Pd in the presence of Cu(I) iodide and triethylamine. The thus formed (2-phenyl-ethynyl)pyridine-3-aldimine **9** is not isolated but cleanly undergoes cyclization to give the 7-phenyl-substituted 1,6-naphthyridine **1** upon thermal reaction with Cu(I) iodide in DMF. It can be assumed that the reaction proceeds by intramolecular nucleophilic addition of the imine function to the C≡C triple bond (→ **10**) with loss of the *t*Bu group under formation of isobutene (**10** → **1**).

Thus, the target molecule **1** is prepared in a three-step sequence with an overall yield of 20% (based on 2-bromopyridine **6**).

(c) Experimental procedures for the synthesis of 1

3.4.3.1   **   2-Bromopyridine-3-carbaldehyde [4,5]

n-Butyllithium (4.30 mL, 1.6 M in hexane, 6.88 mmol) is added by means of a syringe to a stirred solution of diisopropylamine (810 mg, 8.00 mmol) in anhydrous THF (20 mL) under an argon atmosphere at −78 °C and stirring is continued for 60 min at this temperature. A solution of 2-bromopyridine (950 mg, 6.00 mmol) in anhydrous THF (5 mL) is then added dropwise and stirring is continued for 4 h at −78 °C. Finally, DMF (2.0 mL, 26.0 mmol) is added at −78 °C. After 30 min at −78 °C, the reaction mixture is allowed to warm to room temperature, whereupon stirring is continued for an additional 2 h.

A saturated solution of NH$_4$Cl (50 mL) is then added, the phases are separated, and the aqueous phase is extracted with Et$_2$O (2 × 40 mL). The combined organic layers are washed with brine (50 mL) and dried over MgSO$_4$. Removal of the solvent *in vacuo* affords a brown oil, which is purified by flash chromatography (EtOAc/n-hexane, 1:10) to give colorless needles; 360 mg (32%), mp 70–71 °C.

> **^{1}H NMR** (400 MHz, CDCl$_3$): δ (ppm) = 10.36 (d, J = 0.8 Hz, 1 H, CHO), 8.58 (dd, J = 4.6, 2.1 Hz, 1 H, H-6), 8.18 (dd, J = 7.6, 2.1 Hz, 1 H, H-4), 7.44 (ddd, J = 7.6, 4.6, 0.8 Hz, 1 H, H-5).
>
> **^{13}C NMR** (100 MHz, CDCl$_3$): δ (ppm) = 191.1, 154.5, 145.4, 138.0, 130.6, 123.5.

3.4.3.2   *   *N*-(2-Bromopyridin-3-ylmethylene)-*tert*-butylamine [2b]

A mixture of the aldehyde **3.4.3.1** (200 mg, 1.08 mmol) and *tert*-butylamine (236 mg, 3.22 mmol) is stirred at room temperature for 15 h.

The excess amine is then removed *in vacuo*, H$_2$O (3 mL) is added, and the mixture is extracted with Et$_2$O (3 × 5 mL). The combined organic layers are dried (Na$_2$SO$_4$) and concentrated to afford the imine as a pale-yellow oil; 248 mg (95%).

> **^{1}H NMR** (400 MHz, CDCl$_3$): δ (ppm) = 8.52 (d, J = 0.8 Hz, 1 H, CH=N), 8.38 (dd, J = 4.5/2.0 Hz, 1 H, H-6), 8.30 (dd, J = 7.5/2.0 Hz, 1 H, H-4), 7.30 (ddd, J = 7.5/4.5/0.8 Hz, 1 H, H-5), 1.32 (s, 9 H).
>
> **^{13}C NMR** (100 MHz, CDCl$_3$): δ (ppm) = 153.2, 151.1, 143.9, 136.9, 132.9, 123.2, 58.44, 29.56.

3.4.3.3   **   7-Phenyl-1,6-naphthyridine [2b]

A mixture of triethylamine (2 mL), (Ph₃P)₄Pd (11.5 mg, 0.01 mmol, note 1), the imine **3.4.3.2** (121 mg, 0.5 mmol), phenylacetylene (62.0 mg, 0.60 mmol), and CuI (2 mg, 0.01 mmol) is heated at 55 °C for 3 h under an argon atmosphere. The reaction is monitored by TLC (SiO_2; n-hexane/EtOAc, note 2).

The reaction mixture is then cooled and diluted with Et_2O (2 mL). The mixture is filtered, the filter cake is washed with Et_2O (5 mL), and the combined filtrate and washings are concentrated under reduced pressure.

The residue is dissolved in DMF (5 mL) and CuI (10 mg, 0.05 mmol) is added. The mixture is heated at 100 °C for 15 h under an argon atmosphere.

The reaction mixture is then cooled, diluted with Et_2O (25 mL), washed with saturated NH_4Cl solution (30 mL), and dried (Na_2SO_4). The solvent is evaporated under reduced pressure and the residue is purified by chromatography on silica gel (n-hexane/EtOAc, 1:1, $R_f = 0.21$) to afford the naphthyridine 70 mg (67%) as a colorless solid, mp 135–136 °C.

> **¹H NMR** (400 MHz, CDCl₃): δ (ppm) = 9.35 (d, $J = 0.7$ Hz, 1 H, H-5), 9.10 (dd, $J = 4.3/1.8$ Hz, 1 H, H-2), 8.35 (s$_{br}$, 1 H, H-8), 8.30 (ddd, $J = 8.3/1.8/0.7$ Hz, 1 H, H-4), 8.22–8.15 (m, 2 H, phenyl), 7.53 (m$_c$, 2 H, phenyl), 7.48 (dd, $J = 8.3/4.3$ Hz, 1 H, H-3), 7.45 (m$_c$, 1 H, phenyl).
>
> **¹³C NMR** (100 MHz, CDCl₃): δ (ppm) = 155.3, 155.1, 152.7, 151.4, 138.9, 135.6, 129.2, 128.9, 127.2, 122.7, 122.2, 117.8.

Notes: (1) As in the original literature [2b], (Ph₃P)₂PdCl₂ can also be used as Pd catalyst.
           (2) It should be noted that the cyclization may already occur to some extent during the Sonogashira reaction.

[1] a) J. P. Stanforth, in *Comprehensive Heterocyclic Chemistry* (Eds.: A. R. Katritzky, C. W. Rees, E. F. V. Scriven), Vol. 7, p. 527, Pergamon, Oxford, **1996**; b) for a review on advances in the chemistry of naphthyridines, see: V. P. Litvinov, *Adv. Heterocycl. Chem.* **2006**, *91*, 189.

[2] a) K. R. Roesch, R. C. Larock, *J. Org. Chem.* **1998**, *63*, 5306; b) *J. Org. Chem.* **2002**, *67*, 86.

[3] It should be mentioned that reaction of *o*-halogenobenzaldehydes and *di*substituted alkynes in the presence of a Pd(0) catalyst leads to 2,3-disubstituted indenones: R. C. Larock, M. J. Doty, *J. Org. Chem.* **1993**, *58*, 4579.

[4] A. Numata, Y. Kondo, T. Sakamoto, *Synthesis* **1999**, 306.

[5] F. Bracher, *J. Heterocycl. Chem.* **1993**, *30*, 157.

[6] For the relative acidities of pyridines, see: Th. Eicher, S. Hauptmann, *The Chemistry of Heterocycles*, 2ⁿᵈ ed., p. 273, Wiley-VCH, Weinheim, **2003**.

3.4.4 Caffeine

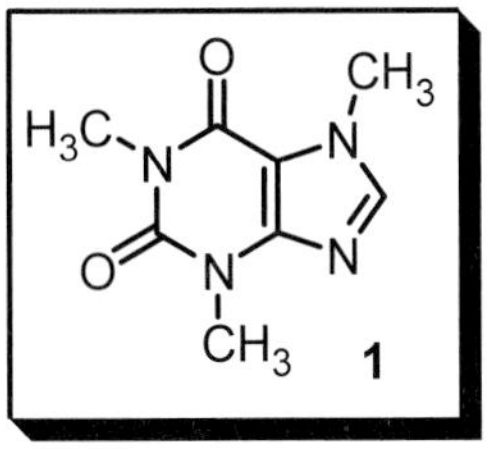

Topics:
- Cyanoacetylation of a urea derivative
- Ring closure of an (*N*-cyanoacetyl)urea to a 4-aminouracil
- Traube synthesis of a purine derivative according to the Bredereck modification

(a) General

Caffeine (**1**, 1,3,7-trimethylxanthine) is a derivative of xanthine **2**, which exists as an equilibrium of the imide **2a** and the energetically less favored lactim **2b**. The underlying heterocycle is purine (**3**).

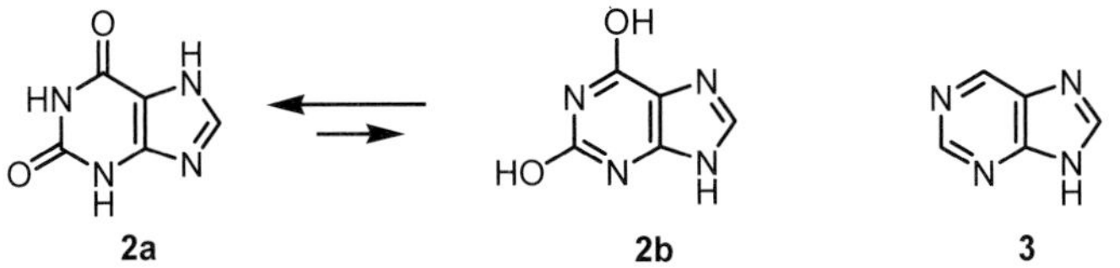

Other alkaloids of the xanthine type are theophylline (**4**) and theobromine (**5**):

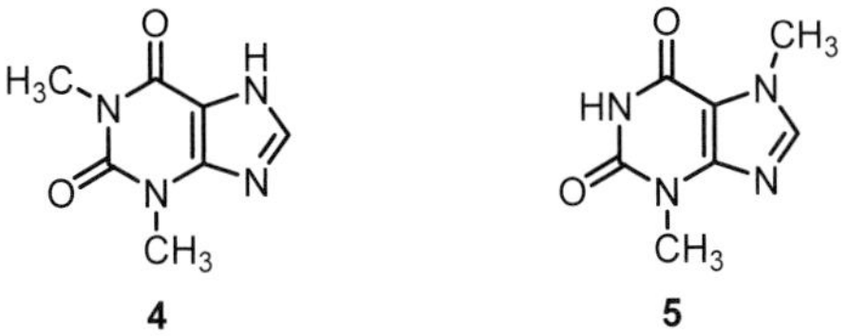

Caffeine occurs in coffee beans and tea leaves; it exhibits stimulating effects on the central nervous system and is used therapeutically as an analeptic [1].

Two different retrosyntheses of purine (**3**), i.e. A and B, have to be considered, according to the heterocyclic substrates **6** and **7** used as starting materials.

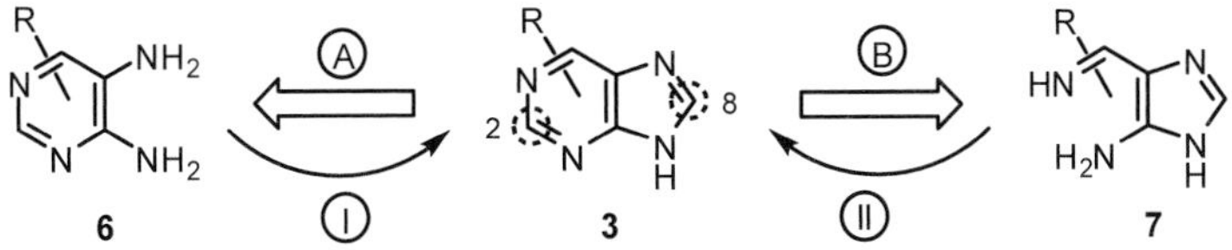

It should be stressed that carbon atoms C-2 and C-8 in purine (**3**) are at the oxidation level of formic acid. Thus, the standard method for purine formation is the classical Traube synthesis [2], in which, according to strategy **I**, 4,5-diaminopyrimidines of type **6** are subjected to cyclocondensation with formic acid or a formic acid derivative (formamide, formamidine, orthoformates, etc.). The 4,5-diaminopyrimidines **6** are generally obtained from 4-aminopyrimidines **8** by nitrosation and reduction of the intermediately formed 5-nitroso compounds **9**:

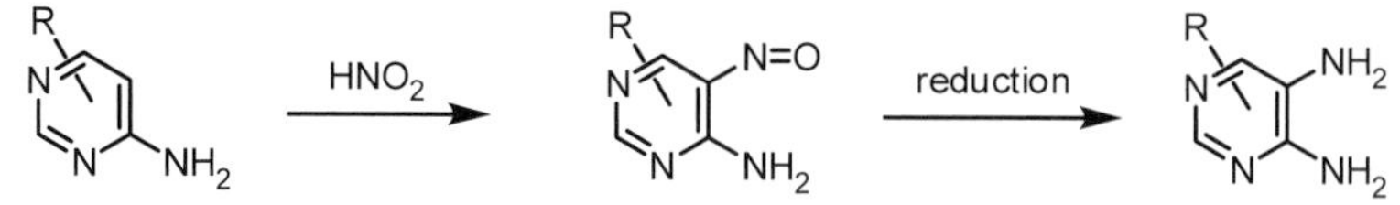

8 9 6

On the other hand, according to strategy **II**, 4,5-disubstituted imidazoles of type **7** can also be used as substrates. However, this approach has only limited scope. A recent and instructive example of the use of strategy **II** is the synthesis of 9-benzyladenine **13** [3]:

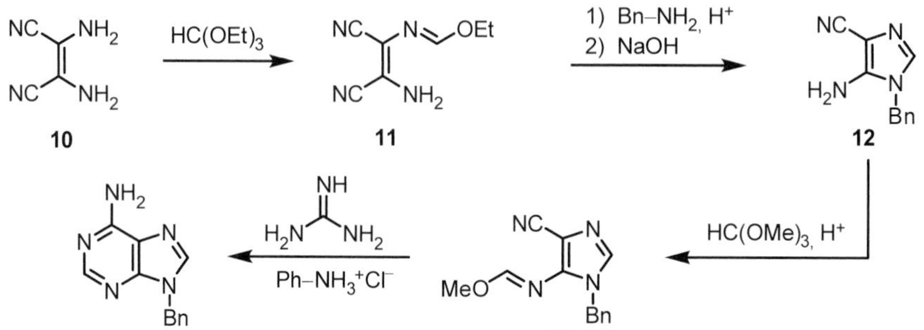

10 11 12 13 14

5-Amino-1-benzyl-4-cyanoimidazole (**12**), readily available from diaminomaleodinitrile (**10**) via the imidate **11** and cyclization with benzylamine, is transformed into the formimidate **14** by reaction with trimethyl orthoformate; **14** undergoes cyclocondensation with guanidine to give the adenine derivative **13**.

The synthesis of caffeine, however, was performed using the Traube method (**I**), which requires 4,5-diaminouracil **15** as intermediate and which is described in detail in section (b).

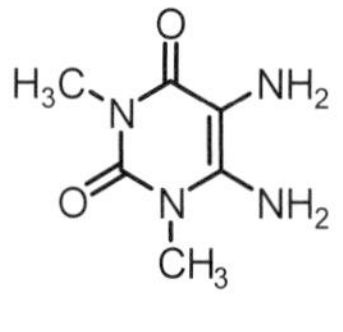

15

(b) Synthesis of 1

N,N'-Dimethylurea (**16**) is acylated with cyanoacetic acid in the presence of acetic anhydride, probably via formation of its mixed anhydride with acetic acid. The formed *N*-(cyanoacetyl)urea **17** readily cyclizes in an aqueous medium in the presence of potassium acetate as a base to provide the 4-aminouracil **19**:

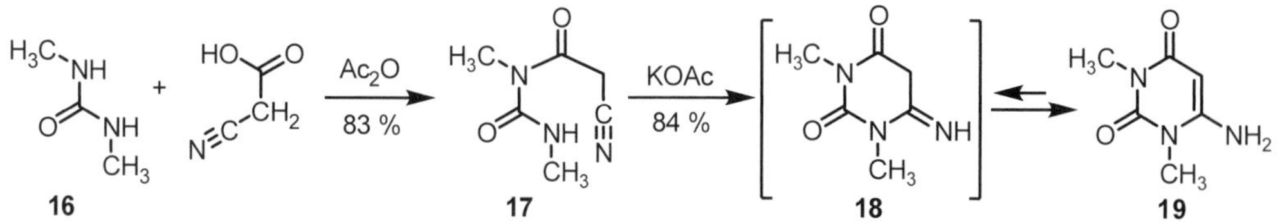

16 17 18 19

This two-step procedure formally represents the basic principle of the Pinner synthesis of pyrimidine derivatives, which consists of the cyclocondensation of a 1,3-bis-electrophile (here: cyanoacetic acid) with an N–C–N system (here: a urea) (cf. **3.3.3**).

In the cyclization step, nucleophilic addition of the urea nitrogen to the nitrile function leads to the imine **18** as intermediate, which tautomerizes to **19** containing a more stable enamide moiety.

To attach the imidazole ring to the uracil system, **19** is first nitrosated at the 5-position. The nitroso compound is then reduced and the 4,5-diaminouracil **15** thus obtained is cyclocondensed with formic acid. This sequence can be carried out in a stepwise manner [4], but from a preparative point of view it is advantageous to perform it as a one-pot procedure (Bredereck protocol [5]), using formamide as solvent for the nitrosation with HNO_2 (→ **20**) and reduction with $Na_2S_2O_4$ (→ **15**). The formamide used as solvent then reacts with **15** to give the xanthine derivative theophylline **4**.

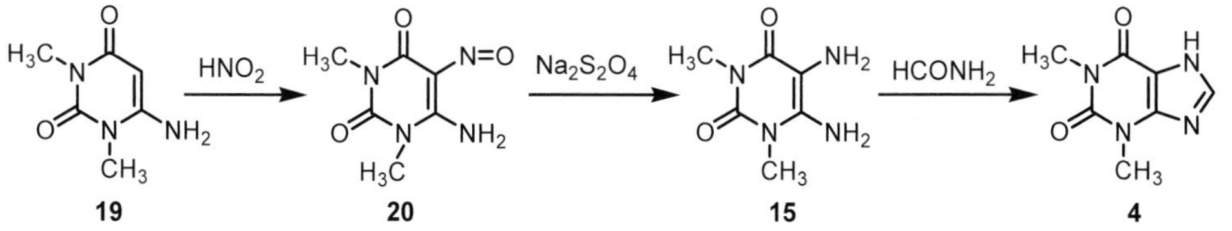

Using this approach, **4** is obtained in 55% yield (over three steps). For the synthesis of caffeine **1**, theophylline (**4**) is methylated at N-7 by reaction with methyl iodide in the presence of sodium ethoxide.

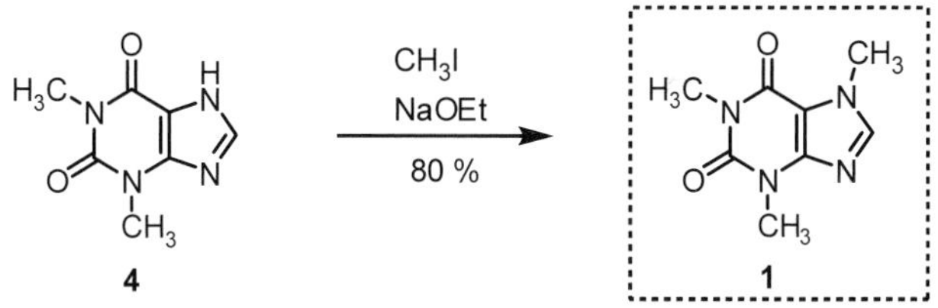

In this way, the target molecule **1** is synthesized in a four-step sequence with an overall yield of 30% (based on **16**) [6].

(c)   Experimental procedures for the synthesis of 1

3.4.4.1   *   N-Cyanoacetyl-*N*,*N'*-dimethylurea [7]

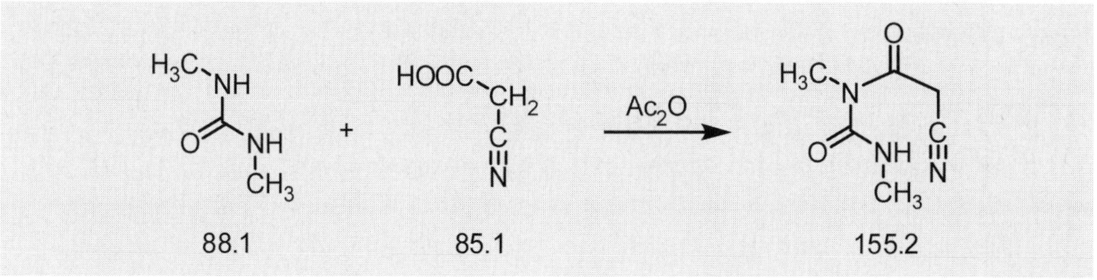

A solution of *N,N'*-dimethylurea (30.0 g, 0.34 mol) and cyanoacetic acid (30.0 g, 0.35 mol) in acetic anhydride (60 mL, note 1) is heated at 100–110 °C (external temperature) for 1.5 h.

The excess acetic anhydride is then distilled off *in vacuo* (~100 mbar) and the remaining dark-brown oil is dissolved in an EtOH/Et$_2$O mixture (60 mL/20 mL). After keeping the solution at 5 °C for 2 h (refrigerator), the yellowish crystals that are formed (note 2) are filtered off and washed with Et$_2$O (2 × 5 mL). The filtrate is concentrated to ca. 40 mL, Et$_2$O (40 mL) is added, and the solution is kept in a refrigerator for 12 h to give another crop of crystals. The total yield is 44.0 g (83%), mp 77–79 °C. Recrystallization from acetone/Et$_2$O (1:2) gives yellowish cubes, mp 82–83 °C, TLC (SiO$_2$; EtOAc): $R_f = 0.60$.

IR (film): $\tilde{v}$ (cm^{-1}) = 3300 (NH), 2260 (C≡N), 1700/1670 (C=O).

^{1}H NMR (CDCl$_3$): δ (ppm) = 8.2 (s$_{br}$, 1 H, NH; exchangeable with D$_2$O), 3.81 (s, 2 H, CH$_2$), 3.29 (s, 3 H, N–CH$_3$), 2.88 (d, J = 4.5 Hz, 3 H, NH–CH$_3$).

Notes: (1) Acetic anhydride should be freshly distilled before use, bp$_{760}$ 139–140 °C.
(2) It is recommended that crystallization is induced by scratching with a glass rod.

3.4.4.2   *   4-Amino-1,3-dimethyluracil [8]

155.2   98.2   155.2

Cyanoacetylurea **3.4.4.1** (31.0 g, 0.20 mol) is added portionwise to a stirred solution of potassium acetate (7.50 g, 76.4 mmol) in water (250 mL) at room temperature. The mixture is heated to reflux. The solid eventually dissolves and then the product begins to crystallize; heating is continued for 30 min.

After cooling to room temperature, the mixture is kept in a refrigerator for 12 h. The crystals that are formed are collected by filtration and washed with ice-cold water; the filtrate is concentrated to about one-third of its original volume, whereupon another crop of crystals is obtained. The total yield of the aminouracil is 26.0 g (84%) as faintly yellow needles, mp 296–297 °C. Recrystallization from H$_2$O with the addition of decolorizing charcoal yields colorless needles, mp 299–300 °C, TLC (SiO$_2$; EtOH): $R_f = 0.70$.

IR (film): $\tilde{v}$ (cm^{-1}) = 3410/3360/3240 (NH$_2$), 1650/1575 (C=O).

^{1}H NMR ([D$_6$]DMSO): δ (ppm) = 6.79 (s, 2 H, NH$_2$; exchangeable with D$_2$O), 4.73 (s, 1 H, 5-H), 3.29, 3.12 (s, 3 H, N–CH$_3$).

Derivative: **4-Amino-5-nitroso-1,3-dimethyluracil** [4]:

Formic acid (2.0 mL) is added dropwise to a stirred solution of uracil **3.4.4.2** (2.00 g, 12.9 mmol) and NaNO$_2$ (0.89 g, 12.9 mmol) in hot H$_2$O (40 mL). Crystals of the nitroso compound are formed immediately. The mixture is kept at 0 °C for 4 h and then the crystals are collected by filtration, washed with ice-cold water, and dried over P$_4$O$_{10}$ *in vacuo*: 2.37 g (100%) of red-violet crystals, mp 260–261 °C (dec.), TLC (SiO$_2$; EtOH): R_f = 0.70.

3.4.4.3 ** Theophylline (1,3-dimethylxanthine) [5]

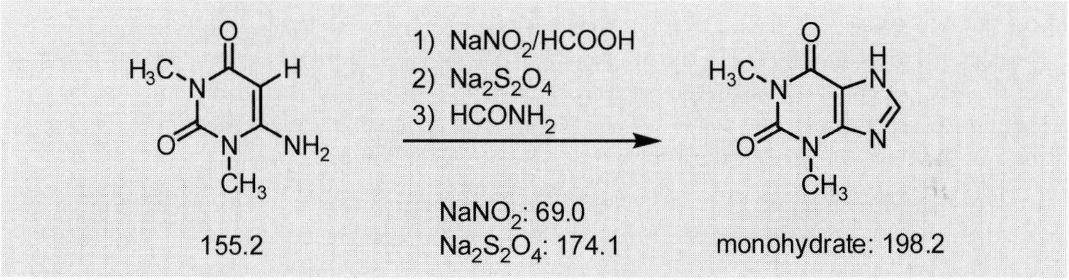

A stirred solution of the aminouracil **3.4.4.2** (23.3 g, 0.15 mol) and sodium nitrite (14.7 g, 0.15 mol) in formamide (120 mL) is heated to 60 °C (internal temperature). With vigorous stirring, formic acid (24.0 mL) is added dropwise over 10 min; the 5-nitroso compound separates as a red-violet precipitate (cf. **3.4.4.2**, *derivative*).

The suspension is then heated to 100 °C and sodium dithionite (4.66 g, 26.8 mmol) is added in small portions over a period of 10 min with stirring; the internal temperature increases to 130–140 °C and a yellow solution is obtained. When the addition of the reducing agent is complete, the reaction mixture is heated to 180–200 °C and kept at this temperature for 30 min.

On cooling to room temperature, the product precipitates in part; it is collected by filtration and washed with water (3 × 20 mL). The combined filtrate and washings are diluted with H$_2$O (ca. 300 mL) and extracted with CHCl$_3$ (3 × 100 mL). The CHCl$_3$ extracts are combined, dried (Na$_2$SO$_4$), and concentrated *in vacuo* to obtain a further crop of the product. The combined fractions of crude product are recrystallized from EtOH/H$_2$O (1:1); theophylline monohydrate is obtained as a faintly yellow, microcrystalline powder; 16.2 g (55%), mp 272–273 °C; TLC (SiO$_2$; EtOH): R_f = 0.60.

IR (KBr): $\tilde{\nu}$ (cm^{-1}) = 3140 (NH), 1710/1665 (C=O).

^{1}H NMR (CDCl$_3$/[D$_6$]DMSO): δ (ppm) = 13.35 (s$_{br}$, 1 H, NH; exchangeable with D$_2$O), 7.73 (s, 1 H, 8-H), 3.54, 3.34 (s, 3 H, N–CH$_3$).

3.4.4.4   **   Caffeine (1,3,7-trimethylxanthine) [4]

monohydrate: 198.2

1) NaOEt
2) CH_3I

Na: 23.0
CH_3I: 174.1

194.2

Sodium (0.60 g, 26.0 mmol) is added in small pieces to anhydrous EtOH (40 mL). Theophylline monohydrate **3.4.4.3** (3.60 g, 18.2 mmol) is then added to the thus obtained solution of NaOEt and the suspension is heated to reflux with stirring for 1.5 h. After cooling to room temperature, a solution of methyl iodide (1.90 mL, 30.0 mmol) in anhydrous EtOH (10 mL) is added dropwise with stirring over a period of 30 min. During the addition of methyl iodide, the external temperature should be kept at 50–55 °C. A suspension results, which is stirred at 50–55 °C for 3–4 h.

The solvent is then removed *in vacuo*; the colorless residue is taken up in H_2O (50 mL) and the aqueous solution is extracted with CH_2Cl_2 (10 × 50 mL). The combined extracts are dried ($MgSO_4$) and the solvent is removed *in vacuo*. Caffeine is obtained as a colorless to faintly ochre-colored microcrystalline powder (without water of crystallization); 2.82 g (80%), mp 227–228 °C. Recrystallization from $EtOH/H_2O$ (1:1) raises the mp to 234–235 °C (monohydrate); TLC (SiO_2; EtOAc): $R_f = 0.30$.

IR (KBr): $\tilde{\nu}$ (cm^{-1}) = 1695/1655 (C=O).

^{1}H NMR ($CDCl_3$): δ (ppm) = 7.50 (s, 1 H, 8-H), 3.99 (s, 3 H, N-7-CH_3), 3.58, 3.40 (s, 3 H, N-1/N-3-CH_3).

---

[1]   Römpp, *Lexikon "Naturstoffe"* (Eds.: W. Steglich, B. Fugmann, S. Lang-Fugmann), p. 144, Georg Thieme Verlag, Stuttgart, **1997**.

[2]   a) J. A. Joule, K. Mills, *Heterocyclic Chemistry*, 4th ed., p. 461, Blackwell Science, Oxford, **2000**; b) Th. Eicher, S. Hauptmann, *The Chemistry of Heterocycles*, 2nd ed., p. 411, Wiley-VCH, Weinheim, **2003**.

[3]   Z. Sun, R. S. Hosmane, *Synth. Commun.* **2001**, *31*, 549.

[4]   W. Traube, *Ber. Dtsch. Chem. Ges.* **1900**, *33*, 3035.

[5]   H. Bredereck, R. Gompper, H. G. v. Schuh, G. Theilig, *Angew. Chem.* **1959**, *71*, 753.

[6]   It should be noted that caffeine has been prepared in a novel six-step sequence starting from uracil: M. A. Zajac, A. G. Zakrzewski, M. G. Kowal, S. Narayan, *Synth. Commun.* **2003**, *33*, 3291.

[7]   F. Baum, *Ber. Dtsch. Chem. Ges.* **1908**, *41*, 525.

[8]   F. Baum, *Ber. Dtsch. Chem. Ges.* **1908**, *41*, 532.

3.4.5   Nedocromil analogon

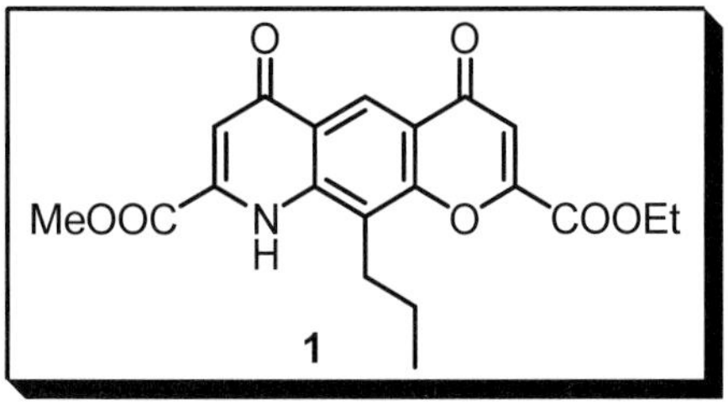

1

Topics:
- Formation of a chromone derivative
- Formation of a 4-quinolone derivative
- Claisen rearrangement of an aryl allyl ether
- Michael addition to dimethyl acetylene-dicarboxylate
- Catalytic hydrogenation of a C=C bond
- *N*-acylation and deacylation
- Ester hydrolysis

(a) General

A number of pyranoquinoline dicarboxylic acids of types **2/3** are pharmaceutically relevant as anti-allergics for the topical treatment of asthma. Among them, nedocromil (**3**, used as its Na salt) shows the strongest therapeutic effects [1]:

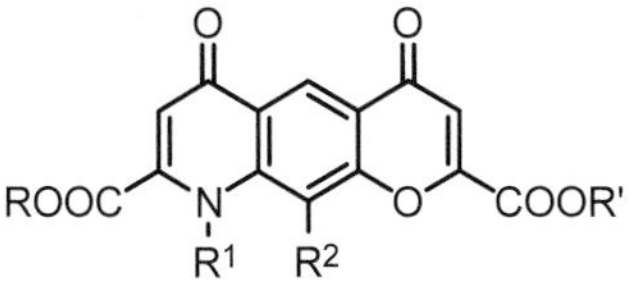

1: R^1= H, R^2= Pr, R = Me, R' = Et
2: R^1= H, R^2= Pr, R = R' = H
3: R^1= Et, R^2= Pr, R = R' = H

Retrosynthesis of the linear condensed heterotricyclic compounds **1–3** can be conducted in two directions starting with disconnection either **A** at the quinolone site or **B** at the chromone site, thus proceeding via either the chromone **4** or the quinolone **5**:

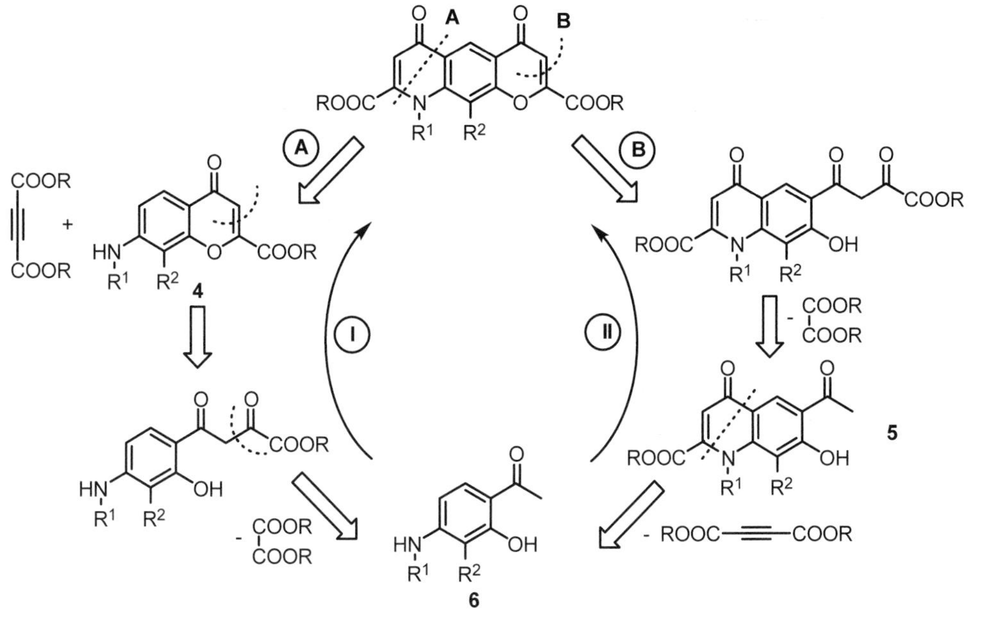

Both retroanalytical pathways lead to the substrate **6**.

The substrate **6** (R^1 = H, R^2 = Pr) required for the synthesis of **1** or **2** – in contrast to the substrate **6** (R^1 = Et, R^2 = Pr) for the synthesis of **3** – is accessible from inexpensive starting materials. Therefore, the preparation of the nedocromil analogon **1** [2] is described here, as presented in section (b). As can be seen from the retrosynthesis according to strategy **I**, the chromone part of **1** is formed first and this is followed by annelation of the 4-quinolone moiety. This has the advantage that the protection/deprotection steps required can be kept to a minimum.

(b) Synthesis of 1

For the formation of chromones and flavones, 2-hydroxyacetophenones are preferentially used as starting materials [3]. Therefore, for the synthesis of **1**, a 2-hydroxyacetophenone of type **6** is employed, which bears a propyl substituent in the 3-position and a protected amino group in the 4-position; it is prepared by conventional means starting from *m*-anisidine. Acetylation of *m*-anisidine using acetic anhydride gives the *N*-acetylated product **7** [4], which is subjected to AlCl₃-catalyzed Friedel–Crafts acylation with acetyl chloride and concomitant cleavage of the methyl ether to yield 4-acetylamino-2-hydroxyacetophenone (**8**) [5]. The phenolic OH group in **8** is then transformed to the *O*-allyl ether **9**, which isomerizes to the C-3-allyl phenol **10** upon thermolysis by [3,3]-sigmatropic Claisen rearrangement [6]. Finally, the 3-allyl group is subjected to catalytic hydrogenation to produce the desired 4-acetylamino-2-hydroxy-3-propylacetophenone (**11**):

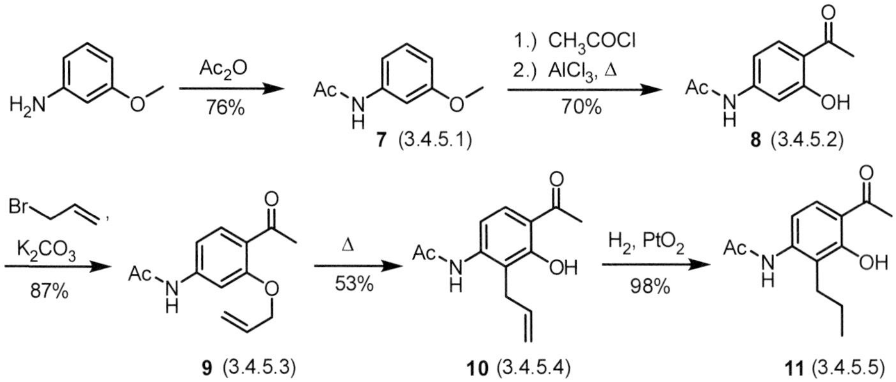

For the formation of the chromone moiety, the 2-hydroxyacetophenone **11** is reacted with diethyl oxalate in the presence of sodium ethoxide. The initially formed product of the Claisen condensation, the β-keto ester **12**, is not isolated, but is directly cyclized by treatment with acid to give the chromone carboxylic ester **13**. In addition, the *N*-acetyl group is removed under the reaction conditions.

Annelation of the 4-quinolone moiety to the chromonone **13** is achieved in a two-step sequence. First, the β-enamino ester **14** is formed by Michael addition of the NH₂ function in **13** to dimethyl acetylenedicarboxylate, and then **14** is cyclized to afford the 4-quinolone diester **1** by heating it in diphenyl ether.

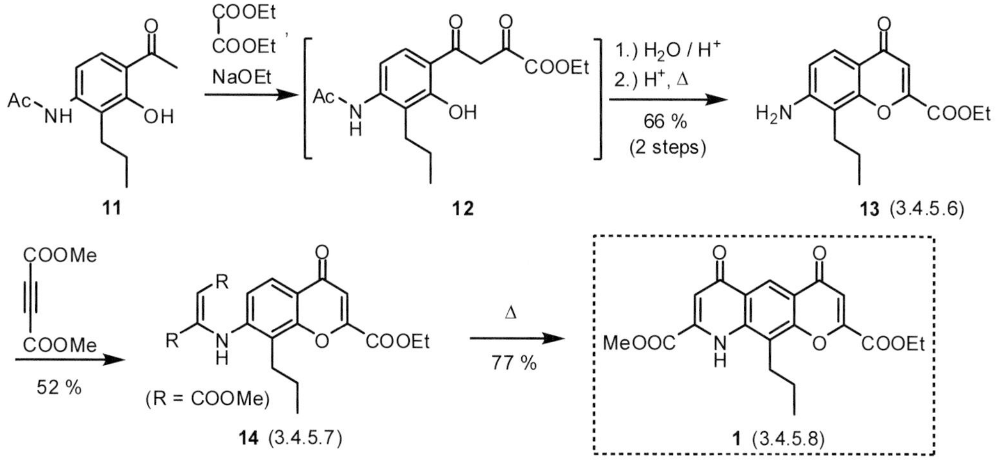

11   **12**   **13** (3.4.5.6)

14 (3.4.5.7)   **1** (3.4.5.8)

The thermal cyclization of β-anilinoacrylic esters (like **14**) to give 4-quinolones is referred to as the Konrad–Limpach synthesis [7]; as a non-catalyzed thermal process, it is likely to proceed as a 6π-$(4n + 2)$-electrocyclization, as formulated for the transformation **14 → 1**:

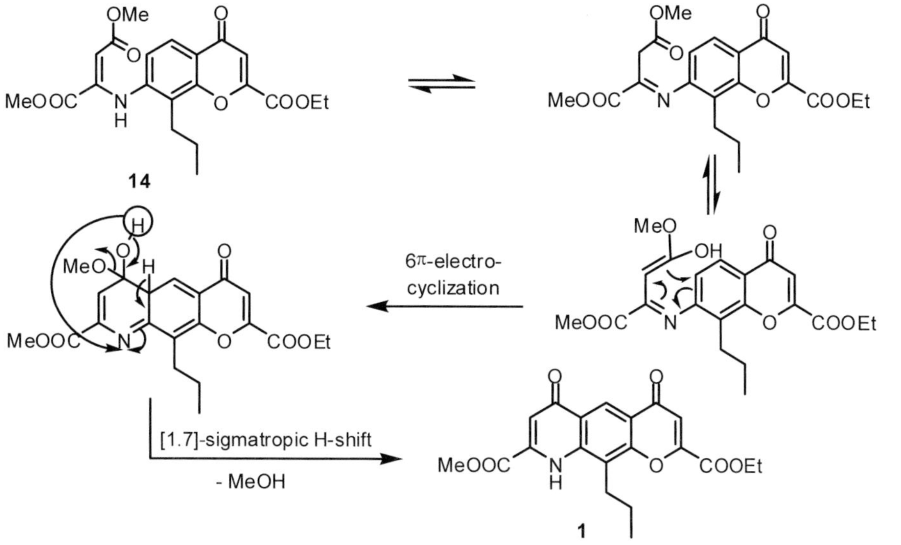

14

6π-electro-
cyclization

[1.7]-sigmatropic H-shift
- MeOH

1

In contrast, the formation of 2-quinolones by cyclization of β-keto anilides is catalyzed by strong acids (Knorr synthesis [7]) and is interpreted as an S_EAr process, as shown in the following example:

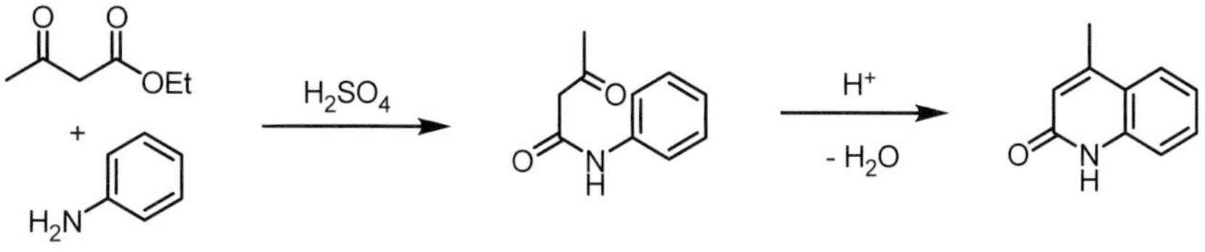

Using the described procedure, the target molecule **1** is obtained by a linear nine-step sequence in an overall yield of 6% (based on *m*-aminophenol).

(c) Experimental procedures for the synthesis of 1

3.4.5.1   *   *N*-(3-Methoxyphenyl)acetamide [4]

$$H_2N \quad\quad OMe \quad\quad\quad \xrightarrow{Ac_2O,\ HOAc} \quad\quad\quad OMe$$

123.2   102.1   165.2

Acetic anhydride (30.0 mL, 0.32 mol) (note) is added dropwise to a stirred solution of *m*-anisidine (30.0 g, 0.24 mol) in glacial acetic acid (30 mL) at 0 °C. The solution is stirred for 15 h at room temperature and then poured onto crushed ice (150 g) in water (150 mL).

A precipitate forms, which is collected by filtration, washed with water, and dried *in vacuo* (50 °C/20–30 mbar) over $CaCl_2$ to give a colorless solid; 30.7 g (76%), mp 78–79 °C, TLC (SiO_2; 5% MeOH in CH_2Cl_2): $R_f = 0.36$.

> **FT-IR** (solid): $\tilde{v}$ (cm^{-1}) = 3255 (NH), 2843, 1662 (C=O), 1601 (C=O), 1482, 1415, 1280, 1152, 1048, 858, 761.
>
> **^{1}H NMR** ($CDCl_3$): δ (ppm) = 7.63 (s_{br}, 1 H, NH), 7.27 (m_c, 1 H, Ar-H), 7.19 (t, J = 8.0 Hz, 1 H, Ar-H), 6.98 (dd, not resolved, J = 7.9 Hz, 1 H, Ar-H), 6.65 (dd, J = 8.2, 2.0 Hz, 1 H, Ar-H), 3.77 (s, 3 H, OCH_3), 2.15 (s, 3 H, CH_3C=O).
>
> **^{13}C NMR** ($CDCl_3$): δ (ppm) = 168.62, 160.14, 139.21, 129.64, 112.08, 110.04, 105.78, 55.29, 24.61.

Note: Acetic anhydride has to be distilled before use, bp_{760} 140–141 °C.

3.4.5.2   **   4-Acetylamino-2-hydroxyacetophenone [5]

165.2   78.5   133.3   193.2

Acetyl chloride (4.72 g, 4.30 mL, 60.0 mmol) (note 1) is added dropwise to a stirred solution of the acetamide **3.4.5.1** (4.00 g, 24.2 mmol) in anhydrous 1,2-dichloroethane (20 mL) under a nitrogen atmosphere. After complete addition, the solution is cooled to 0 °C and, with vigorous stirring, anhydrous $AlCl_3$ (10.2 g, 76.0 mmol) (note 2) is added at such a rate that the internal temperature is kept below 15 °C. After completion of the addition, the dark reaction mixture is heated to reflux for 2 h (release of HCl gas) and then allowed to cool to room temperature.

The viscous brown oil obtained is poured onto crushed ice (~100 g) and the resulting mixture is stirred for 30 min. The yellow precipitate is collected, washed with water, dried *in vacuo* (50 °C/20–30 mbar), and recrystallized from cyclohexane/EtOAc (2:1, 140 mL). The insoluble material is filtered from the hot suspension; 3.30 g (70%) of a light-yellow crystalline solid, mp 138–140 °C.

FT-IR (solid): $\tilde{v}$ (cm^{-1}) = 3179 (NH), 3105, 3045 (OH), 1602 (C=O), 1407, 1362, 1250, 788.

^{1}**H NMR** (CDCl$_3$): δ (ppm) = 12.47 (s, 1 H, OH), 7.67 (d, J = 8.5 Hz, 1 H, Ar-H), 7.57 (s$_{br}$, 1 H, NH), 7.17 (dd, J = 8.5, 1.6 Hz, 1 H, Ar-H), 7.08 (d, J = 1.6 Hz, 1 H, Ar-H), 2.58 (s, 3 H, CH$_3$COAr), 2.21 (s, 3 H, CH$_3$CON).

^{13}**C NMR** (CDCl$_3$): δ (ppm) = 203.03, 168.69, 163.75, 145.00, 132.08, 116.23, 110.25, 107.18, 26.35, 24.86.

Notes: (1) Acetyl chloride has to be distilled before use, bp$_{760}$ 51–52 °C.
          (2) It is recommended that the addition is performed under constant N$_2$ gas flow.

3.4.5.3   *   *N*-[4-Acetyl-3-(2-propenyloxy)phenyl]acetamide [2]

Allyl bromide (2.14 g, 1.53 mL, 17.7 mmol) is added dropwise to a stirred suspension of the hydroxyacetophenone **3.4.5.2** (2.44 g, 12.6 mmol) and potassium carbonate (note 1) (2.70 g, 19.5 mmol) in anhydrous DMF (25 mL). The yellow reaction mixture is stirred at room temperature for 5 h.

It is then poured into water (100 mL) and the resulting aqueous mixture is extracted with EtOAc (5 × 30 mL). The combined organic extracts are washed with 10% NaOH (3 × 30 mL) and H$_2$O (3 × 30 mL), dried (MgSO$_4$), and concentrated *in vacuo*. The yellow residue is dried *in vacuo* (50 °C/20–30 mbar); 2.56 g (87%) (note 2), mp 107–108 °C, TLC (SiO$_2$; EtOAc/hexane, 4:1): R_f = 0.50.

FT-IR (solid): $\tilde{v}$ (cm^{-1}) = 3319 (NH), 1697 (C=O), 1586, 1263, 1187, 931, 830.

^{1}**H NMR** (CDCl$_3$): δ (ppm) = 7.98 (s$_{br}$, 1 H, NH), 7.74–7.75 (combined signals, 2 H, Ar-H), 6.77 (dd, J = 8.5, 1.9 Hz, 1 H, Ar-H), 6.06 (m$_c$, 1 H, =CH), 5.43 (dt, J = 17.3, 1.6 Hz, 1 H, =CH), 5.32 (dt, J = 10.4, 1.3 Hz, 1 H, =CH), 4.63 (ddd, not resolved, J = 5.7, 1.3, 1.6 Hz, 2 H, OCH$_2$), 2.63 (s, 3 H, CH$_3$COAr), 2.20 (s, 3 H, CH$_3$CON).

^{13}**C NMR** (CDCl$_3$): δ (ppm) = 198.52, 168.96, 159.47, 143.43, 132.39, 131.44, 123.52, 118.56, 110.96, 103.75, 69.57, 32.10, 24.82.

Notes: (1) It is recommended that K$_2$CO$_3$ is dried for 24 h at 80 °C.
          (2) The product can be used in the next step without further purification.

3.4.5.4 * N-[4-Acetyl-3-hydroxy-2-(2-propenyl)phenyl]acetamide [2]

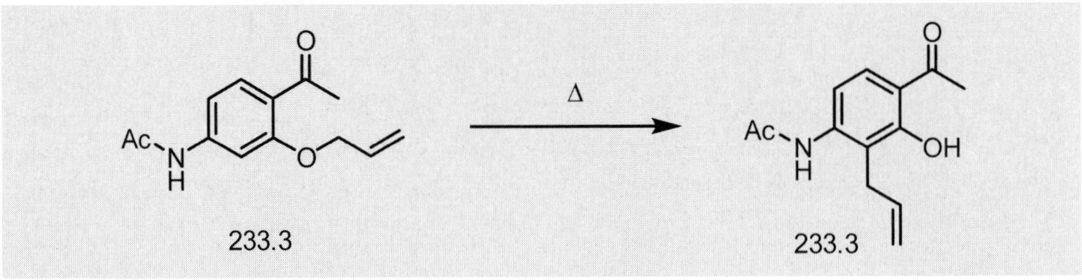

233.3   233.3

A solution of the aryl allyl ether **3.4.5.3** (12.2 g, 52.3 mmol) in *N,N*-dimethylaniline (60 mL) is heated under reflux (230 °C external temperature) for 4 h.

The solution is slowly cooled to ambient temperature and the precipitate formed is collected by filtration, washed with petroleum ether (bp 40–60 °C, ~250 mL), and dried *in vacuo* (50 °C/20–30 mbar) to give a crystalline grey solid; 6.49 g (53%), mp 177–179 °C, TLC (SiO$_2$; EtOAc/hexane, 4:1): R_f = 0.66.

FT-IR (solid): $\tilde{\nu}$ (cm^{-1}) = 3256 (NH), 3002 (OH), 1659 (C=O), 1625, 1515, 1355, 1276, 810, 668.

^{1}H NMR (CDCl$_3$): δ (ppm) = 12.91 (s, 1 H, OH), 7.73 (s$_{br}$, 1 H, Ar-H), 7.65 (d, *J* = 8.8 Hz, 1 H, Ar-H), 7.50 (s$_{br}$, 1 H, NH), 5.94 (ddt, *J* = 6.0, 17.3, 10.1 Hz, 1 H, =CH), 5.19 (dt, not resolved, *J* = 10.1 Hz, 1 H, =CH), 5.13 (dt, not resolved, *J* = 17.3 Hz, 1 H, =CH), 3.51 (ddd, not resolved, *J* = 6.0, 1.6 Hz, 2 H, ArCH$_2$), 2.17 (s, 3 H, CH$_3$CON), 2.60 (s, 3 H, CH$_3$COAr).

^{13}C NMR (CDCl$_3$): δ (ppm) = 203.62, 168.35, 161.04, 143.71, 135.48, 130.04, 116.19, 116.07, 112.13, 27.79, 26.47, 24.76.

3.4.5.5 * N-[4-Acetyl-3-hydroxy-2-propylphenyl]acetamide [2]

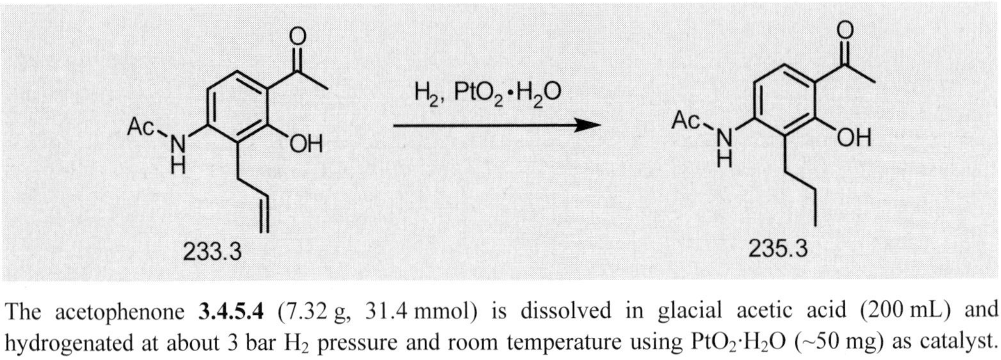

233.3   235.3

The acetophenone **3.4.5.4** (7.32 g, 31.4 mmol) is dissolved in glacial acetic acid (200 mL) and hydrogenated at about 3 bar H$_2$ pressure and room temperature using PtO$_2$·H$_2$O (~50 mg) as catalyst. The H$_2$ uptake is complete after 14 h.

The reaction mixture is then moderately heated to redissolve the suspended product and the catalyst is removed from the warm solution by filtration. The filtrate is concentrated and the residue is dried *in vacuo* (50 °C/20–30 mbar) over CaCl$_2$; 7.23 g (98%), mp 190–191 °C, TLC (SiO$_2$; EtOAc/hexane, 4:1): R_f = 0.40.

FT-IR (solid): $\tilde{\nu}$ (cm^{-1}) = 3294 (NH, OH), 2958 (CH), 2871 (CH), 1659 (C=O), 1625, 1512, 1362, 1278, 1110, 807, 666.

^{1}H NMR (CDCl$_3$): δ (ppm) = 12.8 (s$_{br}$, 1 H, OH), 7.68 (s$_{br}$, 1 H, Ar-H), 7.59 (d, J = 8.8 Hz, 1 H, Ar-H), 7.30 (s$_{br}$, 1 H, NH), 2.62 (t, J = 7.5 Hz, 2 H, ArCH$_2$), 2.59 (s, 3 H, CH$_3$COAr), 2.23 (s, 3 H, CH$_3$CON), 1.56 (sextet, J = 7.5 Hz, 2 H, CH$_2$CH$_2$CH$_3$), 0.99 (t, J = 7.5 Hz, 3 H, CH$_2$CH$_3$).

^{13}C NMR (CDCl$_3$): δ (ppm) = 203.71, 168.49, 161.40, 142.36, 129.18, 119.42, 116.09, 112.36, 26.50, 25.47, 24.89, 21.80, 14.17.

3.4.5.6 **   Ethyl 7-amino-4-oxo-8-propyl-4*H*-1-benzopyran-2-carboxylate [2]

235.3   Na: 23.0   146.1   275.3

Sodium (1.00 g, 4.25 mmol) is added in small pieces to anhydrous EtOH (25 mL). A suspension of the acetophenone **3.4.5.5** (2.00 g, 8.50 mmol) and diethyl oxalate (3.05 g, 20.8 mmol) in anhydrous EtOH (60 mL) is then added to the thus formed solution of EtONa with intense stirring (note). The yellow solution is heated under reflux for 2 h and then stirred for 1 h at room temperature.

The mixture is poured into water (100 mL) and, after acidification with 7% HCl, a yellow precipitate appears. The mixture is extracted with CHCl$_3$ (5 × 50 mL). The combined extracts are washed with brine (3 × 50 mL), dried (MgSO$_4$), and concentrated *in vacuo*, and the residue is suspended in anhydrous EtOH (30 mL) containing concentrated HCl (0.3 mL). The reaction mixture is heated under reflux for 15 h.

It is then poured into water (100 mL) and the resulting mixture is extracted with EtOAc (5 × 80 mL). The combined organic extracts are washed with water (4 × 80 mL), dried over MgSO$_4$, and concentrated *in vacuo* to give a dark-colored sticky gum. On triturating with a small amount of petroleum ether (bp 40–60 °C), the product is obtained as a dark-yellow solid; 1.54 g (66%), mp 86–89 °C.

^{1}H NMR (CDCl$_3$): δ (ppm) = 7.90 (d, J = 8.6 Hz, 1 H, Ar-H), 7.04 (s, 1 H, Ar-H), 6.87 (d, J = 8.6 Hz, 1 H, Ar-H), 4.43 (q, J = 7.3 Hz, 2 H, OCH$_2$), 2.82 (t, J = 7.5 Hz, 2 H, ArCH$_2$), 1.68 (sextet, J = 7.5 Hz, 2 H, CH$_2$CH$_2$CH$_3$), 1.43 (t, J = 7.3 Hz, 3 H, OCH$_2$CH$_3$), 1.03 (t, J = 7.5 Hz, 3 H, CH$_2$CH$_3$).

^{13}C NMR (CDCl$_3$): δ (ppm) = 177.95, 160.75, 155.66, 151.58, 148.36, 124.45, 117.42, 115.38, 114.31, 62.64, 25.87, 21.41, 14.15, 14.08.

Note: It is recommended that the reaction is performed under an N$_2$ atmosphere.

3.4.5.7 * **(Z)-Dimethyl N-[2-(ethoxycarbonyl)-4-oxo-8-propyl-4H-1-benzopyran-7-yl]-2-amino-2-butene-1,4-dioate** [2]

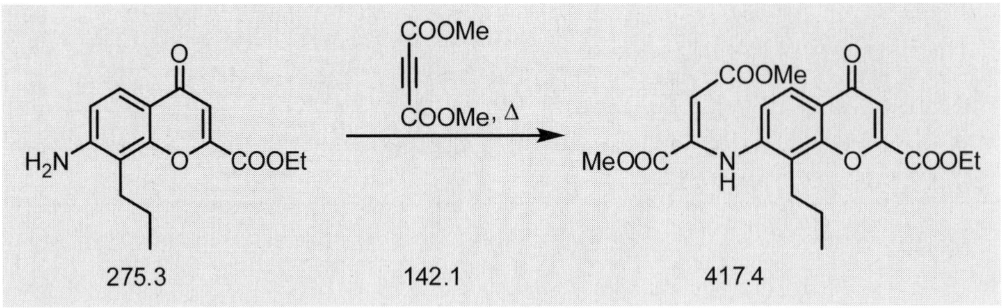

275.3   142.1   417.4

A solution of the amine **3.4.5.6** (1.10 g, 4.00 mmol) and dimethyl acetylenedicarboxylate (0.66 g, 4.70 mmol) in anhydrous EtOH (5 mL) is heated under reflux for 17 h.

On cooling (refrigerator) the product precipitates; it is collected by filtration and dried to give a yellow solid; 860 mg (52%), mp 138–139 °C.

FT-IR (solid): $\tilde{v}$ (cm^{-1}) = 3438, 3387, 3095, 2960, 1725, 1660, 1642, 1593, 1338.

^{1}H NMR (500 MHz, CDCl$_3$): δ (ppm) = 9.86 (s$_{br}$, 1 H, NH), 7.93 (d, J = 8.5 Hz, 1 H, Ar-H), 7.07 (s, 1 H, Ar-H), 6.75 (d, J = 8.5 Hz, 1 H, Ar-H), 5.67 (s, 1 H, =CH), 4.43 (q, J = 6.9 Hz, 2 H, OCH$_2$), 3.79 (s, 3 H, COOCH$_3$), 3.71 (s, 3 H, COOCH$_3$), 3.00 (t, J = 7.3 Hz, 2 H, ArCH$_2$), 1.77 (sextet, J = 7.3 Hz, 2 H, CH$_2$C$\underline{H}_2$CH$_3$), 1.44 (t, J = 6.9 Hz, 3 H, OCH$_2$C$\underline{H}_3$). 1.07 (t, J = 7.3 Hz, 3 H, CH$_2$CH$_2$C$\underline{H}_3$).

^{13}C NMR (125 MHz, CDCl$_3$): δ (ppm) = 178.16, 169.71, 164.21, 160.59, 155.01, 152.17, 146.56, 144.41, 123.55, 122.34, 120.66, 118.52, 114.43, 97.76, 62.80, 53.02, 51.62, 26.53, 22.08, 14.09, 14.00.

3.4.5.8 * **2-Ethyl 8-methyl 6,9-dihydro-4,6-dioxo-10-propyl-4H-pyrano[3,2-g]quinoline-2,8-dicarboxylate** [2]

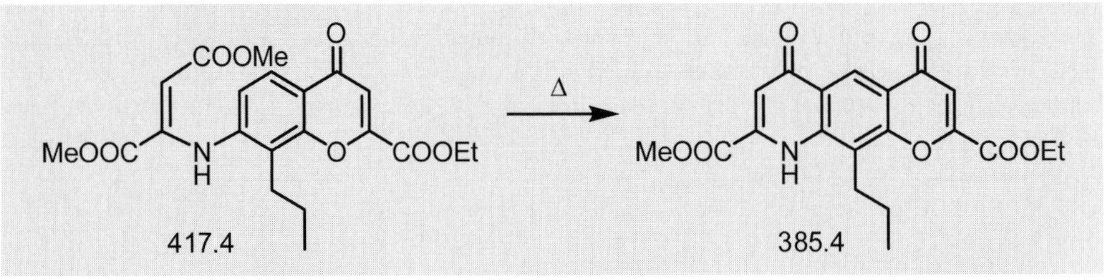

417.4   385.4

The triester **3.4.5.7** (500 mg, 1.20 mmol) is added in one portion to refluxing diphenyl ether (12.5 mL) with stirring and the mixture is heated for 10 min.

The solution is cooled, poured into petroleum ether (bp 60–80 °C, 50 mL), and the precipitated product is collected and dried in vacuo over P$_2$O$_5$. Recrystallization from EtOAc affords the diester as a yellow solid; 355 mg (77%), mp 177–178 °C.

FT-IR (solid): $\tilde{v}$ (cm^{-1}) = 3370, 3094, 2870, 2577, 2465, 1740, 1731, 1637, 1614.

^{1}H NMR (CDCl$_3$): δ (ppm) = 9.00 (s, 1 H, Ar-H), 8.95 (s$_{br}$, 1 H, NH), 7.05 (s, 1 H, Ar-H), 6.87 (s, 1 H, Ar-H), 4.49 (q, J = 7.3 Hz, 2 H, OCH$_2$), 4.08 (s, 3 H, COOCH$_3$), 3.13 (t, J = 7.6 Hz, 2 H, ArCH$_2$), 1.81 (sextet, J = 7.6 Hz, 2 H, CH$_2$C$\underline{H}_2$CH$_3$), 1.47 (t, J = 7.3 Hz, 3 H, OCH$_2$C$\underline{H}_3$), 1.0 (t, J = 7.6 Hz, 3 H, CH$_2$CH$_2$C$\underline{H}_3$).

^{13}C NMR (125 MHz, CDCl$_3$): δ (ppm) = 179.51, 178.04, 163.32, 130.32, 155.00, 152.32, 140.61, 136.89, 124.41, 123.61, 120.67, 118.37, 113.96, 111.44, 63.04, 54.16, 25.69, 22.05, 14.10.

[1] a) A. Kleemann, J. Engel, *Pharmaceutical Substances*, 3rd ed., p. 1313, Georg Thieme Verlag, Stuttgart, **1999**; b) H. Auterhoff, J. Knabe, H.-D. Höltje, *Lehrbuch der Pharmazeutischen Chemie*, 13th ed., p. 586, Wissenschaftliche Verlagsgesellschaft mbH, Stuttgart, **1994**.

[2] H. Cairns, D. Cox, K. J. Gould, A. H. Ingall, J. L. Suschitzky, *J. Med. Chem.* **1985**, *28*, 1832.

[3] Th. Eicher, S. Hauptmann, *The Chemistry of Heterocycles*, 2nd ed., p. 263, Wiley-VCH, Weinheim, **2003**.

[4] a) H. Akhavan-Tafti, R. DeSilva, Z. Arghavani, *J. Org. Chem.* **1998**, *63*, 930; b) Z. Zhang, L. M. V. Tillekerante, R. A. Hudson, *Synthesis* **1996**, 377.

[5] M. Julia, *Bull. Soc. Chim. Fr.* **1952**, 639.

[6] For a review on Claisen rearrangement, see: A. M. Martin Castro, *Chem. Rev.* **2004**, *104*, 3037.

[7] Ref. [3], p. 330.

3.4.6　High-pressure reaction

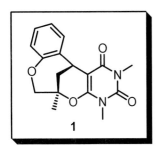

1

Topics:
- Mesylation of a primary homoallylic alcohol
- Nucleophilic substitution (Williamson ether synthesis)
- Knoevenagel condensation
- Intramolecular hetero-Diels–Alder reaction under high pressure

(a) General

The reaction rate and the equilibrium position of many chemical transformations can be strongly influenced by applying high pressure, usually in the range up to 1.5 GPa (15 kbar) (pressure units: 1 kbar = 100 MPa = 0.1 GPa = 14503.8 psi = 986.92 atm) [1]. Of synthetic value is the application of high pressure to transformations with a large negative volume of activation ($\Delta V^{\ddagger}$), since this will increase the reaction rate, allowing the process to be run at a lower temperature; examples include Diels–Alder reactions, 1,3-dipolar cycloadditions, [2+2]-cycloadditions, sigmatropic rearrangements, and radical polymerizations (Table 1).

	$\Delta V^{\ddagger}$ [cm^3 mol^{-1}]
Free radical bond cleavage	0 to 13
S_N2 reaction	0 to –20
Formation of acetals	–5 to –10
Claisen, Cope rearrangements	–8 to –15
Free-radical polymerization	–10 to –25
Diels–Alder reaction	–25 to –50
[2+2]-cycloaddition	–35 to –50

Table 1. Typical $\Delta V^{\ddagger}$ values of organic reactions

Mathematical correlations between reaction rate and applied pressure for different values of volume of activation are indicated in Table 2.

$k(p)/k(0.1\ \text{MPa}) = \exp[-\Delta V^{\ddagger}/RT(p-1)]$

Pressure [MPa]	$\Delta\Delta V^{\ddagger}$ [cm^3 mol^{-1}]			
	+10	–10	–20	–30
100	0.67	1.5	2.2	3.4
300	0.30	3.4	11	38
500	0.13	7.5	56	420
700	0.06	17	280	4800
1000	0.02	56	3200	180000

Table 2. Influence of pressure on rates of reaction at 25 °C

Theoretically, a transformation with a $\Delta V^{\ddagger}$ of -30 cm^3 mol^{-1} can be accelerated by a factor of 2.0×10^6 at 1.5 GPa compared to the reaction at atmospheric pressure; however, the calculated rates are usually only accurate for pressures up to 0.2 GPa. At higher pressures, the influence of increasing viscosity on dynamic effects must be taken into consideration, which would lead to a retardation of any given process [2, 3].

One of the first examples of the usefulness of the application of high pressure was the total synthesis of ($\pm$)-cantharidin (**5**, an ingredient of Spanish Fly) by Dauben and co-workers, using a Diels–Alder reaction of **2** and **3** to give **4** [4]. The reaction does not take place at ambient pressure. Compound **4** can be transformed into ($\pm$)-cantharidin by hydrogenation using Raney nickel as catalyst.

Besides increasing the reaction rate, high pressure can also be employed to improve the chemo-, regio-, diastereo-, and/or enantioselectivity of a chemical transformation [5, 6]. This can be attributed to a temperature effect on transformations for which there is a large difference in reaction enthalpies for the pathways leading to the different isomers. Thus, lowering the temperature usually has a strong effect on the observed selectivity. On the other hand, pure pressure effects on selectivity are also known. The latter are observed for reactions with a pronounced $\Delta\Delta V^{\ddagger}$ value of the different reaction channels leading to the isomers. A $\Delta\Delta V^{\ddagger}$ value of 10 cm^3 mol^{-1} at 1000 MPa corresponds to an isomer ratio of $C_1/C_2 = 1:56.6$ in the product mixture.

An example of a change in the mechanism of a chemoselective reaction upon the application of high pressure is the transformation of the benzylidene-1,3-dicarbonyl compound **6** into **1** by way of an intramolecular hetero-Diels–Alder reaction and to **7** by way of an intramolecular ene reaction [7]. At 110 °C and 100 MPa in CH$_2$Cl$_2$, a ratio of **1** to **7** of 11:1 was observed, whereas at 90 °C and 550 MPa the ratio was found to be 76.3:1. Thus, higher pressure and lower temperature favor the formation of cycloadduct **1**. The $\Delta\Delta V^{\ddagger}$ value amounts to $-(10.7 \pm 1.9)$ cm^3 mol^{-1} and the $\Delta\Delta H^{\ddagger}$ value to $-(32.4 \pm 7.2)$ kJ mol^{-1}.

$\Delta\Delta V^{\ddagger} = -(10.7 \pm 1.9)$ cm^3 mol^{-1}
$\Delta\Delta H^{\ddagger} = -(32.4 \pm 7.2)$ kJ mol^{-1}

Pressure [MPa]	Selectivity (1:7)
75	19.5:1
100	23.5:1
320	40.7:1
550	76.3:1

Influence of pressure on the chemoselectivity of the reaction of **6** in CH$_2$Cl$_2$ at 90°C.

The large difference between the volumes of activation for the two reaction pathways can be correlated with the intrinsic contribution of $\Delta V^{\ddagger}$ for the formation of a covalent bond. In the Diels–Alder reaction to give **1** two single bonds are formed, whereas in the ene reaction to give **7** only one single bond (a C–H bond is not counted) is produced. It should also be noted that the $\Delta\Delta V^{\ddagger}$ value strongly depends on the solvent.

(b) Synthesis of 1

The precursor **13** for the hetero-Diels–Alder reaction is prepared in three steps starting from the commercially available homoallyl alcohol **8**.

Mesylation of **8** gives the sulfonate **9**, which is used for *O*-alkylation of salicylaldehyde (**10**) to afford the ether **11** in an S_N process. Knoevenagel condensation of **11** with *N,N*-dimethylbarbituric acid (**12**) in the presence of a catalytic amount of ethylene diammonium diacetate (EDDA) leads to the benzylidene compound **13**. Intramolecular hetero-Diels–Alder reaction of **13** is carried out under a pressure of 9 kbar and leads to the cycloadduct **1** as the main product.

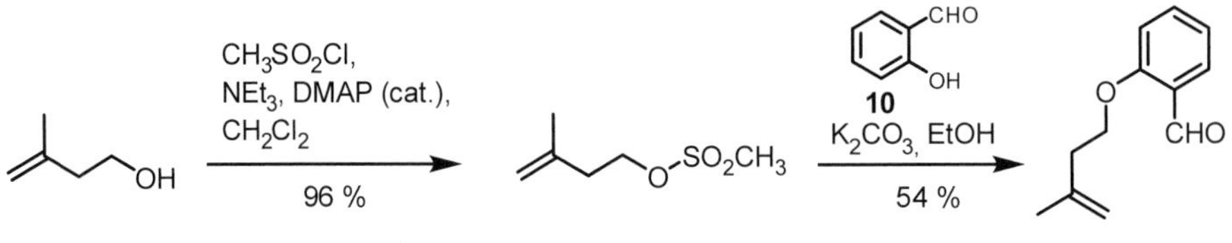

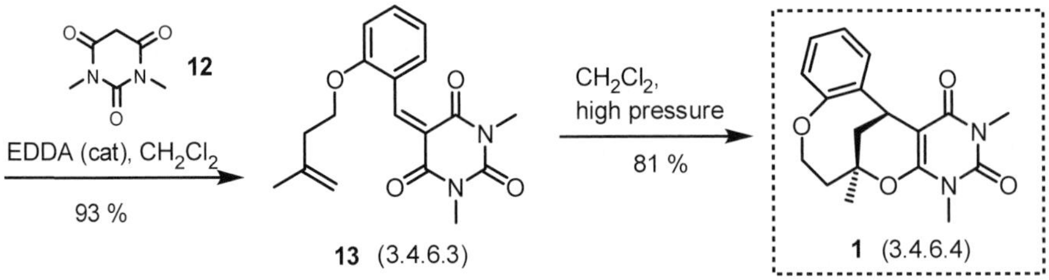

(c) Experimental procedures for the synthesis of 1

3.4.6.1   **   3-Methyl-3-butenyl-methanesulfonate [8]

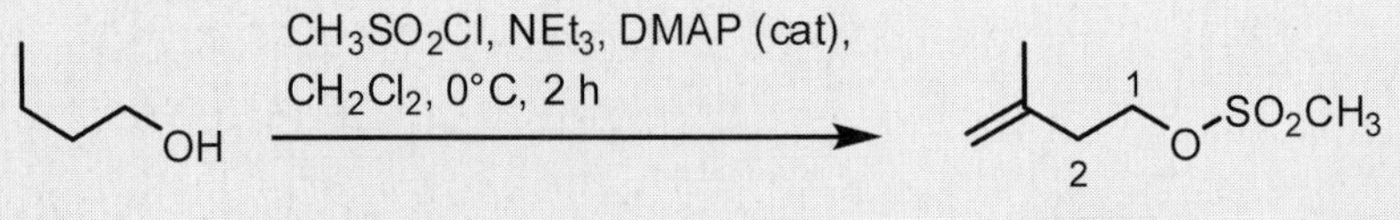

NEt₃ (7.08 g, 70.0 mmol) and a catalytic amount of DMAP are added to a stirred solution of 3-methyl-3-buten-1-ol (5.17 g, 60.0 mmol) in CH_2Cl_2 (120 mL) at 0 °C. Stirring is continued for 15 min, methanesulfonyl chloride (7.56 g, 66.0 mmol) is then added dropwise, and stirring is continued for 2 h at 0 °C.

The reaction is then quenched by the addition of H_2O (150 mL); the organic layer is separated and the aqueous phase is extracted with CH_2Cl_2 (3 × 50 mL). The combined organic layers are washed with saturated NH_4Cl solution (100 mL), saturated $NaHCO_3$ solution (100 mL), and brine (100 mL), and dried over Na_2SO_4. After evaporation of the solvent, the residue is purified by column chromatography (tBuOMe/petroleum ether, 1:3) to yield the sulfonate; 9.45 g (96%), $R_f = 0.28$ (tBuOMe/petroleum ether, 1:3).

IR (KBr): $\tilde{v}$ (cm^{-1}) = 2972, 2942, 2920 (C–H), 1652 (C=C), 1354, 1174 (RSO$_2$OR').

^{1}H NMR (200 MHz, CDCl$_3$): δ (ppm) = 4.88 (m$_c$, 1 H, 4-H), 4.79 (m$_c$, 1 H, 4-H), 4.34 (t, J = 7.0 Hz, 2 H, 1-H$_2$), 3.00 (s, 3 H, S–CH$_3$), 2.46 (t, J = 7.0 Hz, 2 H, 2-H$_2$), 1.77 (s, 3 H, 3-CH$_3$).

3.4.6.2   **   2-(3-Methyl-3-butenyloxy)-benzaldehyde [8]

164.2   122.1   190.2

A stirred suspension of salicylaldehyde (2.50 g, 20.5 mmol), anhydrous K_2CO_3 (3.11 g, 22.5 mmol), and the mesylate **3.4.6.1** (3.03 g, 18.4 mmol) in anhydrous EtOH (40 mL) is heated under reflux for 6 h.

The dark-yellow reaction mixture is then concentrated *in vacuo* and H_2O (60 mL) is added. Extraction with Et$_2$O (3 × 50 mL), washing of the combined organic layers with NaOH (2 M, 50 mL) and brine (50 mL), drying over Na_2CO_3, and removal of the solvent *in vacuo* provides a yellow oil. After column chromatography (tBuOMe/petroleum ether, 1:20), 1.88 g (54%) of the O-alkylated salicylaldehyde is obtained, $R_f = 0.41$ (tBuOMe/petroleum ether, 1:10).

UV (CH$_3$CN): λ_{max} (nm) (lg ε) = 318 (3.6752), 251 (3.9983), 215 (4.3349).

IR (KBr): $\tilde{v}$ (cm^{-1}) = 3042, 2970, 2940, 2882 (C–H), 1690 (C=O), 1600 (C=C), 1458 (CH$_2$, CH$_3$).

^{1}H NMR (200 MHz, CDCl$_3$): δ (ppm) = 10.49 (s, 1 H, CHO), 7.89 (dd, J = 8.0, 2.0 Hz, 1 H, 6-H), 7.60–7.46 (m, 1 H, 4-H), 7.06–6.96 (m, 2 H, 3-H, 5-H), 4.84 (m$_c$, 2 H, 4'-H$_2$), 4.20 (t, J = 6.5 Hz, 2 H, 1'-H$_2$), 2.57 (t, J = 6.5 Hz, 2 H, 2'-H$_2$), 1.81 (s, 3 H, 3'-CH$_3$).

^{13}C NMR (50.3 MHz, CDCl$_3$): δ (ppm) = 189.8 (CHO), 161.3 (C-2), 141.6 (C-3'), 135.9 (C-4), 128.1 (C-6), 125.0 (C-1), 120.6 (C-5), 112.6 (C-4'), 112.5 (C-3), 66.76 (C-1'), 37.11 (C-2'), 22.59 (3'-CH$_3$).

MS (EI, 70 eV): m/z (%) = 190 (11) [M]$^+$, 122 (100) [M – C$_5$H$_8$]$^+$, 69 (68) [C$_5$H$_9$]$^+$, 41 (89) [C$_3$H$_5$]$^+$.

3.4.6.3 * 5-[2-(3-Methyl-3-butenyloxy)-benzylidene]-1,3-dimethyl-pyrimidine-2,4,6-trione [8]

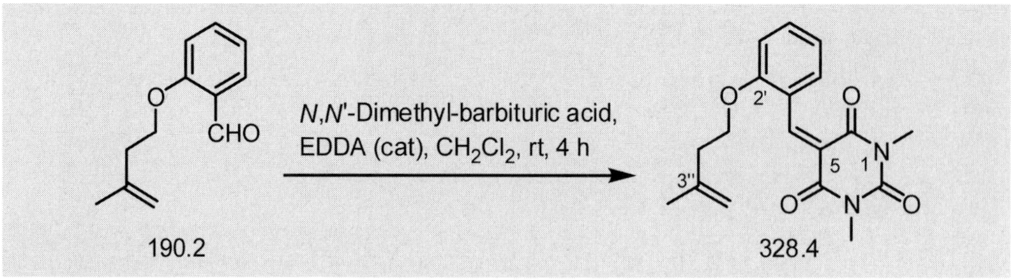

The aldehyde **3.4.6.2** (1.00 g, 5.27 mmol) is added to a mixture of *N,N*-dimethylbarbituric acid (0.78 g, 5.00 mmol) and ethylene diammonium diacetate (EDDA; 10.0 mg, 0.056 mmol) in anhydrous CH_2Cl_2 (40 mL) and the reaction mixture is stirred for 4 h at room temperature.

After evaporation of the solvent under reduced pressure, the resulting yellow oil crystallizes on standing at –20 °C. Recrystallization from MeOH affords the pyrimidine-2,4,6-trione as yellow crystals; 1.53 g (93%), mp 128–129 °C, $R_f = 0.28$ (*t*BuOMe/petroleum ether, 1:3).

UV (CH_3CN): λ_{max} (nm) (lg ε) = 373 (4.0078), 315 (3.8353), 245 (4.0054), 221 (4.0440).

IR (KBr): $\tilde{\nu}$ (cm^{-1}) = 3046, 2966, 2942, 1666, 1574, 1462.

^{1}H NMR (200 MHz, CDCl$_3$): δ (ppm) = 8.90 (s, 1 H, α-H), 8.03 (dd, J = 8.0, 2.0 Hz, 1 H, 6'-H), 7.47 (m$_c$, 1 H, 4'-H), 7.05–6.89 (m, 2 H, 3'-, 5'-H), 4.88–4.78 (m, 2 H, 4''-H$_2$), 4.16 (t, J = 7.0 Hz, 2 H, 1''-H), 3.42 (s, 3 H, N-CH$_3$), 3.34 (s, 3 H, N-CH$_3$), 2.53 (t, J = 7.0 Hz, 2 H, 2''-H$_2$), 1.80 (s, 3 H, 3''-CH$_3$).

^{13}C NMR (50.3 MHz, CDCl$_3$): δ (ppm) = 162.4 (C=O), 160.4 (C=O), 122.3 (C-2'), 119.7 (C-5'), 117.2 (C-5), 112.5 (C-4''), 111.3 (C-3'), 67.11 (C-1''), 37.12 (C-2''), 28.88 (N-CH$_3$), 28.31 (N-CH$_3$), 22.74 (C-3'').

MS (70 eV): m/z (%) = 328 (10) $[M]^+$, 243 (100) $[M - OC_5H_9]^+$, 41 (43) $[C_3H_5]^+$.

3.4.6.4 ** (6R,14S)-(±)-6,14-Methano-2,4,6-trimethyl-6,7,8,14-tetrahydro-4H-5,9-dioxa-2,4-diaza-dibenzo[a,d]cyclodecene-1,3-dione [8]

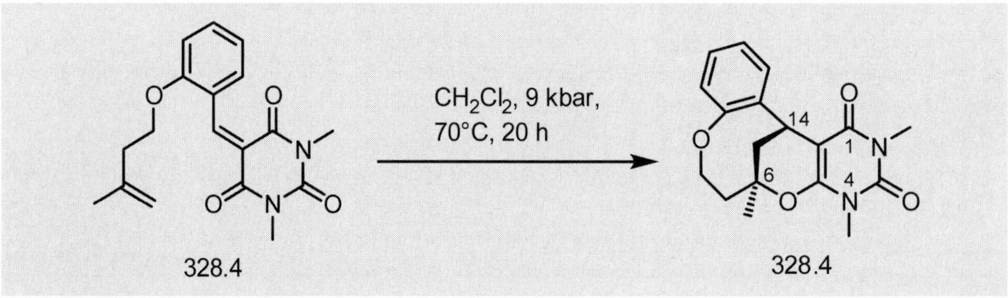

A solution of the benzylidene compound **3.4.6.3** (55.0 mg, 0.17 mmol) in CH_2Cl_2 (4 mL) is placed in a Teflon tube, one end of which is closed. The tube is sealed under argon (heating pliers), placed in a high-pressure device, and kept for 20 h at 9 kbar and 70 °C.

The tube is then opened, the solvent is evaporated under reduced pressure, and the remaining yellow-orange oil is purified by flash chromatography (EtOAc) to yield the title compound; 44.6 mg (81%), mp 160–161 °C, R_f = 0.49 (EtOAc).

UV (CH_3CN): λ_{max} (nm) (lg ε) = 226 (3.9800).

IR (KBr): $\tilde{v}$ (cm^{-1}) = 3016, 2966, 2930, 1700, 1646, 1634, 1612, 1456.

^{1}H NMR (300 MHz, $CDCl_3$): δ (ppm) = 7.51 (d_{br}, J = 8.0 Hz, 1 H, 13-H), 7.22–7.08 (m, 2 H, 11-H, 12-H), 6.90 (dd, J = 8.0, 1.5 Hz, 1 H, 10-H), 4.25–4.08 (m, 2 H, 8-H_{eq}, 14-H), 4.00 (dt, J = 12.0, 4.5 Hz, 1 H, 8-H_{ax}), 3.41 (s, 3 H, N-CH_3), 3.29 (s, 3 H, N-CH_3), 2.39 (s_{br}, 1 H, 7-H_{eq}), 2.13 (dd, J = 15.0, 6.0 Hz, 1 H, 15-H), 1.97 (dd, J = 10.0, 4.5 Hz, 1 H, 7-H_{ax}), 1.88 (dt, J = 15.0, 4.5 Hz, 1 H, 15-H), 1.56 (s, 3 H, 6-CH_3).

^{13}C NMR (50.3 MHz, $CDCl_3$): δ (ppm) = 162.6 (C-1), 156.5 (C-9a), 154.4 (C-4a), 151.1 (C-3), 136.7 (C-13a), 130.9 (C-13), 128.6 (C-11), 125.2 (C-12), 122.8 (C-10), 89.89 (C-14a), 82.23 (C-6), 69.42 (C-8), 41.20 (C-7), 37.72 (C-15), 32.81 (6-CH_3), 31.80 (C-14), 28.63 (N-CH_3), 27.90 (N-CH_3).

MS (70 eV): m/z (%) = 328 (100) $[M]^+$, 243 (51) $[M - OC_5H_9]^+$, 69 (8) $[C_5H_9]^+$, 41 (10) $[C_3H_5]^+$.

---

[1] R. van Eldik, F.-G. Klärner (Eds.), *High-Pressure Chemistry*, Wiley-VCH, Weinheim, **2002**.

[2] L. Nikowa, D. Schwarzer, J. Troe, J. Schroeder, *J. Chem. Phys.* **1992**, *97*, 4827.

[3] T. Asano, K. Cosstick, H. Furuta, K. Matsuo, H. Sumi, *Bull. Chem. Soc. Jpn.* **1996**, *69*, 551.

[4] N. G. Dauben, C. R. Kressel, K. H. Takemura, *J. Am. Chem. Soc.* **1980**, *102*, 6893.

[5] Reviews: a) L. F. Tietze, P. L. Steck, in *High-Pressure Chemistry* (Eds.: R. van Eldik, F.-G. Klärner), p. 239, Wiley-VCH, Weinheim, **2002**; b) F.-G. Klärner, F. Wurche, *J. Prakt. Chem.* **2000**, *342*, 609; c) A. Drljaca, C. D. Hubbard, R. van Eldik, T. Asano, M. V. Basilevski, W. J. le Noble, *Chem. Rev.* **1998**, *98*, 2167; d) K. Matsumoto, M. Kaneko, H. Katsura, N. Hayashi, T. Uchida, R. M. Acheson, *Heterocycles* **1998**, *47*, 1135; e) G. Jenner, *Tetrahedron* **1997**, *53*, 2669.

[6] Monographs: a) N. S. Isaacs, *Liquid High-Pressure Chemistry*, Wiley, New York, **1991**; b) W. J. le Noble (Ed.), *Organic High-Pressure Chemistry*, Elsevier, Amsterdam, **1988**; c) R. M. Acheson, E. K. Matsumoto, *Organic Synthesis at High Pressure*, Wiley, New York, **1991**; d) R. van Eldik, C. D. Hubbard, *Chemistry under Extreme and Non-Classic Conditions*, Wiley, New York, **1997**.

[7] L. F. Tietze, C. Ott, unpublished results.

[8] C. Ott, *Ph.D. Thesis*, University of Göttingen, **1994**.

3.5   Other heterocyclic systems, heterocyclic dyes

3.5.1   (±)-Samin

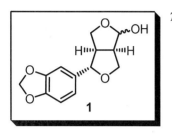

1

Topics:
- Synthesis of a natural product of the furofuran type
- Wittig reaction
- Regioselective halogenation of a C=C double bond with interception of an external nucleophile
- *exo-trig* Radical cyclization
- Reduction COOR → CH$_2$OH, OH silylation, desilylation
- Hydroboration/oxidation
- Swern oxidation R–CH$_2$OH → R–CH=O
- Lactol formation

(a) General

The furofuran lignans are one of the largest groups of naturally occurring lignans, the members of which show a variety of biological activities, e.g. as anti-tumor and anti-fungal agents or as growth inhibitors. Samin (**1**) is a constituent of sesame oil and a central intermediate for the biogenesis of furofuran lignans [1].

Retrosynthesis of **1** starts with FGI of the lactol functionality to a 4-hydroxyaldehyde unit in the 2,3,4-trisubstituted tetrahydrofuran **2**, the substituents on which have a 2,3-*trans*- and a 3,4-*cis*-relationship:

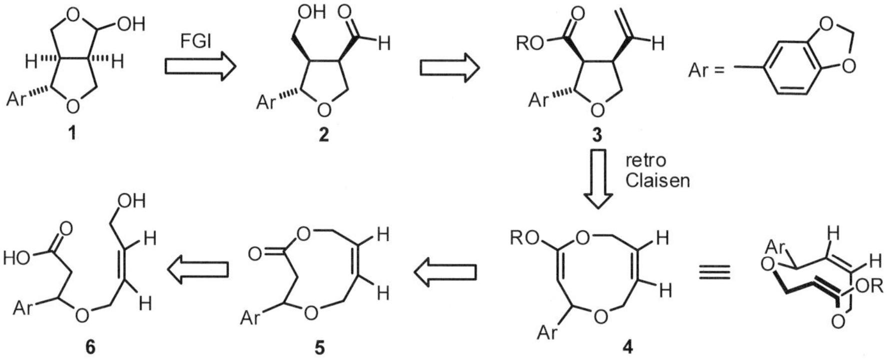

Further retroanalysis may proceed, for example, with FGI of CH$_2$OH to a carboxylic ester, reduction of which requires protection of the 4-aldehyde function, e.g. as a C=C double bond, and may lead to the key intermediate **3**. From **3**, a retro-Claisen process provides the ketene acetal **4**, which in turn leads to the nine-membered ring lactone **5** and finally to the carboxylic acid **6** containing an allylic alcohol moiety with (*Z*)-configuration. Thus, according to the presented retroanalytical approach, the carboxylic acid **6** would serve as a starting material.

In fact, a synthesis of **1** from **6** based on a Claisen rearrangement has been reported [2]. The carboxylic acid **6** is prepared by 1,4-addition of the TBDMS-monoprotected (*Z*)-2-butene-1,4-diol (**8**) to the α-sulfonylcinnamate **7** (→ **9**), reductive removal of the sulfonyl group, ester hydrolysis, and desilylation (**9** → **6**):

Macrolactonization of the hydroxy acid **6** to give **5** is accomplished using the Mukaiyama reagent **10** in the presence of NEt₃ (cf. **3.1.3**) by activation of the carboxylic acid. It can be assumed that a 2-acyloxypyridinium ion is formed, which facilitates intramolecular acyl transfer to the hydroxyl group [3]. Treatment of lactone **5** with LDA generates an ester enolate, which is trapped by silylation with (CH₃)₃SiCl. The formed *O*-silyl ketene acetal **4** (R = (CH₃)₃Si) undergoes a Claisen [3,3]-sigmatropic rearrangement on warming from −78 °C to room temperature, thus leading to the trisubstituted tetrahydrofuran derivative **11** with the correct stereochemistry at the centers C-3/C-4/C-5. In **11**, the ester functionality is reduced to the primary alcohol with LiAlH₄, the vinyl group is dihydroxylated with OsO₄, and the 1,2-diol is cleaved by oxidation with NaIO₄ to give the hydroxyaldehyde **2**, which spontaneously forms samin **1**.

In section (b), however, another approach to (±)-samin is presented [4], which employs a synthetic strategy based on a different retroanalysis utilizing a radical cyclization process as a key transformation.

(b) Synthesis of 1

The starting material for the synthesis of **1** is the cinnamate **13**, which is readily available by a Wittig reaction of piperonal with the (ethoxycarbonylmethylene)phosphorane **12**. Reaction of the α,β-unsaturated ester **13** with NBS in the presence of propargyl alcohol affords the α-bromoester **15** bearing a propargyloxy residue in the β-position.

The transformation **13 → 15** can be explained by the initial formation of a bromonium ion **14**, which is attacked by propargyl alcohol as an external nucleophile at the benzylic position (either directly or via equilibration **14a ⇌ 14b** favored by stabilization of the positive charge through the aryl donor system).

When the bromoester **15** is treated with tributyltin hydride/AIBN, the 4-methylenetetrahydrofuran **18** is obtained in an *endo-trig* radical cyclization process [5]. Initially, dehalogenation of **15** leads to a radical intermediate **16**, in which the aryl and COOEt substituents may occupy pseudo-equatorial positions, this arrangement being responsible for the *trans* orientation of the C-3/C-4 substituents in the cyclized radical **17** and the product **18** [6]:

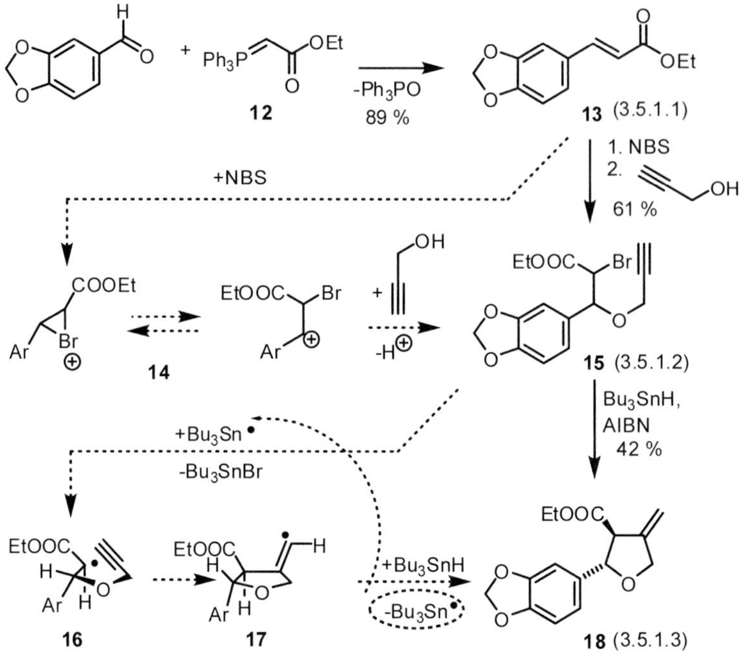

As is usually the case in dehalogenations with Bu₃SnH, tributylstannyl radicals resulting from initiation with AIBN are further produced in the chain propagation sequence (S• = starter radical, cf. **1.8.1**):

$$\text{Initiation:} \quad \text{S} \bullet \quad + \quad \text{Bu}_3\text{SnH} \rightarrow \text{S–H} \quad + \quad \text{Bu}_3\text{Sn} \bullet$$

$$\text{Propagation: Bu}_3\text{Sn} \bullet + \quad \text{R–Br} \quad \rightarrow \text{Bu}_3\text{Sn–Br} \quad + \quad \text{R} \bullet \quad (= \mathbf{16})$$

$$\text{cyclization:} \quad \text{R} \bullet \ (= \mathbf{16}) \quad \rightarrow \quad \text{R'} \bullet \ (= \mathbf{17})$$

$$\text{R'} \bullet \ (= \mathbf{17}) + \quad \text{Bu}_3\text{SnH} \rightarrow \text{R'-H} \ (= \mathbf{18}) \quad + \quad \text{Bu}_3\text{Sn} \bullet$$

To complete the synthesis, the COOEt group in the key intermediate **18** is transformed into a CH₂OH group and the *exo*-methylene moiety is converted into an aldehyde function. Thus, reduction of **18** with LiAlH₄ leads to the primary alcohol **19**, which is protected using TBDMS-Cl in the presence of imidazole, thereby affording the silyl ether **20**. Hydroboration of **20** with B₂H₆/H₂O₂ followed by reaction with (COCl)₂/DMSO (Swern oxidation, cf. **2.3.1**) gives the aldehyde **21** as a mixture of diastereomers. However, on reaction of **21** with TBAF, the silyl protecting group is removed and the *cis*-diastereomer of the formed hydroxyaldehyde, which is in equilibrium with the *trans*-diastereomer, spontaneously forms the lactol function of (±)-samin by addition of the hydroxyl group to the aldehyde function. In this way, the diastereomeric mixture of **21** is entirely transformed into **1**.

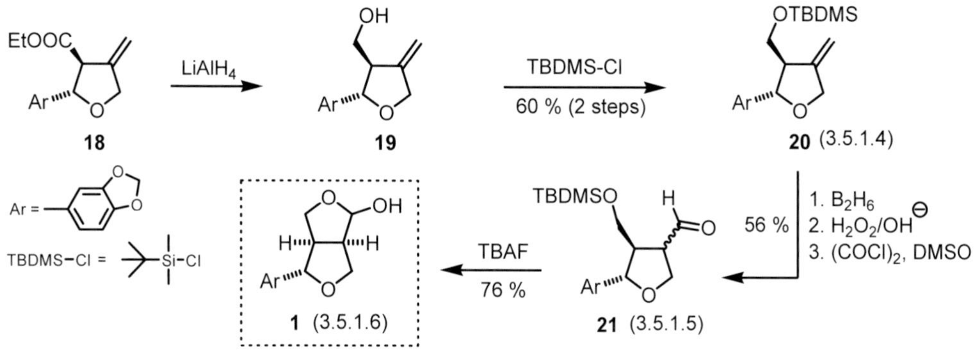

18 → LiAlH₄ → **19** → TBDMS-Cl, 60 % (2 steps) → **20** (3.5.1.4)

Ar = piperonyl group

TBDMS-Cl = chlorotrimethylsilane structure

1 (3.5.1.6) ← TBAF, 76 % ← **21** (3.5.1.5) ← 56 %, 1. B₂H₆, 2. H₂O₂/OH⁻, 3. (COCl)₂, DMSO

Thus, the target molecule **1** is obtained in a sequence of six individual steps in an overall yield of 8% (based on piperonal).

(c) Experimental procedures for the synthesis of 1

3.5.1.1   *   (*E*)-Ethyl 3-(benzo[*d*][1,3]dioxol-5-yl)acrylate [7]

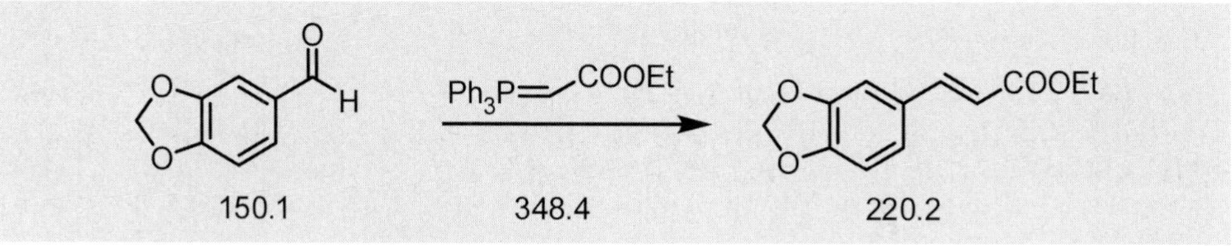

150.1   Ph₃P=CHCOOEt   348.4   220.2

A stirred solution of piperonal (1.18 g, 7.80 mmol) and (ethoxycarbonylmethylene)-triphenyl-phosphorane (3.00 g, 8.60 mmol) in THF (100 mL) is heated under reflux under a nitrogen atmosphere for 15 h.

The solution is then cooled to room temperature, diluted with Et₂O (100 mL), and washed with brine (80 mL). The organic phase is dried over Na₂SO₄ and concentrated *in vacuo*. The remaining light-yellow solid is purified by flash chromatography (CH₂Cl₂) to give a colorless solid; 1.53 g (89%), mp 64–65 °C; R_f = 0.68 (CH₂Cl₂).

UV: λ_{max} (nm) = 324, 216.

IR (KBr): $\tilde{v}$ (cm⁻¹) = 2990, 2904, 1702, 1641, 1504, 1490, 1244, 1173, 927.

¹H NMR (CDCl₃): δ (ppm) = 7.38 (d, *J* = 15.9 Hz, 1 H, =CH), 6.98–7.01 (m, 2 H, Ar-H), 6.79 (d, *J* = 7.9 Hz, 1 H, Ar-H), 6.22 (d, *J* = 15.9 Hz, Ar-CH), 5.99 (s, 2 H, O–CH₂–O), 4.21 (q, *J* = 7.1 Hz, 2 H, O–CH₂), 1.24 (t, *J* = 7.1 Hz, 3 H, CH₃).

¹³C NMR (CDCl₃): δ (ppm) = 166.9, 149.4, 148.2, 144.0, 128.7, 124.2, 116.0, 108.3, 106.3, 101.4, 60.1, 14.2.

HRMS: [*M* + Na]⁺ = 243.0628 (calcd. 243.0628), [*M* + H]⁺ = 221.0808 (calcd. 221.0808).

3.5.1.2   **   Ethyl 3-(benzo[d][1,3]dioxol-5-yl)-2-bromo-3-(prop-2-ynyloxy)propanoate [8]

220.2   56.1   178.0   355.2

A solution of N-bromosuccinimide (1.46 g, 8.23 mmol) in propargyl alcohol (14 mL) is cooled to $-30\,°C$ whereupon a solution of the cinnamate **3.5.1.1** (1.51 g, 6.86 mmol) in anhydrous CH_2Cl_2 (8 mL) is added dropwise. The solution is stirred for 2 h at $-15\,°C$, then slowly heated to $40\,°C$ and stirred for 12 h at this temperature.

The reaction mixture is then diluted with NaOH solution (1 M, 20 mL). The resulting mixture is extracted with CH_2Cl_2 (3 × 30 mL). The combined organic phases are washed with brine, dried over Na_2SO_4, and concentrated *in vacuo*. The crude product is purified by flash chromatography (CH_2Cl_2) to give a colorless oil; 1.77 g (61%); $R_f = 0.38$ (CH_2Cl_2).

> **UV:** λ_{max} (nm) = 287, 237, 202.
>
> **IR** (film): $\tilde{v}$ (cm^{-1}) = 3292, 1741, 1489, 1445, 1371, 1295, 1248, 1182, 1149, 1068, 1038, 640.
>
> **^{1}H NMR** ($CDCl_3$): δ (ppm) = 6.79–6.86 (m, 3 H, Ar-H), 5.96 (m, 2 H, O–CH_2–O), 4.82 (d, $J = 10.2$ Hz, 1 H, Br–CH), 4.23–4.37 (m, $J = 7.2$ Hz, 2 H, CH_3–$\underline{CH_2}$), 4.80 (d, $J = 10.2$ Hz, 1 H, O–CH), 4.12–4.06 (dd, $J = 15.6/2.1$ Hz, 1 H, HC≡C–$\underline{CH_2}$), 3.87–3.81 (dd, $J = 15.6/2.1$ Hz, 2 H, HC≡C–$\underline{CH_2}$), 2.41 (t, $J = 2.4$ Hz, 1 H, HC≡), 1.31 (t, $J = 7.2$ Hz, 3 H, CH_3).
>
> **^{13}C NMR** ($CDCl_3$) δ (ppm) = 168.4, 148.3, 148.0, 129.7, 122.6, 108.0, 107.6, 101.3, 80.6, 76.6, 75.0, 62.1, 56.2, 47.5, 13.9.
>
> **HRMS:** $[M + K]^+$ = 392.9735 (calcd. 392.9734), $[M + Na]^+$ = 376.9995 (calcd. 376.9995).

3.5.1.3   ***   Ethyl 2-(benzo[d][1,3]dioxol-5-yl)-4-methylenetetrahydrofuran-3-carboxylate [9]

355.2   291.1   276.3

nBu₃SnH (321 μL, 1.24 mmol) is added by means of a syringe to a stirred solution of the propargyl ether **3.5.1.2** (400 mg, 1.13 mmol) and a catalytic amount of AIBN in anhydrous, degassed toluene (70 mL) under nitrogen atmosphere. The solution is heated under reflux for 4.5 h.

The reaction mixture is then diluted with 10% KF solution (40 mL). The organic phase is separated, dried over Na_2SO_4, and concentrated *in vacuo*. The crude product obtained is purified by flash chromatography (CH_2Cl_2) to give a colorless oil, 131 mg (42%), $R_f = 0.18$ (CH_2Cl_2).

UV: λ_{max} (nm) = 287, 238, 201.

IR (KBr): $\tilde{\nu}$ (cm^{-1}) = 3291, 1740, 1489, 1445, 1248, 1068, 1038.

^{1}H NMR (CDCl$_3$): δ (ppm) = 6.77–6.90 (m, 3 H, Ar-H), 5.94 (s, 2 H, O–CH$_2$–O), 5.08–5.19 (m, 2 H, O–CH$_2$–C=), 5.13 (d, J = 8.0 Hz, 1 H, Ar–CH), 4.41–4.63 (m, 2 H, =CH$_2$), 4.09–4.29 (m, 2 H, C$\underline{H}_2$–CH$_3$), 3.41 (dd, J = 8.0/3.0 Hz, O–CH–C$\underline{H}$), 1.22 (t, J = 8.0 Hz, 3 H, CH$_2$–C$\underline{H}_3$).

^{13}C NMR (CDCl$_3$): δ (ppm) = 170.6, 147.8, 147.4, 146.4, 133.7, 119.8, 108.2, 106.6, 106.4, 101.1, 83.3, 71.5, 61.2, 57.2, 14.2.

3.5.1.4   *   ((2-(Benzo[*d*][1,3]dioxol-5-yl)-4-methylenetetrahydrofuran-3-yl)methoxy)(*tert*-butyl)dimethylsilane [4]

A stirred mixture of the ester **3.5.1.3** (121.0 mg, 0.43 mmol) and LiAlH$_4$ (33.7 mg, 0.89 mmol) in Et$_2$O (10 mL) under an argon atmosphere is heated under reflux for 3 h.

The suspension is then cooled to 0 °C and diluted with saturated aqueous Na$_2$SO$_4$ solution. The organic layer is separated, dried over Na$_2$SO$_4$, and concentrated. The crude product is dissolved in anhydrous DMF (2 mL) under an argon atmosphere and imidazole (73.5 mg, 1.08 mmol) is added. The solution is cooled to 0 °C and *tert*-butyldimethylchlorosilane (97.2 mg, 0.65 mmol) is added (3 × 32.4 mg) over 60 min. The solution is allowed to warm to room temperature overnight.

The reaction mixture is then diluted with H$_2$O (25 mL) and extracted with EtOAc (30 mL). The aqueous phase is extracted with EtOAc (3 × 30 mL) and the combined organic layers are dried over Na$_2$SO$_4$ and concentrated. The crude product obtained is purified by flash chromatography (*n*-pentane/EtOAc, 95:5) to give a colorless oil, 90.8 mg (60%); R_f = 0.55 (*n*-pentane/EtOAc, 95:5).

UV: λ_{max} (nm) = 286, 236, 201.

IR (film): $\tilde{\nu}$ (cm^{-1}) = 2954, 2929, 2857, 1504, 1490, 1445, 1252, 1105, 1041, 837, 777.

^{1}H NMR (CDCl$_3$): δ (ppm) = 6.87–6.72 (m, 3 H, Ar-H), 5.91 (s, 2 H, O–CH$_2$–O), 5.03–4.97 (m, 2 H, O–CH$_2$–C=), 4.80 (d, J = 6.6 Hz, 1 H, Ar–CH), 4.56–4.33 (m, 2 H, =CH$_2$), 3.75–3.65 (m, 1 H, Si–O–CH$_2$), 2.77–2.67 (m, 1 H, O–CH–C$\underline{H}$), 0.85 (s, 9 H, C(CH$_3$)$_3$), 0.02 (d, 6 H, J = 2.1 Hz, Si(CH$_3$)$_2$).

^{13}C NMR (CDCl$_3$): δ (ppm) = 148.9, 147.7, 146.9, 136.0, 119.7, 108.0, 106.8, 104.7, 100.9, 83.5, 71.4, 63.5, 54.2, 25.8, 18.2, –5.5.

3.5.1.5 ** 2-(Benzo[*d*][1,3]dioxol-5-yl)-3-((*tert*-butyldimethylsilyloxy)methyl)-tetrahydrofuran-4-carbaldehyde [10]

BH$_3$·THF complex (1 M, 427 µL, 0.427 mmol) is added dropwise to a stirred solution of the alkene **3.5.1.4** (67.0 mg, 0.192 mmol) in THF (2 mL) at 0 °C under an argon atmosphere. The reaction mixture is allowed to warm to room temperature overnight. EtOH (0.28 mL), NaOH (3 M, 0.83 mL), and 30% H$_2$O$_2$ solution (0.83 mL) are then added at 0 °C and the mixture is stirred for 12 h at room temperature.

Subsequently, the mixture is diluted with EtOAc (25 mL) and the organic phase is separated. The aqueous phase is extracted with EtOAc (3 × 30 mL) and the combined organic phases are washed with 10% Na$_2$SO$_3$ solution and brine, dried over Na$_2$SO$_4$, and concentrated. The residue is dissolved in CH$_2$Cl$_2$ and added dropwise to a solution of oxalyl chloride (19.7 µL, 0.230 mmol) and DMSO (34.0 µL, 0.480 mmol) in CH$_2$Cl$_2$ (3 mL) at –60 °C under an argon atmosphere. After 1 h, triethylamine (138 µL, 1.00 mmol) is added and the mixture is allowed to warm to room temperature, whereupon it is stirred for a further 1 h.

The reaction is then quenched by the addition of H$_2$O (1 mL) and the mixture is diluted with CH$_2$Cl$_2$ (20 mL). The organic layer is washed with HCl solution (1 M, 10 mL) and brine, dried over Na$_2$SO$_4$, and concentrated *in vacuo*. The crude product obtained is purified by filtration through a short column of silica gel (EtOAc/*n*-pentane, 25:75) to give a colorless oil; 39.4 mg (56%), R_f = 0.62 (EtOAc).

UV: λ_{max} (nm) = 286, 236, 201.

IR (film): $\tilde{v}$ (cm^{-1}) = 2956, 2928, 2856, 1722, 1490, 1445, 1098, 1040, 809.

^{1}H NMR (CDCl$_3$): δ (ppm) = 9.85 (d, *J* = 2.8 Hz, 1 H, CHO), 6.83–6.71 (m, 3 H, Ar-H), 5.93 (m, 2 H, O–CH$_2$–O), 4.72 (d, *J* = 8.3 Hz, 1 H, Ar–CH–CH), 4.23 (d, *J* = 6.9 Hz, 2 H, CH$_2$–CH–CHO), 3.84–3.58 (m, 2 H, Si–O–CH$_2$), 3.24–3.13 (m, 1 H, CH–CHO), 2.56–2.45 (Ar–CH–CH), 0.87 (s, 9 H, C(CH$_3$)$_3$), 0.02 (d, 6 H, *J* = 3.1 Hz, Si(CH$_3$)$_2$).

^{13}C NMR (CDCl$_3$): δ (ppm) = 200.9, 147.9, 147.2, 134.8, 119.6, 108.1, 106.3, 101.0, 81.9, 67.3, 59.2, 53.7, 53.4, 25.7, 18.0, –5.8.

DCI-MS: *m/z* = 746.6 [2*M* + NH$_4$]$^+$, 399.3 [*M* + NH$_3$ + NH$_4$]$^+$, 382.3 [*M* + NH$_4$]$^+$.

3.5.1.6 * 4-(Benzo[*d*][1,3]dioxol-5-yl)hexahydrofuro[3,4-*c*]furan-1-ol (*rac*-samin) [4]

A solution of tetrabutylammonium fluoride in THF (1 M, 0.294 mL, 0.294 mmol) is added to a stirred solution of the aldehyde **3.5.1.5** (35.6 mg, 0.098 mmol) in THF (1 mL) at 0 °C and stirring is continued for 6 h at room temperature.

The solvent is then removed *in vacuo* and H$_2$O (5 mL) and EtOAc (10 mL) are added. The organic layer is separated, washed with 10% aqueous NaOH solution (20 mL) and brine (20 mL), dried over Na$_2$SO$_4$, and the solvent is evaporated. The residue is purified by flash chromatography (*n*-pentane/EtOAc, 60:40) to give *rac*-samin as a crystalline solid; 17.8 mg (73%), mp 106–107 °C, R_f = 0.31 (*n*-pentane/EtOAc, 60:40).

UV: λ_{max} (nm) = 286, 235, 201.

IR (film): $\tilde{v}$ (cm^{-1}) = 3406, 2961, 2926, 2854, 1495, 1451, 1263, 1245, 1040, 1023, 808.

^{1}H NMR (CDCl$_3$): δ (ppm) = 6.80–6.67 (m, 3 H, Ar-H), 5.89 (m, 2 H, O–CH$_2$–O), 5.32 (s, 1 H, CH–OH), 4.35–4.25 (m, 2 H, Ar–CH–O–CH, Ar–CH), 4.13–4.07 (m, 1 H, HO–CH–O–CH), 3.85 (d, *J* = 9.7 Hz, 1 H, HO–CH–O–CH), 3.50 (m, 1 H, Ar–CH–O–CH), 3.06–2.96 (m, 1 H, HO–CH), 2.84–2.75 (m, 1 H, Ar–CH–CH).

^{13}C NMR (CDCl$_3$): δ (ppm) = 148.0, 147.3, 134.5, 119.6, 108.1, 106.5, 102.2, 101.1, 86.9, 71.2, 69.3, 53.6, 52.8.

DCI-MS: *m/z* = 267.1 [*M* + NH$_3$]$^+$, 250.1 [*M* – H$_2$O + NH$_4$]$^+$.

---

[1] For an instructive overview on samin and its synthesis, see: T. Wirth, *Syntheseplanung – aber wie?*, p. 89, Spektrum Akademischer Verlag, Heidelberg, **1998**.

[2] H. M. Bradley, D. W. Knight, *J. Chem. Soc., Chem. Commun.* **1991**, 1641.

[3] Th. Eicher, S. Hauptmann, *The Chemistry of Heterocycles*, 2nd ed., p. 308, Wiley-VCH, Weinheim, **2003**.

[4] G. Maiti, S. Adhikari, S. C. Roy, *Tetrahedron* **1995**, *51*, 8389.

[5] For the classification of ring-closure reactions (Baldwin rules), see: M. B. Smith, J. March, *March's Advanced Organic Chemistry*, p. 282, John Wiley & Sons, Inc., New York, **2001**.

[6] In an analogous natural product synthesis, stereoselective formation of trisubstituted tetrahydrofurans by radical cyclization has also been performed using hypophosphite: S. C. Roy, C. Guin, K. K. Rana, G. Maiti, *Tetrahedron* **2002**, *58*, 2435.

[7] J .C. Galland, S. Dias, M. Savignac, J.-P. Genêt, *Tetrahedron* **2001**, *57*, 5137.

[8] M. Okabe, M. Abe, M. Tada, *J. Org. Chem.* **1982**, *47*, 1775.

[9] S. C. Roy, S. Adhikari, *Tetrahedron* **1993**, *49*, 8415.

[10] L. C. Dias, G. Diaz, A. A. Ferreira, P. R. R. Meira, E. Ferreira, *Synthesis* **2003**, 603.

3.5.2 Dibenzopyridino[18]crown-6

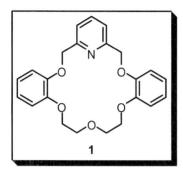

1

Topics:
- Synthesis of a crown ether
- Alkyl chlorides from alcohols and $SOCl_2$
- Monoalkylation of a bisphenol
- Carboxylic esters by alcoholysis of carboxylic acid chlorides
- Reduction of an ester to give a primary alcohol
- Formation of a macrocycle by cycloalkylation of a bisphenol with a bishalide
- Use of the Ziegler–Ruggli high-dilution principle

(a) General

Podands, coronands, and cryptands are ring-open, cyclic, and bi(poly)cyclic receptor molecules, respectively (e.g. **2–4**), which bind other molecules or ions by electrostatic, van der Waals, coordinative or donor–acceptor interactions in a "host–guest" relationship according to basic features of supramolecular chemistry [1]:

2	**3**	**4**
EDTA, a podand	[18]-crown-6, a coronand	Bicyclo[8.8.8]-1,10-diaza hexaoxahexaeicosane, a cryptand

Macrocyclic polyethers of type **3**, usually named crown ethers [2], and their analogues containing sulfur or nitrogen atoms instead of oxygenatoms, exhibit unusual potentiality and specificity in the complexation of cations through ion–dipole interactions. For example, in the presence of the crown ether **3**, $KMnO_4$ shows solubility in benzene due to the formation of a stable **3**·[K^+] complex, thus allowing oxidations with $KMnO_4$ in an organic medium that would otherwise be impossible. Many other applications of crown-ether complexation with cations are known in preparative chemistry, e.g. use in phase-transfer catalysis, acceleration of S_N reactions, enhancement of ester hydrolysis, among others [2].

The target molecule **1** is a crown ether in which one oxygen atom is replaced by a nitrogen atom. The compound can be traced back to [18]crown-6 (**3**) by exchanging one CH_2–O–CH_2 moiety by pyridine and two lateral O–CH_2–CH_2–O groups by catechol. Its synthesis utilizing typical reactions of crown ether synthesis is described in detail in section (b).

(b) Synthesis of 1

For the synthesis of **1**, a convergent approach [3] is used. First, the separate synthesis of the two building blocks **7** and **10** is performed. Then, an intramolecular bisalkylation of the heterobenzylic bishalide **10** and the bisphenol **7** is performed to give **1**, applying the high-dilution principle of Ziegler/Ruggli [4].

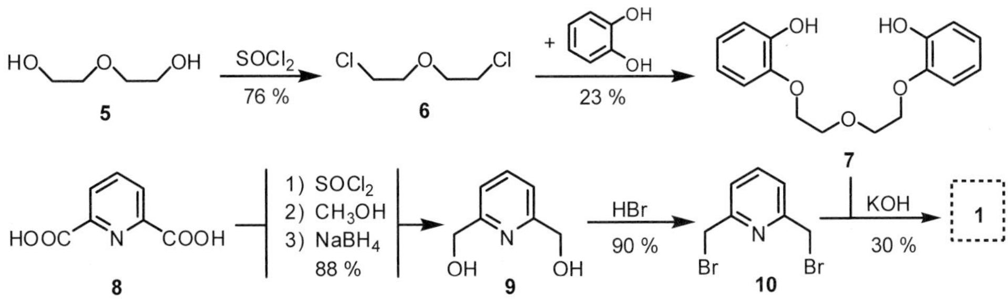

Building block **7** is prepared by way of a two-step procedure starting from diethylene glycol (**5**). By reaction with thionyl chloride, the OH groups of **5** are replaced by chloro substituents to yield 1,5-dichloro-3-oxapentane (**6**). S_N reactions at both electrophilic sites of **6** with two molecules of catechol then lead to the bisphenol **7**. The yield of the alkylation step **6 → 7** is rather low; however, the substrates for the formation of **7** are inexpensive and the product can be easily separated. It might be possible to improve the yields by the use of monoprotected catechol, but this would prolongate the synthesis of **7** by four steps [5]:

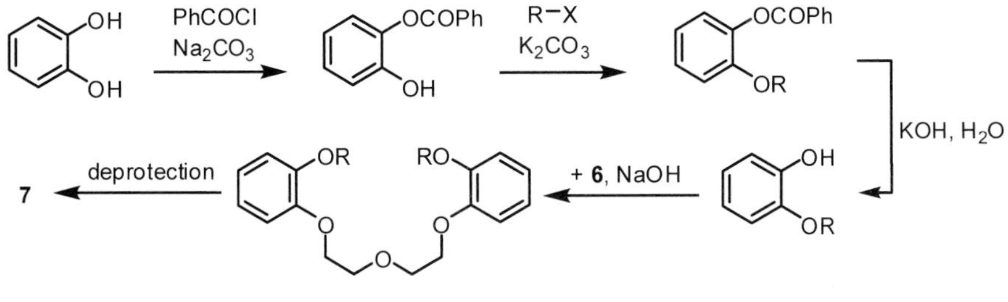

Building block **10** (2,6-bis(bromomethyl)pyridine) is prepared starting from pyridine-2,6-dicarboxylic acid (**8**) by esterification (alcoholysis of the acid chloride of **8**), reduction of the ester with sodium borohydride to 2,6-bis(hydroxymethyl)pyridine (**9**), and reaction of **9** with HBr.

The final step is the combination of the two building blocks **7** and **10** in a cycloalkylation reaction using a benzene/DMF/EtOH/H$_2$O mixture in the presence of KOH as a base to give the macrocycle **1** in 30% yield after chromatographic purification.

The product forms a well-defined crystalline 1:1 complex with KSCN, showing K$^+$ ion specificity, as expected for an analogue of the crown ether **3**.

Thus, the synthesis of the target molecule **1** can be performed in six steps, where the building block **7** is obtained in two steps in 17% yield and the building block **10** is obtained in three steps in 72% yield. The combination of **7** and **10** is accomplished in 30% yield.

(c) Experimental procedures for the synthesis of 1

3.5.2.1 * 1,5-Dichloro-3-oxapentane [3]

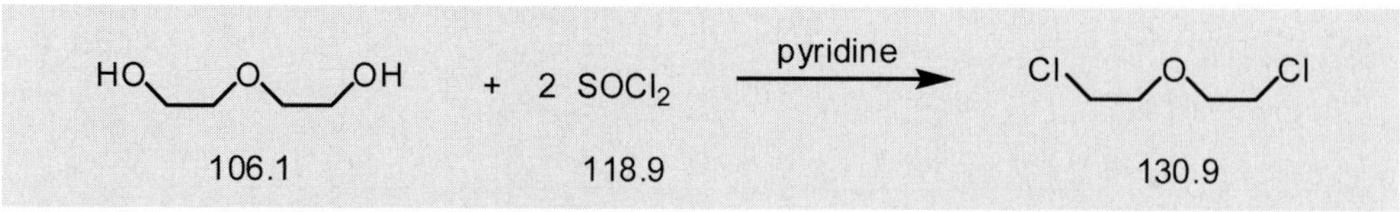

A stirred solution of diethylene glycol (106 g, 1.00 mol), benzene (900 mL, Caution!), and pyridine (180 mL, Caution!) is heated to 86 °C. Thionyl chloride (264 g, 1.40 mol, ca. 162 mL) is then added dropwise and stirring is continued for 16 h at 86 °C.

The solution is then cooled to room temperature and a mixture of concentrated HCl (50 mL) and H_2O (200 mL) is added dropwise over 15 min. The phases are separated, the aqueous phase is extracted several times with benzene (Caution!), and the combined organic phases are washed with ice-cold brine and dried over Na_2SO_4. The solvent is evaporated and the residue is distilled *in vacuo* to give the product as a colorless oil; 100 g (76%), bp$_{11}$ 60–62 °C, n_D^{20} = 1.4570.

3.5.2.2 * 1,5-Bis(2-hydroxyphenoxy)-3-oxapentane [3]

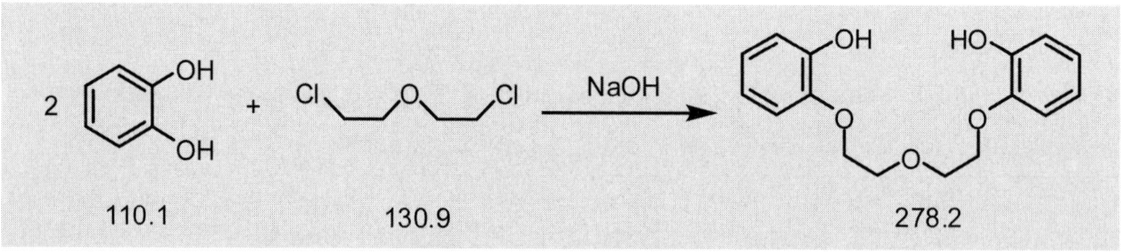

1,5-Dichloro-3-oxapentane **3.5.2.1** (32.8 g, 0.25 mol) is added in one portion to a solution of catechol (55.0 g, 0.50 mol) and NaOH (20.0 g, 0.50 mol) in water (500 mL) under a nitrogen atmosphere. The biphasic system is vigorously stirred to form an emulsion and then heated under reflux for 24 h with stirring.

The mixture is then acidified with concentrated HCl and concentrated *in vacuo*. The dark-brown, tar-like residue is triturated with hot MeOH (500 mL) and filtered to remove salts. The MeOH extract is concentrated to ca. 1/4 of its original volume, giving an impure precipitate mixed with brown particles. Two recrystallizations of the crude product from MeOH give colorless crystals; 32.0 g (23%), mp 86–88 °C.

> **^{1}H NMR** (CDCl$_3$): δ (ppm) = 7.5 (s, 2 H, OH), 7.1–6.75 (m, 8 H, Ar-H), 4.35–5.05, 4.0–3.75 (m, 4 H, CH$_2$–CH$_2$).

3.5.2.3   *   2,6-Bis(hydroxymethyl)pyridine [3]

167.1   195.2   139.1

(a) A stirred solution of pyridine-2,6-dicarboxylic acid (31.0 g, 186 mmol) in thionyl chloride (200 mL) is heated under reflux for 10 h. Excess $SOCl_2$ is then distilled off and the residue (acid chloride) is cooled in an ice bath; anhydrous MeOH (250 mL) is added dropwise with stirring. The resulting solution is heated under reflux for 30 min.

The MeOH is partially distilled off (150 mL) and the remaining solution is cooled in an ice bath to allow crystallization of the formed methyl diester of pyridine-2,6-dicarboxylic acid. The solid is collected by filtration and washed with ice-cold MeOH; 34.6 g (95%), mp 115–120 °C. The product is sufficiently pure to be used in the next step; the pure product has mp 120–121 °C (from CH_3OH).

(b) Sodium borohydride (26.0 g, 0.70 mol) is added portionwise over 15 min to a stirred suspension of the methyl diester prepared in (a) (29.0 g, 0.15 mol) in anhydrous EtOH (400 mL) with ice cooling. The mixture is stirred for 1 h at 0 °C; the ice bath is then removed and an exothermic reaction starts, which brings the solution to reflux. The solution is stirred at room temperature for 3 h and heated under reflux for 10 h.

The solvent is then distilled off *in vacuo*, the residue is dissolved in acetone (100 mL), the solution is filtered, and the filtrate is concentrated *in vacuo*. The residue is taken up in saturated aqueous K_2CO_3 solution (100 mL) and the mixture is heated on a steam bath for 1 h. Continuous extraction of the mixture with chloroform for 10 h followed by evaporation of the solvent *in vacuo* gives 19.3 g (93%) of the diol, mp 112–114 °C (the pure product has mp 114–115 °C).

^{1}H NMR (CDCl$_3$): δ (ppm) = 8.4–7.6 (m, 3 H, pyridine-H), 5.45 (t, J = 6.0 Hz, 2 H, OH), 4.95 (d, J = 6.0 Hz, 4 H, CH$_2$).

3.5.2.4   *   2,6-Bis(bromomethyl)pyridine [3]

139.1   264.9

2,6-Bis(hydroxymethyl)pyridine 3.5.2.3 (30.0 g, 0.22 mol) is dissolved in 48% hydrobromic acid (300 mL) with stirring and the solution is heated under reflux for 2 h.

On cooling the solution to room temperature, a colorless precipitate forms. The mixture is neutralized with concentrated aqueous NaOH solution, keeping the temperature at 0 °C (dry-ice cooling bath). The amorphous residue is collected by filtration, washed with H_2O, and dried over P_4O_{10} *in vacuo*. Recrystallization from petroleum ether (50–70 °C, ca. 750 mL) gives colorless needles of the dibromide; 46.6 g (82%), mp 86–89 °C (Caution: the product is a lachrymator!).

¹H NMR (CDCl₃): δ (ppm) = 8.1–7.4 (m, 3 H, pyridine-H), 4.80 (s, 4 H, CH₂).

3.5.2.5 * Dibenzopyridino[18]crown-6 [3]

264.9   278.2   381.3

A solution of the dibromide **3.5.2.4** (4.3 g, 20.0 mmol) in benzene (250 mL, Caution!), a solution of the bisphenol **3.5.2.2** (5.81 g, 20.0 mmol) in DMF (250 mL), and a solution of KOH (3.24 g, 40.0 mmol) in an ethanol/water mixture 50:1 (250 mL) are simultaneously added dropwise with stirring to refluxing *n*-butanol (1000 mL) over a period of 8–10 h. After the addition is complete, the solution is heated under reflux for an additional 2 h.

The solvents are then evaporated and the stirred oily residue is triturated with H₂O to remove DMF. The solidified crude product is taken up in hot CHCl₃, the solution is filtered, and the filtrate is dried (MgSO₄). After evaporation of the solvent *in vacuo*, the residue is purified by chromatography on basic aluminum oxide eluting with CHCl₃. The product migrates ahead of a yellow fraction. After evaporation of the solvent, the residue is recrystallized from EtOAc/*n*-hexane to yield the crown ether as colorless crystals; 2.28 g (30%), mp 131–132 °C (dec.) (note).

IR (KBr): ṽ (cm⁻¹) = 1600, 1510, 1255, 1130, 1055, 1010.

¹H NMR (CDCl₃): δ (ppm) = 7.9–7.6 (m, 3 H, pyridine-H), 7.2–6.8 (m, 8 H, Ar-H), 5.16 (s, 4 H, CH₂), 4.25–3.65 (m, 8 H, CH₂–CH₂).

Note: The crown ether can be characterized as its 1:1 complex with KSCN according to ref. [3]; colorless platelets, mp 212–213 °C.

[1]   a) J.-M. Lehn, *Naturwiss. Rundschau* **1997**, *50*, 421; b) J.-M. Lehn, *Supramolecular Chemistry*, VCH, Weinheim, **1995**; c) J. F. Stoddart, *Acc. Chem. Res.* **1997**, *30*, 393; d) F. Vögtle, *Supramolekulare Chemie*, Teubner, Stuttgart, **1989**; e) J. L. Atwood, D. D. MacNicol, J. E. D. Davies, *Comprehensive Supramolecular Chemistry*, Vols. 1–11, Pergamon, Oxford, **1996**.

[2]   C. J. Pedersen, *Angew. Chem.* **1988**, *100*, 1053; *Angew. Chem. Int. Ed. Engl.* **1988**, *27*, 1021.

[3]   a) E. Weber, F. Vögtle, *Chem. Ber.* **1976**, *10*, 1803; b) E. Weber, F. Vögtle, *Angew. Chem.* **1980**, *92*, 1067; *Angew. Chem. Int. Ed. Engl.* **1980**, *19*, 1030; c) F. Vögtle, private communication, 1981.

[4]   a) F. Vögtle, *Chem. Ztg.* **1972**, *96*, 396; b) L. Rossa, F. Vögtle, *Top. Curr. Chem.* **1983**, *113*, 1.

[5]   In analogy to the synthesis of guaiacol: L. F. Tietze, Th. Eicher, *Reaktionen und Synthesen im Organisch-chemischen Praktikum und Forschungslaboratorium*, 2ⁿᵈ ed., p. 409, Georg Thieme Verlag, Stuttgart, **1991**.

3.5.3   Indigo

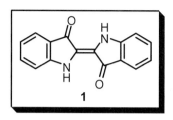

1

Topics: • Nitroaldol addition (Henry reaction)
          • Oxidative dimerization of a 3*H*-indol-3-one precursor
            to indigo

(a) General

Indigo (**1**) and other members of the group of indigoid vat dyes, e.g. thioindigo (**3**), possess a doubly-cross-conjugated, two-fold donor–acceptor substituted olefinic double bond as chromophoric system (with NH: **2**):

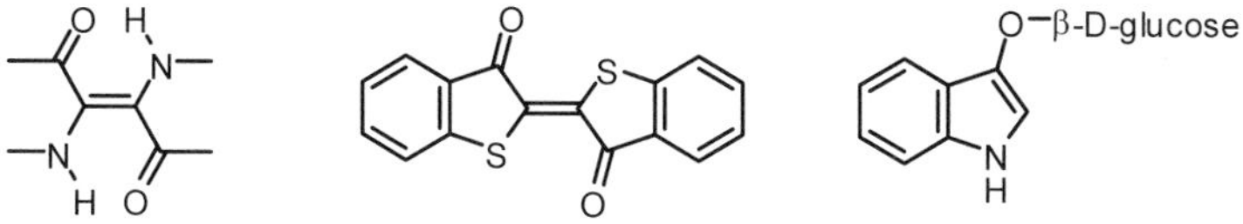

2   **3**   **4**

Since antiquity, indigo was obtained from indican (**4**) by enzymatic hydrolysis to give indoxyl (**7**), which undergoes oxidative dimerization (**7 → 1**). Indican (**4**) is the β-glucoside of indoxyl (**7**) found in the tropical indigo plant (*indigofera tinctoria*) and in European woad (*isatis tinctoria*; dyer's woad). However, since the beginning of the last century, indigo of natural origin has been completely replaced by indigo produced by industrial synthesis [1–3]:

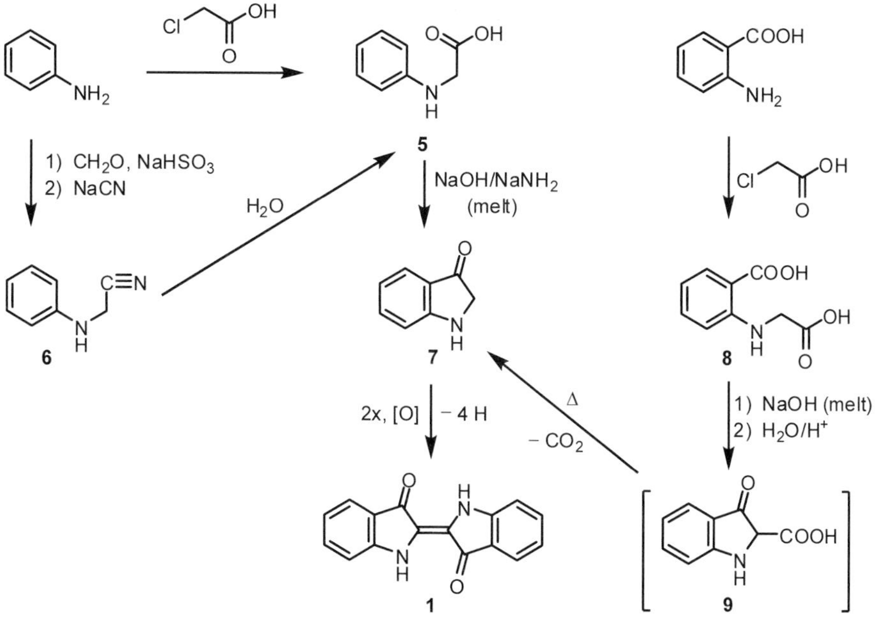

5

1) CH$_2$O, NaHSO$_3$
2) NaCN

H$_2$O

NaOH/NaNH$_2$
(melt)

1) NaOH (melt)
2) H$_2$O/H$^+$

6   **7**   **8**

2x, [O]   – 4 H

Δ
– CO$_2$

1   **9**

The technically relevant syntheses of indigo [1] start either from aniline or from anthranilic acid. In the first Heumann synthesis, aniline is *N*-alkylated by chloroacetic acid to give *N*-phenylglycine (**5**), which is cyclized to indoxyl (**7**) in an NaOH/NaNH₂ melt. Alternatively (and with higher yields), *N*-phenylglycine is synthesized by alkaline hydrolysis of *N*-phenylglycinonitrile (**6**), obtained by reaction of aniline with formaldehyde/NaHSO₃ followed by NaCN.

In the second Heumann synthesis, *N*-phenylglycine-*o*-carboxylic acid (**8**), accessible from anthranilic acid and chloroacetic acid, cyclizes in an alkali melt to give indoxyl-2-carboxylic acid (**9**), thermal decarboxylation of which also yields indoxyl (**7**). As the final step in these syntheses, indoxyl is oxidized by aerial oxygen to afford indigo (**1**).

For the laboratory synthesis of indigo, a preparatively more convenient procedure [4] that avoids the high-temperature alkali melt formation of indoxyl is presented in section (b).

(b) Synthesis of 1

The substrate for the synthesis of **1** [4] is *o*-nitrobenzaldehyde, which is subjected to an aldol addition with nitromethane (Henry reaction) in the presence of sodium methoxide to give the nitroaldol, which is isolated as the sodium salt **10**. On reduction of this nitronate salt with sodium dithionite in aqueous NaOH and subsequent oxidation with air, indigo (**1**) is obtained in high yield as a blue crystalline powder:

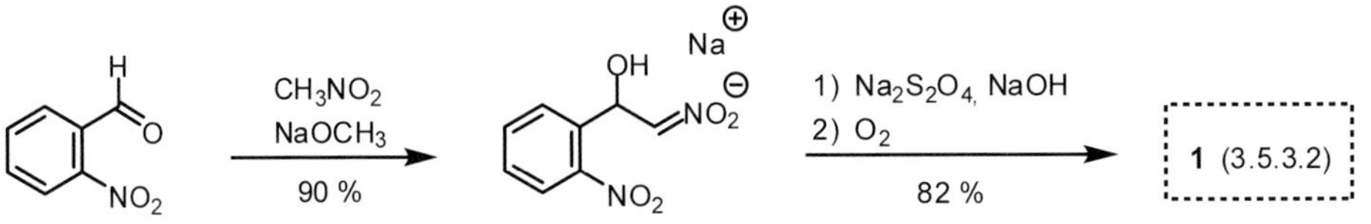

The mechanism of this indigo formation (**10 → 1**) remains speculative.

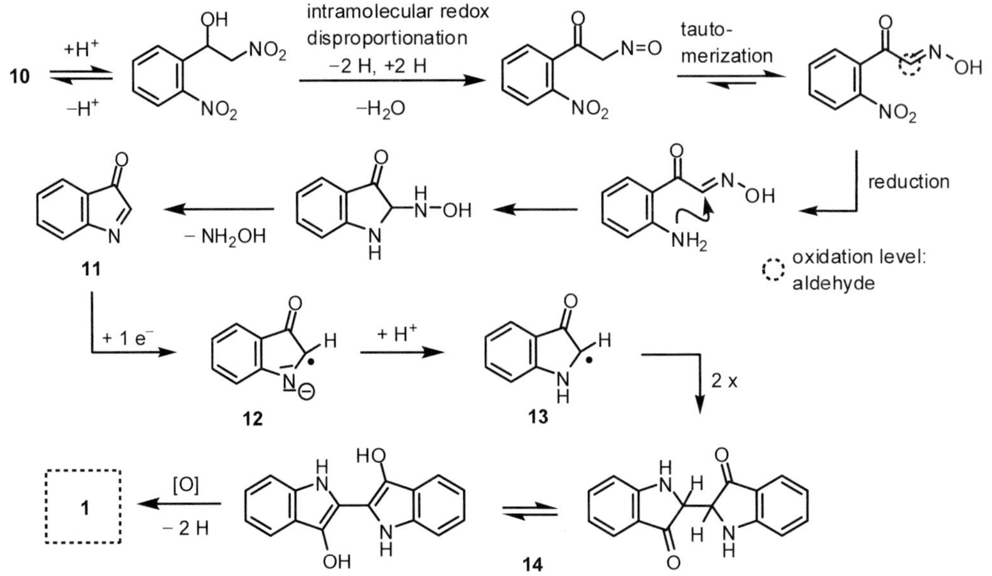

It has been assumed [5] that the elusive 3*H*-indol-3-one (**11**) is the primary intermediate, from which an SET process via radical anion **12** and radical **13** followed by dimerization may lead to the leuco form of indigo, **14**. This is finally dehydrogenated by aerial oxygen to give indigo (**1**). The sequence **13** → **14** → **1** corresponds to indigo formation from indoxyl (**7**) [6].

The proposed intermediacy of 3*H*-indol-3-one (**11**) might be rationalized by a working hypothesis that includes intramolecular redox disproportionation [7] of the nitronate **10** (or of the hydroxynitro compound obtained by protonation), tautomerization of the resulting α-nitroso ketone to afford an α-oximino ketone, reduction of NO_2 → NH_2, and eliminative cyclization to form the indolone system.

According to the described procedure, the target molecule is prepared in two steps in an overall yield of 73% (based on *o*-nitrobenzaldehyde).

(c) Experimental procedures for the synthesis of 1

3.5.3.1 * 1-(*o*-Nitrophenyl)-2-nitroethanol, Na salt [8]

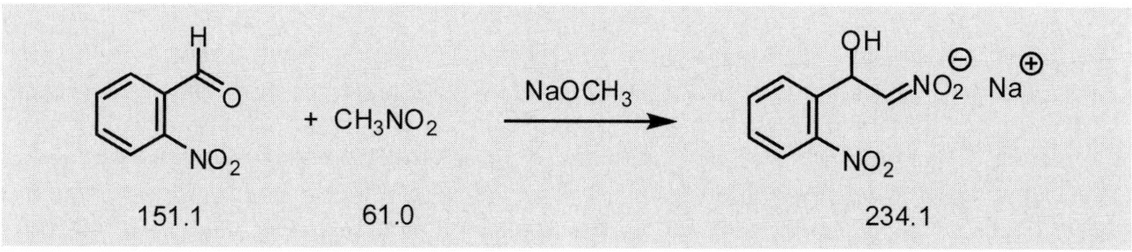

A solution of sodium methoxide is first prepared by portionwise addition of sodium (1.80 g, 78.3 mmol) to anhydrous MeOH (30 mL). This solution is then added dropwise over 20 min to a stirred solution of *o*-nitrobenzaldehyde (10.0 g, 66.2 mmol) and anhydrous nitromethane (4.60 g, 75.4 mmol) in MeOH (50 mL) at 0–5 °C. Towards the end of the addition, the yellow product begins to crystallize out. The mixture is left at 0 °C for 15 h and can be used directly for the next step.

The nitronate salt can be isolated by filtering it off, washing it with MeOH (2 × 10 mL) and Et$_2$O (3 × 10 mL), and drying it over P_4O_{10} *in vacuo*; 14.0 g (90%), yellow, air-sensitive powder.

> IR (KBr): $\tilde{v}$ (cm^{-1}) = 3120 (br), 1570, 1530, 1345 (NO$_2$).

3.5.3.2 * Indigo [8]

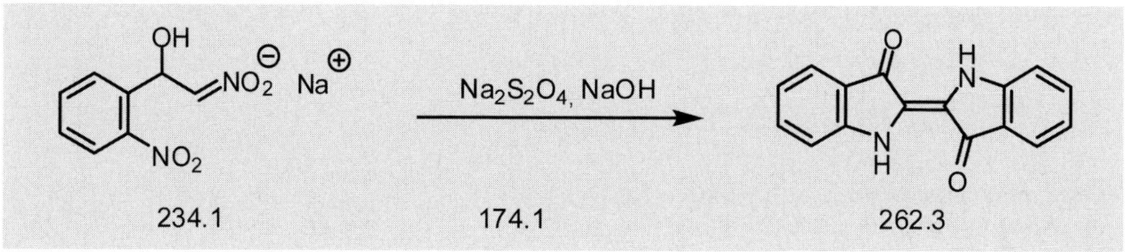

The product from **3.5.3.1** (note) is dissolved in water (200 mL), aqueous NaOH (2 M, 60 mL) is added, and the yellow solution is cooled to 6 °C. With vigorous stirring, sodium dithionite (33.6 g, 193 mmol) is added in small portions at such a rate that the temperature remains well below 15 °C; the addition time is ca. 15 min. The solution rapidly darkens and indigo begins to precipitate as a blue-black solid. When the dithionite addition is complete, air is rapidly bubbled through the reaction mixture for ca. 30 min.

The solid is collected by filtration, washed with water until alkaline-free, and then with EtOH (3 × 20 mL) and Et$_2$O (3 × 20 mL). The product is dried at 120 °C for 3 h to give a dark-blue, crystalline powder with a metallic sheen; 7.13 g (82%) of indigo, mp 390–393 °C (dec.).

UV/Vis (DMSO): λ_{max} (nm)/(log ε) = 619 (4.20), 580 (sh), 330 (sh), 287 (4.41) [9].

Note: When the reaction mixture (slurry) of **3.5.3.1** is used directly, the methanol is removed *in vacuo* at 25 °C and the residue is dissolved in water (200 mL).

[1]   Beyer-Walter, *Lehrbuch der Organischen Chemie*, 23rd ed., p. 779, S. Hirzel Verlag, Stuttgart, **1998**.

[2]   Römpp *Lexikon Naturstoffe* (Eds.: W. Steglich, B. Fugmann, S. Lang-Fugmann), p. 313, Thieme Verlag, Stuttgart, **1997**.

[3]   Th. Eicher, S. Hauptmann, *The Chemistry of Heterocycles*, 2nd ed., p. 81/109, Wiley-VCH, Weinheim, **2003**.

[4]   J. Harley-Mason, *J. Chem. Soc.* **1950**, 2907.

[5]   a) J. Gosteli, *Helv. Chim. Acta* **1977**, *60*, 1980; b) S. P. Hiremath, M. Hooper, *Adv. Heterocycl. Chem.* **1978**, *22*, 123.

[6]   Ref. [3], p. 110.

[7]   A long-known example of such an intramolecular redox reaction is the formation of anthranilic acid from *o*-nitrotoluene in NaOH/EtOH: S. Hauptmann, *Organische Chemie*, 2nd ed., p. 510, VEB Verlag, Leipzig, **1988**.

[8]   Modification of the method reported in ref. [4].

[9]   W. Lüttke, M. Klessinger, *Chem. Ber.* **1964**, *9*, 2342.

3.5.4   Pyrvinium iodide

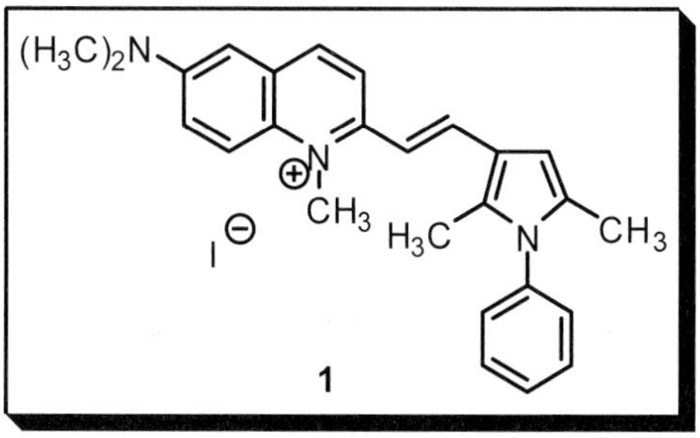

Topics:
- Synthesis of an unsymmetrical cyanine dye
- Paal–Knorr synthesis of a pyrrole derivative
- Doebner–Miller synthesis of a quinoline derivative
- Aldol condensation with C–H-acidic substituted heterocycles
- Vilsmeier reaction of a hetarene

(a) General

The pyrvinium salt **1** (6-dimethylamino-2-[2-(2,5-dimethyl-1-phenyl-3-pyrrolyl)vinyl]-1-methyl-quinolinium iodide) is used pharmaceutically in the form of the embonate as an anthelmintic [1]. It is structurally related to the cyanine dyes, which represent an important class of dyestuffs and are technically relevant as sensitizers in color photography [2].

Cyanine dyes contain as chromophore a polymethine chain with an odd number of methine CH groups, which bears at its terminal positions an (uncharged) amino nitrogen and a (charged) iminium nitrogen, thus allowing symmetrical charge delocalization over the chain, as exemplified in the case of the pentamethine cyanine system **2**:

The terminal nitrogens can be incorporated into heterocycles, thus giving rise to cyanine dyes with heterocyclic end groups, which can be arranged symmetrically (as in **3**) or unsymmetrically (as in **1**); both cyanine systems **1** and **3** contain a pentamethine cyanine structural unit.

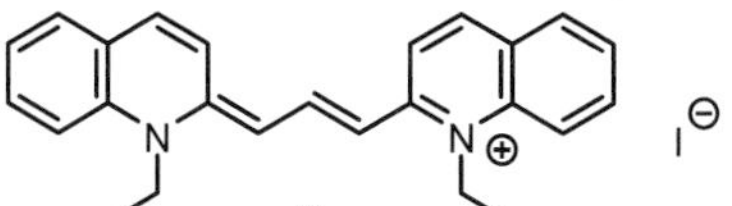

For the retrosynthesis of the target molecule **1**, the CH=CH group connecting the two heterocyclic moieties is of strategic relevance, since its disconnection according to a retro aldol mode leads to the two building blocks **4** and **5**. Thus, **1** should be accessible by an aldol condensation of **4** and **5**.

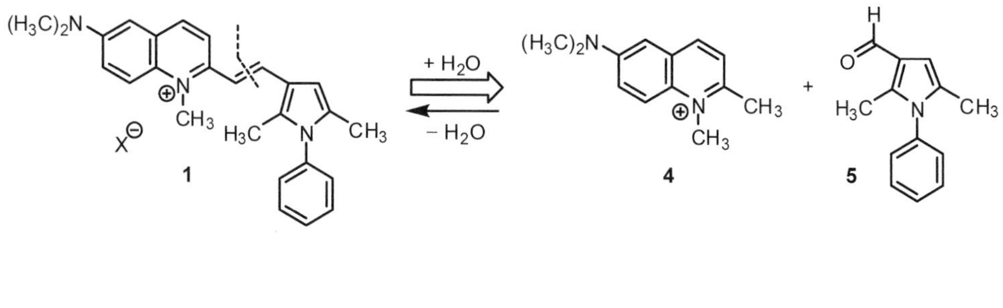

The basis for the retroanalytical approach is the well-known CH-acidity of heterobenzylic C–H bonds, preferentially of CH_3 groups. The CH-acidity is strongly enhanced by *N*-quaternation. Thus, methyl groups in the 2- and/or 4-position of azines and benzazines [3] can be deprotonated with a base to give a carbanion, which can undergo C–C bond-forming transformations such as alkylations, acylations, or aldol reactions, e.g.:

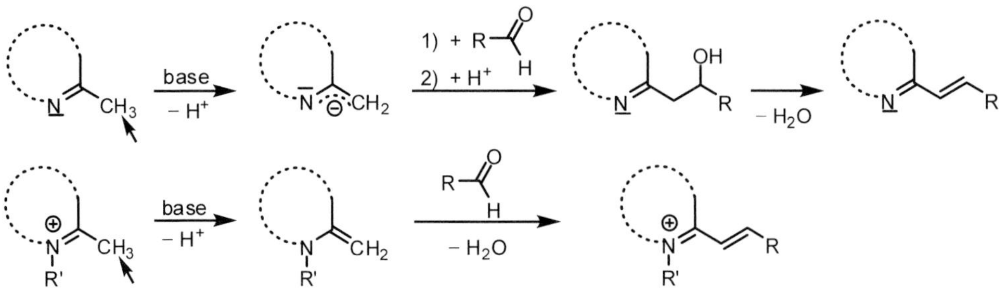

For the synthesis of the two building blocks **4** and **5**, two universally applicable methods of heterocycle synthesis are applied, namely the Paal–Knorr synthesis of pyrroles and the Doebner–Miller synthesis of quinolines [4].

In the Paal–Knorr synthesis, 1,4-dicarbonyl compounds **6** are cyclocondensed with ammonia or primary amines thus producing 2,5-disubstituted pyrroles **7**:

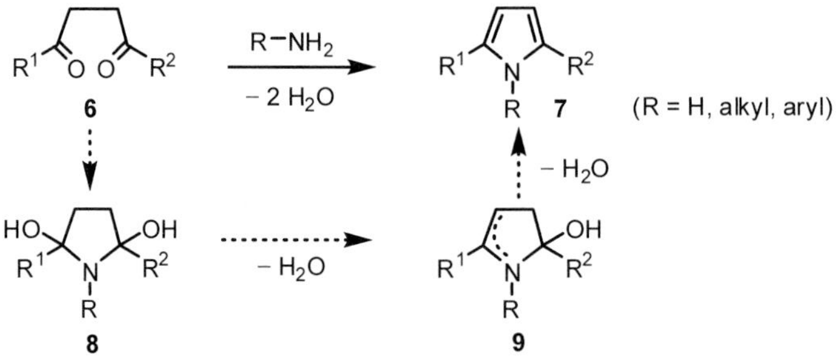

The initial reaction step leads to twofold hemiaminals **8**, which give the pyrroles **7** by stepwise H_2O elimination via imine (R = H) or enamine (R ≠ H) intermediates **9** [5].

In the Doebner–Miller synthesis, primary arylamines with an unsubstituted *ortho*-position are reacted with α,β-unsaturated carbonyl compounds in the presence of a proton acid and an oxidant (nitroarene, As_2O_5, etc.) [6] to give quinoline derivatives **11**:

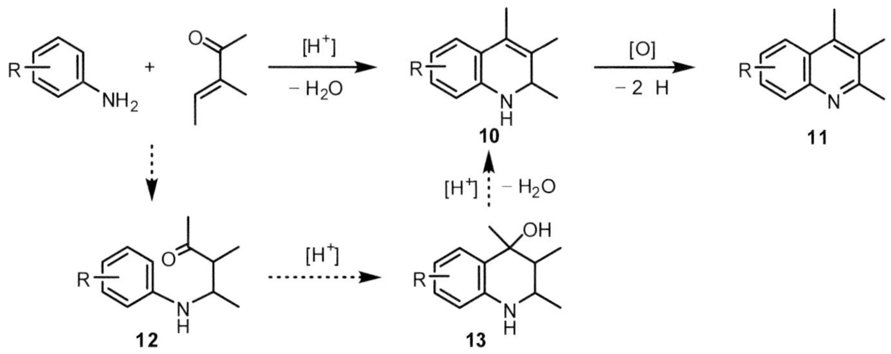

For this synthesis, a complex multi-step sequence has been established, which includes Michael addition of the arylamine to the enone system (→ 12), ring closure of intermediate 12 by H⁺-catalyzed intramolecular hydroxyalkylation (→ 13), and dehydration leading to a 1,2-dihydroquinoline 10, which is dehydrogenated (by the oxidant) to give the quinoline 11.

(b) Synthesis of 1

Based on the retrosynthetic considerations in section (a), a convergent approach [7] for the construction of 1 is presented, in which the building blocks 4 and 5 are first prepared separately.

The synthesis of pyrrolecarbaldehyde 5 starts with the cyclocondensation of hexane-2,5-dione (14) with aniline according to the Paal–Knorr procedure to give the pyrrole 15. Introduction of an aldehyde function at the 3-position of the activated heterocycle is achieved by means of Vilsmeier formylation (→ 5):

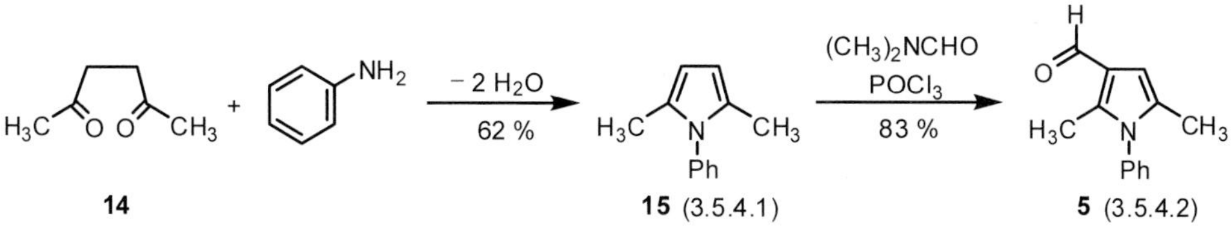

The synthesis of the quaternized quinaldinium salt 4 follows the pattern of the Doebner–Miller method. p-(Dimethylamino)aniline (16) is cyclocondensed with crotonaldehyde in 6 N aqueous hydrochloric acid in the presence of ZnCl₂ to give the 2-methylquinoline 17:

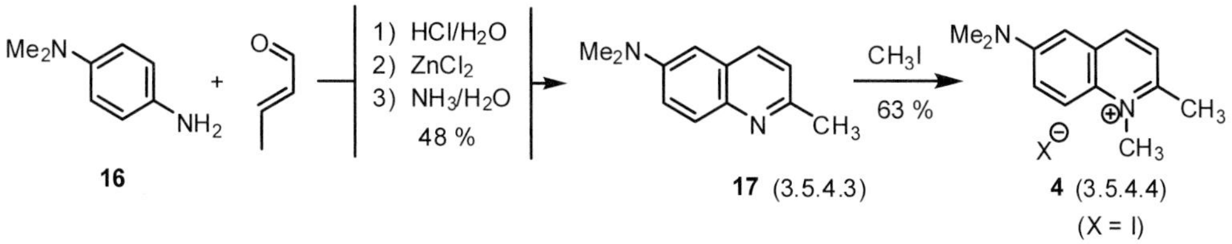

In this variant of the Doebner–Miller synthesis, a Zn complex of the cyclization product 17 is isolated first, which is then decomposed using ammonia. In this way, a process that is usually accompanied by side reactions can be greatly improved.

The quinaldine **17** is transformed to the quaternary salt **4** (X = I) by alkylation with methyl iodide. Initially, a mixture of products methylated at the azine nitrogen and at the $(CH_3)_2N$ group results; however, the product isomerizes thermally to the *N*-methylquinaldinium salt (**4**, X = I) [8].

Finally, the building blocks **4** and **5** are combined by aldol condensation in the presence of piperidine as base to provide pyrvinium iodide (**1**):

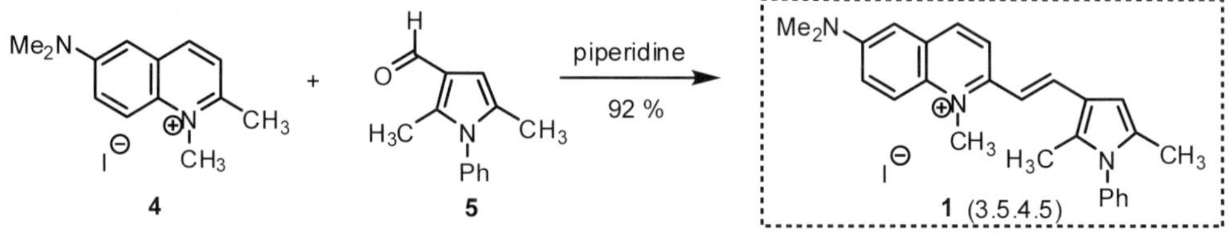

Thus, the target molecule **1** is obtained in a convergent synthesis in one step in 92% yield from **4** and **5**, which are formed in two steps each in yields of 51% and 30%, respectively.

(c) Experimental procedures for the synthesis of 1

3.5.4.1 * 2,5-Dimethyl-1-phenylpyrrole [9]

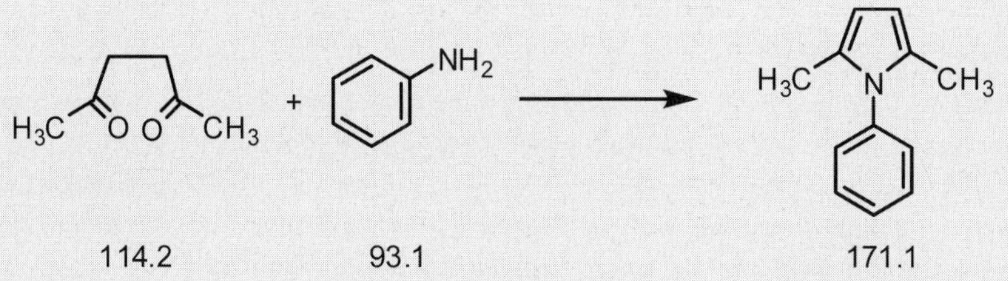

Aniline (27.9 g, 0.30 mol) (note 1) and hexane-2,5-dione (34.2 g, 0.30 mol) are heated under reflux for 1 h.

The reaction mixture is then cooled to room temperature and poured into a mixture of H_2O (100 mL) and concentrated HCl (10 mL). The precipitate is collected by suction filtration, washed with iced water, and recrystallized from a mixture of MeOH (150 mL) and H_2O (15 mL) (note 2). The product is obtained as colorless crystals, 31.8 g (62%), mp 50–51 °C; TLC (SiO$_2$; cyclohexane): $R_f = 0.75$.

IR (KBr): $\tilde{v}$ (cm^{-1}) = 1595, 1490, 1400, 1315.

^{1}H NMR (CDCl$_3$): δ (ppm) = 7.6–7.1 (m, 5 H, phenyl-H), 5.93 (s, 2 H, 3-H/4-H), 1.98 (s, 6 H,

---

Notes: (1) Aniline (bp$_{20}$ 84–85 °C) has to be distilled before use.

(2) If the product separates as an oil on cooling, it has to be redissolved by the addition of a small amount of MeOH (ca. 3 mL).

3.5.4.2   *   2,5-Dimethyl-1-phenylpyrrole-3-carbaldehyde [10]

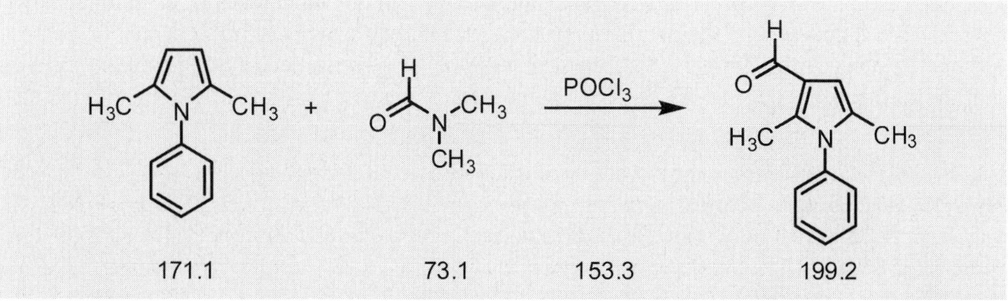

171.1   73.1   153.3   199.2

The pyrrole **3.5.4.1** (25.0 g, 146 mmol) and anhydrous dimethylformamide (16.0 g, 219 mmol) are dissolved in anhydrous toluene (100 mL). With vigorous stirring, POCl₃ (27.0 g, 219 mmol) (note) is added dropwise over 30 min, the temperature of the solution rises to ca. 80 °C, and a dark color develops. When the addition of POCl₃ is complete, the solution is heated with stirring to 100 °C for 6 h.

The reaction mixture is then cooled to room temperature, poured into saturated aqueous NaOAc solution (300 mL), and the resulting mixture is vigorously stirred for 30 min. The organic phase is separated and the aqueous phase is extracted with toluene (2 × 200 mL). The organic phases are combined and washed successively with 10% aqueous Na₂CO₃ solution (200 mL) and H₂O (200 mL). The toluene is distilled off *in vacuo* and the residue is purified by fractionating distillation *in vacuo*. The pyrrole carbaldehyde is obtained as a yellowish oil, which solidifies on cooling, 24.0 g (83%), bp₁₂ 190–191 °C, mp 90–91 °C; TLC (SiO₂; Et₂O): R_f = 0.70.

> **IR** (KBr): $\tilde{\nu}$ (cm⁻¹) = 1650 (C=O), 1600, 1540.
>
> **¹H NMR** (CDCl₃): δ (ppm) = 9.88 (s, 1 H, OC-H), 7.6–7.4, 7.3–7.1 (m, 5 H, phenyl-H), 6.39 (d, *J* = 1 Hz, 1 H, 4-H), 2.28 (s, 3 H, CH₃), 1.99 (d, *J* = 1 Hz, 3 H, CH₃).

Note: POCl₃ is a lachrymator and should be handled only in a hood; it has to be distilled before use (bp₇₆₀ 105–106 °C).

3.5.4.3   *   6-Dimethylamino-2-methylquinoline [8]

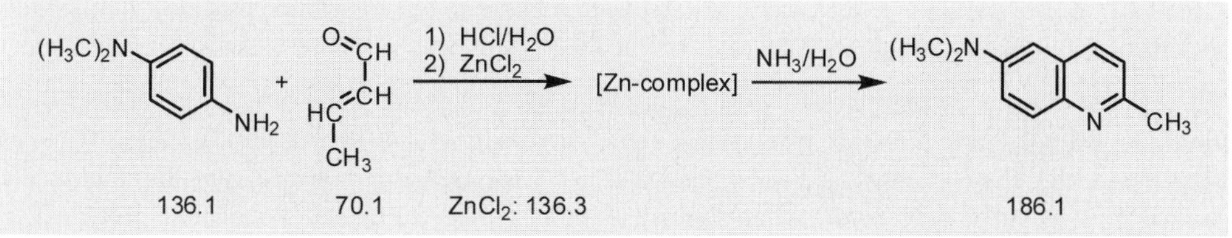

136.1   70.1   ZnCl₂: 136.3   186.1

A solution of *p*-(dimethylamino)aniline (35.0 g, 257 mmol) in hydrochloric acid (6 N, 130 mL) is heated to reflux. With intense stirring, crotonaldehyde (25.0 g, 357 mmol) is added dropwise over 30 min. When the addition is complete, the dark solution is heated to reflux for 1 h.

The reaction mixture is cooled to room temperature and extracted with Et₂O (100 mL) to remove undissolved dark impurities; anhydrous zinc chloride (35.4 g, 0.26 mol) is then added to the clear brown-red solution with stirring. Concentrated NH_3 is then added with stirring, until the pH of the solution reaches 5–5.5. An orange-red zinc complex of the product (note 1) is formed, which crystallizes. It is collected by suction filtration, suspended in isopropanol (200 mL), and the suspension is stirred for 5 min. After filtration, the Zn complex is washed with isopropanol (in portions of 50 mL) until the washings are only faintly colored; the product is then washed with Et_2O (100 mL) and air-dried.

The Zn complex is decomposed by portionwise addition to concentrated aqueous NH_3 (150 mL) and the quinoline derivative formed is extracted with CH_2Cl_2 (4 × 200 mL). The CH_2Cl_2 extracts are combined, dried over K_2CO_3, filtered, and the solvent is distilled off; the residue (ca. 28 g) is then fractionated *in vacuo*. The quinoline derivative is obtained as a yellow oil, which solidifies on cooling (note 2), 23.0 g (48%), bp$_{0.01}$ 120–121 °C, mp 92–93 °C; TLC (SiO_2; Et_2O): $R_f = 0.50$.

IR (KBr): $\tilde{v}$ (cm^{-1}) = 1630 (C=N), 1605, 1515.

^{1}H NMR (CDCl$_3$): δ (ppm) = 7.91 (d, J = 3 Hz, 1 H, 3-H), 7.81 (d, J = 3 Hz, 1 H, 4-H), 7.31 (dd, J = 8.5/3.0 Hz, 1 H, 7-H), 7.13 (d, J = 8.5 Hz, 1 H, 8-H), 6.78 (d, J = 3 Hz, 1 H, 5-H), 3.03 (s, 6 H, N(CH$_3$)$_2$), 2.66 (s, 3 H, CH$_3$).

Notes: (1) The complex contains $ZnCl_2$ and two molecules of the quinoline [8].
(2) The product is air-sensitive. Cleavage of the Zn complex should be rapid, the distillation should be performed under an N_2 atmosphere, and the product kept in a refrigerator under N_2. Due to the sensitivity of the quinoline, the next step (alkylation) should follow immediately.

3.5.4.4   *   6-Dimethylamino-1,2-dimethylquinolinium iodide [8]

186.1   141.9   328.0

A solution of the quinoline derivative **3.5.4.3** (22.0 g, 118 mmol) and methyl iodide (33.5 g, 236 mmol) in anhydrous isopropanol (130 mL) is heated under reflux for 2 h with stirring. Orange-red crystals of the quinolinium salt are formed.

The reaction mixture is then cooled to room temperature and the crystals formed are collected by filtration, washed with ice-cold isopropanol, and air-dried: 34.4 g (89%), mp 253–258 °C. For purification, the crude product is heated to 200–210 °C (external temperature) for 10–15 min; the product must attain a dark color. After cooling to room temperature, the crystals are dissolved in boiling H_2O (ca. 220 mL) and the hot solution is filtered. On cooling to room temperature, the methoiodide crystallizes in brown-red needles, which are collected by suction filtration, washed with H_2O, and dried *in vacuo* over P_4O_{10}, 24.5 g (63%), mp 265–267 °C.

IR (KBr): $\tilde{v}$ (cm^{-1}) = 3050, 2940 (CH), 1625 (C=N), 1610, 1525 (C=C).

^{1}H NMR ([D$_6$]DMSO): δ (ppm) = 8.73 (d, J = 8 Hz, 1 H, 3-H), 8.30 (d, J = 8 Hz, 1 H, 4-H), 7.85 (d, J = 8 Hz, 1 H, 8-H), 7.68 (dd, J = 8/3 Hz, 1 H, 7-H), 7.23 (d, J = 3 Hz, 1 H, 5-H), 4.40 (s, 3 H, $^+$N–CH$_3$), 3.14 (s, 6 H, N(CH$_3$)$_2$), 3.03 (s, 3 H, CH$_3$).

3.5.4.5   *   Pyrvinium iodide [7]

| 328.0 | 199.2 | 509.1 |

Freshly distilled piperidine (2.20 g, bp$_{760}$ 105–106 °C) is added to a stirred solution of the methoiodide **3.5.4.4** (8.80 g, 26.8 mmol) and the pyrrole carbaldehyde **3.5.4.2** (5.34 g, 26.8 mmol) in anhydrous MeOH (100 mL). On heating to reflux, the solution becomes intensely red and after some minutes the product precipitates as red crystals. Heating to reflux is continued for 30 min.

The reaction mixture is cooled to room temperature, and the product is collected by suction filtration, washed with MeOH, and dried over P$_4$O$_{10}$ *in vacuo*; 12.5 g (92%), reddish-brown crystalline powder, mp 286–287 °C.

IR (KBr): $\tilde{v}$ (cm^{-1}) = 1620 (C=N), 1575, 1530 (C=C).

^{1}H NMR ([D$_6$]DMSO): δ (ppm) = 8.45–6.9 (m, 12 H, arom. H + vinyl H), 6.50 (s, 1 H, pyrrole 4-H), 4.40 (s, 3 H, $^+$N–CH$_3$), 3.20 (s, 6 H, N(CH$_3$)$_2$), 2.25, 2.08 (s, 3 H, CH$_3$).

[1]   A. Kleemann, J. Engel, *Pharmaceutical Substances*, 3rd ed., p. 1641, Thieme Verlag, Stuttgart, **1999**.

[2]   See textbooks on organic chemistry, e.g. Beyer-Walter, *Lehrbuch der Organischen Chemie*, 23rd ed., p. 806, S. Hirzel Verlag, Stuttgart, **1998**.

[3]   Th. Eicher, S. Hauptmann, *The Chemistry of Heterocycles*, 2nd ed., p. 281 and 322, Wiley-VCH, Weinheim, **2003**.

[4]   a) Ref. [3], p. 94 (pyrrole), p. 325 (quinoline); see also: *Houben-Weyl*, Vol. E 6a and Vol. E 7°; b) a related reaction principle operates in the MCR synthesis of quinaldates from aromatic amines, aliphatic aldehydes, and glyoxylate: T. Inada, T. Nakajima, I. Shimizu, *Heterocycles* **2005**, *66*, 611.

[5]   V. F. Ferreira, M. C. B. V. De Souza, A. C. Cunha, L. O. R. Pereira, M. L. G. Ferreira, *Org. Prep. Proced. Int.* **2001**, *33*, 411.

[6]   Originally, for the reaction of anilines and acrolein to give 2,3,4-unsubstituted quinolines the name "Skraup synthesis" and for the reaction of anilines and crotonaldehyde to give quinaldines the name "Doebner–Miller synthesis" was used. Today, the reaction of enones with arylamines in general is listed as "Doebner–Miller synthesis".

[7]   USP 2 252 912 (*Chem. Abstr.* **1951**, *45*, 3567f).

[8]   a) W. Cocker, D. G. Turner, *J. Chem. Soc.* **1941**, 143; see also b) C. Leir, *J. Org. Chem.* **1977**, *42*, 911.

[9]   E. Wolthuis, *J. Chem. Educ.* **1979**, *56*, 343.

[10] R. Rips, N. Ph. Buu-Hoi, *J. Org. Chem.* **1959**, *24*, 372.

3.5.5   2,3,7,8,12,13,17,18-Octamethylporphyrin

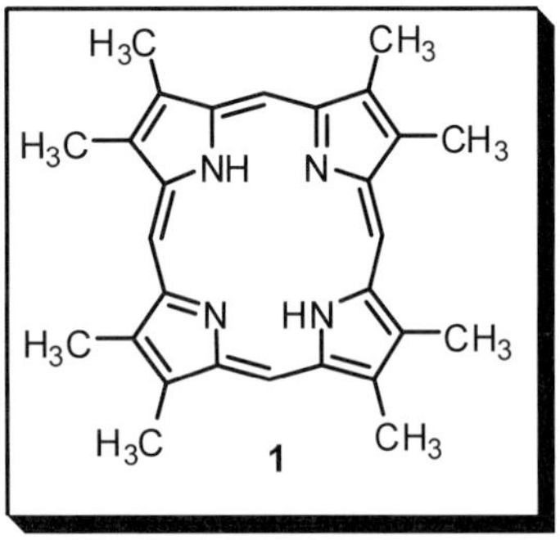

Topic: • Synthesis of a symmetrically substituted porphyrin by oxidative cyclotetramerization of a pyrrole with an aldehyde

(a) General

Porphyrin (**2**) is the parent compound of the natural product family of tetrapyrroles (e.g. haemine, chlorophyll). In porphyrin, four pyrrole-derived units are linked together at their two α-positions by four methine (sp^2-C) bridges. They form a planar C_{20} macroheterocycle with a conjugated delocalized aromatic π-system of 18 π-electrons ($4n + 2$ with $n = 4$; altogether there are 22 π-electrons) [1].

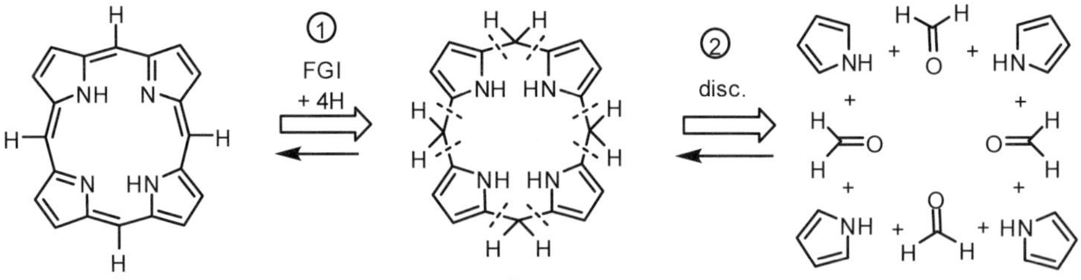

Among the numerous approaches for the synthesis of porphyrins [2], the most simple and straightforward strategy stems from the retroanalytical considerations that: (1) in a reductive FGI, the four sp^2 methine bridges of **2** can be transformed into sp^3 methane bridges as in **3** (porphyrinogen), and (2) successive disconnection of the methane bridges leads to four molecules of pyrrole and four molecules of formaldehyde.

The retrosynthesis step (2) is based on the reversal of the well-known formation of dipyrrolylmethanes **4** by H^+-catalyzed hydroxyalkylation/alkylation of pyrroles with free α-positions by carbonyl compounds, which is one of the most important electrophilic reactions of pyrroles [3]:

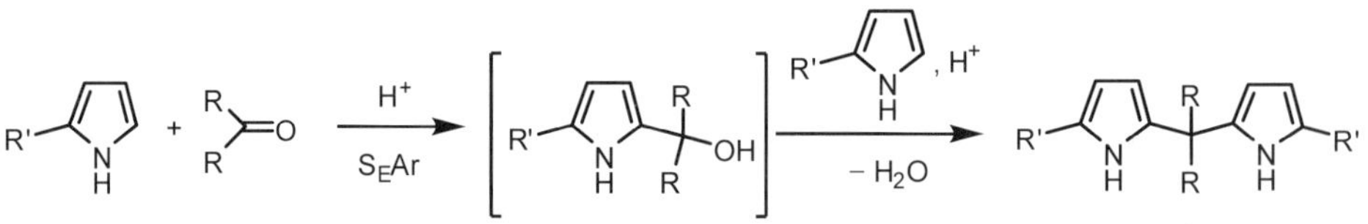

Accordingly, as shown in section (b), the cyclotetramerization of pyrrole or pyrroles with identical substituents at C-3 and C-4 with aldehydes in the presence of a proton acid or Lewis acid followed by

dehydrogenation represents the method of choice for the synthesis of symmetrically substituted porphyrins like **1**; it has been realized for pyrrole and aryl aldehydes [4] as well as for 3,4-dialkyl-pyrroles and formaldehyde or aryl aldehydes [5–7].

The procedure corresponds very well with the biosynthesis of natural tetrapyrroles starting from the pyrrole derivative porphobilinogen (**5**) [8]. In an enzymatic linear condensation, the acyclic tetramer hydroxymethylbilane **6** is formed, which cyclizes to give uroporphyrinogen III (**7**) with inversion of ring D [9]. **7** is the substrate of other pigments essential to life, such as the hemes, chlorophylls, corrins, and factor 43 [10].

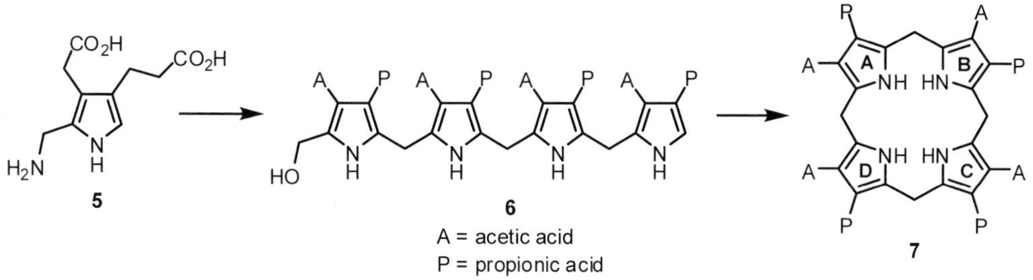

A = acetic acid
P = propionic acid

(b) Synthesis of 1

3,4-Dimethylpyrrole (**8**, cf. **3.2.2**) is subjected to cyclotetramerization with formaldehyde in the presence of *p*-toluenesulfonic acid as catalyst in benzene solution. The octamethylporphyrinogen **9** initially formed by azeotropic removal of H$_2$O is not isolated, but is dehydrogenated *in situ* by reaction with oxygen to give the octamethylporphyrin (**1**) in a one-pot procedure [11]:

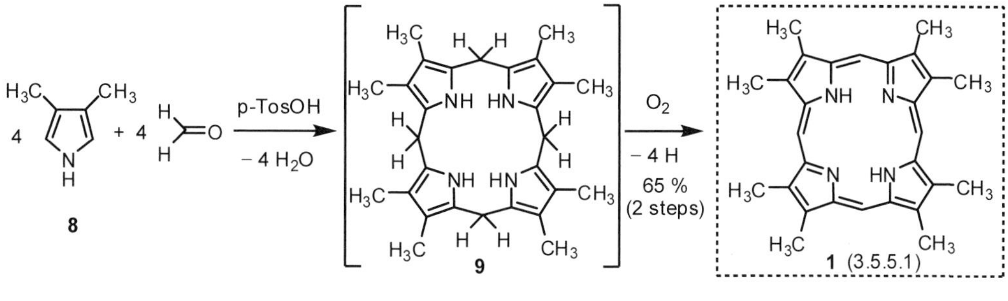

(c) Experimental procedure for the synthesis of 1

3.5.5.1   **   2,3,7,8,12,13,17,18-Octamethylporphyrin [7, 8]

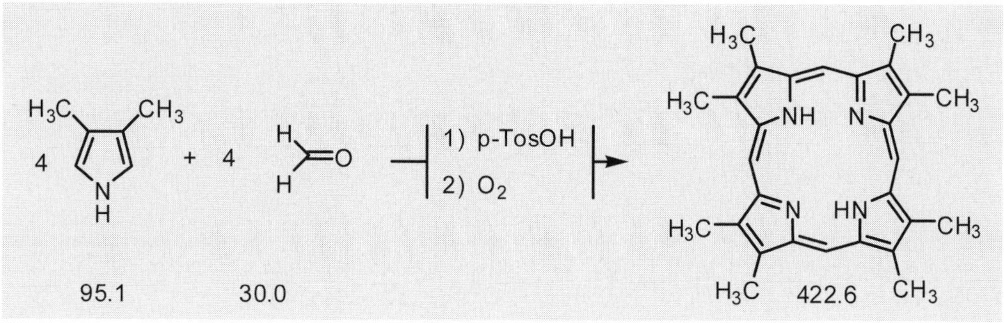

95.1   30.0   422.6

Under a nitrogen atmosphere, a 500-mL round-bottomed flask, wrapped with aluminum foil (for light protection) and equipped with a Dean–Stark trap and a reflux condenser, is charged with 3,4-dimethyl-pyrrole (cf. **3.2.2**; 0.77 g, 8.10 mmol), benzene (300 mL, Caution!), aqueous formaldehyde solution (37%, 0.73 mL, 8.9 mmol), and *p*-toluenesulfonic acid (0.03 g, 1.7 mmol). The mixture is heated to reflux with stirring and removal of water for 8 h.

The brown reaction mixture is then cooled to room temperature and oxygen is bubbled through it (frit) at room temperature for 12 h with stirring to give a black suspension.

The solvent is then evaporated *in vacuo*, and the residue is washed with CHCl₃ (5 mL) and MeOH (5 mL) and dried *in vacuo* to afford an amorphous, purple-black powder; 0.52 g (61%) (note).

---

¹H NMR (CDCl₃/CF₃COOH, 400 MHz): δ (ppm) = 10.57 (s, 4 H, methine-CH), 3.55 (s, 24 H, CH₃) (note).

¹³C NMR (CDCl₃/CF₃COOH, 400 MHz): δ (ppm) = 142.06 (pyrrole-C$_\alpha$), 138.74 (pyrrole-C$_\beta$), 98.26 (methine-CH), 11.93 (CH₃).

Note: The product is insoluble in most common solvents; however, it can be recrystallized from nitrobenzene [7]. It is also soluble in CHCl₃ on addition of a small amount of TFA, producing a deep purple-red color by formation of the porphyrin dication [2]. For NMR measurements, a solution of 7 mg of the above product in 1 mL of CDCl₃ and 2 drops of TFA is used. According to ¹H NMR, the product is of >98% purity.

[1]   M. K. Cyrański, T. M. Krygowski, M. Wisiorowski, N. J. R. van Eikema Hommes, M. K. Cyrański, T. M. Krygowski, M. Wisiorowski, N. J. R. van Eikema Hommes, P. von Ragué Schleyer, *Angew. Chem.* **1998**, *110*, 187; *Angew. Chem. Int. Ed.* **1998**, *37*, 177.

[2]   Th. Eicher, S. Hauptmann, *The Chemistry of Heterocycles*, 2ⁿᵈ ed., p. 485, Wiley-VCH, Weinheim, **2003**.

[3]   Ref. [2], p. 487.

[4]   R. A. W. Johnstone, M. L. P. G. Nunes, M. M. Pereira, A. M. d'A. Gonsalves, A. C. Serra, *Heterocycles* **1996**, *43*, 1423.

[5]   K. M. Barkigia, M. D. Berber, J. Fajer, C. J. Medforth, M. W. Renner, K. M. Smith, *J. Am. Chem. Soc.* **1990**, *112*, 8851.

[6]   J. L. Sessler, A. Mozaffari, M. R. Johnson, *Org. Synth.* **1992**, *70*, 68.

[7]   D. L. Boger, R. S. Coleman, J. S. Panek, D. Yohannes, *J. Org. Chem.* **1984**, *49*, 4405.

[8]   F. P. Montforts, M. Glasenapp-Breiling, *Fortschr. Chem. Org. Naturst.* **2002**, *84*, 1–51.

[9]   a) L. F. Tietze, H. Geissler, *Angew. Chem.* **1993**, *105*, 1087–1089; *Angew. Chem. Int. Ed. Engl.* **1993**, *32*, 1038–1040; b) L. F. Tietze, H. Geissler, *Angew. Chem.* **1993**, *105*, 1090–1091; *Angew. Chem. Int. Ed. Engl.* **1993**, *32*, 1040–1042; c) L. F. Tietze, H. Geissler, G. Schulz, *Pure. Appl. Chem.* **1994**, *66*, 2303–2306; d) L. F. Tietze, G. Schulz, *Chem. Eur. J.* **1997**, *3*, 523–529.

[10] A. R. Battersby, *Nat. Prod. Rep.* **2000**, *17*, 507–526.

[11] The synthesis of **1** by cyclotetramerization of 3,4-dimethylpyrrole with formaldehyde described in ref. [7] was modified, adapting the conditions used in ref. [6] for the preparation of the corresponding octaethylporphyrin.

3.5.6 Synthesis of a rotaxane

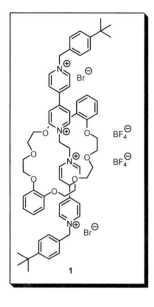

Topics:
- Preparation of a rotaxane consisting of a crown ether and a blocked linear bis(4,4'-dipyridylium)ethane unit
- Crown ethers
- Dipyridylium salts
- Electrostatic interactions
- π–π Stacking

(a) General

Rather unusual types of compounds are the catenanes [1], consisting of two interlocking rings, and the rotaxanes [2], which consist of an assembly of one ring and a dumbbell. The striking feature of these substances is the fact that the two parts of the molecules are not connected to each other by a covalent bond, but by a so-called mechanical bond, which in the case of the catenanes is also a topological bond. Cleavage of catenanes requires breaking of one of the two rings, whereas in the case of rotaxanes a deformation of one of the two parts is necessary to dissociate the ring from the dumbbell, which is normally prevented from unthreading by large stoppers at its two ends.

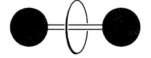

General structures of a catenane (left) and a rotaxane (right).

Such compounds, especially the rotaxanes, have attracted much attention in recent years due to their photophysical and electronic properties as well as their dynamic behavior [3]; thus, rotaxanes can be considered as wheels. This allows the design of molecular motors [4] with an axle rotating inside a stator. For a long time it was thought that the wheel and molecular motors were inventions of humankind, but recently it has been demonstrated that Nature also exploits this type of concept in ATP synthases [5]. These enzymes generate ATP from ADP and are responsible for the supply of chemical energy in all living organisms.

In the early work in this area, the synthesis of catenanes and rotaxanes completely depended on statistical approaches leading to the desired compounds usually in only very low yields [6]. Nowadays, directed and template methods are used. In the case of rotaxanes, in the directed method a pre-rotaxane is synthesized, in which the cyclic part and the dumbbell are linked together. Cleavage of the connections between the two parts then leads to the rotaxanes. The most efficient method – the template method, of which one example is described in the following – uses non-covalent interactions (ionic, van der Waals, hydrogen bonding, $\pi-\pi$ stacking, metal–ligand interactions) to assemble a pre-rotaxane, which is then transformed into the rotaxane by blocking the ends of the linear axle to avoid its slipping out.

(b) Synthesis of 1

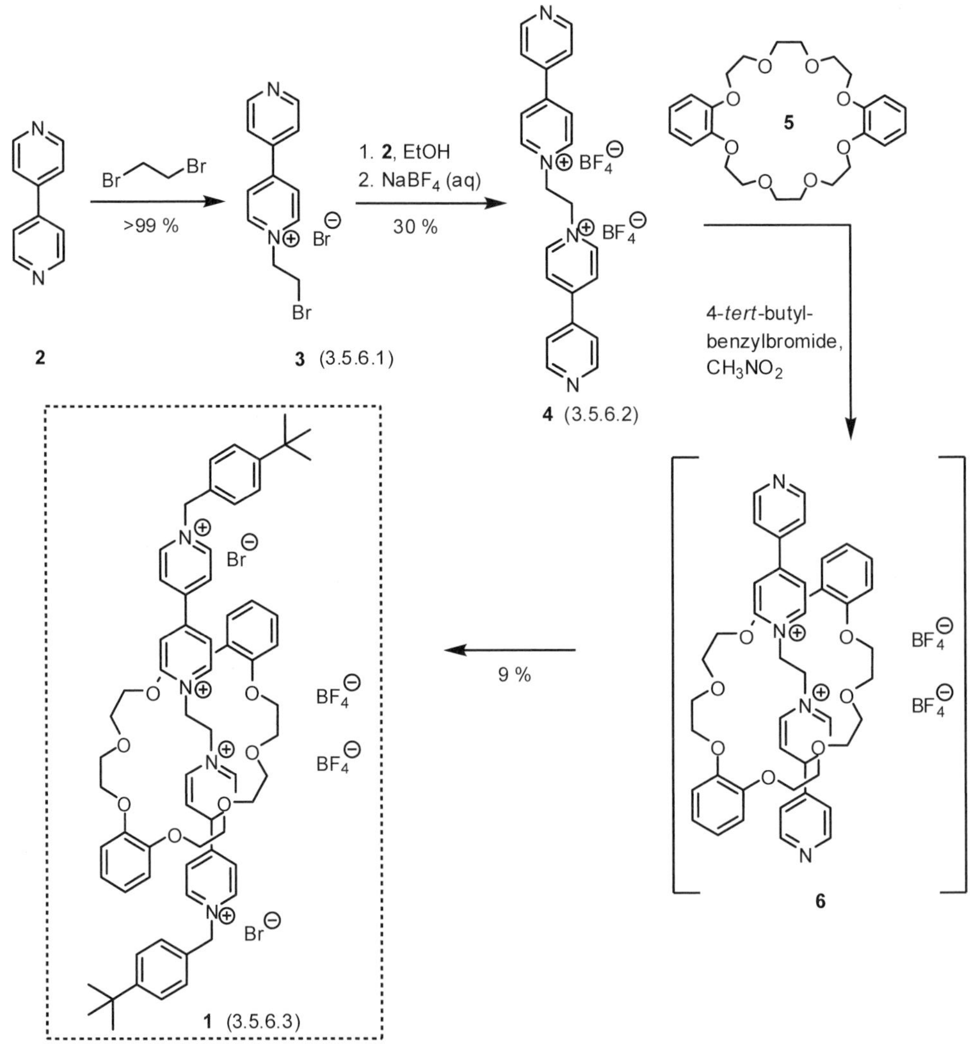

Stoddart and co-workers have shown that *ortho*- and *meta*-substituted crown ethers of appropriate size are excellent receptors for bipyridinium dications [7]. As non-covalent binding forces, electrostatic interactions and π–π stacking can be assumed. Based on this pre-organization, Wisner [8] prepared the rotaxane **1** using the dipyridinium salt **4** and the crown ether **5**. First, a pre-rotaxane **6** is formed by insertion of **4** into the crown ether **5**; then, **6** is stabilized by blocking the ends of the dipyridinium system through alkylation with 4-*tert*-butyl-benzyl bromide. The success of the reaction can be recognized from a strong bathochromic effect on the UV/Vis absorbance of **1** compared to that of **4**. The obtained yields are moderate, but the procedure is simple and illustrative.

(c) Experimental procedures for the synthesis of 1

3.5.6.1   *   1-(2-Bromoethyl)-[4,4']bipyridyl-1-ium bromide [8]

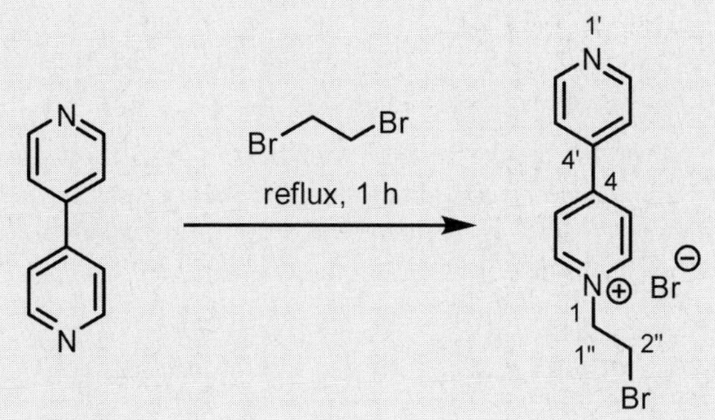

156.2   187.9   344.1

A stirred solution of 4,4'-dipyridyl (5.00 g, 32.0 mmol) in 1,2-dibromoethane (86.9 g, 463 mmol, 40.0 mL) is heated to reflux for 1 h.

After cooling to room temperature, the resulting salt is filtered off, washed with Et$_2$O, and dried *in vacuo* to yield the bromide as a beige solid; 10.9 g (99%).

UV (MeOH): λ_{max} (nm) (lg ε) = 267.5 (4.27), 201.0 (4.40).

IR (KBr): $\tilde{v}$ (cm^{-1}) = 2999, 1643, 1599, 1548, 1531, 1494, 1469, 1409, 1366, 1225, 1175, 1071, 995, 889, 814, 748, 714, 661, 477.

^{1}H NMR (300 MHz, D$_2$O): δ (ppm) = 9.09 (d, J = 1.9, 5.2 Hz, 2 H, 2-H), 8.84 (dd, J = 1.9, 4.7 Hz, 2 H, 2'-H), 8.51 (dd, J = 1.9, 5.2 Hz, 2 H, 3-H), 7.98 (dd, J = 1.9, 4.7 Hz, 2 H, 3'-H), 5.17 (t, J = 5.6 Hz, 2 H, 2"-H), 4.11 (t, J = 5.6 Hz, 2 H, 1"-H).

^{13}C NMR (50 MHz, D$_2$O/MeOH): δ (ppm) = 155.70 (C-4), 150.99 (C-2), 146.14 (C-2'), 143.53 (C-4'), 127.05 (C-3), 123.53 (C-3'), 62.73 (C-2"), 31.21 (C-1").

MS (ESI): *m/z* (%) = 265.1 (53) [(*M* − Br)]$^+$, 263.0 (52) [(*M* − Br)]$^+$, 184.3 (13) [(*M* − 2Br)]$^+$, 183.2 (100) [(*M* − 2Br)]$^+$.

3.5.6.2   **   1,2-Di-(4,4'-dipyridylium tetrafluoroborate)-ethane [8]

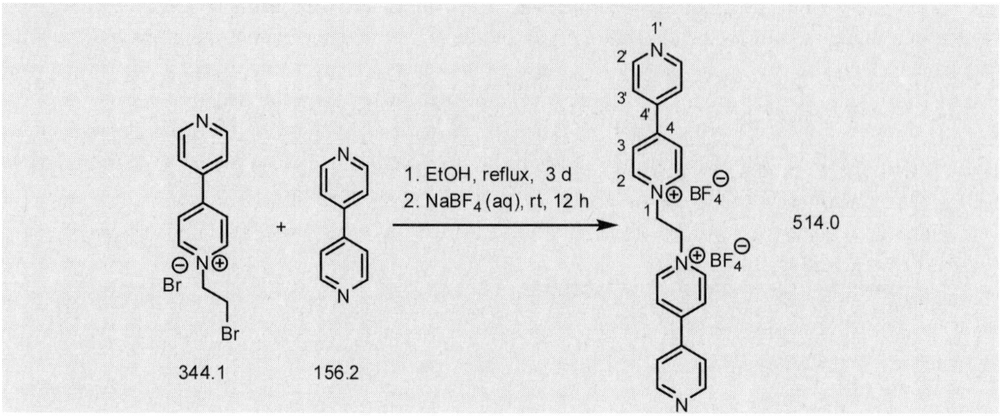

A solution of the bromide **3.5.6.1** (640 mg, 1.86 mmol) and 4,4'-dipyridyl (1.30 g, 8.32 mmol) in anhydrous EtOH (50 mL) is heated to reflux for 3 d.

After cooling to room temperature, the precipitate formed is filtered off, dried *in vacuo*, and dissolved in boiling H_2O (3 mL). Saturated aqueous sodium tetrafluoroborate solution is added dropwise to the boiling solution. The mixture is left at room temperature for 12 h, whereupon the dipyridium salt is obtained as a beige solid, which is collected by filtration; 287 mg, (30%).

^{1}H NMR (300 MHz, D_2O): δ (ppm) = 9.20 (d, J = 7.2 Hz, 4 H, 2-H), 9.01 (dd, J = 1.5, 5.3 Hz, 4 H, 2'-H), 8.65 (d, J = 7.2 Hz, 4 H, 3-H), 8.36 (dd, J = 1.5, 4.9 Hz, 4 H, 3'-H), 5.56 (s, 4 H, 2 × CH$_2$).

3.5.6.3   ***   Rotaxane [9]

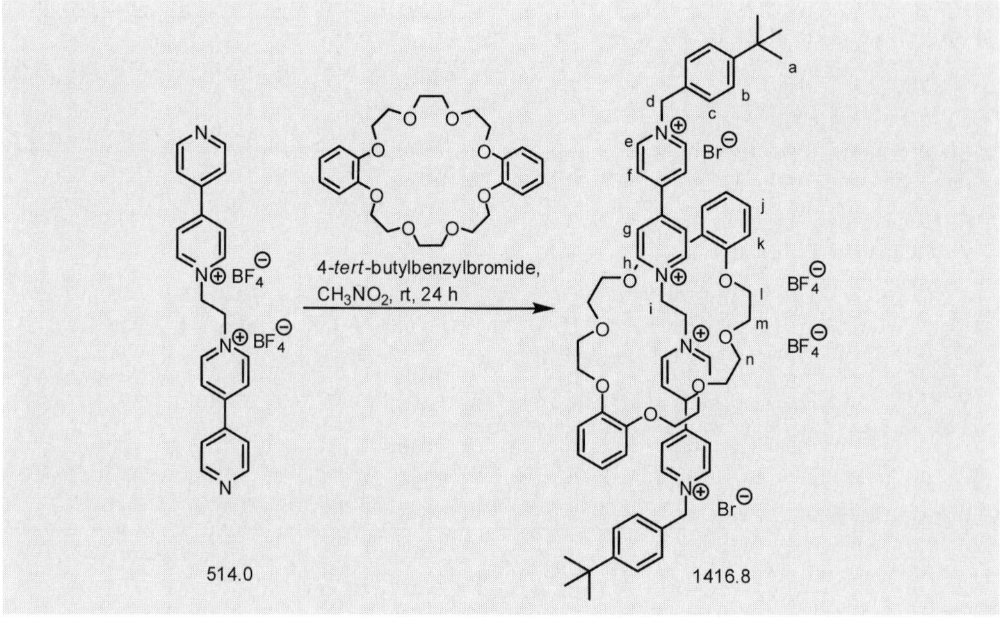

A solution of the dipyridinium salt **3.5.6.2** (50 mg, 97 μmol) and dibenzo-24-crown-8 (131 mg, 0.292 mmol) in nitromethane (5.00 mL) is stirred for 30 min at room temperature; 4-(*tert*-butyl)benzyl bromide (133 mg, 0.58 mmol) is then added dropwise over 10 min and stirring is continued for 24 h.

The resulting mixture is then filtered, the filtrate is concentrated *in vacuo*, and the residue is recrystallized from CH_2Cl_2/Et_2O. The rotaxane is obtained as a dark-red solid; 12.8 mg (9%).

^{1}H NMR (300 MHz, [D$_6$]DMSO): δ (ppm) = 9.45 (d, J = 6.8 Hz, 4 H, h), 9.20 (d, J = 6.8 Hz, 4 H, e), 8.59 (d, J = 6.8 Hz, 4 H, g), 8.48 (d, J = 6.8 Hz, 4 H, f), 7.60 (m$_c$, 8 H, b, c), 6.63 (dd, J = 3.4, 5.7 Hz, 4 H, k), 6.21 (dd, J = 3.4, 6.2 Hz, 4 H, j), 5.93 (s, 4 H, d), 5.48 (s, 4 H, i), 4.01-3.90 (m, 24 H, l, m, n), 1.28 (s, 18 H, *t*Bu).

^{13}C NMR (126 MHz, [D$_6$]DMSO): δ (ppm) = 146.2 (e), 145.4 (h), 126.5 (f), 128.6, 125.8 (b, c), 125.5 (g), 120.7 (j), 112.2 (k), 70.30, 69.91, 67.36 (l, m, n), 63.34 (d), 57.90 (i), 30.86 (a).

MS (ESI-HRMS):
calcd.:   628.30900 [$(M + 2BF_4)^{2+}$]
found: 628.30919 [$(M + 2BF_4)^{2+}$]

[1]   J.-C. Chambron, J.-P. Collin, V. Heitz, D. Jouvenot, J.-M. Kern, P. Mobian, D. Pomeranc, J.-P. Sauvage, *Eur. J. Org. Chem.* **2004**, 1627.

[2]   a) G. Wenz, B.-H. Han, A. Müller, *Chem. Rev.* **2006**, *106*, 782; b) J.-Cl. Chambron, J.-P. Sauvage, *Chem. Eur. J.* **1998**, *4*, 1362 ; c) for a review on molecular knots, see: C. Dietrich-Buchecker, B. X. Colasson, J.-P. Sauvage, *Top. Curr. Chem.* **2005**, *249*, 261.

[3]   a) M. J. Gunter, *Eur. J. Org. Chem.* **2004**, 1655; b) M.-J. Blanco, M. Consuelo Jiménez, J.-Cl. Chambron, V. Heitz, M. Linke, J.-P. Sauvage, *Chem. Soc. Rev.* **1999**, *28*, 293.

[4]   V. Balzani, A. Credi, F. M. Raymo, J. F. Stoddart, *Angew. Chem.* **2000**, *112*, 3484; *Angew. Chem. Int. Ed.* **2000**, *39*, 3348.

[5]   C. A. Schalley, K. Beizai, F. Vögtle, *Acc. Chem. Res.* **2001**, *34*, 465.

[6]   a) J. S. Siegel, *Science* **2004**, *304*, 1256; b) Th. Dünnwald, Th. Schmidt, F. Vögtle, *Acc. Chem. Res.* **1996**, *29*, 451.

[7]   a) B. L. Allwood, F. H. Kohnke, J. F. Stoddart, D. J. Williams, *Angew. Chem.* **1985**, *97*, 584; *Angew. Chem. Int. Ed. Engl.* **1985**, *24*, 581; b) B. L. Allwood, H. Shahriari-Zarvareh, J. F. Stoddart, D. J. Williams, *J. Chem. Soc., Chem Commun.* **1987**, 1058.

[8]   J. A. Wisner, Ph.D. Thesis, University of Windsor, **1999**.

[9]   Modification of the procedure given in ref. [8].

4   Selected Natural Products

4.1   Alkaloids

Introduction

Alkaloids are nitrogen-containing natural products usually having a complex cyclic structure; simple amines, amino acids, and proteins, as well as nucleosides and nucleic acids, which also contain nitrogen, are not included in this group [1]. However, heterocyclic compounds such as caffeine (cf. **3.4.4**) are alkaloids. Originally, the term alkaloid was used for amines from plants, indicating that these compounds have a basic (= alkaline) character, but nowadays the definition has been broadened to also cover non-basic nitrogen compounds such as ammonium salts and amides from all natural sources.

The vast majority of the over 20000 known natural alkaloids have been isolated from plants, such as morphine (**1**) from *Papaver somniferum* [2], but alkaloids are also found in animals, for example pumiliotoxin C (**2**) from the frog *Dendrobates pumilio* [3], and in mushrooms, for example muscarine (**3**) from *Amanita muscaria* [4].

Morphine	Pumiliotoxin	Muscarine
1	**2**	**3**

There are several ways of classifying alkaloids, either according to their origin, e.g. ergot alkaloids, or according to their heterocyclic core structure, e.g. indole alkaloids. However, the best classification of alkaloids is based on their biosynthesis. Despite a few exceptions [5] – see also later – alkaloids are formed from amino acids and biogenic amines, respectively. The most common precursors of aliphatic alkaloids are *L*-ornithine and *L*-lysine. Thus, *L*-ornithine is the precursor of the pyrrolidine and pyrrolizidine alkaloids. An example of a pyrrolidine alkaloid is the well known narcotic cocaine (**4**). On the other hand, *L*-lysine serves as the precursor of the piperidine alkaloids, such as piperine (**5**) from *Piper nigrum* (black pepper) (cf. **1.5.1**). Here, a piperidine unit, which is formed from *L*-lysine, is acylated by a phenylpropanoic compound after C$_2$-chain extension.

Interestingly, the alkaloid coniine (**6**), isolated from hemlock, which also contains a piperidine unit, is formed from acetate via 5-oxooctanoic acid [6].

Cocaine	Piperine	Coniine
4	**5**	**6**

Phenylalanine and tyrosine are precursors of aromatic alkaloids such as the *Amaryllidaceae* alkaloid buflavine (**7**) (cf. **4.1.3**), the alkaloid 2,3-dimethoxyberbine (**8**) (cf. **4.1.2**), and morphine (**1**).

Reactions and Syntheses in the Organic Chemistry Laboratory. L. F. Tietze, Th. Eicher, U. Diederichsen, A. Speicher
Copyright © 2007 WILEY-VCH Verlag GmbH & Co. KGaA, Weinheim
ISBN: 978-3-527-31223-8

In the formation of aromatic alkaloids such as morphine (**1**) and buflavine (**7**), so-called phenol oxidation (cf. **1.7.6**) is a very important transformation, whereby a phenolate is oxidized to give a radical that can undergo C–O or C–C bond formation [7]. Thus, in the biosynthesis of morphine (**1**), the benzyltetrahydroisoquinoline reticulin (**9**) is transformed via the proposed diradical **10** into salutaridine (**11**), which is further converted into **1** via several intermediates [8]. The alkaloid 2,3-dimethoxyberbine (**8**) is also formed from a benzyltetrahydroisoquinoline akin to **9**, but in this case oxidation of the *N*-methyl group takes place to form an iminium ion, which undergoes an electrophilic aromatic substitution.

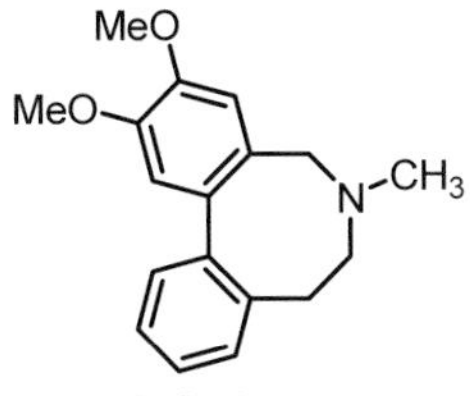

Buflavine
7

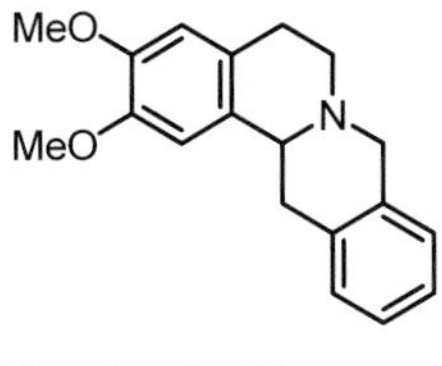

Dimethoxyberbine
8

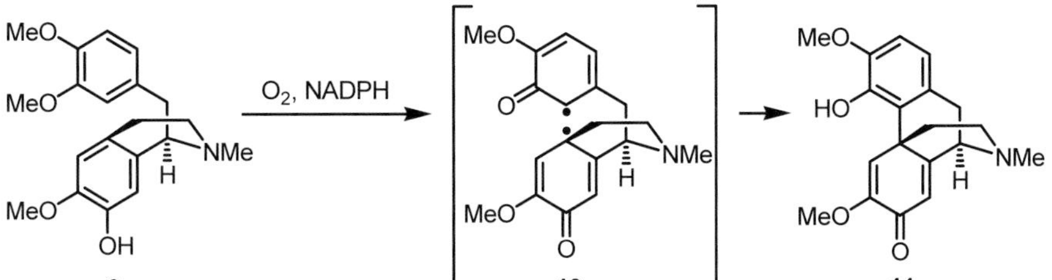

O₂, NADPH

9   **10**   **11**

One of the biggest groups of alkaloids are the indole alkaloids, which are formed from tryptophan and tryptamine. Simple compounds belonging to this class of natural products are melatonin (**12**) (cf. **3.2.5**) and (*R*)-salsolidine (**13**) (cf. **3.3.2**). However, very complex structures are found in the group of so-called monoterpenoid indole alkaloids and alkaloids derived from them, such as the cinchona alkaloids and the pyrroloquinoline alkaloids [9]. Several highly bioactive compounds originate from this class. In the biosynthesis of these compounds, tryptamine (**14**) is first condensed with the monoterpene secologanin (**15**) to give strictosidine (**16**), which, after cleavage of the glucose moiety, forms indole alkaloids of the corynanthe family such as geissoschizine (**17**).

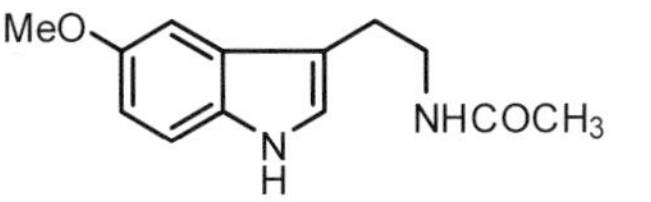

12 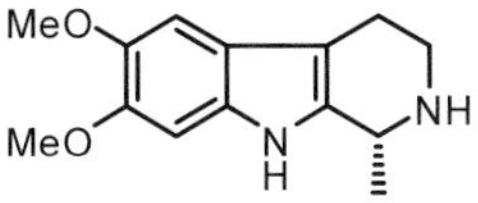   **13**

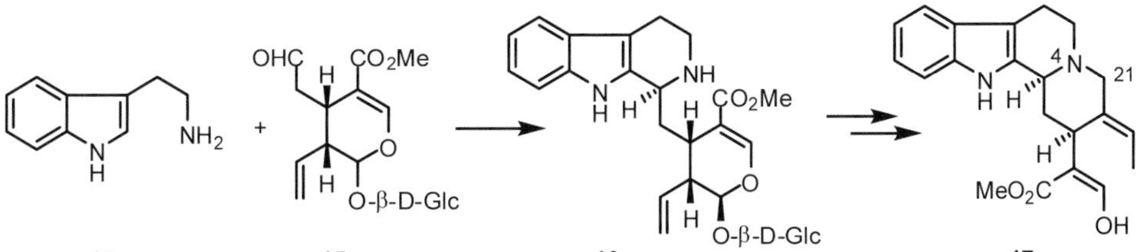

14   +   **15**   **16**   **17**

The alkaloid hirsutine (**18**) (cf. **4.1.1**) also belongs to the corynanthe family. Further transformations of **17** and its 4,21-didehydro derivative, respectively, lead to indole alkaloids of the aspidosperma and iboga families. Moreover, oxidative cleavage of the indole moiety followed by aldol condensation furnishes cinchona alkaloids such as quinine (**19**), which contains a quinoline moiety as one of the heterocyclic core structures, as well as pyrroloquinoline alkaloids.

Nearly all alkaloids have strong biological activities. They can act as poisons, such as coniine (**6**), which was used by the people of Athens to kill Socrates nearly 2500 years ago [6], or tetrodotoxin (**20**), which is found in the puffer fish [10]. However, at appropriate concentrations, they can also be used as drugs, such as morphine (**1**), a very strong pain reliever, or the dimeric indole alkaloid vincristine (**21**), which is used for the treatment of pediatric leukemia with a success rate of almost 70% [12].

Hirsutine
18

Quinine
19

Tetrodotoxin
20

Vincristine
21

(−)-Sparteine
22

On the other hand, some alkaloids are used as catalysts, for example sparteine (**22**) is used for some enantioselective addition reactions (cf. **1.1.3**), and derivatives of quinine (**19**) and quinidine are used for Sharpless bishydroxylations (cf. **2.1.1**).

[1]   a) *Modern Alkaloids* (Eds.: E. Fattorusso, O. Taglialatela-Scafati), Wiley-VCH, Weinheim, **2007**; b) M. Hesse, *Alkaloids – Nature's Curse or Blessing?*, Wiley-VCH, Weinheim, **2002**.

[2]   S. Benyhe, *Life. Sci.* **1994**, *55*, 969.

[3]   a) C. Macfoy, D. Danosus, R. Sandit, T. H. Jones, H. M. Garraffo, T. F. Spande, J. W. Daly, *Z. Naturforsch. C* **2005**, *60*, 932; b) J. E. Warnick, P. J. Jessup, L. E. Overman, M. E. Eldefrawi, Y. Nimit, J. W. Daly, E. X. Albuquerque, *Mol. Pharm.* **1982**, *22*, 565.

[4]   Z. Jin, *Nat. Prod. Rep.* **2005**, *22*, 196.

[5]   G. Bringmann, J. Mutanyatta-Comar, M. Greb, S. Rüdenauer, T. F. Noll, A. Irmer, *Tetrahedron* **2007**, *63*, 1755.

[6]   T. Reynolds, *Phytochemistry* **2005**, *66*, 1399.

[7]   M. Schmittel, A. Haeuseler, *J. Organomet. Chem.* **2002**, *661*, 169.

[8]   G. W. Kirby, *Science* **1967**, *155*, 170.

[9]   a) S. E. O'Connor, J. J. Maresh, *Nat. Prod. Rep.* **2006**, *23*, 532; b) T. Kawasaki, K. Higuchi, *Nat. Prod. Rep.* **2005**, *22*, 761.

[10] U. Koert, *Angew. Chem.* **2004**, *116*, 5690; *Angew. Chem. Int. Ed.* **2004**, *43*, 5572.

4.1.1 Hirsutine

1

Topics:
- Pictet–Spengler reaction
- Enantioselective transfer hydrogenation of an imine
- Protection as carbamates (Boc and Cbz)
- Reduction of an ethyl ester to an aldehyde
- Domino Knoevenagel–hetero-Diels–Alder reaction
- Formation of an enamine
- Diastereoselective hydrogenation of a cyclic enamine

(a) General

Hirsutine (**1**) [1] is an indole alkaloid of the corynanthe subgroup, which has been isolated from the plants *Mitragyna hirsuta* and *Uncaria rhynchophylla* MIQ [2]. The extract of the latter plant is used in the old Chinese folk medicine "Kampo". Nowadays, hirsutine is of great medicinal interest, since it shows a strong inhibition of the influenza A virus of the subtype H3N2 [3, 4]. Its activity, with an ED_{50} value (effective dosage, at which 50% of the maximum activity occurs or at which 50% of the test candidates show a particular reaction) of 0.40 to 0.57 µg/mL, is 10 to 20 times higher than that of the clinically used drug ribavirine. In addition, it shows anti-hypertensive and anti-arrhythmic properties [5].

In 1967, the absolute configuration of hirsutine was established [6] and since then several syntheses have been developed. The first enantioselective synthesis of hirsutine (**1**) was published by the Tietze group [7] in 1999, which, moreover, is an excellent example of a highly efficient domino reaction [8]. This type of transformation is defined as a process in which two or more bonds are formed in one sequence under the same reaction conditions without isolating any intermediates. In this process, the ensuing bond-forming reactions take place at the functionalities obtained in the preceding bond-forming steps. The approach allows the efficient synthesis of complex molecules starting from simple substrates. Moreover, it has great economical and ecological advantages, since it saves time, reduces the amount of waste, and is beneficial in terms of conserving natural resources.

The synthesis of hirsutine partly follows the biosynthesis of the monoterpenoid indole alkaloids, in which, as already mentioned in the introduction to this chapter, the monoterpene secologanin (**3**) [9] undergoes a Pictet–Spengler-type reaction with tryptamine (**2**) to give strictosidine (**4**).

In the first step, the sugar moiety in strictosidine (**4**), used by Nature as a protective group, is removed and one of the two aldehyde functionalities formed then undergoes a condensation with N-4 to give either the corynanthe (N-4 to C-21) or the vallesiachotamine (N-4 to C-17) indole alkaloids.

Accordingly, the retrosynthesis of hirsutine (**1**) leads to **5**, where the transformation of **5** into **1** can be achieved by an ester condensation with methyl formate followed by a methylation of the enol formed; these two steps will not be described here.

Returning to the retrosynthesis of **1**, the bond between C-21 and N-4 in **5** is obtained as in the biosynthesis by a reductive amination of the aldehyde functionality at C-21, which is protected as an acetal in **6**. Compound **6** can be synthesized by a domino Knoevenagel–hetero-Diels–Alder reaction of the chiral tetrahydro-β-carboline **8** bearing an aldehyde moiety, Meldrum's acid (**9**), and the enol ether **10**, via the initially formed cycloadduct **7**, which loses acetone and CO_2.

(b) Synthesis of 5

The β-carboline **8** is prepared by starting with a so-called Pictet–Spengler reaction of tryptamine (**2**) and carbethoxypyruvic acid (**11**), obtained from diethyl oxalacetate by partial hydrolysis, to give a racemic mixture of **12**. This can be oxidized with $KMnO_4$ to afford the achiral imine **13**.

Enantioselective transfer hydrogenation using the chiral Ru catalyst **14** developed by Noyori and co-workers [10], in the presence of formic acid and NEt$_3$, allows face-selective reduction of the imine **13** to give the almost enantiopure tetrahydro-β-carboline (*R*)-**12**. The catalyst **14** is prepared *in situ* by reaction of [RuCl$_2$(C$_6$H$_6$)]$_2$, obtained from RuCl$_3$·3H$_2$O and 1,4-cyclohexadiene, with the mono-toluenesulfonamide of a chiral diamine (cf. **3.3.2.1**). Using the enantiomeric catalyst *ent*-**14**, (*S*)-**12** is also accessible. The subsequent transformations can be performed either using (*R*)-**12**, leading to enantiopure **5**, or with racemic (*R,S*)-**12**, which would give racemic **5**. Reaction of (*R*)-**12** with CbzCl in the presence of NEt$_3$ affords **15a**, which is transformed into **15b** using Boc$_2$O in the presence of a catalytic amount of DMAP. Reduction of **15b** with DIBAH in CH$_2$Cl$_2$ directly affords the aldehyde **8**.

Aldehyde **8** is then used for the domino Knoevenagel–hetero-Diels–Alder reaction with Meldrum's acid (**9**) and a diastereomeric mixture of the enol ether **10** under sonication in benzene in the presence of a few crystals of ethylenediammonium diacetate (EDDA). After formation of the Knoevenagel adduct **16**, the cycloadduct **7** is obtained, which, however, is unstable under the reaction conditions and loses CO$_2$ and acetone upon reaction with H$_2$O formed in the condensation step to give the lactone **6** with an induced 1,3-diastereoselectivity of >24:1 with reference to C-15.

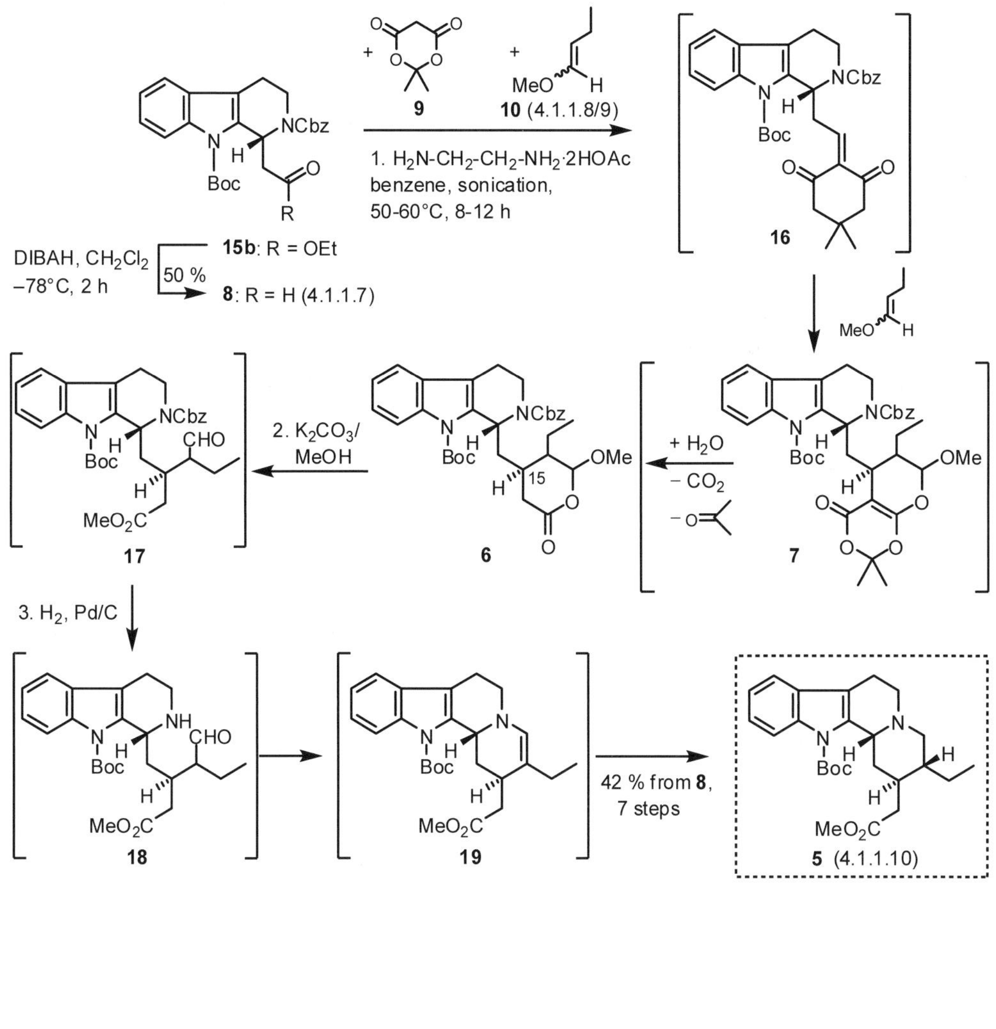

The other two stereogenic centers are formed unselectively since the enol ether **10** is used as a diastereomeric mixture and the *endo/exo* selectivity is low. However, this is of no concern since these two stereogenic centers are lost during the further transformations. Thus, without isolation, **6** is treated with MeOH and a catalytic amount of K_2CO_3 after removal of excess **10**, which is followed by hydrogenation in the presence of Pd/C to afford **5** as a single diastereomer. In this process, the lactone moiety in **6** is attacked by methoxide to give **17** containing a methyl ester and an aldehyde moiety, with elimination of methoxide. The subsequent hydrogenolysis liberates the secondary amino function (**17** → **18**), which attacks the aldehyde function to yield the enamine **19**. The final step is hydrogenation of the enamine **19** under stereoelectronic control to give **5** as a single enantiopure diastereomer.

An unusual feature of the synthesis of **5** that merits discussion is the high 1,3-diastereoselectivity of the hetero-Diels–Alder reaction. It is assumed that the 1-oxa-1,3-butadiene **16** adopts the conformation K-1 rather than K-2 due to the steric hindrance caused by the Boc protecting group at the indole nitrogen. Furthermore, the enol ether should attack the (*E*)-1-oxa-1,3-butadiene moiety and not the less reactive (*Z*)-1-oxa-1,3-butadiene, which also exists in the molecule **16**. Finally, the attack should come from above, *syn* to the hydrogen as the more accessible site of the molecule, to give **6**. Interestingly, employing an aldehyde of type **8** without the Boc protecting group at the indole nitrogen leads to the (*R,S*)-diastereomer of **6** as the main product, albeit with lower diastereoselectivity.

16-K-1   **16-K-2**

The second interesting stereochemical aspect in the formation of **5** is the high selectivity in the formation of the stereogenic center C-20. One can assume that the enamine **19** is the intermediate, which may exist in the conformations K-1 and K-2. However, K-1 should be preferred due to a favorable pseudo equatorial orientation of the acetate moiety at C-15. Assuming that the attack of the hydrogen should come from the β-face due to a favorable chair-like transition state, the stereoselective formation of **5** can be easily explained.

19-K-1   **19-K-2**

(c) Experimental procedures for the synthesis of 5

4.1.1.1   *   Carbethoxypyruvic acid [11]

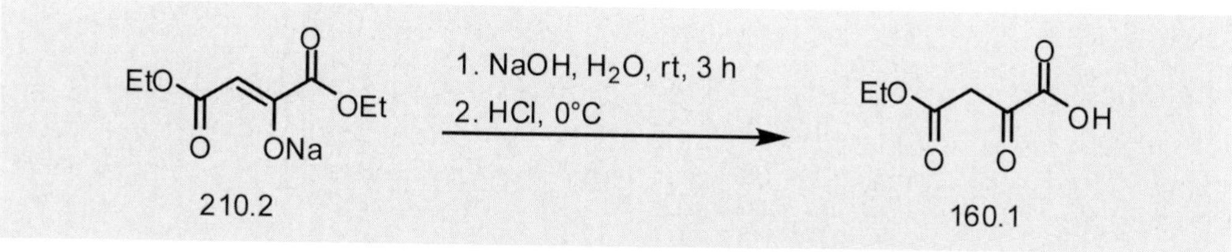

210.2

1. NaOH, H$_2$O, rt, 3 h
2. HCl, 0°C

160.1

Aqueous NaOH solution (6 M, 33.4 mL) is added dropwise to a stirred solution of the sodium salt of diethyl oxalacetate (42 g, 200 mmol) in H$_2$O (400 mL) at room temperature. The resulting mixture is vigorously stirred for 3 h, and then HCl (6 M, 70 mL) is added at 0 °C.

The acidic solution is extracted with Et$_2$O (8 × 140 mL) and EtOAc (3 × 140 mL) and the combined organic fractions are dried over MgSO$_4$. Removal of the solvents under reduced pressure affords the acid as an orange suspension, consisting of the crystalline enolic form and the oily ketonic form; 28.0 g (88%).

UV (CH$_3$CN): λ_{max} (nm) (lg ε) = 257.0 (0.353), 211.5 (0.126), 194.0 (0.196).

IR (KBr): $\tilde{v}$ (cm^{-1}) = 2986, 1729, 1415, 1217, 1026, 853, 787.

^{1}H NMR (300 MHz, CDCl$_3$): δ (ppm) = 5.93 (s, enolic form, 0.6 H, CH), 4.25 (q, ketonic form, J = 7.0 Hz, 2 H, CH$_2$), 4.09 (q, enolic form, J = 7.0 Hz, 1.5 H, CH$_2$), 1.29 (t, ketonic form, J = 7.0 Hz, 3.0 H, CH$_3$), 1.22 (t, enolic form, J = 7.0 Hz, 2.4 H, CH$_3$).

^{13}C NMR (50.3 MHz, CDCl$_3$): δ (ppm) = 172.4 (CO-CO$_2$H), 170.7 (CO-OCH$_2$CH$_3$), 170.0 (CO$_2$H), 96.89 (CH), 61.90, 61.74 (CH$_2$), 47.72 (OCH$_2$CH$_3$), 14.36, 14.30 (OCH$_2$CH$_3$).

MS (DCI-NH$_3$): m/z (%) = 178 (29) [M + NH$_4$]$^+$, 195 (11) [M + NH$_3$ + NH$_4$]$^+$.

4.1.1.2   *   1-Carbethoxymethyl-1,2,3,4-tetrahydro-β-carboline [11]

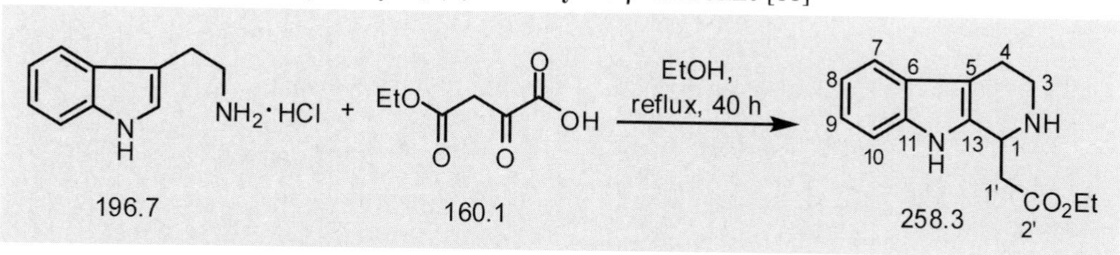

196.7

160.1

EtOH, reflux, 40 h

258.3

A solution of the acid 4.1.1.1 (28.4 g, 178 mmol) in EtOH (85 mL) is added over 30 min to a stirred solution of tryptamine hydrochloride (25.0 g, 127 mmol) in EtOH (400 mL) at reflux. The mixture is refluxed for an additional 40 h and is then kept overnight at 0 to 5 °C to allow crystallization of the hydrochloride of the formed β-carboline.

The precipitate is filtered off, added to a saturated NaHCO$_3$ solution, and the suspension is stirred until the solid has dissolved. The liberated free amine is extracted with EtOAc. The organic layer is washed with brine (2 × 20 mL), dried over Na_2SO_4, filtered, and concentrated under reduced pressure to afford the β-carboline as a light-brown oil; 36.7 g (80%), R_f = 0.55 (MeOH/EtOAc, 1:3, + 1% (v/v) NEt$_3$).

UV (CH$_3$CN): λ_{max} (nm) (lg ε) = 289.0 (0.314), 280.5 (0.390), 226.5 (1.659), 195.0 (1.183).

IR (KBr): $\tilde{v}$ (cm^{-1}) = 3400, 2930, 1722, 1451, 1373, 1157, 1112, 1023, 742.

^{1}H NMR (300 MHz, CDCl$_3$): δ (ppm) = 8.64 (s$_{br}$, 1 H, N-H indole), 7.48 (d, J = 10.0 Hz, 1 H, 7-H), 7.28 (d, J = 10.0 Hz, 1 H, 10-H), 7.10 (2 × ddd, J = 7.0, 7.0, 1.0 Hz, 2 H, 8-H, 9-H), 4.44 (t, J = 7.0 Hz, 1 H, 1-H), 4.20 (q, J = 7.0 Hz, 2 H, 3'-H$_2$), 3.21 (dt, J = 13.0, 5.5 Hz, 1 H, 3-H$_b$), 3.07 (dt, J = 13.0, 5.5 Hz, 1 H, 3-H$_a$), 2.79 (d, J = 7.0 Hz, 2 H, 1'-H$_2$), 2.72 (m, 2 H, 4-H$_2$), 1.90 (s$_{br}$, 1 H, NH), 1.28 (t, J = 7.0 Hz, 3 H, CH$_3$).

^{13}C NMR (50.3 MHz, CDCl$_3$): δ (ppm) = 173.0 (CO), 135.4 (C-11), 134.9 (C-13), 127.0 (C-6), 121.6 (C-8), 119.1 (C-9), 118.0 (C-7), 110.8 (C-10), 108.9 (C-5), 60.95 (C-3'), 48.72 (C-1), 41.79 (C-3), 40.64 (C-1'), 22.45 (C-4), 14.07 (C-4').

MS (DCI-NH$_3$): m/z (%) = 259 (100) [M + H]$^+$, 517 (20) [2M + H]$^+$.

4.1.1.3 ** Dichloro(η^6-benzene)ruthenium(II) dimer [12]

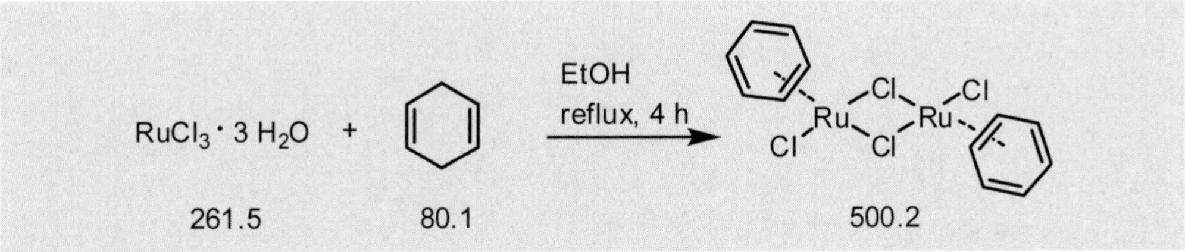

A mixture of ruthenium trichloride monohydrate (300 mg, 41% Ru), EtOH (4 mL), and 1,4-cyclohexadiene (3 mL) is refluxed for 4 h, filtered, and concentrated to dryness under reduced pressure to give [RuCl$_2$(C$_6$H$_6$)]$_2$ as a brown powder; 300 mg (98%).

MS (DCI): m/z (%) = 285 [M + NH$_3$ + NH$_4$]$^+$.

4.1.1.4 *** (1R)-1-Carbethoxymethyl-1,2,3,4-tetrahydro-β-carboline [12]

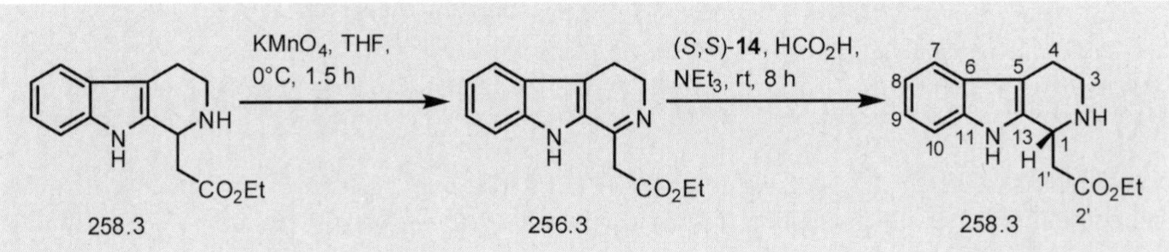

Powdered KMnO$_4$ (1.0 g) is added in small portions over a period of 0.5 h to a vigorously stirred solution of the β-carboline **4.1.1.2** (260 mg, 1.00 μmol) in dry THF (20 mL) at 0 °C. The reaction mixture is kept at 0 °C and stirred for 1 h (TLC control). The precipitate is then filtered off, washed with THF (2 × 10 mL), and the combined filtrate and washings are concentrated to leave the crude imine as a pale-yellow solid; 250 mg (97%).

The catalyst for the enantioselective hydrogenation is prepared *in situ* by stirring [RuCl$_2$(C$_6$H$_6$)]$_2$ (**4.1.1.3**, 6.0 mg, 24 μmol) and the diamine **3.3.2.1** (7.3 mg, 20 μmol) in CH$_3$CN (2.0 mL) for 5 min.

Formic acid/NEt$_3$ (5:2, 2.0 mL) and the preformed catalyst are added to a solution of the imine (120 mg, 500 μmol) in CH$_3$CN (5.0 mL) at 0 °C and the mixture is stirred at room temperature for 6 h. Purification of the crude product by flash chromatography, eluting with EtOAc/EtOH/NEt$_3$ (100:10:1) gives the (1*R*)-β-carboline as a brown oil; 230 mg (90%, 93% *ee*), [α]$_D^{20}$ = +61.9 (*c* = 0.5, CHCl$_3$).

The spectra are identical to those of the racemic compound.

4.1.1.5   *   1-Carbethoxymethyl-2-benzyloxycarbonyl-1,2,3,4-tetrahydro-β-carboline [13]

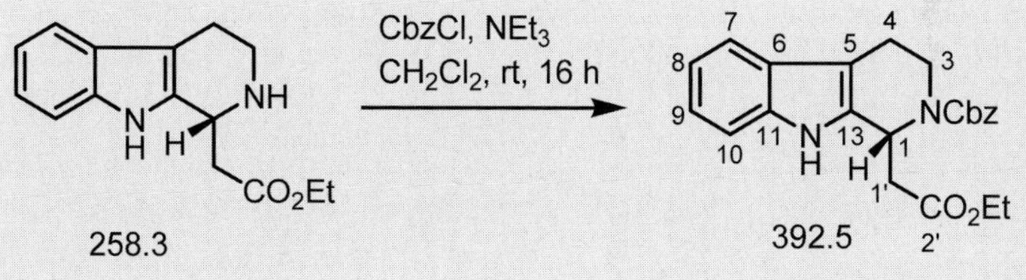

A solution of benzyl chloroformate (4.78 mL, 33.6 mmol) in CH$_2$Cl$_2$ (17 mL) is added to a stirred solution of the β-carboline **4.1.1.4** (7.28 g, 28.0 mmol), NEt$_3$ (11.6 mL, 84 mmol), and a catalytic amount of DMAP in dry CH$_2$Cl$_2$ (28 mL) at 0 °C and stirring is continued at room temperature for 12 h.

The organic layer is then washed with H$_2$O (45 mL), HCl (2 M, 45 mL), further H$_2$O (45 mL), saturated Na$_2$CO$_3$ solution (45 mL), and brine (45 mL). After drying over Na$_2$SO$_4$, the solvent is evaporated under reduced pressure. The brown oily residue obtained is used for the next reaction without further purification; 9.8 g (89%), R_f = 0.69 (EtOAc).

UV (CH$_3$CN): λ$_{max}$ (nm) (lg ε) = 289.0 (0.176), 278.5 (0.235), 273.0 (0.237), 224.0 (1.107).

IR (KBr): $\tilde{\nu}$ (cm^{-1}) = 3395, 2981, 1699, 1424, 1361, 1266, 1218, 1100, 1020, 742, 699.

^{1}H NMR (300 MHz, CDCl$_3$): δ (ppm) = 8.91, 8.75 (2 × s$_{br}$, 1 H, indole NH), 7.47 (dd, *J* = 7.1, 7.1 Hz, 1 H, 10-H), 7.42–7.32 (m, 6 H, 7-H, Ph-H), 7.16 (ddd, *J* = 7.1, 7.1, 1.0 Hz, 1 H, 9-H), 7.08 (dd, *J* = 7.1, 7.1 Hz, 1 H, 8-H), 5.67 (ddd, *J* = 25.0, 8.0, 4.0 Hz, 1 H, 3-H$_b$), 5.19 (s, 2 H, Ph-CH$_2$), 4.51 (ddd, *J* = 39.0, 13.0, 5.0 Hz, 1 H, 1-H), 4.20 (m, 2 H, CH$_2$), 3.14 (m$_c$, 1 H, 3-H$_a$), 2.99–2.67 (m, 4 H, 4-H$_2$, 1'-H$_2$), 1.26 (td, *J* = 7.0, 3.5 Hz, 3 H, CH$_3$).

^{13}C NMR (50.3 MHz, CDCl$_3$): δ (ppm) = 172.9 (C-2'), 155.0 (C-1"), 136.5 (C-13), 135.6 (C-11), 133.0, 128.5, 128.0, 127.8 (Ph-C), 126.3 (C-6), 121.9 (C-8), 119.3 (C-9), 118.0 (C-7), 111.1 (C-10), 108.1 (C-5), 61.16 (CH$_2$), 47.52 (C-1), 39.29 (C-3), 39.06 (C-2"), 39.00 (C-1'), 21.05 (C-4), 14.05 (CH$_3$).

MS (DCI-NH$_3$): *m/z* (%) = 393 (70) [*M* + H]$^+$, 410 (62) [*M* + NH$_4$]$^+$.

4.1.1.6 * 1-Carbethoxymethyl-2-benzyloxycarbonyl-12-*tert*-butyloxycarbonyl-1,2,3,4-tetrahydro-β-carboline [13]

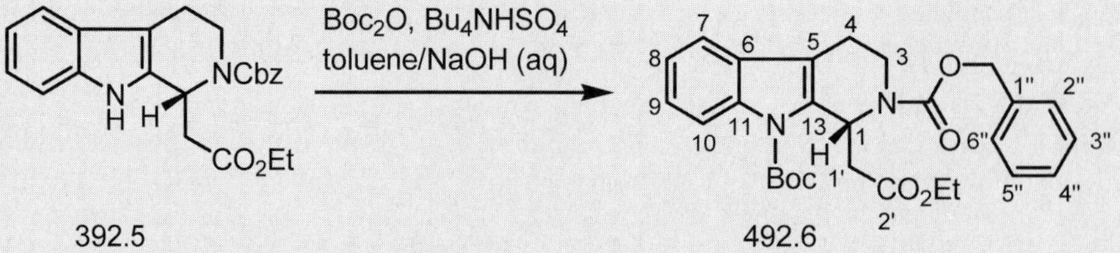

392.5

Boc$_2$O, Bu$_4$NHSO$_4$
toluene/NaOH (aq)

492.6

A solution of 2,2-dimethylpropionic acid anhydride (6.8 g, 31.2 mmol) in CH$_3$CN (80 mL) and DMAP (826 mg, 6.8 mmol) are added to a stirred solution of the β-carboline **4.1.1.5** (9.6 g, 24.5 mmol) in CH$_3$CN (80 mL) at room temperature and stirring is continued for 12 h.

The reaction is then quenched with HCl (0.4 M, 160 mL). After separation of the organic layer, the aqueous layer is extracted with CH$_2$Cl$_2$ (2 × 160 mL). The combined organic layers are washed with saturated NaHCO$_3$ solution and brine, dried over Na$_2$SO$_4$, filtered, and concentrated under reduced pressure to afford the protected β-carboline as a light-brown oil, which is used for the next reaction without further purification; 11.2 g (95%), R_f = 0.41 (EtOAc/*n*-pentane, 1:4).

UV (CH$_3$CN): λ_{max} (nm) (lg ε) = 293.0 (0.093), 265.5 (0.336), 228.5 (0.557).

IR (KBr): $\tilde{\nu}$ (cm^{-1}) = 2980, 1732, 1457, 1424, 1323, 1141, 1116, 1036, 855, 749, 699.

^{1}H NMR (300 MHz, CDCl$_3$) (ratio of rotamers: 3:2): δ (ppm) = 8.18 (d, J = 7.0 Hz, 1 H, 10-H), 7.44–7.18 (m, 8 H, 7-H, 8-H, 9-H, Ph-H), 6.37, 6.27 (2 × dd, J = 10.2, 3.5 Hz, 1 H, 1-H), 5.18, 5.14 (2 × s, 2 H, Ph-CH$_2$), 4.52, 4.40 (m, dd, J = 14.4, 6.0 Hz, 1 H, 3-H$_b$), 4.15–3.99 (m, 2 H, CH$_2$), 3.32 (m$_c$, 1 H, 3-H$_a$), 3.10–2.60 (m, 4 H, 4-H$_2$, 1'-H$_2$), 1.73, 1.60 (2 × s, 9 H, C(CH$_3$)$_3$), 1.16 (m, 3 H, CH$_3$).

^{13}C NMR (50.3 MHz, CDCl$_3$): δ (ppm) = 170.2, 169.9 (C-2'), 155.5 (N-2-CO), 149.8 (N-12-CO), 136.7 (Ph-C), 136.7 (C-13), 136.3 (C-11), 133.5 (C-11), 128.6 (C-6), 128.4 (2 × Ph-C), 128.2, 128.0, 124.7 (3 × Ph-C), 124.6 (C-8), 122.9 (C-7), 118.1, 118.0 (C-9), 115.9 (C-10), 115.4 (C-5), 84.77 (C(CH$_3$)$_3$), 67.75, 67.34 (Ph-CH$_2$), 60.66 (CH$_2$), 49.86 (C-1), 39.10, 38.96 (C-1'), 36.80, 36.35 (C-3), 28.21, 28.08 (C(CH$_3$)$_3$), 21.26, 20.59 (C-4), 14.11 (CH$_3$).

MS (DCI-NH$_3$): m/z (%) = 510 (74) [M + NH$_4$]$^+$, 493 (38) [M + H]$^+$.

4.1.1.7 ** 1-(2-Formylmethyl)-2-benzyloxycarbonyl-12-*tert*-butyloxycarbonyl-1,2,3,4-tetrahydro-β-carboline [13]

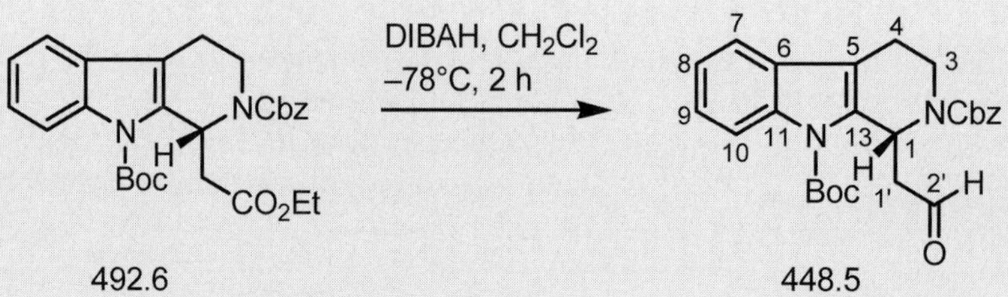

492.6

DIBAH, CH$_2$Cl$_2$
−78°C, 2 h

448.5

A solution of diisobutylaluminum hydride (DIBAH) (1 M in *n*-hexane, 22.7 mL, 22.7 mmol), precooled to –78 °C, is added to a stirred solution of the ethyl ester **4.1.1.6** (11.2 g, 22.7 mmol) in anhydrous CH$_2$Cl$_2$ (230 mL) at –78 °C via a transfer cannula. Stirring is continued at the same temperature for 2 h, and then the reaction is carefully quenched by the dropwise addition of a 9:1 mixture of MeOH and 2 M aqueous HCl (68 mL).

The reaction mixture is quickly warmed to room temperature and washed with saturated NH$_4$Cl solution. After separation of the organic layer, the aqueous layer is extracted with CH$_2$Cl$_2$ (2 × 300 mL). The combined organic layers are washed with saturated NaHCO$_3$ solution, dried over Na$_2$SO$_4$, filtered, and concentrated under reduced pressure. The residue obtained is purified by flash chromatography on silica gel (EtOAc/*n*-pentane, 1:4) to give the aldehyde as a colorless foam; 5.07 g (50%), (1*R*): $[\alpha]_D^{20}$ = –88.8 (*c* = 0.16, CH$_2$Cl$_2$), R_f = 0.32 (EtOAc/*n*-pentane, 1:2).

UV (CH$_3$CN): λ_{max} (nm) (lg ε) = 293.0 (0.102), 264.5 (0.361), 228.5 (0.621).

IR (KBr): $\tilde{v}$ (cm^{-1}) = 2979, 1725, 1456, 1423, 1369, 1235, 1143, 1117, 1016, 853, 751, 699.

^{1}H NMR (300 MHz, CDCl$_3$) (ratio of rotamers: 1:1): δ (ppm) = 9.90, 9.77 (2 × d, J = 4.0 Hz, 1 H, CHO), 8.07 (d, J = 7.5 Hz, 1 H, 10-H), 7.38–7.17 (m, 8 H, 7-H, 8-H, 9-H, Ph-H), 6.53, 6.38 (2 × d, J = 8.9 Hz, 1 H, 1-H), 5.16 (dd, J = 15.0 Hz, 2 H, Ph-CH$_2$), 4.51, 4.37 (2 × dd, J = 13.8, 5.0 Hz, 1 H, 3-H$_b$), 3.30–2.50 (m, 5 H, 3-H$_a$, 4-H$_2$, 1'-H$_2$), 1.69, 1.58 (2 × s, 9 H, C(CH$_3$)$_3$).

^{13}C NMR (50.3 MHz, CDCl$_3$): δ (ppm) = 200.9, 200.1 (C-2'), 155.5 (N-2-CO), 149.9 (N-12-CO), 136.2 (Ph-C), 135.6 (C-11), 133.5 (C-13), 128.5 (C-6), 128.4 (2 × Ph-C), 128.3, 128.0, 127.7 (3 × Ph-C), 124.6, 124.5 (C-8), 122.9 (C-7), 118.0, 117.9 (C-9), 115.8 (C-10), 116.0, 115.4 (C-5), 84.60 (C(CH$_3$)$_3$), 67.76, 67.46 (Ph-CH$_2$), 48.16 (C-1), 47.67 (C-1'), 36.78, 36.39 (C-3), 28.08 (C(CH$_3$)$_3$), 20.41, 21.00 (C-4).

MS (DCI-NH$_3$): *m/z* (%) = 466 (100) [*M* + NH$_4$]$^+$, 914 (3) [2*M* + NH$_4$]$^+$.

4.1.1.8   **   1,1-Dimethoxybutane [11]

A suspension of trimethyl orthoformate (16.0 mL), methanol (16.0 mL), and Montmorillonite K-10 clay (10.0 g) is stirred for 10 min at room temperature under an argon atmosphere. The mixture is then diluted with CH$_2$Cl$_2$ (50.0 mL), cooled to 0 °C, and treated with butanal (8.00 mL, 6.42 g, 89.0 mmol, note). The resulting suspension is stirred for 15 h at room temperature and then filtered through Celite. The filter cake is washed with CH$_2$Cl$_2$ (150 mL) and the combined filtrates are washed with saturated NaHCO$_3$ solution (100 mL), water (100 mL), and brine (100 mL), dried over MgSO$_4$, and concentrated under reduced pressure (not below 400 mbar at 30 °C). The residue is distilled over a 10 cm Vigreux column at ambient pressure to afford the acetal as a colorless liquid (bp 113 °C); 7.64 g (73%), n_D^{20} = 1.3882.

IR (NaCl): $\bar{v}$ (cm^{-1}) = 2960, 2940, 2880, 2830, 1190, 1135, 1125.

^{1}H NMR (200 MHz, CDCl$_3$): δ (ppm) = 4.33 (t, J = 5.0 Hz, 1 H, CH), 3.28 (s, 6 H, OCH$_3$), 1.70–0.80 (m, 7 H, CH$_2$, CH$_3$).

Note: Distilled, bp 75–76 °C.

4.1.1.9   **   (*E*/*Z*)-1-Methoxy-1-butene [11]

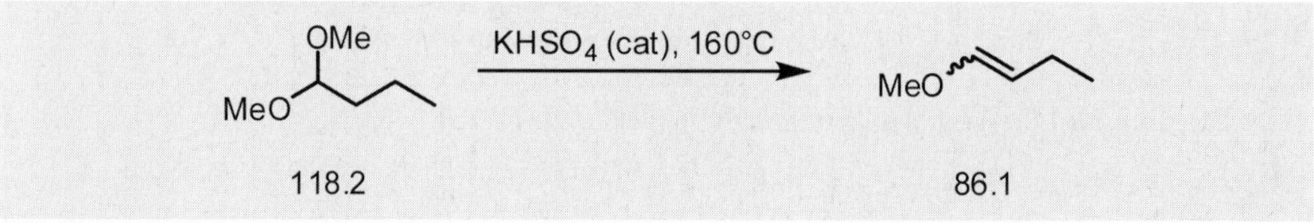

A distillation apparatus consisting of a 25 mL two-necked round-bottomed flask and a 10 cm Vigreux column with a cooler is charged with KHSO$_4$ (15.0 mg, 110 μmol) and 1,1-dimethoxybutane **4.1.1.8** (7.00 g, 59.2 mmol). The stirred mixture is heated to 160 °C (bath temperature) while the enol ether and MeOH formed are continuously distilled off into a flask containing 10 mL of 1% K$_2$CO$_3$ solution. The temperature of the column should be no higher than 50–55 °C. The distillate is washed with 1% K$_2$CO$_3$ solution (3 × 10 mL) to remove MeOH. After distillation of the crude product over a 10 cm Vigreux column, the enol ether is obtained as a 1.8:1 mixture of the *Z*- and *E*-isomers (bp 70–72°C); 2.04 g (40%).

---

^{1}H NMR (200 MHz, CDCl$_3$): δ (ppm) = 6.28 (dt, J = 12.5, 1.0 Hz, 1 H, 1-H (*E*-isomer)), 5.82 (J = 6.5, 1.5 Hz, 1 H, 1-H (*Z*-isomer)), 4.76 (dt, J = 12.5, 7.0 Hz, 1 H, 2-H (*E*-isomer)), 4.32 (dt, J = 6.5, 6.0 Hz, 1 H, 2-H (*Z*-isomer)), 3.60 (s, 3 H, OCH$_3$ (*Z*-isomer)), 3.50 (s, 3 H, OCH$_3$ (*E*-isomer)), 2.06 (dqd, J = 7.5, 6.5, 1.5 Hz, 2 H, 3-H$_2$ (*Z*-isomer)), 1.94 (dqd, J = 7.0, 7.0, 1.0 Hz, 2 H, 3-H$_2$ (*E*-isomer)), 0.97 (t, J = 7.0 Hz, 3 H, CH$_3$ (*E*-isomer)), 0.94 (t, J = 7.5 Hz, 3 H, CH$_3$ (*Z*-isomer)).

^{13}C NMR (50.3 MHz, CDCl$_3$): δ (ppm) = 146.4 (C-1 (*E*-isomer)), 145.5 (C-1 (*Z*-isomer)), 108.8 (C-2 (*Z*-isomer)), 104.9 (C-2 (*E*-isomer)), 59.42 (OCH$_3$ (*Z*-isomer)), 55.79 (OCH$_3$ (*E*-isomer)), 21.03 (C-3 (*E*-isomer)), 17.23 (C-3 (*Z*-isomer)), 15.38 (CH$_3$ (*E*-isomer)), 14.47 (CH$_3$ (*Z*-isomer)).

MS (EI, 70 eV): m/z (%) = 86 (32) $[M]^+$, 71 (100) $[M - CH_3]^+$.

4.1.1.10   ***   (3RS,15RS,20RS)-15-(Methoxycarbonylmethyl)-20-ethyl-1-(*tert*-butyloxycarbonyl)-3,4,5,6,14,15,20,21-octahydro-indolo[2,3a]quinolizine [11]

A few crystals of ethylenediammonium diacetate (EDDA) are added to a solution of the aldehyde **4.1.1.7** (50 mg, 0.11 mmol), the *E/Z* mixture of the enol ether **4.2.1.9** (28.4 mg, 0.33 mmol), and Meldrum's acid (19.0 mg, 0.13 mmol) in benzene (0.5 mL). The reaction vessel is sealed under an argon atmosphere and the red solution obtained is sonicated at 50–60 °C for 8–12 h (TLC monitoring). After complete conversion, the solution is subjected to a column filtration through silica gel (gradient: petroleum ether to petroleum ether/EtOAc, 2:1). The solvents are evaporated and the crude product is dissolved in MeOH (5 mL). K_2CO_3 (10 mg) and Pd/C (50 mg) are added and the resulting mixture is stirred at room temperature for 30 min. The suspension is then stirred vigorously under a hydrogen atmosphere for 4 h, and thereafter is filtered through silica gel (MeOH/EtOAc/NEt$_3$, 1:3:0.05). The solvent is evaporated to give the desired product as a yellow foam; 15.6 mg (42%), $R_f = 0.29$ (petroleum ether/EtOAc, 3:1), (+)-enantiomer: $[\alpha]_D^{20} = +93.0$ ($c = 0.3$, CH$_2$Cl$_2$).

UV (CH$_3$CN): λ_{max} (nm) (lg ε) = 293 (3.452), 268 (4.066), 229 (4.304).

^{1}H NMR (500 MHz, CDCl$_3$): δ (ppm) = 7.92 (dd, $J = 8.0$, 1.5 Hz, 1 H, 12-H), 7.39 (dd, $J = 8.0$, 1.5 Hz, 1 H, 9-H), 7.24, 7.20 (2 × ddd, $J = 8.0$, 8.0, 1.5 Hz, 2 H, 10-H, 11-H), 4.03 (s$_{br}$, 1 H, 3-H), 3.71 (s, 3 H, OCH$_3$), 3.00–2.52 (m, 8 H, 5-H$_2$, 6-H$_2$, 21-H$_2$, 16-H$_2$), 2.10 (m$_c$, 1 H, 15-H), 1.95 (ddd, $J = 13.0$, 3.5, 3.0 Hz, 1 H, 14-H$_{eq}$), 1.88 (ddd, $J = 13.0$, 9.0, 5.0 Hz, 1 H, 14-H$_{ax}$), 1.66 (s, 9 H, *t*-butyl-H), 1.58–1.49 (m, 2 H, C$\underline{H}_2$CH$_3$), 0.91 (t, $J = 7.5$ Hz, 3 H, CH$_3$).

^{13}C NMR (50.3 MHz, CDCl$_3$): δ (ppm) = 173.6, 150.5 (CO), 136.6, 133.2, 129.4, 123.6, 122.5, 117.9, 117.0, 115.1 (Ar-C), 83.56 ($\underline{C}$(CH$_3$)$_3$), 55.97 (C-3), 53.99 (C-), 51.39 (OCH$_3$), 50.93 (C-21), 40.45 (C-20), 37.47 (C-16), 33.45 (C-15), 31.57 (C-14), 28.14 (C($\underline{C}$H$_3$)$_3$), 25.76 (C-19), 21.42 (C-6), 12.17 (CH$_3$).

MS (70 eV, EI): *m/z* (%) = 426 (20) [*M*]$^+$, 369 (100) [*M* – *t*-butyl]$^+$, 325 (20) [*M* – Boc]$^+$, 57 (15) [*t*-butyl]$^+$.

[1] The name hirsutine was also used for a sesquiterpene of the triquinone type. Nowadays, these compounds are named as hirsutane and hirsutene.

[2] a) E. J. Shellard, A. H. Becket, P. Tantivatana, J. D. Phillipson, C. M. Lee, *J. Pharm. Pharmacol.* **1966**, *18*, 553; b) G. Laus, H. Teppner, *Phyton* **1996**, *36*, 185.

[3] H. Takayama, Y. Limura, M. Kitajima, N. Aimi, K. Konno, *Bioorg. Med. Chem. Lett.* **1997**, *7*, 3145.

[4] L. F. Tietze, A. Modi, *Med. Res. Rev.* **2000**, *20*, 304.

[5] a) H. Masumiya, T. Saitoh, Y. Tanaka, S. Horie, N. Aimi, H. Takayama, H. Tanaka, K. Shigenobu, *Life Sci.* **1999**, *65*, 2333; b) S. Horie, S. Yano, N. Aimi, S. Sakai, K. Watanabe, *Life Sci.* **1992**, *50*, 491.

[6] W. F. Trager, C. M. Lee, *Tetrahedron* **1967**, *23*, 1043.

[7] L. F. Tietze, Y. Zhou, *Angew. Chem.* **1999**, *111*, 2076; *Angew. Chem. Int. Ed.* **1999**, *38*, 2045.

[8] a) L. F. Tietze, U. Beifuss, *Angew. Chem.* **1993**, *105*, 137; *Angew. Chem. Int. Ed. Engl.* **1993**, *32*, 131; b) L. F. Tietze, *Chem. Rev.* **1996**, *96*, 195; c) L. F. Tietze, G. Brasche, K. Gericke, *Domino Reactions in Organic Synthesis*, Wiley-VCH, Weinheim, **2006**.

[9] L. F. Tietze, *Angew. Chem.* **1983**, *95*, 840; *Angew. Chem. Int. Ed. Engl.* **1983**, *22*, 828.

[10] a) N. Uematsu, A. Fujii, S. Hashiguchi, T. Ikariya, R. Noyori, *J. Am. Chem. Soc.* **1996**, *118*, 4916; b) R. Noyori, S. Hashiguchi, *Acc. Chem. Res.* **1997**, *30*, 97; c) K. J. Haack, S. Hashiguchi, A. Fujii, T. Ikariya, R. Noyori, *Angew. Chem.* **1997**, *109*, 297; *Angew. Chem. Int. Ed. Engl.* **1997**, *36*, 285.

[11] Y. Zhou, *Ph.D. Thesis*, University of Göttingen, **1998**.

[12] L. F. Tietze, Y. Zhou, E. Töpken, *Eur. J. Org. Chem.* **2000**, 2247.

[13] L. F. Tietze, K. Klapa, unpublished results.

4.1.2   *rac*-2,3-Dimethoxyberbine

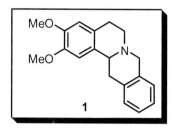

1

Topics:
- Baeyer–Villiger oxidation of a cycloalkanone to a lactone
- Aminolysis of a lactone
- Domino process:
  Bischler–Napieralski isoquinoline synthesis / transformation $CH_2OH \rightarrow CH_2Cl$ / cycloalkylation
- Reduction of iminium salts $R_2C={}^+NR'_2 \rightarrow HR_2C-NR'_2$

(a) General

The target molecule **1** is a protoberberine alkaloid with interesting physiological and pharmacological properties, including cytostatic activity [1, 2]. In Nature, the 1-benzylisoquinoline-based protoberberine alkaloids mainly occur in two structural varieties, namely as protoberberinium salts (e.g. **2**, palmatine) and tetrahydroprotoberberine (e.g. **3**, tetrahydropalmatine). They usually bear hydroxy, methoxy, or methylenedioxy groups on the terminal aromatic rings (A/D). The unsubstituted tetracyclic system **4** has been named berbine [3].

2   **3**   **4**

Retrosynthesis of **1** can be conducted in two ways. Disconnections A and B lead to 1-benzylisoquinoline derivatives **5** and **10**, respectively, the latter via **6**, which might be accessible by Bischler–Napieralski reactions of the amides **8** and **11**, respectively. A different approach could be followed according to disconnection path C, leading to the isoquinolinium salts **7**, which should be accessible by alkylation of isoquinoline with the halide **9**. A problem could arise from the need for regioselective cyclization of **7** at C-3 (**7 → 1**) since, in general, the more reactive position of isoquinolinium salts is C-1.

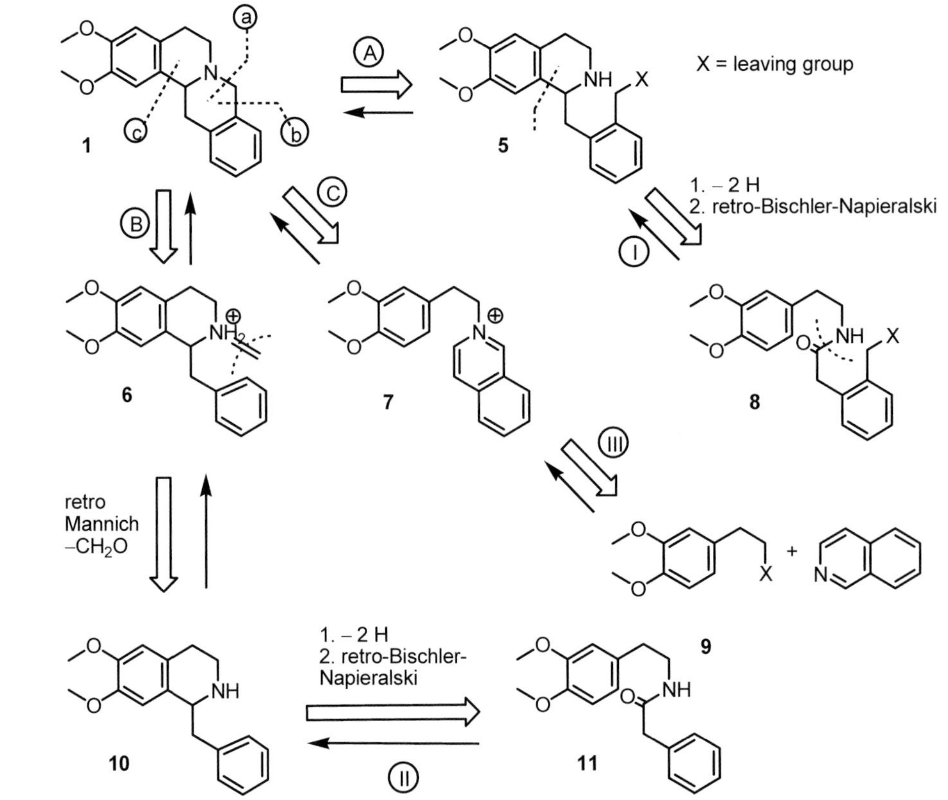

X = leaving group

1. – 2 H
2. retro-Bischler-Napieralski

1

5

6

7

8

retro
Mannich
–CH₂O

9

1. – 2 H
2. retro-Bischler-
Napieralski

10

11

Approach I allows the most straightforward access to **1** and is presented in detail in section (b) [4].

Approach II is the basis of a reported synthesis of **1** [5] by intramolecular aminoalkylation of the 1-benzyltetrahydroisoquinoline **10** with formaldehyde (**11** → **10** → **6** → **1**) [6].

The aforementioned problem associated with approach III is elegantly solved by a radical cyclization strategy [2], which starts from the *ortho*-brominated (β-arylethyl)halide **12** and thus involves the corresponding isoquinolinium salt **13**:

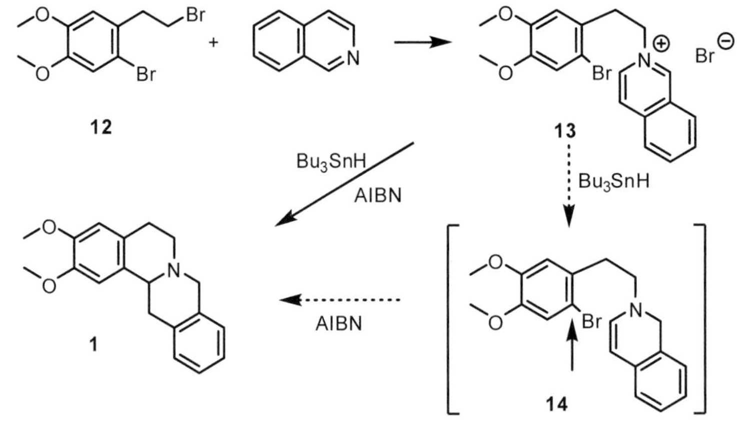

12

13

Bu₃SnH
AIBN

Bu₃SnH

1

AIBN

14

Reaction of **13** with *n*Bu₃SnH/AIBN (3:1) induces a domino process to give **1**, which comprises initial reduction of the more reactive C1–N double bond in **13** followed by formation of an aryl radical at the bromoaryl position and subsequent cyclization by addition to the C-3/C-4 double bond of the intermediate 1,2-dihydroisoquinoline **14**.

A concept completely different from approaches I–III is based on a cycloaddition of the diyne **15** with acetylene to give the tetrahydroisoquinoline moiety in **1** in one step.

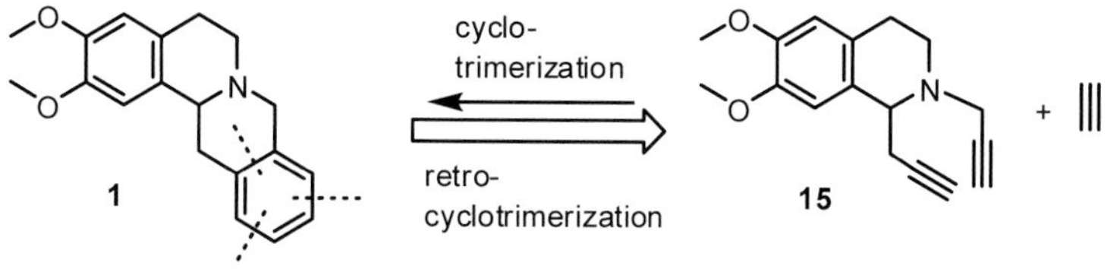

The cyclooligomerization of alkynes [7] is preferentially mediated by Co(I) complexes and its potency is demonstrated by a further synthesis of **1** [8]:

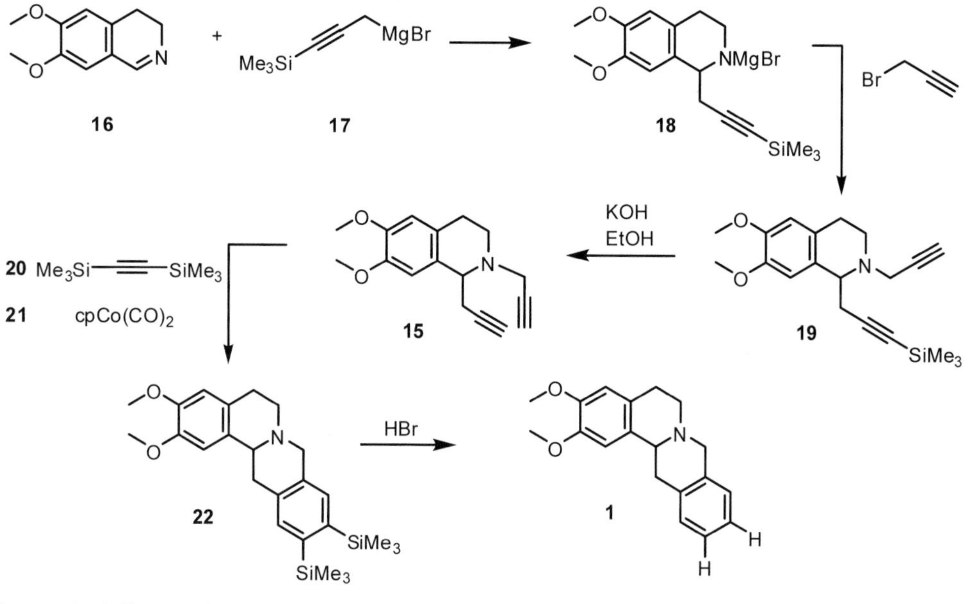

The required diyne **15** is obtained from 3,4-dihydro-6,7-dimethoxyisoquinoline (**16**) by addition of the TMS-protected propargylic Grignard compound **17** at C-1 of **16** and subsequent *N*-propargylation of the intermediately formed Mg amide **18** to afford **19**. After removal of the TMS group (KOH in EtOH), the diyne **15** is subjected to a co-cyclooligomerization with bis(trimethylsilyl)acetylene (**20**) catalyzed by the commercially available η⁵-cyclopentadienyl-Co(I)-dicarbonyl (**21**), leading to the cycloadduct **22**. The use of the bis-TMS-acetylene (**20**) is superior to that of acetylene itself, since the bulky TMS substituents prevent the otherwise inevitable acetylene autotrimerization. The synthesis is completed by protodesilylation of **22** with HBr to give the HBr salt of **1**.

(b) Synthesis of 1

The starting material for the synthesis of **1** according to approach I is 2-indanone (**23**), which is first subjected to a Baeyer–Villiger oxidation [9] with *m*-chloroperbenzoic acid. This oxidative *O*-sextet rearrangement effects ring-expansion of the cyclopentanone derivative **23** to the lactone **24**, which is ring-opened by reaction with homoveratrylamine (**25**) to afford the amide **26**:

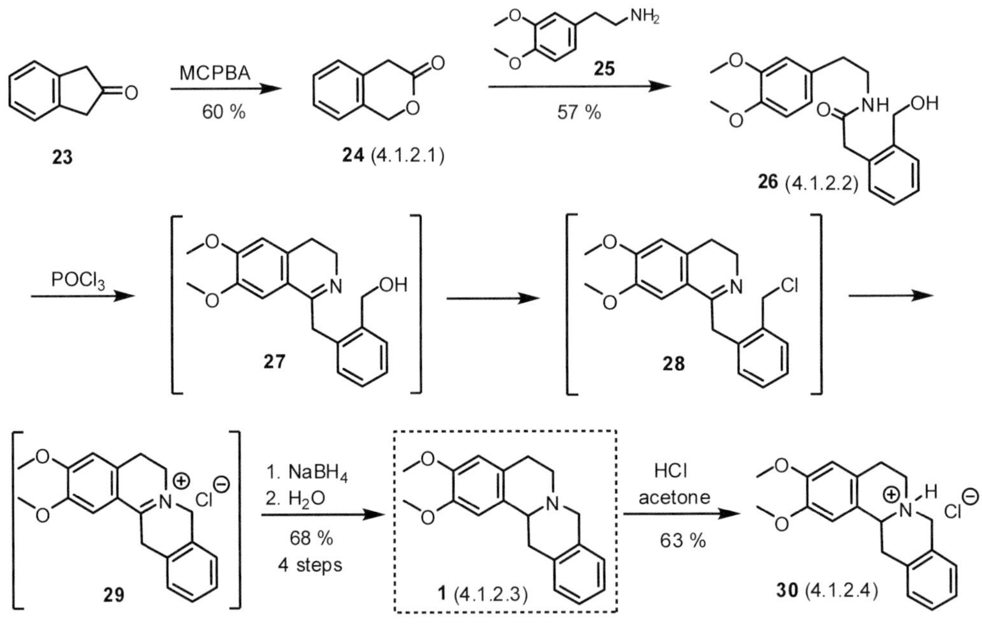

On treatment with POCl$_3$, the amide **26** undergoes two-fold ring-closure, which is followed by reduction with NaBH$_4$ to give the tetracyclic alkaloid **1**. This is transformed into the hydrochloride salt **30**.

The sequence **26 → 1** can be reasonably understood in terms of (1) a domino process consisting of Bischler–Napieralski cyclization (**26 → 27**, cf. **3.3.2**), transformation of the benzylic alcohol into the corresponding chloride (**26 → 28**), and ring closure by an intramolecular *N*-alkylation (cycloalkylation **28 → 29**); (2) hydride reduction of the iminium moiety in **29** to give the tertiary amine functionality of **1** (R$_2$C=NR'$_2$$^+$ → HR$_2$C–NR'$_2$).

Thus, the target molecule **1** is obtained by a three-step sequence in an overall yield of 21% based on 2-indanone **23**.

(c) Experimental procedures for the synthesis of 1

4.1.2.1   **   3-Isochromanone [4]

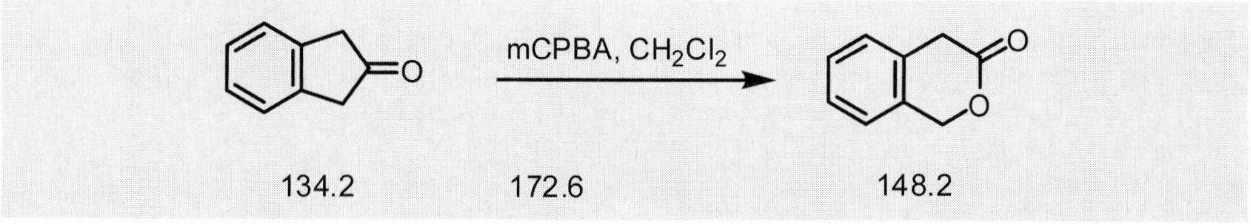

$$134.2 \qquad\qquad 172.6 \qquad\qquad\qquad 148.2$$

An ice-cold solution of 2-indanone (5.00 g, 36.4 mmol) in dry CH_2Cl_2 (10 mL) is added to an ice-cold solution of m-chloroperbenzoic acid (7.82 g, 45.5 mmol) in dry CH_2Cl_2 (50 mL) over 20 min and the reaction mixture is kept at 0 °C for 10 d.

The precipitated m-chloroperbenzoic acid is filtered off and washed with CH_2Cl_2. The combined organic layers are washed with 1% sodium hydrogencarbonate solution (100 mL) and H_2O (100 mL) and dried over Na_2SO_4. After removal of the solvent $in\ vacuo$, the residue is crystallized from MeOH to give a white solid: 3.23 g (60%), mp 81 °C; R_f (SiO₂; EtOAc/toluene, 95:5) = 0.76.

UV: λ_{max} (nm) = 202.

IR (KBr): $\tilde{\nu}$ (cm^{-1}) = 3459, 3019, 2989, 2893, 1853, 1747, 1606, 1495, 1486, 1458, 1407, 1392, 1298, 1252, 1224, 1187, 1148, 1110, 1037, 1028, 992, 959, 928, 897, 867, 819, 777, 761, 743, 693, 633, 551, 498, 458, 430.

^{1}H NMR (CDCl₃): δ (ppm) = 7.36–7.16 (m, 4 H, Ph), 5.30 (s, 2 H, δ-CH₂), 3.70 (s, 2 H, γ-CH₂).

^{13}C NMR (CDCl₃): δ (ppm) = 170.7, 128.8, 127.3, 127.1, 124.7, 36.2.

EI HRMS: m/z = 148.0524.

4.1.2.2   *   N-[2-(3,4-Dimethoxyphenyl)ethyl]-2-(hydroxymethyl)phenylacetamide [4]

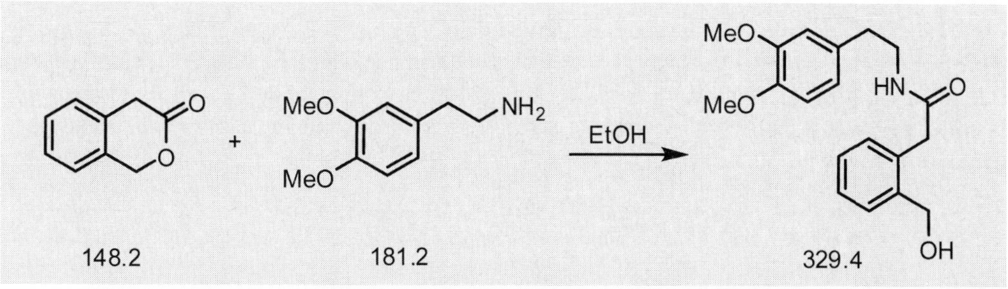

$$148.2 \qquad\qquad\qquad 181.2 \qquad\qquad\qquad\qquad 329.4$$

A solution of 3-isochromanone **4.1.2.1** (2.20 g, 14.9 mmol) and 2-(3,4-dimethoxyphenyl)ethylamine (3.20 mL, 19.3 mmol) in EtOH (50 mL) is refluxed for 20 h.

After evaporation of the solvent $in\ vacuo$, the residue is purified by column chromatography on silica gel (toluene/EtOAc, 1:1) to give a light-yellow solid: 2.80 g (57%), mp 98–99 °C; R_f (SiO₂; EtOAc/toluene, 95:5) = 0.30.

UV: λ_{max} (nm) = 279, 204, 203.

IR (KBr): $\tilde{v}$ (cm^{-1}) = 3288, 3190, 3099, 3015, 3002, 2929, 2867, 2831, 1829, 1639, 1603, 1576, 1519, 1493, 1469, 1453, 1438, 1420, 1366, 1343, 1333, 1293, 1262, 1238, 1201, 1157, 1140, 1067, 1031, 1013, 948, 855, 800, 767, 748, 702, 639, 605, 562, 506, 474, 452, 412.

^{1}H NMR (CDCl$_3$): δ (ppm) = 7.36–7.11 (m, 4 H, Ph$_D$), 6.63–6.50 (m, 3 H, Ph$_A$), 6.23–6.18 (m, 1 H, NH), 4.58 (s, 2 H, CH$_2$OH), 3.83 (s, 3 H, OCH$_3$), 3.78 (s, 3 H, OCH$_3$), 3.53 (s, 2 H, CH$_2$C(O)), 3.40 (q, J = 6 Hz, 2 H, CH$_2$CH$_2$NH), 2.66 (t, J = 6 Hz, 2 H, CH$_2$CH$_2$NH).

^{13}C NMR (CDCl$_3$): δ (ppm) = 171.5, 148.9, 147.6, 139.4, 134.0, 131.1, 130.3, 130.2, 128.4, 127.7, 120.6, 111.8, 111.2, 63.6, 55.9, 55.8, 40.9, 34.8.

HRMS: calcd. for C$_{19}$H$_{23}$NO$_4$ 329.1627; found 330.1700 ([M + H]$^+$), 352.1519 ([M + Na]$^+$).

4.1.2.3   **   *rac*-2,3-Dimethoxyberbine [4]

$$\begin{array}{ccc} & \text{1. POCl}_3 & \\ & \text{2. NaBH}_4 & \end{array}$$

329.4   OH   POCl$_3$: 153.0   295.4
NaBH$_4$: 37.9

A solution of phosphoryl chloride (4.16 mL, 44.6 mmol) in dry toluene (15 mL) is added at room temperature over 15 min to a stirred solution of amide **4.1.2.2** (978 mg, 2.97 mmol) in dry toluene (15 mL) under a nitrogen atmosphere and the mixture is heated under reflux for 1.5 h. The reaction mixture is then cooled to about 50 °C and excess phosphoryl chloride and toluene are evaporated *in vacuo*. The residue obtained is dissolved in MeOH (20 mL) and, after cooling the solution to 0 °C, sodium borohydride (3.20 g, 84.4 mmol) is added in small portions over a period of 1 h. The solution is kept overnight at 0 °C.

Excess sodium borohydride is then decomposed by the addition of ice-cooled water (20 mL) and the mixture is extracted with chloroform (3 × 20 mL). The combined chloroform layers are washed with H$_2$O (20 mL) and brine (20 mL) and dried over Na$_2$SO$_4$. After removal of the solvent, a yellowish gum is obtained, which is purified by column chromatography on neutral alumina (toluene/EtOAc, 4:1) to give a yellow solid, 600 mg (68%); mp 193–194 °C, R_f (SiO$_2$; EtOAc/toluene, 1:1) = 0.83; R_f (neutral alumina; EtOAc/toluene, 1:1) = 0.50.

^{1}H NMR (CDCl$_3$): δ (ppm) = 7.26–7.02 (m, 4 H, Ph$_D$), 6.74 (s, 1 H, Ph$_A$), 6.60 (s, 1 H, Ph$_A$), 4.03 (d, J = 15 Hz, 1 H, NCH$_{2C}$), 3.92–3.81 (m, 9 H, OCH$_3$), 3.74 (d, J = 15 Hz, 1 H, NCH$_{2C}$), 3.62 (dd, J = 10, 5 Hz, 1 H, CH), 3.33 (dd, J = 15, 6 Hz, 1 H, NCH$_{2B}$), 3.20–3.06 (m, 2 H, CH$_{2B}$), 2.96–2.84 (m, 1 H, NCH$_{2B}$), 2.72–2.54 (m, 2 H, CH$_{2C}$).

^{13}C NMR (CDCl$_3$): δ (ppm) = 147.4, 134.3, 128.6, 126.2, 126.1, 125.8, 111.3, 108.4, 59.5, 58.5, 56.0, 55.8, 51.4, 36.7, 29.0.

ESI MS: [M + H]$^+$ = 296.

4.1.2.4   *   *rac*-2,3-Dimethoxyberbine hydrochloride [4]

A 2 N solution of HCl in acetone (2 mL) is added to a stirred solution of dimethoxyberbine **4.1.2.3** (600 mg, 2.03 mmol) in acetone (20 mL). The precipitate formed is collected by filtration and crystallized from MeOH to afford the desired product as a yellow solid: 425 mg (63%), mp 205–206 °C.

UV: λ_{max} (nm) = 285, 205, 202.

IR (KBr): $\tilde{v}$ (cm^{-1}) = 3424, 3071, 3003, 2937, 2909, 2836, 2689, 2497, 1613, 1591, 1525, 1455, 1411, 1364, 1342, 1277, 1264, 1252, 1237, 1216, 1185, 1157, 1129, 1109, 1081, 1053, 1036, 1019, 993, 962, 889, 858, 832, 793, 773, 739, 722, 550, 528, 472, 436.

^{1}H NMR (CDCl$_3$): δ (ppm) = 7.35–7.22 (m, 4 H, Ph$_D$), 7.06 (s, 1 H, Ph$_A$), 6.82 (s, 1 H, Ph$_A$), 4.82–4.69 (m, 1 H), 4.66–4.52 (m, 2 H), 4.00–3.88 (m, 1 H), 3.84–3.70 (m, 6 H, OCH$_3$), 3.45–3.06 (m, 4 H), 2.94–2.81 (m, 1 H).

^{13}C NMR (CDCl$_3$): δ (ppm) = 147.4, 128.6, 126.2, 126.1, 125.8, 111.3, 108.4, 59.5, 58.5, 56.0, 55.8, 51.4, 36.7, 29.0.

HRMS: calcd. for C$_{21}$H$_{23}$N 295.1572; found 296.1645 ([M + H]$^+$).

---

[1] Römpp, *Lexikon "Naturstoffe"* (Eds.: W. Steglich, B. Fugmann, S. Lang-Fugmann), p. 519, Georg Thieme Verlag, Stuttgart, **1997**.

[2] K. Orito, Y. Satoh, H. Nishizawa, R. Harada, M. Tokuda, *Org. Lett.* **2000**, *2*, 2535.

[3] W. Meise, F. Zymalkowski, *Arch. Pharm.* **1971**, *304*, 175; *ibid.* **1971**, *304*, 182.

[4] A. Chatterjee, S. Ghosh, *Synthesis* **1981**, 818.

[5] Smith, Kline & French, Engl. Patent 1004077 (*Chem. Abstr.* **1965**, *63*, 18054d); this synthesis, however, could not be verified according to ref. [3].

[6] For other syntheses of tetrahydroprotoberberines, see: S. C. Pakrashi, R. Mukhopadhyay, P. P. Ghosh Dastidar, A. Bhattacharya, *J. Indian Chem. Soc.* **1985**, *62*, 1003.

[7] L. S. Hegedus, *Organische Synthese mit Übergangsmetallen*, p. 223, VCH, Weinheim, **1995**.

[8] R. L. Hillard III, C. A. Parnell, K. P. C. Vollhardt, *Tetrahedron* **1983**, *39*, 905; compare the remarkable synthesis of estrone: R. L. Funk, K. P. C. Vollhardt, *J. Am. Chem. Soc.* **1980**, *102*, 5253.

[9] M. B. Smith, J. March, *March's Advanced Organic Chemistry*, 5th ed., p. 1417, John Wiley & Sons, Inc., New York, **2001**.

4.1.3 Buflavine

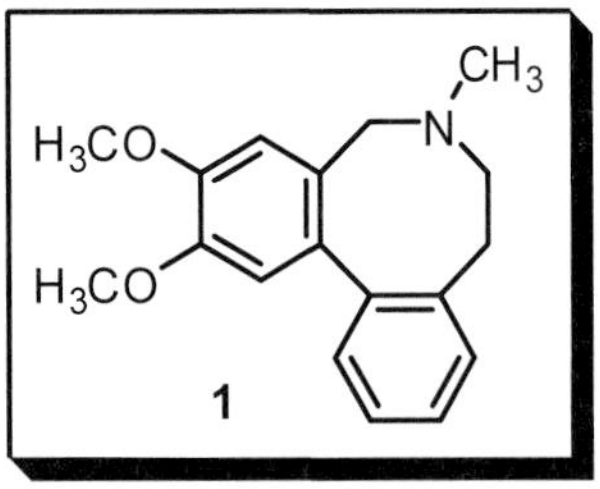

Topics:
- Synthesis of an *Amaryllidaceae* alkaloid
- Suzuki–Miyaura cross-coupling reaction
- Reduction R–CN → R–CH₂–NH₂
- Domino process: Pictet–Spengler synthesis / Eschweiler–Clarke reaction

(a) General

Buflavine (**1**) belongs to the group of *Amaryllidaceae* alkaloids and possesses the very rare 5,6,7,8-tetrahydrodibenz[*c,e*]azocine skeleton, which is composed of a biaryl unit integrated at its *o,o'*-positions in an eight-membered *N*-heterocyclic ring system. Compounds of this type have been shown to exhibit potential α-renolytic and anti-serotonin activities [1].

For the retrosynthesis of buflavine (**1**), after *N*-demethylation to give the secondary amine **2**, three reasonable disconnections (a/b/c) can be envisaged. Two of them (a and c) involve iminium ion formation (**2 → 3**, **2 → 6**) as a favorable retroanalytical operation, while the third (b) involves the retro-process of a reductive amination (**2 → 5**):

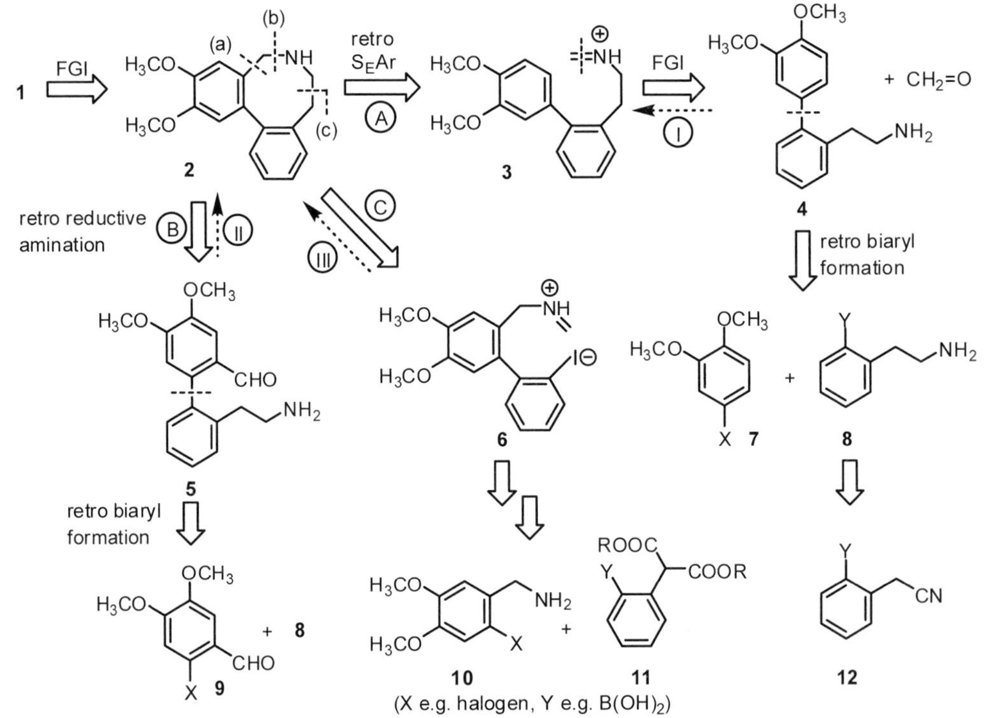

(X e.g. halogen, Y e.g. B(OH)₂)

The retroanalytical pathways **A** and **B** straightforwardly lead to the substrates **7**, **8**, **12** and **5**, **9**, **8**, respectively. Accordingly, two relatively simple approaches I/II for the synthesis of **1** can be envisaged.

In I, after biaryl coupling of **7** and **12** and reduction of the coupling product to give the primary amine **4**, the eight-membered ring of **2** should result from a Pictet–Spengler-like cyclization of the iminium ion **3**, which is formed *in situ* by reaction of **4** with formaldehyde.

In II, after biaryl coupling of the (protected) aldehyde **9** with **12** and reduction/deprotection to give the amino aldehyde **5**, intramolecular reductive amination should provide the dibenzazocine system **2**.

Retrosynthetic pathway **C** leads to the synthon **6** with the functionalities of an iminium cation and a benzyl anion. Suitable substrates might be **10** and **11**. However, route III to the target molecule will be more complex than routes I and II and will therefore not be discussed in detail here.

Nevertheless, all three approaches, I–III, have been used in actual syntheses of **1**.

(1): The *N*-methyl-*N'*-bis(trimethylsilyl)methyl amide **13** is regioselectively metalated in the *ortho* position with respect to the amide function (cf. **3.2.3**) and then the aryllithium is treated with trimethyl borate to give the boronic acid **14** after acid hydrolysis. Suzuki–Miyaura cross-coupling of **14** with 2-bromobenzaldehyde gives the highly functionalized biaryl **16**, which is subjected to an intramolecular Peterson olefination (cf. **1.1.7**) under high-dilution conditions to give the cyclization product **15**. Catalytic hydrogenation, LiAlH$_4$ reduction of the amide moiety (C=O → CH$_2$), and selective cleavage of the isopropyl ether leads to 8-*O*-demethylbuflavine (**17**, another *Amaryllidaceae* alkaloid), which is methylated to yield buflavine (**1**) [2]. This synthesis covers the essentials of approach III.

(2): Oxazoline **18** (derived from 2,4,5-trimethoxybenzoic acid) undergoes a Pd(0)-catalyzed unsymmetrical biaryl coupling according to Meyers' protocol [3] with the Grignard compound **20** (derived from 2-bromobenzaldehyde) leading to the biaryl **19**. Cleavage of the dioxolane moiety using FeCl$_3$·6H$_2$O leads to the aldehyde **21**, which is subjected to a PO-activated olefination with the Boc-protected phosphine oxide **22** to yield the enecarbamate **24**. Subsequent hydrogenation of **24** affords the oxazoline **23**, which is reductively cleaved by NaBH$_4$ after *N*-alkylation [4] to yield the aldehyde

25. After Boc-deprotection (TFA), intermolecular reductive amination of the corresponding secondary amine with Na[HB(OAc)$_3$] results in the formation of the azocine **1** [5] according to approach II:

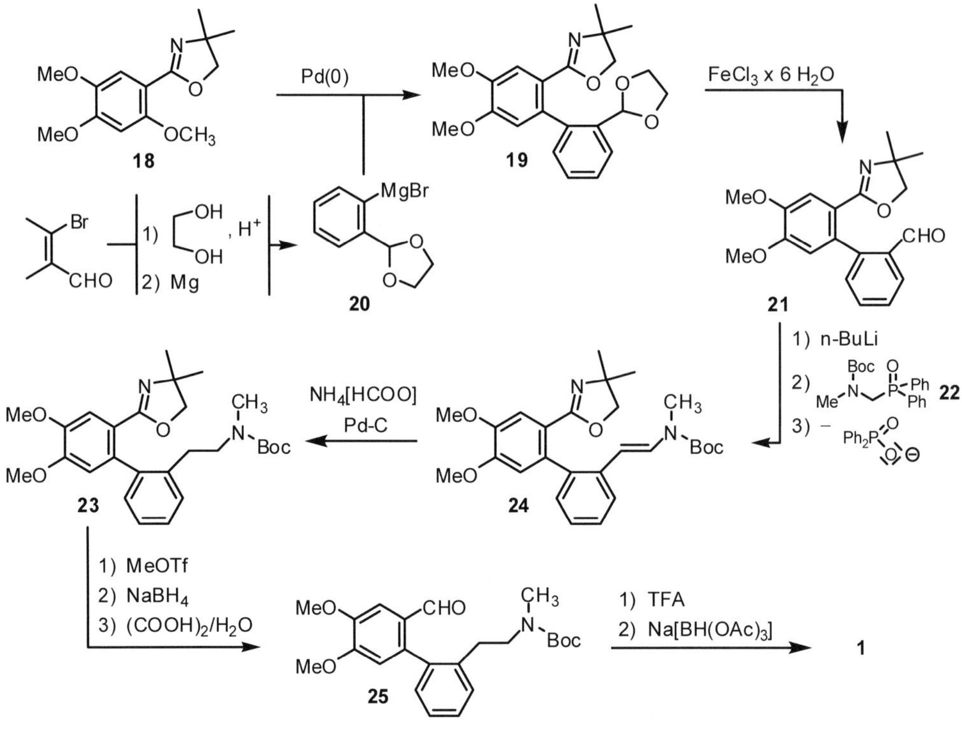

Realization of the retrosynthesis-based approach I leads to a short and efficient synthesis of **1** [1], which is presented in section (b).

(b) Synthesis of 1

Substrates for the key intermediate **28** are the readily accessible arylboronic acid **26** and arylacetonitrile **27**, which are connected (cf. **1.6.2**) by a Suzuki–Miyaura cross-coupling reaction in the presence of Pd(PPh$_3$)$_4$ and K$_2$CO$_3$ in almost quantitative yield to give the biaryl **28** [1].

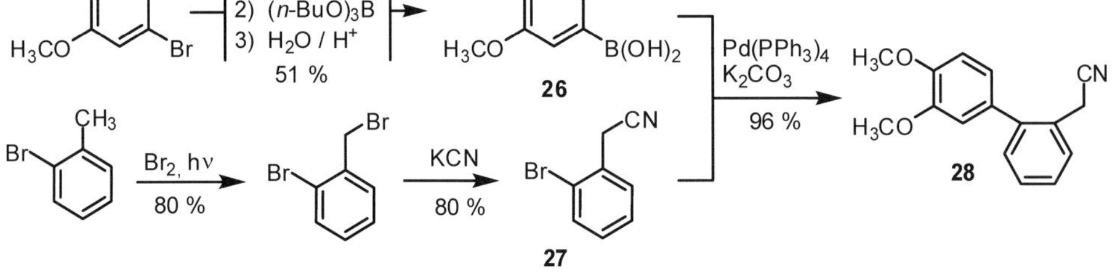

In the concluding steps, the arylacetonitrile **28** is transformed into the β-arylethylamine **4** by CoCl$_2$-assisted reduction with NaBH$_4$. The amine **4** is then treated with a fivefold excess of paraformaldehyde in formic acid to give the target molecule **1** in a yield of 55%. In addition, according to ref. [1], 5% of a side-product **29** can be isolated by preparative TLC. This product is not described in the experimental part.

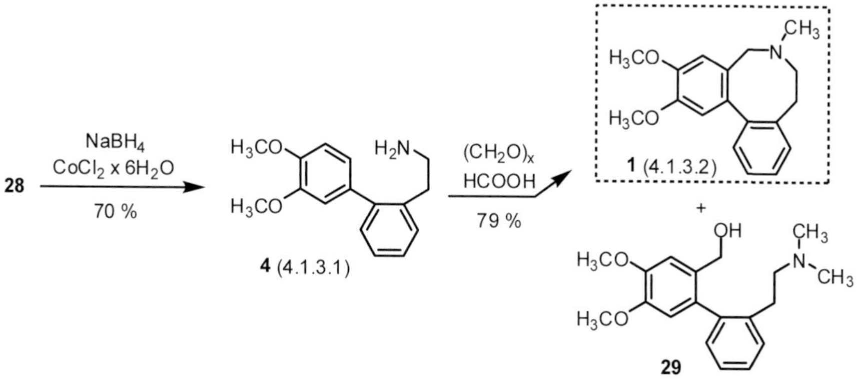

Apparently, in the key step (**4 → 1**), the eight-membered ring is closed in a similar fashion to the formation of 1,2,3,4-tetrahydroisoquinolines according to a Pictet–Spengler synthesis [6] by iminium ion formation followed by an intramolecular S_EAr reaction with ring closure and *N*-methylation according to the Eschweiler–Clarke reaction [7]. For this domino process [8], two alternative routes (either **4 → 3 → 2 → 1** or **4 → 30 → 31 → 1**) are possible:

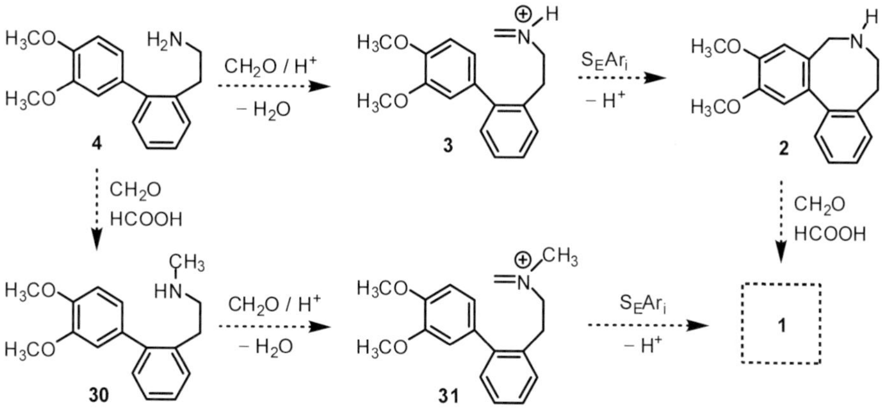

The side-product **29** is presumably derived from **4** by a hydroxymethylation (S_EAr) of the donor-activated A-ring in **4** with (protonated) formaldehyde (**4 → 32**). This is then followed by an Eschweiler–Clarke *N,N*-dimethylation (**32 → 29**) of the primary amine function in **32**.

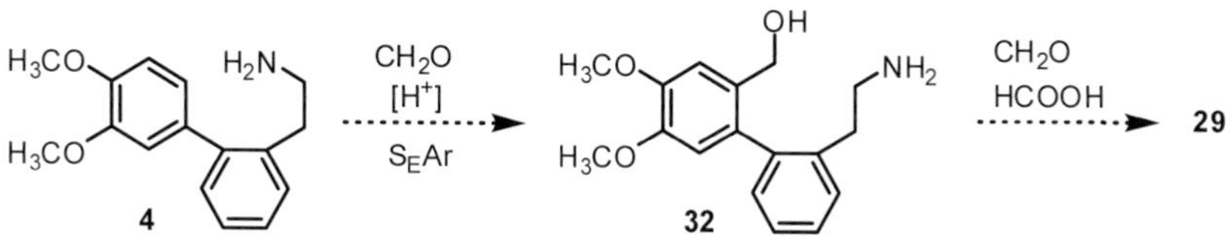

Since the substrates **26** and **27** are commercially available, the target molecule **1** is prepared in a three-step sequence with an overall yield of 53%.

(c) Experimental procedures for the synthesis of 1

4.1.3.1   *   2-(2-Aminoethyl)-3',4'-dimethoxybiphenyl [1]

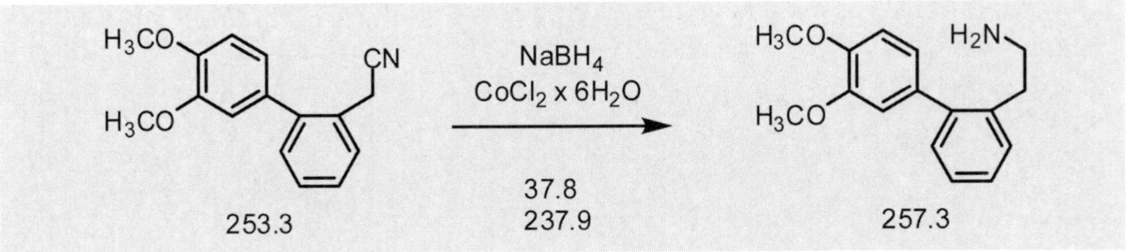

Cobalt chloride hexahydrate (3.76 g, 15.8 mmol) is added to a stirred solution of the nitrile **1.6.2.3** (2.00 g, 7.90 mmol) in methanol/benzene (4:1, 100 mL). The mixture is then cooled to 0 °C and sodium borohydride (2.99 g, 79.0 mmol) is added. The resulting dark solution is allowed to warm to room temperature and stirred for 3.5 h, after which HCl (3 M, 100 mL) is added and stirring is continued for an additional 1.5 h.

The mixture is then concentrated (rotary evaporator) and basified with concentrated ammonia, and the resulting slurry is extracted with Et_2O (3 × 100 mL). The ethereal layers are combined, washed with water and brine, dried over $MgSO_4$, and concentrated; 1.42 g (70%), yellow viscous oil.

^{1}H NMR (500 MHz, CDCl$_3$): δ (ppm) = 7.32–7.20 (comb. m, 4 H, Ar-H), 6.95–6.80 (comb. m, 4 H, Ar-H), 3.92 (s, 3 H, OCH$_3$), 3.88 (s, 3 H, OCH$_3$), 2.78 (comb. m, 4 H, Ar–CH$_2$–CH$_2$), 1.21 (s$_{br}$, 2 H, NH$_2$).

^{13}C NMR (125 MHz, CDCl$_3$): δ (ppm) = 148.55, 148.08, 142.15, 137.39, 134.57, 130.35, 129.61, 127.32, 126.07, 121.40, 112.77, 111.00, 55.95, 43.40, 37.55.

4.1.3.2   *   2,3-Dimethoxy-6-methyl-5,6,7,8-tetrahydrodibenzo[c,e]azocine (buflavine) [1]

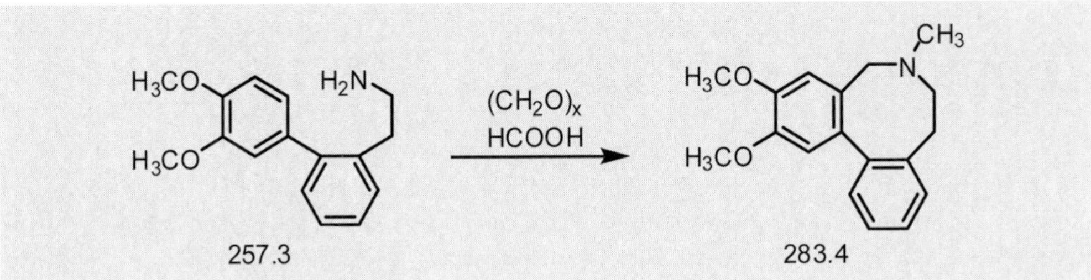

A mixture of the β-arylethylamine **4.1.3.1** (800 mg, 3.11 mmol), paraformaldehyde (516 mg, 15.5 mmol, 90–92%), and formic acid (20 mL) is stirred at room temperature for 24 h. Additional paraformaldehyde (516 mg, 15.5 mmol) is then added and the solution is heated to reflux for 24 h.

The mixture is then concentrated *in vacuo*, made alkaline with aqueous Na_2CO_3 (2 M, 100 mL), and extracted with CHCl$_3$ (3 × 80 mL). The combined organic phases are washed with water (3 × 100 mL), dried (MgSO$_4$), and concentrated. The viscous oil obtained is purified by column chromatography on silica gel using MeOH/CH$_2$Cl$_2$ (1:9) as eluent; 700 mg (79%), yellow viscous oil.

IR (KBr): $\tilde{v}$ (cm^{-1}) = 3050, 2930, 2840, 2785, 1605, 1515, 1440, 1210, 1145, 1020, 860, 750.

^{1}H NMR (500 MHz, CDCl$_3$): δ (ppm) = 7.36–7.27 (comb. m, 3 H, Ar-H), 7.25–7.22 (comb. m, 1 H, Ar-H), 6.90 (s, 1 H, Ar-H), 6.80 (s, 1 H, Ar-H), 3.95 (s, 3 H, OCH$_3$), 3.89 (s, 3 H, OCH$_3$), 3.52 (d, J = 13.6 Hz, 1 H), 3.26 (comb. m, 1 H), 3.06 (d, J = 13.6 Hz, 1 H), 2.75–2.67 (comb. m, 1 H), 2.55–2.49 (comb. m, 2 H), 2.49 (s, 3 H, CH$_3$).

^{13}C NMR (125 MHz, CDCl$_3$): δ (ppm) = 148.54, 147.58, 141.40, 140.12, 133.04, 130.06, 129.49, 129.07, 127.90, 126.08, 113.74, 112.34, 58.81, 58.43, 56.01, 55.95, 45.93, 32.68.

---

[1]   P. Sahakitpichan, S. Ruchirawat, *Tetrahedron. Lett.* **2003**, *44*, 5239.

[2]   P. A. Patil, V. Snieckus, *Tetrahedron Lett.* **1998**, *39*, 1325.

[3]   R. H. Hutchings, A. I. Meyers, *J. Org. Chem.* **1996**, *61*, 1004.

[4]   For comparison, see: L. F. Tietze, Th. Eicher, *Reaktionen und Synthesen im organisch-chemischen Praktikum und Forschungslaboratorium*, 2nd ed., p. 384, Georg Thieme Verlag, Stuttgart, **1991**.

[5]   C. Hoarau, A. Couture, E. Deniau, P. Grandclaudon, *J. Org. Chem.* **2002**, *67*, 5846.

[6]   Th. Eicher, S. Hauptmann, *The Chemistry of Heterocycles*, 2nd ed., p. 346, Wiley-VCH, Weinheim, **2003**.

[7]   *Organikum*, 21st ed., p. 578, Wiley-VCH, Weinheim, **2001** (Leuckart–Wallach reaction).

[8]   a) L. F. Tietze, *Chem. Rev.* **1996**, *96*, 115; b) L. F. Tietze, G. Brasche, K. Gericke, *Domino Reactions in Organic Synthesis*, Wiley-VCH, Weinheim, **2006**.

4.2   Isoprenoids

Introduction

The terpenes are a huge group of natural products, which can be formally deduced from the hydrocarbon isoprene (**1**). According to the number of carbon atoms, which is always a multiple of five (= isoprene), one distinguishes between monoterpenes (C_{10}), sesquiterpenes (C_{15}), diterpenes (C_{20}), sesterterpenes (C_{25}), triterpenes (C_{30}), and tetraterpenes (C_{40}), as well as polyterpenes [1]. In this book, syntheses of the following monoterpenes are described: α-terpineol (**2**) (**1.7.5**), *trans*-chrysanthemic acid (**4**) (**4.2.1**), nerol (**5**) (**4.2.2**), (–)-menthol (**6**) (**4.2.3**), and artemisia ketone (**7**) (**4.2.4**). In addition, preparations of the sesquiterpene veticadinol (**8**) (**4.2.5**) and the diterpene all-*trans*-vitamin A (**9**) (**4.2.6**) are also presented.

Isoprene	α-Terpineol	Multistriatin	Chrysanthemic acid	Nerol
1	(+)-**2**	**3**	**4**	**5**

Menthol	Artemisia ketone	Veticadinol	all-*trans*-vitamin A (all-*trans*-retinol)
(–)-**6**	**7**	**8**	**9**

To date, nearly 50000 different terpenes have been isolated from natural sources as primary ingredients of the essential oils of many types of flowers and plants. Some terpenes are also found in animals, where they act, for example, as pheromones, such as multistriatin (**3**), and in marine organisms, where they are commonly shighly halogenated. A large amount of terpenes is emitted from conifers (over 1,000,000,000 tons per annum), which can sometimes lead to the development of a haze in the summertime, as in the Smoky Mountains in the USA. Terpenes can form linear chains and cyclic compounds of different sizes, as well as annulated and bridged compounds.

In 1910, Professor Otto Wallach of the Georg-August-University in Göttingen was awarded the Nobel prize for his excellent work in this field.

Monoterpenes and sesquiterpenes are mostly used as fragrances in perfumery [2]. There are several important diterpenes, such as the plant growth factor gibberilinic acid (**10**), the anticancer agent taxol (**11**) [3], and the sight purpur retinal (**9**, CHO instead of CH_2OH). A notable triterpene is the tetracyclic compound lanosterol (**12**), the precursor of steroids [7–9] in animals.

Important tetraterpenes are the acyclic lycopene, the dye of tomatoes, and the carotenes. The best known polyterpene is natural rubber, in which all of the double bonds are of (Z)-configuration. Moreover, many compounds found in Nature are degradation products of terpenes, such as β-ionone (**13**) (cf. **1.5.3**), which is formed by oxidative cleavage of a tetraterpene.

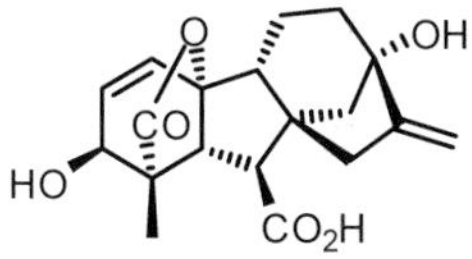

Gibberilinic acid GA$_3$
10

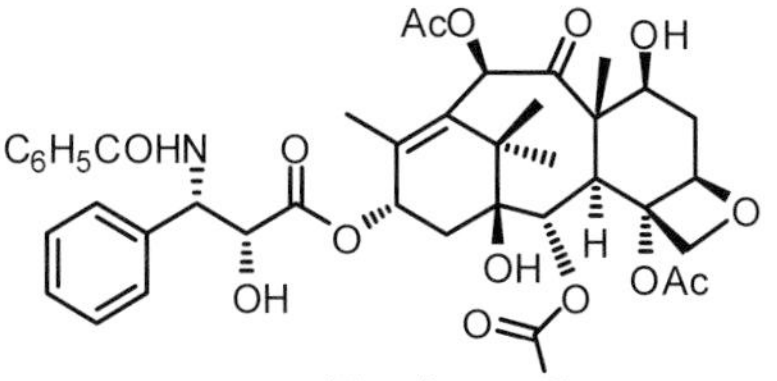

Taxol   Ph
11

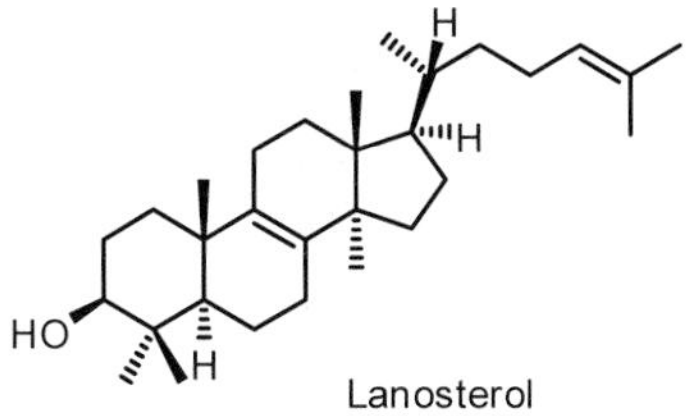

Lanosterol
12

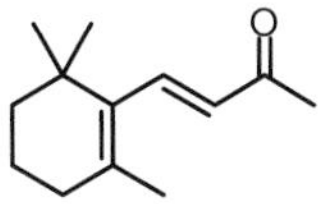

β-Ionone
13

Terpenes are biosynthesized from acetyl-CoA via 3-hydroxy-3-methylglutaryl-CoA (**14**) (HMG-CoA) and mevalonic acid (**15**) (MVA) to afford isopentenyl diphosphate (**16**) (IDP) and dimethylallyl diphosphate (**17**) (DMADP), which undergo a head-to-tail condensation to give the monoterpene geranyl diphosphate (**18**) [4]. This is the natural substrate for nearly all other terpenes. However, in a few cases, a head-to-head connection also takes place. This type of condensation is found in the biosynthesis of chrysanthemic acid (**4**) (**4.2.1**) [5] and in a way also in that of artemisia ketone (**7**) (**4.2.4**).

Recently, it has been shown that in some bacteria and plastids of plants another pathway is operative, which starts from a C$_4$-sugar, namely 2-methyl-D-erythritol-4-phosphate (MEP pathway), to again give IDP and DMADP [6].

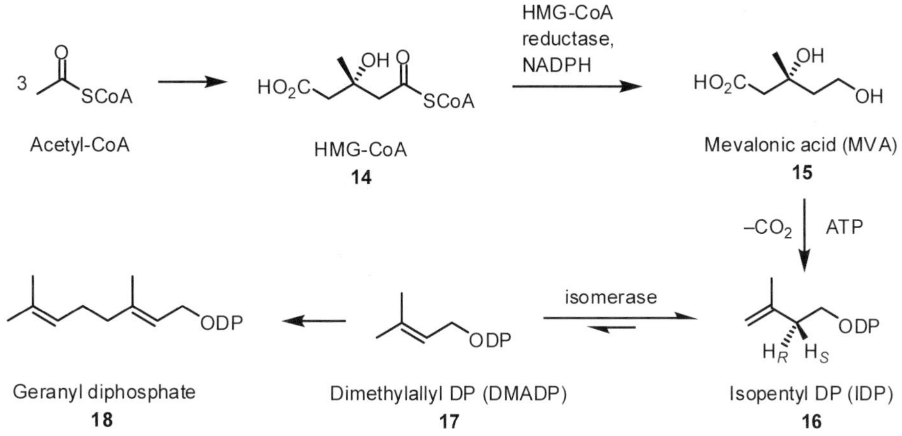

Acetyl-CoA

HMG-CoA
14

Mevalonic acid (MVA)
15

Geranyl diphosphate
18

Dimethylallyl DP (DMADP)
17

Isopentyl DP (IDP)
16

Closely related to the terpenes are the steroids, since they are formed from lanosterol (**12**) in animals as well as in fungi, and from cycloartenol (**24**) in plants as well as in algae. They all contain a perhydrogenated cyclopenta[c]phenanthrene carbon skeleton **19**.

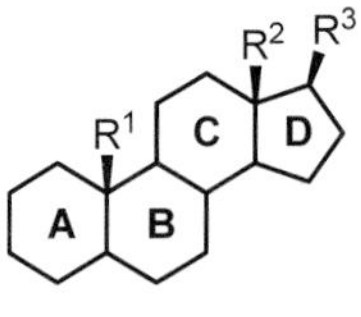

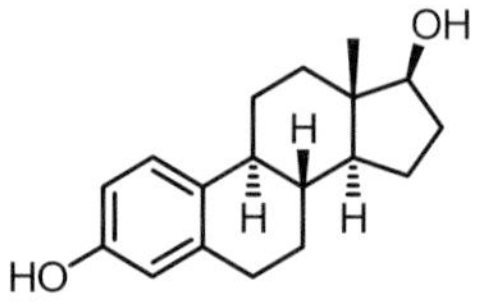

Estradiol
20

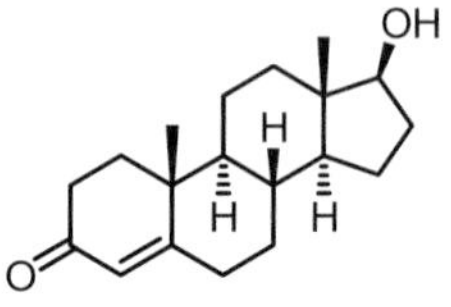

Testosterone
21

Fundamental skeleton	R^1	R^2	R^3
Gonane	H	H	H
Estrane	H	CH$_3$	H
Androstane	CH$_3$	CH$_3$	H
Pregnane	CH$_3$	CH$_3$	CH$_3$

The steroids are by far the best investigated group of natural products due to their pronounced biological activity as hormones, such as estradiol (**20**) and testosterone (**21**). They are formed in Nature from the acyclic triterpene squalene by cyclization of its epoxide **22** to give either lanosterol (**12**) or cycloartenol (**24**) via the intermediate carbocation **23** [7].

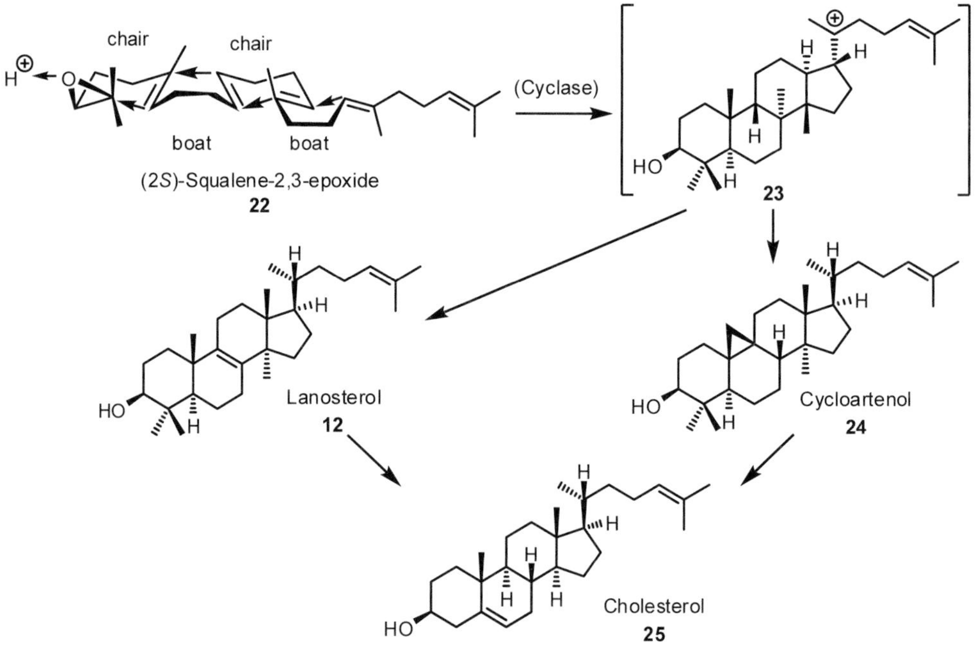

(2S)-Squalene-2,3-epoxide
22

(Cyclase)

23

Lanosterol
12

Cycloartenol
24

Cholesterol
25

In animals, lanosterol (**12**) is transformed into cholesterol (**25**), which is the principal animal steroid. It is present in the membranes of all animal cells and, furthermore, is the substrate for the formation of other steroid hormones.

Moreover, vitamin D (**27**) is also a steroid derivative since it is formed from a steroid, i.e. **26**, by an interesting photochemical ring-opening of the cyclohexadiene moiety followed by a thermal 1,7-sigmatropic hydrogen shift [8]. In 1928, Prof. A. Windaus of the Georg-August-University in Göttingen was awarded the Nobel prize for his outstanding work on vitamin D.

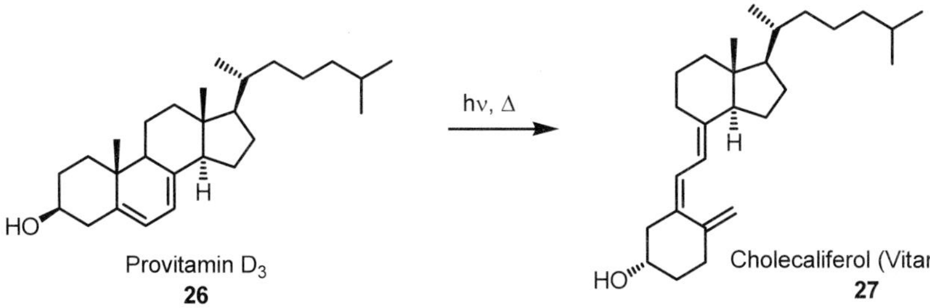

Provitamin D$_3$
26

Cholecaliferol (Vitamin D$_3$)
27

[1]   a) Monoterpenes: D. H. Grayson, *Nat. Prod. Rep.* **2000**, *17*, 385; b) sesquiterpenes: B. M. Fraga, *Nat. Prod. Rep.* **2005**, *22*, 465; c) diterpenes: J. R. Hanson, *Nat. Prod. Rep.* **2005**, *22*, 594; d) sesterterpenes: J. R. Hanson, *Nat. Prod. Rep.* **1996**, *13*, 529; e) triterpenes: J. D. Connolly, R. A. Hill, *Nat. Prod. Rep.* **2007**, *24*, 465.

[2]   C. C. C. R. de Cavalho, M. M. R. da Fonseca, *Biotech. Adv.* **2006**, *24*, 134.

[3]   K. C. Nicolaou, W.-M. Dai, R. K. Guy, *Angew. Chem.* **1994**, *106*, 38; *Angew. Chem. Int. Ed. Engl.* **1994**, *33*, 15.

[4]   P. M. Dewick, *Nat. Prod. Rep.* **2002**, *19*, 181.

[5]   M. P. Crowley, P. J. Godin, H. S. Inglis, M. Snarey, E. M. Thain, *Biochim. Biophys. Acta* **1962**, *60*, 312.

[6]   W. Eisenreich, A. Bacher, D. Arigoni, F. Rohdich, *Cell. Mol. Life. Sci.* **2004**, *61*, 1401.

[7]   a) K. U. Wendt, G. E. Schulz, E. J. Corey, D. R. Liu, *Angew. Chem.* **2000**, *112*, 2930; *Angew. Chem. Int. Ed.* **2000**, *39*, 2812; b) D. M. Harrison, *Nat. Prod. Rep.* **1985**, *2*, 525; c) W. S. Johnson, *Angew. Chem.* **1976**, *88*, 33; *Angew. Chem. Int. Ed. Engl.* **1976**, *15*, 9.

[8]   G.-D. Zhu, W. H. Okamura, *Chem. Rev.* **1995**, *95*, 1877.

4.2.1   (±)-*trans*-Chrysanthemic acid

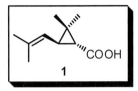

Topics:
- Synthesis of a cyclic monoterpene carboxylic acid
- Allylic halides from allylic alcohols and HX with allyl inversion
- Esterification/ester hydrolysis
- Sulfinates by reduction of sulfochlorides
- Sulfones by rearrangement of sulfinic esters
- Domino process: Michael addition/cyclopropane formation by 1,3-elimination

(a) General

Chrysanthemic acid (**1**) is a *gem*-dimethyl-substituted cyclopropane carboxylic acid, in which a β-dimethylvinyl moiety is attached *trans* to the COOH group. It belongs to the monoterpene natural product family and occurs in pyrethrums (e.g. *Chrysanthemum cinnerariaefolium Vis.*) as an ester of a hydroxycyclopentenone (pyrethrolones). These esters (pyrethrins), such as Pyrethrin I/II (**2**), are important insecticides. They are only slightly toxic for warm-blooded animals, but possess effective and rapid activity (knock-down effect) against insects.

Pyrethrin I: R = CH₃
Pyrethrin II: R = COOCH₃

By structural modifications of chrysanthemic acid and of the alcohol component, the insecticidal activity can be improved. Thus, the ester **3** of (β-dichlorovinyl) dimethyl cyclopropane carboxylic acid with (3-phenoxy)benzyl alcohol finds wide application in plant protection under the name Permethrin [1]. In comparison to other established insecticides, Permethrin shows low toxicity in warm-blooded animals and reasonable stability under light and in air and it is easily metabolized. For its application, only relatively small amounts of substance are required. However, recently some disadvantages have emerged, such as potentiality as an allergen or toxicity for bees and fishes [2]. The introduction of a cyano group at the benzyl position (↑) of the alcohol component and the replacement of chlorine by bromine in the dihalovinyl group further increases the activity of the insecticide **3** [3].

Retrosynthesis of the cyclopropane-containing target molecule conventionally entails the following considerations:

- [2+1]-cycloreversion of the three-membered ring leading to olefinic substrates and carbenes (and their precursors),

- opening of a cyclopropane C–C bond leading to cyclizable 1,3-functionalized open-chain compounds.

Accordingly, chrysanthemic acid (**1**) and its ester **6**, respectively, offer two principal disconnection modes, **A** (leading to synthons **4** and **5**) and **B** (leading to synthon **7**):

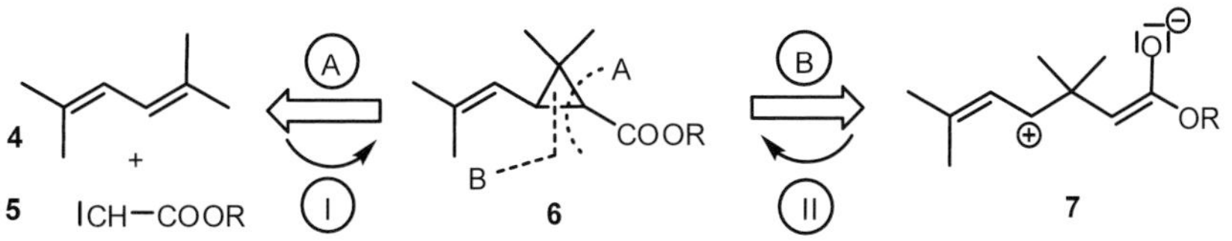

Retrosynthesis according to **A** directly discloses a pattern of synthesis for **6** (strategy **I**), which is realized by the [2+1]-cycloaddition of carbalkoxy carbene **5** – generated by Cu- or Rh-catalyzed thermolysis of diazoacetic ester **8** – to the easily accessible 2,5-dimethylhexa-2,4-diene (**4**) [4]. Mono-cyclopropanation and subsequent base-induced equilibration leads to the *trans*-ester **6** with >95% stereoselectivity [5]. Using the (dimeric) $Rh_2(OAc)_4$ as catalyst, cyclopropanation with diazoacetate is likely to proceed via a rhodium(II) carbene complex (after elimination of N_2) and its addition to one of the double bonds of diene **4** [6]:

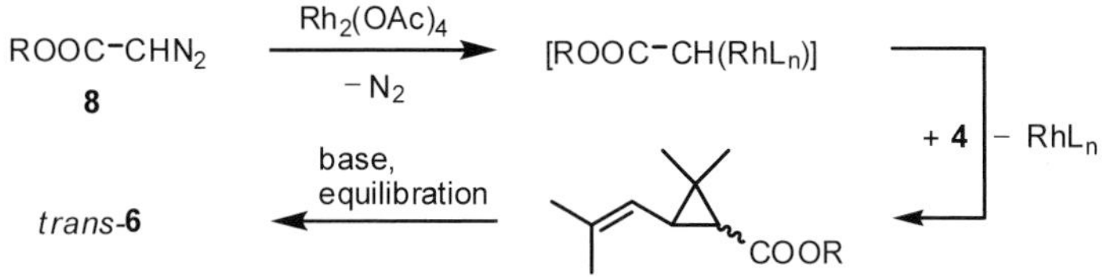

The asymmetric synthesis of *trans*-chrysanthemate **6** is achieved by the use of chiral ester components in the diazoacetate and/or catalysis by chiral Cu catalysts based on amino acids (Aratani catalysts, *ee* values > 90%) [7, 8]. Particularly effective is the chiral Cu(I) catalyst **10**, which facilitates the generation of the chrysanthemate **9** in a *trans*/*cis* ratio of 95:5 with 94% *ee* [9, 10].

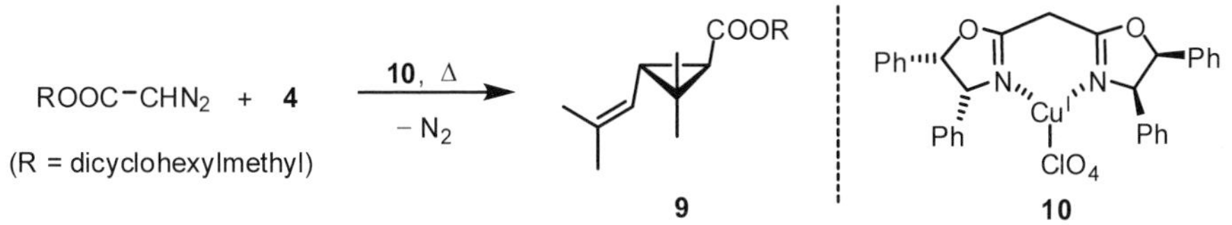

Retrosynthesis according to **B** leads to synthon **7**, which might be represented by a building block **11**, a ring-open ester bearing a suitable functionality **X** in the γ-position and retrosynthetically deduced from the anionic synthon **12** and β,β-dimethyl acrylate **13** (acting as a Michael acceptor):

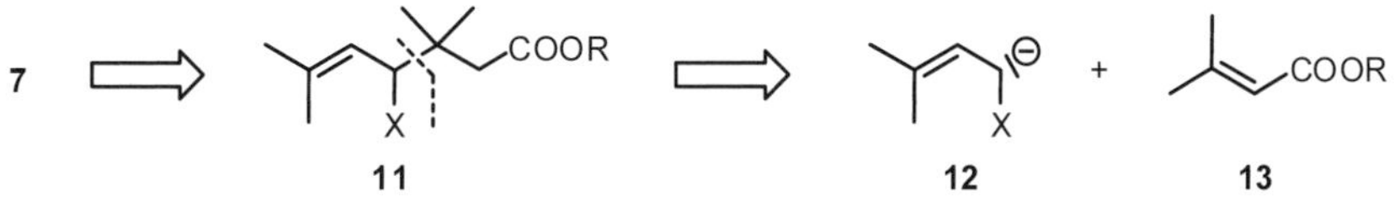

In the γ-functionalized system **11**, it is necessary that the functionality **X** not only increases the CH-acidity of the C–H bonds in the α-position, but also serves as a good leaving group. This prerequisite is fulfilled, for example, by the group SO_2R. Thus, a key intermediate for strategy **II** is compound **14**, which is suitably predisposed for a sequence of base-induced deprotonation to give **12**, Michael addition of **12** to **13** to give **15**, cyclopropane ring closure of the ester enolate **15** by intramolecular nucleophilic substitution (S_Ni, 1,3-elimination), and finally equilibration in favor of the thermo-dynamically more stable *trans*-product **6**. For reasons of efficiency, it is desirable to conduct the series of base-induced reactions (leading from **14** to **6**) as a domino process [11]. Indeed, a synthesis of **1** following this strategy has been documented in the literature [see section (b)].

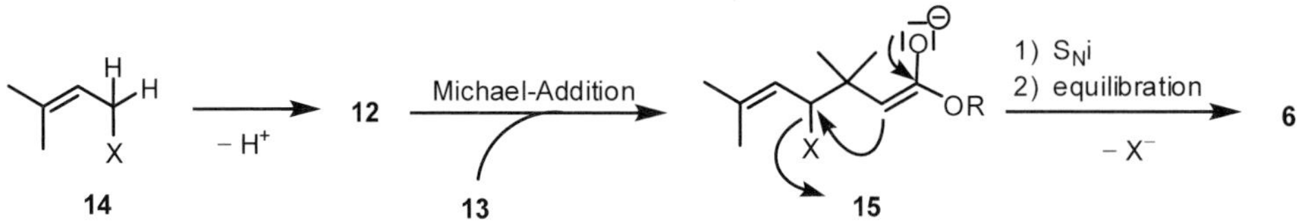

It should be noted that due to the unsymmetrical cyclopropane structure of the target molecule **1**, further possibilities of syntheses according to **A** and **B** can be deduced following the [2+1]-cycloaddition or 1,3-elimination protocols [12]. A third principle for the formation of cyclopropane carboxylic acids is the ring contraction of α-halogeno cyclobutanones by the Favorskii reaction, which has been successfully applied to the synthesis of β,β-dihalovinyl analogues of chrysanthemic acid [13].

(b) Synthesis of 1

For reasons of handling and practicability, a synthesis of **1** based on strategy **II** is chosen, which was specifically designed for laboratory use and which uses easily accessible substrates [14].

The key intermediate is (*p*-tolyl)-3,3-dimethylsulfone **18**, which is obtained from sodium *p*-toluenesulfinate **16** and 3,3-dimethylallyl bromide **17**. Initial *O*-alkylation of the sulfinate anion leads to an allyl sulfinate, which rearranges under the reaction conditions to give the allyl sulfone. For this rearrangement, the mechanism of an electrophilic 1,2-*O,S* migration is discussed, since it is restricted to sulfinates with ester residues representing resonance-stabilized carbenium ions such as benzyl or allyl [15]:

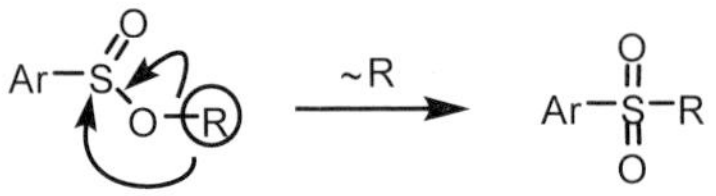

Sodium *p*-toluenesulfinate **16** is obtained by reduction of *p*-toluenesulfonyl chloride with Zn in aqueous NaOH, and the allyl bromide **17** is derived from dimethyl vinyl carbinol and 48% HBr by way of an S_N' reaction with allyl inversion [16].

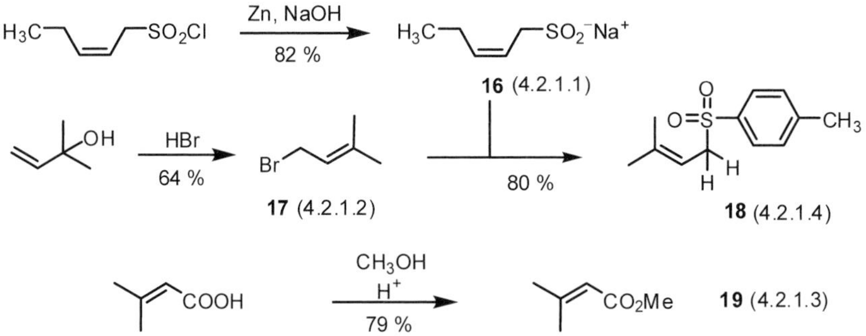

Methyl 3-methyl-2-butenoate **19** is obtained by esterification of 3-methyl-2-butenoic acid (senecic acid) with methanol according to the Fischer method. The α,β-unsaturated ester **19** is reacted with the sulfone **18** in the presence of sodium methoxide to furnish racemic methyl chrysanthemate **20**, which comprises > 95% of the desired *trans*-diastereomer (according to ^{1}H NMR, see **4.2.1.5**). The final step of the synthesis is saponification of the ester **20** by KOH in methanol to give (±)-*trans*-chrysanthemic acid **1**.

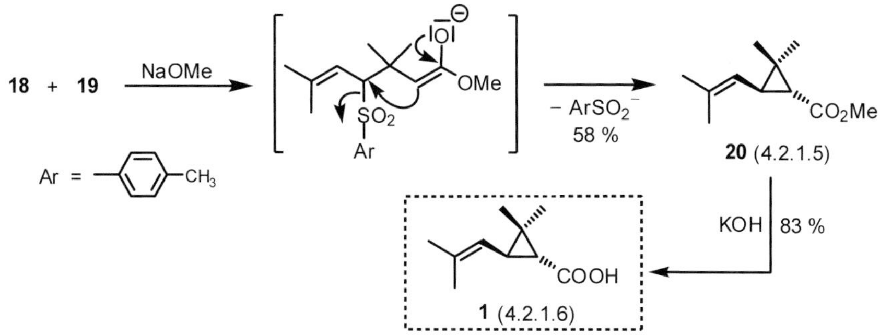

18 + **19** → NaOMe → [intermediate] → − ArSO$_2^-$ 58 % → **20** (4.2.1.5)

Ar = (p-tolyl)—CH$_3$

1 (4.2.1.6)

KOH | 83 %

By this doubly-convergent approach, the target molecule **1** is obtained in six steps with an overall yield of 32% (based on **16**).

(c) Experimental procedures for the synthesis of 1

4.2.1.1   *   Sodium *p*-toluenesulfinate [16]

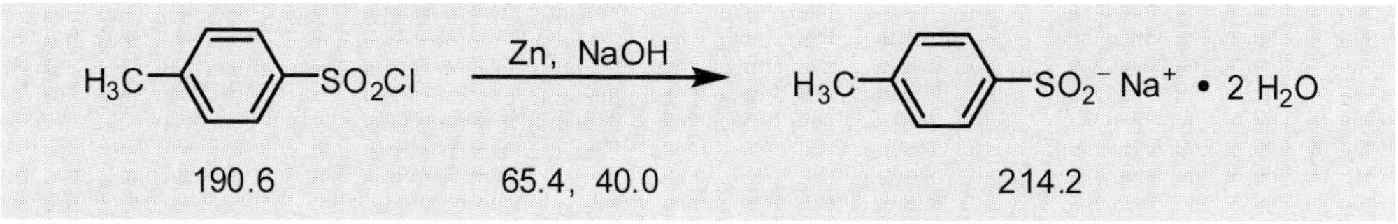

$$H_3C\!-\!\!\langle\rangle\!-\!SO_2Cl \xrightarrow{\text{Zn, NaOH}} H_3C\!-\!\!\langle\rangle\!-\!SO_2^-\;Na^+ \cdot 2\,H_2O$$

190.6   65.4, 40.0   214.2

Finely powdered *p*-toluenesulfonyl chloride (50.0 g, 0.26 mol) is added in portions over 10 min to a stirred suspension of zinc dust (45.0 g, 0.69 mol) in water (500 mL) maintained at 70–75 °C. The temperature rises slightly. After stirring for 10 min, a solution of sodium hydroxide (12.0 g, 0.30 mol) in water (25 mL) is added dropwise over 3 min at 70–75 °C. The reaction mixture becomes lighter in color and reaches pH ≈ 7. Sodium carbonate (20.0 g, 0.20 mol) is then added to reach pH 9–10.

The hot suspension is rapidly filtered through a large Büchner funnel and the solid is triturated twice with hot H$_2$O (250 mL). The combined filtrates are concentrated to a volume of 130 mL in an open flask. On cooling, sodium *p*-toluenesulfinate dihydrate crystallizes in large needles. It is collected by filtration and dried in air to constant weight. A second crop can be obtained by reducing the volume of the mother liquor to ca. one-third. The total yield is 46.0 g (82%). The salt does not have a sharp melting point, but starts to decompose at ca. 340 °C.

IR (KBr): $\tilde{\nu}$ (cm^{-1}) = 1010, 970 (SO$_2^-$, very intense).

4.2.1.2   *   1-Bromo-3-methyl-2-butene [17]

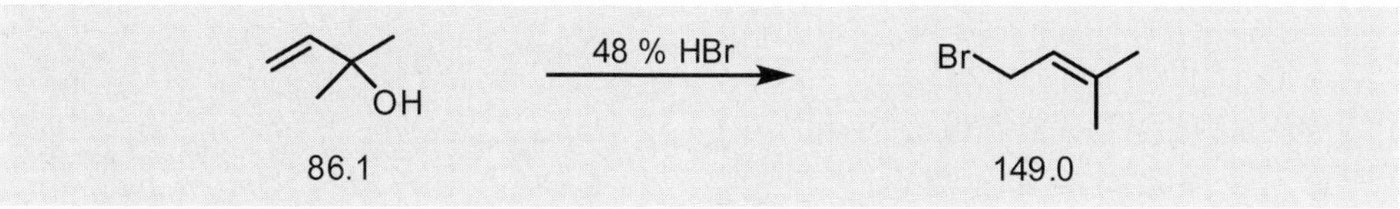

48 % HBr

86.1   149.0

2-Methyl-3-buten-2-ol (86.1 g, 1.00 mol) in 48% hydrobromic acid (400 mL) is vigorously stirred at room temperature for 15 min.

The aqueous phase is separated from the oil formed and extracted with benzene (250 mL) (Caution: carcinogenic!). The combined organic phases are rapidly washed with ice-cold, dilute $NaHCO_3$ solution (100 mL) and then dried ($CaCl_2$). The solution is fractionally distilled at 150 mbar to give 1-bromo-3-methyl-2-butene as a slightly yellow oil. The yield is 95.0 g (64%), bp_{150} 82–83 °C.

IR (film): $\tilde{v}$ (cm^{-1}) = 1670 (C=C).

^{1}H NMR (CDCl$_3$): δ (ppm) = 5.49 (t, J = 8 Hz, 1 H, =C–H), 3.95 (d, J = 8 Hz, 2 H, =C–CH$_2$), 1.80, 1.83 (s$_{br}$, 3 H, CH$_3$).

4.2.1.3   *   Methyl 3-methyl-2-butenoate [15]

	100.1				114.1

Concentrated sulfuric acid (5.0 mL by pipette) is carefully added to a solution of 3-methyl-2-butenoic acid (3,3-dimethylacrylic acid; senecic acid)) (18.0 g, 0.18 mol) in anhydrous MeOH (100 mL) and the mixture is heated under reflux for 2 h.

The solution is then cooled, poured into iced water (100 mL), and extracted with Et$_2$O (3 × 75 mL). The combined ethereal extracts are washed with brine (100 mL) and dried over MgSO$_4$. The solution is distilled at atmospheric pressure to give Et$_2$O, a small forerun, and then the ester as a colorless oil with a fruit-like odor. The yield is 16.2 g (79%), bp_{760} 133–134 °C, n_D^{20} = 1.4375.

IR (film): $\tilde{v}$ (cm^{-1}) = 1720 (C=O), 1660 (C=C).

^{1}H NMR (CDCl$_3$): δ (ppm) = 5.64 (s, 1 H, vinyl-H), 3.65 (s, 3 H, OCH$_3$), 2.14 (s, 3 H, =C–CH$_3$ *cis* to CO$_2$Me), 1.87 (s, 3 H, =C–CH$_3$ *trans* to CO$_2$Me).

4.2.1.4   *   (3-Methyl-2-butenyl)-(p-tolyl) sulfone [15]

	214.2				149.0				224.3

1-Bromo-3-methyl-2-butene **4.2.1.2** (16.0 g, 0.10 mol) is added dropwise to a stirred suspension of sodium p-toluenesulfinate dihydrate **4.2.1.1** (27.0 g, 0.13 mol) in DMF (100 mL) over a period of 15 min at room temperature. The temperature rises by ca. 8 °C and a clear solution forms in about 10 min. After the addition, the temperature is held at 85 °C for 1.5 h.

The colorless solution is then cooled to room temperature and poured into H_2O (500 mL), forming a flocculent precipitate. If an oil forms, the mixture is stirred for 14 h. The solid is collected by filtration, washed with H_2O, and recrystallized from isopropyl alcohol (20 mL) to give colorless prisms; 18.6 g (80%), mp 80–81 °C.

> **IR** (KBr): $\tilde{v}$ (cm^{-1}) = 1665 (C=C, weak).
>
> **^{1}H NMR** (CDCl$_3$): δ (ppm) = 7.73, 7.30 (d, J = 8.5 Hz, 2 H, tolyl-H), 5.16 (t, J = 8 Hz, 1 H, vinyl-H), 3.75 (d, J = 8 Hz, 2 H, allyl CH$_2$), 2.42 (s, 3 H, *p*-tolyl CH$_3$), 1.68, 1.33 (s, 3 H, =C–CH$_3$).

4.2.1.5   **   Methyl (±)-*trans*-chrysanthemate [15]

| 224.3 | 114.1 | 54.0 | 182.3 |

Sodium methoxide (10.0 g, 185 mmol) is added in one portion to a well-stirred solution of ester **4.2.1.3** (9.00 g, 79.0 mmol) and dimethylallyl sulfone **4.2.1.4** (15.0 g, 67.0 mmol) in anhydrous DMF (75 mL) under a nitrogen atmosphere. The suspension turns brown and is stirred at room temperature for 72 h.

The mixture is then poured into a mixture of concentrated HCl (25 mL), H_2O (50 mL), and ice (50 g). An orange oil separates. The resulting mixture is extracted with *n*-pentane (5 × 50 mL). A small amount of brown oil that forms between the layers is separated with the aqueous layer and discarded. The combined *n*-pentane extracts are washed with saturated NaHCO$_3$ solution and brine (each 100 mL), dried over MgSO$_4$, and concentrated *in vacuo*. The residue is fractionally distilled *in vacuo* to give a colorless oil with a refreshing odor; 7.40 g (58%), bp$_1$ 49–50 °C, n$_D^{20}$ = 1.4645.

> **IR** (film): $\tilde{v}$ (cm^{-1}) = 1730 (C=O), 1650 (C=C, weak).
>
> **^{1}H NMR** (CCl$_4$): δ (ppm) = 4.86 (d, J = 8 Hz, 1 H, vinyl-H; note), 3.60 (s, 3 H, OCH$_3$), 1.95 (m, 1 H, CH–CO$_2$Me), 1.73 (s, 6 H, =C(CH$_3$)$_2$), 1.2 (m, 1 H, CH vinyl), 1.24, 1.13 (s, 3 H C(CH$_3$)$_2$).

Note: This peak is assigned to the *trans*-ester. A vinyl proton signal due to the *cis*-isomer (ca. 5%) appears at δ = 5.30 ppm. A long reaction time favors formation of the *trans* form.

4.2.1.6   *   (±)-*trans*-Chrysanthemic acid [15]

| 182.3 | 56.1 | 168.2 |

A mixture of the ester **4.2.1.5** (5.00 g, 27.4 mmol) and potassium hydroxide (5.00 g, 90.0 mmol) in 95% EtOH (75 mL) is heated under reflux for 2 h.

The solvent is then evaporated *in vacuo* and the dark, oily residue is dissolved in H_2O (100 mL). The solution is extracted with Et_2O (50 mL), and the reddish aqueous phase is acidified to pH 1–2 with concentrated HCl and extracted with Et_2O (3 × 40 mL). The chrysanthemic acid separates from the acidic solution in the form of dark oily drops, which dissolve in the Et_2O. The combined ethereal phases are dried over $MgSO_4$ and concentrated. The residue is vacuum distilled in a microdistillation apparatus. The product distils at a bath temperature of 105–120 °C. The yield is 3.80 g (83 %), $bp_{0.4}$ 83–85 °C, n_D^{20} = 1.4782; the product solidifies on standing at 4 °C for 14 h; mp 45–47 °C.

IR (KBr): $\tilde{v}$ (cm^{-1}) = 1685 (C=O).

^{1}H NMR ($CDCl_3$): δ (ppm) = 11.73 (s, 1 H, OH), 4.90 (d, J = 8 Hz, 1 H, vinyl-H; note), 2.1 (m, 1 H, CH–COOH), 1.72 (s, 6 H, =C(CH$_3$)$_2$), 1.25 (m, 1 H, CH vinyl), 1.30, 1.19 (s, 3 H, C(CH$_3$)$_2$).

Note: An additional peak appears at δ = 5.30 ppm, which is assigned to the *cis*-acid formed as a by-product (< 5%).

[1]   a) K. H. Büchel, *Pflanzenschutz und Schädlingsbekämpfung*, Georg Thieme Verlag, Stuttgart, **1977**; b) K. Naumann, *Nachr. Chem. Tech. Lab.* **1978**, *26*, 120.

[2]   H. Marquart, S. G. Schäfer, *Lehrbuch der Toxikologie*, p. 475, BI Wissenschaftsverlag, Mannheim, **1994**.

[3]   M. Elliot, *Nature* **1973**, *246*, 169; **1974**, *248*, 710.

[4]   The diene **3** can be synthesized starting from acetone either by Wittig reaction with Ph$_3$P=CH–CH=C(CH)$_2$ (B. Bogdanovic, S. Konstantinovic, *Synthesis* **1972**, 481) or by ethynylation via 2,5-dimethylhex-3-yne-2,5-diol (H. J. Sanders, A. W. Taff, *Ind. Eng. Chem.* **1954**, *46*, 414).

[5]   A. J. Hubert, A. F. Noels, A. J. Anciaux, P. Teyssie, *Synthesis* **1976**, 600.

[6]   L. S. Hegedus, *Organische Synthese mit Übergangsmetallen*, p. 165, VCH, Weinheim, **1995**.

[7]   T. Aratani, Y. Yoneyoshi, T. Nagase, *Tetrahedron Lett.* **1977**, 2599.

[8]   a) D. Arlt, M. Jautelat, R. Lantzsch, *Angew. Chem.* **1981**, *93*, 719; *Angew. Chem. Int. Ed. Engl.* **1981**, *20*, 703. For enantioselective syntheses of cyclopropanes, see: b) H. Lebel, J.-F. Marcoux, C. Molinaro, A. B. Charette, *Chem. Rev.* **2003**, *103*, 977; c) H.-U. Reißig, *Angew. Chem.* **1996**, *108*, 1049; *Angew. Chem. Int. Ed. Engl.* **1996**, *35*, 971.

[9]   R. E. Lowenthal, S. Masamune, *Tetrahedron Lett.* **1991**, *32*, 7373.

[10] Accordingly, BINOL-based chiral iodomethylzinc phosphates have been found to catalyze asymmetric Simmons–Smith cyclopropanations with Zn(CH$_2$I)$_2$: M.-C. Lacasse, C. Poulard, A. B. Charette, *J. Am. Chem. Soc.* **2005**, *127*, 12440.

[11] a) L. F. Tietze, *Chem. Rev.* **1996**, *96*, 115; b) L. F. Tietze, G. Brasche, K. Gericke, *Domino Reactions in Organic Synthesis*, Wiley-VCH, Weinheim, **2006**.

[12] For an enantioselective synthesis based on strategy **B**, see: A. Krief, W. Dumont, D. Baillicul, *Synthesis* **2002**, 2019.

[13] a) P. Martin, H. Kreuter, D. Bellus, *J. Am. Chem. Soc.* **1979**, *101*, 5853; b) for a review on chrysanthemic acid syntheses, see: S. Jeanmart, *Aust. J. Chem.* **2003**, *56*, 559.

[14] a) P. F. Schatz, *J. Chem. Educ.* **1978**, *55*, 468; b) J. Martell, C. Hauynh, *Bull. Soc. Chim. Fr.* **1967**, 985.

[15] S. Oae, *Organic Chemistry of Sulfur*, p. 639, Plenum Press, New York, London, **1977**.

[16] L. F. Tietze, Th. Eicher, *Reaktionen und Synthesen im Organisch-Chemischen Praktikum und Forschungslaboratorium*, 2nd ed., p. 489, Georg Thieme Verlag, Stuttgart, **1991**.

[17] a) Brit. Pat. 735,428, 31.8.1955 (*Chem. Abstr.* **1956**, *50*, 88706a); b) prenyl bromide has alternatively been prepared by surface-mediated hydrobromination of isoprene: M. C. S. De Mattos, A. M. Sanseverino, *Synth. Commun.* **2003**, *33*, 2181.

4.2.2   Nerol

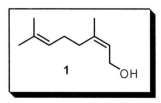

Topics:
- Synthesis of a monoterpene alcohol
- Stereoselective dimerization (telomerization) of isoprene induced by LDA
- Transformation of a tertiary allylamine to an allyl chloride
- Crown ether-catalyzed S_N reaction
- Saponification of a carboxylic ester

(a) General

Nerol ((2*Z*)-3,7-dimethylocta-2,6-dien-1-ol) (**1**) and its (2*E*)-stereoisomer geraniol (**2**) belong to the group of unsaturated acyclic monoterpene alcohols most widespread in Nature. Nerol (**1**) and geraniol (**2**) are found as free alcohols and in esterified form in palmarosa oil and geranium oil; nerol (**1**) is also found in the etheric oil of the straw flower *Helichrysum italicum* (*Helichrysum angustifolium*, Asteraceae). Together with the allyl-isomeric linalool (**3**), geraniol (**2**) and nerol (**1**) are used in perfumery and cosmetics [1]. Nerol (**1**) is technically produced along with geraniol (**2**) and linalool (**3**) from β-pinene via myrcene (**4**) [2].

Geranyl and neryl diphosphates are important intermediates in the biosynthesis of acyclic and cyclic monoterpenes according to the mevalonate pathway [3].

Several retrosynthetic approaches for nerol (**1**) are possible:

According to (A), FGI leads to the α,β-unsaturated ester **5**, which can be disconnected at the C-2/C-3 double bond in a retro-Wittig fashion leading to methylheptenone **6** and a P-ylide. **6** can be further disconnected to give acetoacetate and a prenyl halide. As might be expected, synthesis of **6** by alkylation of acetoacetate followed by "ketone cleavage" of the alkylated β-keto ester is straightforward [4], but the carbonyl olefination of **6**, e.g. by the Wittig–Horner method [5], leads to (E)/(Z) mixtures of the unsaturated ester **5** and thus lacks the stereoselectivity required for a concise synthesis of **1**.

According to (B), the alkyne **7** would be a good precursor of **1** since the addition of an organometallic M–CH$_3$ species (M = metal) would allow (Z)-stereoselective construction of nerol (**1**). Similar considerations (C) would lead to the acetylenic ester **8**, its transformation to **5** by *syn*-addition of an M–CH$_3$ species, such as a cuprate [6], and finally reduction of the (Z)-α,β-unsaturated ester moiety in **5** to give nerol (**1**).

In fact, strategy (II) serves as a basis for several stereoselective syntheses of nerol (**1**) [7], as shown by the following examples.

Treatment of the propargylic alcohol **7** with isobutyl-MgCl catalyzed by Cp$_2$TiCl$_2$ leads to hydromagnesiation of the triple bond, which proceeds readily in a *syn*-fashion and leads regioselectively to the vinylmagnesium compound **9**; its C-methylation by CH$_3$I with retention of configuration followed by hydrolysis gives nerol (**1**) in high yield [8].

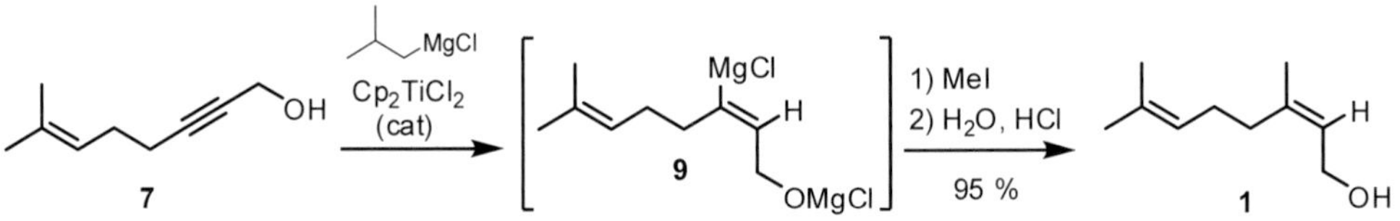

Likewise, addition of the alkenylcopper reagent **10** to propyne occurs with complete *syn*-selectivity at low temperature to yield the vinylcopper intermediate **11**, which is trapped as the vinyl iodide **12** by reaction with iodine and transformed to the vinyllithium compound **13** by halogen–metal exchange with *n*-BuLi; since the double-bond configuration is preserved in the transmetalation sequence (**11** → **12** → **13**), addition of **13** to formaldehyde and subsequent hydrolysis leads to nerol (**1**) [9]:

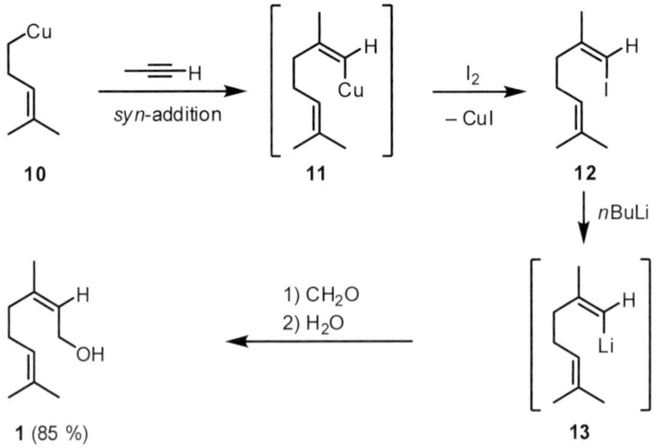

It should be noted that while *syn*-addition of alkyl cuprates to acetylenic esters of type **8** is documented in numerous examples [10], the transformation of **8** to afford **5** followed by reduction to give nerol (**1**)

has not yet been realized. This is probably due to the instability of **5**, since (*Z*)-α,β-unsaturated esters easily isomerize to give the corresponding (*E*)-compounds.

For reasons of preparative viability, the synthesis of nerol (**1**) described here follows a different reaction principle, namely the LDA-induced stereoselective telomerization of isoprene.

(b) Synthesis of (1)

Reaction of isoprene with lithium diethylamide (LDA) in anhydrous benzene leads to *N,N*-diethylnerylamine (**20**) in 65% yield after hydrolytic work-up [11]. Obviously, LDA effects head-to-tail coupling of two isoprene units – in analogy to the biosynthesis of isoprenoids [2] – but directs this process to proceed stereoselectively in a (*Z*)-fashion with respect to the C-2/C-3 double bond formed in the product **20**:

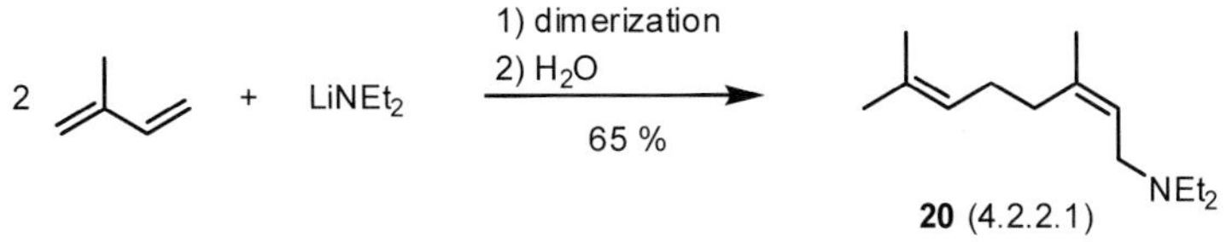

20 (4.2.2.1)

This remarkable finding that dimerization (telomerization, cf. **1.8.1**) is favored over the expected anionic polymerization of the 1,3-diene (isoprene → polymer **18**) can be rationalized on the basis of the following mechanistic considerations [11].

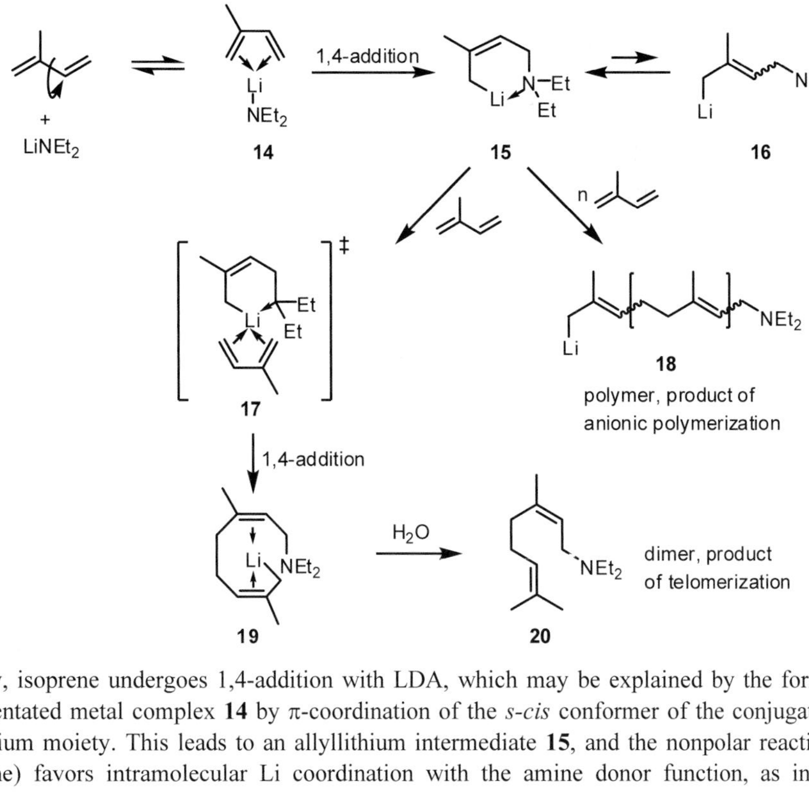

Initially, isoprene undergoes 1,4-addition with LDA, which may be explained by the formation of a pre-orientated metal complex **14** by π-coordination of the *s-cis* conformer of the conjugated diene to the lithium moiety. This leads to an allyllithium intermediate **15**, and the nonpolar reaction medium (benzene) favors intramolecular Li coordination with the amine donor function, as in **15** with a

(Z)-configuration at the C-2/C-3 double bond. As a consequence, addition of a second isoprene unit may proceed via π-coordination to the Li center of **15** in a highly ordered transition state **17**. The product of this second 1,4-addition, **19**, which is probably stabilized by internal N- and π-coordination, leads to nerylamine (**20**) on hydrolysis.

For the transformation of nerylamine (**20**) into nerol (**1**), the amine function is first replaced by a chloro substituent by reacting **20** with ethyl chloroformate to give the allylic chloride **22**. This substitution process is generally applicable to tertiary allylic amines [12] and is assumed to occur via N-acylation (**20** → **21**) and elimination of a urethane moiety by attack of chloride at the allylic ammonium position to produce the chloride **22**.

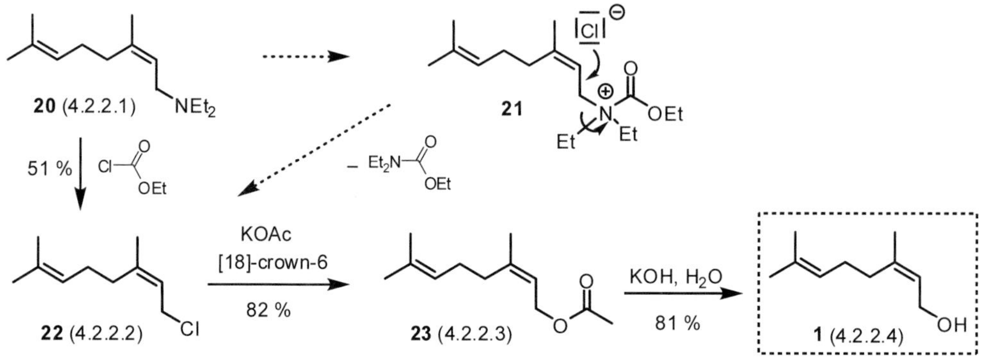

Finally, the chloride **22** is subjected to an S_N displacement in the presence of a crown ether to increase the nucleophilicity of the anion (cf. **3.5.3**) using KOAc. The resulting allylic acetate **23**, which is formed without allylic inversion, is saponified with aqueous KOH to provide nerol (**1**).

Thus, the target molecule **1** is obtained by a four-step sequence in an overall yield of 23% based on isoprene.

(c) Experimental procedures for the synthesis of 1

4.2.2.1   ***   *N,N*-Diethylnerylamine [11]

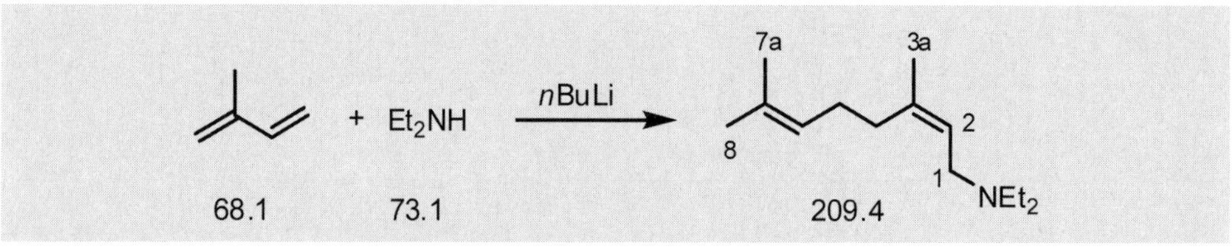

Under an argon atmosphere, isoprene (34.1 g, 500 mmol) and diethylamine (7.31 g, 100 mmol) are dissolved in anhydrous benzene (40 mL) (Caution: carcinogenic!) in a 250 mL three-necked round-bottomed flask, fitted with a reflux condenser and an inert gas inlet. *n*BuLi (0.75 M in *n*-hexane, 27.0 mL, 20 mmol) is added with stirring, and the solution is heated to 51 °C for 30 h. During this time, a precipitate is formed, which redissolves after several hours of stirring and the solution becomes yellow as the internal temperature rises to 67 °C.

The solution is then cooled, EtOH (20 mL) is added dropwise, and the resulting solution is extracted with H_2O (70 mL). The aqueous phase is saturated with NaCl and extracted with benzene (3 × 50 mL) (Caution!). The combined organic phases are washed with brine, dried over Na_2SO_4, and the solvent is evaporated under reduced pressure. The residue is distilled to give the product as a colorless oil; 13.5 g (65%), bp_{19} 135–138 °C, n_D^{20} = 1.4669.

IR (NaCl): $\tilde{v}$ (cm^{-1}) = 2960, 2920, 2865, 2800, 1670.

^{1}H NMR (300 MHz, $CDCl_3$): δ (ppm) = 5.22 (t, J = 6.8 Hz, 1 H, 6-H), 5.08 (m_c, 1 H, 2-H), 3.01 (dd, J = 6.8 Hz, 2 H, 1-H_2), 2.47 (q, J = 7.1 Hz, 4 H, 2 × C$\underline{H}_2$CH$_3$), 2.00–2.05 (m, 4 H, 4-H_2, 5-H_2), 1.68–1.71 (m, 3 H, 3a-H_3), 1.65 (s, 3 H, 8-H_3), 1.58 (s, 3 H, 7a-H_3), 0.99 (t, J = 7.1 Hz, 6 H, 2 × CH$_2$C$\underline{H}_3$).

^{13}C NMR (75 MHz, $CDCl_3$): δ (ppm) = 137.7 (C-3), 131.7 (C-7), 124.1 (C-2), 122.7 (C-6), 50.36 (C-1), 46.62 (2 × C$\underline{H}_2$CH$_3$), 32.15 (C-4), 26.54 (C-5), 25.68 (C-8), 23.52 (C-3a), 17.59 (C-7a), 11.82 (2 × CH$_2$C$\underline{H}_3$).

Note: Isoprene (bp 34–35 °C) and diethylamine (bp 56–57 °C, dried over KOH) are distilled before use.

4.2.2.2   *   Neryl chloride [13]

$$\text{209.4} \qquad\qquad \text{108.5} \qquad\qquad\qquad \text{172.7}$$

N,N-Diethylnerylamine **4.2.2.1** (11.5 g, 55.2 mmol) is added dropwise with stirring to ethyl chloroformate (11.9 g, 110 mmol) (note 1) at 0 °C (inner temperature) over a period of 15 min. The mixture is stirred at room temperature for 15 h as the amine odor changes to a fruity odor.

The product is distilled directly from the reaction mixture under reduced pressure; the excess ethyl chloroformate distils first at room temperature (Caution: foaming!), followed by *N,N*-diethylethoxyformamide (bp_{15} 67–73 °C, 7.50 g) and neryl chloride as a colorless oil; 5.37 g (note 2), bp_{15} 98–99°C, n_D^{20} = 1.4728.

IR (film): $\tilde{v}$ (cm^{-1}) = 1665, 675.

^{1}H NMR (200 MHz neat, TMS$_{ext}$): δ (ppm) = 5.30–4.85 (m, 2 H, 2 × vinyl-H), 3.76 (d, J = 13 Hz, 2 H, CH$_2$–Cl), 1.89 (2 × d, J = 3.2 Hz, 2 × 2 H, 2 × allyl-CH$_2$), 1.53, 1.47, 1.40 (3 × s, 3 × 3 H, 3 × CH$_3$).

Notes: (1) Ethyl chloroformate has to be distilled before use, bp 94–95 °C.

(2) According to ^{1}H NMR, the neryl chloride obtained is 90% pure, corresponding to a 51% yield. The contaminating Et_2N–CO_2Et can be removed by spinning-band distillation; however, it does not interfere with the following reaction.

4.2.2.3   *   Neryl acetate [13]

KOAc, [18]-crown-6,
CH$_3$CN, 60°C, 4 h

172.7   196.3

Potassium acetate (dried over P$_4$O$_{10}$ under vacuum; 2.93 g, 30.0 mmol) is added to a solution of neryl chloride **4.2.2.2** (4.41 g, 32.2 mmol) and [18]-crown-6 (0.53 g, 2.00 mmol) in anhydrous CH$_3$CN (25 mL) and the mixture is stirred for 4 h at 60 °C.

The mixture is then cooled to room temperature, filtered, and the filtrate is concentrated. The brown residue is distilled *in vacuo* in a microdistillation apparatus. The product is 90% pure with a contamination of Et$_2$N–CO$_2$Et (^{1}H NMR); 4.15 g (82%), bp$_{15}$ 119–122 °C, n$_D^{20}$ = 1.4602.

> **IR** (film): $\tilde{v}$ (cm^{-1}) = 1740, 1230, 1020.
>
> **^{1}H NMR** (200 MHz neat, TMS$_{ext}$): δ (ppm) = 5.25–4.85 (m, 2 H, 2 × vinyl-H), 4.25 (d, J = 12 Hz, 2 H, CH$_2$–OAc), 1.85 (2 × d, J = 3.2 Hz, 4 H, 2 × allyl-CH$_2$), 1.70 (s, 3 H, OCO–CH$_3$), 1.49, 1.43, 1.37 (3 × s, 3 × 3 H, 3 × CH$_3$).

4.2.2.4   *   Nerol (3,7-dimethylocta-2(*Z*)-6-dien-1-ol) [13]

KOH, H$_2$O, rt, 19 h

196.3   154.2

Neryl acetate **4.2.2.3** (3.55 g, 16.0 mmol) is dissolved in methanolic KOH (1.50 g, 27.0 mmol of KOH in 10.7 mL of MeOH). The solution is stirred at room temperature for 19 h, whereupon potassium acetate is precipitated.

H$_2$O (40 mL) is then added and the mixture is extracted with CHCl$_3$ (1 × 40 mL, then 3 × 25 mL). The combined organic phases are washed with H$_2$O (30 mL), dried over Na$_2$SO$_4$, and the solvent is evaporated at atmospheric pressure. The yellow, oily residue is fractionally distilled *in vacuo* to give nerol as a colorless oil; 2.28 g, bp$_{15}$ 111–113 °C, n$_D^{20}$ = 1.4730.

The product is 95% pure (^{1}H NMR) (still contaminated with Et$_2$N–CO$_2$Et), which corresponds to a yield of 88%. The product shows a *Z/E* ratio of 99.5:0.5 and is uniform by TLC (SiO$_2$; CH$_2$Cl$_2$) with R_f = 0.9 for nerol (as compared to R_f = 0.65 for geraniol) (note).

> **IR** (NaCl): $\tilde{v}$ (cm^{-1}) = 3320, 1675, 1000.
>
> **^{1}H NMR** (neat, TMS$_{ext}$): δ (ppm) = 5.25–4.90 (m, 2 H, 2 × vinyl-H), 4.33 (s, 1 H, OH), 3.80 (d, J = 12 Hz, 2 H, C$\underline{H}_2$-OH), 1.85 (2 × d, J = 3.2 Hz, 4 H, 2 × allyl-CH$_2$), 1.52, 1.41 (3 × s, 3 × 3 H, 3 × CH$_3$).

Note: Crystalline derivatives of nerol are:
          the tetrabromide, mp 118–119 °C; the diphenylurethane, mp 52–53 °C.
          Compare with the corresponding derivatives of geraniol:
          the tetrabromide, mp 70–71 °C; the diphenylurethane, mp 81–82 °C.

[1]   Römpp, Lexikon *"Naturstoffe"* (Eds.: W. Steglich, B. Fugmann, S. Lang-Fugmann), p. 255, Georg Thieme Verlag, Stuttgart, **1997**.

[2]   Ullmann's *Encyclopedia of Industrial Chemistry*, 6th ed., Vol. 14, p. 85, Wiley-VCH, Weinheim, **2003**.

[3]   P. Nuhn, *Naturstoffchemie*, 3rd ed., p. 513, S. Hirzel Verlag, Stuttgart, **1997**.

[4]   L. F. Tietze, Th. Eicher, *Reaktionen und Synthesen im organisch-chemischen Praktikum und Forschungslaboratorium*, 2nd ed., p. 497, Georg Thieme Verlag, Stuttgart, **1991**.

[5]   K. Tanaka, N. Yamagishi, R. Tanikaga, A. Kaji, *Bull. Chem. Soc. Jpn.* **1979**, *51*, 3619.

[6]   N. Krause, *Metallorganische Chemie*, Spektrum Akademischer Verlag, p. 183, Heidelberg, **1996**.

[7]   For a summary on the earlier syntheses of geraniol and nerol, see: A. F. Thomas, in: *The total synthesis of natural products* (Ed.: J. ApSimon), p. 1, John Wiley & Sons, Inc., New York, **1973**.

[8]   S. Inoue, H. Takaya, K. Tani, S. Otsuka, T. Sato, R. Nojori, *J. Am. Chem. Soc.* **1990**, *112*, 4897.

[9]   G. Cahiez, D. Bernard, J. F. Normant, *Synthesis* **1976**, 245.

[10] For a summary, see: R. C. Larock, *Comprehensive Organic Transformations*, 2nd ed., p. 452, Wiley-VCH, New York, **1999**.

[11] K. Takabe, T. Katagiri, J. Tanaka, *Tetrahedron Lett.* **1972**, *13*, 4009.

[12] J. H. Cooley, E. J. Evain, *Synthesis* **1989**, 1.

[13] K. Takabe, T. Katagiri, J. Tanaka, *Chem. Lett.* **1977**, 1025.

4.2.3 (–)-Menthol

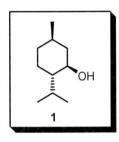

1

Topics:
- Lewis acid-mediated ene reaction
- Catalytic hydrogenation of an olefinic double bond

(a) General

(–)-Menthol ((1R,3R,4S)-4-isopropyl-1-methylcyclohexan-3-ol or (1R,3R,4S)-p-menthan-3-ol) (**1**) possesses three defined stereogenic centers within its cyclohexane core. It is one of the eight possible stereoisomers with this skeleton, of which four occur in Nature ((–)-menthol (**1**), (+)-neomenthol (**2**), (+)-isomenthol (**3**), and (+)-neoisomenthol (**4**)) [1]:

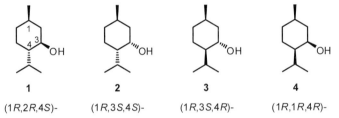

1	**2**	**3**	**4**
(1R,2R,4S)-	(1R,3S,4S)-	(1R,3S,4R)-	(1R,1R,4R)-

(–)-Menthol (**1**) is widespread in Nature as the main component of peppermint and cornmint oils, which are obtained from the species *Mentha piperita* and *Mentha arvensis* in free and esterified forms (e.g. acetate, isovalerate). (–)-Menthol (**1**) is used as a cooling and refreshing ingredient in cigarettes, cosmetics, toothpastes, sweets, and medicines [2].

In the industrial production of (–)-menthol (**1**), isolation from natural sources competes with partial or total synthesis. Among the numerous syntheses [3], two methods are worthy of mention.

(1) (–)-Menthol has been synthesized by catalytic hydrogenation of thymol (**5**) [2]:

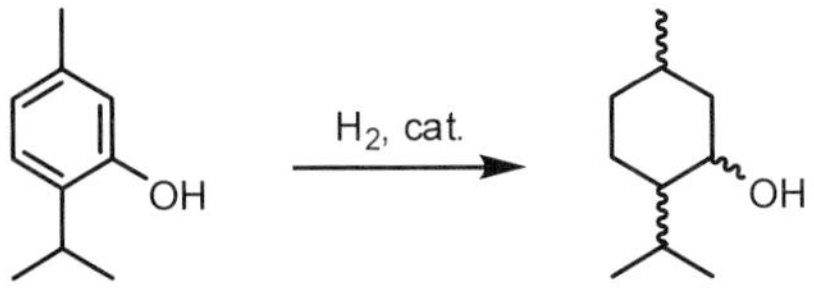

$$ \text{5} \xrightarrow{\text{H}_2,\ \text{cat.}} $$

5

This process yields a mixture of the four diastereomers in various proportions, from which *rac*-menthol is separated by fractional distillation and resolved into its enantiomers by selective crystallization of the benzoates.

(2) (–)-Menthol is obtained on an industrial scale (Takasago) [4] by an enantioselective asymmetric hydrogen shift of *N,N*-diethylgeranylamine (**6**) catalyzed by a chiral rhodium(I)-(*S*)-BINAP complex to give the (*E*)-enamine **7** of (*R*)-citronellal:

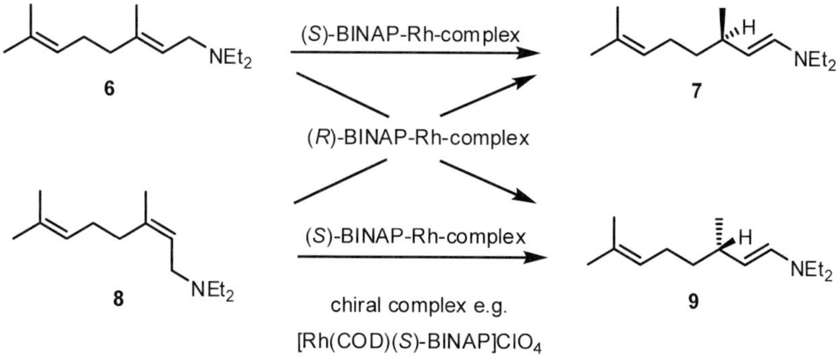

Interestingly, *N,N*-diethylnerylamine (**8**) (cf. **4.2.2**) can also serve as a substrate for the formation of **7** if the Rh-(*R*)-BINAP complex is used as catalyst, while the enantiomeric enamine **9** arises from **6** by catalysis with the Rh-(*R*)-BINAP complex and from **8** by catalysis with the Rh-(*S*)-BINAP complex. The origin of the enantioselection has been discussed in detail in terms of chiral recognition caused by the effects of the environments about the chiral Rh-BINAP complexes [3–5].

The chiral enamine **7** is hydrolyzed in an acidic medium (AcOH/H_2O) to give (*R*)-citronellal (**10**). This is then subjected to a Lewis acid-catalyzed intramolecular carbonyl ene reaction (see section (b)), which proceeds stereoselectively to form (–)-isopulegol (**11**). Hydrogenation of **11** then leads to the desired (–)-menthol with an *ee* value > 98%. Alternatively, (*R*)-citronellol (**12**) can be obtained from the aldehyde **10** by catalytic hydrogenation [4].

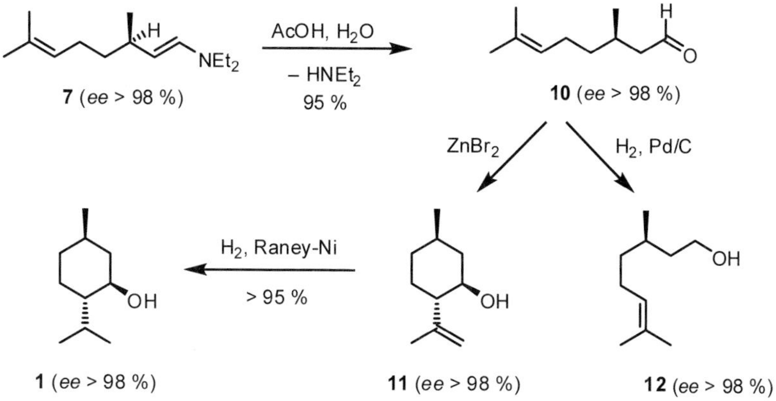

In the Takasago process, all transformations are reported to occur with high chemical yields (95–100%) and excellent enantioselectivities (*ee* > 98%).

Since the chiral Rh complexes applied for the enantioselective H-shift in **6** to give **7** are very expensive, only the concluding transformations of the technical synthesis of (–)-menthol are presented in section (b).

(b) Synthesis of 1

In the first step, (R)-citronellal (**10**) is cyclized to (–)-isopulegol (**11**) in benzene solution at 5–10 °C in the presence of ZnBr$_2$. It can be assumed that an oxenium ion is initially formed by coordination of the Lewis acid to the carbonyl moiety. This then undergoes a carbonyl ene reaction via a chair-like transition state **13**, with an equatorial orientation of the methyl group controlling the stereochemistry of the newly formed stereogenic centers in **11**.

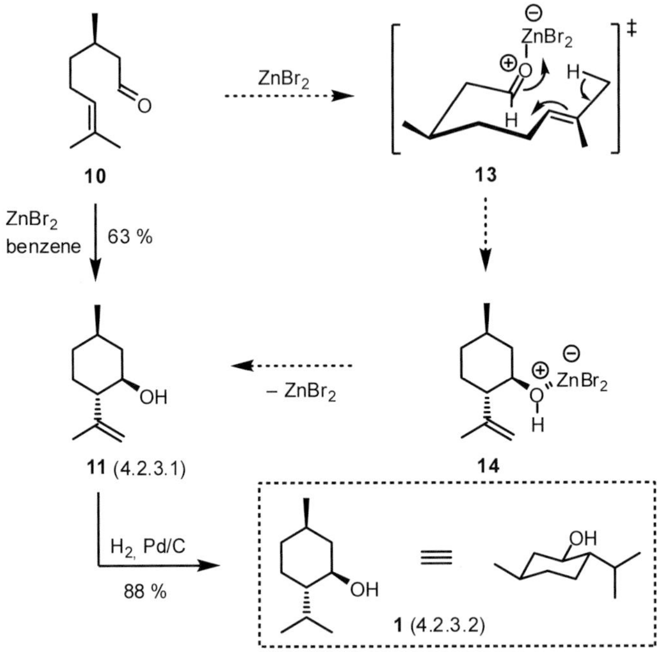

Ene reactions take place between olefinic substrates bearing an allylic hydrogen and either a second olefin bearing electron-withdrawing groups (one or two) or a carbonyl moiety as the "enophile". "All-carbon" ene reactions usually follow a concerted mechanism as pericyclic reactions, whereas carbonyl ene reactions proceed via a carbocation as an intermediate in a two-step mechanism [6]. In cases in which transfer of chirality or high stereoselection in product formation is observed, a highly ordered transition state can be assumed.

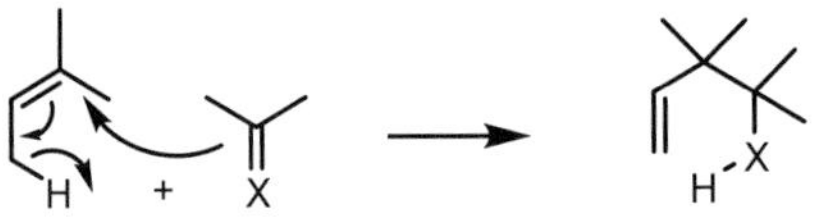

X = CR$_2$: ene reaction

X = O   : carbonyl ene reaction

In the second step, (–)-isopulegol (**11**) is subjected to catalytic hydrogenation to yield (–)-menthol (**1**) in high yield (88%) and with high enantiomeric purity ($ee > 98\%$).

(c) Experimental procedures for the synthesis of 1

4.2.3.1 * Isopulegol [7]

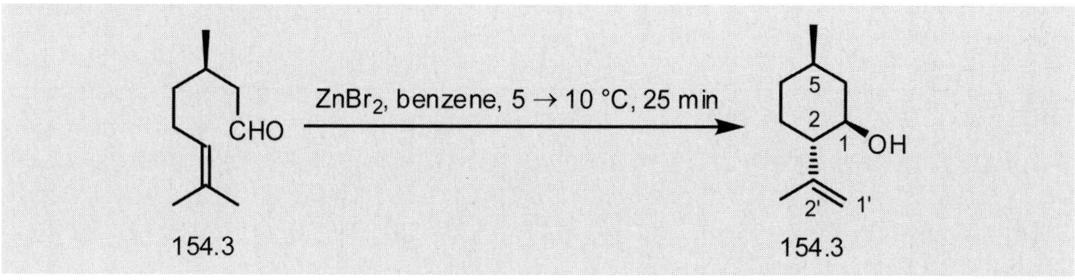

154.3   154.3

ZnBr$_2$ (219 mg, 972 µmol) is added portionwise to a stirred solution of (R)-citronellal (150 mg, 972 µmol) in anhydrous benzene (2 mL) (Caution: carcinogenic!) at 5 °C under an argon atmosphere and stirring is continued at 5–10 °C for 25 min.

After filtration, the ZnBr$_2$ is rinsed with Et$_2$O (10 mL), and the filtrate is washed with H$_2$O (10 mL) and saturated aqueous NaHCO$_3$ solution (10 mL) and dried over Na$_2$SO$_4$. The solvent is removed *in vacuo* and the crude product is purified by flash column chromatography on silica gel (*n*-pentane/Et$_2$O, 10:1) to afford (–)-isopulegol as a colorless oil; 94.5 mg (63%), bp$_{2.6}$ 50–60 °C; n_D^{20} = 1.4695; $[\alpha]_D^{20}$ = –18.8 (c = 1.0, CHCl$_3$), R_f = 0.51 (Et$_2$O/*n*-pentane, 1:1).

> **IR** (NaCl): $\tilde{\nu}$ (cm^{-1}) = 2923, 1645, 1455, 1095, 1027, 886, 846.
>
> **^{1}H NMR** (300 MHz, CDCl$_3$): δ (ppm) = 4.90 (s, 1 H, 1'-H$_b$), 4.86 (s, 1 H, 1'-H$_a$), 3.47 (td, J = 10.4, 4.3 Hz, 1 H, 1-H), 2.04 (m$_c$, 1 H, 2-H), 1.94–1.84 (m, 2 H, alkyl-CH$_2$), 1.71 (s, 3 H, 2'-CH$_3$), 1.70–1.63 (m, 2 H, alkyl-CH$_2$), 1.59–1.42 (m, 1 H, 5-H), 1.40–1.24 (m, 2 H, alkyl-CH$_2$), 0.95 (d, J = 6.6 Hz, 3 H, 5-CH$_3$).
>
> **^{13}C NMR** (50 MHz, CDCl$_3$): δ (ppm) = 146.6 (C-2'), 112.9 (C-1'), 70.30 (C-1), 54.11 (C-2), 42.60 (C-6), 34.29 (C-4), 31.42 (C-5), 29.59 (C-3), 22.22 (5-CH$_3$), 19.17 (C-3').
>
> **MS** (EI, 200 eV): m/z (%) = 154.2 (40) $[M]^+$.

4.2.3.2 ** (–)-Menthol [7]

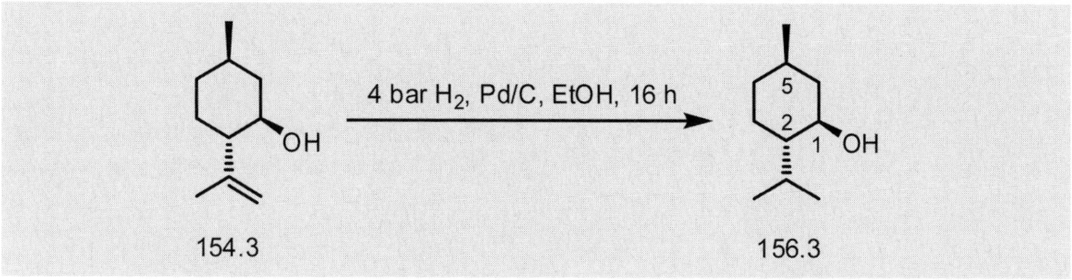

154.3   156.3

A mixture of isopulegol **4.2.3.1** (702 mg, 4.55 mmol) and 10% Pd on charcoal (200 mg) in EtOH (45 mL) is shaken for 18 h at 20 °C under an atmosphere of hydrogen (4 bar) (Caution!).

The catalyst is then filtered off by passing the mixture through a pad of Celite, which is subsequently rinsed with EtOH (50 mL). Evaporation of the solvent *in vacuo* and purification of the crude product by flash column chromatography on silica gel (*n*-pentane/EtOAc, 9:1) affords (–)-menthol as a colorless solid; 627 mg (88%), mp 40–41 °C, $[\alpha]_D^{20} = -37.1$ (*c* = 2.7, EtOH); $R_f = 0.51$ (Et$_2$O/*n*-pentane, 1:1).

IR (NaCl): $\tilde{\nu}$ (cm^{-1}) = 2954, 1455, 1045, 1025.

^{1}H NMR (300 MHz, CDCl$_3$): δ (ppm) = 3.41 (td, *J* = 10.6, 4.4 Hz, 1 H, 1-H), 2.17 (sept of d, *J* = 7.2, 2.8 Hz, 1 H, C<u>H</u>(CH$_3$)$_2$), 1.96 (m$_c$, 1 H, 5-H), 1.71–1.57 (m, 2 H), 1.51–1.34 (m, 2 H), 1.16–1.06 (m, 1 H), 1.05–0.95 (m, 1 H), 0.93 (d, *J* = 4.7 Hz, 3 H, CH(C<u>H</u>$_3$)$_2$), 0.90 (s, 3 H, 5-CH$_3$), 0.81 (d, *J* = 4.4 Hz, 3 H, CH(C<u>H</u>$_3$)$_2$).

^{13}C NMR (50 MHz, CDCl$_3$): δ (ppm) = 71.52 (C-1), 50.13 (C-2), 45.04 (C-6), 34.52 (C-4), 31.62 (C-5), 25.81 (<u>C</u>H(CH$_3$)$_2$), 23.12 (C-3), 22.18 (5-CH$_3$), 20.98, 16.07 (CH(<u>C</u>H$_3$)$_2$).

MS (EI, 200 eV): *m/z* (%) = 156.2 (2) [*M*]$^+$.

---

[1] Römpp, Lexikon "*Naturstoffe*" (Eds.: W. Steglich, B. Fugmann, S. Lang-Fugmann), p. 392, Georg Thieme Verlag, Stuttgart, **1997**.

[2] Ullmann's *Encyclopedia of Industrial Chemistry*, 6th ed., Vol. 14, p. 100, Wiley-VCH, Weinheim, **2003**.

[3] For an instructive overview on menthol synthesis, see: K. C. Nicolaou, E. J. Sorensen, *Classics in Total Synthesis*, p. 343, Wiley-VCH, Weinheim, **1996**.

[4] S. Otsuka, K. Tani, *Synthesis* **1991**, 665.

[5] S. Inoue, H. Takaya, K. Tani, S. Otsuka, T. Sato, R. Noyori, *J. Am. Chem. Soc.* **1990**, *112*, 4897.

[6] M. B. Smith, J. March, *March's Advanced Organic Chemistry*, 5th ed., p. 1021, John Wiley & Sons, Inc., New York, **2001**.

[7] K. Tani, T. Yamagata, S. Akutagawa, H. Kumobayashi, T. Taketomi, H. Takaya, A. Miyashita, R. Noyori, S. Otsuka, *J. Am. Chem. Soc.* **1984**, *106*, 5208.

4.2.4   Artemisia ketone

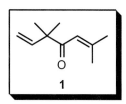

1

Topics:
- Synthesis of a monoterpene ketone
- Formation of an acid chloride from a carboxylic acid
- Formation of an allylsilane from an allyl Grignard compound and trimethylchlorosilane
- Acylation of an allylsilane with C–Si cleavage and allylic inversion

(a) General

Artemisia ketone (**1**, 3,3,6-trimethyl-1,5-heptadien-4-one) has been isolated from the etheric oils of *Artemisia annua* (mugwort) and *Santolina chamaecyparissus* (lavender cotton), which contain mixtures of **1** and its isomer **2** (*iso* artemisia ketone). In terms of the biological activity of **1**, no practical application is known [1].

1   **2**   isoprene

As a C_{10} monoterpene ketone, the structure of **1** can be deduced from two isoprene C_5 units. Notably — and in contrast to the usual head-to-tail orientation (C-1 to C-4) according to the mevalonate pathway of terpene biogenesis (cf. **4.2.2**) — the two building blocks of **1** are in a C-2/C-4 alignment.

Retrosynthesis of **1** offers disconnection at the sp^3 site of the carbonyl group leading to synthons **3** and **4**, which can be assigned to a β,β-dimethylallyl organometalic and β,β-dimethylacrylic acid (**5/6**):

1   **3**   **5** (M = metal)

4   **6**

Synthesis of **1** according to this retroanalytical concept — simply by regioselective acylation of the metalorganic compound **5** with the acid **6** — is associated with the problem that electrophilic attack has to occur at the sterically more hindered site of an appropriate allyl organometallic **5**.

As shown in section (b), the use of an allylsilane [2] offers an elegant solution to this problem, allowing the desired acylation to proceed with complete allylic inversion.

(b) Synthesis of 1

In the first step, the required (β,β-dimethylallyl)silane **9** is prepared by coupling of the Grignard compound **8**, prepared from prenyl bromide (**7**), with trimethylchlorosilane [3]. Notably, the electrophilic attack of Me_3SiCl takes place at the CH_2 site of the allyl Grignard compound **8** bearing the covalent C–Mg bond.

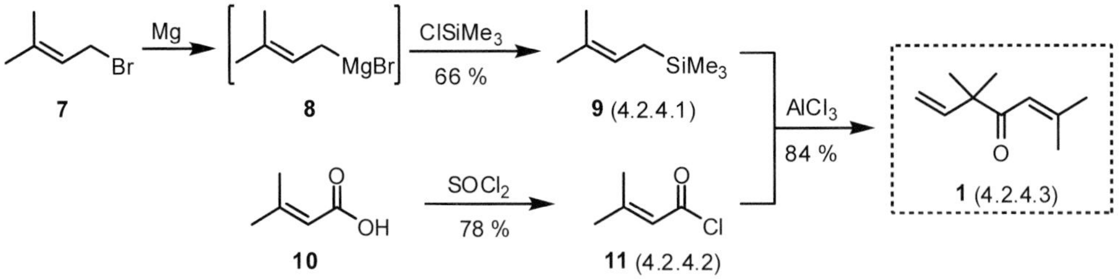

Then, the acid chloride **11** is prepared in the standard manner from senecic acid (**10**) by treatment with thionyl chloride.

In the last step, the allylsilane **9** is reacted with the acid chloride **11** in the presence of AlCl₃. As evidenced by the structure of the product **1**, the electrophilic acylation of **9** apparently occurs in a different mode compared to the reaction of **8 → 9**, that is, with C–Si cleavage accompanied by allylic inversion. It can be assumed that the acylation process (**9 → 1**) involves a cyclic transition state **12**, which (1) is preceded by complexation of AlCl₃ at the C=O group of the acid chloride **11** and (2) is likely to be favored by gaining the high Si–Cl bond energy after C–Si cleavage:

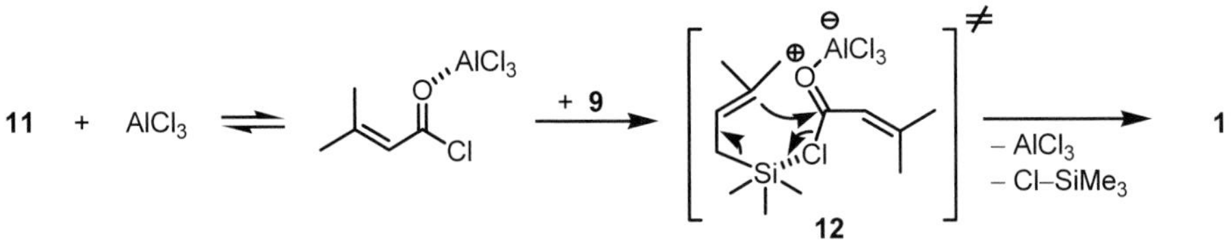

Thus, a three-step convergent synthesis of **1** provides the target molecule in an overall yield of 55% (based on senecic acid (**10**)).

(c) Experimental procedures for the synthesis of 1

4.2.4.1 ** 1-Trimethylsilyl-3-methyl-2-butene [4]

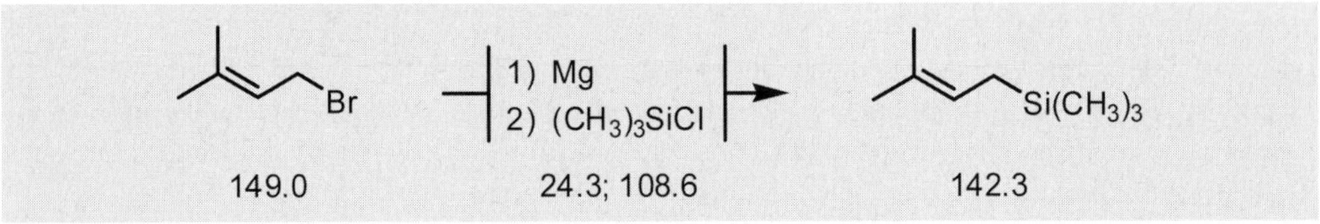

Magnesium turnings (14.6 g, 0.60 mol) in anhydrous THF (160 mL) are treated with one crystal of iodine and prenyl bromide (1 g). As soon as the Grignard reaction starts (disappearance of the iodine color), the mixture is cooled in an ice bath and a solution of prenyl bromide **4.2.1.2** (29.8 g, 0.20 mol, total amount including the initial 1 g) and trimethylchlorosilane (20.6 g, 0.19 mol) in anhydrous THF (60 mL) is added dropwise with stirring over 40 min. When the addition is complete, stirring is continued for 30 min at 0 °C and for 15 h at room temperature.

The excess magnesium is then removed by filtration, the filtrate is cooled to –20 °C, and saturated aqueous NH₄Cl solution (150 mL) is slowly added dropwise with stirring. The phases are separated, the aqueous phase is extracted with Et₂O (50 mL), and the combined organic phases are dried

(Na$_2$SO$_4$). The solvents are evaporated and the residue is fractionally distilled to give the product as a colorless oil; 17.7 g (66%), bp$_{300}$ 100–101 °C; n$_D^{20}$ = 1.4308.

IR (film): $\tilde{\nu}$ (cm^{-1}) = 1675 (weak, C=C), 1250, 865.

^{1}H NMR (CDCl$_3$): δ (ppm) = 5.09 (t$_{br}$, J = 8.5 Hz, 1 H, =CH), 1.67, 1.53 (s$_{br}$, 3 H, =C–CH$_3$), 1.36 (d, J = 8.5 Hz, 2 H, =C–CH$_2$), 0.05 (s, 9 H, Si(CH$_3$)$_3$).

4.2.4.2　*　Senecyl chloride (3,3-dimethylacryloyl chloride) [5]

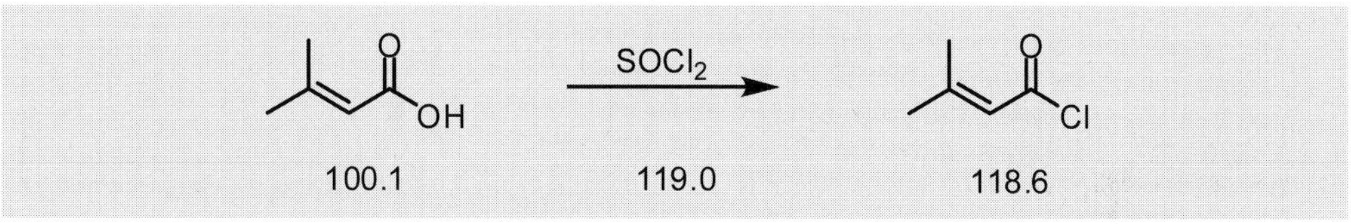

100.1　　　　　　119.0　　　　　　118.6

A stirred mixture of senecic acid (3,3-dimethylacrylic acid, 50.0 g, 0.50 mol), thionyl chloride (89.2 g, 0.75 mol), and anhydrous DMF (1 drop) is heated under reflux until the initially vigorous gas evolution (Hood: SO$_2$ and HCl are formed!) ceases (ca. 2 h).

The excess thionyl chloride is then evaporated *in vacuo* (Hood!) and the residue is distilled *in vacuo* to give the acid chloride as a colorless oil; 64.1 g (78%), bp$_{13}$ 52–53 °C.

IR (film): $\tilde{\nu}$ (cm^{-1}) = 1765, 1730 (C=O), 1600 (C=C).

^{1}H NMR (CDCl$_3$): δ (ppm) = 6.05 (sept, J = 1.5 Hz, 1 H, =C–H), 2.14, 1.97 (d, J = 1.5 Hz, each 3 H, =C–CH$_3$).

4.2.4.3　**　Artemisia ketone (3,3,6-trimethylhepta-1,5-dien-4-one) [3]

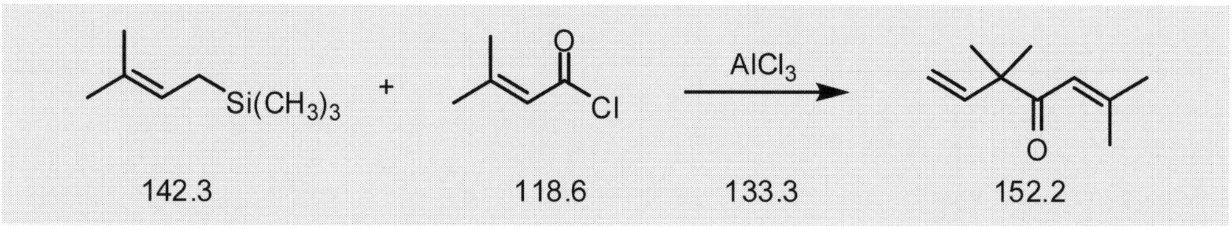

142.3　　　　　　118.6　　　133.3　　　　　152.2

3,3-Dimethylacryloyl chloride **4.2.4.2** (5.93 g, 50.0 mmol) is added to anhydrous aluminum chloride (6.67 g, 50.0 mmol) in anhydrous CH$_2$Cl$_2$ (25 mL) at 0 °C. This solution is added dropwise over 30 min to a stirred solution of the allylsilane **4.2.4.1** (7.83 g, 55.0 mmol) in CH$_2$Cl$_2$ (50 mL) at –65 °C and stirring is continued for 10 min.

The reaction mixture is then poured into a vigorously stirred mixture of NH$_4$Cl (30 g) and crushed ice (100 g). The phases are separated, the aqueous phase is extracted with CH$_2$Cl$_2$ (2 × 50 mL), and the combined organic phases are dried (Na$_2$SO$_4$). The solvent is evaporated and the residue is fractionally distilled *in vacuo* to give the product as a colorless oil with an aromatic odor; 6.35 g (84%), bp$_{20}$ 81–81 °C, n$_D^{20}$ = 1.4670.

IR (film): $\tilde{\nu}$ (cm^{-1}) = 3090, 1670 (C=O), 1635 (C=C).

^{1}H NMR (CDCl$_3$): δ (ppm) = 6.18 (s$_{br}$, 1 H, =C–H), 5.94, 5.20, 4.91 ("q" and "oct", total 3 H, X and AB of the ABX system –CH=CH$_2$; J_{AB} = 1.5 Hz, J_{AX} = 10 Hz, J_{BX} = 18 Hz), 2.10, 1.89 (s, each 3 H, =C–CH$_3$), 1.18 (s, 6 H, C(CH$_3$)$_2$).

Note: The following derivatives may be prepared:
        (1) the 2,4-dinitrophenylhydrazone, mp 66–67 °C;
        (2) the semicarbazone, mp 71–72 °C.

[1] Römpp, *Lexikon "Naturstoffe"* (Eds.: W. Steglich, B. Fugmann, S. Lang-Fugmann), p. 58, Georg Thieme Verlag, Stuttgart, **1997**.

[2] a) A. Hosomi, *Acc. Chem. Res.* **1988**, *21*, 200; b) H. Mayr, M. Patz, *Angew. Chem.* **1994**, *106*, 991; *Angew. Chem. Int. Ed. Engl.* **1994**, *33*, 938.

[3] J. P. Pillot, J. Dunogues, R. Calas, *Tetrahedron Lett.* **1976**, *17*, 1871.

[4] A. Hosomi, H. Sakurai, *Tetrahedron Lett.* **1978**, *19*, 2589.

[5] H. Staudinger, E. Ott, *Ber. Dtsch. Chem. Ges.* **1911**, *44*, 1633.

4.2.5   Veticadinol

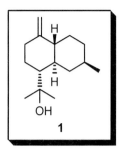

1

Topics:
- Synthesis of the sesquiterpene veticadinol
- Knoevenagel condensation
- Lewis acid-mediated intramolecular ene reaction
- Krapcho decarboxylation
- Prins reaction (oxa-ene reaction)
- Tosylation and nucleophilic substitution
- Intramolecular alkylation
- Grignard reaction

(a) General

Veticadinol (**1**) is a natural product belonging to the sesquiterpene family [1], which is formed in Nature from mevalonic acid. It was first obtained as a mixture with other terpenes from vetiver oil in 1961 [2], which is obtained from the roots of the grass *Vetiveria zizanoides* by steam distillation. Several syntheses have since been published, although again these led only to product mixtures. The first stereoselective synthesis, which gave veticadinol in eight steps in 36% overall yield starting from (*R*)-citronellal (**7**) and dimethyl malonate (**6**), was developed by Tietze and co-workers in 1988 [3].

Retrosynthetic analysis of **1** leads to the ester **2**, treatment of which with methylmagnesium iodide would give **1** with a tertiary alcohol moiety [4]. Cleavage of the decalin skeleton would lead to the alcohol **3**, which in the synthetic approach could undergo an alkylation after transformation of the hydroxyl group into the corresponding iodide. The central retrosynthetic steps are the retro-Prins reaction (**3 → 4**) and the retro-ene reaction to give **5**, which should be easily accessible by a Knoevenagel condensation of the aldehyde **7** and dimethyl malonate (**6**).

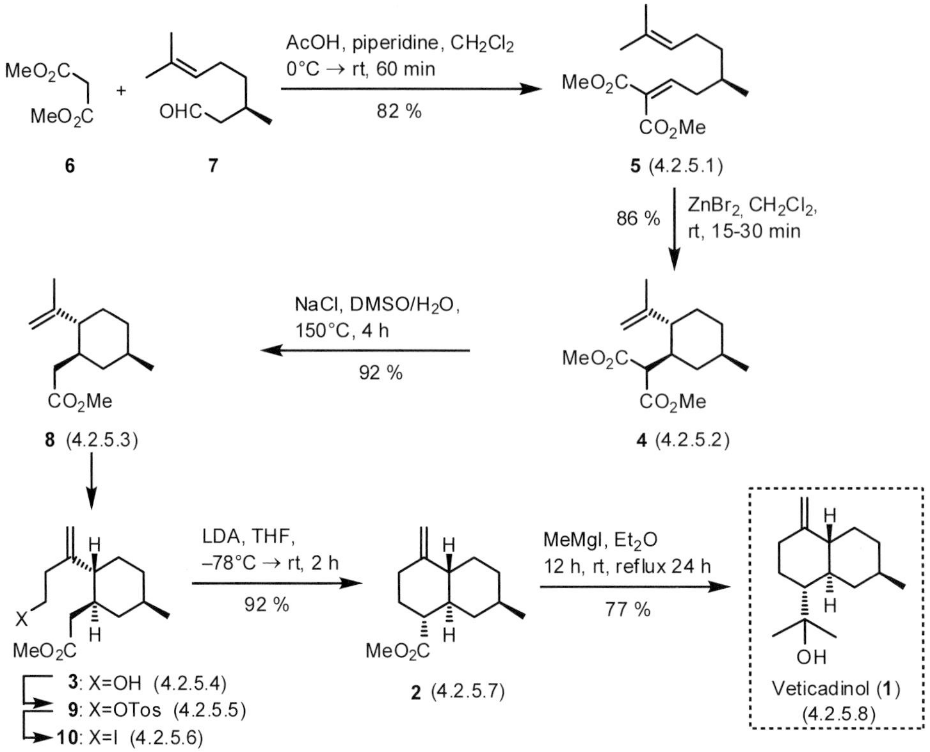

(b) Synthesis of 1

According to the retrosynthetic analysis, citronellal (7) (ee = 97%) is used for the Knoevenagel condensation [5] with dimethyl malonate in the presence of piperidinium acetate to give the alkylidene-1,3-dicarboxylate 5 in 82% yield. The key step in the synthesis of veticadinol (1) is the subsequent Lewis acid-mediated intramolecular ene reaction of 5 [6], which gives the *trans*-1,2-disubstituted cyclohexane 4 in 86% yield and with excellent selectivity (simple and induced diastereoselectivity). Thereafter, one of the ester moieties in 4 is removed using NaCl in DMSO at 150 °C (Krapcho reaction) [7]. In this transformation, the methyl ester is first cleaved by nucleophilic substitution at the methyl group to give the corresponding acid, which then undergoes decarboxylation to yield 8. Usually, esters are cleaved by aqueous hydrolysis; however, in the present case, this procedure gives only low yields.

The next step in the synthesis of 1 is an oxa-ene reaction (Prins reaction) using formaldehyde and dimethylaluminum chloride [8]. This reaction is not a concerted transformation like the normal ene reaction, but proceeds via a carbocation. For the ring closure of 3 to give the decalin 2, the alcohol moiety in 3 is first transformed into the corresponding tosylate 9 by reaction with TosCl; 9 is then subjected to nucleophilic substitution using sodium iodide in acetone to give the iodide 10 in 87% yield over the two steps. Intramolecular alkylation of 10 using LDA for the formation of the ester enolate then leads to 2 in 92% yield. The final step is a two-fold Grignard reaction at the ester moiety in 2 with MeMgI to afford the desired veticadinol (1) in 77% yield. Thus, starting from malonate and almost enantiopure citronellal, veticadinol (1) could be prepared in eight steps in 33% overall yield.

4.2.5.5 **   (1'*R*,2'*R*,5'*R*)-2-{[2'-(4''-*p*-Toluenesulfonyloxy-1''-buten-2''-yl)-5'-methyl]-cyclohex-1'-yl}-acetic acid methyl ester [11]

TosCl, pyridine,
0°C → 4°C, 12 h

240.3   394.5

A solution of the hydroxy ester **4.2.5.4** (800 mg, 3.33 mmol) in pyridine (1.05 g, 13.3 mmol) is cooled to 0 °C and *p*-toluenesulfonyl chloride (630 mg, 3.33 mmol) is added. The mixture is stirred at 0 °C for 1 h and at 4 °C overnight.

It is then partitioned between ice-cold HCl (2 M, 50 mL) and ice-cold Et$_2$O (20 mL). The aqueous layer is extracted with Et$_2$O (3 × 20 mL). The combined organic layers are washed with HCl (2 M) to completely remove the pyridine, then washed with saturated NaHCO$_3$ solution (10 mL) and brine (10 mL), dried over Na$_2$SO$_4$, and concentrated *in vacuo*. The resulting crude product can be used for subsequent reactions without further purification. Analytically pure samples are prepared by chromatography on silica gel using Et$_2$O/petroleum ether (1:4) as eluent to afford the tosylate; 1.22 g (93%), $[\alpha]_D^{20} = -19.6$ (*c* = 1, CH$_3$CN); R_f = 0.44 (Et$_2$O/*n*-hexane, 1:1).

UV (CH$_3$CN): λ_{max} (nm) (lg ε) = 273 (2.76), 267 (2.81), 262 (2.81), 255 (2.76), 225 (4.11).

IR (NaCl): $\tilde{v}$ (cm^{-1}) = 3060, 3020, 2940, 2920, 2850, 1735, 1640, 1595, 1360, 1190, 1175, 965, 905, 815, 770, 660.

^{1}H NMR (CDCl$_3$, 200 MHz): δ (ppm) = 7.88–7.32 (m, 4 H, Ar-H), 4.87 (s, 1 H, 1''-H), 4.74 (m$_c$, 1 H, 1''-H), 4.13 (t, *J* = 7.0 Hz, 2 H, 4''-H), 3.64 (s, 3 H, OCH$_3$), 2.46 (s, 3 H, Ar-H), 2.38 (m$_c$, 1 H), 2.32 (tm, *J* = 7.0 Hz, 2 H, 3''-H), 1.94–1.27 (m, 7 H), 1.20 (dqm, *J* = 13.0, 3.0 Hz, 1 H), 0.98–0.78 (m, 1 H), 0.86 (d, *J* = 6.5 Hz, 3 H, 5'-CH$_3$), 0.66 (dt, *J* = 13.0, 11.5 Hz, 1 H, 6'-H$_{ax}$).

^{13}C NMR (CDCl$_3$, 20 MHz): δ (ppm) = 173.5 (C-1), 146.9 (C-2''), 144.8, 133.2 (Ar-C), 129.9, 127.9, 112.3 (C-1''), 68.83 (C-4''), 51.26, 50.50 (C-2', OCH$_3$), 41.13 (C-2), 39.12 (C-6'), 36.89 (C-1'), 34.87 (C-4'), 32.80, 32.59 (C-3', C-3''), 32.23 (C-5'), 22.43 (5'-CH$_3$), 21.57 (Ar–CH$_3$).

MS (EI, 70 eV): *m/z* (%) = 394 [*M*]$^+$, 363 (3) [*M* – CH$_3$O]$^+$, 362 (3) [*M* – CH$_3$OH]$^+$, 334 (3) [*M* – CH$_3$OH + CO]$^+$, 222 (47) [C$_{14}$H$_{22}$O$_2$]$^+$, 193 (35) [222 – CH$_3$O]$^+$, 148 (84) [222 – C$_3$H$_6$O$_2$]$^+$, 91 (61) [C$_7$H$_7$]$^+$, 74 (100) [C$_3$H$_6$O$_2$]$^+$, 41 (37) [C$_3$H$_5$]$^+$.

4.2.5.6 ** **(1'R,2'R,5'R)-2-{[2'-(4''-Iodo-1''-buten-2''-yl)-5'-methyl]-cyclohex-1'-yl}-acetic acid methyl ester** [11]

NaI, acetone, rt, 12 h

TosO — CO$_2$Me — 394.5

1'' 3' 3'' 2'' 2 1' 5' I ^{1}CO$_2$Me — 350.2

Sodium iodide (2.09 g, 13.9 mmol) is added to a solution of the tosylate **4.2.5.5** (1.10 g, 2.78 mmol) in acetone (11.0 mL) at room temperature. The reaction mixture is stirred for 12 h at room temperature.

Sodium tosylate is then removed by filtration and the filtrate is partitioned between petroleum ether (20 mL) and H$_2$O (10 mL). The organic layer is dried and concentrated (rotary evaporator bath temperature: 20 °C). The crude product can be used for subsequent reactions without further purification. Analytically pure samples are prepared by chromatography on silica gel using Et$_2$O/petroleum ether (1:7) as eluent to afford the iodo ester, which decomposes easily giving rise to a red color; 919 mg (94%), $[\alpha]_D^{20} = -30.7$ ($c = 1$, CH$_3$CN), $R_f = 0.55$ (Et$_2$O/n-hexane, 1:1).

UV (CH$_3$CN): λ_{max} (nm) (lg ε) = 252 (2.83).

IR (NaCl): $\tilde{\nu}$ (cm^{-1}) = 3060, 2940, 2910, 2850, 1735, 1640, 1430, 1360, 1155, 895.

^{1}H NMR (CDCl$_3$, 200 MHz): δ (ppm) = 4.92 (s, 1 H, 1''-H), 4.84 (m$_c$, 1 H, 1''-H), 3.65 (s, 3 H, OCH$_3$), 3.24 (tm, J = 7.5 Hz, 2 H, 4''-H), 2.54 (tm, J = 7.5 Hz, 2 H, 3''-H), 2.50 (m$_c$, 1 H), 2.00–1.16 (m, 8 H), 0.88 (d, J = 6.5 Hz, 3 H, 5'-CH$_3$), 1.04–0.80 (m, 1 H), 0.70 (dt, J = 13.0, 11.5 Hz, 1 H, 6'-H$_{ax}$).

^{13}C NMR (CDCl$_3$, 20 MHz): δ (ppm) = 173.6 (C-1), 151.0 (C-2''), 111.5 (C-1''), 51.32, 50.07 (C-2', OCH$_3$), 41.18 (C-2), 39.23 (C-6'), 38.53 (C-3''), 37.13 (C-1'), 34.99 (C-4'), 33.28 (C-3'), 32.22 (C-5'), 22.45 (5'-CH$_3$), 3.07 (C-4'').

MS (EI, 70 eV): m/z (%) = 319 (5) $[M - CH_3O]^+$, 276 (2) $[M - C_3H_6O_2]^+$, 223 (61) $[M - I]^+$, 149 (80) $[223 - C_3H_6O_2]^+$, 93 (100), 81 (61) $[C_6H_9]^+$, 74 (67) $[C_3H_6O_2]^+$, 67 (41) $[C_5H_7]^+$, 55 (68) $[C_4H_7]^+$, 41 (80) $[C_3H_5]^+$.

4.2.5.7 ** **(1R,5R,6R,8R)-8-Methyl-2-methylene-bicyclo[4.4.0]decane-5-carboxylic acid methyl ester** [11]

LDA, THF, −78°C → rt, 2 h

I — CO$_2$Me — 350.2

H 2 1 6 5 8 MeO$_2$C H — 222.3

A solution of *n*-butyllithium in *n*-hexane (1.6 M, 2.15 mL, 3.43 mmol) is added to a solution of diisopropylamine (389 mg, 3.84 mmol) in THF (29 mL) at 0 °C. After stirring for 5 min, the mixture is cooled to –78 °C. A solution of the iodo ester **4.2.5.6** (800 mg, 2.29 mmol) in THF (2.0 mL) is added dropwise with stirring over 15 min. The mixture is allowed to warm to room temperature over a period of 2 h.

Saturated aqueous NH$_4$Cl solution (2.5 mL) is then added and the mixture is diluted with petroleum ether (60 mL). The organic phase is washed with H$_2$O (20 mL) and brine (20 mL), dried over Na$_2$SO$_4$, and concentrated *in vacuo*. Purification by chromatography on silica gel using Et$_2$O/petroleum ether (1:1) as eluent affords the product; 466 mg (92%), $[\alpha]_D^{20} = -10.4$ ($c = 1$, CH$_3$CN), $R_f = 0.57$ (Et$_2$O/*n*-hexane, 1:1).

IR (NaCl): $\tilde{v}$ (cm^{-1}) = 3090, 3000, 2930, 2875, 2850, 1740, 1650, 1435, 1370, 1160, 895.

^{1}H NMR (CDCl$_3$, 200 MHz): δ (ppm) = 4.71 (m$_c$, 1 H, C=CH$_2$), 4.61 (m$_c$, 1 H, C=CH$_2$), 3.69 (s, 3 H, OCH$_3$), 2.47–2.33 (m, 1 H), 2.24 (ddd, J = 12.3, 11.0, 3.5 Hz, 1 H, 5-H$_{ax}$), 2.10 (dddm, J = 14.0, 4.5, 3.0 Hz, 1 H, 3-H$_{eq}$), 2.03–1.19 (m, 9 H), 1.09–0.84 (m, 1 H), 0.87 (d, J = 6.5 Hz, 3 H, 8-CH$_3$), 0.74 (q, J = 12.0 Hz, 1 H, 7-H$_{ax}$).

^{13}C NMR (CDCl$_3$, 50 MHz): δ (ppm) = 175.8 (5 C-α), 150.6 (C-2), 105.2 (2 C-α), 51.30 (OCH$_3$), 50.51 (C-5), 45.16, 45.03 (C-1, C-6), 40.70 (C-7), 35.54 (C-3), 34.69 (C-9), 32.14 (C-8), 31.41 (C-4), 28.69 (C-10), 22.49 (8-CH$_3$).

MS (EI, 70 eV): *m/z* (%) = 222 (17) [*M*]$^+$, 207 (2) [*M* – CH$_3$]$^+$, 193 (7) [*M* – C$_2$H$_5$]$^+$, 191 (7) [*M* – CH$_3$O]$^+$, 190 (9) [*M* – CH$_3$OH]$^+$, 163 (70) [*M* – C$_2$H$_3$O$_2$]$^+$, 162 (100) [*M* – CH$_3$OH + CO]$^+$, 148 (13) [163 – CH$_3$]$^+$, 147 (20) [162 – CH$_3$]$^+$, 107 (35) [C$_8$H$_{11}$]$^+$, 95 (46) [C$_7$H$_{11}$]$^+$, 93 (34) [C$_7$H$_9$]$^+$, 81 (46) [C$_6$H$_9$]$^+$, 79 (38) [C$_6$H$_7$]$^+$.

4.2.5.8   **   (1*R*,5*R*,6*R*,8*R*)-5-(2-Hydroxyisopropyl)-8-methyl-2-methylene-bicyclo[4.4.0]decane (veticadinol) [11]

222.3   222.4

A solution of the ester **4.2.5.7** (350 mg, 1.58 mmol) in Et$_2$O (8 mL) is added dropwise to a stirred solution of methylmagnesium bromide, prepared by reaction of methyl iodide (741 mg, 5.22 mmol) and magnesium (115 mg, 4.73 mmol) in anhydrous Et$_2$O (16 mL). The mixture is stirred for 12 h at room temperature and thereafter for 12 h at reflux temperature.

After cooling, the mixture is carefully poured into saturated NH_4Cl solution (10 mL) and the aqueous layer is extracted with Et_2O (5 × 5 mL). The combined organic layers are washed with H_2O (10 mL) and brine (10 mL), dried over Na_2SO_4, and concentrated *in vacuo*. Purification by chromatography on silica gel using Et_2O/petroleum ether (1:4) as eluent affords veticadinol as a solid; 270 mg (77%), mp 84–85 °C, $[\alpha]_D^{20} = +11.8$ ($c = 1$, CH_3CN), $R_f = 0.10$ (Et_2O/n-hexane, 1:1).

IR (KBr): $\tilde{\nu}$ (cm^{-1}) = 3330, 3100, 3000, 2980, 2960, 2940, 2880, 1650, 1460, 1382, 1372, 1160, 1140, 892, 885.

^1H NMR (C_6D_6, 200 MHz): δ (ppm) = 4.78 (m_c, 1 H, C=CH$_2$), 4.71 (m_c, 1 H, C=CH$_2$), 2.57 (dddm, $J = 13.0$, 6.0, 3.0 Hz, 1 H, 7-H$_{eq}$), 2.32 (dtm, $J = 13.0$, 3.5 Hz, 1 H, 3-H$_{eq}$), 2.01 (dm, $J = 12.5$ Hz, 1 H, 3-H$_{ax}$), 1.92 (dq, $J = 12.5$, 3.0 Hz, 1 H, 10-H$_{eq}$), 1.78–1.60 (m, 2 H, 4-H$_{eq}$, 9-H$_{eq}$), 1.52 (tm, $J = 11.0$ Hz, 1 H, 1-H$_{ax}$), 1.33 (dq, $J = 12.0$, 3.5 Hz, 1 H, 10-H$_{ax}$), 1.00 (s, 3 H, 5-C-α-CH$_3$), 0.96 (s, 3 H, 5 C-α-CH$_3$), 0.94 (d, 3 H, 8-CH$_3$), 1.40–0.84 (m, 5 H, 4-H$_{ax}$, 5-H$_{ax}$, 6-H$_{ax}$, 8-H$_{ax}$, 9-H$_{ax}$), 0.74 (dt, $J = 13.0$, 11.0 Hz, 1 H, 7 H$_{ax}$), 0.70 (s, 1 H, OH).

^{13}C NMR (C_6D_6, 50 MHz): δ (ppm) = 152.7 (C-2), 103.9 (2 C-α), 73.62 (5 C-α), 53.35 (C-5), 46.75 (C-6), 46.01 (C-1), 42.43 (C-7), 37.00 (C-3), 34.94 (C-9), 32.97 (C-8), 31.83 (5 C-β), 31.26 (C-4), 29.82 (C-10), 24.56 (5 C-β), 23.31 (8-CH$_3$).

MS (EI, 70 eV): m/z (%) = 222 (0.04) $[M]^+$, 204 (28) $[M - H_2O]^+$, 164 (33) $[M - C_3H_6O]^+$, 149 (43) $[164 - CH_3]^+$, 135 (21) $[C_{10}H_{15}]^+$, 121 (23) $[C_9H_{13}]^+$, 93 (21) $[C_7H_9]^+$, 81 (22) $[C_6H_9]^+$, 59 (100) $[C_3H_7O]^+$.

---

[1]   E. Breitmaier, *Terpene*, Wiley-VCH, Weinheim, **2005**.

[2]   G. Chiurdoglu, A. Delsemme, *Bull. Soc. Chim. Belg.* **1961**, *70*, 5.

[3]   L. F. Tietze, U. Beifuss, J. Antel, G. M. Sheldrick, *Angew. Chem.* **1988**, *100*, 739; *Angew. Chem. Int. Ed. Engl.* **1988**, *27*, 703.

[4]   For a discussion of the synthesis, see: J. A. Gewert, J. Görlitzer, S. Götze, J. Looft, P. Menningen, T. Nöbel, H. Schirok, C. Wulff, in *Organic Synthesis Workbook*, Wiley-VCH, Weinheim, **2000**.

[5]   L. F. Tietze, U. Beifuss, in *Comprehensive Organic Synthesis* (Ed.: B. M. Trost), Vol. 2, p. 341, Pergamon, Oxford, **1991**.

[6]   a) L. F. Tietze, U. Beifuss, *Angew. Chem.* **1985**, *97*, 1067; *Angew. Chem. Int. Ed. Engl.* **1985**, *24*, 1042; b) L. F. Tietze, U. Beifuss, *Tetrahedron Lett.* **1986**, *27*, 1767; c) L. F. Tietze, U. Beifuss, M. Ruther, A. Rühlmann, J. Antel, G. M. Sheldrick, *Angew. Chem.* **1988**, *100*, 1200–1201; *Angew. Chem. Int. Ed. Engl.* **1988**, *27*, 1186–1187; d) L. F. Tietze, U. Beifuss, *Liebigs Ann. Chem.* **1988**, 321–329; e) L. F. Tietze, U. Beifuss, M. Ruther, *J. Org. Chem.* **1989**, *54*, 3120–3129; f) L. F. Tietze, C. Schünke, *Angew. Chem.* **1995**, *107*, 1901–1903; *Angew. Chem. Int. Ed. Engl.* **1995**, *34*, 1731–1733.

[7]   A. P. Krapcho, *Synthesis* **1982**, 805.

[8]   B. B. Snider, D. J. Rodini, T. C. Kirk, R. Cordova, *J. Am. Chem. Soc.* **1982**, *104*, 555.

[9]   a) R. W. Hoffmann, *Chem. Rev.* **1989**, *89*, 1841; b) L. F. Tietze, G. Schulz, *Liebigs Ann. Chem.* **1996**, 1575.

[10]   I. Fleming, *Frontier Orbitals and Organic Chemical Reactions*, p. 161, VCH, Weinheim, **1976**.

[11]   L. F. Tietze, U. Beifuss, *Synthesis* **1988**, 359–362.

4.2.6   all-*trans*-Vitamin A acetate

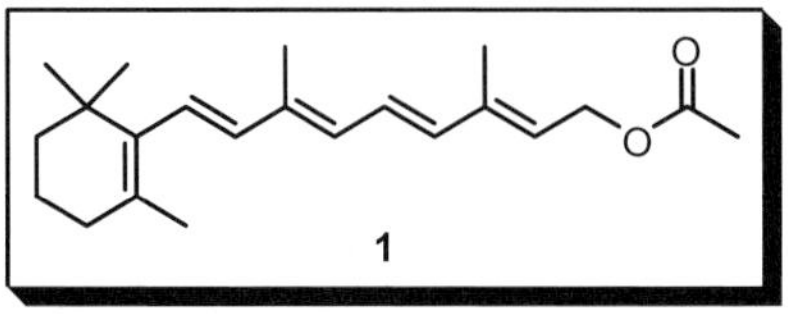

1

Topics:
- Synthesis of a diterpene-derived polyene ester
- Carbonyl olefinations according to Wittig and Horner reactions
- Formation of phosphonium salts
- Reduction of esters to primary alcohols

(a) General

The name vitamin A refers to a number of monocyclic C_{20}-diterpenes (retinoids), in which a trimethylcyclohexene unit is combined with a terminally functionalized polyene side chain. The most prominent member, the polyene alcohol retinol (vitamin A_1), occurs exclusively in animal tissues and is stored, e.g., in the liver in the form of esters of higher fatty acids (e.g., palmitic acid). Retinol plays a number of roles in the organism, the most important being growth, development, and differentiation of epithelial tissue, reproduction, and vision [1].

Retinol is manufactured almost exclusively in the form of the more stable esters, mainly the acetate **1**, but also as propionate and palmitate. Vitamin A is produced (1) by isolation from natural sources, or (2) in increasing amounts by synthesis on an industrial scale. The target molecule of most syntheses is retinol acetate (**1**).

For a retrosynthetic approach to **1**, two restrictions prove to be useful. First, it should be considered that β-ionone (**8**, cf. **1.5.3**) is the key intermediate in practically all syntheses. Second, C–C disconnections in the remaining polyene side chain should focus on the C=C double bonds and therefore on the retro-transformations of carbonyl olefination and/or aldol-type condensation reactions.

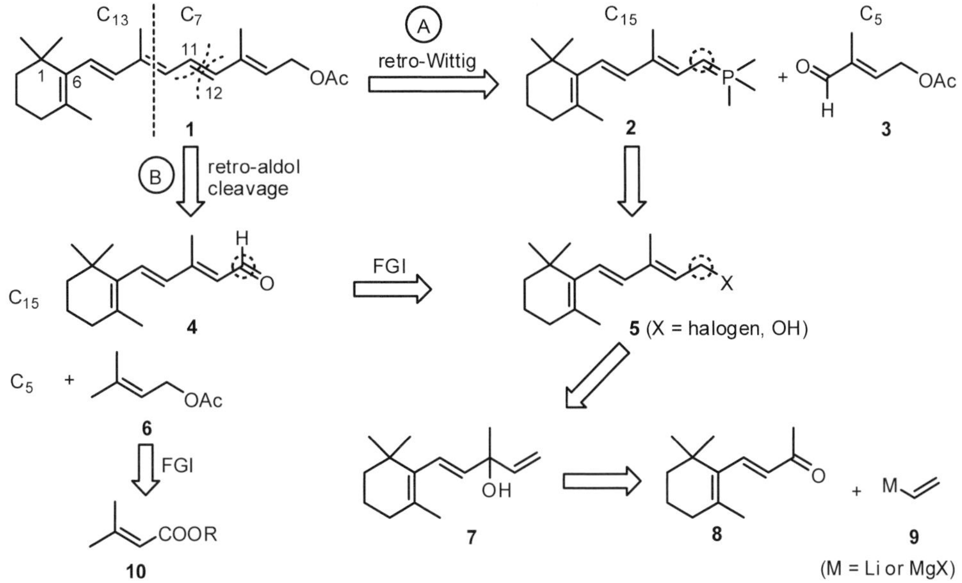

Since C_2-elongation at the carbonyl group of the β-ionone C_{13} unit is simple (**8 → 7**), diconnection at the C-11/C-12 double bond in **1** is most attractive (C_{20} → C_{15} + C_5) and gives rise to a retro-Wittig mode **A** (with the olefination components ylide **2** and aldehyde **3**) and to a retro-aldol reaction **B** with the condensation components aldehyde **4** and acetate **6**, of which the latter is accessible from ester **10**. Ylide **2** and aldehyde **4** can be deduced from the common intermediate **5**, allylic rearrangement of which leads to the isomeric C_{15} alcohol **7**, the product of addition of, e.g., a vinyl Grignard compound **9** (M = MgX) to β-ionone (**8**). Educt **10** is an ester of senecic acid, while aldehyde **3** has been the subject of an earlier synthesis (cf. **1.1.1**).

From these retroanalytical considerations, two of the industrial syntheses of **1** documented in the literature [1, 2] can be deduced.

In the Sumitomo synthesis [3], aldol condensation of aldehyde **4** with 3-methyl-2-butenoic ester **10** using KNH$_2$ in liquid NH$_3$ produces all-*trans* retinoic ester **11**, reduction of which with LiAlH$_4$ and subsequent acetylation yields **1**:

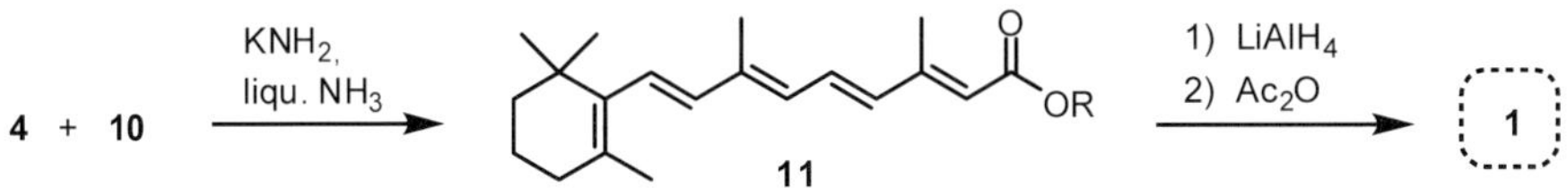

In the BASF synthesis [4], β-ionone (**8**) is transformed to vinyl-β-ionol (**7**) by ethynylation with acetylene followed by partial hydrogenation. Treatment of **7** with Ph$_3$P/HX provides the C_{15} phosphonium salt **12**, the key building block for the synthesis of retinoids [4]. Wittig reaction with the aldehyde **3** leads to a 70:30 mixture of C-11-(*E*)/(*Z*)-isomers, which is transformed to the all-*trans* acetate **1** by treatment with iodine.

In the Rhône-Poulenc synthesis [5], the starting material is the allyl phenyl sulfone **13** obtained from vinyl-β-ionol (**7**) and sodium phenylsulfinate. Sulfone **13** is deprotonated at the α-position and alkylated with the allylic halide **14** (a precursor of aldehyde **3**, cf. **1.1.1**) to give **15**. Finally, the acetate **1** or free retinol can be obtained from **15** by base-induced elimination of sulfinate:

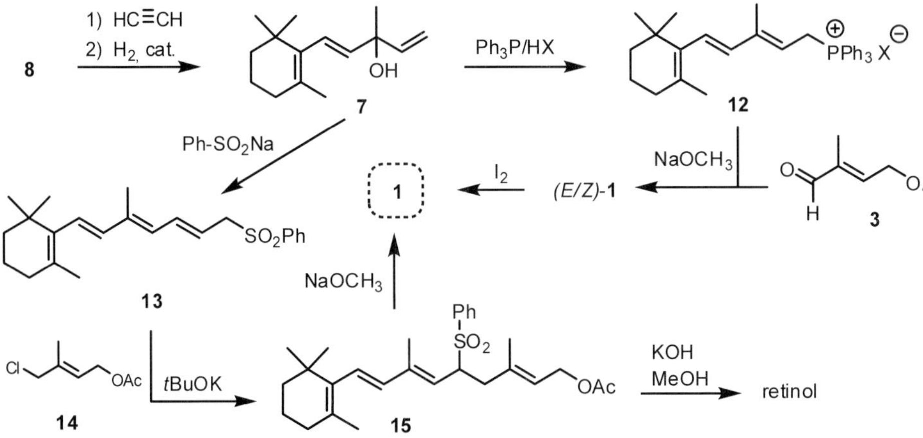

The conceptually different approach of Hoffmann-La Roche [6] starts with the C_{14}-building block **17**, which is obtained from β-ionone (**8**) via **16** by a Darzens synthesis. Aldehyde **17** is then converted to the C_{20}-alkyne diol **21** by Grignard reaction with the bis-MgBr derivative **19** of 3-methyl-2-penten-4-yn-1-ol (accessible from methyl vinyl ketone by ethynylation and H$^+$-catalyzed allylic rearrangement).

cis-Stereoselective partial hydrogenation using a Lindlar catalyst to give the (*E*)/(*Z*)-diol **18**, acetylation of the primary hydroxyl function (→ **20**), and subsequent acid-induced dehydration with double-bond isomerization lead to **1**:

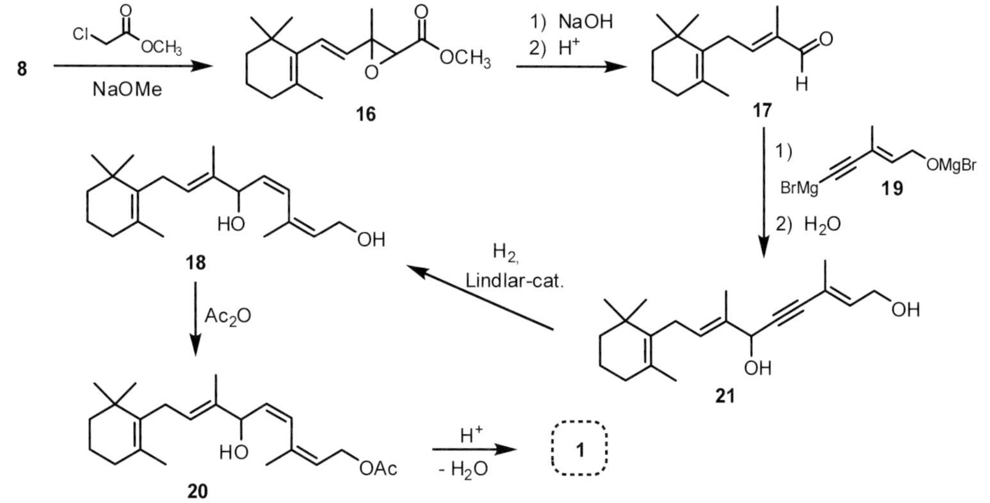

It should be noted that syntheses of the polyene chain by formation of $C(sp^2)-C(sp^2)$ single bonds by means of Heck reactions are also known [7].

(b) Synthesis of 1

For the synthesis of vitamin A acetate **1**, a laboratory procedure is presented, which is based on the retrosynthesis according to **A** and which contains elements of the industrial synthesis developed by BASF [4]. Thus, β-ionone (**8**) is olefinated with diethyl (ethoxycarbonyl)methylphosphonate (**22**) in the presence of NaOCH$_3$ in a Horner reaction (P–O-activated carbonyl olefination, cf. **1.1.7**) with concomitant transesterification to give methyl β-ionylidene acetate (**23**):

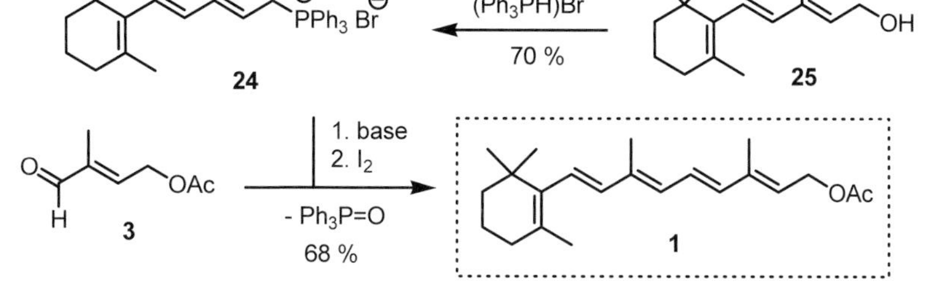

The ester **23** is reduced using LiAlH$_4$ to give β-ionylidene ethanol **25**, which is converted into the C$_{15}$-phosphonium bromide **24** by reaction with triphenylphosphonium hydrobromide.

Thereafter, phosphonium salt **24** is deprotonated to provide (*in situ*) the ylide **2**, which in a Wittig reaction is combined with the aldehyde **3** (cf. **1.1.1**) to give a mixture of (*E*)- and (*Z*)-retinol acetates under elimination of Ph$_3$PO. Isomerization with iodine leads to the all-*trans*-configured retinol acetate **1**.

For the deprotonation of **24**, two specific methods are used, namely (1) base-free ylide formation with an oxirane (1,2-butene oxide) as HX acceptor, and (2) ylide formation in a two-phase system with aqueous NaOH/dichloromethane.

The synthesis of β-ionone **8** (cf. **1.5.3**) as the starting material for the preparation of the C$_{15}$-salt **24** requires five steps starting from acetoacetate (overall yield 30%) and the transformation of **8** into **24** needs an additional three steps with 55% yield. Since the aldehyde **3** is prepared in three steps with 48% yield, the described synthesis of vitamin A acetate synthesis requires 12 steps and has an overall yield of 5%.

(c) Experimental procedures for the synthesis of 1

4.2.6.1   **   Methyl β-ionylidene acetate [9]

192.3   224.2   248.3

A solution of NaOCH$_3$ (prepared from sodium (3.68 g, 160 mmol) in anhydrous MeOH (80 mL) is slowly added dropwise to a stirred solution of β-ionone **1.5.3** (30.0 g, 156 mmol) and diethyl (ethoxycarbonyl)methylphosphonate (36.0 g, 160 mmol) in anhydrous benzene (80 mL, Caution: carcinogen!) at room temperature. Stirring is continued at 40 °C for 15 min.

The solution is then poured onto ice (300 g) and extracted with Et$_2$O (3 × 100 mL). The combined organic phases are washed with H$_2$O (2 × 200 mL), dried over Na$_2$SO$_4$, and concentrated *in vacuo*. The residue is fractionally distilled to give the product as a light-yellow oil; 34.6 g (89%), bp$_{0.3}$ 118–120 °C.

IR (film): $\tilde{v}$ (cm^{-1}) = 1715 (C=O), 1610 (C=C), 1235, 1135 (C–O–C).

^{1}H NMR (CDCl$_3$): δ (ppm) = 6.52, 6.08 (d, *J* = 16 Hz, 1 H, CH=CH), 5.77 (m, 1 H, 2-H), 3.72 (s, 3 H, OCH$_3$), 2.35 (s, 3 H, =C–CH$_3$), 2.2–1.8 (m, 2 H, allyl CH$_2$), 1.70 (s, 3 H, =C–CH$_3$), 1.6–1.1 (m, 4 H, CH$_2$–CH$_2$), 1.02 (s, 6 H, C(CH$_3$)$_2$).

Note: As a derivative, β-ionylidene acetic acid, mp 124–125 °C, can be easily prepared by saponification (KOH in methanol, 24 h, room temperature).

4.2.6.2   **   β-Ionylidene-ethanol [9]

248.3   38.0   220.3

A solution of β-ionylidene acetate **4.2.6.1** (20.0 g, 80.0 mmol) in anhydrous Et$_2$O (80 mL) is added dropwise to a stirred suspension of LiAlH$_4$ (3.40 g, 90.0 mmol) in anhydrous Et$_2$O (60 mL) at 0 °C over 30 min. Stirring is continued at 0 °C for 1 h (note).

A mixture of MeOH/H$_2$O (9:1; 20 mL) is slowly dropped into the solution with stirring, followed by NH$_4$Cl solution (10%). During the addition, external cooling with ice is necessary. The phases are separated and the aqueous phase is extracted with Et$_2$O (3 × 100 mL). The combined organic phases are washed with H$_2$O (50 mL), dried (Na$_2$SO$_4$), and then concentrated *in vacuo*. The residue is fractionally distilled to give the product as a nearly colorless oil; 17.0 g (90%), bp$_{0.3}$ 140–145 °C, n$_D^{20}$= 1.5390.

> **IR** (film): $\tilde{v}$ (cm^{-1}) = 3320 (br, OH), 1460, 1010, 980.
>
> **^{1}H NMR** (CDCl$_3$): δ (ppm) = 6.09 (s, 2 H, vinyl-H), 5.73 ("t", J = 7 Hz, 1 H, vinyl-H), 4.29 (d, J = 7 Hz, 2 H, HO–CH$_2$), 2.1–1.85 (m, 2 H, allyl CH$_2$), 1.91 (s, 1 H, OH), 1.86, 1.70 (s, 3 H, =C–CH$_3$), 1.6–1.2 (m, 4 H, CH$_2$–CH$_2$), 1.02 (s, 6 H, C(CH$_3$)$_2$).

Note: The progress of the reaction can be followed by TLC (SiO$_2$; CH$_2$Cl$_2$).

4.2.6.3   *   (β-Ionylidene-ethyl)triphenylphosphonium bromide [10]

220.3   342.2   545.5

β-Ionylidene-ethanol **4.2.6.2** (11.0 g, 50.0 mmol) and triphenylphosphonium hydrobromide (17.1 g, 50.0 mmol) in MeOH (200 mL) are stirred at room temperature for 48 h. During this time, the phosphonium salt goes into solution, which becomes yellow.

The solvent is then evaporated *in vacuo*, and the yellow crystalline residue is dissolved in the minimum volume of acetone. Addition of Et$_2$O and scratching with a glass rod cause the C$_{15}$-salt to crystallize as pale-yellow prisms; 18.2–18.8 g (67–69%), mp 151–153 °C.

> **IR** (film): $\tilde{v}$ (cm^{-1}) = 1435, 1110, 745, 720, 685.
>
> **^{1}H NMR** (CDCl$_3$): δ (ppm) = 7.80 (m, 15 H, phenyl-H), 6.00 (s, 2 H, vinyl-H), 5.5–5.1 (m, 1 H, vinyl-H), 4.75 (dd, J_{HP} = 15 Hz, J = 8 Hz, 2 H, P–CH$_2$), 2.2–1.8 (m, 2 H, allyl CH$_2$), 1.63, 1.47 (s, 3 H, =C–CH$_3$), 1.6–1.1 (m, 4 H, CH$_2$–CH$_2$), 0.97 (s, 6 H, C(CH$_3$)$_2$).

4.2.6.4   **   all-*trans*-Vitamin A acetate [10, 11]

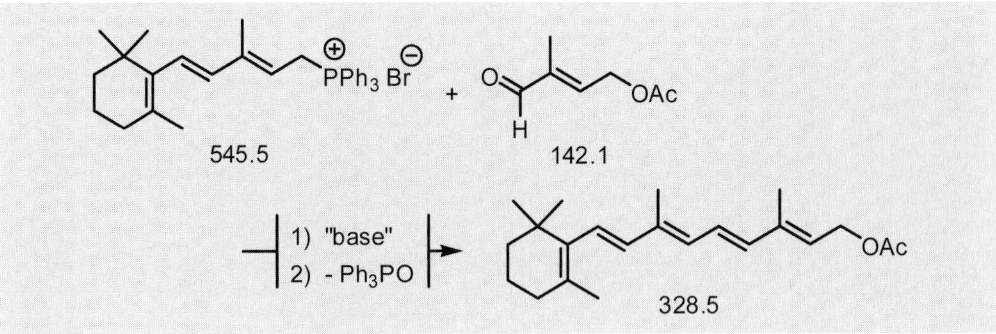

Method 1:

The C_{15}-salt **4.2.6.3** (12.5 g, 23.0 mmol) is dissolved in anhydrous DMF (50 mL) under a nitrogen atmosphere and the solution is cooled to 0 °C. The C_5-aldehyde **1.1.1** (4.00 g, 28.0 mmol) is added with stirring, and then 1,2-butene oxide (4.18 g, 58.0 mmol) is added dropwise. The mixture is stirred for 16 h at room temperature and for 4 h at 60 °C.

Petroleum ether (100 mL, 40–60 °C fraction) is then added and the solution is poured into ice-cold 20% H_2SO_4 (150 mL). The organic phase is separated and the aqueous phase is extracted with petroleum ether (2 × 100 mL). The combined organic phases are dried (Na_2SO_4) and concentrated *in vacuo*. The crude retinol acetate is obtained as a yellow oil, 5.52 g (73%).

The UV spectrum [**UV** (EtOH): λ_{max} (log ε) = 327 nm (4.54)] indicates a 68:32 mixture of the **11(*E*)/11(*Z*)** isomers. The UV spectrum of the pure **11(*Z*)**-isomer shows λ_{max} (log ε) = 327 nm (4.70).

Isomerization:

The crude product (5.00 g) is dissolved in *n*-pentane (10 mL), iodine (2.5 mg) is added, and the solution is left for 2 h in the dark at room temperature.

The solution is then diluted with *n*-pentane (100 mL), washed sequentially with dilute $Na_2S_2O_3$ solution and H_2O, dried (Na_2SO_4), and concentrated *in vacuo*.

The residue contains only all-*trans*-retinol acetate [**UV** (EtOH): λ_{max} (log ε) = 327 nm (4.67)], 0.98 g (93%). The product can be crystallized from *n*-hexane (at –20 to –30 °C) or MeOH/EtOAc (2:1) (at –20 °C) to give yellow prisms, mp 58–59 °C.

Method 2:

A solution of the C_5-aldehyde **1.1.1** (0.52 g, 3.60 mmol) in CH_2Cl_2 (200 mL) is layered with aqueous NaOH (2 M, 200 mL). A solution of the C_{15}-salt **4.2.6.3** (1.96 g, 3.60 mmol) in CH_2Cl_2 (200 mL) is added dropwise with vigorous stirring. The organic phase acquires a red color; the mixture is stirred for 30 min at room temperature.

The organic phase is separated and washed with H_2O until the aqueous washings are neutral. The CH_2Cl_2 phase is dried (Na_2SO_4) and concentrated *in vacuo*. The crude oily product [0.82 g (69%)] is crystallized by triturating with *n*-hexane at –20 to –30 °C to give yellow prisms, mp 57–59 °C. According to the UV spectrum [**UV** (EtOH): λ_{max} (log ε) = 327 nm (4.67)], the product consists of 94% all-*trans*-retinol acetate.

UV (EtOH): λ_{max} (log ϵ) = 327 nm (4.69); 98% all-*trans*-retinol acetate.

IR (KBr): $\tilde{\nu}$ (cm^{-1}) = 1730 (C=O), 1220, 1020, 980, 950.

^{1}H NMR (CDCl$_3$): δ (ppm) = 6.65 (dd, J = 15/11 Hz, 1 H, 11-H), 6.27 (d, J = 15 Hz, 1 H, 12-H), 6.12 (d, J = 16 Hz, 2 H, 7-H/8-H), 6.09 (d, J = 11 Hz, 1 H, 10-H), 5.61 ("t", J = 7 Hz, 1 H, vinyl-H), 4.70 (d, J = 7 Hz, 2 H, allyl–CH$_2$), 2.1–1.8 (m, 2 H, allyl–CH$_2$), 1.95, 1.89, 1.70 (s, 3 H, 9-CH$_3$/13-CH$_3$/5-CH$_3$), 1.6–1.1 (4 H, CH$_2$–CH$_2$), 1.03 (s, 6 H, C(CH$_3$)$_2$).

[1]   Ullmann's *Encyclopedia of Industrial Chemistry*, 6th ed., Vol. A 38, p. 119, Wiley-VCH, Weinheim, **2003**.

[2]   A. Kleemann, J. Engel, *Pharmaceutical Substances*, 3rd ed., p. 1669, Thieme Verlag, Stuttgart, **1999**.

[3]   a) Sumitomo Chem., US Pat. 2 951 853, **1960** (M. Masui); b) M. Masui, *J. Vitaminol.* **1958**, *4*, 178.

[4]   a) W. Reif, H. Grassner, *Chem. Ing. Tech.* **1973**, *45*, 646; b) H. Pommer, A. Nürrenbach, *Pure. Appl. Chem.* **1975**, *43*, 527; c) H. Pommer, *Angew. Chem.* **1977**, *89*, 437; *Angew. Chem. Int. Ed. Engl.* **1977**, *16*, 423.

[5]   Rhône-Poulenc, DE-OS 2 202 689, **1972** (M. Julia).

[6]   a) O. Isler, W. Huber, A. Ronco, M. Kofler, *Helv. Chim. Acta* **1947**, *30*, 1911, 1922; b) O. Isler, A. Ronco, W. Guex, N. C. Hindley, W. Huber, K. Dialer, M. Kofler, *Helv. Chim. Acta* **1949**, *32*, 489; c) U. Schwieter, G. Saucy, M. Montavon, C. v. Planta, R. Rüegg, O. Isler, *Helv. Chim. Acta* **1962**, *45*, 517, 528, 541, 548.

[7]   See ref. [1]; compare: a) E. Negishi, T. Takahashi, S. Baba, D. E. Van Horn, N. Okukado, *J. Am. Chem. Soc.* **1987**, *109*, 2393; b) N. Miyaura, K. Yamada, H. Suginome, A. Suzuki, *J. Am. Chem. Soc.* **1985**, *107*, 972.

[8]   Y. Ishikawa, *Bull. Soc. Chem. Jpn.* **1964**, *37*, 207.

[9]   A. G. Andrewes, S. Liaaen-Jensen, *Acta Chem. Scand.* **1973**, *27*, 1401.

[10] H. Pfander, A. Lachenmeier, M. Hadorn, *Helv. Chim. Acta* **1980**, *63*, 1377.

[11] a) H. Pommer, A. Nürrenbach, private communication, **1981**; b) J. Buddrus, *Chem. Ber.* **1974**, *100*, 2050.

4.3   Carbohydrates

Introduction

Carbohydrates are the most widely distributed natural products, representing about two-thirds of the annually renewable biomass [1]. The majority exists in the form of simple or complex oligo- and polysaccharides in the supporting tissue of plants, microbial cell walls, mammalian membranes, and in the casings of insects and the shells of crabs. New findings in recent decades have revealed an influence of carbohydrates in many biological processes, such as cell recognition [2], signal transduction [3], oncogenesis [4], bacterial infection, and probably even in Alzheimer's disease [5]. In addition, carbohydrates are valuable substrates in organic synthesis.

Monosaccharides, the building blocks of oligo- and polysaccharides, as well as glycosides, are polyhydroxy aldehydes (aldoses) or polyhydroxy ketones (ketoses) of different length and with defined stereochemistry at the stereogenic centers. The diversity is further increased by replacing hydroxyl groups by amino functionalities (amino sugars), hydrogen (deoxy sugars), or other groups. Aldopentoses such as 2-deoxyribose and ketohexoses such as fructose usually exist as tetrahydrofurans, while aldohexoses such as glucose contain a tetrahydropyran moiety, both having a hemiacetal functionality. Further important derivatives of monosaccharides are the cyclic uronic acids, such as glucuronic acid, and the acyclic ucaric acids, such as glucaric acid, and sugar alcohols such as glucitol.

2-Deoxy-β-D-ribose
2-Deoxy-β-D-ribofuranose
Aldopentoses

α/β-D-Glucose
α/β-D-Glucopyranose
Aldohexoses

D-Glucaric acid
Aldaric acids

α/β-D-Fructose
α/β-D-Fructofuranose
Ketohexoses

α/β-D-Glucuronic acid
α/β-D-Glucopyranuronic acid
Alduronic acids

Among the reactions of carbohydrates, two transformations are most commonly encountered, namely the introduction of protecting groups at the hydroxyl groups, and the stereoselective formation of acetals of the hemiacetal moiety (glycosylation or glycosidation).

It is important to bear in mind that the protecting groups have a great influence not only on the reactivity in the glycosylation step, but also on the stereochemistry (α or β) of the reaction. Many glycosylation methods have hitherto been developed, which allow stereoselective acetal formation at the anomeric centers of carbohydrates as donor molecules with either an alcohol, a phenol, or another sugar moiety with a free hydroxyl group as the acceptor molecule. The first glycosylation reactions were reported by Arthur Michael in 1879, followed by Emil Fischer (1893), as well as Wilhelm Koenigs and Eduard Knorr (1901) [6].

Fischer-type glycosylation, in which a monosaccharide is reacted with an alcohol in the presence of a catalytic amount of a strong acid, does not require any protecting group on the carbohydrate moiety. However, this method is strongly limited by the need for a large excess of the alcohol (often used as the solvent) and does not permit stereoselective glycosylation. The procedure is used nowadays on a multi-ton scale for the synthesis of alkyl glucosides of dodecanol as important biocompatible tensides.

The most important versatile stereoselective glycosylation methods of today are the Koenigs–Knorr reaction [6] using acyl-protected glycosyl halides, most often the bromides, and the Schmidt procedure employing a trichloroacetimidate moiety at the anomeric center of the glycosyl donor. However, thioglycosides, N-allylthiocarbamates [7a], phosphites, 1,2-anhydro sugars [7b, c], TMS-glycosides [7d], and several other sugar derivatives have also been used as donors [7e]. In addition, enzymes can also be employed for glycosylations.

L = leaving group,
e.g. Br, Cl, OC(NH)CCl₃,
SR, OC(S)NHAll etc.

"oxocarbenium"-ion

In the Koenigs–Knorr procedure, e.g., the 2,3,4,6-tetra-O-acetyl-α-D-glucopyranosyl bromide (**1**) is reacted with a silver or mercury salt as an activator, which abstracts the halide to give an oxocarbenium cation **2**. This is attacked by the neighboring acetyl group to form a new carbocation **3**, which then reacts with an alcohol from the β-side in a 1,2-*trans* mode to give the β-glucoside **4**. The corresponding mannose derivative would therefore give the α-mannoside. As side products, orthoesters such as **5** may be formed by attack of the alcohol at the cationic center of the 1,3-dioxolane ring in **3**. Orthoesters themselves can also be applied as donors in glycosylations.

For stereoselective glycosylation using the Koenigs–Knorr procedure, an acyl protecting group must be in place at C-2. However, the reactivity of the compounds is decreased compared to alkylated or benzylated sugars. A comparative study, e.g. of thiogalactosides, showed an increasing reactivity with different substituents at C-2 in the order $-N_3 < -OCOCH_2Cl < -NPhth < -OBz < -OAc < -OBn$.

In glycosylations, a peracetylated donor proved to be 1189 times less reactive than the corresponding perbenzylated compound, but due to the lack of a neighboring group effect, the transformation of the latter compound results in the formation of a mixture of α- and β-galactosides. Comparing different sugar types, a reactivity order of fucose > galactose > glucose > mannose was found [8].

The Schmidt glycosylation using α- or β-glucosyl trichloroacetimidates **6** bearing an acetoxy group at C-2 [9] follows a similar route as described for the Koenigs–Knorr reaction. After activation with a catalytic amount of the Lewis acid $BF_3 \cdot Et_2O$ or TMSOTf, an intermediate **7** is formed in the case of $BF_3 \cdot Et_2O$, which usually reacts to give the cation **3**, which then undergoes an attack of the alcohol from the β-side. If one uses α- or β-glucosyl trichloroacetimidates with benzyl protecting groups, the reaction proceeds in a different way. Under mild Lewis acid catalysis, e.g. using $BF_3 \cdot Et_2O$, the β-imidate leads to the α-glucoside, whereas the α-imidate affords the β-glucoside in a type of S_N2 reaction. In contrast, with strong Lewis acids such as TMSOTf, the α-glucoside is usually predominately obtained. Similar results are found for other monosaccharides; an exception is D-mannose with an axial positioned hydroxyl group at C-2, from which the α-mannosides can always be obtained with very good selectivity but the selective formation of β-mannosides is difficult.

A special feature in glycosylation chemistry is the nitrile effect. Thus, when a glycosylation is performed in the presence of a nitrile such as acetonitrile using, e.g., gluco-trichloroacetimidates with non-participating protecting groups at C-2, β-selectivity is observed. It is proposed that in these reactions an α-glucopyranosyl nitrilium ion is formed, which is attacked by the alcohol with inversion of configuration.

The great advantage of the glycosyl trichloroacetimidates is their higher stability compared to the halides and the possibility of preparing the α- and β-glycosyl trichloroacetimidates in a largely selective manner by judicious choice of the base used and the reaction time (see **4.3.1.4**).

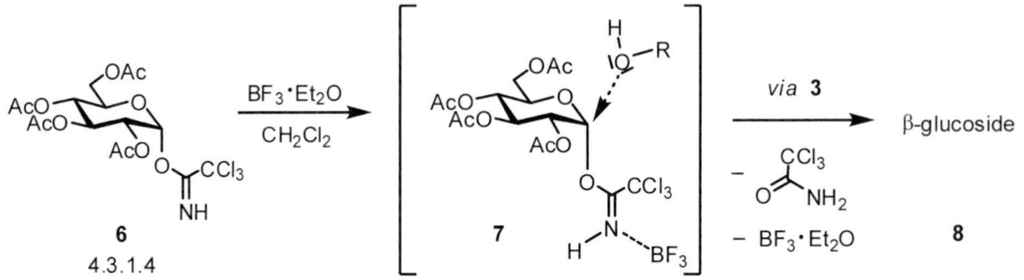

It should be noted that in carbohydrate chemistry there are no universal tools for all types of glycosylation problems. To find the right procedure for a given synthetic problem is not an easy task and one has to apply different conditions. The German carbohydrate chemist Hans Paulsen wrote in his excellent review in *Angewandte Chemie* in 1982 about the state-of-the-art of chemical oligosaccharide synthesis [10]:

"Although we have now learned to synthesize oligosaccharides, it should be emphasized that each oligosaccharide synthesis remains an independent problem, whose resolution requires considerable systematic research and a good deal of know-how. There are no universal reaction conditions for oligosaccharides".

Despite significant advances in the last 20 years, Paulsen's statement has largely remained true.

In the following, different procedures for glycosylations are described, as well as the synthesis of the corresponding glycosyl donors.

4.3.1   Synthesis of glucosyl donors

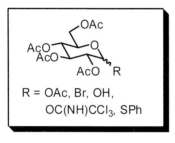

R = OAc, Br, OH,
   OC(NH)CCl₃, SPh

Topics: • Acetylation of free sugars
          • Displacement of the 1-*O*-acetyl group with thiols or bromide
          • Anomeric deprotection
          • Introduction of a trichloroacetimidate moiety

(a) Acetylation of *D*-glucose

Many methods are available for the synthesis of peracetylated carbohydrates. Most commonly, the sugar is treated with acetic anhydride in the presence of NaOAc [11]. One can also use pyridine instead of NaOAc or a mixture of acetic anhydride, acetyl chloride, and a small amount of dimethylaminopyridine (DMAP). Acidic conditions can also be employed, such as the reaction with acetic anhydride in the presence of perchloric acid [12]. Depending on the method and the reaction temperature, one can partly control the ratio of the resulting peracetylated α/β-pyranose and α/β-furanose mixture.

4.3.1.1   *   1,2,3,4,6-Penta-*O*-acetyl-α/β-*D*-glucopyranose

Ac₂O, NaOAc, rt, 20 h

180.2   390.4

A suspension of *D*-glucose (5.00 g, 27.8 mmol) and freshly fused, powdered NaOAc (0.46 g, 5.55 mmol) in acetic anhydride (20 mL, 21.7 g, 213 mmol) is vigorously stirred at room temperature overnight. EtOH (4.5 mL) is then added and stirring is continued for 30 min.

The solution is poured into iced water (~50 mL), diluted with *n*-pentane/CH₂Cl₂ (2:1; 150 mL), and the phases are separated. The organic layer is washed with iced water (50 mL), saturated NaHCO₃ solution (4 × 50 mL), and ice-cold brine (50 mL). The resulting solution is filtered through cotton and the solvents are evaporated under reduced pressure. The residue is co-distilled with toluene/EtOH (5:1; 3 × 25 mL) and the distillate is concentrated under reduced pressure to yield the desired compound as a colorless solid. The pure β-compound can be separated by crystallization from EtOH; 10.7 g (99%), α-compound: $R_f = 0.58$ (EtOAc/*n*-pentane, 1:1); β-compound: mp 130–131 °C, $[\alpha]_D^{20} = +4.2$ ($c = 1.0$, CHCl₃), $R_f = 0.58$ (EtOAc/*n*-pentane, 1:1).

α-compound:

IR (KBr) (α/β): $\tilde{v}$ (cm^{-1}) = 2968, 1746, 1370, 1224, 1038, 912.

^{1}H NMR (300 MHz, CDCl$_3$) (from α/β mixture): δ (ppm) = 6.33 (d, J = 3.7 Hz, 1 H, H-1), 5.46 (dd, J = 9.9, 9.8 Hz, 1 H, H-3), 5.11, 5.12 (m, 2 H, H-2, H-4), 4.27 (dd, J = 11.0, 4.2 Hz, 1 H, H-6$_b$), 4.13, 4.10 (m, 2 H, H-5, H-6$_a$), 2.17, 2.08, 2.03, 2.01, 2.00 (5 × s, 15 H, 5 × COCH$_3$).

^{13}C NMR (75 MHz, CDCl$_3$): δ (ppm) = 170.5, 170.2, 169.3, 169.6, 168.7 (5 × COCH$_3$), 89.02 (C-1), 69.84, 69.14, 67.84 (C-2, C-3, C-4, C-5), 61.40 (C-6), 20.83, 20.77, 20.66, 20.52, 20.40 (5 × COCH$_3$).

MS (ESI): m/z (%) = 802.6 (10) [2M + Na]$^+$, 413.0 (100) [M + Na]$^+$.

β-compound:

^{1}H NMR (300 MHz, CDCl$_3$): δ (ppm) = 5.72 (d, J = 8.2 Hz, 1 H, H-1), 5.48, 5.26 (2 × m$_c$, 2 H, H-3, H-4), 5.12 (dd, J = 9.2, 8.2 Hz, 1 H, H-2), 4.30 (dd, J = 12.5, 4.5 Hz, 1 H, H-6$_b$), 4.11 (dd, J = 12.5, 2.2 Hz, 1 H, H-6$_a$), 3.85 (ddd, J = 9.9, 4.4, 2.2 Hz, 1 H, H-5), 2.12, 2.09, 2.04, 2.02 (5 × s, 15 H, COCH$_3$).

^{13}C NMR (75 MHz, CDCl$_3$): δ (ppm) = 170.5, 170.0, 169.3, 169.2, 168.9 (5 × COCH$_3$), 91.65 (C-1), 72.74 (C-5), 72.67 (C-3), 70.17 (C-2), 67.69 (C-4), 61.40 (C-6), 20.77, 20.66, 20.62, 20.52 (5 × COCH$_3$).

MS (ESI): m/z (%) = 802.6 (10) [2M + Na]$^+$, 413.0 (100) [M + Na]$^+$.

(b) Bromination of D-glucose

The normal protocol for the generation of glycosyl bromides is the reaction of a 1-O-acyl-glycoside with 30% HBr solution in glacial acetic acid and CH$_2$Cl$_2$ as solvent. Other procedures start from the 1-hydroxy derivatives and use oxalyl bromide or TMSBr in CH$_2$Cl$_2$. The method presented here is a two-step, one-pot procedure, in which the free sugar is first peracetylated and then brominated at C-1 with HBr generated *in situ* by reaction of PBr$_3$ with H$_2$O. In general, only the α-bromides are obtained due to the well-known anomeric effect [13].

$$2\,P\ +\ 3\,Br_2\ \longrightarrow\ 2\,PBr_3\ \xrightarrow{\ H_2O\ (6\ eq),\ 0°C\ }\ 2\,P(OH)_3\ +\ 6\,HBr$$

4.3.1.2 ** 2,3,4,6-Tetra-O-acetyl-α-D-glucopyranosyl bromide [11]

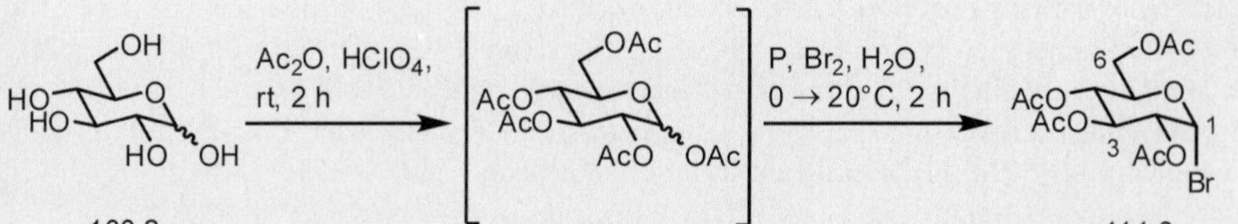

D-Glucose (10.0 g, 55.4 mmol) is added in small portions over 30 min to a stirred mixture of acetic anhydride (40 mL, 43.5 g, 426 mmol) and perchloric acid (0.24 mL), at such a rate that the temperature does not exceed 40 °C. After 2 h, amorphous red phosphorus (3.00 g, 96.9 mmol) is added and the mixture is cooled to 0 °C. At this temperature, bromine (18.0 g, 5.80 mL, 113 mmol) is slowly dropped into the mixture, immediately followed by H_2O (3.62 mL, 200 mmol) over 15 min. A local rise of the internal temperature should be avoided; stirring is continued at room temperature for 2.5 h.

The solution is then diluted with CH_2Cl_2/*n*-pentane (1:2; 100 mL) and poured into iced water (150 mL). The mixture is extracted with CH_2Cl_2/*n*-pentane (1:2; 2 × 50 mL). The combined organic layers are filtered through cotton (removal of phosphorus), washed with saturated ice-cold $NaHCO_3$ solution (2 × 80 mL) (vigorous evolution of carbon dioxide), iced water (80 mL), and ice-cold brine (80 mL), and filtered through cotton once more. Activated charcoal (500 mg) and $NaHCO_3$ (25 mg) are added and the mixture is stirred at room temperature for 1 h. After filtration through Celite® and evaporation of the solvents, a yellow syrup is obtained, which is recrystallized from absolute Et_2O to afford the bromide as white needles; 18.8 g (83%), mp 88–89 °C, $[\alpha]_D^{20} = +195$ (c = 3.0, $CHCl_3$), R_f = 0.69 (EtOAc/*n*-pentane, 1:1).

IR (KBr): $\tilde{v}$ (cm^{-1}) = 2964, 1745, 1384, 1229, 1112, 1041, 923, 555, 486.

^{1}H NMR (300 MHz, $CDCl_3$): δ (ppm) = 6.62 (d, J = 4.0 Hz, 1 H, H-1), 5.56 (t, J = 9.8 Hz, 1 H, H-4), 5.17 (dd, J = 10.1, 9.6 Hz, 1 H, H-3), 4.84 (dd, J = 10.0, 4.0 Hz, 1 H, H-2), 4.34 (dd, J = 13.1, 3.2 Hz, 1 H, H-6$_b$), 4.30 (ddd, J = 9.8, 3.9, 1.4 Hz, 1 H, H-5), 4.13 (m, 1 H, H-6$_a$), 2.11, 2.10, 2.06, 2.04 (4 × s, 12 H, 4 × COCH_3).

^{13}C NMR (75 MHz, $CDCl_3$): δ (ppm) = 170.5, 169.8, 169.7, 169.4 (4 × $\underline{C}$OCH$_3$), 86.53 (C-1), 72.10 (C-2), 70.56 (C-3), 70.12 (C-5), 67.13 (C-4), 60.91 (C-6), 20.63, 20.61, 20.58, 20.51 (4 × COC$\underline{H}_3$).

MS (ESI): *m/z* (%) = 434.9 (57) $[M + Na]^+$.

(c) Anomeric deprotection

A common problem in carbohydrate chemistry is the selective deprotection of the anomeric center in the presence of similar protecting groups. In general, this is possible due to the higher reactivity of the anomeric center, which corresponds to a derivative of a hemiacetal. In the case of an acyl protecting group, the standard reagent for deprotection is hydrazinium acetate, whereas for the cleavage of a 1-*O*-methyl group strong acids such as H_2SO_4 can be used.

Hydrazine attacks the carbonyl group of the 1-*O*-acyl-protected carbohydrate **9** to give the corresponding carbohydrate **10** with a free anomeric center.

4.3.1.3   *   2,3,4,6-Tetra-*O*-acetyl-α/β-*D*-glucopyranose [14]

A solution of the pentaacetate **4.3.1.1** (5.00 g, 12.8 mmol) and hydrazinium acetate (1.47 g, 16.0 mmol) (Caution: carcinogenic!) in DMF (55 mL) is stirred for 60 min at room temperature.

The reaction mixture is then diluted with EtOAc (340 mL), and washed with iced water (120 mL), cold brine (120 mL), cold saturated $NaHCO_3$ solution (2 × 120 mL), and again with iced water (120 mL). The solution is filtered through cotton, which is washed with EtOAc, and the combined filtrate and washings are concentrated under reduced pressure to afford the glucopyranose as a white foam, which solidifies under high vacuum; 3.92 g (88%), $R_f = 0.19$ (EtOAc/*n*-pentane, 1:2).

---

^{1}H NMR (300 MHz, CDCl₃) (from α/β mixture, 2:3): δ (ppm) = 5.54 (dd, *J* = 10.0, 9.7 Hz, 0.4 H, H-3α), 5.47 (d, *J* = 3.4 Hz, 0.4 H, H-1α), 5.26 (t, *J* = 9.5 Hz, 0.6 H, H-3β), 5.10, 5.09 (2 × t, *J* = 9.7 Hz, 1 H, H-4), 4.92 (dd, *J* = 9.9, 3.5 Hz, 0.4 H, H-2α), 4.91 (dd, *J* = 9.6, 8.0 Hz, 0.6 H, H-2β), 4.76 (d, *J* = 8.0 Hz, 0.6 H, H-1β), 4.27, 4.16 (m, 2.4 H, H-5α, H-6a, H-6b), 4.07 (s${}_{br}$, 0.6 H, OHβ), 3.80 (s${}_{br}$, 0.4 H, OHα), 3.77 (ddd, *J* = 10.1, 4.8, 2.4 Hz, 0.6 H, H-5β), 2.11, 2.10, 2.05 (β), 2.04 (α), 2.03 (4 × s, 12 H, COCH₃).

α-compound:

^{13}C NMR (75.5 MHz, CDCl₃): δ (ppm) = 170.8, 170.2, 170.1, 169.6 (4 × COCH₃), 90.0 (C-1), 71.0 (C-2), 69.8 (C-3), 68.4 (C-4), 67.1 (C-5), 61.9 (C-6), 20.71, 20.64, 20.58, 20.54 (4 × COCH₃).

β-compound:

^{13}C NMR (75.5 MHz, CDCl₃): δ (ppm) = 170.8, 170.2, 169.5 (4 × COCH₃), 95.4 (C-1), 73.1 (C-2), 72.2 (C-3), 71.9 (C-5), 68.3 (C-4), 61.9 (C-6), 20.70, 20.67, 20.54 (4 × COCH₃).

MS (ESI): *m/z* (%) = 1088.4 (51) [3*M* + 2Na]⁺, 718.7 (100) [2*M* + Na]⁺, 371.1 (70) [*M* + Na]⁺.

(d) Synthesis of glycosyl trichloroacetimidates

As previously described, glycosyl trichloroacetimidates serve as glycosyl donors in glycosylation reactions. Their syntheses start from the corresponding 1-hydroxy derivatives **12**, which are deprotonated by a base before treatment with trichloroacetonitrile to give the desired product. The choice of base and the reaction time have a high influence on the preferred formation of the α- or β-anomer. Under kinetic control using weak bases, the β-anomer **15** is formed predominately, whereas a long reaction time and strong bases such as NaH lead to the α-anomer **6** as the thermodynamically more stable compound via re-anomerization of the less stable β-anomer **15** [8b].

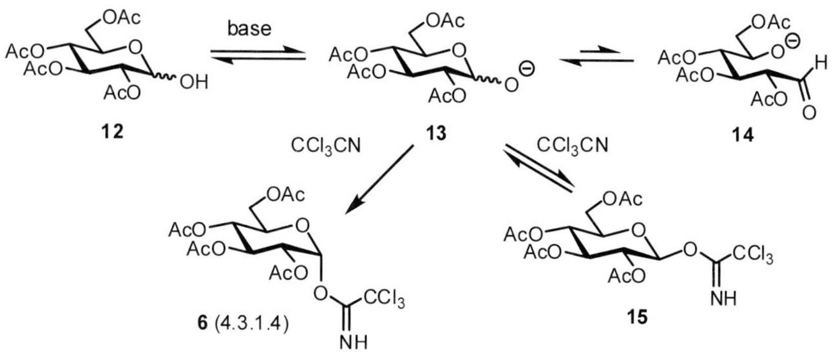

4.3.1.4 *** 2,3,4,6-Tetra-O-acetyl-α-D-glucopyranosyl trichloroacetimidate [15]

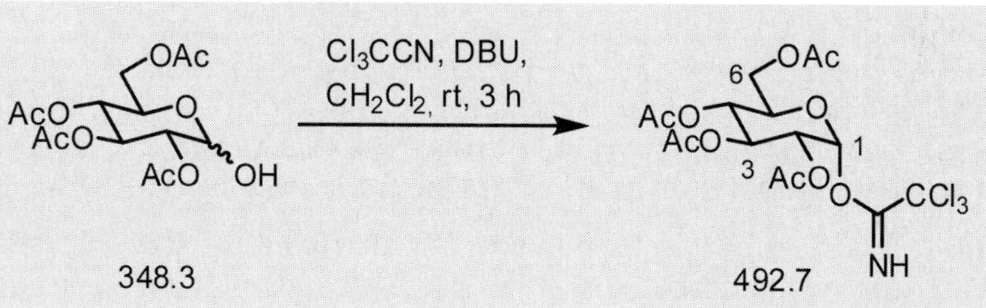

A mixture of *D*-glucopyranose tetraacetate **4.3.1.3** (1.74 g, 5.00 mmol), trichloroacetonitrile (14.4 g, 10.2 mL, 100 mmol), and 1,8-diazabicyclo[5.4.0]undec-7-ene (DBU) (0.38 g, 0.38 mL, 2.50 mmol) in CH$_2$Cl$_2$ (30 mL) is stirred at room temperature for 3 h.

It is then concentrated under reduced pressure at < 30 °C. The residue is purified by column filtration through silica gel with *n*-pentane/EtOAc (3:1; containing 2.5% of NEt$_3$) to afford the trichloroacetimidate as a colorless (or slightly yellow) syrup. For further purification, the product can be crystallized from Et$_2$O/*n*-pentane to give a colorless solid; 1.70 g (69%), mp 153–154 °C, $[\alpha]_D^{20}$ = +8.1 (*c* = 1.0, CHCl$_3$), R_f = 0.73 (EtOAc/*n*-pentane, 1:1).

---

^{1}H NMR (300 MHz, CDCl$_3$): δ (ppm) = 8.70 (s$_{br}$, 1 H, NH), 6.57 (d, *J* = 3.7 Hz, 1 H, H-1), 5.57 (t, *J* = 9.8 Hz, 1 H, H-4), 5.19 (dd, *J* = 10.0, 9.8 Hz, 1 H, H-3), 5.14 (dd, *J* = 10.2, 3.7 Hz, 1 H, H-2), 4.28 (dd, *J* = 12.1, 4.0 Hz, 1 H, H-6$_b$), 4.22 (ddd, *J* = 10.2, 4.1, 1.8 Hz, 1 H, H-5), 4.13 (dd, *J* = 12.1, 2.0 Hz, 1 H, H-6$_a$), 2.08, 2.06, 2.04, 2.02 (4 × s, 12 H, 4 × COCH$_3$).

^{13}C NMR (75.5 MHz, CDCl$_3$): δ (ppm) = 170.6, 170.0, 169.9, 169.5 (4 × $\underline{C}$OCH$_3$), 160.8 (CCl$_3$), 92.92 (C-1), 69.92, 69.71, 69.63, 67.68 (C-2, C-3, C-4, C-5), 61.29 (C-6), 20.63, 20.55, 20.40 (4 × COC$\underline{H}_3$).

(e) Synthesis of thioglycosides

The most widely used approach for the synthesis of thioglycosides is the condensation of a 1-O-acyl-protected sugar with a thiol under Lewis acid catalysis, usually with $BF_3 \cdot Et_2O$, $SnCl_4$, or $ZnCl_2$. After coordination of the Lewis acid to the carbonyl oxygen of the acetyl group at C-1, a displacement with the added thiol, such as thiophenol, takes place.

4.3.1.5 ** **Phenyl 2,3,4,6-tetra-O-acetyl-1-thio-β-D-glucopyranoside** [16]

Thiophenol (6.72 g, 6.26 mL, 61.0 mmol) and tin(IV) chloride (9.25 g, 4.15 mL, 35.5 mmol) are slowly added to a stirred solution of D-glucose pentaacetate **4.3.1.1** (19.8 g, 50.8 mmol) in CH_2Cl_2 (100 mL) at room temperature. The reaction mixture is heated at reflux for 16 h.

It is then cooled to room temperature, and quenched by the addition of saturated $NaHCO_3$ solution (250 mL) (strong odor and formation of carbon dioxide). CH_2Cl_2 (75 mL) and n-pentane (350 mL) are then added. The aqueous layer is separated and extracted with n-pentane/CH_2Cl_2 (2:1; 2 × 150 mL). The combined organic layers are washed with brine and the resulting solution is filtered through cotton, which is subsequently washed with CH_2Cl_2. The solvents are evaporated from the combined filtrate and washings under reduced pressure to provide a white solid. The crude material is recrystallized from EtOAc/n-pentane to afford the thioglucopyranoside as white needles; 16.1 g (72%), mp 122–123 °C, $[\alpha]_D^{20}$ = +17.1 (c = 1.0, $CHCl_3$), R_f = 0.57 (EtOAc/n-pentane, 1:1).

UV (CH_3CN): λ_{max} (nm) (lg ε) = 246 (0.193).

IR (KBr): $\tilde{\nu}$ (cm^{-1}) = 2949, 1746, 1226, 1089, 1043, 913, 745.

^{1}H NMR (300 MHz, $CDCl_3$): δ (ppm) = 7.18–7.54 (m, 5 H, Ph-H), 5.23 (t, J = 9.3 Hz, 1 H, H-3), 5.05 (m, 1 H, H-4), 4.98 (dd, J = 10.0, 9.4 Hz, 1 H, H-2), 4.72 (d, J = 10.1 Hz, 1 H, H-1), 4.24 (dd, J = 12.5, 5.1 Hz, 1 H, H-6$_b$), 4.18 (dd, J = 12.4, 2.9 Hz, 1 H, H-6$_a$), 3.74 (ddd, J = 10.1, 4.8, 2.8 Hz, 1 H, H-5), 2.10, 2.09, 2.03, 2.00 (4 × s, 12 H, 4 × $COCH_3$).

^{13}C NMR (75.5 MHz, $CDCl_3$): δ (ppm) = 170.5, 170.1, 169.4, 169.2 (4 × $\underline{C}OCH_3$), 133.1 (C-8, C-12), 131.6 (C-7), 128.9 (C-9, C-11), 128.4 (C-10), 85.68 (C-1), 75.74 (C-5), 73.90 (C-3), 69.86 (C-2), 68.13 (C-4), 62.09 (C-6), 20.73, 20.57 (4 × $CO\underline{C}H_3$).

MS (ESI): m/z (%) = 902.7 (50) $[2M + Na]^+$, 463.1 (100) $[M + Na]^+$.

4.3.2   Glycosylations of glucosyl donors with cyclopentanol

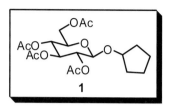

1

Topics: • Glycosylation with trichloroacetimidates
• Koenigs–Knorr glycosylation
• Glycosylation with thioglycosides

General

In this section, β-glycoside formations employing cyclopentanol as the alcohol component and several types of *O*-acetyl-protected glycosyl donors, such as the trichloroacetimidate **2**, the bromide **3**, and the thioglycoside **4**, are described.

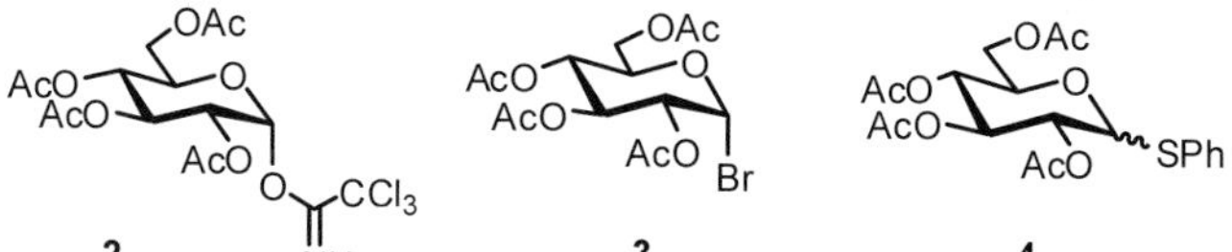

2   **3**   **4**

While glycosylations with **2** and **3** have already been described in detail in the introductory section, for the condensation of the 1-thio-β-*D*-glucopyranoside **4** the following aspects are relevant.

Activation of thioglycosides is usually achieved by the addition of halonium or formal methyl cations. Besides the use of iodonium-*sym*-dicollidine perchlorate [18a] (IDCP), MeI, and dimethyl-(methylthio)sulfonium triflate (DMTST) [17b], a mixture of *N*-iodosuccinimide **5** (NIS) [17c] and trifluoromethanesulfonic acid (TfOH) is most commonly applied. The acid serves to generate an iodonium cation **6**, which attacks the sulfur atom in the thioglycoside to give oxocarbenium ion **7** (see above) and iodo-thiophenol (I–S–Ph). The subsequent attack of the nucleophilic alcohol corresponds to the Koenigs–Knorr reaction and the stereochemistry is once again influenced by the protecting group at C-2.

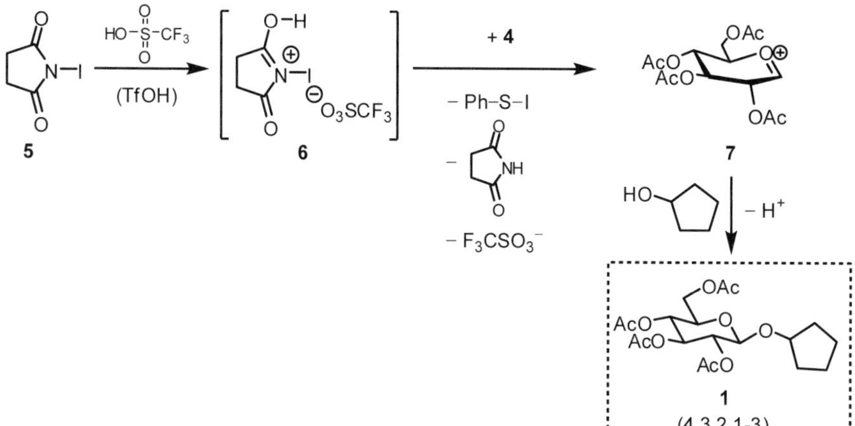

5   **6**   + **4**   **7**

− Ph–S–I

− F$_3$CSO$_3^-$

− H$^+$

1

(4.3.2.1-3)

4.3.2.1   ***   Cyclopentyl-2,3,4,6-tetraacetyl-β-D-glucopyranoside I [17]

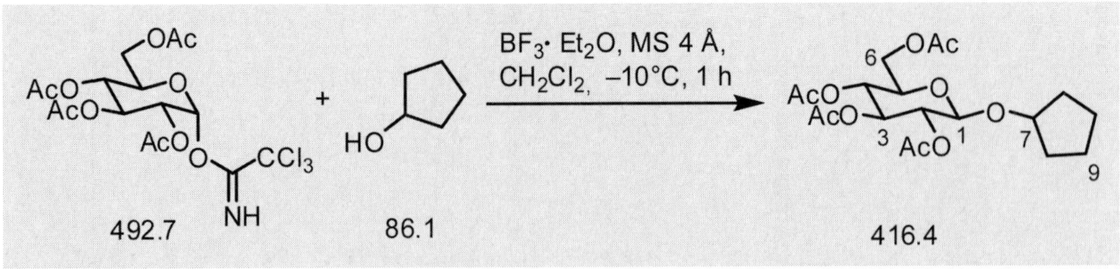

A mixture of the trichloroacetimidate **4.3.1.4** (160 mg, 0.32 mmol), cyclopentanol (32 μL, 30.0 mg, 0.36 mmol), and powdered 4 Å molecular sieves (200 mg) in CH_2Cl_2 (1.5 mL) is stirred at room temperature for 30 min. The solution is then cooled to –10 °C, boron trifluoride etherate (10.3 μL, 11.5 mg, 0.08 mmol) in CH_2Cl_2 (0.5 mL) is added dropwise, and stirring is continued for 1 h.

Thereafter, the reaction is terminated by adding NEt_3 (50 μL). The solvent is removed under reduced pressure and the residue is purified by column chromatography on silica gel (7 g) with n-pentane/EtOAc (3:1) as eluent to provide the β-glucoside as a white solid; 98 mg (73%), mp 118–120 °C, $[\alpha]_D^{20}$ = –34.2 (c = 1.0, $CHCl_3$), R_f = 0.71 (EtOAc/n-pentane, 1:1).

^{1}H NMR (300 MHz, $CDCl_3$): δ (ppm) = 5.21 (t, J = 9.5 Hz, 1 H, H-3), 5.07 (t, J = 9.6 Hz, 1 H, H-4), 4.94 (dd, J = 9.6, 8.0 Hz, 1 H, H-2), 4.53 (d, J = 8.0 Hz, 1 H, H-1), 4.28 (m, 1 H, H-7), 4.27 (dd, J = 12.2, 4.9 Hz, 1 H, H-6$_b$), 4.13 (dd, J = 12.2, 2.5 Hz, 1 H, H-6$_a$), 3.69 (ddd, J = 9.9, 4.8, 2.5 Hz, 1 H, H-5), 2.09, 2.04, 2.03, 2.01 (4 × s, 12 H, 4 × COCH_3), 1.45–1.83 (m, 8 H, cyclopentyl-H).

^{13}C NMR (75.5 MHz, $CDCl_3$): δ (ppm) = 170.7, 170.3, 169.4, 169.2 (4 × $\underline{C}$OCH$_3$), 99.48 (C-1), 81.51 (C-7), 72.83 (C-3), 71.66 (C-5), 71.36 (C-2), 68.48 (C-4), 62.06 (C-6), 33.07, 32.06 (C-8, C-11), 23.31, 23.04 (C-9, C-10), 20.73, 20.61, 20.59 (4 × CO$\underline{C}$H$_3$).

MS (ESI): m/z (%) = 854.8 (39) $[2M + Na]^+$, 439.2 (100) $[M + Na]^+$.

4.3.2.2   **   Cyclopentyl-2,3,4,6-tetraacetyl-β-D-glucopyranoside II [17]

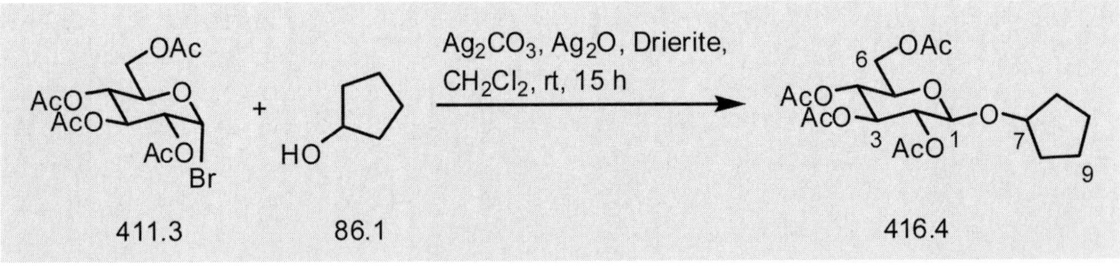

Under exclusion of light, silver(I) oxide (295 mg, 1.27 mmol) and silver(I) carbonate (68 mg, 0.25 mmol) are added to a stirred solution of the D-glucopyranosyl bromide **4.3.1.2** (133 mg, 0.32 mmol) and cyclopentanol (32 μL, 30.0 mg, 0.36 mmol) in CH_2Cl_2 (10 mL) containing Drierite (540 mg) and the mixture is stirred at room temperature for 15 h.

combination with a characteristic *m*-RNA nucleotide sequence. Amino acids with *D*-configuration are available by isomerization from all naturally occurring amino acids. Enzymatic modifications of proteins after translation also lead to non-proteinogenic amino acids. Well-known post-translational modifications are functionalizations with recognition elements such as carbohydrates or lipids. Phosphorylation of serine is used to regulate enzymes by switching enzymatic activity "on" or "off" or to influence receptor fitting. Methylation of arginine or lysine is similarly used to regulate biochemical processes. Post-translational functional group modifications also provide further amino acids, such as citrulline generated from arginine. Additional post-translational modifications at the protein level are the formation of disulfide bridges and the cutting and shortening of peptide chains (as known, e.g., from the insulin production pathway). Moreover, non-standard amino acids found in Nature are the result of metabolism or catabolism; examples are ornithine, homocysteine, dehydroalanine, the inhibitory neurotransmitter γ-aminobutyric acid, α-aminoisobutyric acid, and hydroxyproline.

Homocysteine   Dehydroalanine   γ-Aminobutyric acid   α-Aminoisobutyric acid   Hydroxyproline

Non-chiral β-amino acids such as β-alanine also exist in Nature. Therefore, extensive research is focused on the synthesis of a huge variety of β-amino acids derived by homologation of proteinogenic α-amino acids or β-amino acids containing artificial side chains [6]. The peptide world derived from β-amino acids (see **4.4.8**) is especially interesting since it is possible to mimic the secondary structures known from α-peptides in combination with enormous metabolic and conformational stability. Various helices as well as sheet and turn structures are obtained with even short sequences by design based on the substitution pattern and the configuration of side chains in the α and β positions. As an additional benefit in the β-peptide series, it is possible to design structural properties, such as the directionality or existence of the overall helical dipole.

The synthesis of artificial amino acids with cofactor side chains, functional groups or recognition elements is of pharmacological relevance (see **4.4.1**). In addition, modification of the peptide chain is of value to generate function, increase recognition or bioavailability, and to obtain metabolic stability. Of the numerous possibilities, only the replacement of the peptide bond by isomorphous vinyl fluorides, the retro-inverso-peptides (simultaneous inversion of strand orientation and configuration of all amino acids, thereby maintaining the spatial orientation of side chains), and the peptoides, with side chains linked to the amide nitrogen, should be mentioned [7].

Vinylfluoride

Retro-inverso-peptide

Peptoide

The following sections deal with the preparation of enantiomerically pure amino acids and the chemical synthesis of peptides on solid supports [8, 9].

[1] H.-D. Jakubke, *Peptide*, Spektrum Verlag, Heidelberg, **1996**.

[2] S. M. Hecht (Ed.), *Bioorganic Chemistry, Peptides and Proteins*, University Press, Oxford, **1998**.

[3] G. E. Schulz, R. H. Schirmer, *Principles of Protein Structure*, Springer, New York, **1979**.

[4] C. Branden, J. Tooze, *Introduction to Protein Structure*, Garland Publishing, New York, **1998**.

[5] a) D. M. Driscoll, P. R. Copeland, *Annu. Rev. Nutr.* **2003**, *23*, 17; b) J. A. Krzycki, *Curr. Opin. Microbiol.* **2005**, *8*, 706.

[6] D. Seebach, A. K. Beck, D. J. Bierbaum, *Chemistry Biodiversity* **2004**, *1*, 1111.

[7] J. Gante, *Angew. Chem.* **1994**, *106*, 1780; *Angew. Chem. Int. Ed. Engl.* **1994**, *33*, 1699.

[8] J. Jones, *Amino Acid and Peptide Synthesis*, University Press, Oxford, **1992**.

[9] M. Bodanzki, *Principles of Peptide Synthesis*, Springer, Heidelberg, **1993**.

4.4.1   *N*-Boc-*N*-methyl-(*S*)-alanyl nucleo amino acid

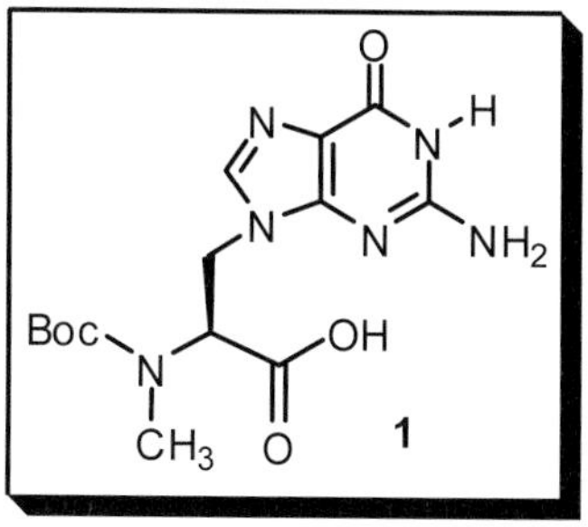

Topics:
- Mitsunobu lactonization
- *N*-Methylation of α-amino acids
- Serine lactone opening with nucleophiles
- Nucleophilic aromatic substitution
- *N*-Boc protection

(a)   General

N-Methyl-α-amino acids are common components of natural products such as cyclosporine, dolastatins, and didemnins, which show a wide range of biological effects [1]. Thus, antibiotic, anticancer, antiviral, and immunosuppressive activities have been reported for *N*-methylated peptides and proteins [2]. *N*-Methylation of amino acids, peptides or proteins has a major impact on pharmacological parameters such as lipophilicity, bioavailability, proteolytic stability, and conformational rigidity, since – *inter alia* – the potential for hydrogen bond formation is usually lost and the amide bond of *N*-methylated amino acids is more prone to *cis/trans* isomerization [3]. Both parameters strongly influence the overall backbone conformation of peptides and proteins containing *N*-methylated amino acids.

For a long time, the synthesis of *N*-methylated amino acids was quite severely restricted due to the harsh reaction conditions of the known synthetic procedures, which were likely to cause racemization [4]. Moreover, these procedures were limited to amino acids without any additional nucleophilic functional groups. However, quite recently, new general methods for the synthesis of *N*-methylated α-amino acids have been developed [5–9]. In the following, the hitherto established general strategies are presented:

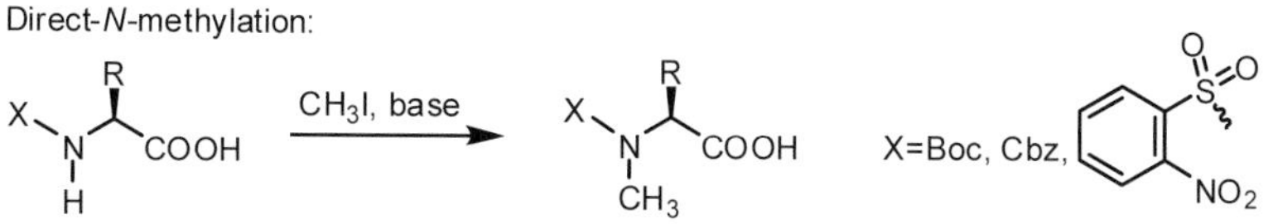

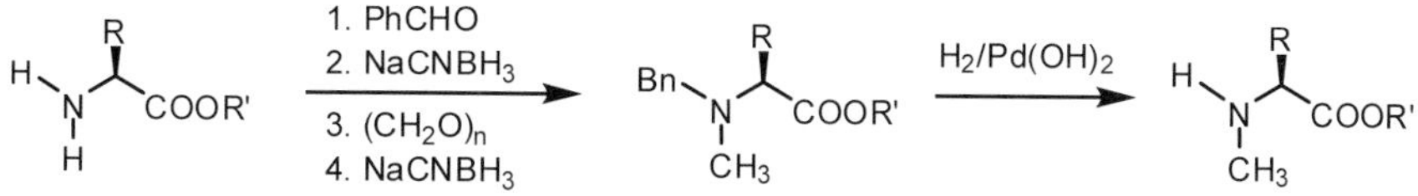

5-Oxazolidinone strategy:

Fmoc, R, N, H, COOH → (CH$_2$O)$_n$ CSA, toluene → Fmoc, R, N, O, O → TFA, Et$_3$SiH, CHCl$_3$ → Fmoc, R, N, CH$_3$, COOH

Serine lactone strategy:

Boc, N, CH$_3$, OH, COOH → Ph$_3$P, DEAD, CH$_3$CN → Boc, N, CH$_3$, O, O → nucleophile, DBU, DMSO → Boc, N, CH$_3$, nucleophile, COOH

Direct N-methylation can be performed by nucleophilic substitution of N-protected α-amino acids and α-amino esters with, e.g., CH$_3$I [4]; in addition, treatment with diazomethane is possible [5]. To avoid strongly basic conditions, the N-protection can be combined with N-activation using the o-nitrobenzenesulfonyl (o-NBS) group [10] to also allow the synthesis of N-methylamino acids with acidic or basic side chains that can be directly applied to Fmoc solid-phase peptide synthesis [6]. Basic N-methylated amino acids can be obtained without side-chain protection by successive reductive amination of the amino acids, first using benzaldehyde/cyanoborohydride and then paraformaldehyde/ cyanoborohydride [7]. A third strategy is based on conversion of Fmoc-protected amino acids into 5-oxazolidinones using paraformaldehyde under acidic conditions [8]; reductive opening with triethylsilane and TFA then provides N-methylated amino acids. Another method is based on serine lactone, which is readily available from Boc-protected N-methyl serine [9] under Mitsunobu conditions [11]. Starting from this substrate, various N-methylated amino acids can be obtained by introducing the respective side chain by nucleophilic opening of the lactone moiety. This method is especially valuable for the preparation of non-proteogenic amino acids such as the alanyl nucleo amino acids, which contain a nucleobase covalently linked at the β-position [12].

Some of the described methylation strategies can also be applied for the synthesis of N-methylated peptides or proteins directly on solid support [6, 13].

(b) Synthesis of 1

Starting from commercially available N-Boc-N-methyl serine (2), the serine lactone 3 can be obtained under Mitsunobu conditions in analogy to the Boc-serine lactone formation described by Vederas [11]. As also known from serine lactone, the N-methyl derivative 3 can be opened by various nucleophiles to yield amino acids with the nucleophile covalently linked at the β-position in the amino acid chain. The introduction of nucleobases is especially interesting due to their recognition potential [12]. However, purines are difficult nucleophiles because of competitive nucleophilicity at positions N-7 and N-9, low solubility and nucleophilicity, as well as aggregation behavior. In order to obtain the guaninyl nucleo amino acid 1, guanine itself cannot be used as the nucleophile. Instead, 6-chloro-2-aminopurine 4 can be employed, providing much better nucleophilicity and regioselectivity. The compound 5 thus obtained is then treated with TFA to provide 6 by nucleophilic aromatic substitution of the chloro substituent. However, under the strongly acidic reaction conditions used, the Boc protecting group is lost and needs to be re-established with di-tert-butyl dicarboxylate to yield the target nucleo amino acid 1.

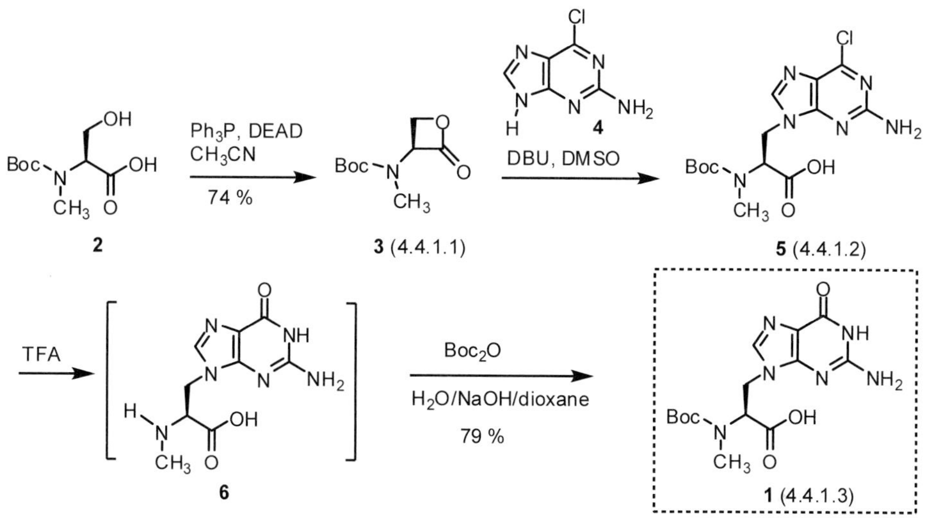

Using the described procedure, the nucleo amino acid **1** is available in three steps starting from *N*-Boc-*N*-methyl serine **2** in an overall yield of 45%. The otherwise comparable reaction with *N*-Boc-serine lactone gives an overall yield of only 21%, hence the use of the *N*-methylated substrate **3** results in a remarkable improvement in yield. The better result using **3** is mainly due to its lower polarity.

(c) Experimental procedures for the synthesis of 1

4.4.1.1   **   (*S*)-*N*-Butoxycarbonyl-*N*-methyl-serine lactone [9]

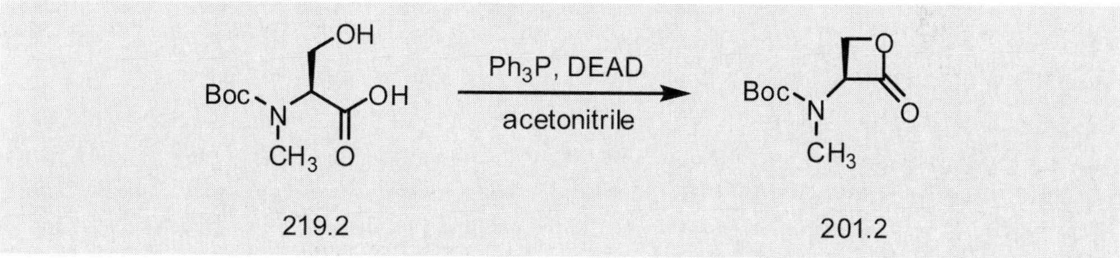

DEAD (3.58 mL, 23.03 mmol) is added to a stirred solution of triphenylphosphine (6.04 g, 23.03 mmol) in anhydrous CH_3CN (100 mL) at –40 °C over a period of 15 min. The reaction mixture is kept at this temperature for about 20 min. A suspension of Boc-*N*-methyl-*L*-serine (5.00 g, 22.81 mmol) in anhydrous CH_3CN (50 mL) is then added and stirring is continued at –35 °C for 2 h and at room temperature for a further 5 h.

After purification by flash chromatography (180 g silica gel; EtOAc/hexane, 1:3), the product is obtained as a colorless solid: 3.37 g (74%), mp 30–35 °C, *ee* > 98 %; R_f (*n*-hexane/EtOAc, 3:1) = 0.37; $[\alpha]_D^{20} = -44$ (*c* = 0.75, MeOH).

IR (film): $\tilde{v}$ (cm^{-1}) = 2980, 2935, 1833, 1762, 1697, 1482, 1455, 1401, 1370, 1352, 1329, 1302, 1254, 1111, 1051, 968, 941.

^{1}H NMR (300 MHz, CDCl$_3$): δ (ppm) = 4.43 (d, 2 H, C$\underline{H}_2$), 2.98 (s, 3 H, C$\underline{H}_3$), 1.47 (s, 9 H, tBu; rotamer).

^{13}C NMR (75 MHz, [D$_6$]DMSO): δ (ppm) = 77.2 (CO lactone), 77.0 (CO), 76.7 ($\underline{C}$(CH$_3$)$_3$), 64.4 (C-2), 58.7 (C-3), 30.3 (NCH$_3$), 27.8 (C($\underline{C}$H$_3$)$_3$).

MS HRS: m/z = 224 [M + Na]$^+$.

4.4.1.2   ***   (*S*)-*N*-*tert*-Butoxycarbonyl-*N*-methyl-β-(2-amino-6-chloro-9-purinyl)-alanine [9]

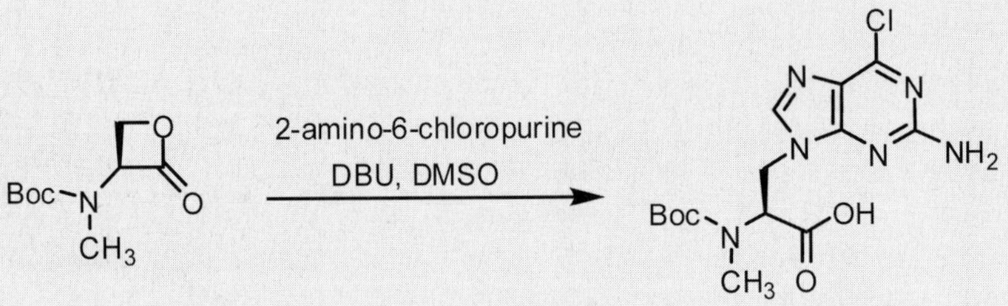

Under an argon atmosphere, DBU (513 µL, 3.43 mmol, 1.1 equiv.) is added to a suspension of 2-amino-6-chloropurine (688 mg, 4.06 mmol, 1.3 equiv.) in anhydrous DMSO (2 mL) and the mixture is stirred for 15 min. A solution of (*S*)-*N*-Boc-*N*-methyl-L-serine lactone **4.4.1.1** (628 mg, 3.12 mmol) in anhydrous DMSO (2 mL) is then added over 15 min and the reaction mixture is kept at room temperature for 210 min.

The reaction is then quenched by the addition of AcOH (127 µL, 2.22 mmol) and the solvent is evaporated. Purification of the residue by flash chromatography (silica gel), eluting with EtOAc/MeOH (8:2, + 0.2–1% AcOH), gives the desired nucleo amino acid as a colorless solid; 889 mg (77%), R_f (isopropanol/H$_2$O/AcOH, 5:2:1, saturated with NaCl) = 0.52; $[\alpha]_D^{20}$ = −109.3 (c = 0.30, DMSO).

IR (KBr): $\tilde{v}$ (cm^{-1}) = 3347, 2977, 1693, 1616, 1565, 1520, 1469, 1393, 1368, 1155, 916, 785, 643.

^{1}H NMR (300 MHz, [D$_6$]DMSO): δ (ppm) = 8.01 (s, 1 H, H-8), 6.85 (s, 2 H, NH$_2$), 4.89 (m, 2 H, α-H, β-H), 4.45 (m, 1 H, β-H), 2.74 (s, 0.5 H, N–CH$_3$), 2.70 (s, 2.5 H, N–CH$_3$), 1.20 (s, 1.5 H, tBu), 1.06 (s, 7.5 H, tBu).

^{13}C NMR (75 MHz, [D$_6$]DMSO): δ (ppm) = 170.2 (COOH), 160.0 (C-4), 155.0 (CONH), 154.2 (C-2), 149.5 (C-6), 142.3 (C-8), 123.2 (C-5), 78.8 ($\underline{C}$(CH$_3$)), 57.4 (α-C), 41.6 (NCH$_3$), 39.9 (β-C), 27.3 (C($\underline{C}$H$_3$)$_3$).

MS HRS: m/z (%): 371.12 [M + H]$^+$.

4.4.1.3 ** (*S*)-*N*-*tert*-Butoxycarbonyl-*N*-methyl-β-(9-guaninyl)-alanine [9]

A solution of amino acid **4.4.1.2** (869 mg, 2.34 mmol) in TFA/H₂O (3:1; 12 mL) is stirred at room temperature overnight. Toluene is then added and the mixture is concentrated to dryness. The formed guaninyl amino acid is dissolved in a mixture of H₂O/1 N NaOH/dioxane (1:1:2; 10 mL) and further 1 N NaOH is added to maintain a pH of 9.5. The reaction mixture is cooled to 0 °C, treated with *N*-di-*tert*-butyl dicarboxylate (563 mg, 2.58 mmol), and kept for 45 min at 0 °C and for 60 h at room temperature. It is then adjusted to pH 6 by the addition of ice-cold 1 N HCl. After evaporation of the solvents *in vacuo*, the residue is purified by RP-column chromatography using MeOH (8%)/water to give the product as a white solid; 522.6 mg (79%); mp >198 °C (decomposition) (93% *ee*); R_f (EtOAc/MeOH/H₂O/AcOH, 6:2:2:1, saturated with NaCl) = 0.69; $[\alpha]_D^{20}$ = –73.4 (*c* = 0.50, DMSO).

---

¹H NMR (300 MHz, [D₆]DMSO): δ (ppm) = 1.11 (s, 9 H, *t*Bu; rotamer), 2.71 (s, 3 H, CH₃–N), 4.95 (m, 1 H, β-H), 4.42 (dd, J_1 = –14.5, J_2 = –3.5 Hz, 1 H, β-H), 4.59 (dd, J_1 = –11.3, J_2 = –3.9 Hz, 1 H, α-H), 6.47 (s_br, 1 H, N*H*Boc), 7.38 (s, 1 H, H-6).

¹³C NMR (75 MHz, [D₆]DMSO): δ (ppm) = 171.2 (COOH), 155.2 (C-2), 154.2 (CONH), 151.2 (C-2), 137.2 (C-6), 116.5 (C-5), 77.8 (*C*(CH₃)₃), 60.4 (α-C), 42.7 (β-C), 27.5 (C*C*H₃), 27.9 (C(*C*H₃)₃).

ESI MS: *m/z* (%): 704.8 [2*M* + H]⁺ (20), 353.1 [*M* + H]⁺ (100).

---

[1] a) J. M. Humphrey, R. A. Chamberlin, *Chem. Rev.* **1997**, *97*, 2243; b) J. Chatterjee, D. Mierke, H. Kessler, *J. Am. Chem. Soc.* **2006**, *128*, 15164.

[2] a) M. Ebata, Y. Takahashi, H. Otsuka, *Bull. Chem. Soc. Jpn.* **1996**, *39*, 2535; b) P. Jouin, J. Poncet, M.-L. Difour, A. Pantaloni, B. Castro, *J. Org. Chem.* **1989**, *54*, 617; c) G. R. Pettit, Y. Kamano, C. L. Herald, Y. Fujii, H. Kizu, M. R. Boyd, F. E. Boettner, D. L. Doubek, J. M. Schmidt, J. C. Chapuis, C. Michel, *Tetrahedron* **1993**, *49*, 9151; d) R. M. Wenger, *Helv. Chim. Acta* **1984**, *67*, 502.

[3] D. P. Fairlie, G. Abbenante, D. R. March, *Curr. Med. Chem.* **1995**, *2*, 654.

[4] L. Aurelio, R. T. C. Brownlee, A. B. Hughes, *Chem. Rev.* **2004**, *104*, 5823.

[5] M. L. Di Gioia, A. Leggio, A. Le Pera, A. Liguori, A. Napoli, C. Siciliano, G. Sindona, *J. Org. Chem.* **2003**, *68*, 7416.

[6] a) E. Biron, H. Kessler, *J. Org. Chem.* **2005**, *70*, 5183; b) E. Biron, J. Chatterjee, H. Kessler, *J. Peptide Sci.* **2006**, *12*, 213.

[7] K. N. White, J. P. Konopelski, *Org. Lett.* **2005**, *7*, 4111.

[8] S. Zhang, T. Govender, T. Norström, P. I. Arvidsson, *J. Org. Chem.* **2005**, *70*, 6918.

[9] R. Ranevski, Synthese und Untersuchung von Alanyl-PNA Oligomeren und deren Einfluß auf β-Faltblatt Strukturen, Dissertation, Universität Göttingen, 2006.

[10] T. Fukuyama, C.-K. Jow, M. Cheung, *Tetrahedron Lett.* **1995**, *36*, 6373.

[11] a) L. D. Arnold, T. H. Kalantar, J. C. Vederas, *J. Am. Chem. Soc.* **1985**, *107*, 7105; b) L. D. Arnold, R. G. May, J. C. Vederas, *J. Am. Chem. Soc.* **1988**, *110*, 2237.

[13] U. Diederichsen, D. Weicherding, N. Diezemann, *Org. Biomol. Chem.* **2005**, *3*, 1058.

[14] M. Teixido, F. Albericio, E. Giralt, *J. Peptide Res.* **2005**, *65*, 153.

4.4.2　(S)-Homoproline

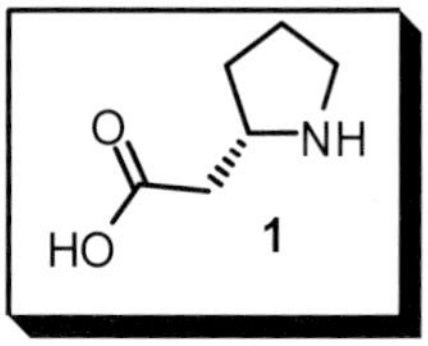

Topics:
- Ring-closing olefin metathesis (RCM)
- Diastereoselective Michael addition of a chiral amide to a 2,4-dienoate
- Hydrogenation with Wilkinson's catalyst
- Debenzylation with Adam's catalyst
- Formation and hydrolysis of a *tert*-butyl ester

(a) General

The cyclic β-amino acid (S)-homoproline (**1**, (S)-(2-pyrrolidine)acetic acid) has attracted interest as a starting material for the synthesis of organo catalysts, e.g. for the Hajos–Parrish–Sauer–Wiechert reaction (cf. **1.3.2**) [1]. It was initially synthesized by classical Arndt–Eistert homologation of (S)-proline [2], which proceeded with retention of configuration at C-2:

In a novel approach [3] – presented in detail in section (b) – the stereogenic center (C-2) of **1** is constructed first by asymmetric synthesis of an acyclic precursor, which is subsequently cyclized by means of olefin metathesis.

Olefin metathesis [4, 5] is the (reversible) reaction of two olefinic substrates

$$A{=}B \ + \ C{=}D \ \rightarrow \ A{=}C \ + \ B{=}C$$

in which the sp^2 moieties constituting the olefinic double bonds are exchanged by mediation of transition metal-derived carbene complexes, as exemplified by the metathesis of two monosubstituted olefins (cross-metathesis, CM):

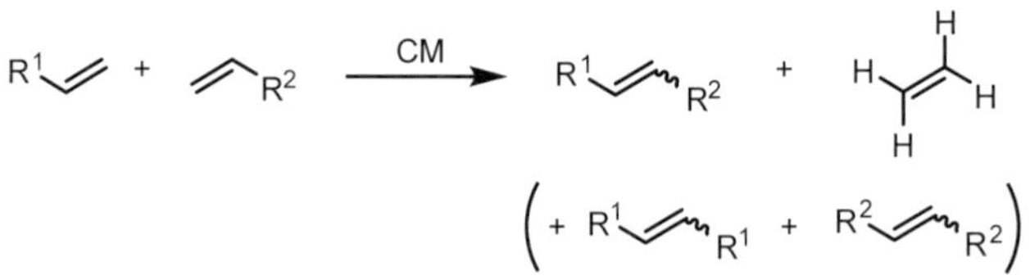

Two main types of catalysts are utilized in olefin metathesis: (1) Schrock carbene complexes [6] of molybdenum and tungsten, e.g. **2**, and (2) Grubbs carbene complexes [3] of ruthenium, e.g. **3a** and **3b**; all of these catalysts of types **2** and **3** are commercially available:

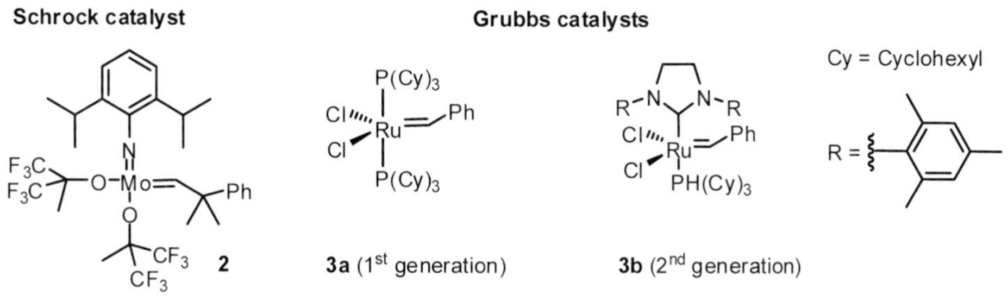

Schrock catalyst   **Grubbs catalysts**

Cy = Cyclohexyl

2   **3a** (1st generation)   **3b** (2nd generation)

With air-sensitive catalysts of the Schrock type, olefin metathesis is compatible with ether, ketone, ester, and amide functionalities; with air-stable catalysts of the Grubbs type, reactions can be performed even in the presence of hydroxyl and carboxylic acid groups in the substrate; moreover, water can serve as solvent.

The mechanism of olefin metathesis can be described as a catalytic cycle (Chauvin mechanism), in which – as generalized for the above process – metallacycles play a central role:

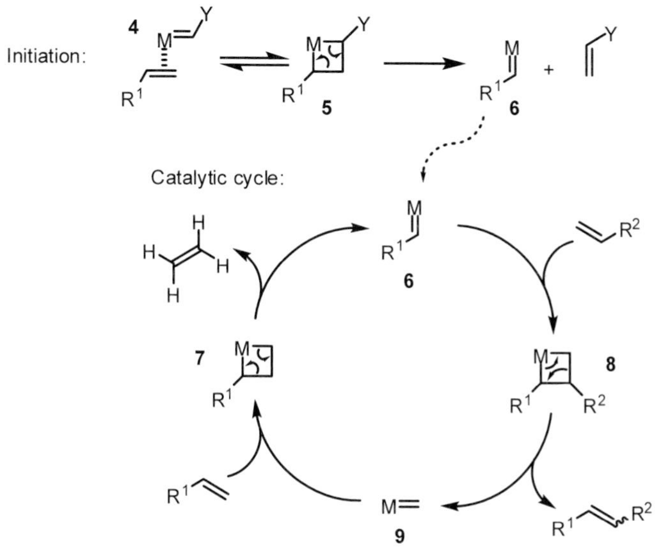

Initiation:

Catalytic cycle:

In an initiation step, the catalysts of the Grubbs or Schrock type (e.g. **2/3**, simplified as **4**) operate as pre-catalysts, i.e. they generate the catalytically active carbene complex **6** via (formal) [2+2]-cycloaddition to the olefin R^1–CH=CH$_2$ (giving rise to the metallacyclobutane **5**) and subsequent cycloreversion of **5**.

The carbene complex **6** then enters a catalytic cycle through cycloaddition to the second olefin moiety R^2–CH=CH$_2$ to give another metallacyclobutane **8**, [2+2]-cycloreversion of which effects formation of the product R^1–CH=CH–R^2 and a new carbene complex **9**. Cycloaddition of **9** to R^1–CH=CH$_2$ produces a third metallacycle **7**, cycloreversion of which affords ethylene and concomitantly regenerates the starting carbene complex **6**, which re-enters the catalytic cycle for further olefin conversion. All partial steps of this cycle are reversible.

On the basis of this mechanism, several competing types of metathesis reactions are possible for α,ω-bisolefinic substrates **10**, which are differentiated as ring-closing metathesis (RCM, **10 → 11**), ring-opening metathesis (ROM, **11 → 10**), acyclic diene metathesis polymerization (ADMET, **10 → 12**), and ring-opening metathesis polymerization (ROMP, **11 → 12**):

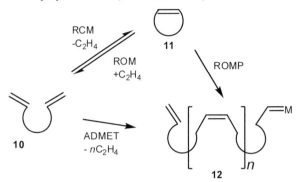

Particularly attractive for synthesis is the RCM process, which as intramolecular olefin metathesis can be used to afford cycloalkenes **11** of conventional ring sizes (five-, six-, and seven-membered rings) as well as macrocyclic ring systems. Irrespective of thermodynamic stability criteria, ring-closing metathesis (**→ 11**) is favored over ADMET polymerization (**→ 12**) when high-dilution conditions are applied for the reaction.

The synthesis described in section (b) utilizes an RCM process as the key transformation.

Of additional synthetic relevance are alkyne metathesis reactions and ring-closing metathesis reactions of diynes. Alkyne metathesis is mediated by Mo or W carbene complexes of the Schrock alkylidene type **13** [4], or a molybdenum amido complex described by Fürstner [7]:

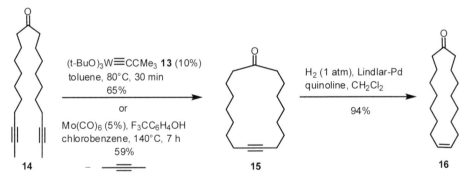

RCM of diyne **14** (chain length > C_{15}) leads to cyclic acetylene **15**, which can be stereoselectively transformed into the macrocyclic alkene with a (Z)-configuration of the C-C double bond, in the illustrated case the (Z)-C_{17}-cycloalkenone civetone (**16**), a natural fragrance of the musc type [8]. This procedure has a great advantage over the normal metathesis of alkenes, which in general leads to mixtures of (Z)- and (E)-diastereomers.

(b) Synthesis of 1

The starting material for the synthesis of **1** is *tert*-butyl (*E*,*E*)-hexa-2,4-dienoate (**17**), which is obtained from the readily available ethyl ester by transesterification with potassium *tert*-butoxide in *tert*-butanol [9]. Asymmetric 1,4-addition (cf. **1.1.4**) of the lithium salt of the commercially available (*S*)-*N*-allyl-*N*-α-methylbenzylamine (**18**) — prepared by *in situ* *N*-metalation using *n*-BuLi — to the α,β-unsaturated acceptor system in **17** leads to the β-amino ester **19** with high diastereoselectivity (*de* > 95%), thus establishing the required configuration at C-2 in the target molecule **1**:

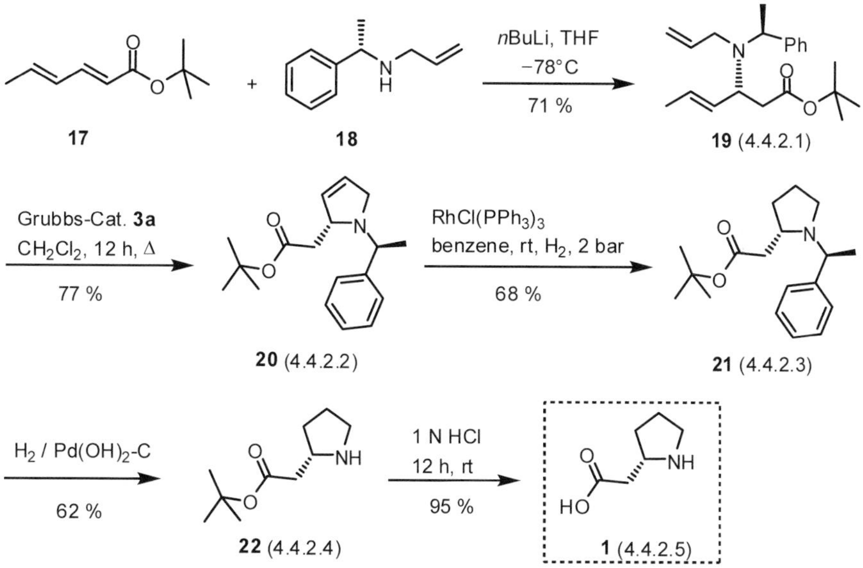

For cyclization by ring-closing metathesis (RCM), the bisallylic β-amino ester **19** is treated with Grubbs benzylidene ruthenium catalyst **3a** in CH$_2$Cl$_2$ solution to give the *N*-α-methylbenzyl-protected pyrroline β-amino ester **20** as a single diastereomer in good yield (77%, *de* > 95%). This proves that no epimerization had taken place during the RCM process. Hydrogenation of the pyrroline double bond in **20** is achieved without affecting the *N*-benzyl and *tert*-butyl protecting groups by applying Wilkinson's rhodium catalyst RhCl(PPh$_3$)$_3$ [10] in a homogeneous system, providing the pyrrolidine β-amino ester **21** in high diastereomeric purity (*de* > 95%). Removal of the auxiliary *N*-protecting group in **21** by catalytic debenzylation using Adam's catalyst (Pd(OH)$_2$ on carbon) [11] leads to **22** in 62% yield; finally, the *tert*-butyl ester moiety in **22** is cleaved by acidic hydrolysis to afford (*S*)-homoproline **1** in high chemical yield (95%) after purification by ion-exchange chromatography.

Thus, the target molecule **1** is obtained by a five-step sequence starting from *tert*-butyl ester **17** in an overall yield of 45%.

The mechanism of the RCM process (simplified as **10** → **11** + ethylene) is analogous to the CM scheme in section (a). In an initiation step, the catalytically active carbene complex **24** is generated from the diene **10** via metallacycle **23** and its cycloreversion. The carbene species **24** enters a catalytic cycle proceeding via intramolecular [2+2]-cycloaddition to the second C=C bond in the substrate **10** (**24** → **25**), cycloreversion with cycloalkene formation (**25** → **11**), and capture of the newly formed carbene complex **26** by a C=C bond of the α,ω-diene **10** (**26** → **27**). The catalytic cycle is closed by cycloreversion of the metallacycle **27**, leading to ethylene and the starting carbene species **24**.

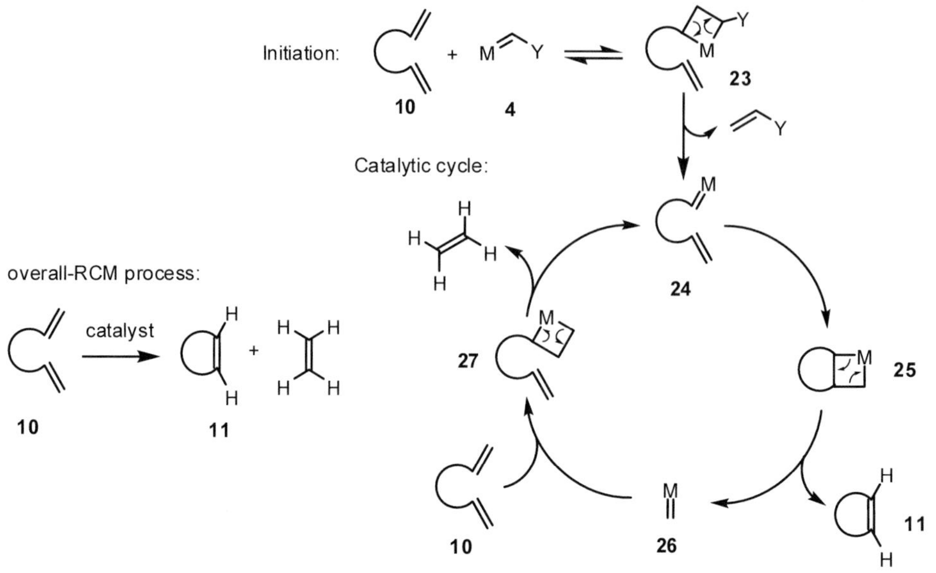

Initiation:

overall-RCM process:

Catalytic cycle:

(c) Experimental procedures for the synthesis of 1

4.4.2.1   ***   3-[Allyl-(1-(S)-phenylethyl)amino]-hex-4-enoic acid *tert*-butyl ester [3]

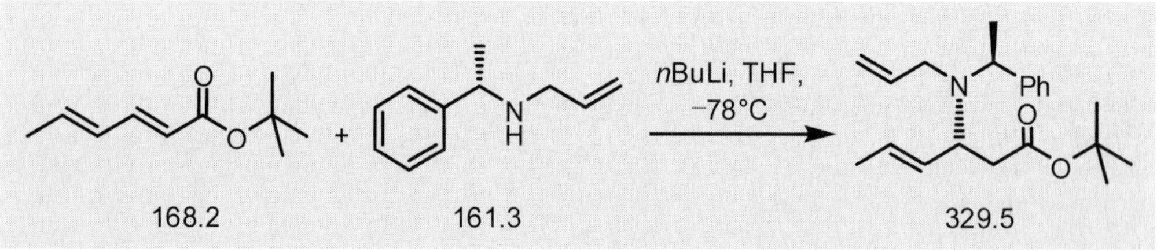

n-Butyllithium (6.90 mmol, 1.60 M, 4.31 mL) is added dropwise over a period of 10 min to a stirred solution of (S)-N-allyl-N-α-methylbenzylamine (1.21 g, 7.50 mmol) in anhydrous THF (12 mL) at –78 °C under an argon atmosphere. Stirring is continued for 30 min, then a precooled (–78 °C) solution of *tert*-butyl sorbate (0.97 g, 5.77 mmol) in THF (5.0 mL) is added dropwise over a period of 15 min.

The mixture is stirred for 16 h at –78 °C, and then the reaction is quenched by the addition of saturated aqueous NH_4Cl solution (5.0 mL). The resulting solution is allowed to warm to room temperature, whereupon brine (15 mL) is added. The layers are separated, and the aqueous layer is extracted with Et_2O (3 × 15 mL). The combined organic extracts are washed with 10% citric acid (1 × 30 mL), dried over $MgSO_4$, and concentrated *in vacuo*. Purification of the residue by column chromatography on silica gel (petroleum ether/Et_2O, 95:5) gives the product as a colorless oil: 1.35 g (71%); $R_f = 0.32$ (petroleum ether/Et_2O, 95:5), $n_D^{20} = 1.6922$, $[\alpha]_D^{20} = -1.0$ ($c = 0.8$, $CHCl_3$).

IR (film): $\tilde{\nu}$ (cm^{-1}) = 2976, 1730, 1453, 1367, 1161.

^{1}H NMR (CDCl$_3$, 300 MHz): δ (ppm) = 7.39–7.15 (m, 5 H, Ph), 5.85–5.69 (m, 1 H, NCH$_2$C**H**CH$_2$), 5.56–5.45 (m, 2 H, 4-H, 5-H), 5.11–4.94 (m, 2 H, NCH$_2$CHC**H**$_2$), 3.99 (q, J = 6.8 Hz, 1 H, 1'-H), 3.86–3.76 (m, 1 H, 3-H), 3.15–3.08 (m, 2 H, NC**H**$_2$CHCH$_2$), 2.40 (dd, J = 6.6, 14.4 Hz, 1 H, 2-CH$_2$), 2.25 (dd, J = 8.5, 14.4 Hz, 1 H, 2-CH$_2$), 1.69 (d, J = 5.0 Hz, 3 H, 6-CH$_3$), 1.39 (s, 9 H, C(CH$_3$)$_3$), 1.36 (d, J = 6.8 Hz, 3 H, 2'-CH$_3$).

^{13}C NMR (CDCl$_3$, 125.7 MHz): δ (ppm) = 171.3 (CO), 145.2 (*ipso*-Ph), 138.9 (NCH$_2$**C**HCH$_2$), 130.6 (5-C), 127.9 (Ph), 127.6 (Ph), 126.8 (4-C), 126.4 (Ph), 115.3 (NCH$_2$CH**C**H$_2$), 79.9 (**C**(CH$_3$)$_3$), 57.2 (3-C), 57.0 (1'-C), 49.6 (N**C**H$_2$CHCH$_2$), 39.9 (2-C), 28.0 (C(**C**H$_3$)$_3$), 18.4 (6-C), 17.9 (2'-C).

DCI (NH$_3$): *m/z* = 330.3 [*M* + H$^+$].

HR MS: (MeOH/NH$_4$OAc): calcd. for C$_{21}$H$_{31}$NO$_2$ [*M* + H$^+$]: 330.24276; found: 330.24261.

4.4.2.2   **   [1-(1-Phenylethyl)-2,5-dihydro-1*H*-pyrrol-2-yl]-acetic acid *tert*-butyl ester [3]

329.5 → Grubbs-cat. 1st gen. CH$_2$Cl$_2$, Δ, 12 h → 287.4

A solution of ester **4.4.2.1** (1.29 g, 3.91 mmol) in anhydrous CH$_2$Cl$_2$ (30 mL) is added to a stirred solution of the Grubbs catalyst (Grubbs 1st, 0.129 g, 0.156 mmol, 4 mol%) in anhydrous CH$_2$Cl$_2$ (100 mL) under an argon atmosphere at room temperature. The mixture is then heated at reflux for 12 h.

Thereafter, the solution is concentrated *in vacuo* and the residue is purified by column chromatography on silica gel (petroleum ether/Et$_2$O, 95:5); 0.86 g (77%), R_f = 0.13 (petroleum ether/Et$_2$O, 95:5), n_D^{20} = 1.6924, $[\alpha]_D^{20}$ = +121 (*c* = 0.8, CHCl$_3$).

UV (CH$_3$CN): λ_{max} (nm) = 205.

IR (film) $\tilde{\nu}$ (cm^{-1}) = 2975, 1726, 1453, 1367, 1143, 702.

^{1}H NMR (CDCl$_3$, 300 MHz): δ (ppm) = 7.35–7.17 (m, 5 H, Ph), 5.76–5.67 (m, 2 H, 3-H, 4-H), 4.19–4.09 (m, 1 H, α-(2)-CH$_2$), 3.85 (q, J = 6.8 Hz, 1 H, 1'-H), 3.64–3.54 (m, 1 H, 5-CH$_2$), 3.37 (dddd, J = 1.8, 4.1, 5.9, 14.4 Hz, 1 H, 5-CH$_2$), 2.59 (dd, J = 4.1, 14.4 Hz, 1 H, β-CH$_2$), 2.33 (dd, J = 9.0, 14.4 Hz, 1 H, β-CH$_2$), 1.45 (s, 9 H, C(C**H**$_3$)$_3$), 1.43 (d, J = 6.8 Hz, 3 H, 2'-CH$_3$).

^{13}C NMR (CDCl$_3$, 125.7 MHz): δ (ppm) = 171.4 (CO), 144.7 (*ipso*-Ph), 130.6 (Ph, 3-C, 4-C), 128.2, 127.4, 126.9, 126.7, 80.1 (**C**(CH$_3$)$_3$), 64.8 (α-(2)-C), 62.1 (1'-C), 58.4 (5-C), 43.0 (β-C), 28.1 (C(**C**H$_3$)$_3$), 22.8 (2'-C).

DCI (NH$_3$): *m/z* = 288.3 [*M* + H$^+$].

HR MS: (MeOH/NH$_4$OAc): calcd. for C$_{18}$H$_{25}$NO$_2$ [*M* + H$^+$]: 288.19581; found: 288.19591.

4.4.2.3   **   [1-(1-Phenylethyl)-pyrrolidin-2-yl]-acetic acid *tert*-butyl ester [3]

287.4

H₂, 2 bar, RhCl(PPh₃)₃
benzene, rt

289.4

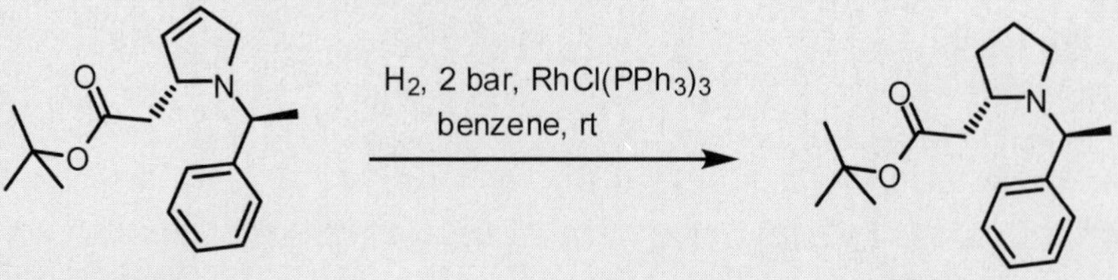

RhCl(PPh₃)₃ (80.0 mg, 86.5 µmol) is added to a degassed solution of ester **4.4.2.2** (784.0 mg, 2.728 mmol) in anhydrous benzene (5.0 mL) and the mixture is stirred under an H₂ atmosphere (2 bar) at room temperature for 20 h.

It is then filtered through Celite, which is washed with MeOH, and the combined filtrate and washings are concentrated *in vacuo*. Purification of the residue by column chromatography (SiO₂; petroleum ether/Et₂O, 8:2) gives the product as a colorless oil; 538 mg (68%), $R_f = 0.13$ (petroleum ether/Et₂O, 8:1), $n_D^{20} = 1.6918$, $[\alpha]_D^{20} = +49.5$ ($c = 1.0$, CHCl₃).

IR (film) $\tilde{v}$ (cm⁻¹) = 2973, 1727, 1367, 1153, 702.

¹H NMR (CDCl₃, 300 MHz): δ (ppm) = 7.39–7.18 (m, 5 H, Ph), 3.76 (q, $J = 6.7$ Hz, 1 H, 1'-H), 3.09–2.99 (m, 1 H, 2-CH₂), 2.83–2.73 (m, 1 H, 5-CH₂), 2.58 (dd, $J = 3.7, 14.4$ Hz, 1 H, β-CH₂), 2.42–2.31 (m, 1 H, 5-CH₂), 2.18 (dd, $J = 9.7, 14.4$ Hz, 1 H, β-CH₂), 1.91–1.53 (m, 4 H, 3-CH₂, 4-CH₂), 1.45–1.41 (m, 12 H, 2'-CH₃, C(CH₃)₃).

¹³C NMR (CDCl₃, 125.7 MHz): δ (ppm) = 171.9 (CO), 142.7 (*ipso*-Ph), 128.0 (Ph), 127.8, 126.8, 80.0 (C(CH₃)₃), 60.4 (1'-C), 57.1 (2-C), 49.7 (5-C), 41.5 (β-C), 30.6 (3-C), 28.1 (C(CH₃)₃), 22.6 (4-C), 22.1 (2'-C).

DCI (NH₃): *m/z* = 289.2 [*M* + H⁺].

HR MS (MeOH/NH₄OAc): calcd. for C₁₈H₂₇NO₂ [*M* + H⁺]: 290.21146; found: 290.21138.

4.4.2.4   *   Pyrrolidin-2-yl-acetic acid *tert*-butyl ester [3]

289.4

H₂ / Pd(OH)₂-C

185.3

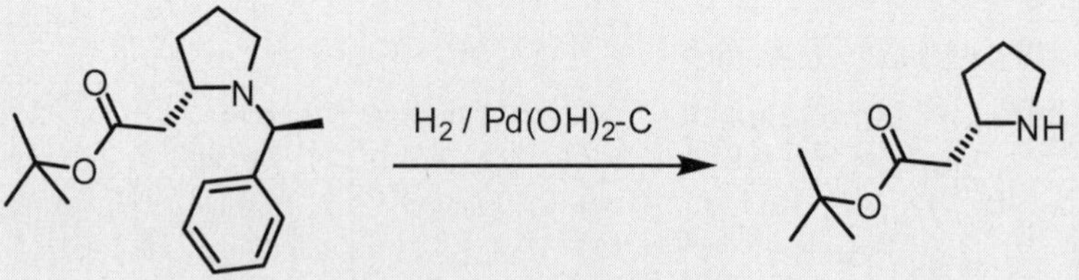

Pd(OH)₂/C (125 mg) is added to a degassed solution of ester **4.4.2.3** (250 mg, 0.863 mmol) in MeOH/H₂O/AcOH (22.51 mL, 40:4:1) and the resulting mixture is stirred under H₂ (1 bar) for 17 h at room temperature.

It is then filtered through Celite, which is washed with MeOH, and the combined filtrate and washings are concentrated *in vacuo*. The residue is dissolved in CH_2Cl_2 (10 mL), and the solution obtained is washed with saturated sodium hydrogencarbonate solution (5 mL), dried over $MgSO_4$, and concentrated. Purification of the residue by column chromatography on silica gel (chloroform/MeOH, 8:2) gives the product as a colorless oil; 98.0 mg (62%), $R_f = 0.14$ (chloroform/MeOH, 8:2), $n_D^{20} = 1.6929–1.6932$, $[\alpha]_D^{20} = -12.7$ ($c = 1.0$, $CHCl_3$).

IR (film): $\tilde{\nu}$ (cm^{-1}) = 2969, 1727, 1367, 1149.

^{1}H NMR ($CDCl_3$, 300 MHz): δ (ppm) = 3.36–3.24 (m, 1 H, 2-CH_2), 3.00–2.76 (m, 2 H, 5-CH_2), 2.36–2.29 (m, 2 H, β-CH_2), 2.21 (s, 1 H, NH), 1.92–1.55 (m, 3 H, 3,4-CH_2), 1.39 (s, 9 H, $C(CH_3)_3$), 1.35–1.16 (m, 1 H, 3,4-CH_2).

^{13}C NMR ($CDCl_3$, 125.7 MHz): δ (ppm) = 171.8 (CO), 80.3 ($\underline{C}(CH_3)_3$), 55.1 (2-C), 46.2 (5-C), 42.0 (β-C), 30.9 (3-C), 28.1 ($C(\underline{C}H_3)_3$), 24.8 (4-C).

DCI (NH_3): m/z = 186.1 [$M + H^+$].

HR MS (MeOH/NH_4OAc): calcd. for $C_{10}H_{19}NO_2$ [$M + Na^+$]: 208.13080; found 208.13070.

4.4.2.5   *   (S)-Homoproline [3]

185.3   129.2

A mixture of HCl (1 N, 7 mL) and ester **4.4.2.4** (70.0 mg, 0.378 mmol) is stirred at room temperature for 12 h and then lyophilized.

Purification by reversed-phase chromatography (RP-gel; H_2O) gives the product as a colorless solid; 41.4 mg (95%), $R_f = 0.13$ (EtOAc/MeOH/H_2O/AcOH, 10:1:1:0.5), mp 194–196 °C, $[\alpha]_D^{20} = -28.4$ ($c = 0.5$, H_2O).

IR (KBr) $\tilde{\nu}$ (cm^{-1}) = 2954, 1733, 1395, 1166, 840.

^{1}H NMR ([D_6]DMSO, 300 MHz): δ (ppm) = 3.77–3.62 (m, 1 H, 2-CH), 3.10–3.08 (m, 2 H, 5-CH_2), 2.87–2.69 (m, 2 H, β-CH_2), 2.13–2.03 (m, 1 H, 3,4-CH_2), 1.99–1.79 (m, 2 H, CH_2), 1.64–1.53 (m, 1 H, 3,4-CH_2).

^{13}C NMR ([D_6]DMSO, 125.7 MHz): δ (ppm) = 171.6 (CO), 55.2 (2-C), 44.2 (5-C), 36.0 (β-C), 29.6 (3-C), 22.8 (4-C).

DCI (NH_3): m/z = 103.1 [$M + H^+$].

HR MS (MeOH, NH_4OAc): calcd. for $C_6H_{11}O_2$: 129.0790; found: 129.0793.

[1] P. Buchschacher, J.-M. Cassal, A. Fürst, W. Meier, *Helv. Chim. Acta* **1977**, *60*, 2747.

[2] J.-M. Cassal, A. Fürst, W. Meier, *Helv. Chim. Acta* **1976**, *59*, 1917.

[3] A. M. Chippindale, S. G. Davies, K. Iwamoto, R. M. Parkin, C. A. P. Smethurst, A. D. Smith, H. Rodriguez-Solla, *Tetrahedron* **2003**, *59*, 3253.

[4] a) A. Fürstner, *Angew. Chem.* **2000**, *112*, 3140, *Angew. Chem. Int. Ed.* **2000**, *39*, 3012; b) M. B. Smith, J. March, *March's Advanced Organic Chemistry*, 5th ed., p. 1457, John Wiley & Sons, Inc., New York, **2001**.

[5] a) K. C. Nicolaou, S. A. Snyder, *Classics in Total Synthesis II*, p. 163, Wiley-VCH, Weinheim, 2003; b) R. H. Grubbs (Ed.), *Handbook of Metathesis*, three volumes, Wiley-VCH, Weinheim, 2003; c) T. J. Katz, *Angew. Chem.* **2005**, *117*, 3070; *Angew. Chem. Int. Ed.* **2005**, *44*, 3010.

[6] For the differentiation of carbene complexes and the chemistry of Schrock carbene complexes, see: L. S. Hegedus, *Organische Synthese mit Übergangsmetallen*, p. 171, VCH, Weinheim, **1995**.

[7] A. Fürstner, C. Mathes, K. Grela, *Chem. Commun.* **2001**, 1057.

[8] A. Fürstner, G. Seidel, *J. Organomet. Chem.* **2000**, *606*, 75.

[9] In analogy to: V. A. Vasin, V. V. Razin, *Synlett* **2001**, 658.

[10] For homogeneous catalytic hydrogenation, see ref. [6], p. 42.

[11] R. Adams, V. Voorhees, R. L. Shriner, *Org. Synth. Coll.* **1941**, *1*, 463.

4.4.3 Amino acid resolution with amino acylase

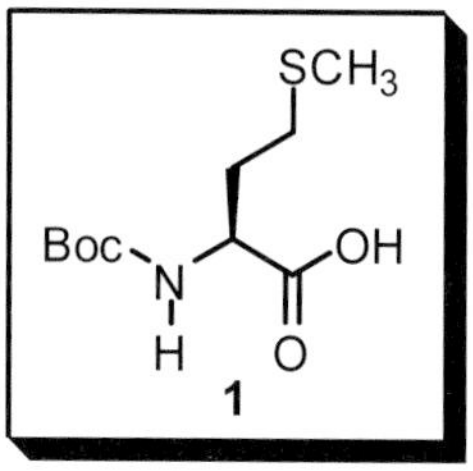

Topics:
- Racemic amino acid synthesis
- Enzymatic reactions
- Kinetic resolution
- Boc-protected amino acids
- Ion-exchange chromatography

(a) General

For industrial and pharmacological purposes, there is huge demand for the synthesis of enantiomerically pure amino acids. Proteinogenic amino acids are needed, e.g., as sweeteners and food additives, and as regulating molecules and inhibitors in medicinal chemistry [1]. Moreover, there is a demand for labeled compounds as well as for modified and non-natural amino acids. The asymmetric synthesis of amino acids and derivatives is well established [2] and, in addition, resolution of racemic mixtures is still an inexpensive and competitive route for obtaining enantiopure amino acids. Co-crystallization with chiral non-racemic amines or acids is often used for resolution. Nevertheless, enzymatic resolution is competitive even on an industrial scale [3]. For such a kinetic resolution, the reaction rates for the two enantiomers must differ strongly, so that the enzyme selectively recognizes, e.g., the L-amino acid, leaving the D-enantiomer unchanged. Enzymes typically used for this purpose are acylase, lipase, esterase, and aminase. After complete conversion of the L-amino acid, separation of the transformed L-amino acid from the remaining D-derivative is required, which is usually achieved by precipitation of one of the compounds or by chromatography. Kinetic resolution is especially useful if both enantiomers are of interest. Otherwise, dynamic kinetic resolution might be a favorable method, combining the enzymatic conversion of amino acid derivatives with *in situ* racemization [4].

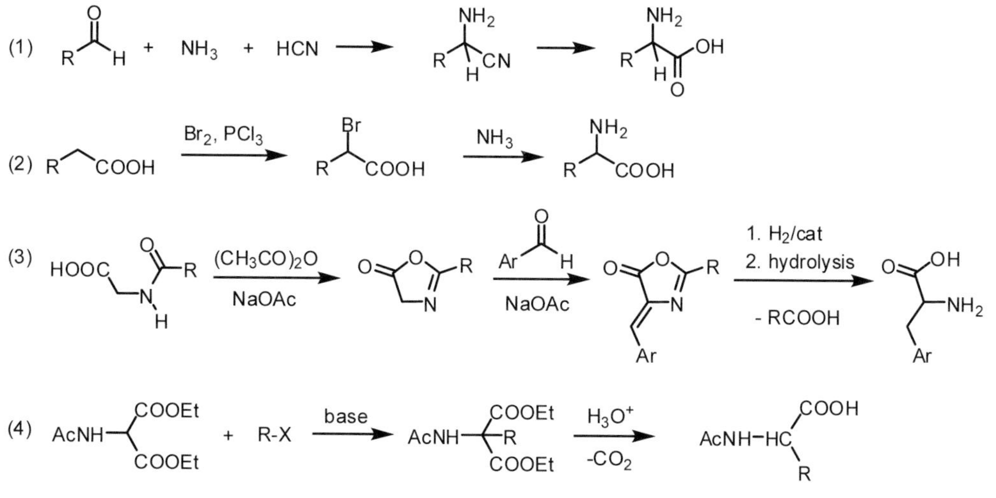

Prior to kinetic resolution, an efficient synthesis of the racemic mixture of the desired amino acid is required. Of the many contemporary methods, the following procedures are probably the most commonly used [5]. The Strecker synthesis (1) allows the introduction of highly diverse side chains by using various aldehydes, which are transformed into iminium ions and these, in turn, are treated with cyanide. The α-aminocyanide thus formed can be hydrolyzed to provide the amino acid. A second method is based on substitution of α-halogenated acids (2) with ammonia. In the Erlenmeyer azlactone synthesis (3), N-benzoylglycine is activated by acetic anhydride to self-condense to give an oxazolone. Further condensation with an aldehyde followed by hydration and hydrolysis provides the respective amino acid. Finally, acylamino malonic acid esters (4) are easily alkylated and the products can then be saponified and decarboxylated.

The following enzymatic transformations are usually applied for kinetic resolution of amino acids. (1) Acylase is used to cleave the amide bond in N-acylated amino acids. (2) Amino acid esters are saponified by lipases or esterases. In both cases, only L-amino acids are enzymatically recognized. Separation of the remaining D-amino acid derivative from the L-amino acid is required after 50% conversion is obtained. (3) Combination of the enzymes hydantoinase, carbamoylase, and racemase serves as an example of dynamic kinetic resolution of amino acids. Hydantoins are readily available, e.g. by conversion of non-protected amino acids with potassium cyanate. The respective hydantoinase is used to generate the enantiomerically pure N-carbamoyl amino acid, which is converted into the amino acid by carbamoylase. Using a racemase for isomerization of the hydantoins, overall, complete conversion to a single enantiomer is obtained.

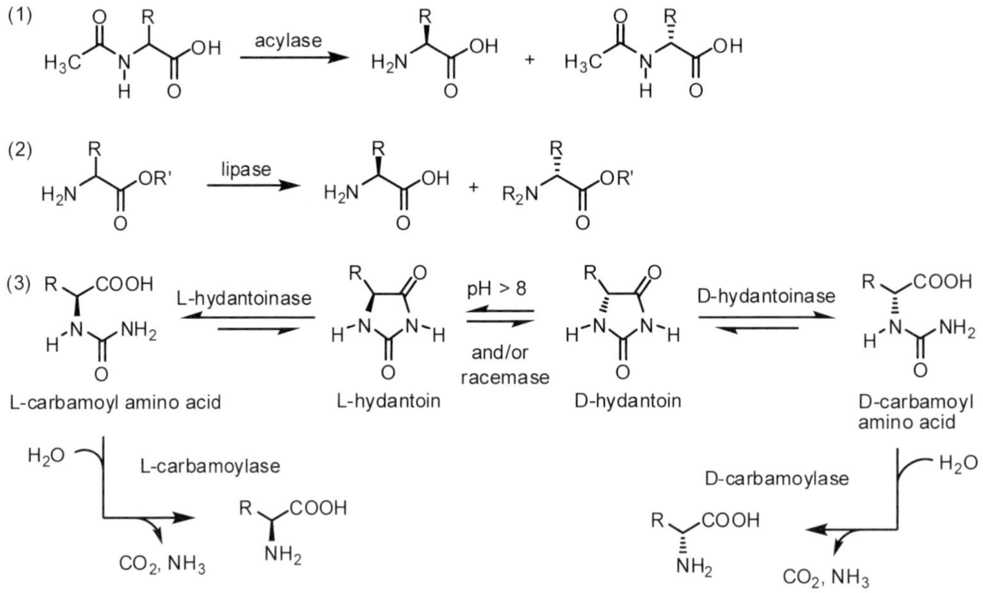

(b) Synthesis of Boc-protected *L*-methionine (1) by kinetic resolution

As an example of the enzymatic resolution of amino acids, the preparation of Boc-protected *L*-methionine (*L*-1) is described. Acylation of racemic methionine *rac*-2 with acetic anhydride under basic conditions readily provides *rac*-3. This is then subjected to enzymatic conversion with acylase (from pork liver) in aqueous solution at neutral pH, which takes two days to fully convert *L*-3 to give *L*-methionine (*L*-2) while *D*-3 remains unaffected. To separate *L*-2 from *D*-3, a cation exchanger is used to extract the liberated amino acid *L*-2. Finally, in order to use the amino acid in solid-phase peptide synthesis according to the Boc strategy, methionine is protected at the amino functionality using butoxycarbonyl anhydride under basic conditions following the standard procedure. Due to the polarity of *L*-1 with a free carboxylic acid moiety, purification on reversed-phase silica gel (C18) is beneficial.

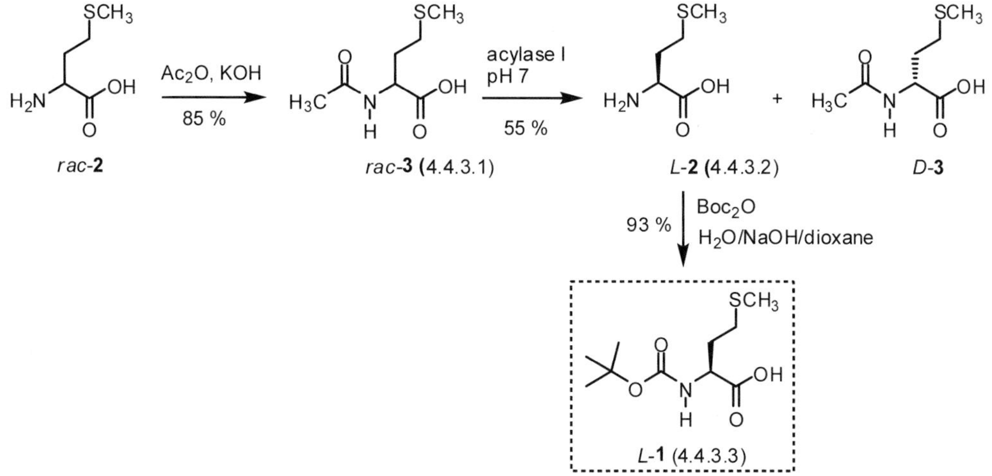

(c) Experimental procedures for the synthesis of *L*-1

4.4.3.1   *   *N*-Acetyl-*rac*-methionine [6]

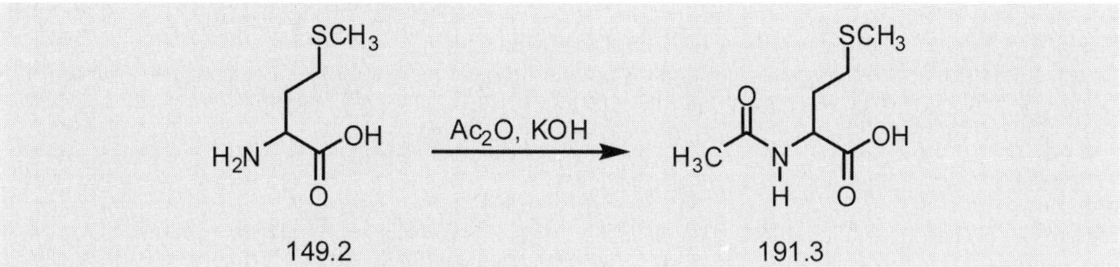

Acetic anhydride (50 mL, 54 g, 530 mmol) is added to a stirred solution of *rac*-methionine (20.0 g, 134 mmol) in 2 M aqueous KOH (70 mL) over 10 min. After stirring for a further 10 min at room temperature, H_2SO_4 (2 N, 70 mL) is added and the solution is concentrated *in vacuo* almost to dryness. The residue is extracted with boiling EtOAc, and the organic solution is dried over Na_2SO_4, filtered, and concentrated. Crystallization from EtOAc affords the product; 21.9 g (85%); mp 114–115 °C; R_f (EtOAc/MeOH/H_2O/AcOH, 10:1:1:0.5) = 0.53.

IR (KBr): $\tilde{v}$ (cm^{-1}) = 3343, 2967, 2918, 2594, 2461, 1694, 1621, 1561, 1446, 1423, 1377, 1339, 1322, 1256, 1235, 1190, 1117, 961, 799, 661, 593, 547.

^{1}H NMR (CD$_3$OD, 300 MHz): δ (ppm) = 4.52 (dd, J = 4.7, 9.6 Hz, 1 H, α-CH), 2.63–2.45 (m, 2 H, γ-CH$_2$), 2.08 (s, 3 H, CH$_3$CO), 2.19–1.85 (m, 2 H, β-CH$_2$), 1.98 (s, 3 H, SCH$_3$).

^{13}C NMR (CD$_3$OD, 75 MHz): δ (ppm) = 175.2 (COOH), 173.5 (CONH), 157.2, 52.8 (α-C), 52.7, 32.1 (β-CH$_2$), 31.2 (γ-CH$_2$), 22.4 (CH$_3$CO), 15.2 (CH$_3$S).

EI HRMS: calcd. for [M + H]$^+$: 192.06888; found: 192.06889.

4.4.3.2   **   *L*-Methionine [6]

In an Erlenmeyer flask, NaOH (1 N, 19 mL) is added to a solution of acetylmethionine **4.4.3.1** (3.82 g, 20 mmol) in distilled H$_2$O (100 mL). The pH is adjusted to 7.2–7.5 by adding further NaOH or acetic acid as required. The overall volume is then made up to 200 mL with distilled H$_2$O. Acylase I (20–30 mg, from pork liver, Sigma, grade II, 500–1500 units/mg) is added and the mixture is kept at 30–35 °C for 2 d.

The reaction is then quenched by the addition of H$_2$SO$_4$ (2 N, 8 mL) and the mixture is filtered through Celite. A cation exchanger (Amberlyst 15, H$^+$-form, Fluka, 30 mL as a suspension in H$_2$O) is added and the mixture is shaken for 2 min. The solution is tested for remaining methionine with ninhydrin. If necessary, an additional ion exchanger is added. The ion exchanger is then filtered off and washed with H$_2$O. An aqueous solution of ammonia diluted with further H$_2$O (1:9, 20–30 mL) is added to the ion exchanger until a pH of 8–9 is obtained. The ion exchanger is separated and washed with H$_2$O. The solution is adjusted to pH 6–7 with glacial AcOH and then concentrated to a volume of about 10 mL *in vacuo*. EtOH (100–200 mL) is then added until the amino acid precipitates. The product is filtered off and dried; 820 mg (55%), mp 280 °C (dec.); [α]$_D^{20}$ = +23.4 (c = 4.11, 5 M HCl).

UV (MeOH): λ$_{max}$ (nm) = 287, 244, 204, 202.

IR (KBr, pellet): $\tilde{v}$ (cm^{-1}) = 2945, 2915, 2724, 2615, 2319, 1656, 1609, 1580, 1515, 1445, 1414, 1340, 1315, 1276, 1220, 1161, 1080, 1046, 930, 878, 780, 719, 685, 553, 439.

^{1}H NMR (D$_2$O, 300 MHz): δ (ppm) = 3.93 (t, J = 6.55 Hz, 1 H, α-CH), 2.72 (t, J = 7.71 Hz, 2 H, γ-CH$_2$), 2.34–2.14 (m, 2 H, β-CH$_2$), 2.21 (s, 3 H, SCH$_3$).

^{13}C NMR (D$_2$O, 125.7 MHz): δ (ppm) = 175.1 (CO), 54.96 (CH), 30.7 (β-CH$_2$), 29.9 (γ-CH$_2$).

EI HRMS: calcd. for C$_5$H$_{11}$NO$_2$S: 149.0510; found: 149.0509.

4.4.3.3   *   *N*-Boc-*L*-methionine [7]

L-Methionine **4.4.3.2** (500 mg, 3.35 mmol) is dissolved in H_2O/1 N NaOH/dioxane (1:1:2; 24 mL) and the solution is cooled to 0 °C. Boc_2O (875 mg, 4.02 mmol) is added and the mixture is stirred for 45 min at 0 °C and for 24 h at room temperature. During the reaction, the pH is adjusted to 9–9.5 by adding further NaOH (1 N).

The reaction is quenched by adding HCl (1 N) until the pH is 6.5 and then the mixture is concentrated *in vacuo*. Purification of the residue by column chromatography on RP-18 (H_2O, gradient MeOH 0 to 30%; 150 mL) and subsequent lyophilization gives the product as a colorless solid; 775 mg (93%); R_f (EtOAc/MeOH, 9:1, +0.5% AcOH) = 0.59; mp 46–48 °C; $[\alpha]_D^{20}$ = –22.5° (c = 0.5 in MeOH).

UV (CH_3CN): λ_{max} (nm) = 212.

IR (KBr, pellet): $\tilde{\nu}$ (cm^{-1}) = 3400, 2977, 2919, 1688, 1592, 1519, 1394, 1367, 1252, 1170, 1050, 1027, 860, 779.

^{1}H NMR ($CDCl_3$, 300 MHz): δ (ppm) = 6.14 (d, J = 7.8 Hz, 1 H, NH), 3.99 (m, 1 H, α-CH), 2.52 (t, J = 7.4 Hz, 2 H, γ-CH_2), 2.06 (s, 1 H, CH_3S), 1.86 (m, 2 H, β-CH_2), 1.42 (s, 9 H, H_{Boc}).

^{13}C NMR ($CDCl_3$, 75.5 MHz): δ (ppm) = 179.3 (COOH), 156.7 (CONH), 79.6 ($\underline{C}(CH_3)_3$), 55.6 (CH), 32.2 (CH_2), 30.7 (CH_2), 28.5 (C($\underline{C}H_3$)$_3$), 15.4 (CH_3S).

EI HRMS: calcd. for $C_{10}H_{19}NO_4S$ [M + Na^+]: 272.09270; found: 272.09285.

---

[1]   G. C. Barrett (Ed.), *Chemistry and Biochemistry of the Amino Acids*, Chapman and Hall, London, **1985**.

[2]   H. Dugas, *Bioorganic Chemistry, A Chemical Approach to Enzyme Action*, pp. 52–77, Springer, New York, **1989**.

[3]   H. Gröger, K. Drauz, in *Asymmetric Catalysis on Industrial Scales*, pp. 131–145 (Eds.: H.-U. Blaser, E. Schmidt), Wiley-VCH, Weinheim, **2004**.

[4]   B. Schnell, K. Faber, W. Kroutil, *Adv. Synth. Catal.* **2003**, *345*, 653.

[5]   J. Jones, *Amino Acid and Peptide Synthesis*, pp. 8–12, Oxford University Press, New York, **1992**.

[6]   J. A. Moore, D. L. Dalrymple, *Experimental Methods in Organic Chemistry*, pp. 253–258, Saunders Company, Philadelphia, 2nd ed., **1976**.

[7]   M. Bodanszky, A. Bodanszky, *The Practice of Peptide Synthesis*, pp. 15–18, Springer, Berlin, **1995**.

4.4.4 γ,δ-Unsaturated α-amino acids

PG−N̈H   COOH

PG = protective group

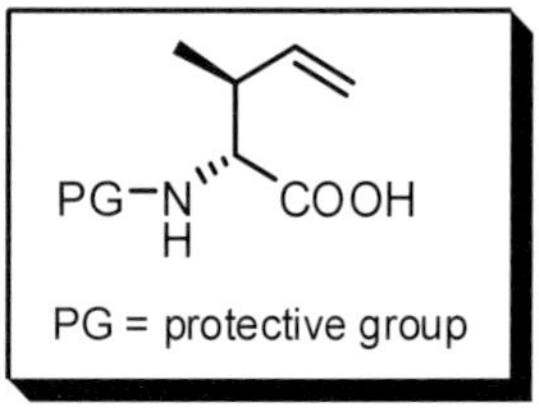

Topics:
- Synthesis of *N*-protected γ,δ-unsaturated α-amino acids
  (a) by [3,3]-sigmatropic rearrangement of chelate-bridged glycine ester enolates,
  (b) by asymmetric glycine ester enolate Claisen rearrangement
- Esterification of Z- and TFA-protected glycine
- Z- and TFA-protection of glycine

(a) General

γ,δ-Unsaturated α-amino acids have attracted considerable attention. Some of these unsaturated α-amino acids occur in Nature, exhibit pronounced activities as, e.g., antibiotics and enzyme inhibitors, and have proved useful for the synthesis of more complex compounds by functionalization of the double bond [1].

The method of choice for the synthesis of γ,δ-unsaturated α-amino acids is the [3,3]-sigmatropic rearrangement of the enolates of *N*-protected glycine allyl esters **1** mediated by chelating metal salts, preferentially ZnCl₂, to give the diastereomeric C-allyl-α-amino acids **2** [1–3]:

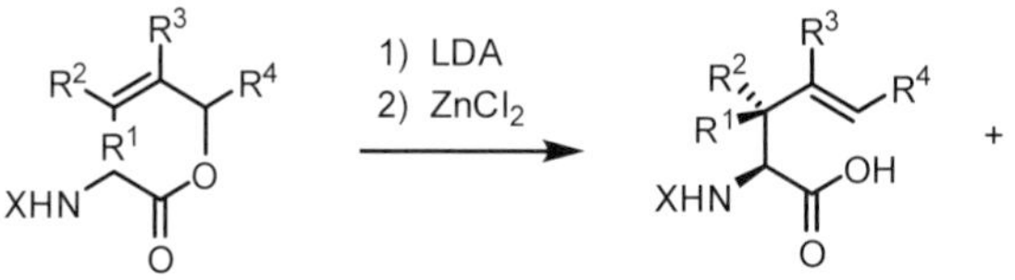

The allyl esters **1** show high diastereoselectivities (*de* = 93–96%) in this rearrangement, in which the formation of *syn*-products from *trans*-substituted allyl esters and of *anti*-products from *cis*-substituted allyl esters is generally favored. This is illustrated for the Z-protected crotyl ester **3** (as presented in section (b)), which is rearranged to the acids **4a/4b** in a diastereomeric ratio of 95:5 in favor of the *syn*-product **4a**:

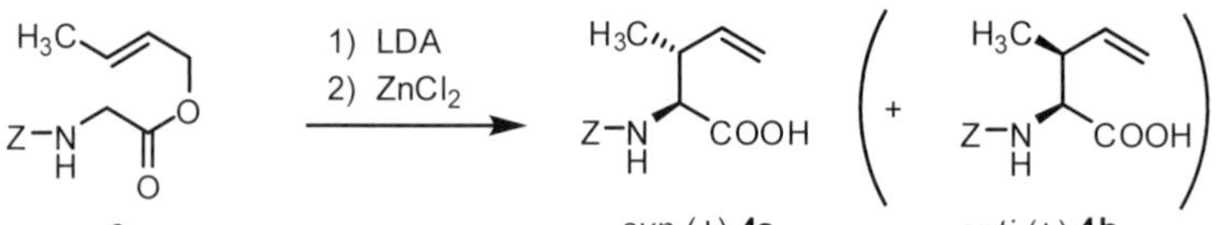

As illustrated for the rearrangement **3 → 4**, a mechanism is proposed [1] in which a Zn-chelated ester enolate **6** (formed from the Li enolate **5** after α-deprotonation of the ester **3** with LDA) serves as the key intermediate. The Zn-enolate **6** is transformed in a Claisen-analogous [3,3]-sigmatropic process to a chelate-bridged (stabilized) carboxylate **7** via a chair-like transition state **8**, the highly ordered geometry of which is likely to account for the high *syn*-diastereoselectivity observed:

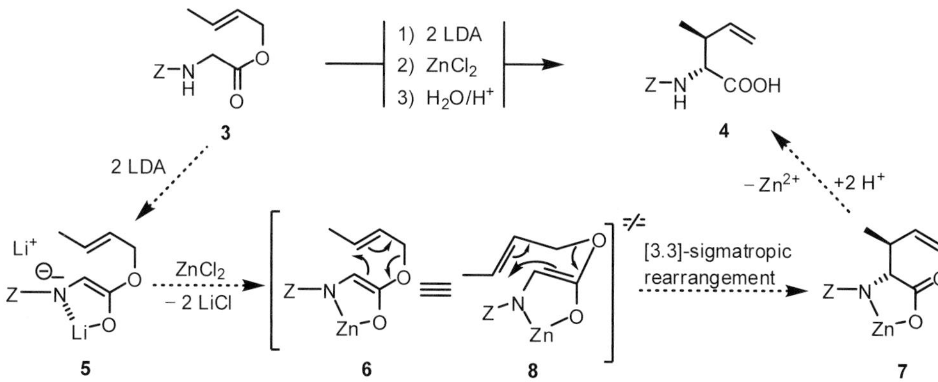

When this glycine allyl ester enolate rearrangement is performed in the presence of a chiral ligand, the [3,3]-sigmatropic reaction proceeds in an asymmetric fashion with high diastereo- and enantio-selectivity [4, 5]. As verified by the example in section (b), the best results are obtained with electron-withdrawing *N*-protecting groups (e.g. trifluoroacetyl), the isopropoxides of Mg and Al as metal components, and the cinchona alkaloids as chiral ligands. Interestingly, the use of quinine gives rise to formation of the (2*R*)-configured amino acid **10** from the TFA-protected glycine ester **9**, whilst quinidine provides the opposite enantiomer (**11**):

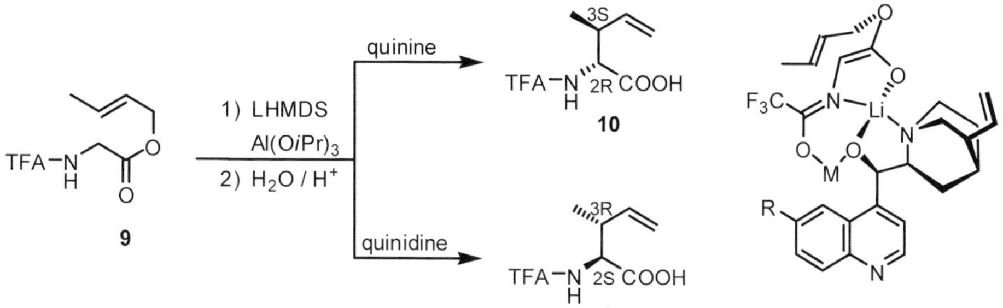

As a mechanistic rationale, a bimetallic complex **12** is postulated [4] to be formed with the bidentate ligand quinine (or quinidine, respectively), which coordinates to the lithium enolate. The incorporation of a second metal ion (Li$^+$, Al^{3+}, Mg^{2+}) should stabilize the complex **12** by imparting a rigid structure, in which one face of the enolate is shielded by the bicyclic substructure of the cinchona alkaloid, thus explaining the high stereoselectivity obtained.

(b) Syntheses of *syn*-(±)-4a and (2*R*,3*S*)-10

(1) Synthesis of *syn*-(±)-4a [1]:

N-(Benzoyloxycarbonyl)glycine (**13**) is prepared by acylation of glycine with carbobenzoxy chloride (Z-chloride) in the presence of a base, e.g. NaOH [6]. The Z-protected glycine **13** is esterified with crotyl alcohol using the DCC/DMAP method to give the Z-protected ester **3**.

The crotyl ester **3** is treated with LDA in THF at −78 °C followed by anhydrous ZnCl$_2$. After hydrolysis with aqueous HCl, the acid **4** is obtained in 90% yield with a *syn/anti* ratio of 95:5 (determined by HPLC analysis of the readily available methyl ester **15**). Recrystallization from diethyl ether/petroleum ether affords the pure *syn*-diastereomer *syn*-(±)-**4a** in 78% yield.

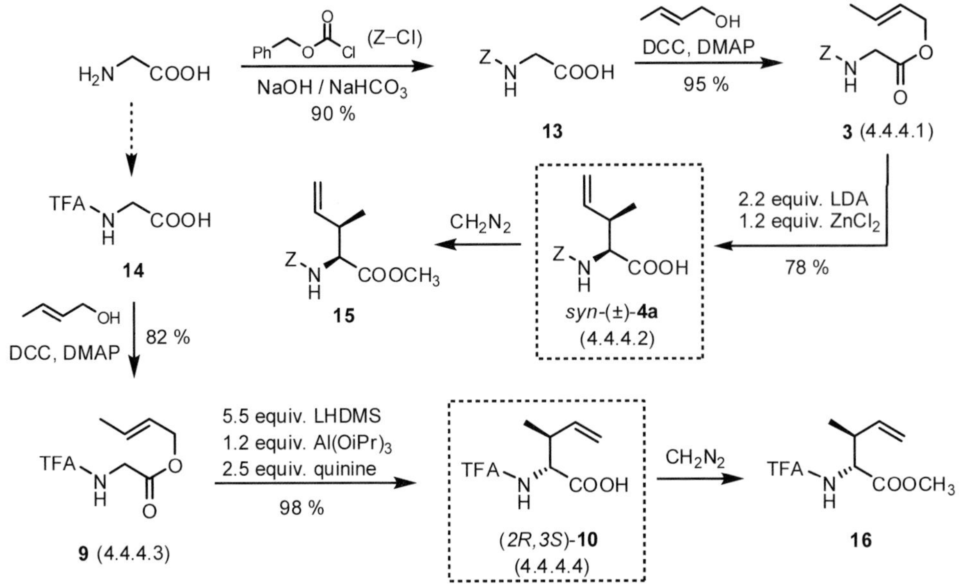

(2) Synthesis of (2R,3S)-**10** [4, 5]:

N-(Trifluoroacetyl)glycine (**14**) can be prepared by trifluoroacetylation of glycine with trifluoroacetic anhydride [7]. The trifluoroacetyl-protected glycine **14** is esterified with crotyl alcohol as above in (1) to give the ester **9**.

The crotyl ester **9** is treated with excess Li-HMDS in THF at −78 °C followed by Al(OiPr)₃ and quinine. After hydrolytic work-up, the acid **10** is obtained in practically quantitative yield; for determination of the diastereomeric and enantiomeric ratios of the product, **10** may be converted to the methyl ester by reaction with diazomethane, which affords the (2R,3S)-ester **16** in 98% yield with de = 98% and ee = 87%.

(c) Experimental procedures for the synthesis of syn-(±)-4a and (2R,3S)-10

4.4.4.1   **   N-(Benzyloxycarbonyl)glycine crotyl ester [4]

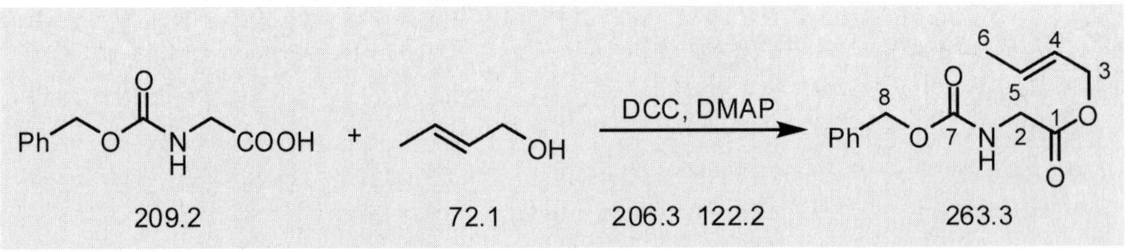

(a) N-(Benzyloxycarbonyl)glycine: Z-Glycine is available commercially or can be prepared according to ref. [7] from glycine and Z-chloride; colorless prisms, mp 120–121 °C (CHCl₃).

(b) DCC (5.88 g, 28.5 mmol) and DMAP (375 mg, 3.0 mmol) are added to a solution of (Z)-crotyl alcohol (3.07 g, 28.5 mmol) in CH$_2$Cl$_2$ (90 mL) at 0 °C. The clear solution is cooled to –20 °C, and after 5 min Z-glycine (5.53 g, 28.5 mmol) is added with stirring. Stirring is continued for 12 h, during which the mixture is allowed to warm to room temperature.

The precipitate of dicyclohexylurea is filtered off and the filtrate is extracted with KHSO$_4$ solution (1 N), saturated NaHCO$_3$ solution, and brine (50 mL each). The organic phase is dried (Na$_2$SO$_4$) and filtered and the solvent is evaporated. The crude ester is purified by flash chromatography (SiO$_2$; petroleum ether/EtOAc, 7:3); 6.82 g (97%), colorless oil, R_f = 0.40.

^{1}H NMR (CDCl$_3$, 300 MHz): δ (ppm) = 7.30–7.35 (m, 5 H, phenyl-H), 5.80 (dq, J = 15.1, 5.7 Hz, 1 H, H-5), 5.55 (dt, J = 15.1, 6.9 Hz, 1 H, H-4), 5.36 (s$_{br}$, 1 H, NH), 5.11 (s, 2 H, H-8), 4.56 (d, J = 5.7 Hz, 2 H, H-3), 3.96 (d, J = 5.5 Hz, 2 H, H-2), 1.72 (d, J = 6.9 Hz, 3 H, H-6).

^{13}C NMR (CDCl$_3$, 75 MHz): δ (ppm) = 169.8 (C-1), 156.2 (C-7), 136.3 (phenyl-C-1), 132.6 (C-5), 128.5 (phenyl-C), 128.1, 124.4 (C-4), 67.0 (C-8), 66.1 (C-3), 42.8 (C-2), 17.7 (C-6).

4.4.4.2   ***   syn-(±)-N-(Benzyloxycarbonyl)-2-amino-3-methyl-4-pentenoic acid [1]

A 1.65 M solution of n-butyllithium in n-hexane (20 mL, 33.0 mmol) is added to a stirred solution of diisopropylamine (5.60 g, 40.0 mmol) in anhydrous THF (30 mL) under an argon atmosphere at –20 °C. After 20 min at this temperature, the solution is cooled to –78 °C and a solution of the ester **4.4.4.1** (3.95 g 15.0 mmol) in THF (15 mL) is added dropwise. A solution of ZnCl$_2$ (2.32 g, 17.0 mmol; note) in THF (20 mL) is then added and the reaction mixture is allowed to warm to room temperature over 12 h.

The clear, faintly yellow solution is hydrolyzed by the addition of aqueous HCl (1 N, ice bath), the organic solvents are removed *in vacuo*, and the residue is dissolved in Et$_2$O. The ethereal phase is washed with HCl (1 N) and then extracted twice with NaOH (1 N). The combined NaOH phases are acidified with concentrated HCl (ice bath) and the product is extracted with Et$_2$O (twice). The ethereal solution is dried (Na$_2$SO$_4$), the solvent is removed *in vacuo*, and the residue (3.55 g (90%), 95:5 mixture of diastereomers) is recrystallized from Et$_2$O/petroleum ether; colorless needles, 3.08 g (78 %), mp 81–82 °C; diastereomerically pure *syn*-(±) acid.

To determine the diastereomeric ratio and the enantiomeric purity, a small sample of the acid is treated with ethereal diazomethane solution [8] and the methyl ester formed is purified by flash chromatography on silica gel (quantitative yield) and examined by HPLC (Chiralcel OD-H, n-hexane/iPrOH, 85:15, 2 mL min^{-1}).

---

^{1}**H NMR** (CDCl$_3$, 300 MHz): δ (ppm) = 7.28–7.40 (m, 5 H, phenyl-H), 5.68 (ddd, J = 17.1, 9.8, 1.4 Hz, 1 H, H-4), 5.28 (d, J = 8.4 Hz, 1 H, NH), 5.08 (s, 2 H, H-9), 5.06 (dd, J = 9.8, 1.4 Hz, 1 H, H-5$_t$), 5.05 (dd, J = 17.1, 1.4 Hz, 1 H, H-5$_c$), 4.35 (dd, J = 9.0, 5.2 Hz, 1 H, H-2), 3.71 (s, 3 H, H-7), 2.64 (dd, J =12.8, 6.7 Hz, 1 H, H-3), 1.03 (d, J = 7.0 Hz, 3 H, H-6).

^{13}**C NMR** (CDCl$_3$, 75 MHz): δ (ppm) = 171.78 (C-1), 155.89 (C-8), 138.41 (C-4), 136.22 (phenyl-C-1), 128.48 (phenyl-C), 128.07, 116.32 (C-5), 67.03 (C-7), 57.86 (C-9), 52.01 (C-2), 40.78 (C-3), 16.38 (C-6),

---

Note: ZnCl$_2$ has to be dried *in vacuo* with a heat gun before use, until it is in the form of a white powder.

4.4.4.3   *   *N*-(Trifluoroacetyl)glycine crotyl ester [5]

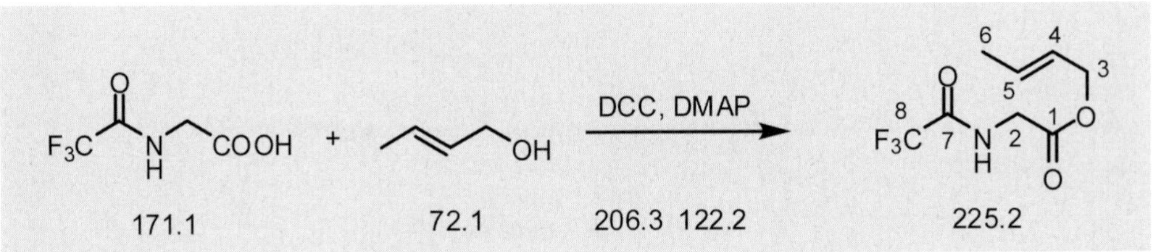

(a) *N*-(Trifluoroacetyl)glycine: *N*-TFA-glycine is commercially available or can be prepared from glycine and trifluoroacetic anhydride [7]; colorless crystals, mp 118–119 °C (Et$_2$O/petroleum ether).

(b) The esterification is performed according to the procedure used in **4.4.4.1** (b) with *N*-(trifluoroacetyl)glycine (7.58 g, 44.4 mmol), (*Z*)-crotyl alcohol (3.85 g, 53.3 mmol), DCC (10.1 g, 48.8 mmol), and DMAP (4.9 mmol). The crude product is purified by flash chromatography on SiO$_2$ (eluent: petroleum ether/Et$_2$O, 4:1, R_f = 0.55) and subsequent recrystallization from CH$_2$Cl$_2$/petroleum ether; 9.80 g (82%), colorless needles, mp 48–49 °C.

---

^{1}**H NMR** (CDCl$_3$, 300 MHz): δ (ppm) = 6.96 (s$_{br}$, 1 H, NH), 5.84 (dt, J = 17.4, 6.9 Hz, 1 H, H-4), 5.58 (dq, J = 17.4, 6.8 Hz, 1 H, H-5), 4.61 (d, J = 6.9 Hz, 2 H, H-3), 4.11 (d, J = 5.1 Hz, 2 H, H-2), 1.72 (d, J = 6.8 Hz, 3 H, H-6).

^{13}**C NMR** (CDCl$_3$, 75 MHz): δ (ppm) = 168.03 (C-1), 157.17 (q, J = 38.0 Hz, C-7), 133.01 (C-5), 123.97 (C-4), 115.62 (q, J = 286.5 Hz, C-8), 66.80 (C-3), 41.38 (C-2), 17.66 (C-6).

4.4.4.4   ***   (2*R*,3*S*)-3-Methyl-2-(trifluoroacetylamino)-4-pentenoic acid [4]

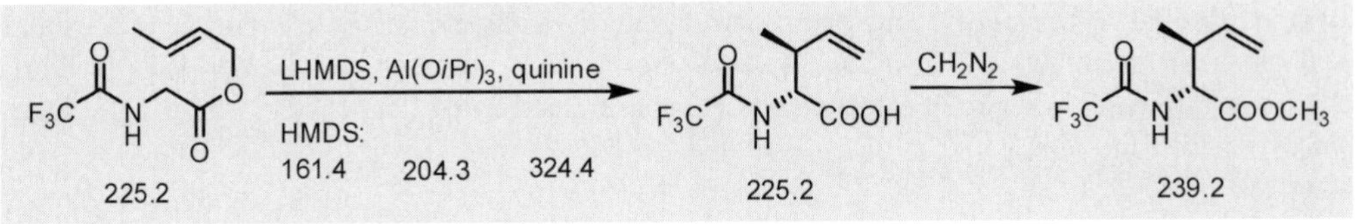

A solution of LHMDS is prepared by adding *n*-butyllithium (1.55 M in *n*-hexane, 1.6 mL, 2.5 mmol) to a stirred solution of hexamethyldisilazane (470 mg, 2.9 mmol) in anhydrous THF (1.5 mL) at –20 °C under an argon atmosphere. After stirring for 20 min, the freshly prepared LHMDS solution is added dropwise to a solution of the ester **4.4.4.3** (124 mg, 0.55 mmol), Al(O*i*Pr)₃ (114 mg, 0.55 mmol), and quinine (405 mg, 1.25 mmol) in anhydrous THF at –78 °C. The reaction mixture is allowed to warm to room temperature over 12 h.

After dilution with Et₂O (50 mL), the reaction mixture is hydrolyzed by the addition of KHSO₄ solution (1 M, 25 mL). The organic layer is washed with KHSO₄ solution (1 M) and the product is extracted into saturated NaHCO₃ solution (3 × 25 mL). The combined basic extracts are subsequently acidified to pH 1 by the careful addition of solid KHSO₄ and then extracted with Et₂O (3 × 25 mL). The combined ethereal extracts are dried (Na₂SO₄) and the solvent is evaporated under reduced pressure.

To determine the enantiomeric and diastereomeric ratios, the residue (121 mg (98%)) is esterified with diazomethane [8]; 125 mg (98%), colorless oil, $[\alpha]_D^{20} = -54.4°$ (c = 2.0, CHCl₃), *ee* = 87%, *de* = 98% (GC on Chirasil-Val, 80 °C, isothermal).

---

¹H NMR (CDCl₃, 300 MHz): δ (ppm) = 6.84 (s_br, 1 H, NH), 5.65 (ddd, *J* = 17.1, 10.5, 6.9 Hz, 1 H, C**H**=CH₂), 5.14 (dd, *J* = 10.5, 1.1 Hz, 1 H, CH=C**H₂**), 5.09 (dd, *J* = 17.1, 1.1 Hz, 1 H, CH=C**H₂**), 4.63 (dd, *J* = 8.5, 5.0 Hz, 1 H, NCH), 3.77 (s, 3 H, OCH₃), 2.73 (ddq, *J* = 8.5, 7.0, 6.9 Hz, C**H**CH₃), 1.09 (d, *J* = 7.0 Hz, 3 H, CHC**H₃**).

¹³C NMR (CDCl₃, 75 MHz): δ (ppm) = 15.4 (CH**C**H₃), 40.6 (**C**HCH₃), 52.6, 56.6 (OCH₃), 115.6 (q, *J* = 287 Hz, CF₃), 117.3 (CH**C**H₂), 137.3 (**C**HCH₂), 156.6 (q, *J* = 38 Hz, CON), 170.3 (COO).

For spectroscopic data of the *anti*-diastereomer, see ref. [5].

[1]   U. Kazmaier, *Angew. Chem.* **1994**, *106*, 1046; *Angew. Chem. Int. Ed. Engl.* **1994**, *33*, 998, and literature cited therein.

[2]   For the transformation 1 → 2, the Claisen–Ireland protocol via silyl ketene acetals (R. E. Ireland, R. H. Mueller, A. K. Willard, *J. Am. Chem. Soc.* **1976**, *98*, 2868) has alternatively been used. The Kazmaier methodology [1] shows clear advantages in terms of reactivity and selectivity.

[3]   For applications of amino acid ester enolates in synthesis, see: a) U. Kazmaier, F. L. Zumpe, *Eur. J. Org. Chem.* **2001**, 1; b) M. Pohlmann, U. Kazmaier, T. Lindner, *J. Org. Chem.* **2004**, *69*, 6909.

[4]   U. Kazmaier, A. Krebs, *Angew. Chem.* **1995**, *107*, 2213; *Angew. Chem. Int. Ed. Engl.* **1995**, *34*, 2012.

[5]   U. Kazmaier, H. Mues, A. Krebs, *Chem. Eur. J.* **2002**, *8*, 1850.

[6]   a) M. Bergmann, L. Zervas, *Chem. Ber.* **1932**, *65*, 1192; b) for Z-protection of α-amino acids, see also: Houben-Weyl, Vol. 15/1, p. 47.

[7]   For alternative methods for trifluoroacetylation of α-amino acids, see: a) F. Weygand, A. Röpsch, *Chem. Ber.* **1959**, *92*, 2095; b) Houben-Weyl, Vol. 15/1, p. 171.

[8]   Prepared according to *Organikum*, 21st ed., p. 647, Wiley-VCH, Weinheim, **2001**.

4.4.5   Passerini hydroxyamide

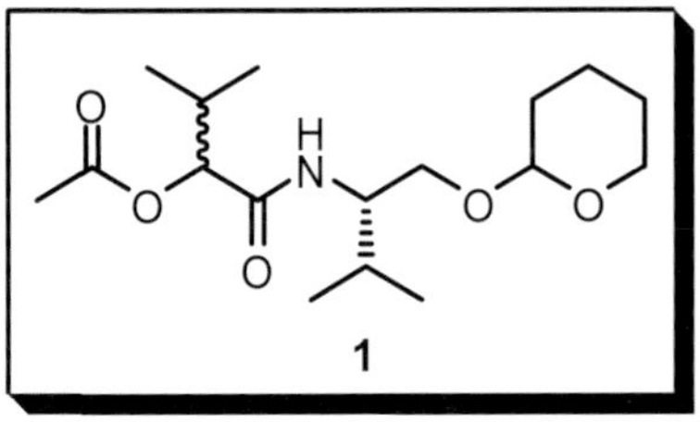

1

Topics:
- Reduction of an amino acid to the corresponding β-amino alcohol
- Chemoselective N-formylation of an amino alcohol
- OH protection by formation of a THP ether
- Isonitrile formation from a primary N-formyl amine
- Passerini reaction

(a) General

Isonitriles, such as **2**, show two basic modes of reactivity due to the specific electronic property of the N=C group. First, they exhibit pronounced C-H acidity of the C–H bonds in the position α- to the N=C function, thus allowing their deprotonation by base to give α-carbanions **3** (or metalation by Li organyls to give α-lithioisonitriles). This opens the possibility of attack by an electrophile, which corresponds to an overall α-substitution (→ **5**). Second, isonitriles are susceptible to stepwise attack of nucleophiles *and* electrophiles at the terminal carbon atom of the N=C function, leading to the geminal introduction of two new substituents; if the electrophile adds first, nitrilium ions **4** are the primary intermediates (→ **6**) [1].

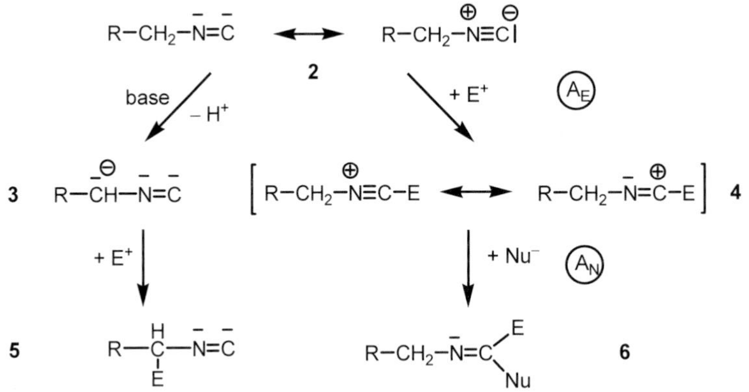

These are the underlying principles of a number of synthetically useful reactions of isonitriles, as illustrated by the following examples.

(1) In the van Leusen synthesis of 1,3-oxazoles [2], the tosyl-substituted isonitrile **2** (R = Tos) (TosMIC) cyclocondenses with an aldehyde in the presence of a base to yield a 5-substituted 1,3-oxazole **8**. In the oxazole formation, the isonitrile group N=C has a threefold function. It facilitates α-deprotonation to give the carbanion **3**, which adds to the aldehyde C=O group to give intermediate **9**. Then, it allows intramolecular interception of the nucleophilic center of **9** (A$_N$), leading to oxazoline **7** by protonation of the 2-anion (A$_E$) formed in the ring-closure **9** → **10**. In the final base-induced elimination step, the oxazoline **7** is aromatized to afford the oxazole **8** by loss of sulfinic acid.

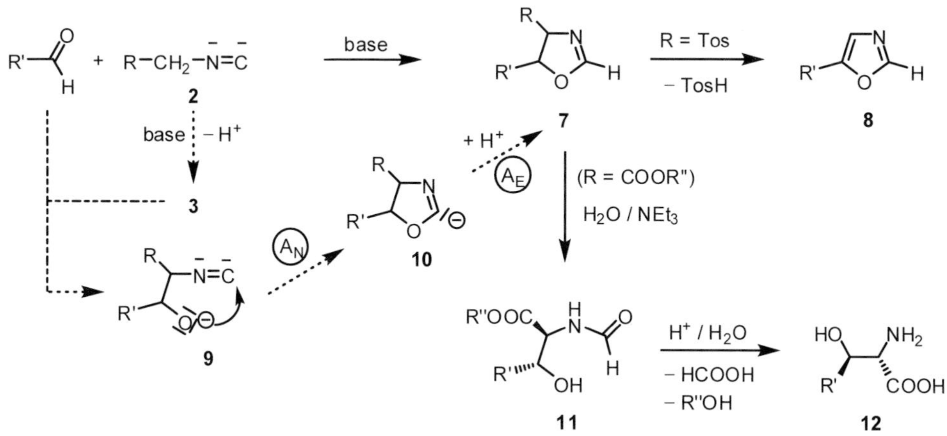

An analogous reaction is observed with the ester-substituted isocyanide **2** (R = COOR'') [3], in which the oxazoline **7** (with *trans*-configuration of R and R') can be isolated and transformed into β-hydroxy-α-amino acids **11/12** by hydrolysis.

(2) In the Passerini reaction [1], isonitriles, aldehydes, and carboxylic acids react in a three-component process (MCR [4]) to give *O*-acylated α-hydroxyamides **13**:

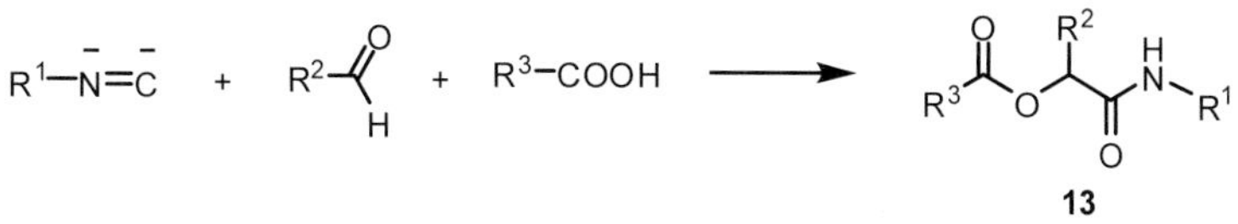

The Passerini reaction is rationalized by the following mechanism [1]:

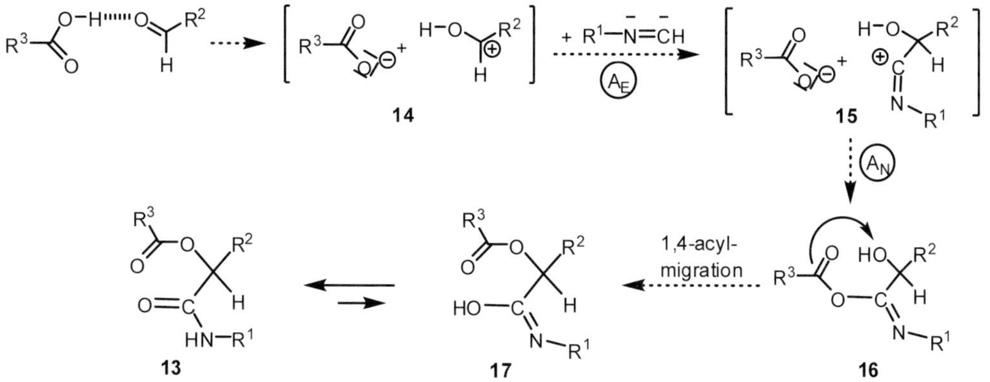

First, the aldehyde carbonyl group is activated by proton transfer from the carboxylic acid. The protonated carbonyl group then undergoes electrophilic addition (A_E) to the isonitrile carbon (**14 → 15**), which is followed by nucleophilic addition (A_N) of the carboxylate to the nitrilium ion in the ion pair **15**; 1,4-acyl migration in the *O*-acylimidate **16** (**16 → 17**) and tautomerization complete the sequence leading to product **13**.

The Passerini reaction can be extended to a four-component process by addition of ammonia or a primary amine (Ugi reaction, cf. **4.4.7**).

(3) As a consequence of the mechanism of the Passerini reaction postulated in (2), β-hydroxy-isonitriles 18 and aldehydes lead to (2-hydroxyalkyl)oxazolines 19 on treatment with a weakly nucleophilic acid such as PPTS [5]:

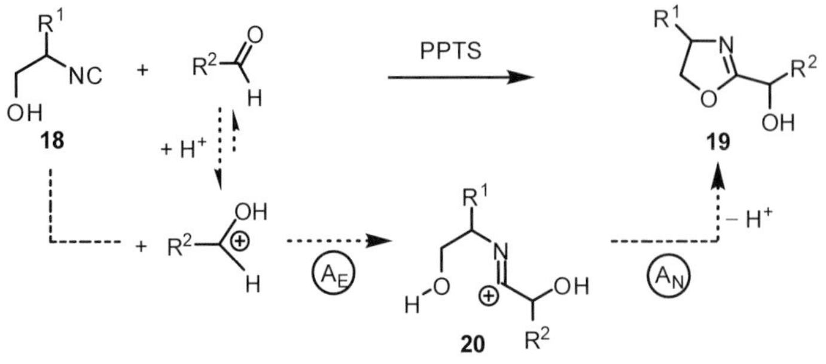

Apparently, the nitrilium ion 20 resulting from addition of the protonated aldehyde to the N=C function is intramolecularly trapped by interception of the β-hydroxy group and is thus cyclized to the oxazoline 19.

Chiral oxazolines of type 19 are of interest as ligands for asymmetric synthesis [5]; the required chiral β-hydroxyisonitriles of type 18 can be conveniently obtained from enantiopure α-amino acids. Thus, the synthesis of a potential precursor and its utility in a Passerini reaction [6] is presented in (b).

(b) Synthesis of 1

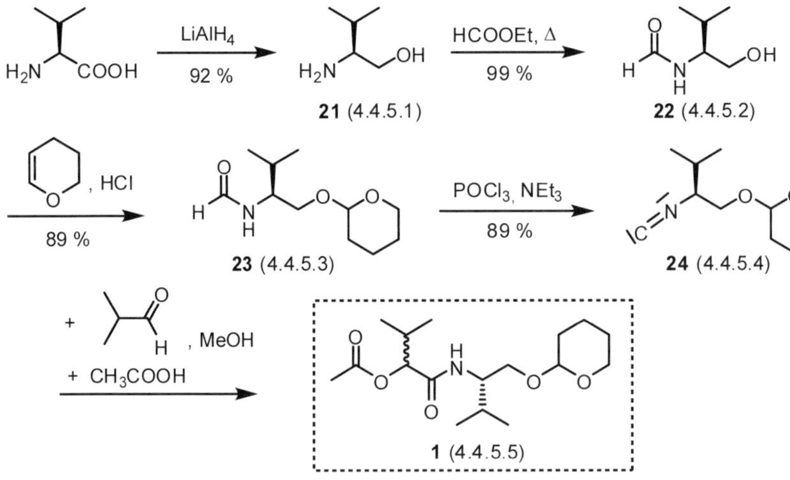

L-Valine is reduced with LiAlH$_4$ and the resulting (S)-valinol (21) is chemoselectively N-formylated by reaction with ethyl formate to give (S)-N-(formyl)-valinol (22). Due to the conditions of the isonitrile-forming dehydration of the formylamino function in the next step, the free OH group in 22 requires protection; this is achieved by proton-catalyzed addition to dihydropyran (DHP) leading to the tetrahydropyranyl ether 23.

As cyclic acetals, tetrahydropyranyl ethers are stable to base, but are sensitive to hydrolysis even with dilute acids. Thus, OH functions in alcohols or phenols can be reversibly blocked by the THP protecting group [7]:

R–OH + [structure] —[H⁺]→ [structure RO–O] —H⁺/H₂O→ [structure HO–O]

Accordingly, treatment of the THP ether **23** with POCl₃ in the presence of triethylamine cleanly affords the β-OH-protected chiral isonitrile **24**.

In the last step, the Passerini reaction is conducted as a three-component domino process with isonitrile **24**, isobutyraldehyde, and acetic acid to give the dihydroxyamide **1** as a mixture of diastereomers (see experimental section), in which the OH groups are protected as an acetate and a tetrahydropyranyl ether, respectively. Thus, the Passerini product **1** is obtained in a five-step sequence with an overall yield of 66% (based on *L*-valine).

(c) Experimental procedures for the synthesis of 1

4.4.5.1 **   (2*S*)-2-Amino-3-methyl-1-butanol (*L*-valinol) [8]

[reaction scheme: H₂N–COOH (117.2) —LiAlH₄ (38.0)→ H₂N–OH (103.2)]

L-Valine (35.1 g, 0.30 mol) is added in small portions to a stirred suspension of LiAlH₄ (25.0 g, 0.66 mol) in anhydrous THF (400 mL) at 0 °C under a nitrogen atmosphere (Caution: H₂ is evolved!). The mixture is then heated under reflux for 15 h.

After cooling to 0 °C (ice bath), the mixture is hydrolyzed by adding crushed ice (ca. 60 g) and filtered through a Buchner funnel. The filter cake is extracted with THF/water (4:1) by heating under reflux for 1 h (three extractions with 50 mL each time, followed by filtration). The combined filtrates are concentrated *in vacuo* and the oily residue is dissolved in CHCl₃ (250 mL). This solution is heated in an apparatus fitted with a Dean–Stark trap until no more water is separated. The solvent is then removed and the crude product is fractionated *in vacuo* to give (*S*)-valinol as a colorless oil with a somewhat unpleasant odor; 28.5 g (92%), bp₁₂ 85–86 °C; the product crystallizes on standing, mp 55–56 °C; $[\alpha]_D^{20} = +25.7$ (*c* = 1.00 in CHCl₃); TLC (SiO₂; acetone): $R_f = 0.54$.

> **IR** (KBr): $\tilde{\nu}$ (cm⁻¹) = 3200 (NH, OH), 2960, 2860 (CH), 1590 (NH), 1050 (OH).
>
> **¹H NMR** (CDCl₃, 80 MHz): δ (ppm) = 0.90 (d, *J* = 7 Hz, 6 H, 2 × CH₃), 1.50 (m_c, 1 H, 3-H), 1.95 (s, 3 H, OH, NH₂; exchangeable with D₂O), 2.55 (ddd, *J* = 9/6/4 Hz, 1 H, 2-H), 3.25 (dd, *J* = 11/9 Hz, 1 H, 1-H) 3.65 (dd, *J* = 11/4 Hz, 1 H, 1-H).

4.4.5.2 *   (2*S*)-(–)-(2-Formylamino)-3-methyl-1-butanol [6]

[reaction scheme: formylamino butanol (131.2) + dihydropyran (84.1) —HCl (catal.)→ product (215.3)]

L-Valinol (**4.4.5.1**; 10.3 g, 0.10 mol) is dissolved in ethyl formate (9.66 g, 120 mmol) and the solution is heated under reflux for 2 h.

The excess formate is then removed *in vacuo* to leave the product as a colorless oil, which crystallizes on cooling or on treatment with Et_2O; 13.0 g (99%), mp 78–80 °C; $[\alpha]_D^{20} = -36.0$ ($c = 1.00$, $CHCl_3$), TLC (SiO_2; EtOAc/EtOH, 4:1): $R_f = 0.49$.

According to ^1H and ^{13}C NMR, the product consists of a 2:1 mixture of the *Z/E* rotamers.

(*Z*)-Isomer:

^1H NMR ($CDCl_3$, 500 MHz): δ (ppm) = 8.19 (d, *J* = 1.9 Hz, 1 H, 5-H), 6.53 (d, *J* = 6.3 Hz, 1 H, N–H), 3.78 (dddd, *J* = 9.9, 6.3, 6.3, 3.6 Hz, 1 H, 2-H), 3.74 (t, *J* = 5.0 Hz, 1 H, O–H), 3.68–3.53 (m, 2 H, 1-H), 1.85 (m, 1 H, 3-H), 0.94, 0.91 (2d, *J* = 6.8 Hz, 6 H, 4-H).

^{13}C NMR ($CDCl_3$, 125 MHz): δ (ppm) = 162.3 (d, C-5), 62.9 (t, C-1), 55.8 (d, C-2), 28.9 (d, C-3), 19.4 (2q, C-4), 18.6.

(*E*)-Isomer:

^1H NMR ($CDCl_3$, 500 MHz): δ (ppm) = 7.95 (d, *J* = 11.8 Hz, 1 H, 5-H), 6.85 (dd, *J* = 11.8, 7.8 Hz, 1 H, N–H), 4.04 (t, *J* = 5.6 Hz, 1 H, O–H), 3.68–3.53 (m, 2 H, 1-H), 1.77 (m, 1 H, 3-H), 3.10 (dddd, *J* = 10.1, 7.8, 6.6 Hz, 3.6 Hz, 1 H, 2-H), 0.93, 0.89 (2d, *J* = 6.8 Hz, 6 H, 4-H).

^{13}C NMR ($CDCl_3$, 125 MHz): δ (ppm) = 165.8 (d, C-5), 63.2 (t, C-1), 60.8 (d, C-2), 29.3 (d, C-3), 19.6 (2q, C-4), 18.3.

4.4.5.3 * (2*S*)-(2-Formylamino)-3-methyl-1-tetrahydropyranyloxybutane [6]

131.2   84.1   215.3

Two drops of concentrated HCl are added to a stirred suspension of the alcohol **4.4.5.2** (5.91 g, 45.0 mmol) in dihydropyran (6.15 g, 67.5 mmol; *d* = 0.992 g mL^{-1}) at 0 °C. The reaction mixture is allowed to slowly warm to room temperature over 12 h, as a clear solution is formed.

The mixture is then diluted with CH_2Cl_2 (50 mL), washed with saturated $NaHCO_3$ solution and brine, dried (Na_2SO_4), and the solvent is removed *in vacuo*. The residue is purified by chromatography on silica gel (eluent: EtOAc) to afford the THP ether as a colorless oil; 8.58 g (89%), 7:3 mixture of *cis/trans* rotamers.

(*Z*)-Isomer:

^1H NMR ($CDCl_3$, 500 MHz): δ (ppm) = 8.20, 8.19 (2d, *J* = 1.4 Hz, 1 H, 5-H), 5.94 (m, 1 H, N–H), 4.53, 4.49 (2m, 1 H, 6-H), 3.94, 3.89 (2m, 1 H, 2-H), 3.83 (dd, *J* = 10.2, 4.4 Hz, 1 H, 1-Hb), 3.79 (m, 1 H, 10-H$_{eq}$), 3.48 (m, 1 H, 10-H$_{ax}$), 3.38 (2dd, *J* = 10.2, 3.8 Hz, 1 H, 1-Ha), 1.90 (m, 1 H, 3-H), 1.76 (m, 1 H, 7-Hb), 1.67 (m, 1 H, 7-Ha), 1.59–1.47 (m, 4 H, 8-/9-H), 0.95, 0.93, 0.92, 0.91 (4d, *J* = 6.8 Hz, 6 H, 4-H).

^{13}C NMR (CDCl$_3$, 125 MHz): δ (ppm) = 160.85, 160.80 (2d, C-5), 99.8, 98.9 (2d, C-6), 68.1, 67.3 (2t, C-1), 62.9, 62.3 (2t, C-10), 52.9, 52.6 (2d, C-2), 30.6, 30.4 (2t, C-7), 29.4, 29.2 (2d, C-3), 25.3, 25.2 (2t, C-9), 19.4, 19.3 (2t, C-8), 19.8, 19.4, 18.9, 18.8 (4q, C-4).

(E)-Isomer:

^{1}H NMR (CDCl$_3$, 500 MHz): δ (ppm) = 8.03, 8.01 (2d, J = 11.9 Hz, 1 H, 5-H), 6.11 (m, 1 H, N-H), 4.58, 4.53 (2m, 1 H, 6-H), 3.79 (m, 1 H, 10-H$_{eq}$), 3.69 (dd, J = 10.4, 4.0 Hz, 1 H, 1-Hb), 3.57 (dd, J = 10.4, 3.8 Hz, 1 H, 1-Ha), 3.48 (m, 1 H, 10-H$_{ax}$), 3.23 (m, 1 H, 2-H), 1.86 (m, 1 H, 3-H), 1.76 (m, 1 H, 7-Hb), 1.67 (m, 1 H, 7-Ha), 1.59–1.46 (m, 4 H, 8-/9-H), 0.96–0.90 (m, 6 H, 4-H).

^{13}C NMR (CDCl$_3$, 125 MHz): δ (ppm) = 164.8, 164.6 (2d, C-5), 99.5, 98.4 (2d, C-6), 68.9, 68.2 (2t, C-1), 62.4, 61.8 (2t, C-10), 58.0, 57.8 (2d, C-2), 30.4, 30.3 (2t, C-7), 29.6, 29.5 (2d, C-3), 25.3, 25.2 (2t, C-9), 19.58, 19.55 (2t, C-8), 19.4, 18.9, 18.3, 18.1 (4q, C-4).

4.4.5.4   **   (2S)-2-Isocyano-3-methyl-1-tetrahydropyranyloxybutane [6]

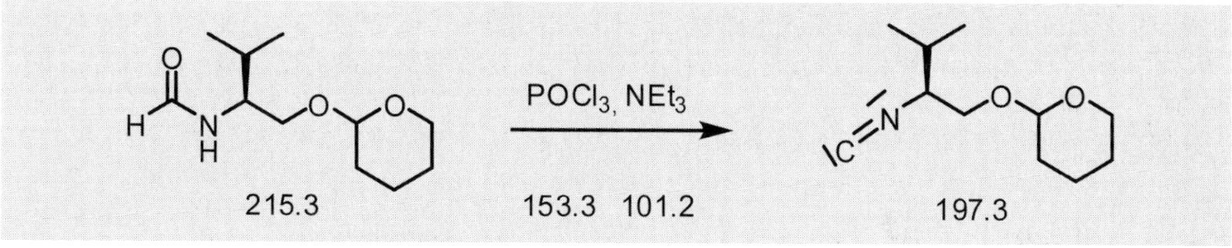

215.3   153.3 101.2   197.3

Triethylamine (10.5 mL, 75.0 mmol; d = 0.72 g mL^{-1}) is added to a stirred solution of the formamide **4.4.5.3** (6.45 g, 30.0 mmol) in anhydrous CH$_2$Cl$_2$ (30 mL) at 0 °C under an N$_2$ atmosphere. Phosphoryl chloride (2.74 mL, 30.0 mmol; d = 1.68 g mL^{-1}) is then added at such a rate that the internal temperature does not exceed 5 °C. For completion of the reaction, stirring is continued at 0 °C for 1 h.

The reaction is then quenched by the dropwise addition of a solution of Na$_2$CO$_3$ (6.0 g) in water (24 mL) with stirring, during which the internal temperature is maintained at 26–28 °C (cooling with an ice bath as necessary). Thereafter, stirring is continued at room temperature for 30 min.

For work-up, water (60 mL) is added, and the aqueous phase is separated and extracted with CH$_2$Cl$_2$ (3 ×). The combined organic phases are washed with brine and dried (K$_2$CO$_3$). The solvent is removed *in vacuo* and the residue is purified by chromatography on silica gel (eluent: EtOAc); 4.86 g (82%), faintly yellow liquid, TLC (Et$_2$O): R_f = 0.74.

^{1}H NMR (CDCl$_3$, 500 MHz): δ (ppm) = 4.64 (t, J = 3.3 Hz, 1 H, 6-H), 3.85 (m, 1 H, 2-H), 3.85, 3.79 (2m, 1 H, 10-H$_{eq}$), 3.60 (m, 1 H, 1-Hb), 3.52 (m, 1 H, 1-Ha), 3.52, 3.46 (2m, 1 H, 10-H$_{ax}$), 1.96 (m, 1 H, 3-H), 1.81 (m, 1 H, 7-Hb), 1.71, (m, 1 H, 7-Ha), 1.64–1.49 (m, 4 H, 8-/9-H), 1.031, 1.028, 1.004, 1.000 (4d, J = 6.8 Hz, 6 H, 4-H).

^{13}C NMR (CDCl$_3$, 125 MHz): δ (ppm) = 156.52, 156.48 (2t, J_{C-N} = 4.7 Hz, C-5), 99.4, 98.5 (2d, C-6), 67.9, 67.2 (2t, C-1), 62.3, 61.9 (2t, C-10), 61.0, 60.7 (2td, J_{C-N} = 5.9 Hz, C-2), 30.4, 30.3 (2t, C-7), 28.91, 28.88 (2d, C-3), 25.3 (t, C-9), 19.6 (t, C-8), 19.2 18.9, 17.0, 16.8 (4q, C-4).

4.4.5.5 * (2S)-2-[(3-Methyl-2-acetoxybutyryl)-amino]-3-methyl-1-tetrahydropyranyloxy-butane [6]

Acetic acid (0.693 mL, 12.0 mmol; $d = 1.04\,\mathrm{g\,mL^{-1}}$) is added to a stirred solution of the isonitrile **4.4.5.4** (1.185 g, 6.00 mmol) in MeOH (1.5 mL). Isobutyraldehyde (1.095 mL, 75.0 mmol; $d = 0.79\,\mathrm{g\,mL^{-1}}$) is then added, leading to an increase in the temperature of the reaction mixture. After the addition is complete, stirring is continued for 30 min at room temperature.

The reaction mixture is then concentrated *in vacuo*, the residue is dissolved in CH_2Cl_2 (45 mL), and this solution is washed with saturated $NaHCO_3$ solution. The aqueous phase is re-extracted with CH_2Cl_2 (3 × 30 mL), and the CH_2Cl_2 phases are combined and dried (Na_2SO_4). After filtration, the solvent is removed *in vacuo* and the residue is purified by chromatography on silica gel (eluent: *n*-hexane/EtOAc, 1:1). The Passerini product is obtained as a mixture of diastereomers; 1.54 g (78%), colorless oil, TLC (*n*-hexane/EtOAc, 1:1): $R_f = 0.54$.

---

^{1}**H NMR** (CDCl$_3$, 500 MHz): δ (ppm) = 6.49, 6.39, 6.16, 6.11 (4d, $J = 9.0/9.3$ Hz, 1 H, N-H), 4.967, 4.959, 4.955, 4.930 (4d, $J = 4.6$ Hz, 1 H, 3-H), 4.46 (m, 1 H, 11-H), 3.84–3.69 (m, 2 H, 7-/15-H$_{eq}$), 3.44 (m, 1 H, 15-H$_{ax}$), 3.76, 3.57, 3.52, 3.33, 3.27 (5m, 2 H, 10-H), 2.22 (m, 1 H, 4-H), 2.091, 2.090, 2.085, 2.075 (4s, 3 H, 1-H), 1.85 (m, 1 H, 8-H), 1.72 (m, 1 H, 12-Hb), 1.63 (m, 1 H, 12-Ha), 1.54–1.43 (m, 4 H, 13-/14-H), 0.91–0.81 (m, 12 H, 5-/9-H).

^{13}**C NMR** (CDCl$_3$, 125 MHz): δ (ppm) = 169.59, 169.53, 169.51, 168.81, 168.73, 168.68 (6s, C-6/-12), 99.6, 99.3, 98.9, 98.8 (4d, C-11), 78.4, 78.3 (2d, C-3), 68.5, 68.0, 67.2 (3t, C-10), 62.53, 62.21, 62.20, 62.15 (4d, C-15), 54.0, 53.8, 53.6, 53.4 (4d, C-7), 30.58, 30.51, 30.41, 30.38 (4t, C-12), 30.26, 30.24, 30.22, 30.21 (4d, C-4), 29.39, 29.35, 29.33, 29.21 (4d, C-8), 25.22, 25.20 (2t, C-14), 20.6 (q, C-1), 19.6–16.7 (m, C-5/-9/-13).

---

[1] M. B. Smith, J. March, *March's Advanced Organic Chemistry*, 5th ed., p. 1251, John Wiley & Sons, Inc., New York, **2001**, and literature cited therein.

[2] Th. Eicher, S. Hauptmann, *The Chemistry of Heterocycles*, 2nd ed., p. 128, Wiley-VCH, Weinheim, **2003**.

[3] D. Hoppe, U. Schöllkopf, *Liebigs Ann. Chem.* **1972**, *763*, 1; see also: L. F. Tietze, Th. Eicher, *Reaktionen und Synthesen im organisch-chemischen Praktikum und Forschungslaboratorium*, p. 525, Georg Thieme Verlag, Stuttgart, **1991**, and literature cited therein.

[4] J. Zhu, H. Bienayme (Eds.), *Multicomponent Reactions*, Wiley-VCH, Weinheim, **2004**.

[5] a) U. Kazmaier, M. Bauer, *J. Organomet. Chem* **2006**, *691*, 2155; b) chiral bis(oxazoline) ligands in asymmetric catalysis: G. Desimoni, G. Faita, K. A. Joergensen, *Chem. Rev.* **2006**, *106*, 3561.

[6] U. Kazmaier, M. Bauer, personal communication.

[7] See ref. [2], p. 240, and P. J. Kocienski, *Protective Groups*, p. 83, Georg Thieme Verlag, Stuttgart, **1994**.

[8] D. A. Evans, M. D. Ennis, D. J. Mathre, *J. Am. Chem. Soc.* **1982**, *104*, 1737.

4.4.6 Aspartame

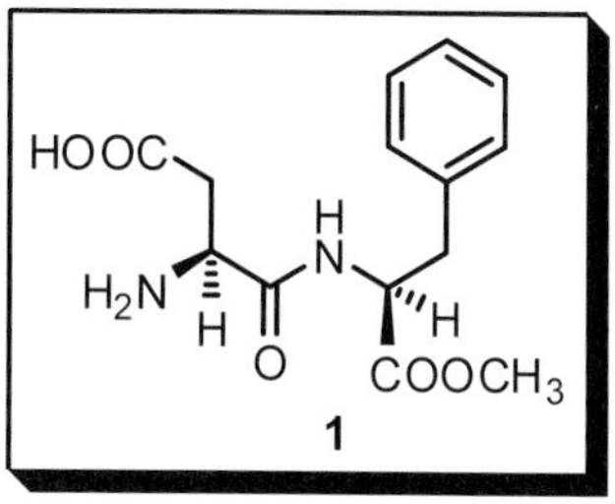

Topics:
- Synthesis of a dipeptide ester
- N-Carboxybenzylation of an α-amino acid
- Esterification of α-amino acids
- Partial hydrolysis of an L-aspartic acid diester
- Peptide formation using the DCC method
- Catalytic debenzylation

(a) General

Aspartame (**1**, N-L-α-aspartyl-L-phenylalanine methyl ester; abridged nomenclature: H-Asp-Phe-OCH$_3$) is a dipeptide ester in which L-phenylalanine methyl ester is attached via its NH$_2$ group to the α-carboxyl function of L-aspartic acid:

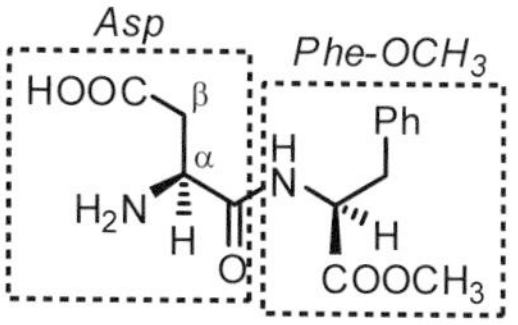

Aspartame is used commercially as a sweetener [1]; it exhibits a very low nutritive value and a sweetness intensity that is 180–200 times higher than that of sucrose. Aspartame is devoid of carcinogenicity, chronic toxicity, or teratogenicity [2].

The synthesis of aspartame is performed according to the fundamentals of peptide synthesis [3]. The key feature is the directed (chemoselective) formation of a CO–NH amide bond ("peptide bond") by condensation of bi- and trifunctional α-amino acids (here: **2/3**), which proceeds according to the following principles:

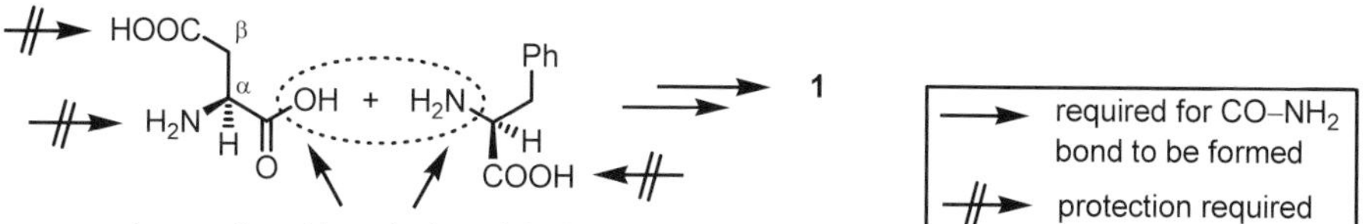

(1) Directed peptide formation demands *protection* of NH$_2$ and COOH functions (as well as other functional groups) that do not participate in the peptide-bond-forming process. Commonly used protecting groups for NH$_2$ are Boc (*tert*-butyloxycarbonyl), Cbz (benzyloxycarbonyl), and Fmoc (fluorenyl-9-methoxycarbonyl); for COOH, various ester groups are applied, often benzyl.

(2) To facilitate amide formation between a COOH and an NH$_2$ group – a reaction that normally requires drastic conditions – the COOH group needs *activation*, in general by formation of a so-called "active ester" through the introduction of a good leaving group for nucleophilic attack of the amine function. Most commonly applied are the DCC (dicyclohexyl carbodiimide), HOBt (1-hydroxy-benzotriazole), and anhydride methods.

(3) After formation of the peptide bond, the protective groups have to be removed (*deprotection*) under reaction conditions that do not affect the amide functionality and that do not lead to isomerization of the stereogenic centers.

(4) The formation of peptide bonds using specific enzymes (proteases) represents a highly economic alternative to the above mentioned strategy (1)–(3) since it can be performed with high selectivity and efficiency, *without* the use of protecting groups [4].

Using the first approach (1)–(3) for aspartame synthesis, *L*-aspartic acid has to be protected at the NH_2 and β-carboxyl groups, and then this bis-protected *L*-aspartic acid **4** has to be condensed with *L*-phenylalanine protected at the COOH group, preferentially as methyl ester **5**. Removal of the protecting groups P^1 and P^2 from the dipeptide ester **6** should then lead to aspartame **1**:

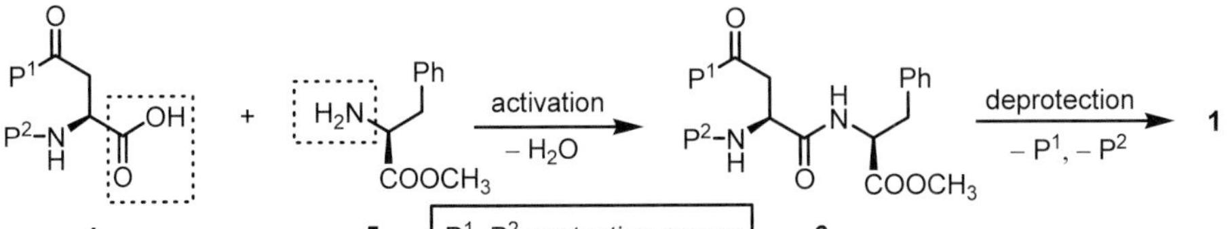

A synthesis of **1** following this strategy [5, 6] is presented in section (b).

For an industrial synthesis of aspartame **1** [7], the anhydride **7** would be a very good starting material, since it represents an *N*-protected and COOH-activated form of *L*-aspartic acid. However, it contains two C=O groups, which in the nucleophilic ring-opening with *L*-phenylalanine would lead to a mixture of *N*-formyl-α- and -β-aspartylphenylalanines **8** and **9**. Fortunately, the reaction can be directed towards the desired α-dipeptide by selecting an appropriate solvent. By treatment with HCl/MeOH, the formyl protecting group is removed and the COOH group of phenylalanine is selectively transformed into the methyl ester to give aspartame hydrochloride **10**, from which aspartame **1** is liberated by reaction with base. The accompanying β-isomer **11** can be efficiently separated by fractional crystallization of the mixture of **10** and **11**.

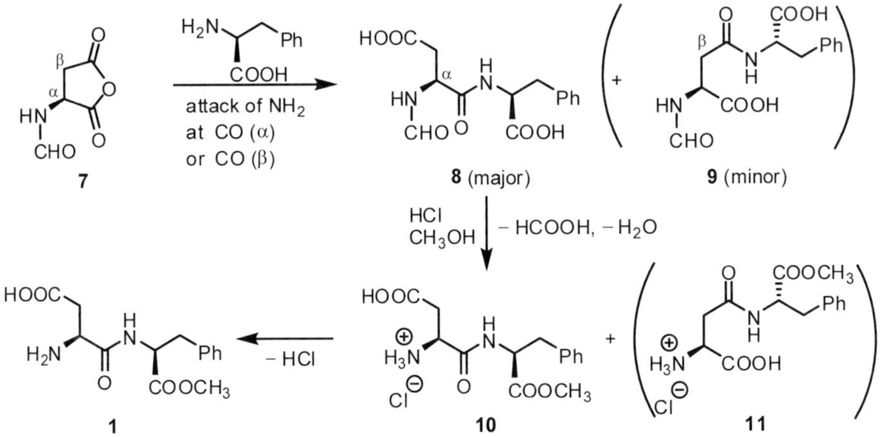

Notably, in enzyme-catalyzed aspartame syntheses [1, 2], protected or unprotected *L*-aspartic anhydride or *L*-aspartic acid itself is linked directly to *L*-phenylalanine or its methyl ester.

(b) Synthesis of 1

L-Aspartic acid (2) is acylated at the NH_2 group using benzyloxycarbonyl chloride (12) in the presence of $NaOH/NaHCO_3$ (→ 13). The formed N-protected aspartic acid 13 is then treated with benzyl alcohol in the presence of a catalytic amount of TosOH to yield dibenzyl N-benzyloxycarbonyl-L-aspartate (14) under azeotropic removal of H_2O:

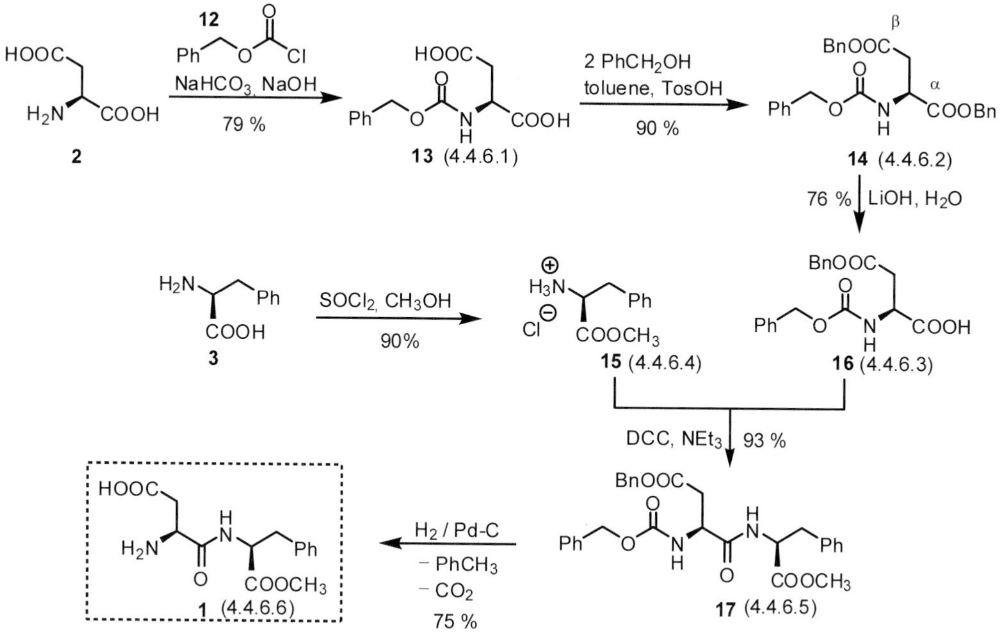

The dibenzyl ester 14 is subjected to partial alkaline hydrolysis ($LiOH/H_2O$) of the α-ester functionality giving rise to the N-protected L-aspartic acid β-benzyl ester 16; the chemoselective saponification 14 → 16 (accelerated reaction of the α-ester vs. the β-ester) might be due to an inductive effect of the amide moiety.

As the second substrate, the hydrochloride 15 of L-phenylalanine methyl ester is prepared by reaction of L-phenylalanine (3) with thionyl chloride in methanol (general method for the preparation of amino esters) [8].

The formation of the dipeptide of the protected aspartic acid derivative 16 and L-phenylalanine methyl ester hydrochloride 15 is then carried out using dicyclohexyl carbodiimide (DCC) and triethylamine to give the all-protected dipeptide ester 17 in almost quantitative yield.

The peptide bond of 17 is created in a complex reaction sequence. First, the NH_2 group of L-phenylalanine methyl ester (5) is liberated from the hydrochloride 15 by deprotonation with NEt_3. Second, the free α-COOH group of the precursor 16 is activated by addition to the heterocumulene moiety of DCC to afford the O-acyl isourea 18 as an "active ester". Third, by addition of the amino component 15 to the "active ester" 18, an acyl transfer takes place in an $S_N t$ process (via 20) with ensuing elimination of dicyclohexylurea (19) as leaving group, to give a CO–NH amide bond between the two reacting amino acid derivatives.

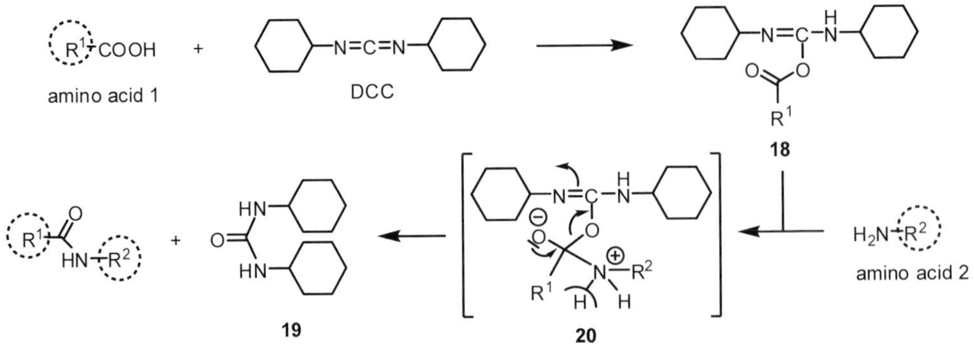

Finally, the protecting groups at the amino and the β-carboxylic acid moieties in the formed dipeptide ester **17** have to be removed. Deprotection at both NH₂ and COOH occurs concomitantly upon catalytic hydrogenation of **17**. Using Pd on charcoal, both functionalities undergo debenzylation with formation of toluene; carbon dioxide is additionally eliminated from the benzyloxycarbonyl group, to afford aspartame (**1**) in a clean-cut reaction.

Thus, the target molecule **1** is obtained in a six-step convergent synthesis, with respect to the five-step linear part in an overall yield of 37% (based on *L*-aspartic acid (**2**)).

(c) Experimental procedures for the synthesis of 1

4.4.6.1   **   *N*-Benzyloxycarbonyl-*L*-aspartic acid [9]

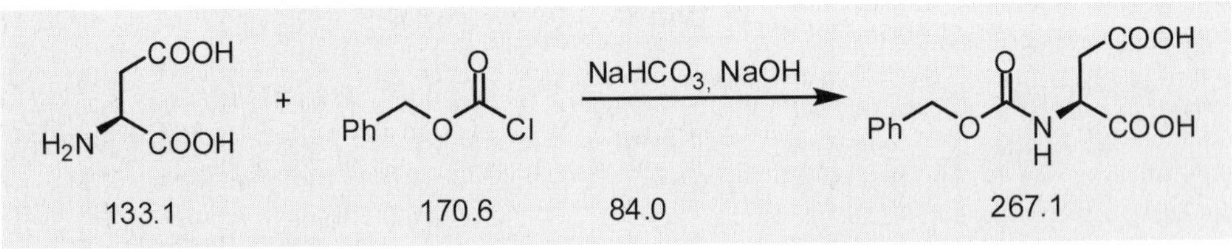

Sodium hydrogencarbonate (67.2 g, 0.80 mol) is added portionwise to a stirred suspension of aspartic acid (53.2 g, 0.40 mol) in water (250 mL); after ca. 15 min, the amino acid has dissolved. Thereafter, benzyloxycarbonyl chloride (75.0 g, 0.44 mol) and NaOH (2 N, 240 mL) are simultaneously added to the well-stirred solution, in such a proportion that the reaction mixture maintains a pH of 8–9 (control with indicator paper). After the addition is complete (ca. 4 h), stirring is continued for a further 1 h.

The reaction mixture is then acidified with concentrated HCl until pH 2 is reached. The (partially precipitated) product is extracted with EtOAc (3 × 200 mL), and the combined extracts are dried (Na₂SO₄), filtered, and concentrated *in vacuo*. The resulting oily residue is dissolved in EtOAc (150 mL), the solution is cooled to 0 °C, and *n*-hexane is slowly added with stirring until turbidity appears; stirring is continued for 15 min and further *n*-hexane (250 mL) is added. The precipitated (partially oily) product crystallizes completely on stirring; it is collected by suction filtration, washed with pre-cooled *n*-hexane, and dried *in vacuo*. The yield is 84.0 g (79%), colorless crystals, mp 105–107 °C; the product can be recrystallized from EtOAc/*n*-hexane, mp 109–110 °C, TLC (SiO₂; Et₂O): R_f = 0.35, $[\alpha]_D^{20}$ = +9.25 (*c* = 2.0, HOAc).

IR (KBr): $\tilde{v}$ (cm^{-1}) = 3340 (NH), 1710 (C=O), 1540, 1420, 1280, 1200.

^{1}H NMR ([D$_6$]DMSO): δ (ppm) = 10.1 (s$_{br}$, 2 H, COOH, exchangeable with D$_2$O), 7.35 (s, 5 H, phenyl-H), 6.43 (d, J = 9 Hz, 1 H, NH, exchangeable with D$_2$O), 5.11 (s, 2 H, benzyl-CH$_2$), 4.6–4.4 (m, 1 H, N–CH), 2.85 (d, J = 6 Hz, 2 H, β-CH$_2$).

4.4.6.2   *   Dibenzyl N-benzyloxycarbonyl-L-aspartate [5]

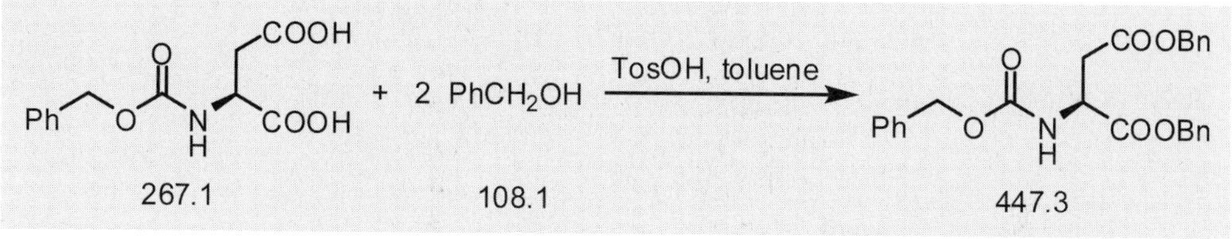

A mixture of the N-protected aspartic acid (Z-Asp-OH) **4.4.6.1** (80.0 g, 0.30 mol), benzyl alcohol (360 mL; note 1), p-toluenesulfonic acid (4.50 g), and anhydrous toluene (360 mL) is heated under reflux with stirring in an apparatus fitted with a Dean–Stark trap. The azeotropic distillation of H$_2$O is complete after 1 h (note 2).

Toluene is removed from the reaction mixture by distillation at 20 mbar, and then the excess benzyl alcohol is removed at 0.5 mbar The oily residue is dissolved in Et$_2$O (150 mL) and the solution is cooled to –30 °C. With vigorous stirring, n-hexane is slowly added, whereupon the dibenzyl ester precipitates in crystalline form. The product is collected by filtration, washed with pre-cooled n-hexane, and dried in vacuo; 120 g (90%), colorless crystals, mp 73–75 °C, TLC (SiO$_2$; CH$_2$Cl$_2$): R_f = 0.45 (note 3).

IR (KBr): $\tilde{v}$ (cm^{-1}) = 3360 (NH), 1745/1705 (C=O), 1420, 1355, 1220, 1195.

^{1}H NMR (CDCl$_3$): δ (ppm) = 7.30 (s, 15 H, phenyl-H), 5.8 (s$_{br}$, 1 H, NH, exchangeable with D$_2$O), 5.12, 5.10, 5.05 (s, 2 H, benzyl-CH$_2$), 4.8–4.5 (m, 1 H, α-CH), 2.98 (t, J = 5 Hz, 2 H, β-CH$_2$).

Notes: (1) Benzyl alcohol has to be distilled before use, bp$_{1013}$ 204–205 °C.
   (2) Theoretically, 10.8 mL of H$_2$O should be formed.
   (3) The product thus obtained is sufficiently pure according to TLC and is used in the next step without further purification.

4.4.6.3   *   N-Benzyloxycarbonyl-L-aspartic acid β-benzyl ester [5]

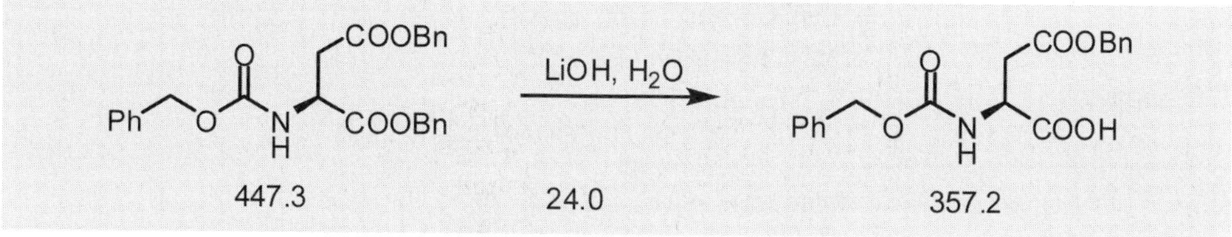

A solution of lithium hydroxide (2.55 g, 106 mmol) in H_2O (100 mL) is added dropwise over 30 min to a stirred solution of diester **4.4.6.2** (45.0 g, 100 mmol) in a mixture of acetone (1900 mL) and H_2O (600 mL) at room temperature; stirring is then continued for 15 min.

After evaporation of the acetone under reduced pressure (bath temperature < 40 °C), the remaining aqueous phase is extracted with Et_2O (3 × 100 mL) (note 1). The aqueous phase is then cooled to 0 °C and acidified by the addition of 6 N HCl with stirring until pH 1 is reached. The oily precipitate of the product crystallizes on further stirring in the ice bath. The product is filtered off, washed with iced water, and dried *in vacuo* over P_4O_{10} to give 17.0 g of the monoester (76% based on transformed dibenzyl ester), colorless crystals, mp 97–99 °C, TLC (SiO_2; EtOH): $R_f = 0.65$ (note 2).

> **IR** (KBr): $\tilde{v}$ (cm^{-1}) = 3330 (NH), 1740/1710/1650 (C=O), 1540, 1290, 1190.
>
> **^{1}H NMR** (CDCl$_3$): δ (ppm) = 7.90 (s$_{br}$, 1 H, COOH, exchangeable with D_2O), 7.35 (s, 10 H, phenyl-H), 5.87 (d, J = 8 Hz, 1 H, NH, exchangeable with D_2O), 5.15 (s, 4 H, benzyl-CH$_2$), 4.75 (q, J = 5 Hz, 1 H, α-CH), 3.02 (t, J = 5 Hz, 2 H, β-CH$_2$).

Notes: (1) The ethereal extract contains unreacted dibenzyl ester; work-up yields 17.0 g of the dibenzyl ester, mp 72–75 °C.
(2) The product is practically pure according to TLC and can be used in the next step without further purification. Recrystallization from benzene gives colorless crystals of mp 106–108 °C, $[\alpha]_D^{20} = -13.1$ (c = 10, in HOAc).

4.4.6.4 * L-Phenylalanine methyl ester hydrochloride [8]

Under an N_2 atmosphere, anhydrous MeOH (175 mL) is cooled to –10 °C and thionyl chloride (19.6 g, 0.27 mol) (note 1) is added with vigorous stirring. Stirring is continued and *L*-phenylalanine (36.3 g, 0.22 mol) is added portionwise. When the addition is complete, the mixture is heated under reflux for 2 h.

The excess MeOH is then removed *in vacuo*, the solid residue is dissolved in the minimum volume of boiling MeOH, and the solution is cooled to –10 °C. The hydrochloride of the product precipitates after slow addition of Et_2O with stirring. The salt is collected by suction filtration, washed with Et_2O, and dried *in vacuo* over P_4O_{10}; 42.5 g (90%), colorless crystals, mp 158–159 °C (note 2).

> **IR** (KBr): $\tilde{v}$ (cm^{-1}) = 1750 (C=O), 1590, 1500, 1455, 1245.
>
> **^{1}H NMR** (CDCl$_3$ + [D$_6$]DMSO): δ (ppm) = 8.90 (s$_{br}$, 2 H, NH$_2$, exchangeable with D_2O), 7.26 (s, 5 H, phenyl-H), 4.4–4.2 (m, 1 H, CH), 3.69 (s, 3 H, COOCH$_3$), 3.5–3.3 (m, 2 H, CH$_2$).

Notes: (1) SOCl$_2$ has to be distilled prior to use, bp$_{1013}$ 78–79 °C.
(2) The hydrochloride is used in the next step without further purification.

4.4.6.5 ** [*N*-Benzyloxycarbonyl-α-*L*-aspartyl(β-benzyl ester)]-*L*-phenylalanine methyl ester
[6]

| 357.2 | 215.6 | 206.3 101.2 | 518.3 |

The monoester **4.4.6.3** (13.9 g, 39.0 mmol) and the hydrochloride **4.4.6.4** (8.40 g, 0.39 mmol) are suspended in anhydrous CH_2Cl_2 (45 mL) and the suspension is cooled to 0 °C. Anhydrous triethylamine (3.96 g, 39.0 mmol) is added with stirring, followed, after 10 min, by dicyclohexyl carbodiimide (DCC) (8.82 g, 42.9 mmol). The reaction mixture is stirred for 1 h at 0 °C and for 12 h at room temperature.

The precipitated dicyclohexylurea is then filtered off and washed with CH_2Cl_2. The combined filtrate and washings are washed successively with HCl (2 N, 60 mL), H_2O (60 mL), aqueous $NaHCO_3$ solution (60 mL), and H_2O (60 mL). The organic phase is then dried ($MgSO_4$) and the solvent is removed *in vacuo*. The residue is recrystallized from EtOAc/petroleum ether (40–65 °C) to yield 18.9 g (93%) of the protected dipeptide as colorless crystals, mp 112–113 °C, TLC (SiO_2; EtOAc): R_f = 0.75.

> **IR** (KBr): $\tilde{v}$ (cm^{-1}) = 3310 (NH), 1740/1720/1650 (C=O), 1535, 1265.
>
> **^{1}H NMR** ($CDCl_3$): δ (ppm) = 7.4–6.8 (m, 15 H, phenyl-H), 5.11 (s, 4 H, benzyl-CH_2), 5.0–4.5 (m, 2 H, benzyl-CH_2), 3.68 (s, 3 H, $COOCH_3$), 3.2–2.6 (m, 4 H, β-CH_2).

4.4.4.6 ** *L*-α-Aspartyl-*L*-phenylalanine methyl ester (aspartame) [5]

| 518.3 | 294.2 |

A mixture of the protected dipeptide **4.4.6.5** (20.0 g, 38.6 mmol), glacial acetic acid (200 mL), water (10 mL), and 10% palladium on charcoal (1.00 g) is hydrogenated at room temperature at a hydrogen pressure of 3–4 bar for 4 h.

The catalyst is then filtered off (Caution: pyrophoric!) and the filtrate is concentrated to dryness *in vacuo*; toluene (10 mL) is added to remove residual H_2O and HOAc by azeotropic distillation *in vacuo*; this procedure is repeated three times. After dissolving the residue in boiling EtOH (ca. 80 mL), the solution is filtered and kept at 0 °C for crystallization. The product is filtered off, washed with a small amount of pre-cooled EtOH, and dried *in vacuo*; 8.50 g (75%); fine, colorless needles with a sweet taste, mp 253–255 °C, TLC (SiO_2; MeOH): R_f = 0.55; $[\alpha]_D^{20}$= –2.3 (c = 1.0 in 1 N HCl).

IR (KBr): $\tilde{\nu}$ (cm^{-1}) = 3330 (NH), 1740, 1670 (C=O), 1545, 1380, 1365, 1230, 700.

^{1}H NMR ([D$_6$]DMSO): δ (ppm) = 8.80 (s$_{br}$, 1 H, CO–NH, exchangeable with D$_2$O), 7.22 (m, 5 H, phenyl-H), 5.30 (s$_{br}$, 2 H, NH$_2$, exchangeable with D$_2$O), 4.54 (t, J = 7 Hz, 1 H, α-aspartyl-CH), 3.60 (s, 3 H, COOCH$_3$), 3.7–3.6 (m, 1 H, phenylalaninyl-CH), 3.2–2.9 (m, 2 H, Ph–CH$_2$), 2.6–2.2 (m, 2 H, β-aspartyl-CH$_2$).

[1]   A. Kleemann, J. Engel, *Pharmaceutical Substances*, 3rd ed., p. 126, Thieme Verlag, Stuttgart, **1999**.

[2]   Ullmann's *Encyclopedia of Industrial Chemistry*, 6th ed., Vol. 35, p. 417, Wiley-VCH, Weinheim, **2003**.

[3]   *Houben-Weyl*, Vols. 15/1 and 15/2; protective groups in peptide synthesis: *Houben-Weyl*, Vol. 15/1, p. 46.

[4]   Compare the difficulties of aspartame synthesis from **2** and **5** without the use of protective groups and enzyme catalysis: Y. Ariyoshi, T. Yamatani, N. Uchiyama, Y. Adachi, N. Sato, *Bull. Chem. Soc. Jpn.* **1973**, *46*, 1893.

[5]   J. M. Davey, A. H. Laird, J. S. Morley, *J. Chem. Soc. (C)* **1966**, 555.

[6]   J. C. Anderson, M. A. Barton, P. M. Hardy, G. W. Kenner, J. Preston, R. C. Sheppard, *J. Chem. Soc. (C)* **1967**, 108.

[7]   See ref. [1] and P. Kuhl, R. Schaaf, *Z. Chem.* **1990**, *30*, 212.

[8]   R. A. Biossonas, S. Guttmann, P.-A. Jaquenoud, J.-P. Weller, *Helv. Chim. Acta* **1956**, *39*, 1421.

[9]   *Houben-Weyl*, Vol. 15/1, p. 321 and 49.

4.4.7   Ugi dipeptide ester

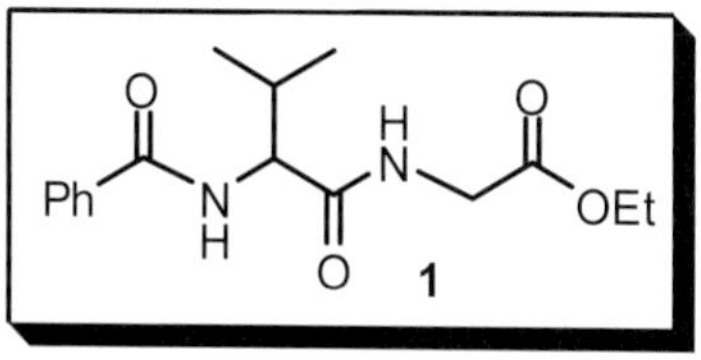

Topics:
- One-pot peptide synthesis
- Four-component Ugi reaction vs. six-component Ugi reaction, combined with aminoacylal hydrolysis

(a) General

Many syntheses of acyclic, carbocyclic, and heterocyclic systems are conducted in the form of "multi-component domino reactions" (MCR) [1], thus creating complex molecules from several simple substrates efficiently and expeditiously in one process. Important examples of MCRs are the Passerini reaction and the Ugi reaction [2].

In the three-component Passerini reaction, isonitrile **2** are combined with a carboxylic acid and an aldehyde (or a ketone) leading to an α-acyloxyamide **3** (cf. **4.4.5**).

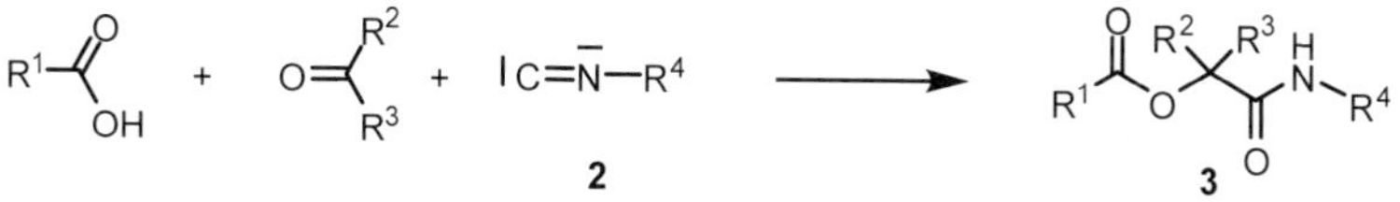

A very similar transformation is the four-component Ugi reaction, in which a mixture of an isonitrile, a carboxylic acid, an aldehyde (or ketone), and a primary amine (or ammonia) is reacted. The products are α-(*N*-acylamino)amides of type **4**:

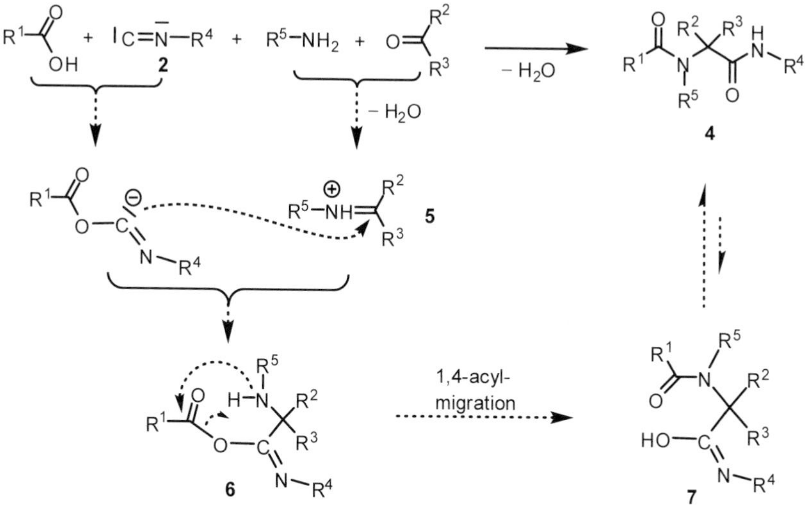

For the Ugi reaction, a mechanism analogous to that for the Passerini reaction is likely to be operative; in the crucial step, a protonated imine **5** – initially formed from the carbonyl compound, the amine (R–NH$_2$ or NH$_3$), and the proton of the carboxylic acid – serves as the electrophilic component (**5** → **6** → **7** → **4**).

The four-component Ugi reaction has remarkable potential in peptide synthesis [3]. In particular, with N-protected amino acids as acid components and isocyanoacetates as isonitrile components, a wide range of di- and tripeptidic fragments of types **8/9/10** are accessible [4]:

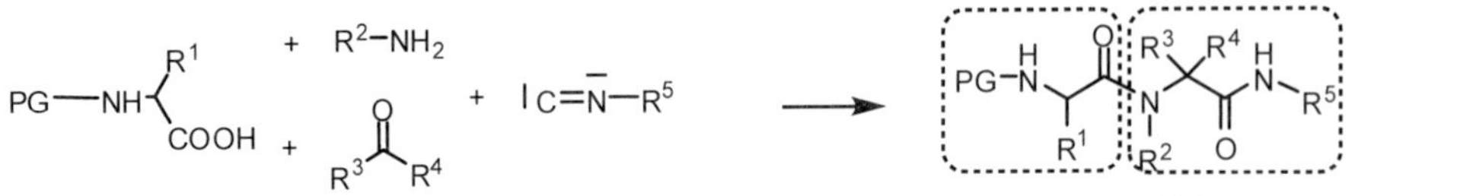

8 (dipeptide)

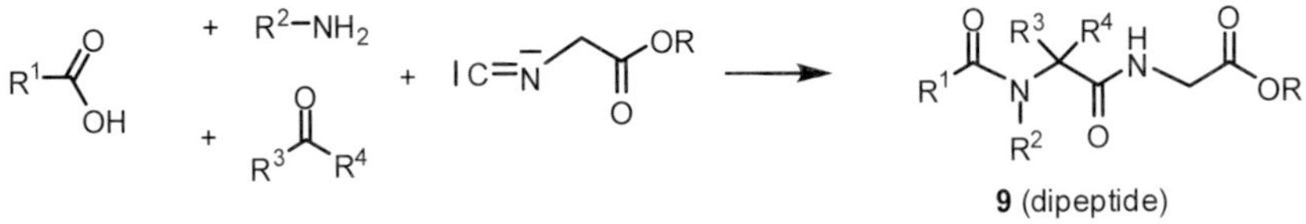

9 (dipeptide)

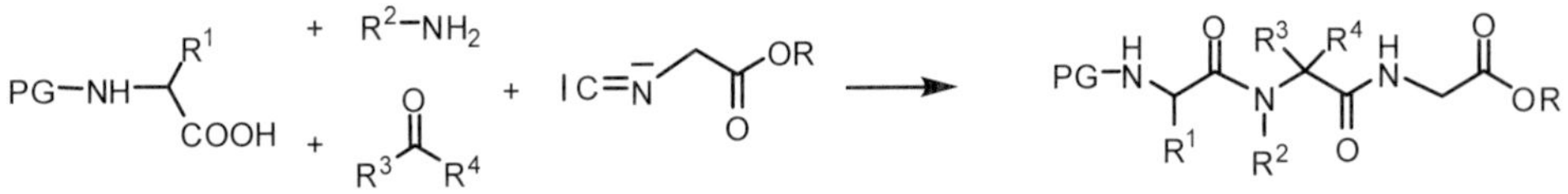

10 (tripeptide)

Since primary alkylamines are the preferred amino substrates, the formed peptidic systems **8–10** contain an "unnatural" N-alkylated peptide bond, –CO–NR–.

For the synthesis of NH peptides by the Ugi principle, ammonia has to be used. However, as shown in a recent investigation [5], the Ugi reaction with NH_3 takes a more complex course, although it may lead straightforwardly to the formation of N-protected dipeptidic systems like **1** under carefully controlled conditions, as presented in section (b).

(b) Synthesis of 1

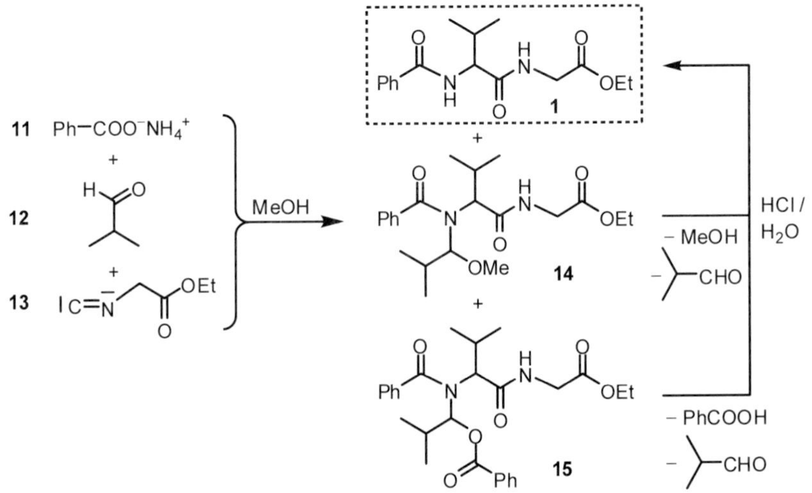

Reaction of ammonium benzoate (**11**, as a source of PhCOOH and NH₃), isobutyraldehyde (**12**), and ethyl isocyanoacetate (**13**, cf. **3.2.2**) in methanol as solvent leads to the dipeptide ester **1**, albeit only in low yield. The major products are the hemiaminals **14** and **15**, indicating that a six-component Ugi reaction with additional incorporation of a second aldehyde moiety together with CH_3OH (**14**), or with a second molecule of PhCOOH (**15**), is favored under these conditions [6]. The amounts of **14** and **15** are even increased by increasing the amount of aldehyde or benzoate; on the other hand, the formation of **14** can be suppressed by using a less nucleophilic solvent such as CF_3CH_2OH. Under optimized conditions (ratio **11**:**12**:**13** = 4.4:2:1, MeOH as solvent), the aminoacylal **15** is the only six-component condensation product formed besides the four-component Ugi product **1** (ratio 3:4).

Selective hydrolysis of the aminoacylal function in **15** is accomplished by treatment with HCl/H_2O at room temperature to quantitatively afford the dipeptide ester **1**. Conveniently, the acidic hydrolysis can be performed directly after the multi-component reaction (i.e. without separation of **1** and **15**) to give **1** in a total yield of 85%.

The following mechanism can be assumed for the six-fold coupling reaction [5]. In the first step, an imine **17** is formed, which obviously does not react directly with the isonitrile (as proposed for the Ugi reaction in (a)), but with another nucleophile such as methanol or benzoate.

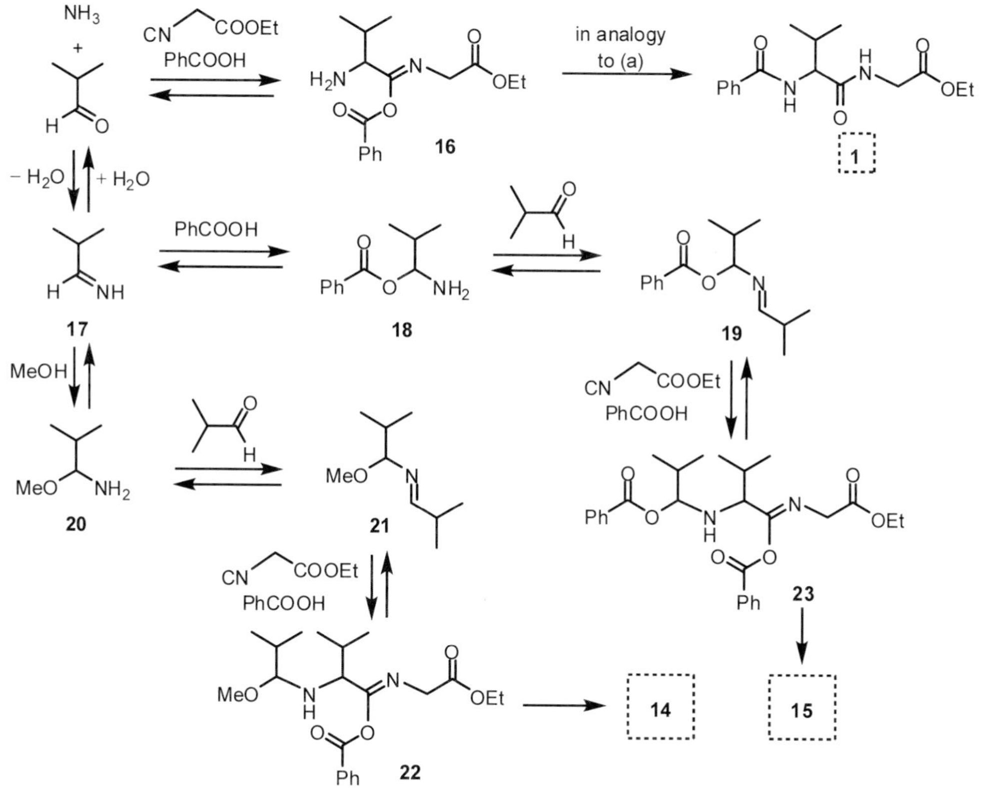

The resulting hemiaminal-type intermediates **18** and **20** can now incorporate a second aldehyde moiety to give the imines **19** and **21**. Addition of the isonitrile and the carboxylate generates the imidates **22** and **23**, which subsequently undergo rearrangement to the amides **14** and **15**.

Thus, the polyfunctional (racemic) target molecule can be obtained by a four-component domino MCR process in high yield (85%).

It should be noted that a synthesis of **1** (racemic or optically active) according to the classical linear strategy outlined in **4.4.5** – preparation of *N*-protected valine and of ethyl glycinate from the corresponding amino acids, peptide coupling of the free COOH and NH$_2$ functions, removal of the protective groups – would require at least four steps.

(c) Experimental procedure for the synthesis of 1

4.4.7.1 ** *rac-N-(N-Benzoylvalinoyl)* glycine ethyl ester (*N*-Bz-Val-Gly-OEt) [5]

Isobutyraldehyde (8.03 mL, 88.0 mmol) is added to a solution of ammonium benzoate (5.56 g, 40.0 mmol) in MeOH (40 mL) at 0 °C. After stirring for 30 min, ethyl isocyanoacetate (2.16 g, 20.0 mmol; cf. **3.2.2**) is added and the mixture is allowed to warm to room temperature over 12 h.

After evaporation of the solvent *in vacuo*, the residue is suspended in a mixture of H$_2$O and CH$_3$CN (30 mL each) and the suspension is acidified to pH 2 by dropwise addition of concentrated HCl and stirred at room temperature for 12 h.

The CH$_3$CN is then evaporated *in vacuo* and the resulting aqueous suspension is treated with H$_2$O and CH$_2$Cl$_2$ (ca. 50 mL each) until two clear layers are formed. The organic layer is separated, washed with saturated NaHCO$_3$ solution, dried (Na$_2$SO$_4$), and the solvent is removed *in vacuo*. The solid residue is triturated with Et$_2$O (50 mL), collected by suction filtration, washed with Et$_2$O and *n*-hexane (10 mL each), and dried *in vacuo*; white powder, 5.20 g (85%), mp 163–164 °C; analytically pure, TLC (SiO$_2$; Et$_2$O): R_f = 0.31.

^{1}H NMR (400 MHz, CDCl$_3$): δ (ppm) = 8.48 (d, *J* = 8.0 Hz, 2 H, Ph-H), 7.49 (d, *J* = 8.0 Hz, 1 H, Ph-H), 7.40 (t, *J* = 8.0 Hz, 2 H, Ph-H), 7.18 (m$_c$, 1 H, NH), 7.08 (d, *J* = 8.8 Hz, 1 H, NH), 4.61 (dd, *J* = 8.8, 7.2 Hz, 1 H, N–CH–CO), 4.17 (q, *J* = 7.1 Hz, 2 H, OCH$_2$), 4.12/3.90 (each dd, *J* = 18.0, 5.2 Hz, each 1 H, N–CH$_2$), 2.23 (m$_c$, 1 H, CH), 1.24 (t, *J* = 7.1 Hz, 3 H, CH$_3$), 1.03/1.01 (2 ×

[1] J. Zhu, H. Bienaymé (Eds.), *Multicomponent Reactions*, Wiley-VCH, Weinheim, **2004**.

[2] M. B. Smith, J. March, *March's Advanced Organic Chemistry*, 5th ed., p. 1251 ff., John Wiley & Sons, Inc., New York, **2001**, and references cited therein.

[3] A. Dömling, I. Ugi, *Angew. Chem.* **2000**, *112*, 3300; *Angew. Chem. Int. Ed.* **2000**, *39*, 3168.

[4] a) I. Ugi, *Angew. Chem.* **1977**, *89*, 267; *Angew. Chem. Int. Ed. Engl.* **1977**, *16*, 259; b) T. Yamada, *J. Chem. Soc. Chem. Commun.* **1990**, 1640.

[5] R. Pick, M. Bauer, U. Kazmaier, C. Hebach, *Synlett* **2005**, 757.

[6] The six-component products **16** and **17** show interesting stereochemical features, cf. ref. [5].

4.4.8   Solid-phase synthesis of β-peptides

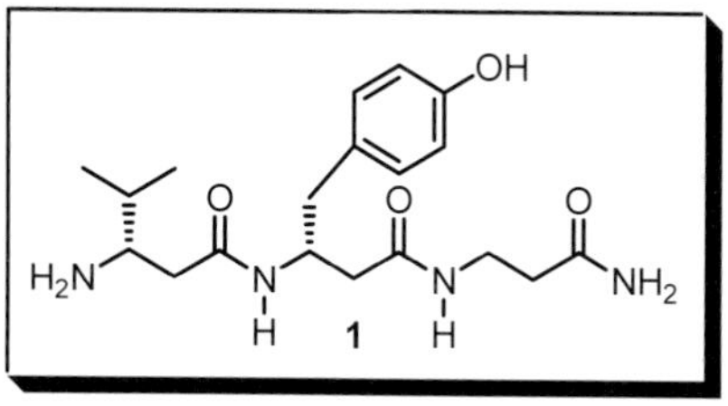

Topics:
- Solid-phase peptide synthesis (SPPS)
- Boc and Fmoc strategies
- Protective groups
- Peptide coupling reagents
- Solid supports
- β-Amino acids

(a) General

Chemical synthesis is used to prepare a variety of small peptides, but proteins with 50–100 amino acids can also be obtained in this way. It is the only method in those cases in which the proteins are difficult to express in biological systems. Moreover, chemical synthesis of peptides and proteins is always necessary when non-natural amino acids are included and backbone modifications are made. Whereas peptide synthesis in solution is only applied in special cases and for very small peptides, solid-phase peptide synthesis (SPPS), first proposed by Merrifield [1], has been developed into a mature automated procedure with coupling yields higher than 99.9% [2–4]. The synthesis is performed on a resin, which allows the use of an excess of reagents, gives high coupling yields, and permits fast purification. A flawless strategy for protecting the functional groups (permanent protecting groups) in the side chain, with individual solutions for the respective amino acids, is required. For chain elongation, which proceeds from the *C* to the *N*-terminal end, temporary protection of the *N*-terminus (Boc or Fmoc strategy) is used, orthogonal with respect to the side-chain protection. This proceeds in combination with *in situ* *C*-terminal activation of the amino acid to be linked. The SPPS methods described in the literature mainly differ in their choice of resin, permanent and temporary protecting group strategies, and coupling reagents. A typical SPPS cycle is illustrated for the *tert*-butyloxycarbonyl (Boc) strategy:

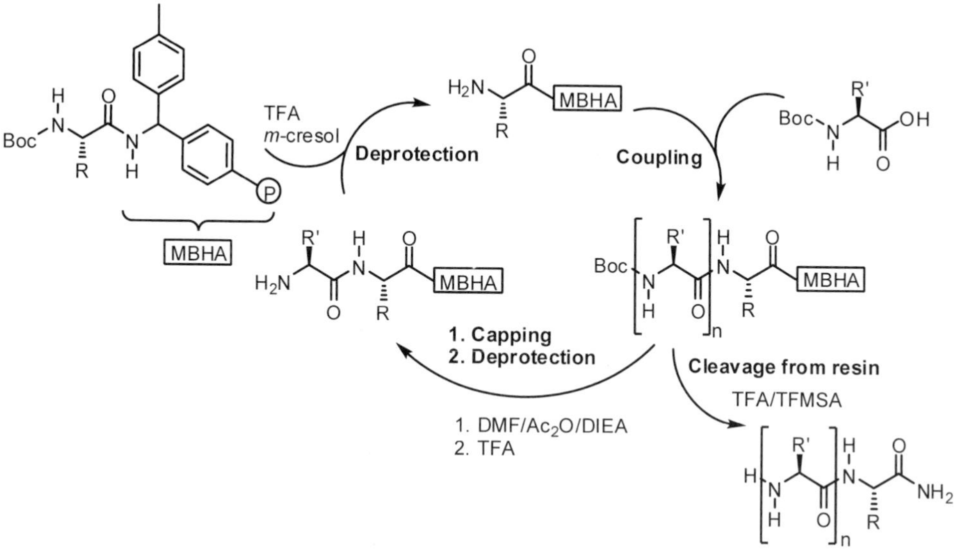

The resin loaded with the first amino acid is Boc-deprotected under acidic conditions. Coupling of the next Boc-protected amino acid takes place with *in situ* activation of the carboxylic acid by coupling reagents. Capping of non-reacted oligomers with acetic anhydride facilitates oligomer separation at a later stage. By Boc-deprotection, the synthesis cycle can be repeated until the desired peptide/protein length is reached. Cleavage from the resin is performed using HF (requiring careful handling) or, more conveniently, under strongly acidic conditions (mixture of trifluoroacetic acid (TFA) and trifluoromethanesulfonic acid (TFMSA)), with simultaneous cleavage of all or most of the permanent side-chain protecting groups.

Fmoc-NHR

The SPPS cycle for the Fmoc strategy is broadly similar to that of the Boc procedure. Nevertheless, the fluorenylmethoxycarbonyl (Fmoc) temporary protecting group for the *N*-terminus, as introduced by Carpino, allows deprotection under basic conditions (20% piperidine in DMF). Side-chain deprotection and cleavage from the resin is usually accomplished with TFA. Therefore, peptide formation using Fmoc chemistry involves milder cleavage conditions. Furthermore, the Fmoc group constitutes a chromophore, which facilitates monitoring of the progress of the synthesis.

Most of the solid supports used for peptide synthesis are based on functionalized polystyrenes, which provide useful physical properties such as swelling, loading, and durability. They differ with respect to the linker, allowing various cleavage mechanisms and different functionalization at the *C*-terminus. Whereas the Merrifield, Wang, and 2-chlorotrityl resins provide a carboxylic acid moiety, cleavage from the MBHA resin results in the formation of an amide. Other resins are known which provide thioesters or esters, as exemplified by the 3-nitro-4-hydroxymethyl benzoyl linker.

Acidic cleavage to give C-terminal carboxylic acid

Wang resin

Mild acidic cleavage to give C-terminal carboxylic acid

2-Chlorotrityl chloride resin

Acidolysis to give a C-terminal amide

Rink amide resin

Base-cleavable linkers

3-Nitro-4-hydroxymethyl benzoyl linker

The most critical aspect of solid-phase peptide synthesis concerns the permanent side-chain protecting groups. They need to be orthogonal to the respective temporary Boc or Fmoc protection used in the chain elongation and should usually be liberated concomitantly upon cleavage from the resin. On the other hand, for some applications, cleavage of the fully protected oligomer from the resin is required. Furthermore, side-chain protection is required for all nucleophiles that might otherwise interfere with chain elongation. The options, advantages, and disadvantages of the individual amino acid side-chain protection protocols are well documented and the different procedures are well established [3].

Carbodiimide   **Aminium salt**   **Phosphonium salt**

EDC

HOBt   HATU   PyBOP

The formation of the amide bond requires activation of the carboxylic acid moiety. In the case of difficult couplings, acid fluorides or active esters such as the pentafluorophenyl ester (OPfp) are applied. In most syntheses, *in situ* activation of the carboxylic acid is preferred. This is exemplified by carbodiimide (DIC) activation followed by treatment with 1-hydroxy-benzotriazole (HOBt) to afford the active ester. Highly reactive coupling reagents, which allow shorter coupling times, are the uronium salts HATU and PyBOP. A major concern in all coupling reactions is to avoid racemization at Cα during active ester formation.

Oligomerization of β-amino acids by means of SPPS leads to β-peptides [5], which are of interest as mimics for α-peptide secondary structures since they provide much higher conformational stability with even short sequences of about six amino acids and are stable with respect to enzymatic degradation. Appropriate choice of the side-chain substituents at the α- and/or β-position in combination with the desired configuration allows the specific design of peptide secondary structures. Since manual and automated syntheses of α-peptides are comprehensively described in the literature, SPPS optimized for the preparation of β-peptides is described here.

(b) Synthesis of β-peptides 1

β-Peptides can be prepared following the Fmoc [6] or Boc protocols [7]. A general method for the coupling of Boc-protected β-amino acids on a 4-methylbenzhydrylamine polystyrene (MBHA) solid support can be applied, which is closely related to SPPS of α-peptides. A major difference, however, lies in the much higher tendency of β-peptides to form aggregates and secondary structures already on the solid support. Therefore, double coupling, higher temperatures, and a reactive coupling reagent are applied, HATU being used in the following synthesis. For β-amino acids with the side chain in the β-position, isomerization of the active esters is not a severe problem as in the α-peptide series since the proton at the stereogenic center is much less acidic.

For the following synthesis of β-peptide **1**, the MBHA solid support is loaded with β-homoglycine (**2**), which is preferentially used at the C-terminus to avoid racemization. TFA-induced deprotection of the resin-bound amino acid yields amine **3**. The coupling step is initiated by activation of Boc-protected β-homotyrosine (**4**) by deprotonation with DIPEA in conjunction with the coupling reagent HATU (**5**).

It is likely that an acyloxy amidinium salt is formed as an intermediate, which immediately reacts with the benzotriazole derivative to give the active ester **6**. The nucleophilic amine is probably coordinated by hydrogen bonding to further improve amide formation. In the case of sterically demanding amino acids, the coupling is repeated to obtain higher coupling yields. The β-dipeptide **7** is elongated by successive deprotection and coupling cycles. The desired β-tripeptide is cleaved from the resin under strongly acidic conditions. Oligomers with a length of four or more amino acids can be isolated by precipitation from cold diethyl ether. Since the tripeptide is too short to be precipitated in this way, the cleavage solution is concentrated *in vacuo* and the residue is directly purified by HPLC to yield the β-peptide **1** as the *N*-terminal amide.

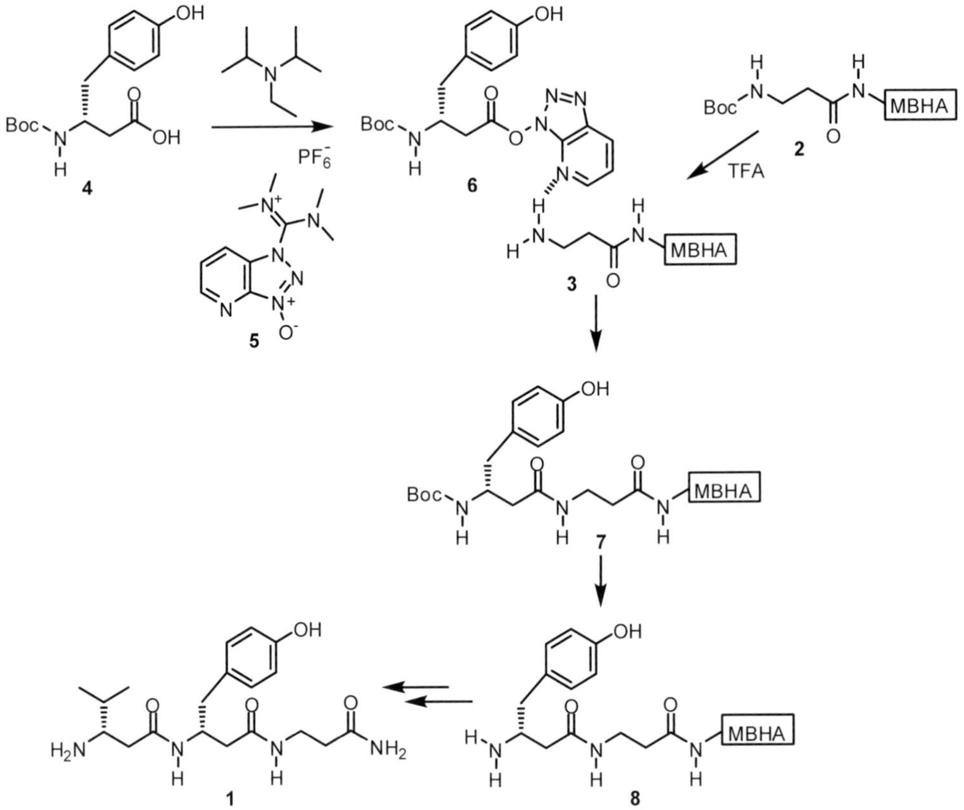

(c) Experimental procedure for the synthesis of 1

4.4.8.1   *   Manual solid-phase β-peptide synthesis [8]**

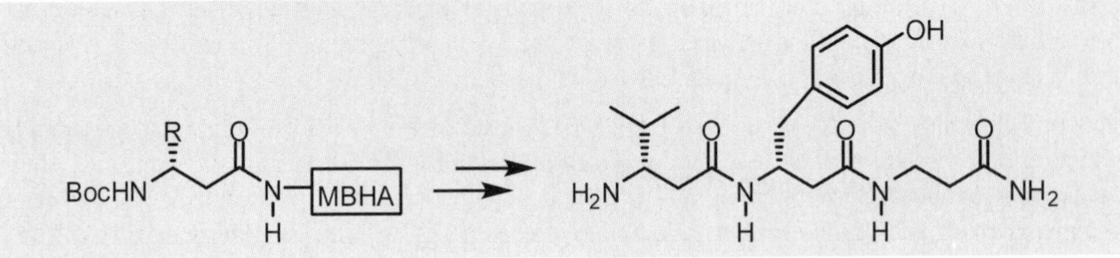

The β-tripeptide **1** is prepared in a small fritted glass column (10 mL) on an MBHA-polystyrene resin with a loading capacity of 0.62 mmol g^{-1}. The resin (48.5 mg) is used preloaded with *N*-Boc-β-homoglycine **2** (19.4 μmol homoglycine amide) and is first swollen by covering it with CH$_2$Cl$_2$ (2 mL) for 2 h. The solvent is then removed by a nitrogen flow and the procedures described for deprotection, coupling, and capping are repeated for each coupling cycle:

(1) Deprotection: The resin is treated with a TFA/*m*-cresol solution (95:5, 2 mL) for 3 min. This deprotection step is repeated, and then the resin is washed three times with CH$_2$Cl$_2$/DMF (1:1; 2 mL) and five times with pyridine (2 mL).

(2) Coupling: Coupling of the β-amino acids is performed in an oven at 50 °C. First, the resin is treated with an excess of Boc-protected amino acid (for the synthesis of **1**, 37.3 mg β-homotyrosine for the first agent and 22.4 mg β-homovaline, 5 equiv., 97.0 μmol, for the second agent), which is activated by *O*-(7-azabenzotriazol-1-yl)-*N,N,N',N'*-tetramethyluronium hexafluorophosphate (HATU) (33.2 mg, 87.3 μmol, 4.5 equiv.), 1-hydroxy-7-azabenzotriazole (HOAt) (194 μL, 97.0 μmol, 5 equiv. of a 0.5 M solution in DMF), and *N,N*-diisopropylethylamine (DIPEA) (46.6 μL, 272 μmol, 14 equiv.) in DMF (400 μL). After gently agitating the resin for 1 h, the reaction mixture is drained.

(3) Capping: Unreacted amines are acylated by treatment with a solution of DMF/Ac$_2$O/DIEA (8:1:1; 2 mL) for 3 min. This capping step is repeated once more, and then the resin is washed with CH$_2$Cl$_2$/DMF (1:1; 2 mL).

The deprotection, coupling, and capping steps are repeated according to the desired peptide. For the synthesis of **1**, after attachment of β-homovaline, the final deprotection step is performed with TFA (3 × 2 mL) and this is followed by washing with CH$_2$Cl$_2$ (5 × 2 mL) and drying of the resin *in vacuo*.

(4) Cleavage from the resin: The resin is transferred into a small glass vessel and suspended in *m*-cresol/thioanisole/ethanedithiol (2:2:1; 500 μL). After stirring for 30 min at room temperature, TFA (2 mL) is added, and the mixture is cooled to −20 °C. Trifluoromethanesulfonic acid (TFMSA) (200 μL) is added dropwise with stirring. The mixture is allowed to warm to room temperature over 1.5 h and stirring is continued for a further 2 h. The mixture is filtered through a fritted glass funnel and the TFA is removed *in vacuo*.

(5) Purification: The crude mixture is concentrated *in vacuo* and the residue is dissolved in H$_2$O/acetonitrile and purified by HPLC on an RP C18 column (150 × 10 mm, 4 μm, 80 Å, flow rate 3 mL min^{-1}) using the eluents: (A) H$_2$O + 0.1% TFA and (B) CH$_3$CN/H$_2$O, 8:2, +0.1% TFA.

β-Peptide **1** is obtained with an HPLC gradient of 5% to 40% B in 30 min; t_R = 15.26 min.

EI HRMS: C$_{19}$H$_{30}$N$_4$O$_4$: calcd. for [M + H$^+$]: 379.23398; found: 379.23408.

---

[1]   R. B. Merrifield, *J. Am. Chem. Soc.* **1963**, *85*, 2149.

[2]   N. L. Benoiton, *Chemistry of Peptide Synthesis*, Taylor and Francis, Boca Raton, **2006**.

[3]   E. Atherton, R. C. Sheppard, *Solid-Phase Peptide Synthesis, A Practical Approach*, IRL Press, Oxford, **1989**.

[4]   J. Jones, *Amino Acid and Peptide Synthesis*, Oxford University Press, New York, **1992**.

[5]   D. Seebach, A. K. Beck, D. J. Bierbaum, *Chemistry Biodiversty* **2004**, *1*, 1111.

[6]   W. C. Chan, P. D. White (Eds.), *Fmoc Solid-Phase Peptide Synthesis, A Practical Approach*, University Press, Oxford, **2000**.

[7]   M. Bodanszky, A. Bodanszky, *The Practice of Peptide Synthesis*, 2nd ed., Springer, Heidelberg, **1994**.

[8]   a) P. Chakraborty, U. Diederichsen, *Chem. Eur. J.* **2005**, *11*, 3207; b) A. M. Brückner, P. Chakraborty, S. H. Gellman, U. Diederichsen, *Angew. Chem.* **2003**, *115*, 4532; *Angew. Chem. Int. Ed.* **2003**, *42*, 4395.

4.5   Nucleotides and oligonucleotides

Introduction

The most important function of deoxyribonucleic acid (DNA) is the storage and replication of information [1]. The complete instruction for the construction of cells and living organisms is coded in genes. Besides coding information, DNA also has structural and regulating functions. Genetic information is encoded in the sequential order of the four nucleobases adenine, guanine, thymine, and cytosine, which are linked together on a linear backbone polymer that consists of deoxyribose units linked by phosphodiesters. Nucleotides are the repeating units of a DNA polymer comprising deoxyribose, a phosphate group at the primary 5'-OH, and the purine or pyrimidine nucleobases linked at the anomeric center (C1') by a β-N-glycosidic bond [2]. DNA polymers made from repeating nucleotides can reach an enormous length, e.g. the largest human chromosome contains 220 million nucleotides.

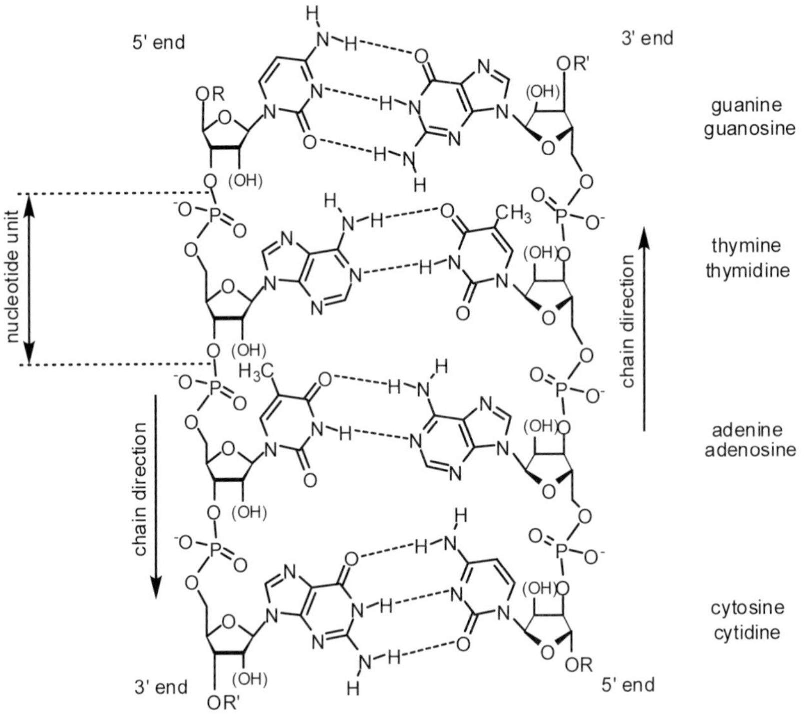

For replication of information, the complementarity of nucleobase recognition is essential. Guanine is specifically recognized by cytosine, forming three hydrogen bonds, whereas adenine and thymine provide the second base pair with only two hydrogen bonds. Both base pairs require an identical size and orientation of the deoxyribose linkages, regardless of purine–pyrimidine or pyrimidine–purine alignment. By this pseudosymmetry of Watson–Crick base pairing, it is ensured that all DNA sequences are possible without disturbing the overall helical structure [3]. The resulting helix topology, on the other hand, defines space and orientation of base pairs in DNA, allowing only the purine–pyrimidine combination and the Watson–Crick pairing mode. Of a number of alternative possibilities for hydrogen-bond recognition between nucleobases, the helix DNA topology leads to a restriction to the Watson–Crick mode and is, therefore, decisive for base pair complementarity and specific replication.

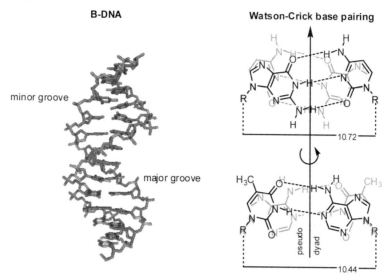

The antiparallel recognition of two complementary DNA strands leads to a right-handed double helix [4]. Besides specific hydrogen bonding between the nucleobases, aromatic stacking interactions and solvent effects contribute nearly equally to the overall stability of a DNA double strand. Stacking of nucleobases occurs with a preferred distance of 3.4 Å, which determines the overall helix structure. The DNA double helix provides a major groove and a smaller minor groove. Small molecules and proteins typically interact with DNA by specifically recognizing the respective hydrogen-bonding pattern of the nucleobase pairs in the grooves, intercalating between base pairs, or non-specifically interacting with the negatively charged DNA backbone. Depending on the environment, DNA helices are able to adopt different conformations. The most prominent secondary structure is the B-form DNA double helix. However, there are also other left-handed helices, such as A, C, and D-DNA, which differ in the thickness of the helix, base pair stacking, and the size and depth of the grooves. In addition, for Z-DNA, a right-handed DNA double helix is also known. The different helix conformations provide an additional recognition motif besides base pair recognition, since there are proteins that require a specific DNA helix secondary structure for interaction and recognition. Furthermore, DNA double strands are highly dynamic structures that, e.g., partially unwind, especially at their termini, where hydrogen-bonded base pairs open and form again. As a consequence of interactions with small molecules, unwinding, bending, base flipping, or other conformational changes are regularly observed. The double helical structure is forced to be extended by repulsion between phosphodiester charges that are uniformly distributed over the helix. Selective charge neutralization at certain points, as provided by cations or cationic amino acids such as lysine or arginine, induces a bend in the DNA double helix. This kind of bending is an important mechanism with regard to the efficient packing of DNA in cells or viruses.

Besides DNA, there is another nucleic acid called ribonucleic acid (RNA), which differs from DNA in that 2-deoxyribose is replaced by ribose as the sugar component and that thymine is replaced by uracil as one of the nucleobases [5]. RNA plays an important role in the translation of genetic information from DNA into proteins. The transcription of the information from DNA into messenger-RNA is followed by a transport of information out of the nucleus as an RNA copy for further processing. Translation into proteins with the help of transfer-RNA is the final step in translating the genetic code into an amino acid sequence synthesizing proteins in the ribosome. With respect to the secondary structure, RNA oligomers form double strands like DNA, but in addition RNA can fold into a variety of motifs. The possibility of adopting various folded forms allows RNA to take over many biological functions and to act as artificial RNA-based enzymes called ribozymes. The 2'-hydroxy group on the sugar component has an influence on the ribosyl conformation, thereby affecting the folding process, but is also responsible for a significant decrease in the stability of RNA oligomers. Intramolecular attack of the 2'-OH on the neighboring phosphorus results in a phosphotriester that can be cleaved by the reverse reaction, by rearrangement to the 2'-linked RNA, or by strand cleavage leading to the cyclophosphate. The shorter lifetime of an RNA oligomer is important from the biological point of view since RNA molecules are intermediate transporters of information or function.

With the accessibility of DNA and RNA by organic synthesis, interest in modifications of oligonucleotides has also grown. Modified oligonucleotides with nuclease resistance and complementarity for a given oligonucleotide sequence are promising for diagnostic purposes as well as in *antigene* (interaction with the DNA double strand) or *antisense* (complementarity to an RNA single strand) therapies, blocking the complementary oligonucleotide sequence of interest and allowing genetic disorders to be translated to the protein level [6]. The recognition of double-stranded DNA by a third oligonucleotide strand is known to proceed in the major groove, especially by interaction with a purine-rich central strand by formation of specific hydrogen bonds on the so-called purine Hoogsteen face [7]. Based on the promising *antigene/antisense* idea, numerous DNA and RNA modifications have been investigated by varying the phosphodiester, the ribose moiety, or the nucleobase. As the first *antigene/antisense* therapeutics, phosphorothioates have been used in clinical trials [8]. In these compounds, one oxygen is replaced by sulfur. Hexose-DNA is chosen as a representative of a modification in which an additional methylene group is introduced in the sugar moiety as compared to DNA [9]. This results in a six-membered sugar with a preferred chair conformation, which has consequences for the overall oligonucleotide topology. Many more conformationally constrained ribosyl derivatives have been investigated, especially with a view to obtaining selective recognition of DNA and RNA, respectively.

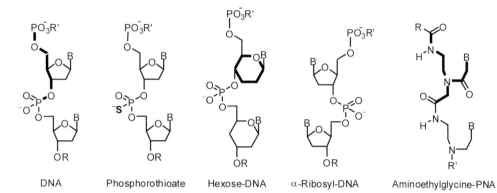

| DNA | Phosphorothioate | Hexose-DNA | α-Ribosyl-DNA | Aminoethylglycine-PNA |

Further modifications of DNA include α-ribosyl DNA, the enantiomer of naturally occurring DNA, and aminoethylglycine peptide nucleic acid (PNA), in which the ribosyl-phosphodiester backbone of DNA is completely substituted by a non-charged, non-chiral polyamide [10]. PNA oligomers are well suited to mimic DNA strands. DNA-PNA, RNA-PNA, and PNA-PNA double strands show high stability and structurally closely resemble the respective DNA-DNA or RNA-DNA double strands.

[1]   G. M. Blackburn, M. J. Gait (Eds.), *Nucleic Acids in Chemistry and Biology*, Oxford University Press, New York, **1996**.

[2]   S. M. Hecht (Ed.), *Bioorganic Chemistry, Nucleic Acids*, Oxford University Press, New York, **1996**.

[3]   W. Saenger, *Principles of Nucleic Acid Structure*, Springer Verlag, New York, **1984**.

[4]   S. Neidle (Ed.), *Nucleic Acid Structure*, Oxford University Press, New York, **1999**.

[5]   R. F. Gesteland, T. R. Cech, J. F. Atkins, *The RNA World*, Cold Spring Harbor, New York, **1999**.

[6]   J. Hunziker, C. Leumann, *Nucleic Acid Analogues, Synthesis and Properties*, in B. Ernst, C. Leumann (Eds.), *Modern Synthetic Methods 1995*, Verlag Chimica Acta, Basel, **1995**.

[7]   V. N. Soyfer, V. N. Potaman, *Triple-Helical Nucleic Acids*, Springer, New York, **1995**.

[8]   F. Eckstein, G. Gish, *Trends Biol. Sci.* **1989**, *14*, 97.

[9]   A. Eschenmoser, M. Dobler, *Helv. Chim. Acta* **1992**, *75*, 218.

[10] P. E. Nielsen, M. Egholm (Eds.), *Peptide Nucleic Acids, Protocols and Applications*, Horizon Scientific Press, Wymondham, **1999**.

4.5.1   2',3'-Dibenzoyl-6'-*O*-DMT-β-*D*-glucopyranosyl-uracil 4'-*O*-phosphoramidite

Topics:
- Vorbrüggen and Hilbert–Johnson nucleosidation
- Phosphoramidite synthesis
- Solid-phase linkage of nucleosides
- Glucopyranosyl protection

(a) General

Deoxyribosyl and ribosyl nucleotides with the canonical nucleobases guanine, adenine, cytosine, as well as thymine and uracil, respectively, are used for solid-phase DNA synthesis by the phosphoramidite strategy and are commercially available [1]. Nevertheless, the preparation of phosphoramidite building blocks, especially for the synthesis of oligonucleotides with nucleobase or sugar modifications, is still a major concern [2]. The key step in the formation of nucleosides and nucleotides is the nucleosidation reaction linking the nucleobase to the anomeric center of the sugar moiety [3]. Besides the nucleosidation reaction, nucleosides can be obtained by an alternative route building up the heterocycle in the presence of the sugar moiety.

General problems associated with the nucleosidation reaction are stereoselectivity and regioselectivity; on the other hand, poor nucleophilicity as well as simultaneous solubility of the nucleobases and sugar building blocks and low stability of the sugar donor are often troublesome. Nucleophilic addition of pyrimidines proceeds with low regioselectivity with respect to N-1 and N-3. Purine nucleosides are formed in comparable amounts as the N-7 and N-9 regioisomers. Furthermore, the stereochemistry at the anomeric center is difficult to control. Neighboring group participation by an ester group in place of the hydroxyl group at C-2 can be used to direct the nucleophilic attack so that exclusively β-isomers are generated. Without neighboring group participation, the anomeric effect is not sufficient to provide α-anomeric products with reasonable selectivity; thus, the formed α/β-anomeric mixtures often need to be separated, which is not always an easy task.

Four fundamentally different methods for nucleosidation reactions are known in the literature: (1) The Fischer–Helferich and Koenigs–Knorr procedures are based on Ag^+ or Hg^{2+} heavy metal salts of the nucleobases substituting a halogen at the anomeric center. Low solubility of the nucleobase salts, fast hydrolysis of the ribosyl halogenides, and harsh reaction conditions limit the utility of these methods [4]. (2) The Hilbert–Johnson nucleosidation employs alkylated nucleobases, which are sufficiently nucleophilic to substitute bromo sugars [5]. The generated quaternary salts offer a gentle means of introducing the nucleobase functionalities. (3) The silyl version of the Hilbert–Johnson method is called Vorbrüggen nucleosidation [6]. Silylated nucleobases are used, which exhibit increased solubility. They are usually reacted with peracylated sugars in the presence of a strong Lewis acid such as $SnCl_4$ or $TfOSiMe_3$. The intermediately generated sugar halogenides or triflates are the true electrophiles, although there is a strong S_N1 contribution due to the possible formation of an intermediate oxocarbenium ion. The silyl groups on the nucleobase are cleaved during the coupling reaction. Silylation of the nucleobases is often performed *in situ* using hexamethyldisilazane with Me$_3$SiCl or bis(trimethylsilyl)acetamide (BSA). (4) Finally, nucleosidation under basic conditions using NaH to deprotonate the nucleobase has been described by Kazimierczuk [7].

Fischer-Helferich and Koenigs-Knorr nucleosidation

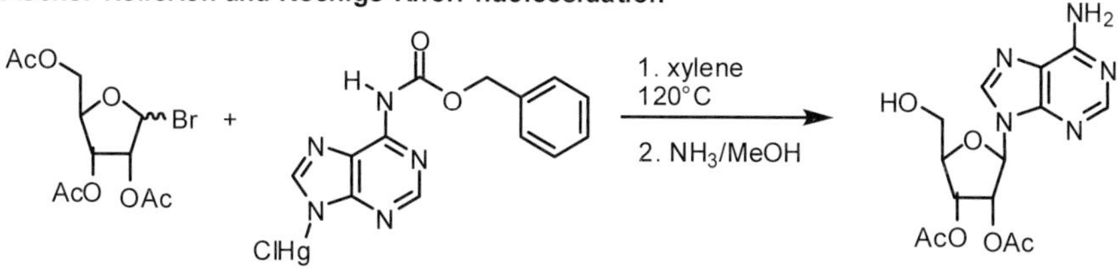

Hilbert-Johnson nucleosidation

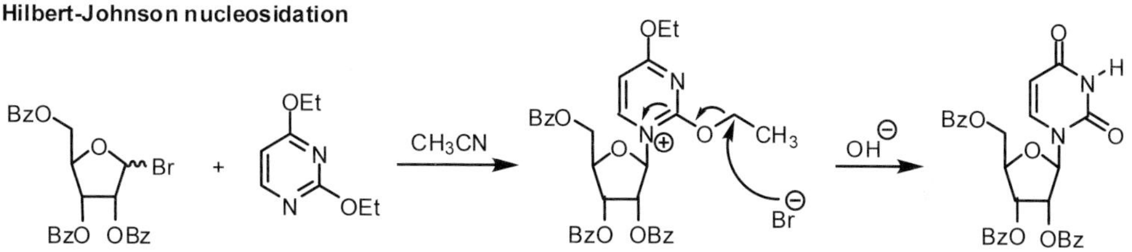

Vorbrüggen nucleosidation

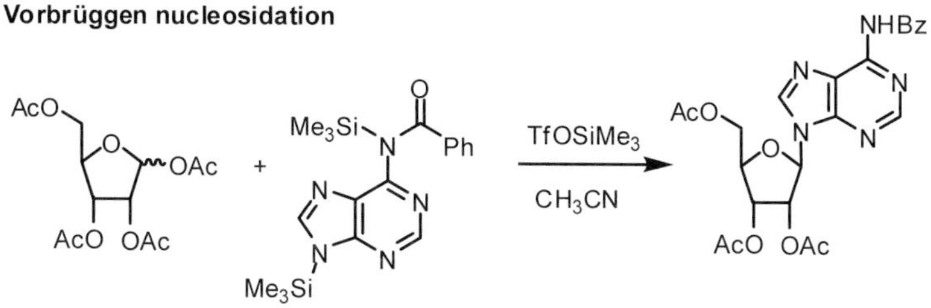

Nucleosidation under basic conditions

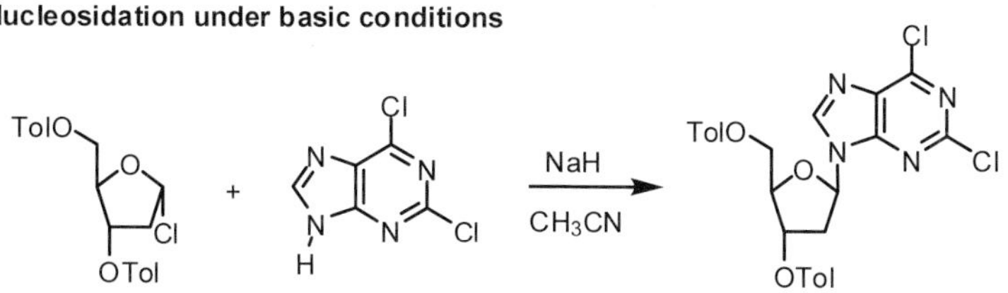

(b) Synthesis of 1

As an example of an artificial DNA analogue, the synthesis of the glucopyranosyl phosphoramidite **1** is described, in which the monomeric unit for an oligonucleotide synthesis is generated on a solid support (cf. **4.5.1.2**). Furthermore, the preparation of a glucopyranosyl nucleoside linked to a controlled pore glass (CPG) support **2** is presented. In general, the well-documented synthesis protocols of ribosyl or deoxyribosyl phosphoramidites can be adapted to the preparation of glucopyranosyl nucleotides by applying harsher reaction conditions [8]. Starting from *D*-glucose (**3**), regioselective acetal protection is followed by benzoylation to give **5** via **4**. Nucleosidation of **5** with uracil under Vorbrüggen conditions is performed with BSA by *in situ* silylation under the assistance of TfOSiMe$_3$. The strongly acidic conditions simultaneously lead to acetal deprotection, resulting in the nucleoside **6**. The β-anomer of the N-1 regioisomer is obtained as the major isomer. The primary hydroxyl group of **6** is selectively protected as the dimethoxytrityl ether (DMT) to form **7**, from which the phosphoramidite **1** is generated by nucleophilic substitution.

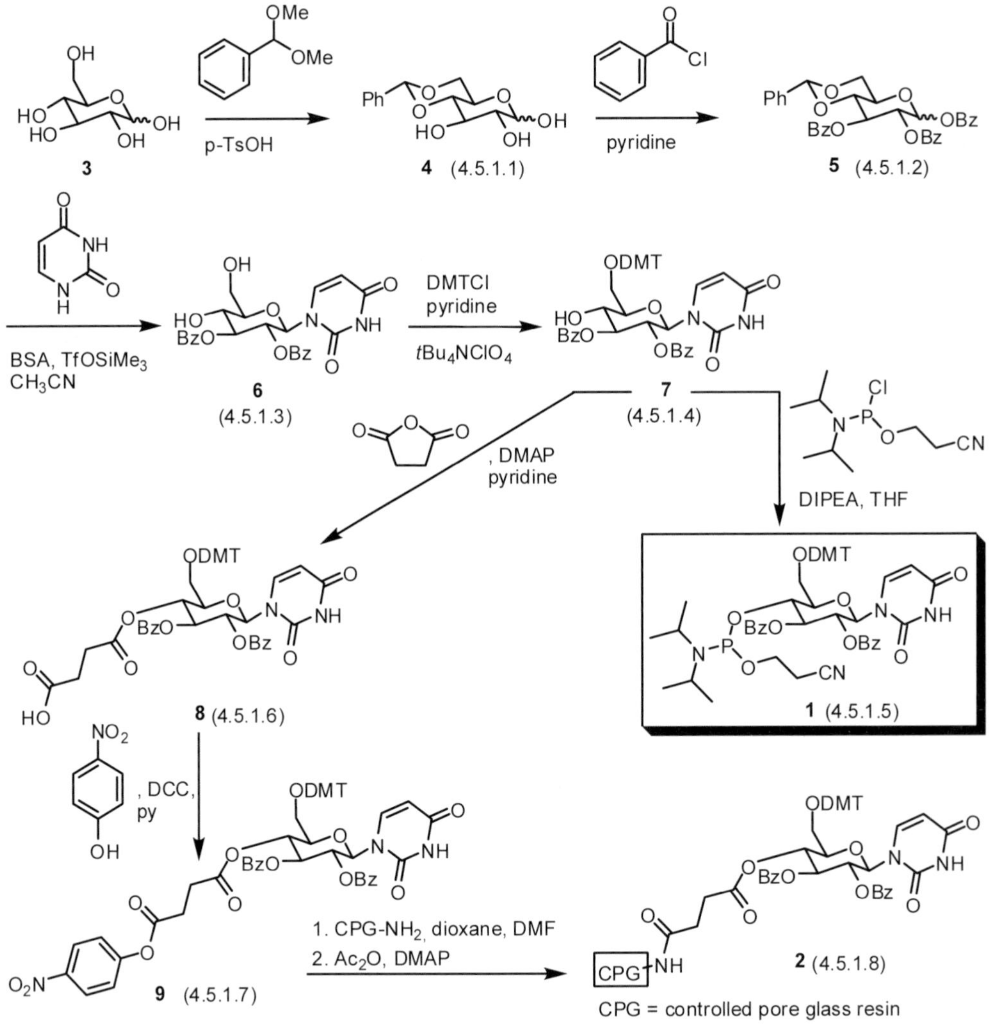

The CPG-bound nucleoside **2** is available from **7** by first using succinic anhydride to establish a linker that can be attached by DMAP activation. The resulting acid **8** is activated as *p*-nitrophenyl ester **9** and bound to the CPG resin as an amide. Finally, acylation of all remaining amino functionalities on the resin with acetic anhydride and DMAP is required as a capping step. The nucleoside-loaded CPG support **2** is used in solid-phase synthesis (cf. **4.5.1.2**).

(c) Experimental procedures for the synthesis of 1

4.5.1.1 * *D*-(4,6-Benzylidene)-glucose [9]

180.2   152.2   268.3

Benzaldehyde dimethyl acetal (3.31 mL, 22.2 mmol) is added to a suspension of *D*-glucose (2.00 g, 11.1 mmol) and *p*-toluenesulfonic acid (422 mg, 2.22 mmol, dried under high vacuum) in anhydrous DMF (32 mL) under an argon atmosphere. The mixture is heated until a solution is obtained, then stirred at room temperature for 2 h, after which Na_2CO_3 (5 g) is added. After stirring for 30 min, the solid is filtered off and washed with DMF. Evaporation of the solvent from the combined filtrate and washings is followed by flash chromatography on silica gel using EtOAc as eluent. The light-yellow oil obtained is dissolved in $CHCl_3$ (4 mL) and precipitated by the addition of ice-cold *n*-hexane (20 mL). The precipitate is crystallized (acetone/MeOH, 10:1) to give the product as a white solid; 1.82 g (61%); the ^1H NMR spectrum shows the existence of two anomers in a ratio of α:β = 3:2; TLC (SiO_2; EtOAc/MeOH, 9:1): R_f = 0.58; $[\alpha]_D^{20}$ = −4.8 (c = 1.83, EtOH).

---

^1H **NMR** (200 MHz, CD_3OD): δ (ppm) = 7.43–7.56 (m, 2 H, Ph), 7.25–7.38 (m, 3 H, Ph), 5.56 (s, 1 H, H-(C7)), 5.13 (d, 0.6 H, *J* = 3.6 Hz, αH-(C1')), 4.60 (d, 0.4 H, *J* = 7.7 Hz, βH-(C1')), 4.23 (dd, 0.4 H, *J* = 11.3, 4.5 Hz, βH-(C6')), 4.18 (dd, 0.6 H, *J* = 9.9, 4.8 Hz, αH-(C6')), 3.97 (dt, 0.6 H, *J* = 11.8, 4.9 Hz, αH-(C5')), 3.87 (t, 0.6 H, *J* = 9.3 Hz), 3.58–3.81 (m, 1.4 H), 3.39–3.50 (m, 2 H), 3.24 (dd, 0.4 H, *J* = 8.8, 7.7 Hz, βH-(C2')).

^{13}C **NMR** (50 MHz, CD_3OD): δ (ppm) = 139.6 (Ph), 139.5, 130.2 (Ph), 129.4, 127.9, 103.3 (C7), 103.2, 99.2, 95.0, 83.4, 82.7, 77.5, 75.0, 74.7, 72.1, 70.5 (C6'), 70.0, 68.0, 63.8.

MS (FAB^+, 3-NOBA): *m/z* = 269 (100, *M*H$^+$).

4.5.1.2 * *D*-(4,6-Benzylidene-1,2,3-tribenzoyl)-glucose [9]

268.3   140.6   580.6

Benzoyl chloride (4.28 g, 30.5 mmol) is added over 5 min to a stirred solution of benzylidene glucose **4.5.1.1** (1.82 g, 6.77 mmol) in anhydrous pyridine (25 mL) at 0 °C under an argon atmosphere. The mixture is stirred for 2 h at room temperature, then cooled to 0 °C, and the reaction is quenched by the addition of MeOH (2 mL). After almost complete removal of the solvent *in vacuo*, the residue is taken up in EtOAc (200 mL) and the solution obtained is extracted with saturated aqueous $NaHCO_3$ solution

(100 mL) and brine (100 mL). The aqueous phases are re-extracted with EtOAc (100 mL) and the combined organic phases are dried over Na_2SO_4. After removal of the solvent, traces of pyridine are removed from the residue by co-evaporation with toluene (2×10 mL). Flash chromatography of the residue on silica gel (*n*-hexane/EtOAc, 4:1) provides the product as a diastereomeric mixture of $\alpha:\beta = 0.56:0.44$ (determined by ^{1}H NMR spectroscopy); 3.39 g (86%); TLC (SiO_2; EtOAc/hexane, 2:1): $R_f = 0.64$; $[\alpha]_D^{20} = -99.6$ ($c = 2.15$, $CHCl_3$).

UV (EtOH): 230 (28700).

^{1}H NMR (300 MHz, $CDCl_3$): δ (ppm) = 8.14 (m, 1 H, Bn, Bz), 8.00 (m, 3 H, Bn, Bz), 7.93 (m, 2 H, Bn, Bz), 7.61 (m, 1 H, Bn, Bz), 7.23–7.59 (m, 13 H, Bn, Bz), 6.77 (d, 0.56 H, $J_{1'/2'} = 3.8$ Hz, αH-(C1')), 6.23 (d, 0.44 H, $J_{1'/2'} = 7.9$ Hz, βH-(C1')), 6.20 (t, 0.56 H, $J_{3'/2'} = J_{3'/4'} = 10.0$ Hz, αH-(C3')), 5.92 (t, 0.44 H, $J = 9.2$ Hz, βH-(C2'), βH-(C3')), 5.80 (t, 0.44 H, $J = 7.9$ Hz, βH-(C2'), βH-(C3')), 5.62 (dd, 0.56 H, $J_{2'/3'} = 10.0$, $J_{2'/1'} = 3.8$ Hz, αH-(C2')), 5.61 (s, 0.56 H, αH-(C7')), 5.58 (s, 0.44 H, βH-(C7')), 4.49 (dd, 0.44 H, $J_{6'S/6'R} = 9.4$, $J_{6'/5'} = 3.8$ Hz, βH-(C6')), 4.40 (dd, 0.56 H, $J_{6'S/6'R} = 10.3$, $J_{6'/5'} = 4.9$ Hz, αH-(C6')), 4.28 (dt, 0.56 H, $J_{5'/4'} = J_{5'/6'} = 9.9$, $J_{5'/6'} = 4.9$ Hz, αH-(C5')), 4.06 (t, 1 H, $J = 9.7$ Hz, H-(C4'), H-(C6')), 4.95 (dt, 0.44 H, $J_{5'/4'} = J_{5'/6'} = 9.6$, $J_{5'/6'} = 4.3$ Hz, βH-(C5')), 3.88 (t, 1 H, $J = 10.1$ Hz, H-(C4'), H-(C6')).

^{13}C NMR (75 MHz, $CDCl_3$): δ (ppm) = 165.7, 165.5, 165.3, 164.6, 136.7, 133.9, 133.6, 133.5, 130.2, 130.1, 129.8, 129.6, 129.4, 129.3, 129.1, 128.9, 128.7, 128.6, 128.4, 128.2, 126.1, 101.7, 93.1, 90.5, 78.9, 78.6, 72.0, 71.5, 70.9, 69.6, 68.6 (C6), 68.5, 67.4, 65.4.

MS (FAB$^+$, 3-NOBA): m/z = 581 (3, MH$^+$), 580 (4, M^+), 579 (9, M^+ – H), 475 (3, M^+ – Bz).

4.5.1.3   ***   (2,3-Dibenzoyl-β-D-glucopyranosyl)-uracil [9]

112.1   580.6   482.5

N,O-Bis(trimethylsilyl)acetamide (7.34 mL, 30 mmol) is added to a stirred suspension of uracil (1.12 g, 10.0 mmol) and the glucose derivative **4.5.1.2** (6.39 g, 11.1 mmol) (both are dried overnight under high vacuum at room temperature) in anhydrous CH_3CN (30 mL) at room temperature under an argon atmosphere. Stirring is continued for 30 min at 100 °C until a homogeneous solution is obtained. After the addition of trimethylsilyl trifluoromethanesulfonate (3.63 mL, 20 mmol), the reaction mixture is stirred for 4.5 h at 100 °C, with an additional portion of trimethylsilyl trifluoromethanesulfonate (3.63 mL, 20 mmol) being added after 1 h. Thereafter, the solvent is evaporated, the yellow oil obtained is dissolved in EtOAc (100 mL), and this solution is washed twice with a saturated $NaHCO_3$ solution and with brine (each 100 mL). The aqueous phases are re-extracted with EtOAc (100 mL) and the combined organic phases are dried over Na_2SO_4. Removal of the solvent and flash chromatography of the residue on silica gel (EtOAc/hexane, 2:1) provides the desired nucleoside after crystallization from acetone; 3.03 g (63%); TLC (SiO_2; EtOAc): $R_f = 0.40$; mp 158 °C; $[\alpha]_D^{20} = 100.0$ ($c = 0.60$, EtOH).

¹H NMR (300 MHz, [D₆]DMSO): δ (ppm) = 11.33 (s$_{br}$, 1 H, NH), 7.97 (d, 1 H, $J_{6/5}$ = 8.2 Hz, H-(C6)), 7.88 (d, 2 H, J = 7.1 Hz, Bz), 7.76 (d, 2 H, J = 7.1 Hz, Bz), 7.58–7.63 (m, 2 H, Bz), 7.42–7.50 (m, 4 H, Bz), 6.10 (d, 1 H, $J_{1'/2'}$ = 9.0 Hz, H-(C1')), 5.74 (m, 2 H, H-(C5), 4'-OH), 5.65 (t, 1 H, $J_{3'/2'} = J_{3'/4'}$ = 9.2 Hz, H-(C3')), 5.53 (t, 1 H, $J_{2'/1'} = J_{2'/3'}$ = 9.3 Hz, H-(C2')), 4.77 (t, 1 H, $J_{OH/6'}$ = 5.6 Hz, 6'-OH), 3.78–3.90 (m, 3 H, H-(C4'), H-(C5'), H-(C6')), 3.58 (m, 1 H, H-(C6')).

¹³C NMR (100 MHz, [D₆]DMSO): δ (ppm) = 165.1, 164.6, 162.6 (C4), 150.3 (C2), 141.1 (C6), 133.8 (Bz), 133.3, 129.4 (Bz), 129.1 (Bz), 129.0, 128.7, 128.5, 128.2 (Bz), 102.3 (C5), 79.6 (C1'), 79.3 (C5'), 75.7 (C3'), 71.2 (C2'), 67.0 (C4'), 60.5 (C6').

MS (FAB⁺, 3-NOBA): m/z = 483 (17, MH^+).

4.5.1.4   **   (2,3-Dibenzoyl-6-O-(dimethoxytrityl)-β-D-glucopyranosyl)-uracil [9]

A mixture of the nucleoside **4.5.1.3** (2.00 g, 4.15 mmol), tetrabutylammonium perchlorate (1.70 g, 4.98 mmol), and 4,4'-dimethoxytrityl chloride (1.69 g, 4.98 mmol) is dried under high vacuum at room temperature and then dissolved under an argon atmosphere in anhydrous pyridine (15 mL). The initially deep-orange, later dark-yellow solution is stirred for 90 min at room temperature. The reaction is then quenched by the addition of MeOH (2 mL) and stirring is continued for 10 min. The mixture is concentrated *in vacuo*, the residue is dissolved in EtOAc (50 mL), and the resulting solution is washed with a saturated NaHCO₃ solution and brine (each 2 × 50 mL). The aqueous phases are re-extracted with EtOAc (50 mL) and the combined organic phases are dried over Na₂SO₄. After removal of the solvent *in vacuo*, further purification is performed by flash chromatography on silica gel (EtOAc/hexane, 2:3) to provide the desired nucleoside as a light-yellow foam; 2.81 g (87%); TLC (SiO₂; EtOAc/hexane, 1:1): R_f = 0.30; mp 142 °C; $[\alpha]_D^{20}$ = 50.3 (c = 1.55, CHCl₃).

¹H NMR (300 MHz, CDCl₃): δ (ppm) = 8.85 (s$_{br}$, 1 H, NH), 7.93 (dd, 2 H, J = 8.4, J = 1.3 Hz, Bz), 7.85 (dd, 2 H, J = 8.4, J = 1.3 Hz, Bz), 7.39–7.60 (m, 5 H, Bz, DMT, H-(C6)), 7.21–7.37 (m, 11 H, Bz, DMT), 6.82 (dd, 4 H, J = 9.0, J = 1.2 Hz, DMT), 6.07 (d, 1 H, $J_{1'/2'}$ = 9.3 Hz, H-(C1')), 5.80 (d, 1 H, $J_{5/6}$ = 9.3 Hz, H-(C5)), 5.66 (t, 1 H, $J_{3'/2'} = J_{3'/4'}$ = 9.4 Hz, H-(C3')), 5.50 (t, 1 H, $J_{2'/1'} = J_{2'/3'}$ = 9.5 Hz, H-(C2')), 4.09 (t, 1 H, $J_{4'/3'} = J_{4'/5'}$ = 9.4 Hz, H-(C4')), 3.85 (m, 1 H, H-(C5')), 3.77 (s, 6 H, OCH₃), 3.49 (m, 2 H, H-(C6')), 3.22 (s$_{br}$, 1 H, 4'-OH).

¹³C NMR (75 MHz, CDCl₃): δ (ppm) = 166.6, 165.5, 162.4, 158.7 (DMT), 150.1 (C2), 144.4 (DMT), 139.2 (C6), 135.5, 133.8 (Bz), 133.6, 130.0, 129.9, 129.6, 129.2, 128.9, 128.4, 128.1, 128.0, 127.1, 113.3, 86.8 (DMT), 80.6 (C1'), 77.8, 76.0, 70.5, 70.1, 63.1 (C6'), 55,2 (OCH₃).

MS (FAB⁺, 3-NOBA): m/z = 785 (1, MH^+), 784 (2, M^+).

4.5.1.5   ***   **(2,3-Dibenzoyl-6-O-(dimethoxytrityl)-β-D-glucopyranosyl)-uracil 4-O-((2-cyanoethyl)-N,N-diisopropylamino-phosphoramidite)** [9]

N,N-Diisopropylethylamine (663 µL, 3.82 mmol) and 2-cyanoethyl N,N-diisopropylchlorophosphoamidite (452 µL, 1.91 mmol) are successively added to a stirred solution of the nucleoside **4.5.1.4** (1.00 g, 1.27 mmol) in anhydrous THF (9 mL) under an argon atmosphere. Stirring is continued for 2.5 h at room temperature. EtOAc (50 mL) is then added and the reaction mixture is washed with a saturated NaHCO$_3$ solution (2 × 50 mL) and brine (2 × 50 mL). After re-extraction of the aqueous phase with EtOAc (50 mL), the combined organic phases are dried over Na$_2$SO$_4$ and concentrated *in vacuo*, and the residue is purified by flash chromatography on silica gel (EtOAc/hexane, 2:3). After removal of the solvent *in vacuo*, the oily product is redissolved in CH$_2$Cl$_2$ and the solvent is evaporated to give the desired phosphoramidite as a white foam (1:1 mixture of diastereomers); 1.09 g (91%); TLC (SiO$_2$; EtOAc/hexane, 2:3): R_{f1} = 0.22, R_{f2} = 0.28; $[\alpha]_D^{20}$ = 73.7 (c = 1.90, CHCl$_3$).

^{1}H NMR (400 MHz, CDCl$_3$): δ (ppm) = 8.42 (s$_{br}$, 1 H, NH), 7.99 (d, 1 H, J = 8.5 Hz, Bz), 7.91 (d, 1 H, J = 8.5 Hz, Bz), 7.89 (d, 1 H, J = 8.5 Hz, Bz), 7.86 (d, 1 H, J = 8.5 Hz, Bz), 7.62 (d, 0.5 H, $J_{6/5}$ = 8.2 Hz, H-(C6)), 7.56 (d, 0.5 H, $J_{6/5}$ = 8.2 Hz, H-(C6)), 7.45–7.54 (m, 3 H, Bz), 7.41 (m, 1 H, Bz), 6.78–6.85 (m, 4 H, DMT), 6.11 (d, 0.5 H, $J_{1'/2'}$ = 9.3 Hz, H-(C1')), 6.09 (d, 0.5 H, $J_{1'/2'}$ = 9.3 Hz, H-(C1')), 5.88 (d, 0.5 H, $J_{5/6}$ = 8.3 Hz, H-(C5)), 5.86 (d, 0.5 H, $J_{5/6}$ = 8.3 Hz, H-(C5)), 5.86 (t, 0.5 H, $J_{3'/2'}$ = $J_{3'/4'}$ = 9.3 Hz, H-(C3')), 5.81 (t, 0.5 H, $J_{3'/2'}$ = $J_{3'/4'}$ = 9.3 Hz, H-(C3')), 5.54 (t, 0.5 H, $J_{2'/1'}$ = $J_{2'/3'}$ = 9.5 Hz, H-(C2')), 5.42 (t, 0.5 H, $J_{2'/1'}$ = $J_{2'/3'}$ = 9.5 Hz, H-(C2')), 4.38 (q, 0.5 H, $J_{4'/3'}$ = $J_{4'/5'}$ = $J_{4'/P}$ = 9.8 Hz, H-(C4')), 4.18 (q, 0.5 H, $J_{4'/3'}$ = $J_{4'/5'}$ = $J_{4'/P}$ = 9.8 Hz, H-(C4')), 3.98 (dd, 0.5 H, J = 9.7, 4 Hz, H-(C5')), 3.87 (dd, 0.5 H, J = 9.7, 2 Hz, H-(C5')), 3.79 (s, 3 H, OCH$_3$), 3.78 (s, 3 H, OCH$_3$), 3.64 (t, 1 H, $J_{6'/5'}$ = $J_{6'S/6'R}$ = 9.7 Hz, H-(C6')), 3.49 (m, 0.5 H, H-(C6')), 3.41 (m, 0.5 H, H-(C6')), 3.35 (m, 2 H, CH$_2$O), 3.29 (m, 1 H, CHiPr), 3.20 (m, 1 H, CHiPr), 2.25 (m, 1 H, CH$_2$CN), 2.08 (m, 0.5 H, CH$_2$CN), 2.02 (m, 0.5 H, CH$_2$CN), 0.86–0.92 (m, 12 H, CH$_3i$Pr).

^{13}C NMR (100 MHz, CDCl$_3$): δ (ppm) = 165.5, 165.4, 162.2 (C4), 158.6 (DMT), 158.5, 150.0 (C2), 144.8 (DMT), 144.7, 139.5 (C6), 139.4, 136.1, 135.9, 135.6, 133.7 (Bz), 133.2, 130.4, 130.3, 130.2, 130.0, 129.8, 129.7, 128.5, 128.4, 128.3, 128.2, 128.1, 128.0, 127.8, 126.9, 117.3 (CN), 113.1 (DMT), 103.4 (C5), 103.3, 86.3 (DMT), 86.1, 80.7 (C1'), 80.6, 78.8 (C5'), 78.7, 75.2 (C3'), 74.4, 70.9 (C2'), 70.6, 70.5 (J_{CP} = 4.7 Hz, C4'), 70.0 (J_{CP} = 6.6 Hz, C4'), 63.2 (C6'), 62.3 (C6'), 57.7 (J_{CP} = 19.6 Hz, CH$_2$O), 57.6 (J_{CP} = 19.6 Hz, CH$_2$O), 55.2 (OCH$_3$), 43.0 (J_{CP} = 8.9 Hz, CHiPr), 42.9 (J_{CP} = 8.9 Hz, CHiPr), 24.4 (CH$_3i$Pr), 24.3 (CH$_3i$Pr), 24.2 (CH$_3i$Pr), 19.9 (J_{CP} = 7.2 Hz, CH$_2$CN), 19.6 (J_{CP} = 7.6 Hz, CH$_2$CN).

^{31}P NMR (162 MHz, CDCl$_3$): δ (ppm) = 151.3, 150.0.

MS (FAB$^+$, 3-NOBA): 985 (0.4, MH$^+$).

4.5.1.6   *   (2,3-Dibenzoyl-6-*O*-(dimethoxytrityl)-β-*D*-glucopyranosyl)-uracil 4-*O*-(succinylacid)-ester [9]

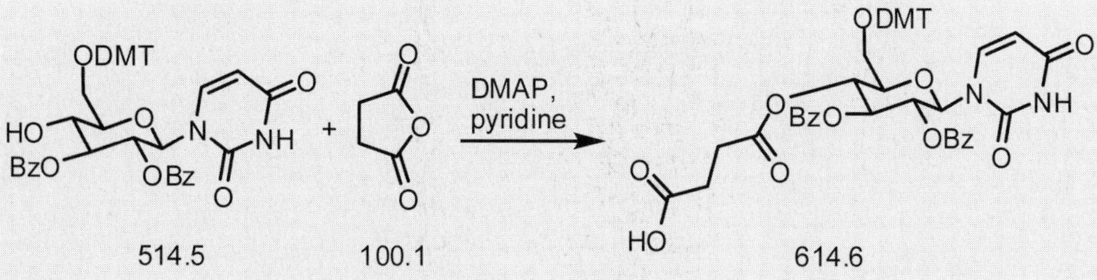

514.5   100.1   614.6

A solution of the nucleoside **4.5.1.4** (392 mg, 500 µmol), DMAP (73 mg, 0.60 mmol), and succinic anhydride (55 mg, 0.55 mmol) in anhydrous pyridine (4 mL) is stirred for 6 h at 60 °C. The pyridine is then distilled off *in vacuo* and any remaining pyridine is removed from the residue by co-evaporation with toluene (2 × 10 mL). The white foam is dissolved in CH_2Cl_2 (20 mL) and this solution is washed with a cooled solution of 10% citric acid and with H_2O. After re-extraction of the aqueous phases with CH_2Cl_2 (10 mL), the combined organic phases are dried over $MgSO_4$ and concentrated *in vacuo* to yield the product as a white foam; 442 mg (99%); TLC (SiO_2; EtOAc/hexane, 2:1): $R_f = 0.31$; mp 136 °C; $[\alpha]_D^{20} = 97.2$ ($c = 1.82$, $CHCl_3$).

---

^{1}H NMR (400 MHz, $CDCl_3$): δ (ppm) = 8.96 (s_{br}, 1 H, NH), 7.84 (m, 4 H, Bz), 7.57 (d, 1 H, $J_{6/5} = 8.3$ Hz, H-(C6)), 7.42 (m, 4 H, Bz, DMT), 7.28 (m, 10 H, Bz, DMT), 7.19 (m, 1 H, Bz, DMT), 6.80 (dd, 4 H, $J = 9.0$, $J = 1.7$ Hz, DMT), 6.13 (d, 1 H, $J_{1'/2'} = 9.2$ Hz, H-(C1')), 5.86 (d, 1 H, $J_{5/6} = 8.2$ Hz, H-(C5)), 5.82 (t, 1 H, $J_{4'/3'} = J_{4'/5'} = 9.7$ Hz, H-(C4')), 5.60 (t, 1 H, $J = 9.8$ Hz, H-(C2'), H-(C3')), 5.55 (t, 1 H, $J = 9.5$ Hz, H-(C2'), H-(C3')), 4.01 (d, 1 H, $J_{5'/4'} = 8.9$ Hz, H-(C5')), 3.75 (s, 3 H, OCH₃), 3.75 (s, 3 H, OCH₃), 3.40 (d, 1 H, $J_{6'S/6'R} = 9.3$ Hz, H-(C6')), 3.17 (dd, 1 H, $J_{6'S/6'R} = 10.9$, $J_{6'/5'} = 3.9$ Hz, H-(C6')), 2.28 (m, 4 H, succinyl).

^{13}C NMR (100 MHz, $CDCl_3$): δ (ppm) = 175.6 (COOH), 170.4 (CO succinyl), 165.7 (CO), 165.3, 162.6 (C4), 158.6 (DMT), 150.1 (C2), 144.2 (DMT), 139.3 (C6), 135.5, 135.4, 133.8 (Bz), 133.5, 130.1, 130.0, 129.8, 129.6, 128.6, 128.5, 128.4, 128.2, 128.0, 127.9, 127.0, 113.2 (DMT), 103.6 (C5), 86.3 (DMT), 80.6 (C1'), 76.5 (C5'), 73.3 (C3'), 70.6 (C2'), 68.3 (C4'), 63.5, 61.5 (C6'), 55.2 (OCH₃), 28.8 (succinyl), 28.7.

MS (FAB⁻, 3-NOBA): $m/z = 884$ (15, M^-), 883 (30, $[M – H]^-$).

4.5.1.7 * (2,3-Dibenzoyl-6-*O*-(dimethoxytrityl)-β-*D*-glucopyranosyl)-uracil 4-*O*-(succinylacid-4-nitrophenyl-ester)-ester [9]

614.6   4-nitrophenol: 108.14   735.7

A solution of *N,N'*-dicyclohexylcarbodiimide (231 mg, 1.12 mol) in anhydrous dioxane (3.9 mL) is added to a stirred solution of the nucleoside **4.5.1.6** (382 mg, 432 µmol) and 4-nitrophenol (60 mg, 430 µmol) (dried overnight under high vacuum at room temperature before use) in anhydrous dioxane (2.5 mL) and anhydrous pyridine (178 µL) at room temperature under an argon atmosphere. After 2 h, the mixture is concentrated *in vacuo* and the residue is purified by flash chromatography on silica gel (EtOAc/hexane, 1:1). Evaporation of the solvent from the appropriate fraction and drying of the residue under high vacuum affords the desired active ester as a white solid, which is used directly for the next step; 355 mg (82%); TLC (SiO$_2$; EtOAc/*n*-hexane, 1:1): $R_f = 0.33$; mp 126 °C.

---

^{1}H NMR (400 MHz, CDCl$_3$): δ (ppm) = 8.18 (s$_{br}$, 1 H, NH), 8.13 (d, 1 H, $J = 9.2$ Hz, NO$_2$Ph), 7.83–7.88 (m, 4 H, Bz, DMT), 7.52 (m, 1 H, H-(C6)), 7.41–7.51 (m, 4 H, Bz, DMT), 7.20–7.36 (m, 13 H, Bz, DMT, NO$_2$Ph), 7.05 (d, 1 H, $J = 9.3$ Hz, NO$_2$Ph), 6.80–6.84 (m, 4 H, Bz, DMT), 6.09 (dd, 1 H, $J = 11.7$, 9.4 Hz, H-(C1')), 5.87 (dd, 1 H, $J = 8.1$, 5.4 Hz, H-(C5)), 5.79 (q, 1 H, $J = 9.7$ Hz, H-(C2'), H-(C3'), H-(C4')), 5.66 (dt, 1 H, $J = 9.9$, 1.3 Hz, H-(C2'), H-(C3'), H-(C4')), 5.57 (dq, 1 H, $J = 9.1$, 3.2 Hz, H-(C2'), H-(C3'), H-(C4')), 3.98 (m, 1 H, H-(C6')), 3.78 (s, 3 H, OCH$_3$), 3.77 (s, 3 H, OCH$_3$), 3.45 (dt, 1 H, $J_{6'/5'} = J_{6'S/6'R} = 10.9$ Hz, $J = 2.1$, H-(C6')), 3.20 (dt, 1 H, $J_{5'/4'} = J_{5'/6'} = 11.3$, $J_{5'/6'} = 4.0$ Hz, H-(C5')), 2.26–2.65 (m, 4 H, succinyl).

^{13}C NMR (100 MHz, CDCl$_3$): δ (ppm) = 171.4 (succinyl CO), 170.1, 165.6 (CO), 165.3, 161.9 (C4), 158.6, 155.1, 153.8, 149.9 (C2), 145.3 (DMT), 144.4, 144.2, 139.1 (C6), 139.0, 135.7, 135.6, 135.5, 135.3, 133.9, 133.4, 130.1, 130.0, 129.9, 129.8, 128.8, 128.7, 128.5, 128.4, 128.2, 128.0, 127.9, 127.0, 125.1, 122.2, 113.2 (DMT), 103.6 (C5), 86.4 (DMT), 80.7 (C1'), 76.5 (C5'), 73.3 (C3'), 70.5 (C2'), 70.4, 68.6 (C4'), 68.3, 61.5 (C6'), 61.4, 55.2 (OCH$_3$), 32.6 (succinyl), 32.5, 30.7, 30.6.

MS (FAB$^+$, 3-NOBA): $m/z = 1006$ (0.2, MH$^+$), 1005 (0.3, M^+).

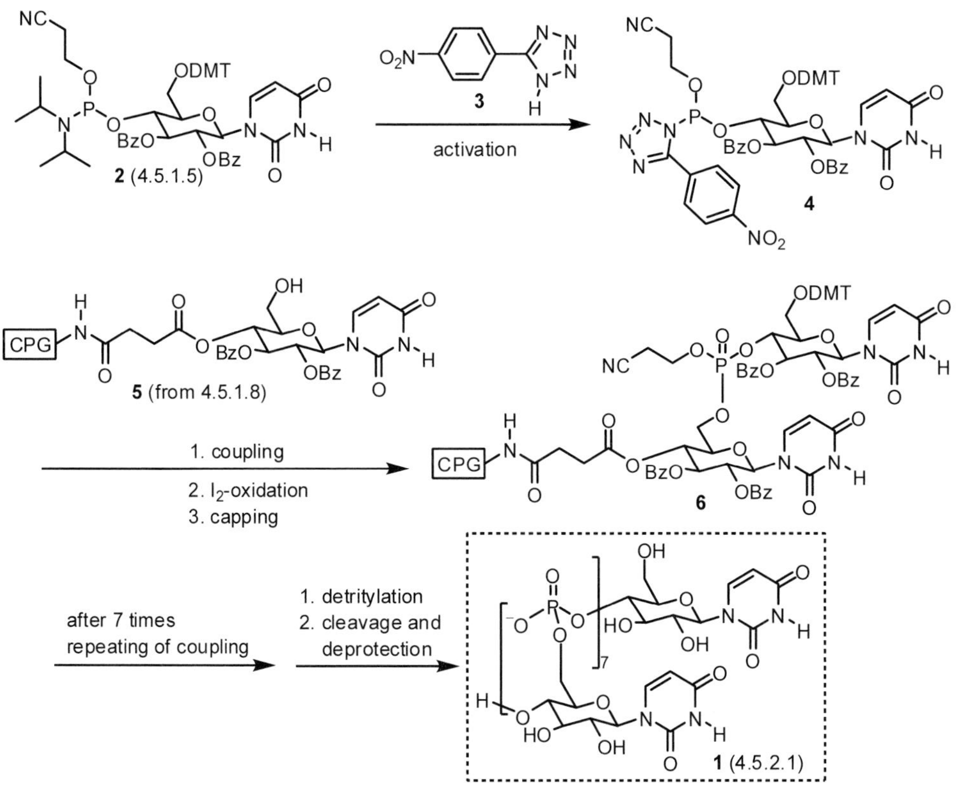

(c)   Experimental procedure for the synthesis of 1

4.5.2.1   * Manual solid-phase synthesis of glucopyranosyl-DNA [11]**

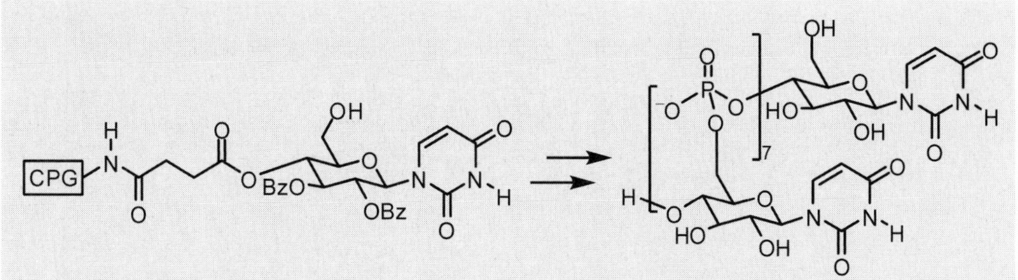

The glucopyranosyl DNA octamer **1** is prepared in a small fritted glass column using a controlled pore glass (CPG) resin (25 mg) loaded with uracil nucleoside (**4.5.1.8**) (25 μmol g^{-1}). The following procedure is repeated for each coupling cycle:

(1) Detritylation: Dichloroacetic acid (2%, 5 mL) is added to the CPG-bound nucleoside or nucleic acid, respectively. After 5 min, the reagent is separated from the resin by filtration using nitrogen pressure and then the resin is washed with CH$_2$Cl$_2$ (5 mL) and Et$_2$O (5 mL). The solid support is dried

in a nitrogen flow. The liquid phase resulting from deprotection is collected in order to determine the yield (> 96%) by analysis of the DMT$^+$ concentration, comparing successive coupling cycles.

(2) Coupling: Phosphoramidite **4.5.1.5** (5 equiv.) and *p*-nitrophenyl tetrazole (20 equiv.) are added to the resin, which is then dried under high vacuum for 1 h. The glass column is flushed with argon, and then CH$_3$CN (150 μL) is added and the mixture is gently shaken for 1 h. Thereafter, the suspension is filtered and the resin is washed with CH$_3$CN (5 mL) and THF (5 mL) and dried in a nitrogen flow.

(3) Oxidation: The solid support is treated with a solution of iodine (0.1 mol) in THF/H$_2$O/2,6-lutidine 2:2:1 (5 mL) for 2 min. The suspension is then filtered and the resin is washed sequentially with THF/H$_2$O/2,6-lutidine (2:2:1; 5 mL), THF (5 mL), MeOH (5 mL), and Et$_2$O (5 mL).

(4) Capping: Under an argon atmosphere, the resin is treated with a solution of DMAP (5.4%) in THF/2,6-lutidine/Ac$_2$O (10:1:1; 6 mL) for 5 min. The suspension is then filtered, and the resin is washed with THF (5 mL), MeOH (5 mL), and Et$_2$O (5 mL) and dried in vacuo.

Steps (1)–(4) are repeated seven times until the desired oligomer length is obtained. A final detritylation step (1) is performed before basic cleavage.

(5) Deprotection and cleavage: The resin-bound oligomer is suspended in concentrated aqueous ammonia (10 mL) for 16 h at 55 °C. After filtration and washing of the resin with H$_2$O, the combined aqueous filtrates are concentrated in vacuo. The residue is redissolved in H$_2$O (10 mL) and lyophilized.

The crude product can be purified by preparative HPLC on an ion-exchange column (Nucleogene DEAE 60-7) with the eluents: A: 20 mM K$_2$HPO$_4$/KH$_2$PO$_4$, pH 6, 20% CH$_3$CN, 80% H$_2$O, and 1 M KCl, B: 20 mM K$_2$HPO$_4$/KH$_2$PO$_4$, pH 6, 20% CH$_3$CN, 80% H$_2$O), gradient 10% to 50% B in 30 min, t_R = 18.6 min, or on an RP C-8 (Aquapore RP-300) column with the eluents: A: 0.1 M tetraethylammonium hydroxide, 0.1 M HOAc in H$_2$O, pH 7, B: 0.1 M tetraethylammonium hydroxide, 0.1 M HOAc in 20% H$_2$O, 80% CH$_3$CN, pH 7) gradient 5% to 20% B in 30 min, t_R = 15.1 min.

[1] H. G. Khorana, K. L. Agarwal, H. Büchi, M. H. Caruthers, N. K. Gupta, K. Kleppe, A. Kumar, E. Ohtsuka, U. L. Raj Bhandary, J. H. Van de Sande, V. Sgaramella, T. Terao, H. Weber, T. Yamada, *J. Mol. Biol.* **1972**, *72*, 209.

[2] M. J. Gait, *Oligonucleotide Synthesis, A Practical Approach*, IRL Press, Oxford, **1984**.

[3] R. L. Letsinger, J. L. Finnan, G. A. Heavner, W. B. Lunsford, *J. Am. Chem. Soc.* **1975**, *97*, 3278.

[4] B. C. Froehler, M. D. Matteucci, *Tetrahedron Lett.* **1986**, *27*, 469.

[5] M. D. Matteucci, M. H. Caruthers, *J. Am. Chem. Soc.* **1981**, *103*, 3185.

[6] P. Herdewijn, *Oligonucleotide Synthesis, Methods and Applications*, Humana Press, Totowa, **2004**.

[7] N. D. Sinha, J. Biernat, J. MacManus, H. Köster, *Nucleic Acids Res.* **1984**, *12*, 4539.

[8] X. Wu, S. Pitsch, *Nucleic Acid Res.* **1998**, *26*, 4315.

[9] S. A. Scaringe, *Methods Enzymol.* **2000**, *317*, 3.

[10] For the synthesis of 5-(*p*-nitrophenyl)tetrazole, see: W. G. Finnegan, R. E. Henry, R. Lofquist, *J. Am. Chem. Soc.* **1958**, *80*, 3908.

[11] U. Diederichsen, A. *Hypoxanthin-Basenpaarungen in HOMO-DNA Oligonucleotiden*; B. *Zur Frage des Paarungsverhaltens von Glucopyranosyl-Oligonucleotiden*, Dissertation ETH Nr. 10122, Zürich, **1993**.

5   Index of Reactions

In this index, reactions and methods are covered, which are performed in the foregoing syntheses in experimental detail. Reaction types, methods and principles only mentioned in the text are not listed here.

A

Acetalization	1.4.3, 1.5.1
Acetylation	
of an alcohol	1.1.1, 1.1.5
of a primary amine	3.2.4, 3.2.5
of carbohydrate OH groups	4.3.1
Acid chloride formation	1.5.1, 4.2.4
Acylation	
electrophilic (Friedel-Crafts)	1.1.6, 1.4.2, 1.4.3, 3.4.1
nucleophilic (Hünig)	1.4.4
of Ar-Li by carboxamide	3.2.3
of C-Si bond	4.2.4
Acid cleavage of acetoacetate	1.1.3
Addition	
of Br_2 to C=C	1.7.2
of R–M to C=O	1.1.1, 1.1.2, 1.1.3, 1.5.2, 1.7.5, 3.2.7
radical, to C=C	1.8.1, 3.5.1
Aldol reaction	1.3.1, 1.3.5, 3.5.4
enantioselective	1.3.2, 1.3.4, 1.3.5
Alkylation	
of acetoacetate	1.2.3, 1.5.3
of an ester	4.2.5
of phenols	1.3.4, 3.5.2
Frater-Seebach	1.4.1
enantioselective, of a carboxylic acid	1.2.2
enantioselective, of a ketone	1.2.1
Alkylhalide formation, from ROH	3.5.2, 4.2.1
Allylsilane formation	4.2.4
Amide formation see: Carboxamide formation	
Amidoxime formation	3.2.6
Amination, reductive	4.1.3
Amino acid resolution see: Enzymatic transformation	
Aminoalkylation	1.3.3
Anhydride formation	1.1.3, 3.4.1

Anomeric deprotection	4.3.1
Auxiliary	
Ephedrine	1.1.4
Evans	1.2.2, 1.7.5
SAMP	1.2.1
Azetidinone formation	3.1.3

B

Baeyer-Villiger oxidation	4.1.2
Baylis-Hillman reaction	1.1.5
Barton-Zard synthesis (pyrrole)	3.2.2
Benzothiophene synthesis	3.2.3
Biginelli reaction	3.3.1
BINAL-H reduction	2.4.1
Birch reduction	1.7.2
Bischler-Napieralski synthesis (isoquinoline)	3.3.2, 4.1.2
tButylester, cleavage of	3.4.1, 4.4.2

C

Carbonyl olefination	1.1.6, 1.1.7, 3.5.1, 4.2.6
Carboxamide formation	
from a carboxylic ester	1.3.5, 4.1.2
from a carboxylic acid	1.5.1, 3.2.3
Carroll reaction	1.5.3
CBS reduction	2.4.2
Chromone synthesis	3.4.5
Claisen condensation	1.4.1
Claisen rearrangement	
of allyl acetoacetates	1.5.3
of allyl arylethers	3.4.5
of glycine ester enolates	4.4.4
of orthoesters	1.8.1
Crown ether	3.5.2, 3.5.6
Cuprate addition	1.6.4

Reactions and Syntheses in the Organic Chemistry Laboratory. L. F. Tietze, Th. Eicher, U. Diederichsen, A. Speicher
Copyright © 2007 WILEY-VCH Verlag GmbH & Co. KGaA, Weinheim
ISBN: 978-3-527-31223-8

7 Subject Index

For a detailed representation of the reactions of the foregoing syntheses, see **Index of Reactions**; for a detailed representation of prepared compounds, see **Index of Products**.

Reactions and Syntheses in the Organic Chemistry Laboratory. L. F. Tietze, Th. Eicher, U. Diederichsen, A. Speicher
Copyright © 2007 WILEY-VCH Verlag GmbH & Co. KGaA, Weinheim
ISBN: 978-3-527-31223-8